19th Process Plant Safety Symposium (PPSS)

Topical Conference at the 2017 AIChE Spring Meeting and 13th Global Congress on Process Safety

San Antonio, Texas, USA
26 - 30 March 2017

ISBN: 978-1-5108-4362-2

Printed from e-media with permission by:

Curran Associates, Inc.
57 Morehouse Lane
Red Hook, NY 12571

Some format issues inherent in the e-media version may also appear in this print version.

Printed by Curran Associates, Inc. (2017)

For permission requests, please contact AIChE
at the address below.

AIChE
120 Wall Street, FL 23
New York, NY 10005-4020

Phone: (800) 242-4363
Fax: (203) 775-5177

www.aiche.org

Additional copies of this publication are available from:

Curran Associates, Inc.
57 Morehouse Lane
Red Hook, NY 12571 USA
Phone: 845-758-0400
Fax: 845-758-2633
Email: curran@proceedings.com
Web: www.proceedings.com

TABLE OF CONTENTS

482030 AGILE Award Keynote Address

Abstract:

Factors in Selecting AGILE Honoree

1. Overall sustained excellent industry leadership in businesses that employ chemical engineers

2. Recognized for efforts in innovation and creating strategies that ensure success

3. Value of the keynote to the meeting attendees and the members of the Institute Schedule

Event	Time	Location
Continental Breakfast	7:00AM - 8:00AM	Henry Gonzalez Convention Center, Lila Cockrell Theatre
AGILE Keynote Address	8:00AM - 9:15AM	Henry Gonzalez Convention Center, Lila Cockrell Theatre

The 2017 AIChE Spring Meeting and 13th GCPS AGILE Keynote Address will be presented by AGILE award recipient John Y. Televantos, Partner, Arsenal Capital Partners.

John Televantos is a Partner at Arsenal Capital Partners, a middle market private equity firm focused on the specialty industrial, healthcare and specialty financial service sectors. Before joining Arsenal in 2006, Dr. Televantos was President of Aqualon Company, owned by Hercules, Inc.; President and Founder of Helios Chemical Company; and Chief Executive Officer for Foamex International.

Dr. Televantos began his career with Union Carbide in 1977 in silicones research. There he developed two key processes for the manufacture of Methyl Chloride and Siloxane polymers that were commercialized. These processes not only improved efficiency and reduced cost but also dramatically reduced waste. He moved through a variety of management positions there and later at ARCO Chemical, where he became vice president of strategy and development.

At ARCO John co-invented and led a team that developed and commercialized a breakthrough technology in polyol manufacture (a key raw material for polyurethanes) eliminating a whole process step along with waste and emissions, while at the same time opening the door to the creation of new higher-performance polyurethanes. The technology has since been licensed by most of the major global polyol manufacturers with significant waste reduction worldwide.

At Arsenal Capital, Dr. Televantos serves as co-head of the specialty industrial practice. During his tenure, he has been part of several acquisitions and successful exits in the chemicals and materials sectors. In addition to serving on the boards of several Arsenal-owned companies, he is lead director on the board of Balchem Corp.

Recipe for a Complete Process Hazard Analysis – Especially Addressing the Key Demands from US CSB

Revonda Tew, Principal Engineer & Instructor
PROCESS IMPROVEMENT INSTITUTE, INC. (PII)
1321 Waterside Lane, Knoxville, TN 37922
e-mail: rtew@piii.com

William G. Bridges, President
PROCESS IMPROVEMENT INSTITUTE, INC. (PII)
1321 Waterside Lane, Knoxville, TN 37922
Phone: (865) 675-3458
Fax: (865) 622-6800
e-mail: wbridges@piii.com

Prepared for Presentation at
13th Global Congress on Process Safety
San Antonio, TX
March 27-29, 2017

UNPUBLISHED

Recipe for a Complete Process Hazard Analysis – Especially Addressing the Key Demands from US CSB

Revonda Tew, Principal Engineer & Instructor
PROCESS IMPROVEMENT INSTITUTE, INC. (PII)

William G. Bridges, President
PROCESS IMPROVEMENT INSTITUTE, INC. (PII)

Keywords: PHAs, Damage Mechanisms, Risk control, Non-routine modes of operations, SOPs, process safety management, PSM

Abstract

Process Hazard Analysis (PHA) is not a new topic. Surprising, more that 80% of PHAs performed today do not comply with the current interpretations by US OSHA, much less the industry best practices. Most PHAs address less than 10% of the hazards during startup, shutdown, and online maintenance[1] and less than about 30% address damage mechanisms such as corrosion, erosion, external impacts, external stresses, vibration, etc. How to address these hazards has been part of the CCPS Guidelines for Hazard Evaluation Procedures since 1991 and the US CSB and US OSHA have noted how these weaknesses have led to many accidents. The citations and comments from US regulators and the CSB are detailed to provide some of the business case for performing more thorough PHAs on these key issues. This paper and presentation also illustrates step-by-step how to address all hazards of the process during ALL modes of operation, during a PHA. The presentation and paper will use examples of process hazards missed and found to illustrate the importance of this step-by-step approach. The results are based on thousands of unit-sized PHAs performed or managed by PII staff.

Background

Process Hazards is a broad term. The relationship between many components define process hazards. For a PHA team, Process Safety Information (information on the chemicals, technology and equipment) and operating procedures are necessary to identify specific components of and analyze a hazard scenario: process deviations (parametric and procedural), their causes, the process safety consequence and various safeguards for detecting deviations and causes, preventing causes, detecting and mitigating consequences and intervening safeguards. A thorough review of incident investigations offers an opportunity to further analyze process hazards.

Over the last twenty years the CSB has brought needed attention, through its investigations of industry incidences, to two significant gaps in many current PHAs; where the teams did not identify and analyze:

- Process hazards unique to non-routine modes of operations resulting in catastrophic consequences and
- Process hazards related to Damage Mechanisms.

Though the intent of the US OSHA PSM regulation has always been for PHA teams to analyze the hazards for all modes of operation, OSHA did not give specific requirements for addressing damage mechanisms and only recently have they increased their enforcement attention on PHA of non-routine modes of operation.

Industry has responded by modifying the *Guidelines for Hazard Evaluation Procedures*, 3rd Edition, 2008,[1] CCPS/AIChE to improve the coverage of these two main issues. API in turn issues guidance on addressing damage mechanisms. Industry best practices now exist for addressing both of these weaknesses. Now, with proper approaches, a team with process safety competencies and process knowledge can thoroughly address the issues during a PHA.

Another less significant gap noted by regulators and by the US CSB has been an incomplete consideration, within the PHA, of previous incidents.

This paper lays out the case from the US CSB and US OSHA on these PHA gaps:

1. Poor or no coverage of hazards during non-routine modes of operations
2. Poor or no coverage of damage mechanisms
3. Poor coverage of prior incidents

This paper also provides proven-in-use best practices for how to close these gaps, while also comparing these best practices to other approaches that have not performed as well.

Although regulatory pressure can be a driver for improvement, data from prior papers on the subject has shown there is near a 1000 to 1 economic return on investment for simply doing a reasonable PHA of non-routine modes of operation (primarily due to the catastrophes avoided).[22]

This paper will **not** cover the other critical aspects of a complete PHA/HAZOP, which can certainly still be problems in PHAs and which PII still finds issues with around the world. The other critical aspects of a fully compliant (excellent) PHA include:

- Excellent team leadership and PHA scribe that are fully competent in all aspects of a Best Practice PHA, including how to efficiently lead and document the PHA
- The most experienced and knowledgeable team members available from operators, operations engineers, and others as needed
- Up-to-date process safety information and access to underlying details, as needed
- Up-to-date, clear, and accurate operating procedures
- Thorough understanding and coverage of facility siting issues and human factors

Increasing Regulatory Pressure Related to PHA/HAZOP of Procedures to Analyzing Process Hazards During All Modes of Operation

Industry initiatives:

Industry has taken some initiatives to help standardize PHA analysis of non-routine modes of operation. One initiative was to improve the focus on PHA of non-routine procedures in *Guidelines for Hazard Evaluation Procedures,* 3rd Ed, 2008, AIChE/CCPS[1]. A new Chapter 9, Section 1, was added that necessitates hazard evaluations of all hazards of the process during all modes of operation. This textbook also explained why, when, and how to perform such analysis of step-by-step procedures.

Many companies have taken the initiatives to do the same, including about 20% of the largest chemical, petrochemical, and refining companies (based on PII data from hundreds of clients) But, the vast majority of companies who should be analyzing step-by-step deviations are not; and the major accidents continue to occur partly because of this. As a result, the US regulators are beginning to bring more pressure for the regulated companies to do a PHA of all modes of operation.

US OSHA Regulation and Enforcement:

The US OSHA PSM regulation requires PHA of all hazards during all modes of operation as well, and several key citations since 1990 have focused on PHA of non-normal modes.

- **Before there was a PSM regulation from US OSHA**, the Agency published CPL 2-2.45 (Systems Safety Evaluation of Operations with Catastrophic Potential).[2] In this guidance document, OSHA stated that a human error analysis should address:
    - *Consequences of failure to perform a task*
    - *Consequences of incorrect performance of a task*

○ *Procedures and controls to minimize errors.*[3]

This approach is still the fundamental analysis method for PHA of non-normal modes of operation.

- **Phillips 66 "PHA" Citation —A citation with 566 instances was issued to Phillips 66 in Pasadena, TX, following their 1989 disaster** that killed 23 workers.[4] **The citation was related to a violation of the General Duty Clause (Section 5(a)(1) of OSH Act of 1970).** US OSHA cited Phillips against the General Duty Clause, since the PSM standard (29 CFR 1910.119) had not yet been issued. OSHA cited Phillips for not protecting its workers from hazards of fire/explosion by, among others, not performing a PHA that should have included an evaluation of the effect of design modifications on operator performance, and the identification of the source of observed human error and the identification of human factors that could result in incident event sequences. The citation stated, **"This review should result in a systematic listing of the (1) types of errors likely to be encountered during normal or emergency operation, (2) factors contributing to such errors, and (3) proposed system modifications to reduce the likelihood of such errors."**

  The settlement agreement[5] between US OSHA and Phillips included the following requirements for process hazard analyses (PHAs) of the rebuilt and surviving units:
  - "Phillips will analyze each process…and will include human factors analysis … [and] will be …led by an independent consultant."
    - William Bridges (of JBF Associates at the time, now with PII) led these PHAs. Before these PHAs began, OSHA, Phillips, and Mr. Bridges decided that the best approach for finding all human error scenarios was to perform a HAZOP of deviations of the steps for the procedures governing activities for startup, shutdown, and particularly online maintenance.
  - "Phillips will provide OSHA an independent consultant's evaluation of the adequacy of its settling leg maintenance procedures performed while the polyethylene reactors are in operation…"
    - As part of the settlement to meet this requirement, it was decided by JBFA, Phillips and OSHA to perform a Human Reliability Analysis (HRA) of the Setting Leg online maintenance procedure, to ensure that the statistical risk of the accident recurring is less than the background risk of driving to work.

  The PHA and HRA resulting from the Phillips settlement agreement are presented as a Case Study later in this paper for sake of clarity.

- **Paragraph (e) of the US OSHA regulation on PSM, 29 CFR 1910.119,[6] and similar requirements in US EPA's rule for risk management programs (RMP), 40 CFR 68.24,[7]** specifically require that PHAs consider and address hazards of the process, i.e., all hazards regardless of the mode of operation (routine or non-routine).
  - 29 CFR 1910.119(e)(1) states that the PHA, "shall identify, evaluate, and

control **the hazards** involved in the process"
- 29 CFR 1910.119(e)(3)(i) states that the process hazard analysis shall address "**the hazards of the process**"
- 29 CFR 1910.119(e)(3)(vi) states that the process hazard analysis shall address human factors
- Appendix C to the OSHA PSM standard states that both routine and non-routine activities need to be addressed by the PHA of the covered process.

*There is no qualifier that limits the OSHA PHA requirement to only routine modes of operation. PSM requires that **all hazards** related to the process be addressed, regardless of the mode of operation or activity (routine or non-routine).*

- **OSHA Inspection No. 103490306 (Nov 2, 1992).**[8] In the first major PSM inspection in 1992 using 29 CFR 1910.119, OSHA assessed a serious violation when the PHAs did not address "human factors such as board operator error, line breaking mistakes, and improper lockout and isolation of process equipment," all of which are errors originating from failure to either perform tasks or perform them incorrectly.

- **US OSHA published an internal document on Program Quality Verification of Process Hazard Analysis in 1993 (by Henry Woodcock, of OSHA).**[3] This document states that a PHA should include analysis of the "procedures for the *operation* and *support* functions" and goes on to define a "procedure analysis" as evaluating the risk of "skipping steps and performing steps wrong." The authors concur and PII has found the same true in PHAs that we have performed using various methods; a 2 Guideword HAZOP approach is normally optimal for PHA of procedures.

- **OSHA Inspection No. 123807828 (Nov 18, 1993)**[9] – Ashland Oil, Catlettsburg, KY. Several operators were preparing to ignite a 2-B-3 crude heater after a two week turnaround. The lead operator had two very inexperienced workers helping him light the heater. A large concentration of fuel gas was allowed to enter the heater before the pilot light was ignited. The resulting explosion killed one employee. The operators bypassed safety shutdown features and did not do a visual check of the emergency shutdown system to ensure that the features were closed. In addition, they did not check the firebox to ensure that it was gas-free before lighting the heater.

  The Kentucky OSHA citation read: *The PHA did not address all hazards of the #2 Crude unit..; The PHA did not address the hazards associated with the startup of the crude unit after a turnaround, ...emergency shutdown..., emergency operations and normal shutdown of the unit. The process hazard analysis that was completed by the PHA team for the #2 Crude unit only evaluated the hazards associated with normal mode of operation of the #2 Crude unit.*

Settlement: All procedures were re-written and all PHAs were redone to include a PHA of deviations from procedural steps for all non-continuous modes of operation.

- **Recent US OSHA PSM National Emphasis Programs** for Chemical Processes[10] and also for Refineries[11] underscore the need for companies to identify potential accident scenarios during non-routine modes, and to reduce the frequency and consequences of such errors as part of an overall process safety management (PSM) program.

OSHA recognizes that CCPS/AIChE has added as Chapter 9.1 in the 3rd edition of *Guidelines for Hazard Evaluation (2008*[1]) to further emphasize the need for a PHA to include hazard evaluations of all modes of operation and that this chapter has added best-practice detail on the approach for doing the hazard evaluation of startup, shutdown, and online maintenance modes of operation. Despite the specific OSHA standard that requires PHAs of covered processes to address all hazards, many PHAs still do not address hazards during all modes of operation. Further, many of the regulated community have stated "Well, OSHA did not tell us to perform a PHA of procedures for non-routine modes of operation." On the other-hand, OSHA did not state to do only a hazard evaluation of normal mode of operation and stop there.

To highlight the importance that PHAs address hazards during all modes of operation and activities (routine and non-routine), OSHA is considering issuing a Hazard Alert that would incorporate the concepts in Chapter 9.1 of *Guidelines for Hazard Evaluation Procedures, 2008, CCPS/AIChE.*[1] Also, as stated above, OSHA has an enforcement initiative, CHEM NEP, that utilizes a list of dynamic questions that OSHA compliance officers use to evaluate compliance at facilities covered by the program. **It is possible that future dynamic list questions could address PHAs of all modes of operation, and is further possible that this CHEM NEP update is drafted and waiting for release.**

Pressure from the US Chemical Safety and Hazard Investigation Board (US CSB)

The CSB has commented on the need for PHAs to address all hazards of the process during all modes of operation. Their clearest statement was in the report **2008-08-I-WV-R1[12] for the Bayer CropScience accident in Institute, WV, 2008.** In that report, CSB asks Bayer to:

- Revise the corporate PHA policies and procedures to require:
    a. Validation of all PHA assumptions to ensure that risk analysis of each PHA scenario specifically examines the risk(s) of intentional bypassing or other nullifications of safeguards
    b. **Addressing all phases of operation and special topics including those cited in chapter 9 of "Guidelines for Hazard Evaluation Procedures" (CCPS, 2008),**
    c. Training all PHA facilitators on the revised policies and procedures prior to assigning the facilitator to a PHA team, and

d. Ensure all PHAs are updated to conform to the revised procedures.

Another strong emphasis on PHA of procedures was in their report: ***Tesoro Anacortes Refinery April 2, 2010 Incident CSB Report, "Catastrophic Rupture of Heat Exchanger," Report 2010-08-I-WA, January, 2014.***[13] This accident resulted in 7 fatalities. In that report, CSB states:

- The startup of the NHT heat exchangers was hazardous non-routine work. Leaks routinely developed that presented hazards to workers conducting the startup activities. Process Hazard Analyses (PHAs) at the refinery repeatedly failed to ensure that these hazards were controlled and that the number of workers exposed to these hazards was minimized.
- None of the Anacortes refinery PHAs effectively evaluated and controlled hazards associated with the non-routine work necessary to periodically clean the NHT heat exchangers. The Washington PSM regulations address the need for non-routine operations to be evaluated and require that at least one member of the PHA team has expertise in non-routine tasks.[14] The CCPS describes the importance of PHA evaluations, as well as the hazardous potential and frequent problems of PHAs that lack sufficient analysis of nonroutine work as follows:[1, 15]
- It is not uncommon for initial PHAs of continuous processes to focus only on normal operations, failing to address non-routine, critical operating modes such as startup, shutdown, preparation for maintenance, emergency operations, emergency shutdown, and other activities whose characteristics may differ considerably from normal operations. Experience indicates that many accidents do not occur during "normal" operation but, rather, during such non-routine modes of operation. Consequently, it is important that a PHA evaluate the hazards of a process during non-routine as well as normal (routine) operating modes. The 1996 Shell Oil PHA for the NHT unit did not evaluate or identify any issues related to non-routine hazardous work associated with the frequent NHT heat exchanger cleaning operations. The 2006 Tesoro NHT unit PHA revalidation identified startup as a non-routine operation but noted that existing procedures were adequately addressing non-routine work.

But there was no PHA on the procedures to determine if there were sufficient protection against errors that could (and did) occur during non-routine operations. Simply stating there are procedures equates to No PHA of non-routine operations.

Another mention was in their report: ***DuPont La Porte, Texas Chemical Facility Toxic Chemical Release Interim Recommendations. Investigation: 2015-01-I-TX. Incident Date: November 15, 2014 Issue Date*: September 30, 2015.**[16] This accident resulted in 4 fatalities. In that report, CSB states:

- DuPont's process hazard analyses (PHAs) and relief system design scenarios do not effectively identify hazards from non-routine operations, such as opening valves to connect the liquid methyl mercaptan piping to the vapor waste gas vent header – the piping connection that provided the pathway for the methyl mercaptan release in this incident. Along the methyl mercaptan feed line there were three locations where it was connected by valves to the waste gas vent header piping. At the time of the incident, one of these valves was fully open and a second valve was slightly open.

US EPA's RMP Regulation

In the Risk Management Program rule (40 CFR 68.24)[7] EPA also recognizes the importance of procedural analysis, by defining the purpose of a PHA to **"examine, in a systematic, step-by-step way, the equipment, systems, and *procedures* (emphasis added) for handling regulated substances."**

> *A well-done PHA should identify all failure scenarios that could lead to significant exposure of workers, the public, or the environment…….For toxics under PSM, however, you may plan to address a loss of containment by venting toxic vapors to the outside air. In each circumstance, a PHA should define how the loss of containment could occur. However, for EPA, the PHA team should reassess venting as an appropriate mitigation measure. (From EPA RMP Guidance, Chapter 7, pgs 7-6 & 7-7; General Risk Management Program Guidance.[26])*

Pressure from the Local Regulator – Contra Costa County (CCC) California

CCC implements the state's Accident Release Prevention (ARP) regulations locally and have supplemented the Cal-ARP with an Industrial Safety Ordinance (ISO)[17] for elements or features of process safety that they deemed were not handled appropriately (or were missing altogether) from the ARP and/or from Cal-OSHA regulations on PSM. One element they added was the requirement that PHA of Procedures, which requires a guideword analysis, step-by-step, through each critical non-routine task. **These requirements closely follow Chapter 9.1 of the Guidelines for Hazard Evaluation Procedures, 3rd Edition.** CCC Hazardous Materials Department enforces these requirements.

CCCHM contracted PII to perform a Safety Evaluation of the Chevron Richmond Refinery following the August 2012 accident in the Crude Unit. The audit was refinery-wide and included the review of safety culture and also reviewed refinery implementation of process safety versus industry best practices. One specific finding listed in the Draft Report[18] was:

- **Recommendation 5 from the Draft Report:** Continue and expand the current Procedural PHA program implementation so that all modes of operation are evaluated. Include a PHA of startup, shutdown, and online maintenance procedures, as described in Chapter 9 of the *Guidelines for Hazard Evaluation Procedures*, 3rd edition, CCPS/AIChE, 2008.[1] This will entail using a 2-guideword HAZOP approach on each step for critical procedures and a What-if (no guideword) approach for less-critical procedures. No procedures should be excluded.[1, 22, 23, 24]

Increasing Regulatory Pressure Related to Addressing Damage Mechanism during PHA/HAZOP

Industry initiatives:

From the second edition of the *Guidelines for Hazard Evaluation Procedures* (1991) until 2007, there has been a good explanation and examples of how to cover damage mechanisms during PHA/HAZOP. However, most PHAs did not follow the examples in this textbook and instead provided no or minimal coverage of damage mechanisms such as corrosion, erosion, wrong materials of construction, external impacts, etc.

At the request of US CSB following the large explosion and losses at *Formosa Plastic*[19], the industry has taken some initiatives to improve the coverage of damage mechanism. In particular, additional text was included in the *Guidelines for Hazard Evaluation Procedures,* 3rd Ed, 2008, AIChE/CCPS[3] that emphasized the earlier examples of how to address damage mechanism during PHA team meetings and in the PHA report. In addition, ***API 571***[20] was issued that discussed various damage mechanisms and how these should be addressed by companies.

Many companies have taken the initiatives to do the same, including many of the largest chemical, petrochemical, and refining companies. But, the vast majority of companies do not provide thorough coverage of damage mechanisms during their PHA/HAZOP.

Pressure from the US Chemical Safety and Hazard Investigation Board (US CSB)

The CSB has made numerous references to the need for PHA/HAZOPs to address all damage mechanisms. Based on interviews of senior staff at CSB by PII staff, they recognize this as the second largest deficiency in current PHAs and have discussed this in presentations at the Global Congress on Process Safety (GCPS) over the past few years. In addition, they have provided in-depth analysis of a few accidents that stemmed from failure to address damage mechanisms, including the 2012 release and fire at Chevron's Richmond, Ca, Refinery and the catastrophic rupture of the heat exchanger at Tesoro Anacortes Refinery, 2010.

One of the first and strongest emphasis by US CSB on considering damage mechanisms during PHAs was in the ***Tesoro Anacortes Refinery April 2, 2010 Incident CSB Report, "Catastrophic Rupture of Heat Exchanger,"*** *Report 2010-08-I-WA, January, 2014.*[13] This accident resulted in 7 fatalities (note that the Anacortes Refinery was owned and operated by the Shell Oil Company prior to 1998). In that report, CSB states:

- The 1996 Shell Oil Naphtha Hydrotreating (NHT) unit PHA simply cited ineffective, nonspecific, judgment-based, qualitative safeguards to prevent equipment failure from high temperature hydrogen attack (HTHA). However, the effectiveness of these safeguards was neither evaluated nor documented; instead the PHA merely listed general safeguards. Had the adequacy of the safeguards been verified, improved safeguards intended to protect against HTHA failure could have been recommended.

- The 2001 and 2006 Tesoro PHA revalidations did not address or modify the analysis performed in the 1996 Shell Oil PHA. The Tesoro 2010 NHT unit PHA failed to identify HTHA as a hazard for the shell of the B and E heat exchangers. For the 15 years before the April 2010 incident, assumptions used by PHA teams at the Anacortes refinery contributed to ineffective safeguards, ineffective hazard identification, and ineffective control of hazards to prevent equipment failures from HTHA damage, such as the E heat exchanger in the NHT unit.
- Shell Oil completed a PHA in 1995 related to process modifications that could increase the hydrogen partial pressure in the NHT heat exchangers. However, when managing this change no consideration, evaluation, or recommendations were made to address the potential for HTHA damage to the NHT heat exchangers.
- Shell Oil and Tesoro periodically performed damage mechanism hazard reviews (DMHRs), called corrosion reviews. However, these reviews did not identify HTHA as a credible failure mechanism for the B and E heat exchangers. These reviews were weakened by primarily relying on design operating data for these heat exchangers rather than data from actual process operating conditions.

Note that the DMHRs were not done as part of a PHA or with a similar PHA team structure.

Another strong emphasis by US CSB on considering damage mechanisms during PHA was in the **U.S. Chemical Safety and Hazard Investigation Board (CSB),** ***Interim Investigation Report: Chevron Richmond Refinery Fire (August 6, 2012).***[21] In that report, CSB states that failure to do damage mechanism hazard reviews and follow-through on recommendations related to know damage mechanisms led in part to the accident.

Pressure from the Local Regulator – Contra Costa County (CCC) California

CCC implements the state's Accident Release Prevention (ARP) regulations locally and have supplemented the Cal-ARP with an Industrial Safety Ordinance (ISO)[17] for elements or features of process safety that they deemed were not handled appropriately (or were missing altogether) from the ARP and/or from Cal-OSHA regulations on PSM. One element they added was the requirement that DM reviews be completed for each unit and that DM be considered in each PHA. CCC Hazardous Materials Department enforces these requirements.

CCCHM contracted PII to perform a Safety Evaluation of the Chevron Richmond Refinery following the August 2012 accident in the Crude Unit. The audit was refinery-wide and included the review of safety culture and also review refinery implementation of process safety versus industry best practices. Recommendation 3 and the background observations from the Draft Report are listed below:

- …since specific/individual damage mechanisms from API 571 "Damage Mechanisms Affecting Fixed Equipment in the Refining Industry"[20] are not considered at each process section (node) within the process, it is likely that unique hazardous scenarios have been missed (best practice, as stated in *Guidelines for Hazard Evaluation Procedures*, 2008[1], is to consider damage mechanisms within each equipment node). It is also likely that some necessary safeguards such as remote isolations, leak detection, etc., for pipe sections, pumps, vessels, and columns have been missed.

- [Refinery] management reports that the Design Engineering Group piloted a specification- break review in 2013 (specification "spec" breaks are where the materials of construction or pipe thickness change). Also, each PHA is intended to identify any potential spec break concerns within the unit (as is typical for PHAs that follow industry best practice). It is understood that a policy and procedures to codify spec-break reviews will be implemented. This approach to damage mechanisms related to materials specification is considered a best practice. Most competent PHA teams in the industry ensure a discussion of each spec break during a PHA.
- **Recommendation 3 from the draft report**: To avoid missing causes and necessary safeguards against loss of containment, be sure hazardous scenarios starting from pertinent damage mechanisms are reviewed during each unit's PHA (pertinent damage mechanisms include corrosion, erosion, seal failure, pump failure, external impact, external fire, material defect, improper maintenance, drains/vents left open, etc.). In particular, review external impact as a damage mechanism for each node. A review of such loss of containment scenarios in each node is recommended, as described in the *Guidelines for Hazard Evaluation Procedures*, 3rd Edition, 2008, CCPS/AIChE.[1] Adding a loss of containment deviation would double-check against missing standard damage mechanisms (such as corrosion and erosion) during unit-wide reviews. It would also improve the current PHAs by providing a review at each node for external impacts and control of drains/vents. Review and incorporate mechanical integrity data during this analysis.

Best Practice for PHA/HAZOP of Procedures to Analyze Process Hazards During All Modes of Operation

Overview of Methodology for Hazard Evaluation of Non-Routine Modes of Operation

The hazard evaluation of non-routine modes of operation involves reviewing procedures using a HAZOP, simplified HAZOP, or What-if analysis to uncover potential accident scenarios associated with non-routine operations, for continuous or batch operations. Human error is more likely and more critical during non-routine operations. By analyzing procedural steps where human error is more likely, and where human error or component failure could lead to a consequence of interest, risk can be reduced. The hazard evaluation team's objective is to evaluate the risk associated with skipping steps and performing steps wrong.

FMEA cannot be applied to procedure-based deviations, unless you create a "human" component, in which case you have simply merged HAZOP deviations for "steps" into FMEA. Pre-Hazard Analysis (P_rHA) and other hazard evaluation methods are not applicable for accomplishing a detailed hazard evaluation of non-routine modes of operations.

Checklist of human factors issues (see an earlier paper[22, 23, 24] and also in the *Guidelines for Hazard Evaluation Procedures*[1]) can be very useful after the detailed hazard evaluation of deviations of steps. Such analysis can indicate where generic weaknesses exist that can make errors during any mode of operation more likely, or that can make errors during maintenance more likely. Such human factors checklists are normally used at the end of the analysis, they can be done piecemeal during an analysis (on breaks from the meetings) by individuals on the team, and then the results of each individual review can be discussed as a team at the end.

Purpose of Hazard Evaluation of Procedures-Based Modes of Operation

Although incorporating human factors considerations into hazard evaluation studies of continuous operation is straightforward by asking why the human might make a mistake that leads to a parametric deviation, this approach only addresses a small fraction of the potential human errors that can affect process safety. Many analysts have tried to find accident scenarios in non-routine modes of operations by adding generic guide words such as "deviations during startup" and "deviations during maintenance/sampling" to the hazard evaluation of equipment nodes/sections. Unfortunately, this only catches a fraction of the accident scenarios that can occur in non-routine modes since a hazard evaluation team is focused on "continuous" mode of operation during HAZOP or What-if of equipment sections/nodes.

From an informal survey of more than 100 companies, most do not currently perform process hazard evaluations of procedures, although many do perform some type of job safety analysis (JSA). The JSA is an excellent starting point for an evaluation of procedures because a JSA identifies the tasks that workers perform and the equipment required to protect workers from typical industrial hazards (slips, falls, cuts, burns, fumes, etc.). Unfortunately, a typical JSA will not usually identify process safety issues or related human factors concerns. For example, from a JSA perspective, it may be perfectly safe for an operator to open a steam valve before opening a feed valve; however, from a process safety perspective, the operator may need to open the feed valve before the steam valve to avoid the potential for overheating the reactor and initiating an exothermic decomposition. The primary purpose of a JSA and other traditional methods for reviewing procedures has been to ensure that the procedures are accurate and complete (which is required of employers in 29 CFR 1910.119(f)(3)).[6]

By contrast, the purpose of a hazard evaluation is ***not*** to ensure the procedures are accurate and acceptable, but instead, to ***evaluate the accident scenarios if the procedures are not followed***. Even the best procedure may not be followed for any number of reasons, and these failures to follow the prescribed instructions can and do result in incidents. In fact, in the chemical industry and most other process industries the chance of an operator or other worker making a mistake in following a procedure is greater than 1/100, and in some cases much greater. When taking into account common human factor deficiencies that accompany non-routine operations, such as fatigue, lack of practice, the rush to restart and return to full production, etc., the probability of errors can climb to 1/10 chances per task (a task being about 1 to 10 detailed steps).[28, 29]

The purpose of a hazard evaluation of non-routine modes of operation (governed by written procedures) is to make sure an organization has enough safeguards for the inevitable instance when a step is either performed wrong or skipped (inadvertently or due to shortcutting or other reasons)

Industry has found that a HAZOP or what-if analysis, structured to address procedures, can be used effectively for finding the great majority of accident scenarios that can occur during non-routine modes of operation.[1, 22, 23, 24] Experience shows that reviews of non-routine procedures have revealed many more hazards than merely trying to address these modes of operation during the P&ID driven hazard evaluations.

To reinforce the need for and to explain the method for analysis of deviations of steps in a procedure, Section 9.1 was included in the 3rd Edition of *Guidelines for Hazard Evaluation Procedures,* 2008[1]; this was one of the major changes to the hazard evaluation procedures.

HAZOP Method for Analyzing Deviations of Procedural Steps

The Hazard and Operability (HAZOP) method has two major variations; one for the continuous mode of operations (where the team brainstorms what would happen if there were deviations of parameters) and procedure-based (where the team brainstorms what would happen when the steps of a procedure are not followed correctly). The procedure-based variation of HAZOP is the oldest form of HAZOP (from ICI in 1960s).[1, 25] It was an expansion of a Hazard Evaluation method based strictly on asking:

- **What happens if the step is skipped?**
- **What happens if the step is performed wrong?**

In turn, the "pre-HAZOP" method for brainstorming accident scenarios from not following procedures (including because the procedure is itself wrong) is based on the understanding that human errors occur by someone not doing a step (errors of omission) or by doing a step incorrectly (errors of commission). So, simply asking what would happen if the operator omitted a step or performed a step wrong is one way to structure a hazard evaluation of a step-by-step procedure. The usefulness of this simple approach to hazard evaluation of steps will be discussed later.

Seven (7) Guide Word Method

In an effort to be more thorough, the inventors of HAZOP (at ICI) broke these two types of errors into subparts and agreed on using the following 7 Guide Words:

Omission:	**Skip (or Step Missing)**
	Part Of
Commission:	**More**

Less
Out of Sequence
As Well As
Other Than
Reverse

In the early 1990s, the guide word Skip was augmented by adding the option of discussing "are there any steps **missing** from the procedure."[24]

Table 1: Definitions of 7-8 Guide Words for HAZOP of Procedure-Based (Non-Continuous Mode) Operation

Guide Word	Meaning When Applied to a Step
Missing (optional guide word)	A step or precaution is missing from the written procedure prior to this step (similar to "Out of Sequence", except the missing step is not written)
Skip (No, Not, Don't)	The specified intent of this step is not performed
Part-of	A portion of the full intent is not performed. Usually only applies to a task that involves two or more nearly simultaneous actions ("Open valves A, B, and C".)
More	Too much of the specified intent is done (does not apply to simple on/off; open/close functions); or it is performed too fast
Less	Too little of the intent is done, or it is performed too slowly
Out of sequence	This step is performed too early in the sequence
As well as	Something happens, or the user does another action, in addition to the specified step being done correctly (could be a short cut)
Other than (or Reverse)	The wrong device is operated, selected, read, etc., or operated in a way other than intended. Or the wrong material is selected or added. "Other than" errors always imply a "Skip" as well.

To apply HAZOP to procedural steps for startup, shutdown, online maintenance, and other modes of operation, the facilitator (or team) first divides the procedure into individual actions. This is already done if there is only one action per step. Then, the set of guide words or questions is systematically applied to each action of the procedure resulting in procedural deviations or what-if questions. The guide words (or procedural deviation phrases) shown in Table 1 above were derived from HAZOP guide words commonly used for analysis of batch processes. The definition of each guide word is carefully chosen to allow universal and thorough application to both routine batch and non-routine continuous and batch procedures. The actual review team structure and meeting progression are nearly identical to that of a process equipment HAZOP or what-if analysis, except that active participation of one or more operators is even more important and usually requires two operators for a thorough review; a senior operator and a junior operator. For each deviation from the intention of the process step (denoted by these guide words applied to the process step or action), the team needs to dig beyond the obvious cause, "operator error," to identify root causes associated with human error such as "inadequate emphasis on this step during training," "responsible for performing two tasks simultaneously," "inadequate labeling of valves," or "instrument display confusing or not readable." The guide word *missing* elicits causes such as "no written procedural step or formal training to obtain a hot work permit before this step," or "no written procedural step or formal training to open the discharge valve before starting the pump."

Two (2) Guide Word Method for Analyzing Deviations of Procedural Steps

A more streamlined guide word approach has also proven very useful for (1) procedures related to less hazardous operations and tasks and/or (2) when the leader has extensive experience in the use of the guide words mentioned previously and can therefore compensate for the weaknesses of a more streamlined approach. The two guide words for this approach (as defined in Table 2 below) encompass the basic human error categories: errors of omission and commission. These guide words are used in an identical way to the guide words introduced earlier. Essentially "omit" includes the errors of omission related to the guide words "skip," "part of," and "missing" mentioned earlier. The guide word "incorrect" incorporates the errors of commission related to the guide words "more," "less," "out of sequence," "as well as," and "other than" mentioned earlier. Note that these two guide words (Table 2) fill the basic requirements for a human error analysis as outlined in OSHA's CPL 2-2.45.[2]

Table 2: 2 Guide Word (Guide Phrases) for Modified-HAZOP of Procedure-Based Operation

Guide Phrase	Meaning When Applied to a Step
Step not performed	The step is not done or part of the step is not done. Some possible reasons include the employee forgot to do the step, did not understand the importance of the step, or the procedures did not include this vital step
Step performed wrong	The employee's intent was to perform the step (not omit the step), however, the step is not performed as intended. Some possible reasons include the employee does too much or too little of stated task, the employee manipulates the wrong process component, or the employee reverses the order of the steps.

Table 3: Example of 2 Guide Word HAZOP of a Critical Step in a Procedure

Drawing or Procedure: SOP-03-002; ***Cooling Water Failure***	**Unit:** HF Alkylation	**Method:** 2 Guide Word Analysis	**Documentation Type:** Cause-by-Cause

Item	Deviation	Causes	Consequences	Safeguards	Recommendation
Node: 23	**Description:** STEP 2: Block in olefin feed to each of the 2 reactors by blocking in feed at flow control valves				
23.1	Step not performed	Operator failing to block in one of the reactors, such as due to miscommunication between control room operator and field operator; or control valve sticking open or leaking through	High pressure due to possible runaway reaction (because cooling is already lost), because of continued feeding of olefin (link to 11.7 - High Rxn Rate; HF Alky Reactor #1/#2) High pressure due to high level in the reactor, because of continued feeding olefin (link to 11.1 - High Level; HF Alky Reactor #1/#2)	High temperature alarm on reactor High pressure alarm on reactor Field operator may notice sound of fluid flow across valve Flow indication (in olefin charge line to reactor that is inadvertently NOT shutdown) Level indicator, high level alarm, and independent high-high level switch/alarm	
		Operator failing to make sure bypass valve is also closed, since this precaution is not listed in the written procedure; or the bypass valve leaks through	High pressure due to possible runaway reaction (because cooling is already lost), because of continued feeding of olefin (link to 11.7 - High Rxn Rate; HF Alky Reactor #1/#2) High pressure due to high level in the reactor, because of continued feeding olefin (link to 11.1 - High Level; HF Alky Reactor #1/#2)	High temperature alarm on reactor High pressure alarm on reactor Operator skill-training requires checking bypasses are closed, when blocking control valves Field operator may notice sound of fluid flow across valve Flow indication in olefin charge line (but likely not sensitive enough for small flows) Level indicator, high level alarm, and independent high-high level switch/alarm	
		Operator failing to close low control valve manually from the DCS because the phrase "block in" is used instead of the word "close"	Valve possibly opens full at restart, allowing too much flow to reactor at restart, resulting in poor quality at startup and/or possibly resulting in runaway reaction and high pressure	Control room skill training requires always manually commanding automatic valves closed before telling field operator to block in control valve	37. Implement best-practice rules for procedure writing, which includes using common terms.
23.2	Step performed wrong	Operator closing the olefin charge flow control valves before shutting down the charge pump, primarily because the steps are written out of the proper sequence	Deadheading of charge pump, leading to possible pump seal damage/failure and/or other leak, resulting in a fire hazard affecting a small area (link to 5.12 - Loss of Containment; Olefin Charge Line/Pump)	Step 3 of procedure that says to shutdown charge pump The step to shut down the charge pump (Step 3) is typically accomplished before Step 2 (in practice)	41. Move Step 3 ahead of Step 2.
		Field operator closing both upstream and downstream block valves	Possible trapping of liquid between block valve and control valve, leading to possible valve damage (due to thermal expansion)	Field operator skill training stresses that only one block valve should be closed	

What-if Method for Analyzing Deviations of Procedural Steps

The What-if method for analyzing procedure-based modes of operations is free brainstorming without the aid (or constraints) of guide words. This method is described in detail in the *Guideline for Hazard Evaluation Procedures (CCPS).*[1] The hazard evaluation team using this method would read the procedure and then answer the question: "What mistakes will lead to our consequences of interest?" The team would list these mistakes and then brainstorm the full consequences, causes, and existing safeguards – the same analysis approach described for the guide word approaches mentioned earlier in this section. What-if brainstorming **is not** applied to each step of the procedure, but rather covers the entire task (procedure) at one time.

Choosing the Right Method for Analysis of Non-Routine Modes of Operation

Obviously the What-if approach takes far less time than the 2-Guide Word method, and the 2 Guide Word method takes much less time than the 7-8 Guide Word method of HAZOP of procedures. Experience has shown that hazard evaluation facilitators, newly trained in the three techniques above, tend to overwork an analysis of non-routine procedures, so a tiered approach is best. In this tiered approach, the first step in choosing the right method of analysis in the hazard evaluation of procedures is to screen the procedures and select only those procedures with extreme hazards. These procedures should be subjected to a detailed HAZOP analysis (7-8 guide word set) presented above. The 2-Guide Word set is efficiently used for less complex tasks or where the consequences are lower. The What-if method is applicable to low hazard, low complexity, or very well understood tasks/hazards.

Experience of the leader or the team plays a major part in selecting the procedures to be analyzed, and then in deciding when to use each guide word set.

Figure 2 shows the typical usage of the three methods described above for a typical set of operations procedures within a complex chemical plant or refinery or other process/ operation. Most of the procedures are simple enough, or have low enough hazards to warrant using the What-if method. Currently, the 7-8 Guide Word approach is used infrequently, since most tasks do not require that level of scrutiny to find the accident scenarios during non-routine modes of operations.

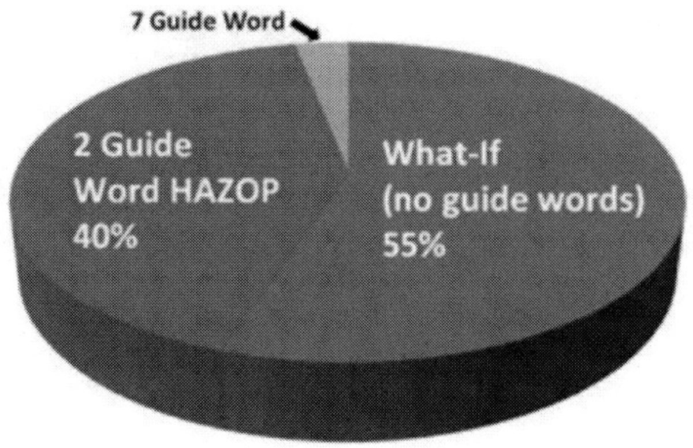

Figure 1. Relative Usage of Techniques for Analysis of Procedure-Based Modes of Operation [1, 22, 23, 24]

The experience of the leader or the team plays a major part in selecting the method to use for each task/procedures to be analyzed. However the first decision will always be "are these procedures ready to be risk reviewed?" If the procedures are up-to-date, complete, clear, and used by operators, then the best approach for completing a complete hazard evaluation of all modes of operation, including routine modes of operation, is shown in Figures 2A and 2B below:

Figure 2A: PHA of a ALL Modes of Operation for a CONTINUOUS Process

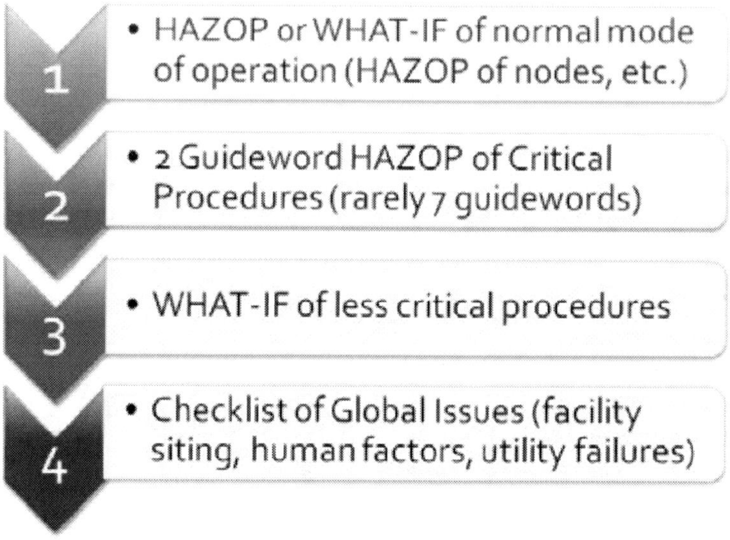

Figure 2B: PHA of a ALL Modes of Operation for a BATCH Process

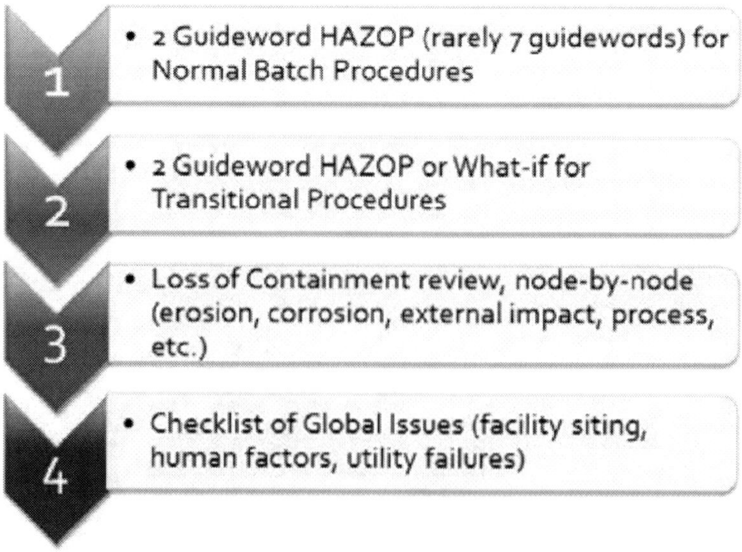

If your procedures are not at least 90% accurate, then the best approach is to develop accurate and up-to-date procedures as quickly as possible and afterwards do a PHA of the newly issued procedures.

Any procedure (even a computer program) can be analyzed using these techniques. Reviews of routine procedures are important, but reviews of non-routine procedures are even more important. As mentioned earlier, the nature of non-routine procedures means that operators have much less experience performing them, and many organizations do not regularly update these procedures [though this should change as companies comply with 29 CFR 1910.119(f)][6]. Also, during non-routine operations, many of the standard equipment safeguards or interlocks are off or bypassed.

Using the approaches above, a company doing a complete hazard evaluation of an existing unit will invest about 65% of their time to evaluate normal (e.g., continuous mode) operation and 35% of their time for evaluating the risks of non-routine modes of operation.[1, 22, 23, 24]

Many companies do **not** perform a thorough analysis of the risk for startup, shutdown, and on-line maintenance modes of operation; the reason normally given is that the analysis of these modes of operation takes "too long." Yet, actually, the hazard evaluation of the normal mode is taking too long and so the organization feels it has no time left for the analysis of procedures for startup and shutdown modes of operation. But, if these hazard evaluations for the normal mode of operation are **optimized** (such as using rules presented elsewhere[23]), the organization will have time for thoroughly analyzing the non-routine modes (typically discontinuous modes) of operation and the organization will still have a net savings overall! This point is critical since 70-80% of catastrophic accidents occur during non-routine modes of operation.[22] Figure 4 illustrates (for a continuous process unit) the typical split of meeting time for analysis of routine mode of operation versus non-routine modes of operation.[1, 22, 23, 24]

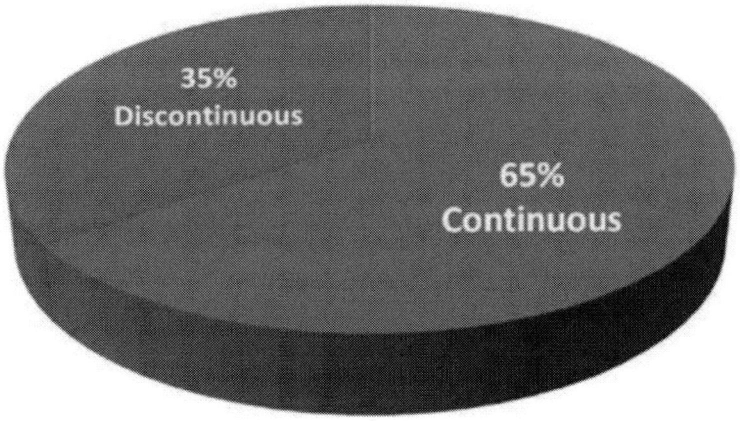

Figure 3. Relative Amount of Meeting Time Spent for Analysis of Routine and Non-routine Modes of Operation for a Continuous Process[1, 22, 23, 24]

General Guidelines for Analyzing Non-routine Modes of Operation or Batch (Step-by-step) Processes

- Define the assumptions about the system's initial status. "What is assumed to be the starting conditions when the user of the procedure begins with Step 1?"
- Define the complete design intention for each step. "Is the step actually 3 or 5 actions instead of one action? If so, what are the individual actions to accomplish this task?"
- Don't analyze safeguard steps that start with ensure, check, verify, inspect, etc., or where the consequence of skip is "loss of one level of safeguard/protection against ….." There is no reason to analyze these steps since they will show up as safeguards of deviations of other steps. This approach is similar to not analyzing a PSV during a HAZOP of continuous mode (parametric deviation analysis); instead the PSV is shown as a safeguard against loss of containment.
- Together with an operator before the meeting, identify the sections of the procedures that warrant use of:
  - 7-8 Guide Words (extremely large consequences can happen if deviations occur)
  - 2 Guide Words (the system is complex, mistakes are costly, or several consequences could occur)
  - On others, use What-If (no guide words or guide phrases; for use on simpler or lower hazard systems)
- Decompose each written step into a sequence of actions (verbs)
- Apply guide words directly to the intentions of each action

The Following Preparation Steps May Also Be Needed:

- Walk through procedure in the plant with one or more operators to see the work situation and verify the accuracy of the written procedure. This is optional and should have also been performed as part of validation of the procedure after it was originally drafted
- Determine if the procedure follows the best practices for "presentation" of the content; the best practices will limit the probability of human error
- Discuss generic issues related to operating procedures, such as:
  - staffing (normal and temporary)
  - human-machine interface
  - worker training, certification, etc.
  - management of change
  - policy enforcement
- Review other related procedures such as lock out/tag out and hot work.

IF the procedures are NOT >90% accurate, then redo procedures first!

Best Practice for Addressing Damage Mechanisms During PHAs

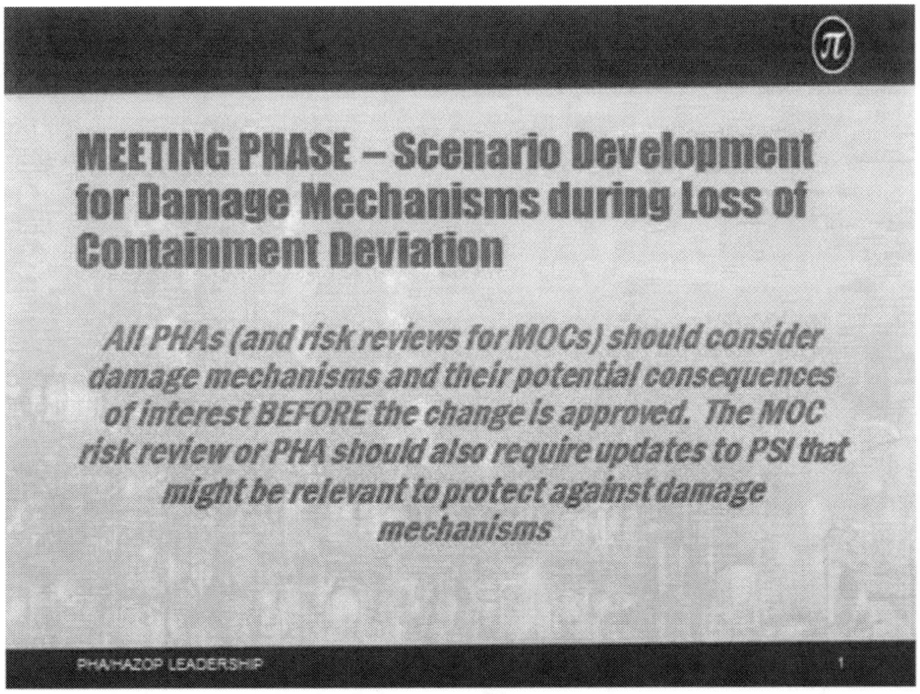

Figure 4. Meeting Phase – Review of Damage Mechanisms

Recently, US CSB and more and more regulators are making analysis of Damage Mechanisms (DMs) a PHA requirement. This requires that potential process damage mechanisms and their potential consequences of interest be identified. Damage mechanisms can be broken down into three main categories. These categories and examples of each are shown in the following table.

Table 4: Types of Damage Mechanisms

Mechanical	• Mechanical loading failure • Mechanical fatigue • Buckling • Cracking • Embrittlement • Ductile fracture • Brittle fracture
Chemical	• Corrosion o Uniform o Localized o Pitting
Physical	• Thermal failures o Creep o Thermal fatigue o Transformation

The team may use aids such as piping specs, established literature and standards, any applicable MOC documents, etc. to help identify DMs. The PHA leader has several options of how to address damage mechanisms:

- Discuss during HAZOP under "Loss of Containment" deviation
    - **This is PII's preferred approach and is one we recommend to all clients, so this is discussed in more detail below.**
- Note damage mechanisms revealed during discussions of other process deviations
    - Pressure
    - Temperature
    - Concentration

  The consequence of these deviations is ultimately loss of containment. Therefore causes of pressure, temperature and concentration deviations can be damage mechanisms and ultimately lead to loss of containment.
- Include Damage Mechanisms as a separate node (for HAZOP technique), *such as proposed by Risk Management Professionals.*[26]
- Perform a "mini" Mechanical Integrity checklist reviews under each process node
- Divide P&ID (or PFD) into separate services for purposes of DM hazard review (DMHR) discussion considering:
    - Process
    - Chemical
    - Spec changes
    - Flow through
    - Operating parameters

  This approach is used by some refineries to accomplish the requirements of a DMHR as recommended in API RP 571.[20] *From PII's experience, this approach has merit and finds some issues that the preferred PII does not always find, BUT overall the PII approach is superior in finding the key scenarios related to damage mechanisms and related safeguards because the review is specific to each Node of process equipment.*

Damage mechanisms can (and sometimes ***should***) be reviewed prior to the PHA. For example, all MOCs should consider damage mechanisms and their potential consequences of interest before the change is approved. The MOC should also require updates to process safety information which might be relevant to protect against DMs.

Coverage of Damage Mechanism for Each Major Section (HAZOP section or What-If node) – as recommended by PII

As mentioned above this is PII's preferred method for thoroughly covering all DM within a PHA/HAZOP of an entire unit. The benefits are that the team can more easily catch changes in DM from section to section and they can more easily identify when unique safeguards, such as Remote Isolation Valves, segregated containments (dike), and unique materials of construction are needed. To facilitate this, a list of generic causes of loss of containment and a generic list of safeguards against loss of containment are covered in the Loss of Containment (LOC) deviation of each section. Below are tables that summarize the typical causes of loss of containment and typical safeguards against loss of containment.

Table 5 Typical Causes of Loss of Containment (and Tube Leak/Rupture) to Discuss in Each Major Process Section (includes Damage Mechanisms)

- Corrosion – internal, including steady corrosion and also stress cracking
- Corrosion - external (including corrosion under insulation
- Erosion
- Fatigue failure (such as due to vibration)
- Improper material of construction (including wrong grade of welding rod/wire)
- Improper maintenance/repair
- Drain valve open
- Vent valve option
- Instrument line breaks
- External impact or external load (crane, load dropped, fork truck, backhoe, failure of structural supports, etc.)
- External fire (flame impingement and heating)
- Thermal expansion (while process is blocked-in)
- Lightning

➢ Any deviation that links into loss of containment as a cause, such as high process temperature or high process pressure

Table 6 Examples of IPLs Protecting Against or Mitigating Damage Mechanisms

RELIEF VALVES

- Relief valve designed for the limiting scenario which vents to a safe location or destruction system (can prevent loss of containment due to high pressure events cause by process upsets or external fire)
- Rupture disks (same criteria and functional protection as relief valve except cannot re-seat)
- Conservation vent (same criteria and functional protection as relief valve)
- Vacuum relief valve (same criteria and functional protection as relief valve, except protects against negative pressure)
- Thermal relief valve (online protects against thermal expansion in a closed system)

ISOLATION

- Remote isolation valves that either operate automatically or can be remotely closed manually and the loop qualifies as a Human IPL
- Locally operated manual valves that can be reached quickly and safely (typically not a valid IPL case)

- Remote shutdown of pumps, blowers, etc. to limit amount of material released

CONTAINMENT & MITIGATION

- Double-walled piping for lines that run under rivers and underground and that carry toxics
- Process contained in enclosed building
- Dikes are provided for secondary containment of spills from major vessels; drain valves on diked areas are kept closed/can be closed quickly
- Process areas are drained to sump
- Rainwater runoff containment
- Fire insulation that protects metal structures for 4 hours

SUPPRESSION AND EXTINGUISHING SYSTEMS

- Deluge system under or over tanks or storage vessels
- Manual pull station fire alarms
- Fire extinguishers in the Power Distribution Center
- Fire extinguishers in the control room rated for electrical fire

Table 7 Examples of Non-IPL Safeguards Protecting Against or Mitigating Damage Mechanisms

PIPING

- Applicable design codes are reflected in equipment design
- Welding done by qualified welders using qualified procedures
- Regular inspection (e.g., ultrasonic, X-ray) for wall thickness
- The wall thickness inspection frequency is increased in critical areas (e.g., areas where erosion or corrosion are likely)
- Systems are leak checked (pressurized with air and soap bubble applied at flanged joints) before starting up after systems are opened for maintenance
- A winterization list is used to check freeze-protection features before the onset of cold weather
- Piping and components that are fabricated on site and installed in the process are pressure tested before they are installed as necessary

PUMPS

- Written specifications in standard maintenance procedures (SMPs) for pumps to help ensure that pumps are maintained, repaired, and replaced in a way that meets original specifications

VALVES

- Fugitive emissions monitoring program that includes monitoring valves for packing leaks and

flange leaks
- Fail-safe settings for control valves upon loss of instrument air, nitrogen, or electric power

CORROSION DETECTION AND PREVENTION
- Periodic inspection of piping, equipment, and structures
- Application of exterior paint and protective coatings when needed on piping, equipment, and structures
- Positive material identification (PMI) of 100% of components and welds in the field/unit

EQUIPMENT MAINTENANCE
- Periodic walkthroughs by personnel inspecting for leaks or other abnormalities, and checking local instrumentation
- Preventive maintenance systems, including:
    - Lubrication schedule
    - Operators inspect pump seals for visible leaks
    - Records of equipment failures
- Periodic inspection of heat exchanger tube bundles

INSTRUMENTATION
- All controllers are indicating controllers
- Testing and calibration of critical process instrumentation

PERSONNEL PROTECTION
- Personal protective equipment (PPE), including:
    - Self-contained breathing apparatus
    - Respirators
    - Protective clothing
- Periodic inspection and testing of PPE and personal safety systems, including:
    - Self-contained breathing apparatus
    - Safety showers
- Availability of:
    - First aid kit
    - Specially trained Emergency Medical Technicians

EMERGENCY RESPONSE
- Remote emergency communication systems, including:
    - Handheld, two-way radios
    - Buddy system for critical or hazardous operations
    - Emergency communications procedures
    - Telephones
    - Emergency reporting stations strategically located in the plant
- Cooperative fire brigade
- Plant wide alarm system
- Facility and unit contingency plan

- Emergency procedures for special incidents
- Trained HAZMAT response teams with emergency equipment
- Spill containment and control guidelines; all operators receive hazardous waste operations and emergency response (HAZWOPER) training

The Leader and Scribe must consciously discuss the typical damage mechanisms listed in Table 5 and discuss at each LOC deviation (so, once in each section).

Documentation of Results of Damage Mechanism Discussions in PHA Meetings

For each node (each line, each vessel, each column, etc.), the PHA team should discuss and document each damage mechanism listed in Table 5 (as a cause of loss of containment), consequence of the failure if the damage mechanism occurs, and the safeguards in place to prevent the damage mechanism, detect the mechanism before failure, prevent the release, detect and response to the release, and mitigate or contain the release.

The documentation style varies between PHAs. Figure 5 show two different styles that have been acceptable in the past.

- One style (Example A) uses a reference to a summary table of typical causes (this summary table is not shown but is similar to Table 5 of this paper) instead of listing each individual damage mechanism. In Example A, the same approach is used for safeguards. This approach saves redundant text and some time, but it requires the leader and scribe to be diligent to cover everything in the Typical tables that were referenced in the Cause and Safeguard columns. A modified approach is to relegate some generic causes and safeguards to Typical tables for reference and then to carefully list the specific cause and/or safeguards of interest for LOC for each specific node.
- Example B does not use or reference a generic or Typical list of causes (damage mechanism) or Typical safeguards, but instead develops a specific listing for the LOC deviation of each node. This style has proven easier to justify to regulators and other outside reviewers but takes more documentation effort.

Regardless of method, the team leader and scribe must ensure the team rigorously discusses all of the damage mechanisms in Table 5 and adequately documents the results of the damage mechanism review in the PHA/HAZOP analysis tables.

Figure 5. Excerpts from Select PHA Reports Showing Various Acceptable Styles for Documenting Damage Mechanism Review Results Within the LOC Deviation

Example A – Referencing Generic Tables for Typical Causes and Typical Safeguards

Dev#	*Deviation*	*Causes*	*Consequences*	*Existing Safeguards*	*Recommendations*
9.10	Loss of containment	Accelerate corrosion External fire High pressure (linked from 9.7) Typical causes of loss of containment (see Table A.1)	Release to atmosphere leading to potential injury of workers and/or community	Pressure Relief Valve's 231A, B, C on Reactor and no valves in line to reactor Most lines / connections are welded construction; only a few flanges Drills conducted each year on evacuation, rescue and isolation Emergency Response personnel are trained at SABIC FTC Generic safeguards protecting against or mitigating process material releases (see Table A.2)	Safety 7. Consider changing the ITPM schedule for managing most PSVs whose inspections are too infrequent based on industry standards and best practice. For instance, the current inspection frequency for PSV-8220 on the Ammonia Receiver (I-2005G) is 9 years, whereas, consensus codes typically recommend testing/inspection every 1-4 years for PSVs in highly toxic services.

Example B – Specific Listing of Causes and Safeguards

1.9	Loss of containment	Corrosion/erosion External fire and/or flame impingement Gasket, packing, or seal failure Improper maintenance Material defect Operator failing to close or inadvertently opening a valve to the atmosphere (e.g., a valve at a hose connection) Railcar inadvertently derailed Valve leaking to the atmosphere High pressure (linked from 1.5) Acid corrosion caused by high concentration of water (linked from 1.8) High ambient temperature External impact (such as from a mini-engine or another railcar)	Catastrophic release of chlorine from a ruptured railcar Steady release of chlorine from a ruptured connection Steady release of chlorine from a leaky connection High pressure caused by thermal expansion of liquid chlorine if railcar is also over-full	Chlorine repair kit Derailer and warning flag to prevent impact by a mini-engine or another railcar Limited vehicular access to area Maintenance/operator response as required, including isolation if needed Operator periodically monitoring the railcar valves while unloading Personal protective equipment in the area Plugs installed in all chlorine valves to the atmosphere when the valves are not in use Relief valve on each railcar for mitigating releases caused by overpressure Supplier maintenance of railcars (per strictly enforced US DOT requirements) Video monitoring of the unloading area Concrete crossties on rail spur Dike preventing any combustibles spilled nearby from reaching the unloading rack area Concrete railroad ties in chlorine unloading area to prevent fires near railcar	10. Consider installing a chlorine detection system in the unloading and vaporizing area to help detect chlorine releases (especially at likely release points) 11. Verify that periodic maintenance and inspections are being performed in accordance with Chlorine Institute recommendations 12. Review the drainage system for the unloading are, and identify the areas that may be affected by a large chlorine release 13. Consider prohibiting the use of heavy equipment (e.g., cranes) in the unloading and vaporizing area unless special precautions to prevent equipment damage are enacted 33. Consider providing a high pressure alarm for each vaporizer 40. Consider providing a water deluge system in the unloading area to help mitigate chlorine releases from the railcar

Best Practices for Addressing Previous Incident

Information on process hazards is also found in incident investigation reports and are often discounted. PHAs predict what can go wrong. Incidents are actual sequences of human errors and equipment failures. Don't miss this opportunity to thoroughly review previous incidents. There is no industry or regulatory guidance regarding how to incorporate previous incidents into HEs. The intent is to review near misses and accidents to ensure that potentially catastrophic events are identified and analyzed.

Typically, this is done by listing and then discussing relevant incidents in the industry and in similar plants to:

- Help identify potential causes or consequences of interest
- Evaluate related safeguards in the process, and
- Make recommendations for additional layers of protection if needed to prevent similar incidents related to the system.

Know Your History

- What caused past incidents? *This includes near misses and accidents and also includes incidents in similar systems. If the company has a good system for getting near misses reported and any incident investigated to root causes, then the PHA team will have a deeper understanding of what can go wrong and why. Root cause analysis of incidents is just as important as PHA. The team should volunteer incidents that may not have been reported formally and these should be kept confidential so that more "unreported" incidents are reported within the PHA team setting.*
- What was done in response? *Were any changes made to the system? Did these improve the system (reduce the error or failure rates)? Has the incident reoccurred?*
- If the incident occurred in a system other than the one under review by the PHA team, should similar actions be applied to the system under evaluation?
- Did the previous countermeasures work? *Did these prevent similar incidents?*

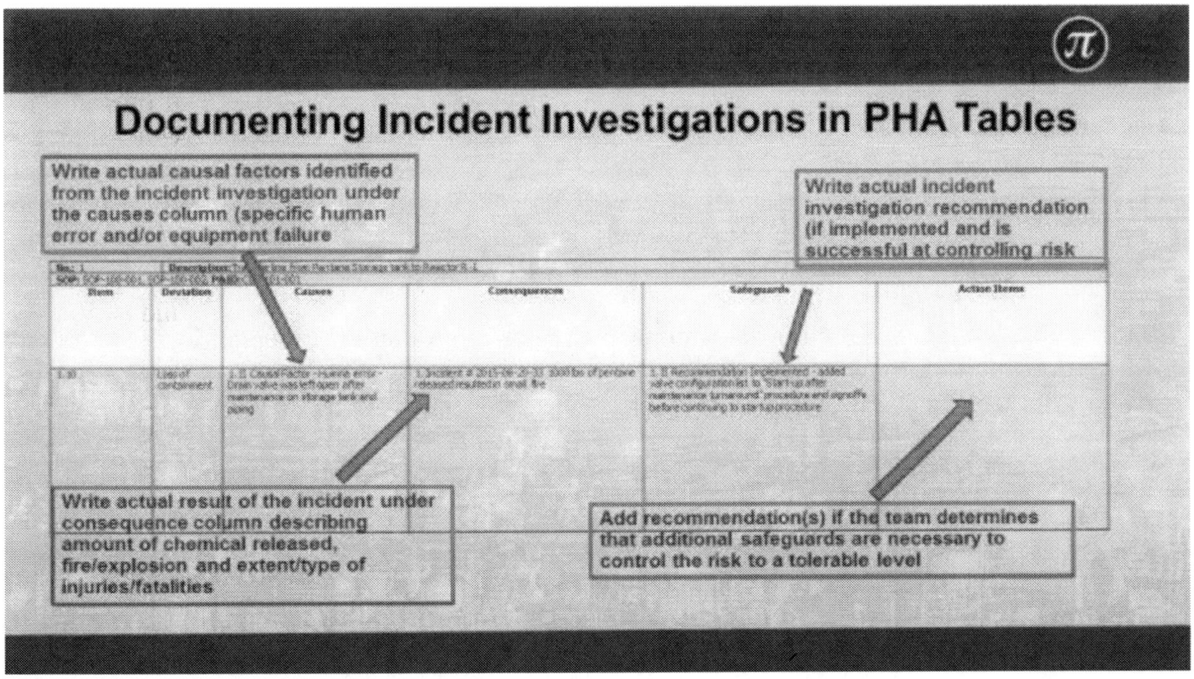

Figure 6. Documenting Incident Investigations in PHA Tables

Conclusions

Qualitative analysis of non-routine operating procedures is an extremely powerful tool for uncovering deficiencies that can lead to human errors and for uncovering accident scenarios during all modes of operation. This approach of step-by-step HAZOP and/or What-If analysis is not new to industry, and regulators have required similar approaches for decades. And regulators continue to note lack of analysis of the risk of non-routine operations and lack of risk review of changes to procedures.

> ***From the Wall Street Journal[27] referencing the presidential commission investigating the Deepwater Horizon accident of April 2010:*** *BP had rules in place governing procedural changes, but its workers didn't consistently follow them, according to BP's September [2010] internal report on the disaster and the report released earlier this month [January 2011] by the presidential commission on the accident. "Such decisions appear to have been made by the BP Macondo team in ad hoc fashion without any formal risk analysis or internal expert review," the commission's report said. "This appears to have been a key causal factor of the blowout."*

> ***From CSB Report on August 2008 Bayer CropScience Explosion:[12]*** *"The accident occurred during the startup of the methomyl unit, following a lengthy period of maintenance ... CSB investigators also found the company failed to perform a thorough Process Hazard Analysis, or PHA, as required by regulation...In particular, for*

operational tasks that depend heavily on task performance and operator decisions, the team should analyze the procedures step-by-step to identify potential incident scenarios and their consequences, and to determine if the protections in place are sufficient."

Since about 80% of major accidents occur during non-routine modes of operation,[1] the PHA of the deviations from the steps of the procedures that govern these modes of operation is certainly critical. But, of the 20% of the major accidents that occur during normal operation, most of these arise from **damage mechanisms** such as corrosion, material defects, gasket failures, external impacts and the like (based on PII data). PHAs can effectively cover DM while discussing LOC.

So, the PHAs must cover both of these concepts (PHA of all modes of operation and addressing DM) thoroughly to fulfill the role of the PHA in controlling risk. **More regulatory pressure is sure to follow on these two issues**

But of course, the PHA must be strong in all aspects, including consideration of lessons learned from previous incidents.

Acronyms Used

AIChE– American Institute of Chemical Engineers
API – American Petroleum Institute
CCPS – Center for Chemical Process Safety (a division of AIChE)
CSB – US Chemical Safety and Hazard Investigation Board
DM – Damage mechanism
DMHR – Damage mechanism Hazard Review
HAZOP – Hazard and Operability Analysis
HAZWOPER – Hazardous and Waste Operater (US OSHA 29 CFR 1910.120)
HTHA – High Temperature Hydrogen Attack
IPL - Independent Protection Layer
LOC – Loss of Containment
LOPA – Layer of Protection Analysis
MOC – Management of Change
NHT – Naphtha Hydrotreater
OSHA – Occupational Safety and Health Administration, US Department of Labor
PHA – Process Hazard Analysis
PII – Process Improvement Institute, Inc.
P&ID – Piping & Instrumentation Diagram
PSI – Process Safety Information
PSM – Process Safety Management
SOP – Standard Operating Procedure

References

1. *Guidelines for Hazard Evaluation Procedures*, 3rd Edition, 2008, CCPS/AIChE.

2. U.S. Department of Labor: Systems Safety Evaluation of Operations with Catastrophic Potential. Occupational Safety and Health Administration Instruction CPL 2-2.45, Directorate of Compliance Programs, September 6, 1988.
3. Woodcock, Henry C., "Program Quality Verification of Process Hazard Analyses (for instructional purposes only)," US OSHA, 1993.
4. OSHA Inspection Number 106612443 - Phillips 66 Company, Houston Chemical Complex, Citations 1-1 through 1-566, Issued 4/19/1990.
5. *Stipulation and Settlement Agreement,* Phillips 66 Company ("Phillips") and Lynn Martin, Secretary of Labor, United States Department of Labor, 8/22/1991.
6. "Process Safety Management of Highly Hazardous Chemicals; Explosives and Blasting Agents," US OSHA Final Rule, 29 CFR 1910.119, February 24, 1992.
7. "Risk Management Programs for Chemical Accidental Release Prevention," US EPA Final Rule, 40 CFR 68, 1994-2000.
8. OSHA Inspection Number 103490306, Issued November 2, 1992.
9. OSHA Inspection Number 123807828; Issued November 18, 1993.
10. U.S. Department of Labor: PSM Covered Chemical Facilities National Emphasis Program. Occupational Safety and Health Administration CPL 03-00-014, Directorate of Enforcement Programs, November 29, 2011.
11. U.S. Department of Labor: Petroleum Refinery Process Safety Management National Emphasis Program. Occupational Safety and Health Administration CPL 03-00-010, Directorate of Enforcement Programs, August 18, 2009.
12. "Investigation Report: Pesticide Chemical Runaway Reaction Pressure Vessel Explosion, at Bayer CropScience, LP, Institute, WV, on August 28, 2008", US Chemical Safety Board, Report No. 2008-08-I-WV, January 2011.
13. U.S. Chemical Safety and Hazard Investigation Board (CSB). *Investigation Report: Catastrophic Rupture of Heat Exchanger (Seven Fatalities), Tesoro Anacortes Refinery, Anacortes, WA, Report No. 2010-08-I-WA, Draft for Public Comment*, January 29, 2014.
14. Washington Administrative Code (WAC) 296-67-291 Appendix C--Compliance guidelines and recommendations for process safety management (nonmandatory) http://www.lni.wa.gov/WISHA/Rules/hazardouschemicals/default.htm#WAC296-67-021 (accessed December 3, 2013).
15. Center for Chemical Process Safety (CCPS). *Revalidating Process Hazard Analyses*. 2001; pp 31-32.
16. U.S. Chemical Safety and Hazard Investigation Board (CSB), *DuPont La Porte, Texas Chemical Facility Toxic Chemical Release Interim Recommendations Investigation: 2015-01-I-TX* Incident Date: November 15, 2014 Issue Date: September 30, 2015
17. Contra Costa County Hazardous Materials Programs, *Contra Costa County Industrial Safety Ordinance (ISO)*, by CCHMP. June 15, 2011.

18. Draft Initial Report "*Safety Evaluation of the Chevron Richmond Refinery*." November, 2015, by Process Improvement Institute, Inc., for Contra Costa County Hazard Materials Department.

19. U.S. Chemical Safety and Hazard Investigation Board (CSB). "CASE STUDY *Fire at Formosa Plastics Corporation: Evaluating Process Hazards*." July 20, 2006.

20. API RP 571. "Damage Mechanisms Affecting Fixed Equipment in the Refining Industry." 2nd ed., April 2011.

21. U.S. Chemical Safety and Hazard Investigation Board (CSB), *Interim Investigation Report: Chevron Richmond Refinery Fire*, U.S. CSB, August 6, 2012.

22. "Necessity of Performing Hazard Evaluations (PHAs) of Non-normal Modes of Operation (Startup, Shutdown, & Online Maintenance)", W. Bridges and Mike Marshall (US OSHA), *18th Annual International Symposium, Mary Kay-O-Connor Process Safety Center, College Station, TX, October 2015 and 12th Global Congress on Process Safety, April, 2016.*

23. "How to Efficiently Perform the Hazard Evaluation (PHA) Required for Non-Routine Modes of Operation (Startup, Shutdown, Online Maintenance)," W. Bridges and T. Clark, *7th Global Congress on Process Safety*, Chicago, AIChE, March 2011.

24. Bridges, W.G., et. al., "Addressing Human Error During Process Hazard Analyses," *Chemical Engineering Progress*, May 1994

25. Rasmussen, B. "Chemical Process Hazard Identification," *Reliability Engineering and System Safety*, Vol. 24, Elsevier Science Publishers Ltd., Great Britain, 1989.

26. "Effectively Addressing New PSM/RMP Damage Mechanism Review Requirements with an Integrated PHA (iPHA)," Nour, Maher, Schutz, all from Risk Management Professionals, 11th GCPS, *April 2015.*

27. Wall Street Journal, January 29, 2011.

28. Gertman, D.; Blackman, H.; Marble, J.; Byers, J. and Smith, C., "The SPAR-H Human Reliability Analysis Method," NUREG/CR-6883, U.S. Nuclear Regulatory Commission, Office of Nuclear Regulatory Research, Washington, DC, August 2005.

29. Swain, A. D., Guttmann, H. E., "Handbook of Human Reliability Analysis with Emphasis on Nuclear Power Plant Applications, Final Report," NUREG/CR-1278, 1983, US Nuclear Regulatory Commission.

When the Fail Open Valve Fails Closed: Lessons from Investigating the "Impossible"

Brenton L. Cox
Exponent Failure Analysis Associates
4580 Weaver Parkway, Suite 100, Warrenville, Illinois 60555
bcox@exponent.com

Sean J. Dee, Russell A. Ogle
Exponent Failure Analysis Associates
4580 Weaver Parkway, Suite 100, Warrenville, Illinois 60555
sdee@exponent.com, rogle@exponent.com

Prepared for Presentation at
American Institute of Chemical Engineers
2017 Spring Meeting and 13th Global Congress on Process Safety
San Antonio, TX
March 26 – 29, 2017

When the Fail Open Valve Fails Closed: Lessons from Investigating the "Impossible"

Brenton L. Cox
Exponent Failure Analysis Associates
4580 Weaver Parkway, Suite 100, Warrenville, Illinois 60555
bcox@exponent.com

Sean J. Dee, Russell A. Ogle
Exponent Failure Analysis Associates

Keywords: Incident investigation, hazard assessment, management of change

Abstract

Process hazard analysis is recognized by many safety professionals and regulatory groups as the key provision of a process safety management system. This analysis evaluates what can go wrong, and what safeguards are needed to prevent incidents. One challenge in process hazard analysis is the reliance on humans to determine reasonable, expected, or anticipated failure modes. The implementation of safeguards may also foster a lost sense of vulnerability to potential process upsets. These factors in combination can lead to a disregard for failure modes considered "unlikely" or even "impossible". Unfortunately, this problem is often revealed when the "impossible" leads to an incident or accident. In this paper, a case study is used to motivate a discussion of incidents involving hazards that were disregarded in process hazard analysis due to their low probability, or supposed impossibility. Tools, techniques, and guidance will be discussed to improve process hazard analysis when evaluating low probability failure modes.

1 Introduction

Process hazard analysis (PHA) is one of the key aspects of a robust process safety management system. PHA is one of the fourteen elements of the Occupational Safety and Health Administration Process Safety Management Standard (OSHA PSM). [1] PHA is also interwoven throughout several elements of a Risk Based Process Safety (RBPS) approach to process safety management. [2] While regulatory compliance is critical for facilities covered under the OSHA PSM standard, PHA is an important best practice for all facilities that aim to identify and manage the risk associated with hazards at their facilities. Therefore, in the context of this manuscript, PHA will be discussed synonymously in the context of the Hazard Identification and Risk Analysis (HIRA) element of RBPS.

The HIRA element focuses on three questions:

- Hazard - What can go wrong?
- Consequence – How bad could it be?
- Likelihood – How often might it happen?

One could argue that the Hazard component of HIRA is almost a prerequisite for the actual analysis. Without an understanding of what can go wrong, it is very challenging to determine a consequence or likelihood for the analysis. Incident investigation teams often revisit pre-incident hazard analyses as a basis for understanding why an accident occurred. Some may argue that facilities feel most vulnerable when faced with responding to a failure that was not identified in their hazard analysis. However, even more concerning are accidents that occur after a facility identified the hazard scenario, but ultimately concluded it was extremely unlikely, or even impossible in their hazard analysis.

This manuscript will discuss the issues facilities face when unexpected failure modes result in accidents. A specific case study involving an unanticipated valve failure is used to guide a more general discussion. This example is particularly illustrative because of the relatively simple function – and failure – of the valve. Lessons learned and opportunities for improvement in PHA will be presented.

2 Process Hazard Analysis

As previously discussed, PHA is intertwined within several elements of a RBPS approach to process safety management. While PHA is part of the foundation of the HIRA element of RBPS, it is also a key component that underlies several other elements including, but in no way limited to, Process Safety Culture, Asset Integrity and Reliability, Management of Change, and Incident Investigation. Thus, facilities that are adept and proficient in PHA can benefit not only in the HIRA element of RBPS, but also in several additional elements. This pushes a facility closer to the goal of implementing a reliable system to manage the process' risk to people, equipment, and the environment.

Hazard identification and analysis has been a focus of process safety management for over 30 years. The first project undertaken by the Center for Chemical Process Safety (CCPS) focused on the development of a comprehensive text to improve process safety performance in the industry. [3] This project resulted in the publication of *Guidelines for Hazard Evaluation Procedures* in 1985, with updates issued in 1992 and 2008. Hazard evaluation is the cornerstone of a process safety management program and is performed throughout the life cycle of a process and facility. Therefore, improvements and advancements in process hazard analysis are an important component of mitigating and preventing process incidents.

Process incidents can be simplified into three components:

1. Existence of a process hazard

2. Determination of a potential initiating cause

3. Occurrence of an incident outcome

Commonly, process hazard analysis is performed to understand and evaluate each of these three elements. First, an analysis aims to identify and understand the potential hazards associated with a process. Next, scenario based or non-scenario based methods are used to identify potential failure modes that could occur. Then, both unmitigated and safeguard mitigated outcomes are considered to determine the potential loss event or impact of a particular failure mode.

Shortcomings in analysis of any of the three elements of a process incident can introduce a vulnerability in the process safety management system. Incidents can occur when facilities either fail to identify or misunderstand a particular hazard in their process. For example, an insufficient understanding of the chemical equilibrium established between elemental sulfur and hydrogen sulfide may result in a failure to identify personnel exposure hazards during material transfer activities. [4] Similarly, hazard evaluations can mistakenly overlook a mechanical failure mode that could initiate a process safety incident. Additionally, the outcome from an initiating event can be much worse than anticipated if its propagation is not characterized, or if controlling safeguards are not properly implemented or maintained.

This paper will focus on a case study related to a valve failure that resulted in an incident at a chemical facility. The facility had identified and understood the hazards involved in the incident, but their hazard analysis and safety system design did not include the valve failure mode that initiated the incident. As a result, several key learnings were developed to improve the facility's hazard analysis, operator response, and safety system design.

3 Case Study

The following case study examines an incident that occurred when a fail open valve failed closed. Heating and cooling were both provided to the facility's polymerization reactor using a jacketed system as shown in Figure 1. When a batch was started, heating was applied to the reactor using steam to initiate the reaction. After the reactor reached the desired temperature, an automatic control chilled water valve would modulate cold makeup water into the jacket circulation loop to maintain the reactor at the appropriate temperature. An operator in the field could also open a manual emergency cooling bypass valve around the automatic control chilled water supply valve in the event that the reactor needed additional cooling.

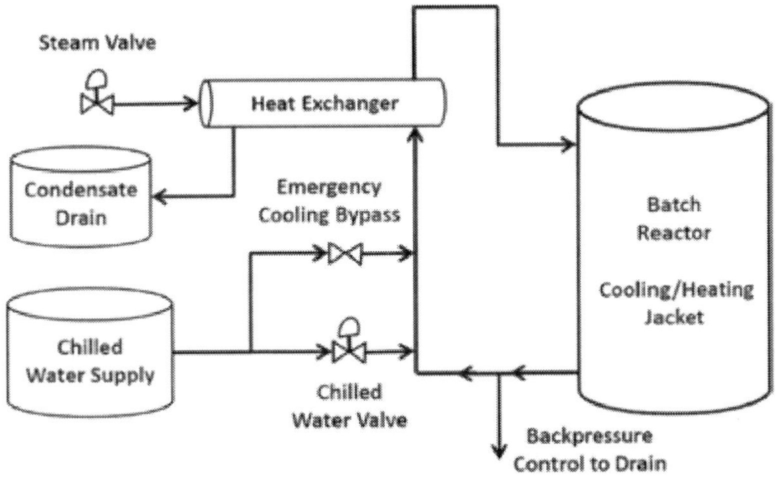

Figure 1. Heating/Cooling Jacket for Batch Polymerization Reactor

On the day of the incident, the temperature within the polymerization reactor began to increase, and triggered the high temperature alarm. The operator at the control station acknowledged the alarm, and noted that the HMI screen indicated the chilled water valve was already 100% open and the temperature of the water supply to the jacket inlet was slowly decreasing. The temperature in the reactor continued to increase, returning an additional error message related to insufficient cooling water flow. The operator confirmed that the chilled water valve was still 100% open on the HMI and that the jacket supply temperature was continuing to decrease indicating that the jacket was receiving cold makeup water. The operator then began to troubleshoot the cooling system. Over the next six to seven minutes, the operator tried to cool the reactor contents by injecting cold water into the reactor. The reactor temperature continued to increase, eventually reaching the high/high alarm set point for the emergency shutdown system (ESS). The ESS then intervened, aborted the batch and dumped the reactor contents to a secondary containment pit. While no one was injured, there was a recordable environmental release, which was reported to the local authority having jurisdiction.

After securing the area and verifying it was safe to enter, operators examined the jacketed system on the reactor. Surprisingly, they found the water supply valve to the reactor water almost completely closed, even though the control station showed the valve as 100% open. An engagement pin from the valve's feedback positioner mechanism was found on the floor next to the chilled water valve. The valve had been designated by the site as a "fail open" valve. However, the valve had failed in a mostly closed position, limiting the flow of cooling water to the jacket circulation loop.

3.1 Chilled water control valve design and operation

The reactor jacket chilled water control valve is shown in Figure 2 and consists of three parts: the valve body, the actuator, and the positioner. As shown in Figure 3, flow through the valve body is regulated by the position of the valve stem, which moves an internal plug up and down in the valve body to open and close the valve. The valve stem is connected to the actuator stem as shown in Figure 4. The entire stem assembly is pushed up to open the valve with an internal actuator spring that pushes against a diaphragm. The stem assembly is pushed down to close the valve by air pressure, which is applied to the top of the diaphragm and overcomes the actuator spring force. The chilled water valve is an air-to-close valve, meaning that the valve requires air pressure to close. Upon loss of air or pneumatic signal, the actuator spring is the sole force acting on the valve stem; in this condition, the valve is forced fully open corresponding to full cooling flow to the reactor jacket.

The positioner shown in Figure 5 controls the valve stem position. An instrument input pressure signal is supplied to the positioner, which expands the bellows. As the bellows pushes up on the "D" shaped beam, the beam rotates about the input axis (vertical) and the flapper assembly moves closer to the nozzle. As the flapper approaches the nozzle, the nozzle pressure increases, which increases the output pressure supplied to the actuator by the relay. As the pressure in the actuator increases and pushes against the diaphragm, the stem assembly and attached travel pin move downward, lowering the rotary shaft arm. The rotary shaft arm then rotates the cam, which pivots the "D" shaped beam about the feedback axis (horizontal) moving the flapper assembly away from the nozzle. This movement decreases the nozzle pressure until it equilibrates for the given bellows position and supply pressure.

Figure 2. Jacket cooling water control valve from reactor.

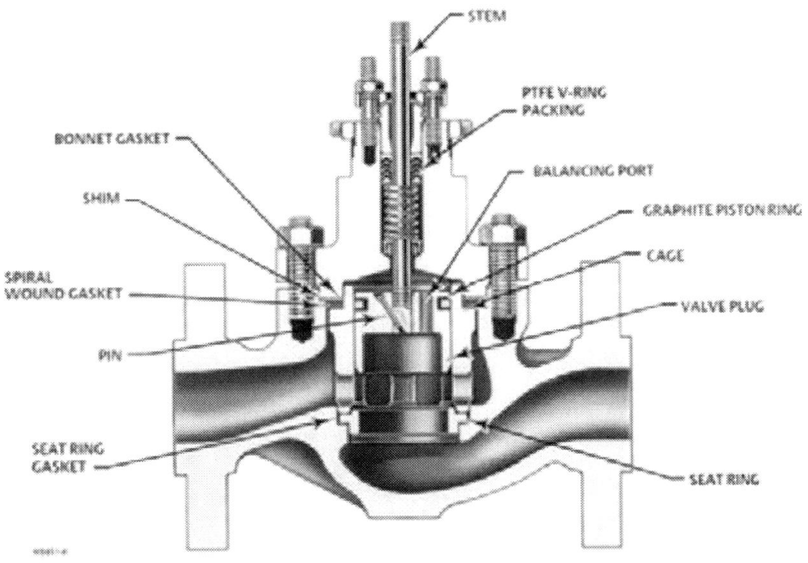

Figure 3. A cutaway diagram for the valve body of the jacket chilled water control valve.

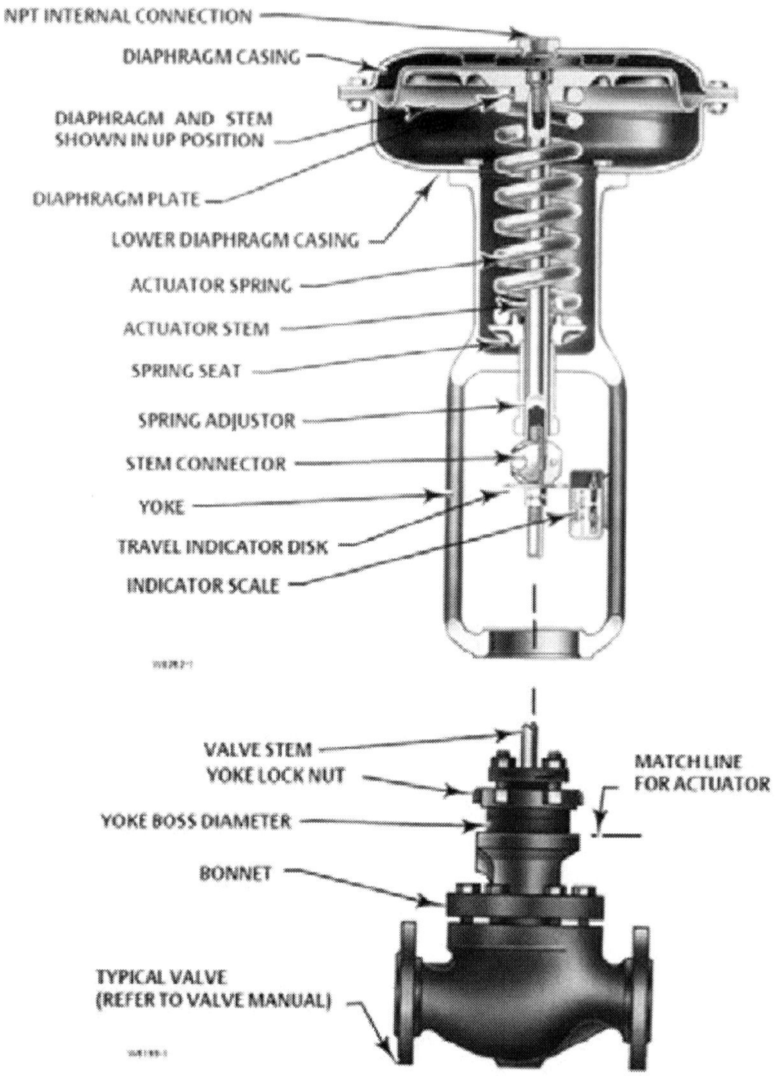

Figure 4. **Valve actuator diagram. The actuator moves the valve stem up and down by counter balancing upward and downward forces on the stem.**

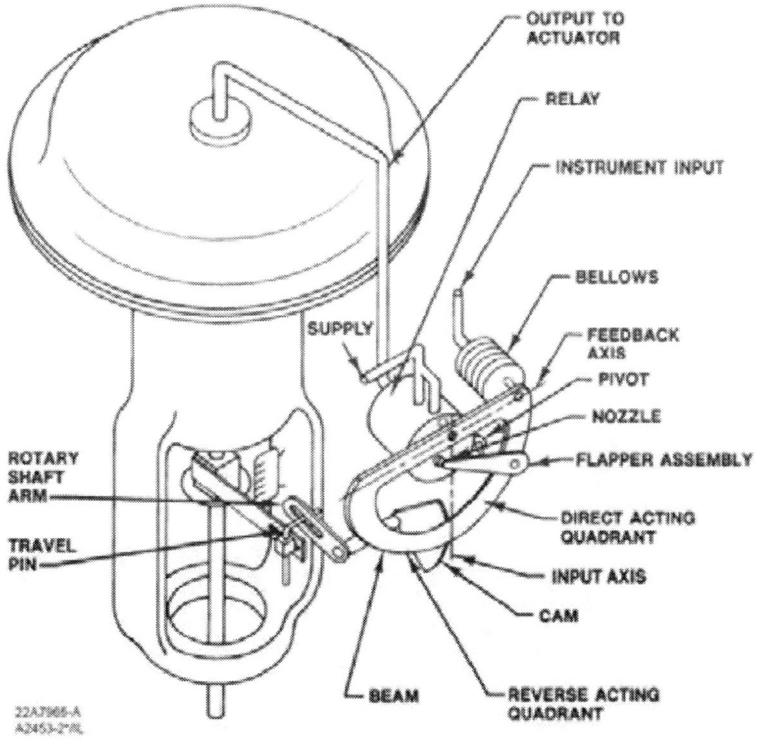

Figure 5. Valve Positioner diagram showing the components of positioner mechanism and the mechanical linkage that attaches to the valve stem.

3.2 Failure mechanism of the chilled water valve.

The incident batch was dumped to the containment pit because the contents of the reactor overheated due to insufficient cooling water introduction into the jacket circulation loop. In the post incident, investigation the travel pin was found disengaged from the rotary shaft arm on the reactor chilled water control valve. With the travel pin disengaged, the rotary shaft arm faulted to a full upward position corresponding to a fully open valve. In fact, the mechanic troubleshooting the issue after the reactor evacuation noted that the input signal pressure gauge on the positioner indicated 3.0 psig (corresponding to the DCS command signal for a 100% open valve). However, with the travel pin disengaged, the valve was found mostly closed. When the travel pin was reinstalled, the valve operation was verified to be within the calibration specified for the valve.

Testing was conducted to examine the effect of disengagement of the travel pin on the open/close state of the chilled water control valve, and is shown in Figure 6. The chilled water valve is calibrated to begin closing from a fully open position at 4.2 psig until it reaches a fully closed position at 9.6 psig. After verifying this calibration, the valve was commanded to a fully closed position (9.6 psig) and the pin was disengaged from the rotary shaft arm. The rotary shaft arm sprung up to the fully open position, but the valve stem did not move. The instrument input pressure was fully released (0.0 psig), and the valve fully opened. Then, a 3.0 psig instrument input pressure was applied and surprisingly, the valve closed. The valve remained closed until the instrument input pressure decreased to approximately 1.0 - 1.5 psig.

Figure 6. Chilled Water Control Valve positions during testing.

Failure of the chilled water control valve with a disengaged travel pin in a mostly closed state at 3.0 psig was consistent with witness statements, process data trends, and independent testing. When the travel pin disengaged from the rotary shaft arm, the arm moved to the fully open position. This movement rotates the cam, such that the flapper assembly is at the maximum distance from the nozzle about the feedback axis. However, the input pressure associated with a fully open command (3 psig) still pushes the "D" shaped beam about the input axis pivoting the flapper assembly toward the nozzle. The resulting failure state supplies almost enough input pressure to close the valve.

Thus, while a 3 psig instrument input signal was insufficient to close the valve with the travel pin engaged, disengagement resulted in movement of the flapper assembly closer to the nozzle. This increased the nozzle pressure and the pressure output to the actuator causing the valve to fail in a mostly closed position. This mostly closed state prevented the cooling water flow to the jacket supply, and inhibited the reactor cooling function. Without sufficient cooling, the reactor overheated and the SIS intervened to evacuate the contents to the emergency containment area.

Upon visual inspection of the control valve, it was determined that three factors were responsible for disengagement of the travel pin:

- Misalignment of the connector arm;
- Wear of the sliding contact surfaces between the travel pin and the rotary shaft arm; and
- Hydraulic forces caused by the transient, turbulent flow through the valve during normal operation

In a discussion with the valve manufacturer, the representative indicated that the disengagement of the travel pin was an extremely rare event. The site had several exemplar valves for additional reactors with jacketed heating/cooling systems. Inspection of the other chilled water control valves in service at the facility revealed that the mechanical components were not identical in appearance and varied in both wear and engagement distances as shown in Figure 7, Figure 8 and Figure 9. In fact, each mechanism had a unique combination of dimensions, orientations, and tolerances. The preventative maintenance for the control valve at the time of the incident did not address any measures to prevent pin disengagement. The manual for the valve did not offer any specific guidance for preventative maintenance procedures that would prevent travel pin disengagement.

Figure 7. Wear pattern observed on bottom side of rotary shaft arm.

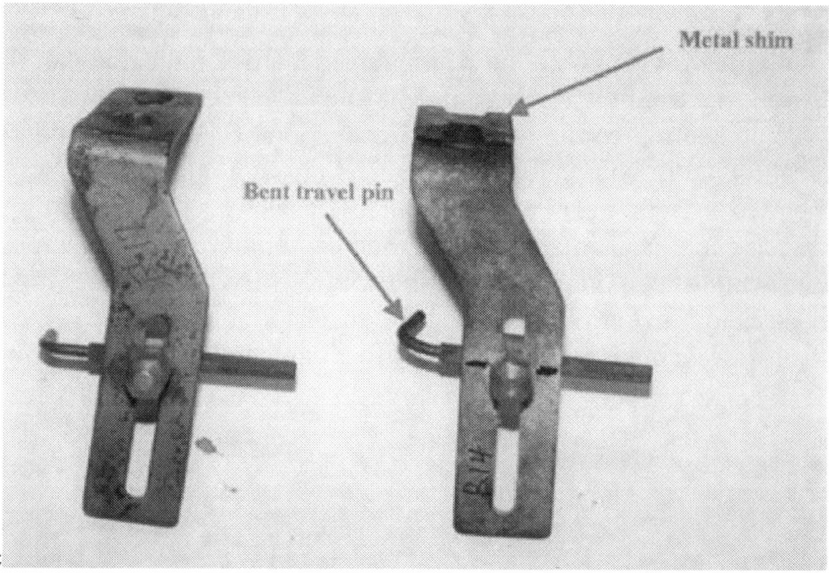

C

Figure 8. Travel pin and bracket assemblies on two Cooling Jacket Control Valves at the facility. The assembly on the right was found to have a bent pin, and a metal shim welded to the bracket.

Figure 9. Pin engagement distance

3.3 Operator intervention based on conflicting information and insufficient time.

The impact due to a failure of the chilled water valve can be mitigated with operator intervention. However, in this incident, conflicting information on the control screen resulted in the wrong course of action in response to the high temperature alarms. The operator screens indicated that the chilled water control valve was fully open, yet the reactor contents continued to increase in temperature. Plant personnel indicated that this situation occurs because the operator screen only displays the command from the system to the control valve, not an actual feedback position from the valve. The operator response was further complicated by the observed decrease in temperature on the jacket supply. The valve did not fail completely closed, and in its impaired state, just enough cooling water was being supplied to decrease the jacket supply water temperature. Operators did not recognize that the decrease in jacket supply temperature was much lower than the decrease in temperature observed with a 100% open valve.

Based on this conflicting information, the operator chose to attempt to manually cool the reactor, rather than order the field operator to open a manual emergency bypass cooling valve. The attempts to cool the reactor were ineffective, and the ESS activated before the operator realized that the manual emergency cooling bypass valve needed to be opened. Given the short response time (6-7 minutes) for the alarm, it was important that the instruction for responding to the high temperature alarms be clear, straight forward, and concise. The timeframe was not sufficient to troubleshoot and diagnose the cooling system performance issues. Instead, opening the emergency cooling bypass valve was viewed as the preferred first response to the high temperature alarm.

4 Lessons Learned

After the incident, the facility identified several key lessons learned regarding their hazard analysis, mechanical integrity/maintenance program, and operational discipline.

4.1 Hazard Analysis

The use of terminology such as "fail open" or "fail close" is common in process engineering. Unfortunately, this terminology does not include the initiating cause for the failure state. As a result, process hazard analysis teams risk misinterpreting a piece of equipment as "fail open" or "fail close" for a wide variety of initiating causes. For example, many components are designated for a specific failure state upon loss of utilities. However, equipment malfunctions, human errors, or external events can also influence a component's operation and cause a failure state that is different than the one observed upon loss of utilities. Therefore, it is important that hazard analysis teams continue to consider additional potential initiating causes and failure states for equipment.

In the presented case study, the facility thought that under all potential failure conditions, the chilled water valve would fail open. As a result, the control valve was also used as a first response emergency cooling valve. When the valve failed, the system not only lost the ability to control temperature in the reactor, but also lost the first response of the emergency cooling system. After the incident, the facility installed a redundant emergency cooling valve. The redundant valve was independent of the control valve and ensured that the failure of the control valve would not impede the first response emergency cooling system.

4.2 Mechanical Integrity and Maintenance

The facility recognized that a valve mechanical linkage failure could be mitigated through preventative maintenance and inspection. After the incident, the facility added inspection of the mechanical linkage on the cooling water control valve to the quarterly preventative maintenance for the process. This inspection included evaluation of the linkage components and the engagement distance of the travel pin into the rotary shaft arm. The goals of the program were to recognize linkage component that were deteriorating and replace them prior to failure, and to make adjustments if parts of the linkage began to move out of alignment.

It is often challenging for a facility to continue to service aging equipment. The original manufacturer of the equipment may go out of business, or elect to stop offering replacement parts. After the incident, the facility reevaluated its procurement process for replacement parts in control valves. It reached out to the valve manufacturer to confirm that the replacement parts it had obtained were approved for use in the valve's mechanical linkage. It also added a dimension verification step to the procure process to confirm that the ordered parts matched the equipment specifications.

4.3 Operational Discipline

As highlighted in the case study, operator intervention is one of the controls that a facility may rely upon to safely navigate process upsets. Using procedures as an administrative control can be particularly challenging because they must be effectively written, implemented, and performed. [5] For this reason, procedures often represent a potential area for improvement in many safety management systems.

After the incident, the facility questioned why the manual emergency bypass line was not opened. The operator in the control room stated that when he responded to the alarm, there were two observations he made: 1. the emergency cooling valve was 100% open; 2. the jacket supply temperature was decreasing. Based on that information, the operator moved to troubleshooting the issue, and did not realize that the reactor temperature was still increasing. The operator also commented that there were several temperature alarms and they tripped infrequently. Given the short response time (~ 6-7 min); it was also possible that a field operator would not have had sufficient time to locate the necessary tools and open the bypass cooling valve prior to the incident.

The facility developed a job aid for the control room operator to improve operator response to the alarms. The job aid was a one-page chart that detailed critical alarms that required opening the bypass cooling valve. The chart also include the alarm set points, and the appropriate response for each alarm. The facility also tied a wrench to the emergency cooling bypass line, to ensure that if the valve needed to be opened, the field operator would have all the necessary tools available at the valve.

5 Conclusions

Process hazard analysis is a fundamental component of several aspects of process safety management systems. When a facility struggles to perform process hazard analysis, there is a potential for risks to go unrecognized, or under mitigated.

A case study was presented that examined an incident related to a "fail open" cooling water valve failing in a partially closed state. The valve failure disabled not only the temperature control function on an exothermic reactor, but also the primary emergency response system. Administrative controls were ineffective as implemented at preventing the incident following failure of the valve. The reactor overheated due to insufficient cooling water flow, and the contents were evacuated to a secondary containment pit resulting in a reportable release. Following the incident, the facility identified several key learnings related to improving their hazard analysis. The facility also implemented corrective actions to mitigate the potential failure of other cooling water control valve, and improve the reliability of the emergency cooling system and operator performance.

6 References

[1] United States Department of Labor, Occupational Safety and Health Administration. Process Safety Management of Highly Hazardous Chemicals, 29 CFR 1910.119.

[2] Center for Chemical Process Safety, Guidelines for Risk Based Process Safety, John Wiley & Sons, Inc., New Jersey, 2007.

[3] Center for Chemical Process Safety, Guidelines for Hazard Evaluation Procedures, John Wiley & Sons, Inc., New Jersey, 2008.

[4] M.J. Viz, R.A. Ogle, S.J. Dee and B.L. Cox. Hydrogen Sulfide Exposure from Molten Sulfur – A Forgotten Hazard? Proceedings of 9th Global Congress on Process Safety, San Antonio, Texas, April 28-May 2, 2013.

[5] S.J. Dee, R.A. Ogle, S.D. Lakhiani. Procedural Safeguard Reliability. Proceedings of 11th Global Congress on Process Safety, Austin, Texas, April 27-April 29, 2015.

13TH GLOBAL CONGRESS ON PROCESS SAFETY

PROCESS HAZARD ANALYSIS (PHA) – ARE STUDIES IN ITSELF GOOD ENOUGH TO REDUCE RISK?

Palaniappan Chidambaram
DuPont Sustainable Solutions
Singapore
Palaniappan.Chidambaram@dupont.com

Prepared for Presentation at
American Institute of Chemical Engineers
2014 Spring Meeting
2017 Spring Meeting and 13th Global Congress on Process Safety
San Antonio, TX
March 26 – 29, 2017

Keywords: Process Hazard Analysis, HAZOP, Risk Recognition, Risk Governance, Operational Risks, Risk Management

Abstract

Despite the advances in processing technologies, construction techniques, inspection methods, and materials, the number of process safety incidents are still increasing, and often the same types of incidents are being repeated. While regulators are enforcing stringent requirements in regards to safety management systems, process safety incident statistics reflects ineffective implementation. Upon closer look at the implementation of safety management systems, it becomes evident that they were not based on robust risk management system. When asked how risks are managed, a common response in the industry is that we have completed the hazard analysis and risk assessment studies. This paper will help understand why studies alone are not sufficient to manage risks effectively. Typical observations based on our client experience will be highlighted in the areas of risk governance, risk thinking culture, risk communication and risk awareness which goes beyond risk recognition/identification and assessment.

Introduction

Operational risks can be defined as the potential loss due to possible or foreseeable worst case consequences with a realistic probability of occurrence during facility operation, where such risks span across the entire value/supply chain. Typical operational risks include employee and public health and safety, community economic impact, environmental damage, asset damage, and business interruption. These risks are managed effectively in an organization when events leading to such risks are identified, well understood, and barriers required to prevent/mitigate them are implemented and maintained. It is essential to recognize risks associated with the process and operations through the lifecycle of the facility and implement systems to manage them effectively.

Organizations that understand the importance of implementing Process Safety Management systems do not always realize the benefits despite significant effort and leadership commitment. This can be primarily attributed to the following factors:

- Ineffective risk governance and lack of risk ownership at different levels;
- Inadequate risk recognition and risk assessment;
- Lack of holistic risk based thinking and risk based implementation;
- Poor risk awareness and risk communication; and
- Lack of robust risk assurance program;

Figure 1 depicts a model to represent operational risk management in a comprehensive manner and is designed to provide organizations with a better understanding and highlight the importance of risk governance and recognition in managing operational risks. In this paper, we will highlight our observations and findings based on our client engagement experience in the areas of risk governance, risk culture, risk recognition and assessment, risk awareness and risk communication. This will help demonstrate that PHA studies alone are not sufficient to manage or reduce risk.

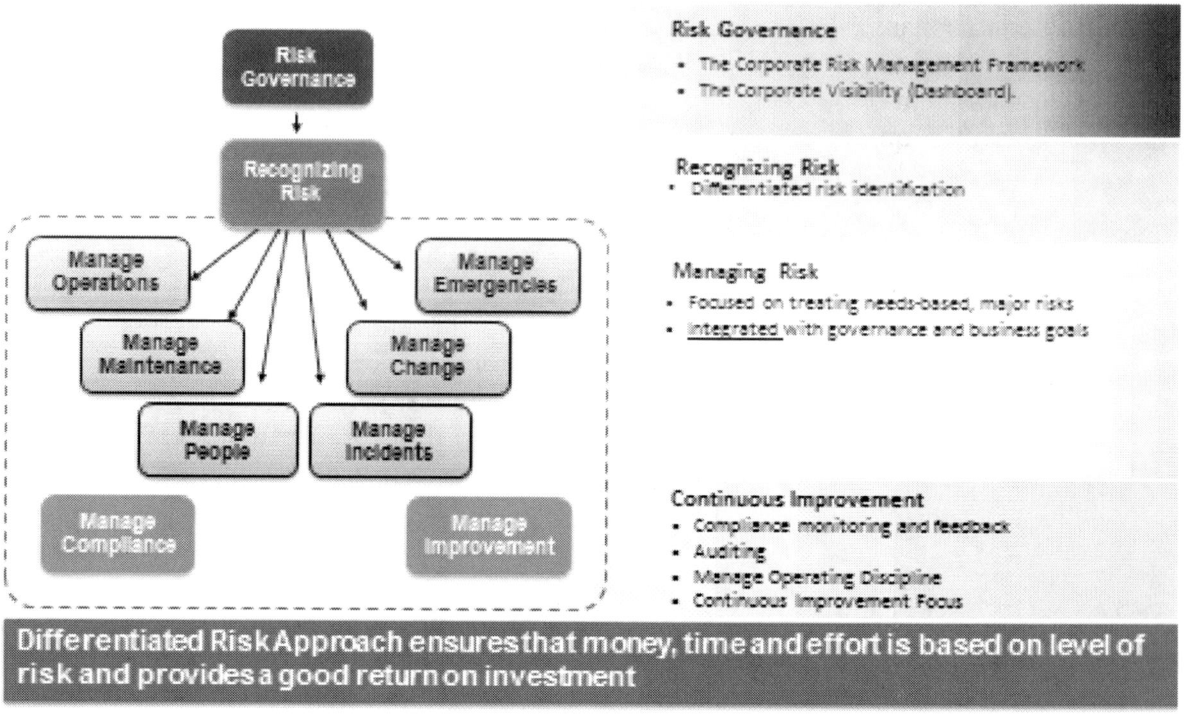

Figure 1 – DuPont Sustainable Solutions' approach to managing Operational Risks
© 2017 DuPont

Risk Governance

To consistently manage risks and make informed decisions, it is essential to have a well codified risk management process and governance structure to execute it. One of the key elements of risk governance is defining the acceptable risk criteria, along with risk ownership and accountability at different levels of the organization. In most cases, risk criteria are represented in the form of a qualitative/semi-quantitative risk matrix. A qualitative or semi-quantitative risk matrix is used during What If/Checklist reviews and Hazard and Operability (HAZOP) reviews. Other common areas where risk based decisions are made include task risk assessments, prioritization of inspection and maintenance activities, as well as recommendations from auditing and incident investigations.

In the absence of a robust risk governance structure and risk thinking culture, risk recognition and assessment efforts are not leveraged effectively (e.g. definition of emergency scenarios, identification of PSM critical elements, training on SOPs, etc.) to drive effective implementation of an operational risk management program. Some of the observations indicating poor risk governance include:

- Inconsistency in the risk matrix utilized for ranking risks at enterprise level and during risk assessment for facilities and sites.
- Identified risks and treatment actions are generic to business and not specific for a given risk event;
- Specific risks identified in hazard evaluation and risk assessment studies are not included in the risk register; and
- Risk escalation and approval mechanism is not in place to address the emerging risks due to changes (for example: impairment/bypass of safety critical devices) or incidents;

This often leads to inconsistent risk based decisions in the organization despite being risk based, leading to underestimation of risk (worst case) and a false sense of security. In addition, efforts and resources are likely to be invested where it is not required.

Risk Culture

It is essential to embed a risk thinking culture into an organization at all levels to achieve effective, systematic and consistent management of operational risks. More often, organizations are task focused with a compliance mindset instead of outcome focused with a commitment towards managing risks. This orientation is often reflected in their behavior and focus on what needs to be done and how, instead of why and how well it needs to be done. Significant effort is invested in activities, whereas outcome of such activities does not result in risk reduction and a shift in risk thinking mindset and behaviors.

Third party consultants are often engaged by organizations to perform their hazard evaluation and risk assessment studies. Organization with task focused mindset tends to complete studies primarily to fulfill insurance, regulatory and corporate requirements. Due to lack of inadequate and competent resources, results and outcome from such studies often resides with a function or department in the organization. Assessment reports are often not reviewed for completeness, understood well, communicated and utilized effectively. Without a strong culture, the outcome of such studies is not reviewed, understood, communicated and utilized in managing risks effectively. Such an organization tends to equate completion of such studies and compliance to local regulations and standards with effective risk management.

Organizations that are unable to respond to the following questions confidently and consistently at all levels of the organization often have difficulty managing risk:

- What are the top risks for the site, sector, plant and section and barriers associated with them?
- Are identified barriers effective and will they function on demand?
- What actions should be taken if barriers do not function?
- What is my role in recognizing and managing risks and consequences of not managing them?

Often responses for each question vary depending on the individuals' risk perception. Based on our client experience, the following paradigms need to be understood and internalized by each employee and contractor:

- Recognition of risks by itself does not result in risk reduction and effective risk management;
- Risks are not static, but dynamic and need continuous and consistent management;
- Not all risks are the same, thus require differentiated risk management;
- Past performance in managing risks does not define future performance; and
- Risk perception varies among individuals (contractors and employees) resulting in varying risk tolerance requiring the need to sensitize everyone to view a given risk consistently.

Cultural traits listed below often lead to risks being overlooked and ineffectively and inconsistently managed in an organization:

- Lack of openness to ideas, opinions and perspectives due to directive & hierarchal culture and lack of "what could go wrong?" and high risk tolerance mindset;
- Inability to foster conviction and commitment to make risk based decisions and help individuals understand the benefit of doing so ("what is in it for them?");
- Silo mentality resulting in risk ownership being limited to their respective area of work or activities;
- Communication rituals & channels being one way monologues; Not recognizing the importance of two-way risk communication internally in raising risk awareness and instilling risk ownership at all levels of organization in the process;

- Race against time (everything is priority, everything is urgent) driving a behavior of ad-hoc quick fixes without understanding of unintended consequences; and
- Lack of ability (will & skill) to say "Positive NO" and challenge at-risk decisions at peer and leadership levels.

Risk Recognition

Process Hazard Analysis (PHA) is a framework for identifying the process hazards, hazardous events and their consequences, evaluating the adequacy of the existing safeguards and determining the need for additional safeguards to reduce the risks to an acceptable/tolerable level of an organization. PHA is often carried out during the design stage by most of the organization and is often limited to HAZOP and SIL assessment studies, in addition to studies required by regulations. Key observations and gaps found based on client experience in diverse sectors are:

1. Initial PHA done by project team without involving operations, resulting in an incomplete assessment;
2. Lack of risk assessment or use of different risk matrices that is not aligned with the organization risk tolerance level without understanding implications;
3. Process Technology or Process Safety Information not handed over by the project team to the operations team or the information is not kept up to date;
4. HAZOP considered as the beginning and end of PHA and lack of augmenting with appropriate quantitative studies, especially for high consequence and low frequency events;
5. PHA outputs not considered in designing the remaining PSM elements;
6. PHA not done periodically or when there are changes in process and/or equipment, resulting in increase of potential unmitigated risks;
7. Lack of Rigor in implementing PHA recommendations;
8. PHA conducted without consideration to internal and external incidents; and
9. Not considering PHA in the early stages of project design.

Similarly, task based risk assessments which are carried out for non-routine activities and as part of permit to work systems are found to be inadequate due to following reasons:

- Viewed as paper exercise and often copied from a previous assessment without detailed review for its validity, correctness and relevance for a given task;
- Assessments are carried from a compliance mindset without giving regards to WHY it is important;
- Not reviewing the tasks with an open mindset and linking to potential consequences beyond the work areas in case of deviation. This limits the severity to the impact on the work crew or individuals directly involved in the task;
- Risk assessments are not communicated effectively (often, one way) to work crew and is assumed that they are familiar with and have understood them;

- Lack of clarity on risk ownership and roles as to what needs to be done, who will do it and how it will be done; and
- Poor risk assurance to ensure that all risks have been identified and barriers are functional and effective during the execution.

Risk Identification

Process Hazard Analysis is often limited to HAZOP and SIL assessment studies in many organizations and jurisdictions. Quantitative Risk assessment and Safety case is developed to satisfy regulatory requirements to obtain the license to operate. The most common method applied in the industry for hazard evaluation is HAZOP.

Based on a review of HAZOP studies in different industry sectors and regions, the quality of HAZOP needs to be improved in terms of application of methodology, level of details and accuracy and completeness based on the below or at organization risk tolerance level. Observations are listed below:

- HAZOP team discussions are mostly limited to few standard parameters.
- "All the causes for each deviation" as well as "all the consequences for each documented cause" are not identified and documented;
- Ultimate consequences, such as impact on people (including occupied plant buildings), environment and business, are not consistently identified resulting in underestimation of consequence severity and risk and in inadequate identification of protection layers;
- Lack of application of organizational risk matrix consistently resulting in identification of protection layers based on team experience;
- Lack of availability and provision of required process safety information to HAZOP team members to enable them to review and provide inputs;
- Not covering non-routine operations such as start-up, shutdown, catalyst change over isolation of equipment for handover to maintenance, etc.;
- HAZOP reviews do not comprehensively capture both the "hazards" and "operability" issues pertaining to process. In some studies, focus is more on "operability" issues, while some are more focused on "hazards".
- Validation of effectiveness, adequacy and independence of layer of protection for which credit is given during qualitative risk assessment. For example,
    - Availability of sufficient time for the operator to intervene upon alarms (e.g. LAH, LAL, TAH, etc.) and specific actions to be undertaken including effectiveness of the action in preventing the consequence; and
    - Review of Safety Valves Sizing Cases (e.g. verification that all overpressure scenarios identified during the HAZOP have been adequately covered in the PSV Datasheet).

Risk Communication & Risk Awareness

Most often, risk awareness among employees is primarily driven by their experience and individual perceptions. Terms such as "Hazard", "Risk", and "Consequence" are often misunderstood and used incorrectly by individuals at different levels of the organization.

Risk communication internally is predominantly absent and ineffective in most organizations. This is reflected in the high degree of variance in risk awareness. Communication is often limited to written email communication or posters and primarily focus on What needs to be done and to some extent on the How. It does not cover Why and How Well it needs to be done, and what are the benefits and consequences of not doing it well.

Absence of two-way communication and ineffective employee engagement results in lack of risk ownership and accountability. Often communication is assumed to have happened in the following areas:

- Policies, principles, standards, procedures, work instructions across the organization;
- Condition of the equipment, unavailability, impairment or bypass of safety critical equipment;
- Changes or modifications to the facilities, personnel, procedures and impact;
- Hazards, consequences, risks and barriers to manage them as part of permit to work system;
- One way tool box sessions and sharing of incident summary and lessons learned;
- Shift change communication;
- Sharing of good practices and lessons learned limited to their respective department, function, unit and sites;

A robust mechanism to assess the quality and effectiveness of risk communication and an attempt to measure is non-existent in most cases. Thus, it is difficult to ensure given risks are understood and will get the same level of treatment and attention across all functions and levels within the organization and that resources are allocated appropriate to the risk level.

CONCLUSION

Technical studies and process safety management systems alone will not help in achieving effective risk management. The organization must embed risk thinking culture at all levels and ensure risks are communicated effectively and seamlessly. This will help manage risks and enable individuals to take ownership for the risks and hold each other accountable for the risks that are prevalent in the organization.

479315 Last Line of Defense - Improving Our Step Back 5x5

Abstract:

Last Line of Defense - Improving Our Step Back 5x5

Step Back 5x5 is our last layer of defense and perhaps our most important. This paper shows how we improved our Step Back 5x5 risk analysis.

Our safety communication has 4 layers:

1. Procedure
2. Permit to Work
3. Job Hazard Analysis
4. Step Back 5x5

Step Back 5x5 is a risk assessment done just before the work begins. Step Back 5x5 is completed by the team leader and their technicians at the job site.

The original intent of Step Back 5x5 was an "informal" risk assessment asking employees to:

- take 5 steps back
- take 5 minutes
- and think about new hazards in the moments before you begin working

Our Step Back 5x5 was taking significantly longer than the intended five minutes. It was dominated by risk questions already asked and answered in other places, such as on a permit or Job Hazard Analysis.

In some cases, following an incident, we would add another question or check box to our Step Back 5x5. Consequently, over time, it grew to 102 questions, 680 words, filling both sides of an A4 page at 10-point type. It was becoming overly bureaucratic and this is not what we wanted.

We formed a working group to make a better Step Back 5x5. Our new Step Back 5x5 has:

- 48 questions (53% reduction)
- 123 words (82% reduction)
- completion time of 10 minutes (50% reduction)
- a graphic orientation - using pictures and graphics instead of words

Most important, we are not asking the same questions covered in the Procedures, Permit to Work, or JHA. We focused our new Step Back 5x5 on a unique task: looking for *new* hazards at the *physical work site* not covered in our other safety documents. We want them on their feet looking at the worksite–not huddled over a questionnaire ticking boxes.

Our paper will show the original Step Back 5x5 and our new version. We are field testing our new Step Back 5x5 now, and we can include these results in our paper END

A Statistical Approach for Using Test and Inspection Results to Adjust Maintenance Strategies

William M. Bradshaw, ABSG Consulting Inc., 10301 Technology Drive, Knoxville, TN, 37932 (865) 671-5829, WMBradshaw@ABSConsulting.com

Jan Wagner, ABSG Consulting Inc., 10301 Technology Drive, and Oklahoma State University (emeritus), (405) 372-8963, jan.wagner@okstate.edu

Prepared for Presentation at
American Institute of Chemical Engineers
2017 Spring Meeting and 13th Global Congress on Process Safety
San Antonio, TX
March 26 – 29, 2017

A Statistical Approach for Using Test and Inspection Results to Adjust Maintenance Strategies

William M. Bradshaw, ABSG Consulting Inc., 10301 Technology Drive, Knoxville, TN, 37932 (865) 671-5829, WMBradshaw@ABSConsulting.com

Jan Wagner, ABSG Consulting Inc., 10301 Technology Drive, and Oklahoma State University (emeritus), (405) 372-8963, jan.wagner@okstate.edu

Keywords: Mechanical integrity; inspection, testing, and preventive maintenance (ITPM); Bayesian; statistical approach

Abstract

Standards, such as American Petroleum Institute (API) standards, provide widely accepted methods and intervals for inspecting fixed equipment such as pressure vessels, piping, and storage tanks. Test methods and test intervals for safety instrumented systems are typically established during the detailed design phase. However, there is little or no guidance regarding inspection, testing, and preventive maintenance (ITPM) activities and intervals for a broad range of equipment that serves a process safety function.

Over the years, most plants have identified ITPM activities and established intervals based on factors such as (1) turnaround intervals, (2) auditors' opinions, (3) historical practices, (4) risk assessments, (5) manufacturers' recommendations, and so forth. Most facilities have not used ITPM results and in-service failure data to determine whether (1) ITPM intervals can be increased while meeting reliability requirements and (2) test and in-service failures of a safety system indicate inadequate performance or are simply the result of statistical noise.

In this paper, we present a statistical method for evaluating the likelihood of an in-service failure given (1) a failure rate distribution, (2) the facility's maximum allowable failure rate (e.g., a risk reduction factor used in a layer of protection analysis), and (3) the user's willingness to get the wrong answer. In addition, we provide a statistically valid method to use ITPM and in-service failure rate data to determine the likelihood that reliability requirements are being met.

1. Introduction

For decades, facilities in the process industries have used time-based test and inspection intervals specified in standards such as the American Petroleum Institute's (API's) standard for inspection of pressure vessels, API 510. More recently, the API has introduced its risk-based approach to fixed equipment inspection, recognizing that in some situations more frequent or enhanced inspections are needed, and in others the intervals may be extended or less intrusive inspection practices may be appropriate (e.g., on-stream rather than an internal visual inspection).

More recent standards, such as the International Society of Automation's (ISA's) ISA 84.01-2004, which applies to safety instrumented systems (SISs) and safety instrumented functions (SIFs), require consideration of maintenance and testing methods and intervals during the design phase [1]. The design and proposed maintenance practices are analyzed to determine if the system meets a target reliability criterion. This becomes an iterative process until an acceptable combination of design and maintenance practices is identified. In practice, tests of SIFs are one of the few instances where the combination of system design and maintenance is considered together at the beginning of the life cycle, with the objective of achieving a specified reliability criterion (e.g., a maximum probability of failure on demand).

In many cases, test and inspection intervals are left to the discretion of facilities. In some cases, manufacturers suggest maintenance methods and intervals, but these tend to be more oriented toward basic maintenance than performance testing. Examples where manufacturers often do not specify test intervals include (1) removal, testing, and reinstallation of pressure safety valves (PSVs), (2) function testing of instrumented systems that are not part of a SIF, (3) condition monitoring activities for pumps and other rotating hardware, (4) maintenance of electrical systems and other utilities, and (5) testing of safety systems such as water curtains or steam snuffing systems. Test and maintenance intervals for these types of systems are often based on turnaround intervals or some other rule-of-thumb. In other cases, these activities and intervals are based on historical plant practices, which are often influenced by major incidents that may have occurred at the facility, within the company, or within the industry. Sometimes, test intervals are influenced by, or based on, insurance company or safety auditor recommendations. After a period of time, test intervals begin to determine turnaround intervals, and the logic becomes very circular.

It is relatively rare that test and inspection intervals are based on a target reliability. When properly used, statistics can predict characteristics of a population based on sample data that represent the entire population. These methods can help answer questions such as "Based on the results we have observed during testing, how confident are we that this safety system will perform as designed at least 99% of the time a demand is placed on it?" Or we might want to know "If we performed the same maintenance activities but extended the interval by a factor of two (or some other number), what will that do to the probability of failure on demand?" We might ask both questions at the same time. For example, "How confident are we that we could achieve a target reliability number if we made some change to our testing program?"

This paper (1) examines several different approaches and statistical methods that can help answer these types of questions, (2) provides three worked examples using the methods that seem most promising, and (3) comments on the pros and cons of the various methods presented. Readers with basic knowledge of statistics should be able to use the discussion, along with the worked examples, to better understand the actual reliability of safety systems and similar equipment at their facility, and have some sense of how changes to maintenance activities or maintenance intervals might affect the reliability of these systems.

2. Background

One of the basic questions when applying statistics to predict attributes of a population is what distribution functions best represent the population. The likelihood of an event can be described in a curve known as a probability density function. Probability density functions are curves that represent the probability of an event or a range of events. For example, there are 11 possible outcomes for the sum of rolling two six-sided dice (2 through 12), and the likelihood of rolling a 7 is greater than rolling a 2 or a 12. We intuitively know this, and the math to support this statement is covered in the discussion of the uniform probability density function and the binomial probability density function in this section. Table 1 summarizes the six probability density functions and Table 2 summarizes three other statistical methods that were investigated or considered, along with a brief summary of how each may apply and a summary of the limitations and key assumptions.

Table 1. Summary of Probability Density Functions Investigated or Considered

Probability Density Function	Possible Applications	Limitations and Key Assumptions
Bathtub curve	Classic model for reliability analysis for failures of electro/mechanical systems such as rotating hardware or control systems	Systems do not always follow this model, and it is mathematically complex Requires user to estimate at least two shape parameters
Uniform	Simple approach for predicting individual failure rates due to random failures/events such as: • Interlock tests • Pressure safety valve tests • Fire deluge system tests	Ignores infant and wear-out failures Only applicable to systems where random failures dominate Requires users to estimate a single parameter – in this case the probability of failure in a single test or demand
Binomial	A means to sum discrete data from multiple tests; for example, a means to evaluate overall performance of a class of assets that are tested in a manner that yields pass/fail results (see the discussion for the uniform distribution for a list of these types of systems)	Same as for the uniform probability density function
Exponential	A means to evaluate failure rates, given initiating event rates and failure probabilities	Requires users to estimate a rate for a demand (on a safety system) and the probability of failure of one or more protection layers Same assumptions as for the uniform probability density function
Normal	A means to model confidence intervals for various test results	Requires relatively large data sets
Poisson	Another classical means to estimate and evaluate discrete test results	No advantages over the binomial probability density function in this case

Table 2. Summary of Other Methods Investigated or Considered

Method	Possible Applications and Limitations
Bayesian methods	A means to predict failure probabilities given a previous estimate and test data Requires a previous estimate and test data
Lot acceptance testing	A means to characterize a population based on sample testing This method requires logical lots, such as all production from a certain line in a specified time period. There is no reason to believe that equipment test results are connected in the same manner as properties of lots of products
Statistical process control charting	A means to categorize a population mean and variance and identify statistically significant changes in these properties over time There is no reason to assume that test results will trend in a certain direction over time

A probability density function can be for a single event or a series of events. For example, the probability density function for rolling a single die would be a flat line, with the numbers 1 through 6 on the x axis and the probability of each outcome being 1/6. The probability density function for rolling two dice at once is a curve. The potential outcomes are 2 through 12, with the maximum at 7 (1/6) and minimums at 2 and 12 (each at 1/36). The cumulative distribution function is the integral of the probability density function. It can be used to estimate the likelihood that an event or group of events will occur over a specified interval of time or a range of individual tries/tests. The following paragraphs discuss several distributions that might describe failure rate data.

The "**bathtub curve,**" which is shown in Figure 1, is the probability density function that is often used to represent failure probability as a function of time. There is a higher than average likelihood of failure when something is first placed in service (sometimes called infant mortality). This is followed by a period where failures are random, relatively rare, and evenly distributed over time. At some point, "wear out" becomes the dominant failure mechanism and the failure rate begins to rise. The advantages of using this probability density function are that it (1) represents failure rates for many systems/equipment items and (2) fits our experience of how they fail over time. The disadvantage is that it requires users to specify at least two parameters, even though in most practical applications it is difficult to know even one parameter with high certainty.

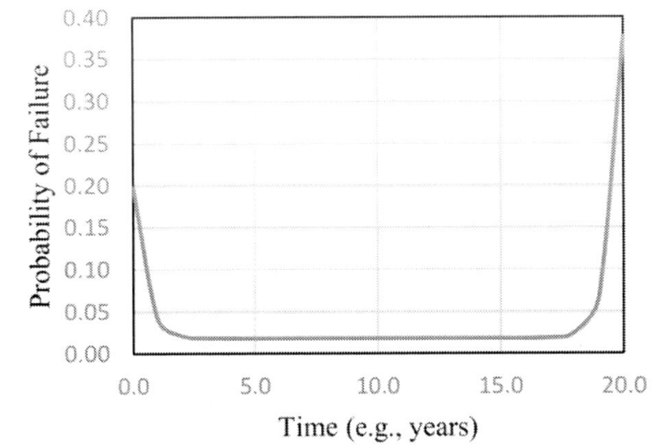

Figure 1. "Bathtub Curve"

In this paper, the **uniform probability density function** is used to estimate the probability of failure in a single test. The *failure probability* does not change from one test to the next, given that the previous test was successful (or the system was repaired if the previous test failed). Consider a low-level interlock that is tested annually and is designed to protect against pump seal failure due to low level in the vessel that supplies the pump. One measure of the probability of failure on demand would be the number of tests where the interlock fails divided by the total number of tests that are performed over a period of time.

The uniform distribution is shown in Figure 2. Note that it matches the middle section of the bathtub curve. Therefore, use of the uniform distribution to model failure rates assumes (1) an effective quality assurance testing program is in place to protect against early life failures and (2) replacement or reconditioning of equipment when condition monitoring activities detect the onset of wear-out. A third assumption that underlies the choice to use the uniform distribution is that time-based degradation, such as corrosion or fouling, is either (1) readily detectable and quickly

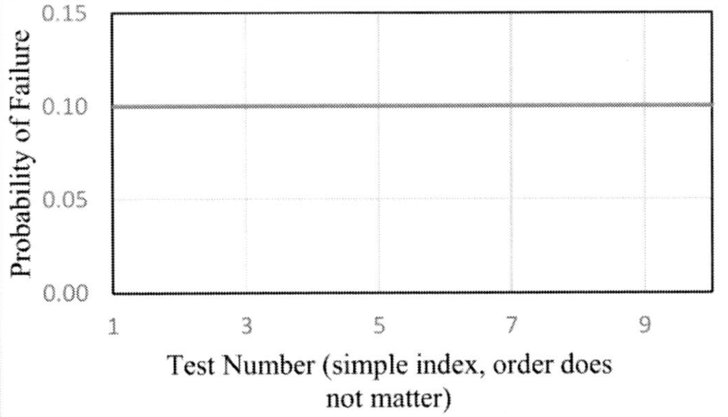

Figure 2. Uniform Probability Density Function

repaired or (2) negligible. If these assumptions are not true, the methods presented in this paper may not apply.

A common example of the uniform distribution is flipping a coin. There is a 50% chance that a fair coin will come up heads. When evaluating the results of multiple coin flips, we obtain a **binomial probability density function**. For example, if we flip a coin 10 times, there are 11 possible outcomes: 0 heads, 1 head, 2 heads, etc., up to 10 heads. The likelihood of exactly zero or exactly 10 heads in 10 flips is quite small (about 1 in 1,000 for either outcome) while the likelihood of exactly 5 heads is 0.25. This distribution is shown in Figure 3. The distribution does not have to be symmetric. For example, if the probability of one outcome is 1 in 6 (with the other outcome being 5 in 6) one could model the results of a series of tests by rolling a 6-sided die with the 1 representing outcome A and 2 through 6 representing outcome B. The probability density function for result A (i.e., probability of N rolls coming up "1" in 10 tries) is again a binomial distribution, but skewed to the left as shown in Figure 4.

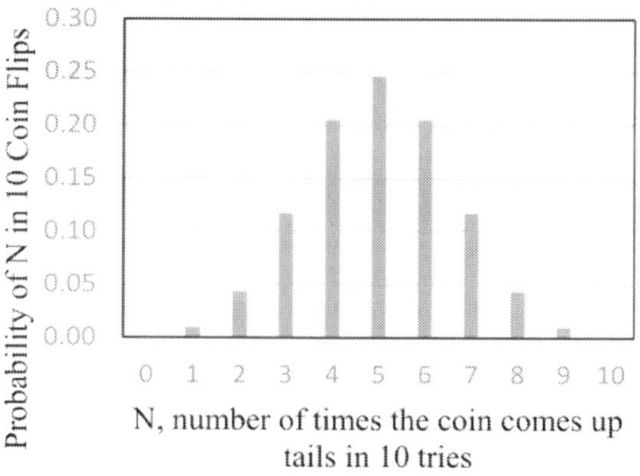

Figure 3. Binomial Probability Density Function, X=0.5 (e.g., a coin flip)

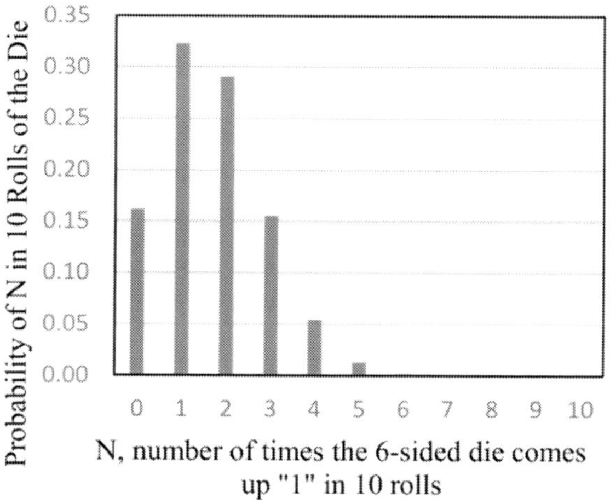

Figure 4. Binomial Probability Density Function, X=0.1667 (e.g., rolling a 6-sided die)

For highly reliable equipment commonly found in the process industries where the failure rate is often below 0.1/year for much of the useful life of a system or equipment, the likelihood of failure is very close to a flat line. Intuitively we know that the reliability of two PSVs that are similar in all respects other than age is nearly the same, assuming the same service and maintenance/testing programs.

Failure rate is the probability of failure per unit time, typically per hour or per year. For the low-level interlock described above, the failure rate is the likelihood the pump seal fails due to low level in the upstream vessel, which is the likelihood that low level occurs times the failure probability for the interlock at that time. Introducing the demand rate (probability of low level in the vessel), requires the use of the **exponential distribution**, which is shown in Figure 5. If the failure rate is λ/year, the probability density function is $\lambda e^{-\lambda t}$ and the cumulative distribution is $F(t;\lambda) = 1 - e^{-\lambda t}$. In other words, the likelihood of failure between time 0 and time T is $1 - e^{-\lambda T}$. The cumulative distribution is shown in Figure 6.

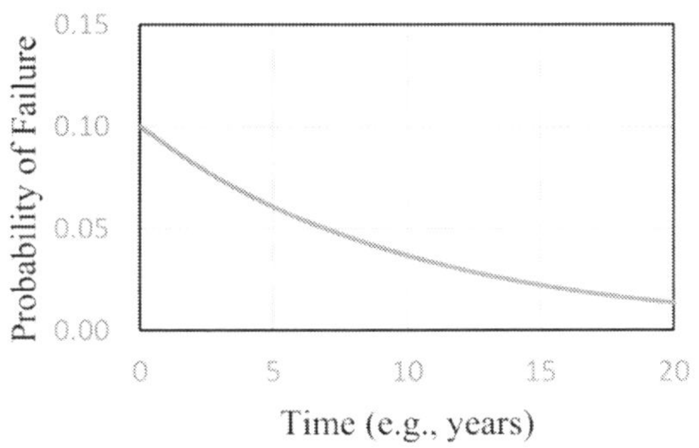

Figure 5. Exponential Probability Density Function, λ=0.1/yr

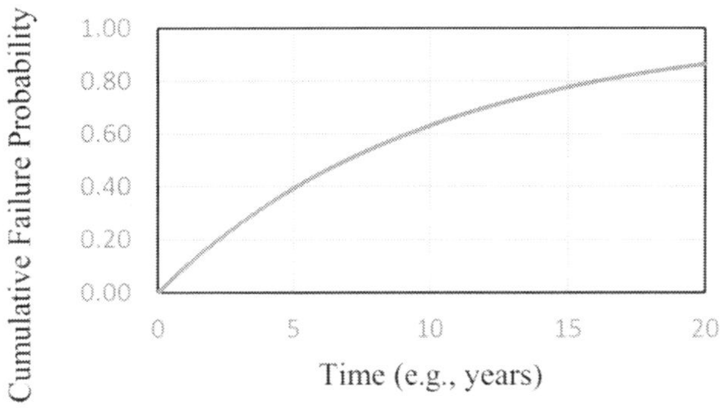

Figure 6. Cumulative Distribution Function Assuming an Exponential Probability Density Function with λ=0.1/yr

There are several other commonly used distributions. The **normal distribution** (i.e., the "bell curve") can be used to estimate a failure rate, particularly if there are a large number of tests. The normal distribution, and its cousin, the Student-t distribution, are used in this paper to determine confidence intervals for failure probabilities and failure rates, and the results will be compared to what would be obtained using the chi-square distribution.

Bayesian methods can also be used to provide a more accurate estimate of a mean and standard deviation. The method requires an estimate of the likelihood of failure for a particular system/equipment (given service conditions and a testing/maintenance strategy) and the variance in the failure rate. The Bayesian method combines this preliminary estimate, which in this case might be based on literature data for the system/equipment under consideration, with the test data collected over time to arrive at the Bayesian estimate of the failure rate and standard deviation of the failure rate. In other words, real-world data are used to adjust generic failure data (from literature data or our initial "best guess") to produce a more accurate, statistically valid estimate.

One other commonly used probability density function, the **Poisson distribution**, could potentially be applied to the problems presented in this paper. It is typically used to model the number of times a particular outcome occurs in a given number of tries. For example, a major league baseball team plays 162 games in a season, and some games are postponed due to rain (i.e., a "rainout"). In a given season, a team could experience between 0 and 162 rainouts, and the Poisson distribution can be used to predict the likelihood of a specific number of rainouts, or a range (e.g., what is the probability of exactly one rainout, or what is the probability of three to seven rainouts?). Likewise if N tests were scheduled during a specific interval (say a quarter, year, or during a turnaround), the number test failures can range from 0 to N, and the Poisson distribution could be used to model the likelihood of a specified number of failures in N tests (say 0 to $0.1 \times N$, rounded to the nearest whole number). The parameters of the Poisson distribution could be used to determine the confidence that the population average is below some specified failure rate. However, for large data sets with a low failure probability, the Poisson distribution yields the same results as the uniform distribution.

Two other approaches commonly used in quality control applications were considered and ultimately rejected: **lot acceptance testing** and **statistical process control charting**. Given a willingness to "get the wrong answer," lot acceptance testing can be used to determine if the items in a production lot meet a certain quality criterion. Lot acceptance testing can be (1) destructive, such as firing ammunition in a standard manner to determine if the round lands within a specified distance of a target, or (2) nondestructive, such as measuring an item using a "go/no-go" feeler gauge. This method can also be used for ongoing surveillance, such as testing to determine if a certain attribute of material in inventory meets a criterion after it has been in storage for 5 years. It seems that this approach might be applied to a plant where PSVs can be removed and tested with equipment in service. For example, if a group of similar PSVs is removed every 5 years (i.e., resulting in a sample each year that represents ~20% of the total population), one might infer something about the condition of all PSVs based on testing a "20% sample" each year. However, over time the sample size will be 100% of the lot because the valves are not randomly selected for testing. Each PSV is scheduled to be tested every 5 years. This is not random sampling, and therefore the lot acceptance testing approach is not valid. The statistical process control approach attempts to detect changes in small lots of samples taken sequentially from a process. This method can be very effective in detecting otherwise hard-to-detect process changes and avoiding reacting to random errors or other random effects in a particular sample or measurement. However, it was rejected because there is no reason to believe that mechanical integrity test results will drift over time in the same manner that a physical property, such as the concentration of a particular impurity, might vary in samples taken from the condensate tank for a particular distillation column. Assuming proper repair, maintenance test results are seldom related to the previous result and can, therefore, be treated as independent.

Finally, we should note that the problems being examined in this paper are not new or unsolved. For example, the commercial nuclear industry has conducted extensive research in the use of probabilistic risk assessment and has developed methods to answer the questions posed in this paper [2]. However, the methods used for the nuclear industry are very complex and would likely be unsuited to most facilities in the process industries.

3. Example Problems

The remainder of this paper focuses on three example problems that are commonly encountered in the process industries.

Example 1: When performing layer of protection analyses (LOPAs), a continuous plant uses a generic probability of failure on demand of 0.01 for PSVs in a particular service. For the scenario under consideration, loss of cooling water is the initiating event with a frequency of 0.1/year.

All of the PSVs are proof tested every 2 years. When proof testing a PSV prior to reconditioning, the ratio of the test pressure (*TP*) at which the valve opened is measured and compared to the valve set pressure (*SP*). If the ratio, $R_p = TP/SP \geq 1.5$, API RP 576 considers the valve to be "stuck shut," meaning the valve would fail to open and relieve excess pressure. (Note that $R_p \geq 1.3$ is designated as a failed test by ASME PCC-3-2007 and API RP 581.) There has only been one failure in the 300 valves tested during the past 5 years. No significant time-based failure modes such as fouling or corrosion have been postulated or observed.

(a) How confident are we that the target probability of failure on demand of 0.01 or lower is being achieved?
(b) What is the impact of increasing the test interval for these valves to every 3 years?

Solution:

(a) The observation that there are no time-related failure modes implies that some random process condition or mechanical failure dominates the failure of this set of PSVs. Therefore, we can use the binomial probability distribution to estimate what we should expect to observe when testing this set of PSVs. The probability density function for multiple independent tests is: [3]

$$P(x) = \frac{n!}{x!(n-x)!} p^x q^{(n-x)}$$

where

p = probability of failure,

q = probability of success = $(1\text{-}p)$,

n = number of tests, and

x = random variable representing the number of test failures.

The cumulative distribution function is:

$$F(x; p, n) = \sum_{i=0}^{x} \frac{n!}{i!(n-i)!} p^i q^{(n-i)}$$

If α represents the fraction of time we are willing to "get the wrong answer" (in decimal format), we can (1) construct a one-sided confidence interval at the $100(1\text{-}\alpha)\%$ confidence level for the true value of the upper limit of probability of failure, p_{ul}, given that we have observed 1 failure in 300 proof tests and (2) compare this value to the target value of $p \leq 0.01$. There is no closed form solution, but we can use numerical techniques to solve the equation [1 (7.2.4.1)]

$$F(1; p_{ul}, 300) - \alpha = 0$$

For $\alpha = 0.20$,

$$\sum_{i=0}^{1} \frac{300!}{i!(300-i)!} p^i (1 - p_{ul})^{(300-i)} - 0.20 = 0$$

and

$p_{ul} = 0.0099$, or very close to 0.01

For $\alpha = 0.10$, $p_{ul} = 0.0129$

This solution is shown graphically in Figures 7 ($\alpha = 0.20$) and 8 ($\alpha = 0.10$) by plotting the cumulative distribution function with

$p = 1/300$

$q = 299/300$

$n = 300$

and determining the point where the area under the curve to the left of 1 failure in 300 tests is $1\text{-}\alpha$.

The target of $p \leq 0.01$ is being achieved at the 80% confidence level, but not at the 90% confidence level based on the results of the 300 proof tests conducted to date.

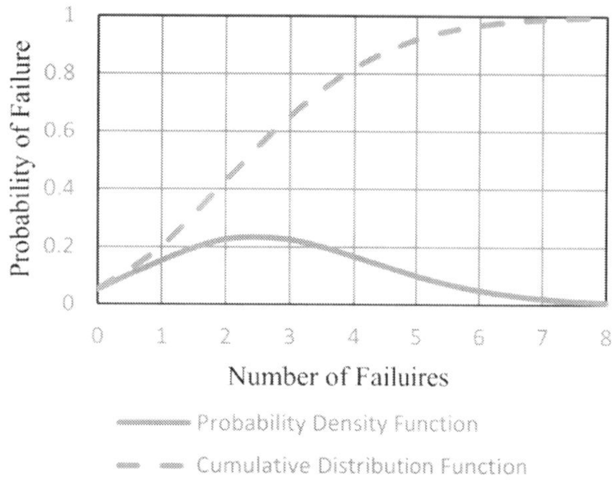

Figure 7. Probability Density Function and Cumulative Distribution Function (failure probability = 0.0099, 300 tests)

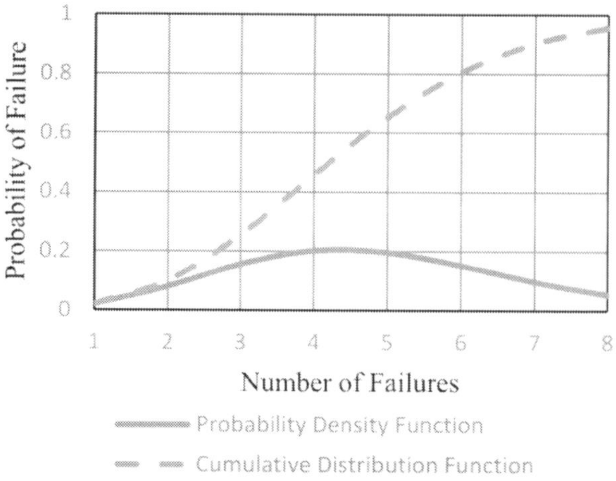

Figure 8. Probability Density Function and Cumulative Distribution Function (failure probability = 0.0129, 300 tests)

The failure rate, λ, is the product of the initiating event frequency (0.1/year for loss of cooling water) and the probability of failure on demand. Therefore, the likelihood that the cooling water will fail and the PSV is "stuck shut" is

$$\lambda = \left(\frac{0.1 \text{ demand}}{\text{year}}\right)\left(\frac{0.01 \text{ failure}}{\text{demand}}\right) = \frac{0.001 \text{ failure}}{\text{year}}$$

Since the failure rate is constant, the exponential distribution can be used to model the time to failure. The probability density and cumulative distributions are [3]

$$f(t;\lambda) = \lambda e^{-\lambda t} \qquad \text{(probability density function)}$$

and by integrating the probability density function we obtain:

$$F(t;\lambda) = 1 - e^{-\lambda t} \text{ (cumulative distribution function)}$$

The cumulative probability of failure for 2-year and 3-year test intervals is shown in Figure 9.

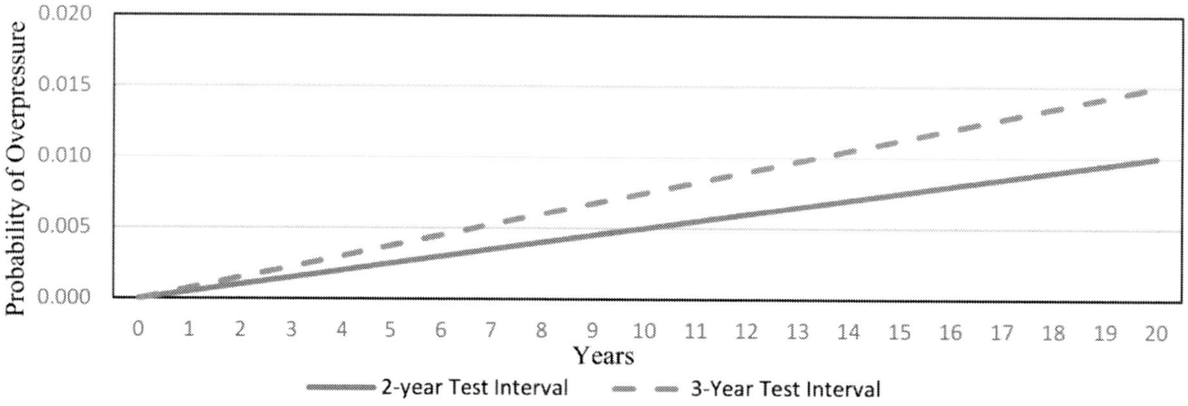

Figure 9. Cumulative Probability of Vessel Overpressure Assuming 0.1 Initiating Events/Year

Changing the proof test interval from 2 years to 3 years causes the probability of vessel overpressure to increase from 0.01 in 20 years to 0.015 in 20 years (assuming the initiating event frequency is 0.1/yr and there are no other safeguards). The probability of failure can also be expressed as the area under the "saw tooth" curves shown in Figure 10, which is $\lambda T/2$, where T is the test interval (in years) and λ is the failure rate of 0.001/year. (While this is a useful model, it should be noted that the probability of failure is never zero and it doesn't ever return to zero, even after each proof test. There is always some baseline of random failures. For example, failure to open a gate valve on the inlet or discharge line for a PSV could render the PSV useless until the error is detected and rectified.) The more general solution to this problem is that the failure rate increases linearly with the test interval, but this is only true if all failures of concern are random (i.e., infant and wearout failures can be ignored and there is no credible time-based failure mode such as fouling or corrosion).

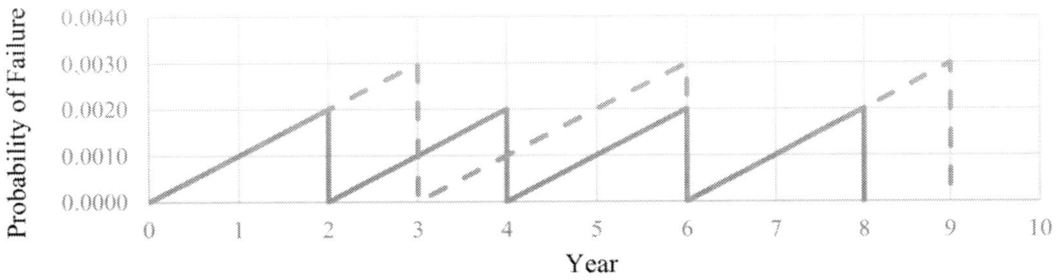

Figure 10. Cumulative Probability of Failure for Proof Test Intervals of 2 and 3 Years

Example 2: A batch chemical plant typically fills and empties 50 mix tanks several times each day. The filling, mixing, and draining steps are controlled by the distributed control system (DCS), and the standard approach is for the DCS to stop the draining step when the tank level reaches 2% on the level transmitter and proceed to make up the next blend. To help prevent a pump seal failure, each tank is fitted with a low level switch in the discharge line with a hard-wired interlock to stop the discharge pump and close the drain valve on the bottom of the tank (and sound an alarm on the DCS) if there is no level in the discharge line for 30 seconds. An in-service failure of this low-level interlock that occurred approximately 10 years ago caused (1) significant quality issues, (2) equipment damage, and (3) a release of flammable process material. In response, the maintenance department began functional testing these low-level interlocks every 6 months. Although the test is relatively simple, it does take time and interrupts production operations. Therefore, the facility is interested in extending the test interval to 1 year. Over the past 10 years, the facility has not had any reported in-service failures of these level switches (i.e., failure to stop a transfer when the tank reaches 2% level concurrent with a low-level interlock failure). However, there have been two low-level interlock test failures observed in slightly over 1,000 tests. What is the expected reliability of these low-level interlocks given the test data that have been accumulated? Assume that no time-based failure mechanisms have been postulated or discovered (i.e., the in-service failure and the two test failures in the past 10 years were determined to be caused by random electromechanical failures or human error).

Solution:

There were 2 failures out of 1000 tests. The maximum likelihood estimate of the proportion of p failures is

$$p^* = \frac{n_f}{n}, \text{ where } n_f \text{ is the number of test failures and } n \text{ is the total number of tests}$$

$$p^* = \frac{2}{1000} = 0.002 \text{ failures per test}$$

This value is well below published failure rates for simple interlocks, which tend to be closer to 0.1. While there seem to be sufficient data to predict that these low-level interlocks fail less than 10% of the time (the observed rate is 0.2%), it is unclear if this low failure rate is the actual failure rate in this service, a random occurrence (i.e., even if the failure rate is 10% there is a slight chance we might observe 2 failures in 1,000 tests), or if the true failure rate is somewhere between 0.002 and 0.1. The Bayesian approach provides a method to (1) make sensible estimates of reliability in cases where few or no failures are observed and (2) update reliability assessments as additional information and data become available [3].

Estimates of the published probability of failure on demand (PFD) for the safety integrity level (SIL) 1 interlocks, which are often simple interlocks such as the one under consideration in this example, range from 0.02 to 0.1 [4]. To obtain a preliminary estimate of the mean and standard deviation of this random variable, assume that the failure rate distribution has an upper tail such that there is less than a 1% chance an interlock will fail more than 10% of the time and only a 5% chance the interlock will fail less than 2% of the time (i.e., designers tend to err on the side of safety; the resulting probability density function for test results is shown in Figure 11).

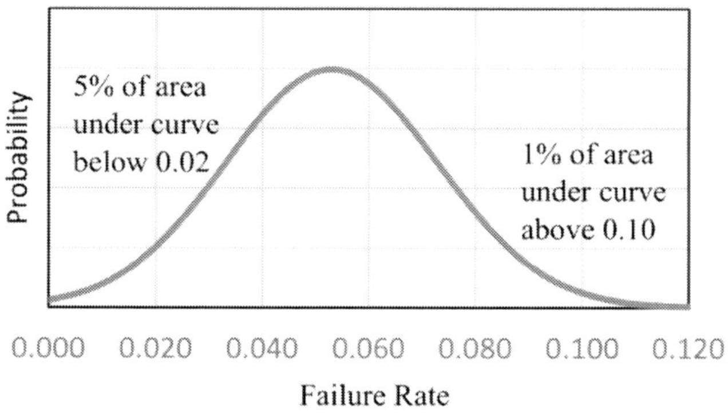

Figure 11. Probability Density Function for the Preliminary Estimate of Failure

The 5% critical value for the Student-t distribution (with 30 or more observations), $t_{0.95,\infty}$, is 1.645, and the 1% critical value for the Student-t distribution, $t_{0.99,\infty}$, is 2.326 (again, with more than 30 observations).

The location of the "lower 5% tail" of the probability density function is $0.02 = \mu_p - t_{0.95,\infty}\,\sigma_p$, or $0.02 = \mu_p - 1.645\,\sigma_p$

and the location of the "upper 1% tail" of the probability density function is $0.1 = \mu_p + t_{0.99,\infty}\,\sigma_p$, or $0.1 = \mu_p + 2.326\,\sigma_p$

Solving these equations, we obtain the following preliminary estimates for the mean and standard deviation of this random variable: $\mu_p = 0.053$ *and* $\sigma_p = 0.020$. These preliminary estimates are based on historical information, engineering judgment, and our assumption that designers "err on the safe side." As we will observe later, these preliminary assumptions are important but not determinant when estimating the actual failure rate.

A natural prior distribution on x (the probability density function for the preliminary estimate of the failure rate) is the beta distribution given by Blischke and Murthy [3]. Depending on the values for α and β, the beta distribution can take on several different forms, but is widely used to describe the probability density function for the preliminary estimates used for the Bayesian method.

$$P(x)=\frac{\Gamma(\alpha+\beta)}{\Gamma(\alpha)\Gamma(\beta)}x^{\alpha-1}(1-x)^{\beta-1}$$

where x is a random variable, α and β are shape parameters, and $\Gamma(x)$ = gamma function of x.

The mean and standard deviation are [3, 5]

$$\mu_p=\frac{\alpha}{\alpha+\beta}$$

and

$$\sigma_p=\sqrt{\frac{\alpha\beta}{(\alpha+\beta+1)(\alpha+\beta)^2}}$$

Using the preliminary estimates of the mean and standard deviation, we can solve for the parameters of the distribution as follows:

Estimate the value of α, $\widehat{\alpha}$.

Calculate the estimated value of $\widehat{\beta}=\widehat{\alpha}\left(\frac{1-\mu_p}{\mu_p}\right)$

Calculate $\widehat{\sigma}=\sqrt{\frac{\widehat{\alpha}\widehat{\beta}}{(\widehat{\alpha}+\widehat{\beta}+1)(\widehat{\alpha}+\widehat{\beta})^2}}$ and iterate until $|\widehat{\sigma}-\sigma_p|\leq$ tolerance

For $\widehat{\alpha}=6.597$, $\widehat{\beta}=117.9$, and $\widehat{\sigma}$=0.020.

These are the prior beta distribution parameters that will be used in the Bayesian update of the interlock reliability using the functional test results. The resulting Bayesian estimate is:

$$p_b=\frac{\alpha+n_f}{\alpha+\beta+n}$$

$$p_b=\frac{6.597+2}{6.597+117.9+1000}=0.0076$$

Based on test data showing 2 failures in 1,000 function tests, and the preliminary estimate of a failure probability of 53 failures in 1,000 tests, the resulting Bayesian estimate of the failure rate is just under 8 failures in 1,000 tests. There is significant movement away from the prior value of 0.053, primarily due to the relatively large number of tests (i.e., $\alpha+\beta+n$ approaches n as the number of tests goes up). [3]. As the number of tests increases, the preliminary estimate has less effect on the resulting Bayesian estimate.

Example 3: For the low-level interlock described in Example 2, what is the expected likelihood of failure on demand if the tests were performed annually instead of every 6 months? What is the one-sided 95% confidence interval for this number (i.e., the estimate of the failure probability that has a 95% chance of not being too low)?

Solution:

The failure rate, λ, is the product of the initiating event frequency and the probability of failure on demand. The estimated frequency of a BCPS failure based on IEC 61511 is $1\text{x}10^{-5}$/hr or 0.0876/year [6]. The estimated failure rate for the interlocks is

$$\lambda = \left(\frac{0.0876 \text{ demand}}{\text{year}}\right)\left(\frac{0.0076 \text{ failure}}{\text{demand}}\right) = \frac{0.0006 \text{ failure}}{\text{year}}$$

The exponential distribution can be used to model the time to failure for a constant failure rate. The cumulative probability distribution is

$$F(t; \lambda) = 1 - e^{-\lambda t}$$

The cumulative probability of failure at 6 months is

$$F(0.5; 0.0006/\text{yr}) = 1 - e^{-(0.0006)(0.5)} = 0.0003$$

and at 1 year

$$F(1.0; 0.0006/\text{yr}) = 1 - e^{-(0.0006)(1.0)} = 0.0006$$

Therefore, doubling the test interval increases the probability of pump seal failure due to low level in the tank supplying the pump from 0.0003/yr to 0.0006/yr.

The upper one-sided confidence interval for the failure rate of an exponential distribution is [3]

$$\lambda_{u} = \frac{\lambda \chi^2_{2n,1-\alpha}}{2n}$$

where $\chi^2_{2n,\alpha}$ is the value of chi-square for $2n$ degrees of freedom (where n is the number of tests) and the significance level is α. For $\alpha = 0.95$ and $n = 1000$, $\chi^2_{2(1000),0.05} = 2105.15$ [7], and

$$\lambda_{u} = \frac{0.0003(2105.15)}{2(1000)} = 0.00032/\text{yr}$$

For $n > 30$, the normal distribution can be used to estimate the confidence limits for the failure rate [3] and

$$\lambda_{u} = \lambda\left(1 + \frac{t_{\alpha,n}}{\sqrt{n}}\right)$$

where t is the Student-t value for n degrees of freedom. For $n = 1{,}000$ (i.e., 1,000 tests were performed) and $\alpha = 0.95$, $t_{0.95,\infty} = 1.645$, and

$$\lambda_{u} = 0.0003\left(1 + \frac{1.645}{\sqrt{1000}}\right) = 0.00032/\text{yr}$$

This is the same result using the chi-square distribution.

As can be seen, the upper limit for the 95% confidence interval is less than 10% greater than the failure rate. Given the uncertainty in other factors that may impact a decision based on this information, the uncertainty in the failure rate is insignificant. Note that the standard deviation, and hence the difference between the mean value and the 95% confidence interval for the upper limit of the failure rate, is linear with the failure rate, so increasing the test interval by a factor of two would result in a predicted failure rate of 0.0006/yr with 95% confidence that the actual failure rate is less than 0.00064/yr. Note also that the difference between the mean and the 95% upper limit for the actual failure rate is a strong function of the number of tests used to arrive at the failure rate and the failure rate itself. As would be expected, this number decreases as the failure rate goes down or the number of tests goes up.

4. Conclusions

The examples used in this paper were intentionally limited to large sets of test results with low failure probability. This is because the uncertainty with small sets is quite high, leading to high uncertainty in the results of any statistical analysis of the data. If there are very few observations (or in this case, tests), particularly if the intent is to achieve very low failure rates, these methods cannot be reliably used. However, this is not without precedent. For example, until recently a widely accepted standard for PSV testing allowed users to extend the test interval by 50% if they observed three consecutive successful "as-found" tests for a particular valve. Consider this: if the true probability of failure of a particular PSV that was being tested every 2 years was 0.1 (which would normally be considered to be unacceptably high), there is a 73% chance that this PSV would pass three consecutive tests (i.e., $0.9^3=0.729$). It seems imprudent to use such a simple approach to extend the test interval from 2 years to 3 years when there is a 3 in 4 chance that a PSV that fails 10% of the time would qualify for the extension. This clause has been removed from the standard. This paper demonstrates that uncertainty in failure rates exist even when there have been hundreds of tests. As shown in Example 1, a data set that includes only 1 failure in 300 tests indicates the failure rate is less than 1%, but this statement cannot be made with 90% confidence.

In Example 2, the test data showed very low likelihood of failure on demand, much lower than published industry data. However, test failures are random; and if they are rare, variability is expected. The Bayesian approach provides a statistically valid means to estimate a failure probability that was well below published data but approximately four times higher than the observed failure probability. In addition, the approach provides an estimate of the failure rate if the test interval was extended from 6 months to 1 year. Depending on the consequence of the predicted increase in the failure rate, it may be acceptable to test these interlocks once per year.

Given sufficient data, the methods described in this paper could be used to adjust nominal failure rates typically used in quantitative risk analyses. However, before doing so the risk analysis teams should carefully consider the assumptions that failures are random and that time-based early-life and wearout failures are negligible.

5. References

1. ANSI/ISA-84.00.01-2004, *Functional Safety: Safety Instrumented Systems for the Process Industries*, International Society of Automation, 2004.
2. *Guidance on the Treatment of Uncertainties Associated with PRAs in Risk-Informed Decision Making – Draft Report for Comment*, U.S. Nuclear Regulatory Commission, NUREG-1844, Rev. 1, March 2013.
3. *Reliability – Modeling, Prediction, and Optimization*, W. R. Blischke and D. N. P. Murthy, John Wiley & Sons, Inc., 2000.
4. *Layer of Protection Analysis, Simplified Process Risk Assessmen*t, Center for Chemical Process Safety, American Institute of Chemical Engineers, 2001.
5. *NIST/SEMATECH e-Handbook of Statistical Methods*, http://www.itl.nist.gov/div898/handbook/
6. IEC 61511, *Functional Safety – Safety Instrumented Systems for the Process Industry Sector*, International Electrotechnical Commission, 2016.
7. Free Statistics Calculators, Version 4.0, Calculator: Critical Chi-Square Value, http://www.danielsoper.com/statcalc/calculator.aspx?id=12 .

Using Data-Based Indicators to Improve Process Safety Performance

Erin P. Collins
JENSEN HUGHES
333 E. 46th Street, #11L
New York, NY 10017
ecollins@jensenhughes.com

Paul J. Amico
JENSEN HUGHES
111 Rockville Pike, Suite 550
Rockville, MD 20850-5109
pamico@jensenhughes.com

Prepared for Presentation at
American Institute of Chemical Engineers
2017 Spring Meeting and 13th Global Congress on Process Safety
San Antonio, TX
March 26 – 29, 2017

Using Data-Based Indicators to Improve Process Safety Performance

Erin P. Collins
JENSEN HUGHES
333 E. 46th Street, #11L
New York, NY 10017
ecollins@jensenhughes.com

Paul J. Amico
JENSEN HUGHES

Keywords: Measurement and Metrics, Performance Indicators, Process Safety Culture

Abstract

Performance indicators have proven to be a key to successful performance-based management. They establish clear goals for objectively measuring achievement of program objectives without constraining the performer with prescriptive requirements. This paper documents an assessment of safety performance indicators as a tool for measuring the safety performance of process facilities. A number of candidate performance indicators covering the broad range of safety aspects of facility operation are identified, including a blend of both leading and lagging indicators.

1 Introduction

Performance indicators (also referred to as performance metrics) have proven to be a key to successful performance-based management. They establish clear goals for objectively measuring achievement of program objectives without constraining the performer with prescriptive requirements, allowing the performer to take ownership of the approach to achieving those goals.

Even if a more prescriptive approach is taken for requirements, performance indicators (PIs) measure whether meeting those requirements is delivering the desired performance. PI programs vary greatly in scope. Initially used almost exclusively for measuring safety performance, they are now often used to measure all aspects of operational performance as part of continuous process improvement programs, such as Six Sigma.

The commercial nuclear power industry has long promoted the development and implementation of safety PIs. The World Association of Nuclear Operators (WANO), which was established in May 1989 after the Chernobyl accident, has been a pioneer in this area, publishing its first PI report in April 1991 [1]. The Institute for Nuclear Power Operations and the US Nuclear

Regulatory Commission (NRC) began identifying performance indicators in the 1985-1986 timeframe [2, 3] and have continued to refine and augment their set of PIs. The Electric Power Research Institute (EPRI) was also involved in the late 1980s in defining means for analyzing and trending nuclear plant safety issues [4] and has continued to evaluate performance indicators to this day, though primarily from the standpoint of organizational factors and human performance.

A series of public meetings was conducted by the NRC in 1998 and 1999 to develop a more objective process for assessing regulatory and safety performance at nuclear plants. The intent of this new process was to use risk-informed insights to focus on issues of safety significance. As a result of those refinements, guidelines for a PI program were issued by the Nuclear Energy Institute at the end of 2001 [5] "to provide the guidance necessary for power reactor licensees to collect and report the data elements that will be used to compute the Performance Indicators." The International Atomic Energy Agency (IAEA) has specifically stated: "It is generally agreed that a comprehensive set of safety performance indicators must be used in combination with other insights such as safety culture and human performance evaluations, inspections and audits, risk analysis, feedback of experience and other external review and self-assessment tools in order to have a complete safety management system."[6]

Initial guidance for the Performance Indicator program used by the US Department of Energy's (DOE) production reactors [7] at Hanford and Savannah River, was issued in 1992. This program has since been developed further by the Training Resources and Data Exchange (TRADE) special interest group for both the DOE and the Assistant Secretary for Defense Programs [8] into a handbook of techniques and tools for performance measurement. This effort was extended still further into a performance metrics manual [9] developed by Westinghouse Savannah River Company for the Energy Facility Contractors Group (EFCOG), which includes a data dictionary with suggested and required performance metrics for submittal to DOE Headquarters.

The application of performance indicators to the safe operation of nuclear power production reactors was initiated in October 1997 by the American Society for Quality (ASQ), leading to a document issued in 1999 [10] that recommended "the implementation...of selected basic leading PIs... in three cross-cutting issues [of] problem identification and resolution, safety conscious(ness in the) work environment, and human performance."

International efforts such as the Canadian National Energy Board with its Safety Performance Indicators Initiative, the United Kingdom Department for Transport Rail Performance Indicators, and the Safety Management System and Transport Safety Performance Indicators in Finland show that the PI concept is well accepted and implemented across a variety of industries. The Nuclear Energy Agency (NEA) Working Group on Inspection Practices (WGIP) of the Organisation for Economic Co-operation and Development (OECD) conducted a survey of international nuclear industry performance indicator programs in 1997 [11] and identified the following advantages and disadvantages of PI programs:

> *"Advantages - Several advantages are cited for a PI programme. In favour of a regulatory set of PIs is the aspect of data control, standardisation, and consistency of interpretation. Having a programme in place provides regulators*

with another means to assess, and to answer questions on performance comparison. PIs provide an additional view of operational performance, enhancing the regulatory body's ability to objectively recognise areas of poor or declining performance.

Disadvantages - Informal use of PIs can lead to wrong conclusions if thresholds are changed and not taken into account properly. Due to their ease of access, PIs are prone to use as a stand-alone means to evaluate and therefore can generate misinformation. They cannot be used for real safety assessment, evaluation, or as a direct tool to regulate due to their limited usefulness. PIs must be used with other tools that a regulator already has in place. Hints may be provided on weaknesses, but they do not give a uniquely clear indication of good or bad performance. The additional benefit of a formal PI programme may not be warranted if the regulatory body already has vigorous supervisory and in-depth evaluation practices in place. In addition to these views, one country noted that no disadvantages exist if the PI programme is used in an appropriate way."

It should be noted, though, that when the WGIP decided to prioritize the issues requiring further study (related to the production of meaningful guidance on effective methods to use in the evaluation of safety performance), the consensus of the Group was that performance indicators (PIs) and how assessments are combined were the two issues that should be addressed first.

The PI programs initiated in the late 1980s and early 1990s continue to this day, indicating that the funding is considered to be well spent and the indicators have proven to be effective and important as part of an industry's safety program.

2 Overview of Performance Indicators

Understanding the elements of a performance indicator program and how they fit together is crucial to the success of its development and implementation.

TRADE [8] provides an overview of the PI program development process, as shown in Figure 1 and through the description of each of the process steps, as follows:

1. Identify the process flow. This is the first and perhaps most important step. If your employees cannot agree on their process(es), how can they effectively measure them or utilize the output of what they have measured?

2. Identify the critical activity to be measured. The critical activity is that culminating activity where it makes the most sense to locate a sensor and define an individual performance measure within a process.

3. Establish performance goal(s) or standards. All performance measures should be tied to a predefined goal or standard, even if the goal is at first somewhat subjective. Having goals and standards is the only way to meaningfully interpret the results of your measurements and gauge the success of your management systems.

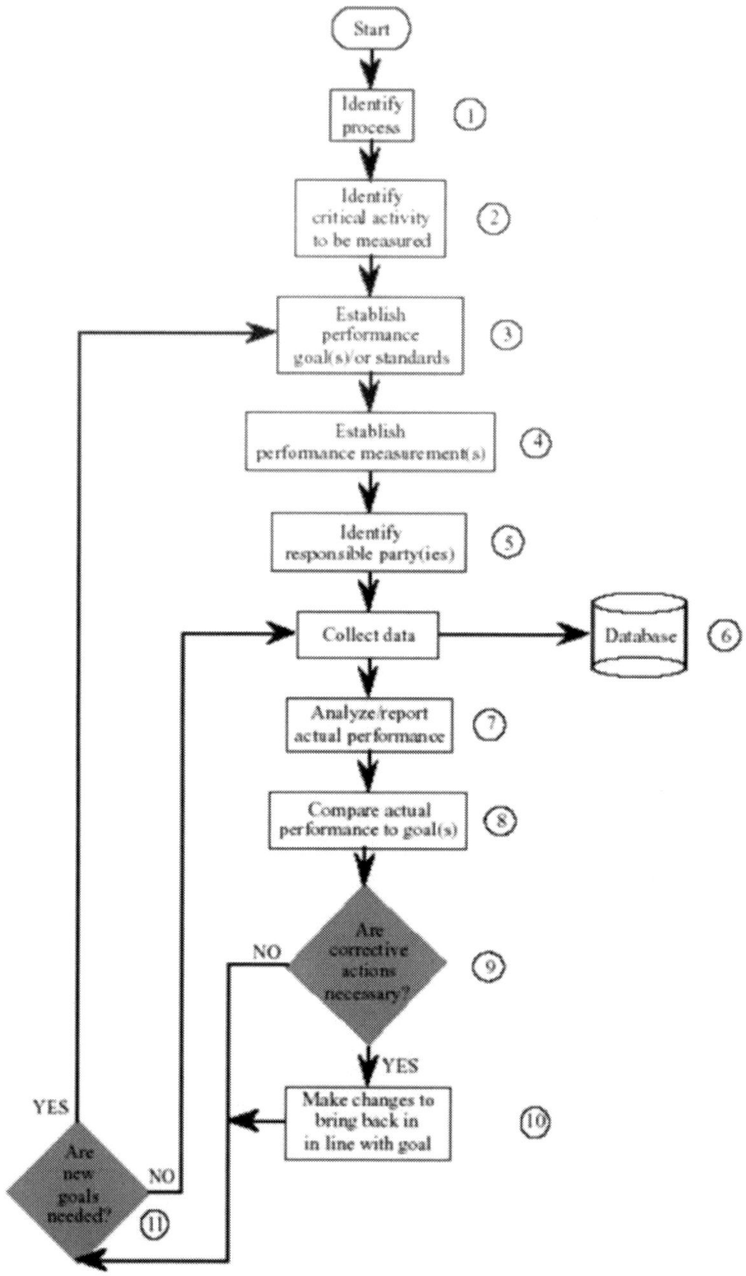

Figure 1. Performance Indicator Development Flowchart (from Ref. [8])

4. Establish performance measurement(s). In this step, you continue to build the performance measurement system by identifying individual measures.

5. Identify responsible party(s). A specific entity (as in a team or an individual) needs to be assigned the responsibilities for each of the steps in the performance measurement process.

6. Collect data. In addition to writing down the numbers, the data need to be pre-analyzed in a timely fashion to observe any early trends and confirm the adequacy of your data collection system.

7. Analyze/report actual performance. In this step, the raw data are formally converted into performance measures, displayed in an understandable form, and disseminated in the form of a report.

8. Compare actual performance to goal(s). In this step, compare performance, as presented in the report, to predetermined goals or standards and determine the variation (if any).

9. Are corrective actions necessary? Depending on the magnitude of the variation between measurements and goals, some form of corrective action may be required.

10. Make changes to bring back in line with goal. This step only occurs if corrective action is expected to be necessary. The actual determination of the corrective action is part of the quality improvement process, not the performance measurement process. This step is primarily concerned with improvement of your management system.

11. Are new goals needed? Even in successful systems, changes may need to be revised in order to establish ones that challenge an organization's resources, but do not overtax them. Goals and standards need periodic evaluation to keep up with the latest organizational processes.

The International Atomic Energy Agency (IAEA) [6] recommends that: "A clear and simple definition of each indicator has to be established as a key part of the programme. The future owners of the data collected, tracking and trend analysis of each indicator, should also be involved in the definition of procedures and methods of calculation of indicators."

When developing indicators that will be used across a variety of sites, such as would likely be the case for the process industry, an even more careful definition and detailed explanation of indicators should be provided, including examples and data collection requirements, such that anyone can understand how to calculate and compare meaningful results among sites.

2.3 Performance Indicator Selection Criteria

According to the IAEA [6]:

> *"The selection and use of PIs require previous consideration of specific criteria and characteristics for their effective use to monitor operational safety performance, among which are the following:*
>
> - *There is a direct relationship between the indicator and safety,*
> - *The necessary data are available or capable of being generated,*
> - *Indicators can be expressed in quantitative terms,*
> - *Indicators are unambiguous,*
> - *Their significance is understood,*
> - *They are not susceptible to manipulation,*

- *They are a manageable set,*
- *They are meaningful,*
- *They can be integrated into normal operational activities,*
- *They can be validated,*
- *They can be linked to the cause of a malfunction,*
- *The accuracy of the data at each level can be subjected to quality control and verification,*
- *Local actions can be taken on the basis of indicators.*

Criteria should be used by management to be alert to indicators containing subjective input or that are subject to manipulations so that abuses do not occur."

The IAEA goes on to state:

"Depending upon their use, for regulatory purposes, operating organization and plant management purposes or plant department supervisor purposes, the number of PIs is going to be different. Most important is that the selection of indicators be done after a complete and thoughtful revision of the key processes and activities by those who have responsibilities and therefore are involved in the processes. It should be taken into account that a reasonable and manageable set of indicators needs to be developed covering the important aspects of the safety management system and following a hierarchical structure. Depending upon their use for decision making they need to be at the appropriate organizational and institutional level."

To paraphrase guidance from the nuclear industry, in selecting safety PIs the following criteria apply: First, degraded performance in the indicated area must be safety-significant. Second, operational conditions (frequency of challenges, etc.) must be such that there is a significant statistical chance that degraded performance will be detected by the indicator within a reasonable time.

2.4 Leading vs. Lagging Indicators

Lagging PIs track the occurrence of undesired impacts (i.e., things have already gone wrong) while leading PIs track the occurrence of "precursors" or "predictors" (i.e., things will go wrong if this continues).

The concept of leading vs. lagging indicators has been well described by the Nuclear Energy Institute [5], as follows:

"According to their use, indicators are generally considered in two groups: leading or proactive and lagging or reactive indicators. Leading indicators are most useful as a precursor to safety degradation for early management reaction. Lagging indicators are most commonly used to drive plant performance, to monitor and for benchmarking against similar plants."

While leading indicators are more desirable, it is important to note that there may be insufficient information about the process facility to recommend the use of leading indicators alone. When a leading indicator is selected, it presupposes that improvements in that indicator will actually improve safety. However, at least initially, this needs to be proven. For example, it would be necessary to prove that if the number of worker entries into hazardous areas in safety gear goes down the recordable injury rate (RIR) or some other direct measure would go down accordingly. If it goes down by a lesser percent, then the program is not capturing sufficient leading information. If the RIR does not go down at all (or, worse, goes up), then the wrong leading indicator is being evaluated. Therefore, a good PI program needs both leading and lagging indicators in order to be effective.

2.5 Performance Levels and Thresholds

Color thresholds are used to characterize performance – such as green, white, yellow, red by the nuclear industry [5] – and signify increasing degradation of safety. The NEI categories of PI metrics indicate trigger levels of "increased regulatory response", "required regulatory response" and "unacceptable performance". The DOE performance reporting process [12] utilizes the following standard colors and definitions:

Blue: "Outstanding" - Performance which significantly exceeds expectations.

Green: "Good" - Performance which meets or exceeds expectations. Generally requires only normal management attention to maintain performance.

Yellow: "Marginal". Yellow can be used to denote either of two conditions:

- Borderline or declining performance, which needs increased management attention and resources to achieve desired performance or to reverse a negative trend.
- Acceptable performance that relies on a set of conditions which could change and quickly send performance into the "Red" category. For example, if events such as the loss of a key manager; the denial of a request for an exemption; or a policy change on the part of a regulator would quickly move performance from acceptable to unacceptable, "Yellow" would be the appropriate designation.

Red: "Poor". Performance which clearly does not meet expectations, and which needs significant management attention and resources to achieve desired level of performance.

White: "No Data".

Regarding the establishment of alert levels, the following guidance is provided by the IAEA [6]:

> *"...when the unacceptable region is not identified by regulatory requirements or by the safety case parameters, an analysis of historical data should be performed to obtain percentile information. The unsatisfactory limit (between warning and unsatisfactory zones) could be set by the 50th percentile. The annual plan (between the warning and operating zones) could be the value set by 40th percentile (if low is good). The vision or strategic goal (between the operating*

and excellent zones) could be the value set by the 20th percentile. In the event enough data is not available, expert judgement would have to be used to select the various limits once the company vision (strategy goal) has been set by the management."

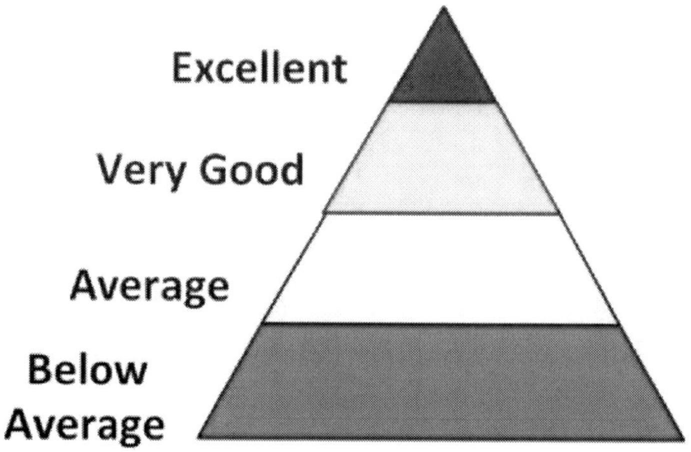

Figure 2. Example Safety PI Performance Level Pyramid

Figure 2 shows the scale of performance levels that could be used in a process industry safety PI program. Preliminary metrics that are applicable to these levels are discussed for some of the proposed safety PIs in the following section. However, it should be recognized that:

- Each PI requires its own specific metrics tailored to the performance to be measured, and
- The metrics provided in this document should be considered preliminary and used for information purposes; metrics specific to certain facilities will have to be developed using data collection and analysis, as discussed further in section 4.0.

3 Candidate Process Safety Performance Indications

The following five key focus areas of safety performance indicators were identified as a result of research conducted into the topic:

- Personnel Safety
- Public Safety
- Emergency Preparedness
- Physical Protection
- Safety Culture

Candidate performance indicators were identified for consideration for a process safety PI program in each of these five focus areas, as summarized in Table 1. Examples of the PIs for the areas of Personnel Safety and Safety Culture are provided below.

Table 1. Candidate Process Safety PI List

Focus Areas	**Candidate PIs**
Personnel Safety	Injuries and Chemical Exposures
	Recordable Injury Rate (RIR)
	Accidents Resulting in Lost Work Time
	Unsafe Acts (Procedure Violations)
Public Safety	Hazmat Releases
	"Near Misses" with Flammable Material
	Public Meeting Effectiveness
	Safety System Performance
	Safety System Challenges
	Safety System Availability
	Percent of Safety Equipment Checked per Schedule
	Emergency Preparedness
	Drill Exercise Performance
	Emergency Response Organization Drill Participation
	Alert and Notification System Reliability
Physical Protection	Security Equipment Performance
	Personnel Screening Program Performance
	Fitness for Duty Program Performance
Safety Culture	Backlog of Corrective Actions
	Safety Culture Training Activities vs. Incidents
	Safety Issues Identified by Management vs Total Safety Issues Identified
	Problem Resolution Implementation Timeline vs. Schedule

	Calls for Re-Work for the Same Problem
	Volume of Waste in Facility

3.1 Personnel Safety

An unsafe work environment translates into low worker morale, decreased productivity and an increased potential for small incidents to become major accidents. Personnel safety can be measured through a variety of indicators, most of which relate to the injuries that have occurred to workers over a period of time. The specific performance indicators that have been identified and are recommended for monitoring personnel safety are discussed individually below.

3.1.1 Injuries and Chemical Exposures

The number of times personnel are injured in a facility can serve as an indication of the level of attention being paid to training and safety briefings, or the number of careless work practices.

According to performance indicator documents from various industries [12], [13], [14], injury-related indicators appear to be generally expressed as injury rates, as discussed further under the recordable injury rate indicator addressed in section 3.1.2. For the process industry, it is recommended that some review of injury and illness reports be conducted to establish a historical basis for a range of injuries and to use that range to set and adjust alert levels accordingly for this proposed PI.

3.1.2 Recordable Injury Rate (RIR)

The US Department of Labor Occupational Safety & Health Administration (OSHA) has a mandatory recordkeeping system for workplace illnesses and injuries that US businesses employing more than 10 workers must complete and submit, unless that business is classified in a specific low hazard retail, service, finance, insurance, or real estate industry according to the Standard Industrial Classification (SIC) Code [15]. The Bureau of Labor Statistics (BLS) also uses injury and illness records as the source data for the Annual Survey of Occupational Injuries and Illnesses that shows safety and health trends nationwide and industrywide. The Recordable Injury Rate (RIR) is defined by the BLS as the rate of injury cases per 100 full time workers[1] in a given year.

3.1.3 Accidents Resulting in Lost Work Time

A review of illness and injury reports, or even the equipment repair record system, quickly reveals that some injuries are serious, while others are quite minor. Lumping all injuries together, therefore, can result in a stilted view of the actual effects of injuries on the worker population and can give an overly pessimistic view of unsafe conditions at a facility. For this reason, some facilities utilize the measure of accidents resulting in lost work time to indicate

[1] Often 200,000 work hours is used as a substitute for 100 full time workers, when calculating injury rates.

those injuries of significance. Of the literature reviewed for this effort, one source from Australian industry [16] used this measure and cited a metric of 10 accidents resulting in lost work time per million hours (or 1E-5 accidents per hour) that was considered to be indicative of "average" performance and 30 such accidents per million hours consistent with "below average" performance. These statistics should be compared against actual facility experience to evaluate their relevance as process industry safety PI trigger levels.

3.1.4 Violations of Procedures (Unsafe Acts)

Safety procedures are used as a means for standardizing practices and behaviors in a way that maximizes safety at a facility. When workers take it upon themselves to circumvent these procedures, accidents and injuries are more likely to occur. For this reason, tracking the number of procedural violations is recommended as part of a process industry safety PI program.

The DOE production site at Savannah River has conducted a pilot program to identify appropriate metrics for performance indicators, including those related to safety [9]. Based on a year of study by the Washington Group involving a variety of subject matter experts, a series of metrics were identified, including a range of values for the number of procedure violations. An average value of <2.5 violations per 200,000 hours of operation was identified, meaning that greater than 2.5 violations in this time-period can be considered as below average performance. The goal stated by DOE for this indicator also cautions that a "healthy balance of operations without suppressing reporting" should be maintained; this is an important concept and requires a delicate balance between encouraging workers to reveal unsafe practices and the effectiveness of management monitoring.

3.2 Safety Culture

As defined by the Idaho National Engineering and Environmental Laboratory (INEEL) [17], in a Total Safety Culture, employees not only feel responsible for their own safety, they feel responsible for their peers' safety, and the organization's formal management systems and leaders' informal management practices facilitate this perspective by encouraging, recognizing, and reinforcing appropriate behaviors.

The IAEA [6] has studied symptoms of a weakening safety culture and has evaluated them in a model that considers stages of organizational decline, each of increasingly severe consequence. Examples they have found of safety culture degradation are: procedures not properly served; incidents not analyzed in-depth and lessons not learned; increasing number of violations; increasing backlog of corrections actions; and failure to deal with findings of external safety reviews. These issues, and others like them, are therefore recommended as safety PIs for the process industry and are discussed further below.

3.2.1 Backlog of Corrective Actions (number)

Backlog refers to the number of corrective actions (or maintenance requests) that are in the queue, waiting to be processed. Metrics for this PI specific to different types of facilities would have to be developed by reviewing the number of maintenance requests that have been issued over several quarters, including the evaluation of historical data from the maintenance record systems. Care would also need to be taken to factor in the effects of non-processing time and the potential for the tendency to hold off on certain maintenance actions until process changeover or system/plant downtimes occur.

3.2.2 Safety Culture Training Activities vs. Incidents

Among the eight key components cited by INEEL [17] to be addressed to achieve an effective safety culture, one element is that employees should have the proper training to perform their jobs in safe ways. However, a clear sign that the training is not having the desired impact is the number of incidents, particularly those related to unsafe actions by staff. This proposed safety PI, identified by the TRADE special interest group among their Security/Safety Performance Measurements [8], would therefore track the number of incidents at the facility versus the safety training activities during the same timeframe. However, no metrics could be found in the literature to recommend performance evaluation levels.

3.2.3 Safety Problems Identified by Management vs. Total Safety Problems Identified

INEEL [17] also rightly states that "a safe and productive working environment is much more likely to be developed from support by key managers" and lists management commitment to safety as one of their eight key safety culture components. The TRADE group [8] included this PI, which would track the percentage of safety problems are identified by management, as a means for identifying the degree of management involvement and awareness of the risk-inducing activities at a facility.

3.2.4 Problem Resolution Implementation Timeliness vs. Schedule (Overdue Corrective Actions)

The DOE EFCOG manual [9] provides an indicator and metrics for the percent of Corrective Action Reports/items completed on time. The proposed indicator discussed here is the inverse of the DOE one and is intended to track the percent of maintenance actions that have yet to be completed and are therefore overdue. Using the inverse of the DOE metric, a trigger level of greater than 15% overdue actions out of the total maintenance records/corrective actions could be used as an initial measure for this candidate safety PI.

3.2.5 Calls for Re-Work for the Same Problem

Since many maintenance activities call for the use of protective suits, which adds risk to workers, it is important to perform maintenance work correctly the first time and not require re-work for the same issue. A review of maintenance records could be performed to identify the number of times requests were made for the same component to be fixed for the same problem within the same week or month.

3.2.6 Volume of Waste in Facility

The issue of the volume of transient waste is based on walkdowns of facilities conducted by the authors and their colleagues as part of fire risk analysis efforts [18], which noted "surprising" amounts of transient waste being temporarily stored in a number of locations.

Housekeeping is a good indicator of safety culture. Given observation of trash storage issues, a performance indicator that tracks the volume of trash that is allowed to accumulate in the facility would focus more attention on the trash issue and formally measure efforts to control that accumulation. Further, the focus on minimizing trash accumulation could also have a beneficial effect on reducing trash generation. Finally, the some facilities have identified fires as the dominant contributor to public risk, and to a large extent this is driven by the possibility that significant amounts of combustibles could be in key areas of the facility. The use of trash volume as a performance indicator would serve as a "leading" indicator of fire risk.

4 Data Collection and Categorization Needs

After the establishment of prospective performance indicators and measurements, a data collection and analysis program must be implemented to provide input to the PI metrics.

According to TRADE [8], to generate useful information, planning for good data collection proceeds along the following lines:

- What question do we need to answer?
- How will we recognize and communicate the answers to the question?
- What data-analysis tools (histogram, bar graph, control charts, etc.) do we envision using? How will we communicate the results?
- What type of data do the data analysis tools require?
- Where in the process can we get these data?
- Who in the process can give us these data?
- How can we collect these data from people with minimum effort and chance of error?
- What additional information do we need to capture for future analysis, reference, and traceability?

The American Society for Quality [10] cites the following guidance for data collection and tracking for performance indicators:

"For each PI, the period for which the data is to be collected should be specifically defined - e.g., monthly. The number of periods for which the data is to be retained and available for trending also should be specifically defined - e.g., for the current month and the immediately preceding eleven months."

"The value of any data point for a PI should be compared to the value of the corresponding regulatory requirement and expectation, industry and plant standard, management requirement and expectation, and industry good practice. Regardless of the absence of a trend, if the value of the data point exceeds the value of a requirement, standard, good practice or expectation (the allowable), self-intervention should be triggered. Self-intervention may include self-assessment, internal independent assessment, or other actions intended to identify the reason(s) for the PI level or trend. Once the reason(s) is understood, appropriate corrective action can be determined."

U.S. Department of Energy (DOE) senior management has asked for regular summary-level reports on the status of DOE sites under their cognizance against a set of top-level performance metrics or "Focus Areas", using an "annunciator panel" approach similar to the one used by the commercial nuclear utility industry. Still, on the matter of data collection and analysis, according to the Department of Energy reporting process [12],

"DOE field offices and their contractors at the various DOE sites already gather, analyze, and track a wide variety of performance information using the specific metrics applicable to the missions. The specific metrics and how they are used vary based on the unique contractual provisions and business necessities of the various sites. It is assumed that no site will start from scratch, and that there will be few cases in which a particular DOE site will have to substantially change its approach to gathering and analyzing information in order to implement the planned performance metrics reporting process. Instead, only the reporting format for the Undersecretary is being standardized."

The DOE findings are echoed by the IAEA [6] in this statement:

"Depending on the starting point of each plant the effort to implement an operational safety indicators programme may be more or less significant. However, because plants already have some kind of data collection and performance monitoring system, it is believed that only small additional resources are required. However, the implementation process implies an additional effort by the plant above all to select and define the indicators and establish the goals."

There is also a point to be made regarding the analysis of so-called "precursors" to failure, meaning, degraded conditions that may provide a preview of complete failures or events to come. A detailed example of such analyses is evidenced by the US Nuclear Regulatory Commission's Accident Sequence Precursor (ASP) program [19], which has been in place since 1979 and identifies nuclear power plant events that are considered precursors to accidents with the potential for severe core damage and uses risk assessment methodologies to determine the quantitative significance of the events. For example, in the equipment sense, a component that has experienced no catastrophic failures will appear to be 100% reliable. However, including degraded and incipient failures in the data analysis provides a more accurate representation of the component's reliability history, particularly in light of the limited data available. Similarly, as the process industry strives to collect data for its safety performance indicator program, caution should be taken since decision-making on the basis of sparse data can be misleading unless all types of failures are included and analyzed.

Table 2 provides a summary of the types of data that should be collected and analyzed to provide metrics for each of the candidate safety PIs.

Table 2. Candidate Safety PI Data Collection Requirements

Candidate Safety PI Focus Area	**Data Needs**
Personnel Safety	• Worker illness and injury reports (including fatalities, agent exposures, accidents resulting in lost work time) • Hazardous area entries, protective suit/glove tears • Post-accident analyses • Number of full-time employees • Employee hours worked • Procedure violations • Hours of facility operation
Public Safety	• Chemical release amounts and frequency • Reportable hazmat release amounts and frequency • Forklift operations
Safety System Performance	• Unscheduled challenges to safety system equipment • Safety system equipment repair dates and durations • Testing and preventive maintenance schedules, actual performance dates and durations by safety system component • Safety system equipment population data
Emergency Preparedness	• Drill/Exercise schedule vs. actual performance frequency • Number of Emergency Response Organization (ERO) personnel • ERO personnel that have participated in a drill, exercise or event • Alert and Notification System repair history, repair duration, and testing/preventive maintenance schedule vs. actual test/PM performance history and duration
Physical Protection	• Protected Area Security Equipment repair history and testing/maintenance schedule • Security violations, infractions and incidents • Personnel reportable events logged by HR related to security violations and to fitness-for-duty
Safety Culture	• Corrective maintenance actions waiting to be processed • Corrective maintenance actions overdue vs. total issued • Corrective maintenance actions issued to address same component and failure mode within same month (re-work) • Safety issues identified by management vs. total identified • Safety culture training program activities per quarter • Incidents/accidents per quarter • Weekly estimate of number of waste bags stored in site areas awaiting disposal

5 Conclusions and Recommendations

This paper has documented an assessment of safety performance indicators as a tool for measuring the safety performance of process industry facilities. A number of candidate performance indicators covering the broad range of safety aspects of facility operation have been identified as summarized in Table 1 and their associated data needs are identified in Table 2. These include a suitable blend of both leading and lagging indicators. Taken as a whole, these indicators would provide a clear picture of safety performance for process industry facilities but should also include quarterly re-evaluation of the success of the program and modifications to the PIs as appropriate. The implementation of such a program would require:

- A further specification of the data requirements for the candidate indicators to determine whether it is practical to collect and analyze the data required to track the indicator;
- For those indicators where numerical metrics have been proposed, further review of the suggested numerical metrics to assure that they are applicable and reasonable;
- For those indicators where numerical metrics have not been proposed, additional research and analysis to establish initial values for the numerical metrics;
- The specification of the approach to be used to summarize and report the program results to process industry management (e.g., DOE annunciator panel approach); and
- Definition of how the results would be used as part of a performance-based management system.

With regard to the first bullet, a determination that one or more of the candidate performance indicators is not practical to use does not necessarily mean that they would need to be replaced with additional indicators. In this case, an evaluation should be conducted to determine if the remaining indicators still adequately cover each of the safety focus areas and provide a balanced mix of both leading and lagging indicators.

6 References

[1] *World Association of Nuclear Operators (WANO) Performance Indicators 2001*, tri-fold brochure, http://www.wano.org.uk/PerformanceIndicators/PI_Trifold/PI_2001_TriFold.pdf

[2] *Performance Objectives and Criteria for Corporate Evaluations*, INPO 85-029, Institute for Nuclear Power Operations, Atlanta, GA, August 1985.

[3] *An Overview of Nuclear Power Plant Operations Experience Feedback Programs*, NRC/Office for Analysis and Evaluation of Operational Data, Report No. AEOD-5602, May 9, 1986.

[4] *Guidelines for Analyzing and Trending Incidents in Nuclear Power Plants*, NSAC-119, Nuclear Safety Analysis Center, Electric Power Research Institute, Palo Alto, CA, June 1988.

[5] "Regulatory Assessment Performance Indicator Guideline", NEI-99-02, Rev.2, Nuclear Energy Institute, Washington, DC, 19 November 2001.

[6] Dahlgren, K., Lederman, L., Palomo, J. and Szikszai, T., *Safety Performance Indicators*, Topical Issue Paper No. 5, International Conference on Topical Issues in Nuclear Safety, International Atomic Energy Agency (IAEA), Vienna, Austria, 3 - 6 September 2001, http://www.iaea.or.at/worldatom/Meetings/2001/infcn82-topical5.pdf.

[7] *Performance Indicators Guidance Document*, US Department of Energy, DOE-STD-1048-92, December 1992.

[8] *How to Measure Performance: A Handbook of Tools and Techniques*, Training Resources and Data Exchange (TRADE) Performance-Based Management Special Interest Group, for Special Project Group – Assistant Secretary for Defense Programs and the Office of Operating Experience, Analysis and Feedback – Assistant Secretary for Environment, Safety & Health, US Dept. of Energy, October 1995, http://www.orau.gov/pbm/.

[9] *Energy Facility Contractors Group Performance Metric Manual*, Rev. 1, WSRC-RP-2002-00252, 2002, http://www.efcog.org/wg/ismwg/ism/pmsg/pmsg.htm.

[10] *Recommendations for the Implementation of Selected Leading Indicators of Performance at Nuclear Power Production Plants*, Nuclear Power Production Committee, Energy and Environmental Division, American Society for Quality (ASQ), EED-99-01, August 1999.

[11] *Performance Indicators and Combining Assessments to Evaluate the Safety Performance of Licensees*, Organisation for Economic Co-operation and Development (OECD), Nuclear Energy Agency (NEA), Committee on Nuclear Regulatory Activities, Working Group on Inspection Practices (WGIP), NEA/CNRA/R(98)3, April 10, 1998.

[12] US Department of Energy, "DOE Performance Metrics Reporting Process", Point Paper, Revision 2, 1 November 2002.

[13] Prevette, Steven S., Fluor Hanford, "Making Business Decisions Using Trend Information", presented to Occurrence Reporting Special Interest Group, 1997 Department of Energy "TRADE" Conference, 2 December 1997.

[14] Prevette, Steven S., Fluor Hanford, "Cleaning Up with SPC", *Quality Progress*, American Society for Quality, September 2001.

[15] *Recordkeeping*, US Department of Labor, Occupational Safety & Health Administration (OSHA), OSHA 3169, 2001, http://www.osha.gov/Publications/osha3169.pdf.

[16] Stacy, Bob, formerly with BHP Engineering,"OH&S Business Case for Change", Key Management Motivators for OHS Forum, hosted by Commonwealth of Australia - National Occupational Health and Safety Commission (NOHSC), Melbourne, Australia, April 2002. http://www.nohsc.gov.au/OHSInformation/OHSSolutions/presentations/BobStacyOHSCostModellingPaper.pdf.

[17] Idaho National Engineering and Environmental Laboratory (INEEL), website on Total Safety Culture, http://www.inel.gov/vpp/total-safety-culture.shtml.

[18] *Tooele Chemical Agent Disposal Facility Quantitative Risk Assessment*, Report No. SAIC-96/2600, prepared by SAIC/Abingdon, MD for PMCD, Aberdeen Proving Ground, MD, December 1996, Section 5.4.1 on Fires.

[19] Minarick, J.W. and Kukielka, C.A., *Precursors to Potential Severe Core Damage Accident Sequences: 1969-1979, A Status Report,* NUREG/CR-2497, Vols.1 and 2, Oak Ridge National Laboratory, Oak Ridge, TN, 1982.

Major Accident Response: Managing the First 96 Hours
By Mark L. Farley

The process industries continue to experience high profile accidents that bring with them intense public scrutiny and the potential for significant reputational harm. These incidents are not limited to any one sector. They have occurred up-stream (exploration and production), mid-stream (transportation), and down-stream (refining and chemical manufacturing).

Whether it is a prolonged loss of well control, the fiery derailment of tank cars carrying volatile crude oil, or a refinery explosion, these incidents present a "perfect storm" of unique challenges that can readily overwhelm even the most sophisticated organizations. Federal criminal investigators are often among the first personnel at an incident scene raising the stakes even higher. A prompt, sure-footed response by counsel experienced in crisis management is essential.

In the United States, multiple federal and state statutes are implicated by a major accident and establish jurisdiction for subsequent investigations by government authorities. It is not unusual to have an alphabet soup of five or more independent agencies, each with its own statutory mandate, investigating an incident.

The Occupational Safety and Health Administration (OSHA) and the U.S. Environmental Protection Agency (EPA) routinely initiate enforcement inspections after major accidents. Independent investigating agencies, such as the National Transportation Safety Board (NTSB) or the U.S. Chemical Safety Board (CSB), also respond depending on the nature of the incident. NTSB investigates transportation incidents while the CSB investigates chemical accidents at stationary facilities that are not incident to transportation. The U.S. Coast Guard (USCG) and the Department of Interior Bureau of Safety and Environmental Enforcement (BSEE) may be involved in off-shore incidents. State and local agencies and authorities also are likely to respond to major events occurring within their jurisdictions.

Crisis Management and the Role of Counsel

Crisis management of a major accident occurs at two levels within an organization: the tactical—the site at which the incident occurred, and the strategic—the corporate organization. The role of a lawyer advising a crisis management team spans these levels with two primary responsibilities: (1) ensuring timely compliance by the company with applicable legal requirements implicated by the crisis (*e.g.*, incident reporting requirements, evidence preservation, etc.); and (2) ensuring reasonable steps to protect the company's legal rights and interests (*e.g.*, preparing for government investigations and third party litigation, safeguarding legal privileges, making certain communications are accurate and do not represent admissions, etc.).

On a more practical level, the lawyer finds herself in the familiar roles of issue spotter, legal advisor, strategic counselor, project manager and communication facilitator, albeit under very demanding circumstances. Fundamentally, the lawyer's job is to consider the legal implications of almost every major decision taken in response to the crisis in an effort to ensure that the company is not unknowingly compromising its interests.

As with every corporate decision, a balancing of interests will occur, and the company may conclude that it must undertake a course of action that adversely affects its legal rights. The lawyer, however, must be knowledgeable and experienced in spotting relevant issues so as to allow appropriate consideration, as circumstances and time permit.

Although the facts and circumstances of any individual crisis may vary, major accidents always implicate similar legal issues and considerations.

Incident Reporting Requirements

Depending upon the type and extent of the incident, there likely will be government notification requirements. Such notification requirements are often fact-specific (*e.g.*, identity of chemical, quantity released, duration of release). The identity of the agencies to whom the release must be reported will depend on the location of the event, the resulting environmental impacts (*e.g.*, whether the release is to air, water or the soil) and the nature and extent of any injuries. Initial notifications usually must occur within hours of the event with follow-up notifications in specified periods of time.

Site Preservation and Evidence Collection

After an incident, a primary concern of any investigating agencies will be the preservation of the incident scene and the collection of relevant evidence. Protecting against subsequent claims of spoliation or obstruction is of critical importance.

Safety and environmental concerns remain paramount, however. Emergency response and site stabilization, including those activities necessary to place operating units in a safe condition, should occur as needed.

As soon as possible under the circumstances, counsel must ensure that appropriate controls are imposed with respect to the incident site. Access should be limited. Changes made during the emergency response must be documented and go-forward changes should be cleared with interested parties, including investigating agencies. In many instances, government agencies such as OSHA or the CSB will seek to enter into an evidence and site control agreement with the employer / site operator. Such an agreement will impose specific restrictions with respect to changes at the incident scene and provide a framework for the development of future protocols regarding documenting the incident scene, as well as evidence collection and subsequent testing.

Litigation is all but assured in the aftermath of major accidents. In anticipation of such litigation and in light of any government investigations, counsel will need to issue typical directives to preserve documents and electronically stored information (ESI). Major accidents implicate unique ESI, however. The energy industry, for example, relies on process control systems that automate and continuously monitor aspects of operations (*e.g.*, Distributed Control Systems (DCS) or Supervisory Control and Data Acquisition (SCADA)). These systems monitor and record key conditions of the process or pipeline systems, including temperature, pressure, and other parameters. DCS or SCADA retain detailed data for set periods of time before a more limited amount of data is recorded by a data historian or some other automated archiving system. After an accident, it is important to back-up the most detailed version of the data before data granularity is lost.

Internal Investigations

A major objective for the company is to independently investigate the causes of the incident. It is important to note that an investigation may be required by law. In the United States, for example, both OSHA's Process Safety Management standard (29 C.F.R. § 1910.119(m)) and the EPA's Clean Air Act Risk Management Program rules (40 C.F.R. § 68.81) require that the employer/facility owner investigate certain process safety incidents and near misses. These rules impose requirements on any such investigations *(e.g.*, deadlines for initiation, team composition, etc.). In addition to these required investigations, the company also will need to investigate the incident and develop facts necessary for the legal defense of the company in follow-on enforcement by the government or litigation by third parties.

The issues related to internal investigations (*e.g.*, single vs. multiple investigations, issues of applicable legal privilege, team composition, use of external experts, etc.) are among the most important and challenging decisions facing the company in the immediate hours after an incident. The meaningful involvement of legal counsel in making these decisions is critical. The initial inclination of counsel unfamiliar with major accidents may be to conduct a single, privileged investigation. Such advice may be ill-advised, especially if an underlying regulation requires the company to conduct an investigation. A company pursuing a single privileged may soon find itself confronting complex issues of waiver when an enforcement or investigating agency ultimately asks to review the legally mandated investigation report.

A company should carefully consider the selection of investigation team leaders and members. Team leaders should have excellent judgment and be accustomed to forging consensus on key issues while appropriately directing a diverse group of potentially opinionated subject matter experts. Previous experience in handling complex investigations and working well with legal counsel on sensitive matters also are important considerations.

Interacting with Government Inspectors

One of the most challenging tasks of on-scene counsel is assuring appropriate interaction with government inspectors. The government investigations may be lengthy and inspectors may be working at the scene for long periods of time. It is not uncommon, for example, to have investigators maintain an almost continuous presence at an incident scene until relevant physical evidence is collected, a process that sometimes may take four to five months. The inspectors' first interactions with company personnel and legal counsel have a significant role in setting the tone of the relationship and can greatly influence the perceptions of the inspectors with respect to the company's commitment to environmental, health and safety issues and its response to the investigation.

At the same time, it is imperative that company counsel establish appropriate controls over the investigative process and safeguard the legal rights of the company and its employees. Finding a balance between these interests can be challenging, especially in the traumatized atmosphere that often accompanies a major accident. The general approach should be to accommodate the reasonable requests of the inspectors, but counsel should be prepared to establish appropriate limits or deny requests if compelling reasons exist for doing so. Cooperation does not require an abdication of responsibility or control. With respect to issues such as changes to the incident

124911888.1

scene that are necessary to further stabilize the situation, conferring with the inspectors prior to such changes and documenting no objection is essential.

Document Retention, Collection and Production

It is imperative that the company promptly develop an effective process for document collection and production. In the case of major accidents, counsel should anticipate receiving requests from multiple agencies within hours or days of the event. NTSB, CSB, and OSHA often make requests immediately upon deployment with an initial production deadline of 24-hours.

It also is unlikely that the agencies will coordinate their requests. The magnitude of produced documents may ultimately reach tens of thousands of pages. If a systematized approach is not adopted, there is a high risk that the overall integrity of custodian files will deteriorate as multiple documents are retrieved, copied, and re-filed.

It is also important to ensure that a system exists that will allow counsel at some point in the future (often years from the event) to identify with specificity the exact file location (which custodian, which file, etc.) from which a document was produced.

Certain original documents may also prove to be of such importance that counsel should collect and secure them immediately so as to ensure a documented (and ideally, limited) chain of custody. For example, the start-up checklist being used by a control room operator in the time immediately preceding an incident may warrant such measures.

Witness Interviews

The inspectors will make requests to interview certain witnesses. It is often in the company's interests to facilitate these interviews, but it is important for the company to make certain that employees understand their individual rights and the company's expectations (*e.g.*, to be professional and polite, to tell the truth, advise against speculation, etc.). It also is important to ensure that individuals understand and acknowledge that company counsel does not represent them personally..

Most agencies will allow observation of their interviews with exempt employees (*e.g.*, supervisors, engineers) by counsel for the company. Agencies, however, often will initially resist efforts to have counsel or others present so counsel should be prepared to push back. Exempt employees should be made aware of their right to have counsel for the company present. In most cases, the exempt employees will condition their interviews on having counsel present.

OSHA, EPA and the CSB almost never allow observation of their interviews with non-exempt, hourly employees. Counsel should make every effort to debrief these employees after their interviews if the employee is willing.

Conclusion

In the aftermath of a catastrophic accident (*e.g.*, an incident with multiple fatalities or significant off-site impacts), it often is difficult for an organization and its leadership to appreciate the true magnitude of the event. A period of denial sometimes exists where management does not fully appreciate the nature and extent of the investigations that are about to commence and the demands that will be placed on the organization. It is important for counsel to begin anticipating and planning for how the organization will promptly provide resources to allow for an effective response to the demands of the agencies.

As discussed above, multiple agencies will respond (some within hours) to the event. It is not uncommon for there to be 15 governmental representatives on site within the first 24 hours and 25-30 within the first 48 hours. Large numbers of other personnel assisting the company (*e.g.*, corporate subject matter experts, in-house and outside attorneys, technical experts, public affairs personnel) also will soon descend on the site with their own demands and needs.

To prepare for the building legal storm, companies should develop reference materials and resources for in-house attorneys. Such major accident guidebooks do not establish procedures that should be strictly followed in every instance. Instead, they are intended to assist in-house counsel in identifying and managing issues likely to arise within the first 48 to 96 hours after an event. These guidebooks should include as appendices examples of relevant documents, including forms that may be used to support the response. The company and its counsel should routinely conduct crisis response drills using these materials. Only through adequate preparation and practice can an organization hope to ride out and properly manage a crisis.

Mark L. Farley co-heads the Environmental and Workplace Safety practice at Katten Muchin Rosenman LLP and routinely works with clients in preparing for and responding to major accidents. He has served in lead attorney roles for some of the most significant workplace incidents in the U.S. in the last 20 years. Mr. Farley also was one of the lead partners supporting the work of the BP U.S. Refineries Independent Safety Review Panel, which investigated the corporate safety culture and oversight at BP's North American refineries on behalf of an independent panel of experts chaired by former U.S. Secretary of State James A. Baker III.

478155 Incident Management System for a Global Company - Lessons Learned

Abstract:

The global rollout of an incident management and action tracking system is no easy task. Although everyone is trying to do the right thing from initial rollout, there are lessons learned along the way that will, in the end, solidify the already existing incident management system. A mid-size, yet global company of about 3,500 employees learned and applied these lessons during a three-phase incident management system and software rollout to three different geographical areas: the Americas, Asia, and Europe. The challenges faced were not only geographical and cultural; they were also a result of the different legacy safety philosophies that come with joint ventures and new partnerships in the ever changing chemical industry.

What was the initial approach? Going back to basics, defining what an incident is, what types of incidents would be tracked, and what the requirements for an investigation were going to be. This paired with the lessons that were learned after each phased rollout resulted in the successful implementation of a global incident management procedure and software system, and the lessons being continuously learned would make for an even more successful rollout in the future.

13TH GLOBAL CONGRESS ON PROCESS SAFETY

Human Error. A Myth Eclipsing Real Causes

Ignacio José Alonso - Mike Broadribb

Consejo de Seguridad de Procesos
Bogotá, Colombia

Prepared for Presentation at
American Institute of Chemical Engineers
2017 Spring Meeting
13th Global Congress on Process Safety
San Antonio, Texas
March 26-30, 2017

Human Error. A myth eclipsing real causes

Ignacio José Alonso
Consejo de Seguridad de Procesos
Bogotá, Colombia
ialonso@csp-la.org

Mike Broadribb
Executive Advisor
Genoa, Nevada, USA
mbroadribb@bakerrisk.com

Keywords: Process Safety (PS), Comprehensive List of Causes (CLC), Management System (MS).

1 Introduction

For years professionals involved in the investigation of industrial accidents have identified "human error" as a common cause of a high number of major events related to process safety. However, incorrect actions performed by workers that lead or contribute to disaster, before or during an event, are associated with factors related to cultural, engineering, situational, psychological, procedural and organizational aspects, just to name a few, and these tend to remain in the background.

The aeronautical industry has developed exhaustive research in causal factors and methodologies to mitigate risks associated with "human error". Pilots work in complex environments and are routinely exposed to high amounts of situational stress in the workplace that can induce pilot error which may result in a threat to flight safety. As in the Oil and Gas industry, catastrophic accidents are infrequent, but are highly visible and often involve massive loss of life.

The environment to which operators and managers of a processing plant can be exposed and that of an aircraft crew can be similar. This paper refers to these similarities and explores factors and causes addressed by both industries to find commonalities and possible root causes.

The CCPS approach with the 20 elements of Risk Based Process Safety has provided in recent years a more comprehensive framework to examine aspects that can determine behaviors of individuals that work in hazardous environments. The paper evaluates typical factors associated to "human error" with possible causes related to the Risk Based Process Safety elements.

2 Some discussion on "human error"

There are several definitions for the word "error". From a dictionary: "a deviation from accuracy or correctness; a mistake, as in action or speech" **[5]**. In the context of Process Safety it could be defined as a failure to obtain an expected or planned result from the execution of one or more actions.

Humans make mistakes often. But error has an intrinsic characteristic associated to it: i.e. it is never intentional. No one really wants to make a mistake. It goes against the natural instinct of a human being.

The term "human error" or "pilot error" is frequently used by the media and general public to refer to the cause of some aviation disasters.

According to the Federal Aviation Administration (FAA) "this is the most common of all accident themes and exists in one form or another on nearly all accidents. It involves humans that, in the course of doing their work, make errors that are later shown to have caused, or substantially contributed to the accident. These are human actions that, if done correctly, results in a safe outcome, but if done incorrectly, can result in an accident. It also represents one of the greatest opportunities for advancing safety by the application of targeted interventions which are intended to reduce the risks for human error". **[1]**

In the aviation context, under normal circumstances, it is not rational that a pilot of a commercial plane would jeopardize the lives of hundreds of people and his own life by intentionally performing poorly. Consequently, it is not rational that a plant operator or manager would put at risk the lives of his colleagues and the safety of his job. However, if there is not adequate control of the conditions that can affect human performance they may end up making an error.

3 Aviation Disasters involving Human Factor / Link to RBPS

The following is a summary of some catastrophic accidents from the airline industry that involved human error as one of the main causes. Listed, at the end of each example, are the suggested process safety elements that according to the investigation should be addressed to resolve deficiencies.

3,1 Flight 254 Varig - September 1989

An old generation 737-200 aircraft was covering the itinerary from Maraba to Belem over the Amazonian jungle. The plane wasn't equipped with Flight Management Computers as most of today's commercial aircraft, and only had a radio navigation system.

When the pilots called air traffic control in Belem to start their approach for landing, they did not receive any answer. Nor did they receive a signal from the VOR (Very High Frequency

Omnidirectional Range) either, and the controller wasn't able to identify their signal on the display of a primary (basic) radar. A recent change that standardized four digits for the input of the flight direction into the autopilot system led to an error when the pilot entered 270° instead of 027° from his interpretation of the 0270 value consigned in the flight plan.

They wandered over the Amazonian trying to find the airport until they ran out of fuel and had to perform an emergency landing in the middle of the jungle. Twelve passengers were killed. Survivors were able to report the crash two days later from a ranch located 700 miles away from the intended destination.

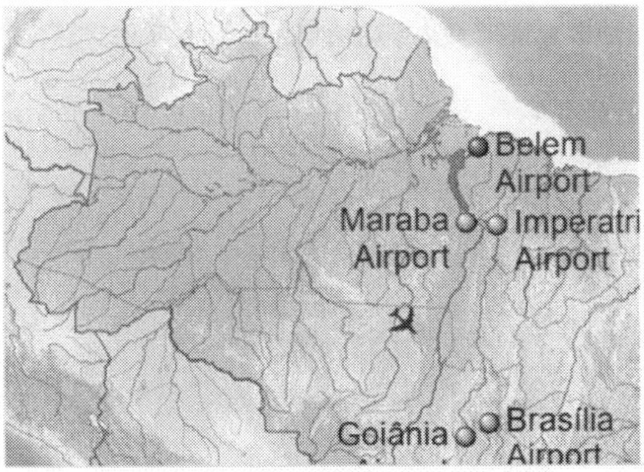

Location of the accident

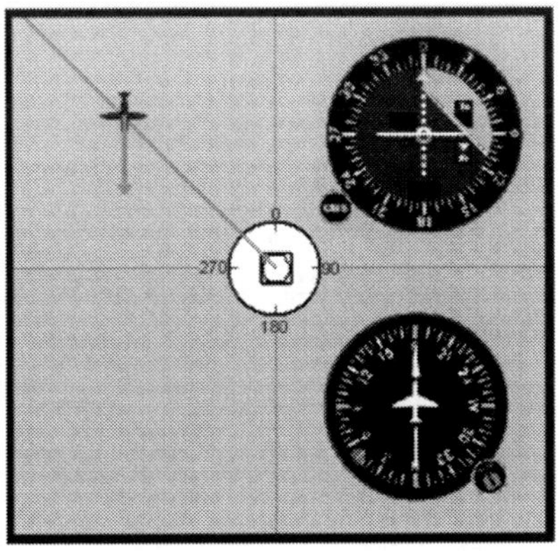

VOR Indication Display

Figure 1. Wrong heading

Suggested PS factors associated to this accident: Culture, Codes and Standards, Management of Change, Training, Operational Discipline, Operating Procedures, Hazard Identification and Risk Analysis.

3.2 Flight 201. Copa Airlines - June 1992

They were flying a Boeing 737 on a scheduled flight from Panama City to Cali, Colombia. During the night and in the middle of a storm, pilots suffered spatial disorientation flying the airplane to a position that was impossible to recover from (pilot error?). The plane stalled, disintegrated in mid-air and fell into the jungle.

The investigation found that the pilot's glideslope indicator malfunction caused by a cable disconnection provided faulty information. The inconsistency between the two instruments resulted in spatial disorientation of the pilots.

The glideslope indicator condition was known and registered in the logbook. The glideslope selector was wrongly positioned activating both instruments. It remains a mystery why they didn't use the reserve glideslope indicator of the aircraft (a third instrument) to have a trustworthy reading.

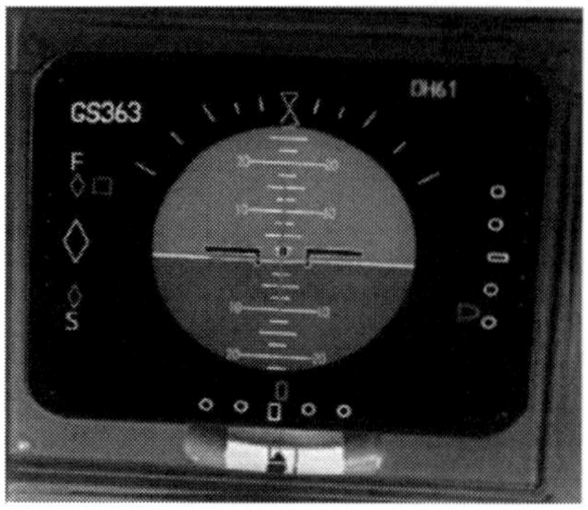

Wrong? or... right?

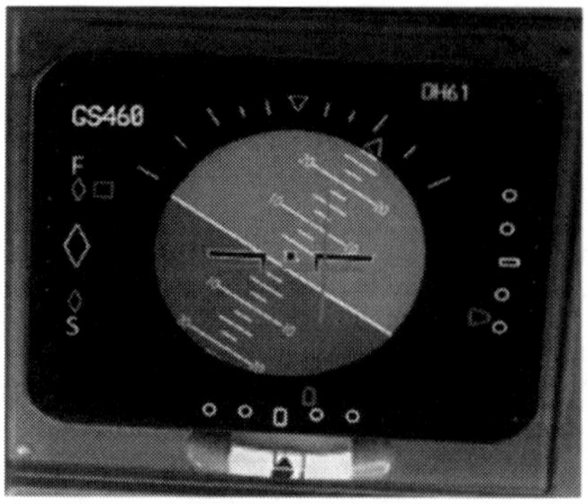

Right? or... Wrong?

Figure 2. Glideslopes

Suggested PS factors associated to this accident: Operational Discipline, Training, Mechanical Integrity, Hazard Identification and Risk Analysis, Emergency Management.

3.3 Private Jet. Mexican Interior Minister - November 2008

During the final approach to Mexico City airport, a Lear Jet used by the Interior Minister lost control when it entered the wake turbulence of a heavy jet and fell in the middle of the city at rush hour. The minister was killed along with another seven people on board and seven on the ground.

The investigation found that the approach speed was too high, the descending angle was inconsistent, and separation to preceding traffic was below the minimum required (4 vs. 5 nautical miles) to avoid wake turbulence due to vortices from a heavy jet.

An essential finding was that the flight certifications of both pilots were fraudulent. Pilots took advantage of a corrupt system to obtain a license to operate an aircraft they were not certified to fly. Also, the flight data register was faulty, as the last data logged was two years old.

Conversation among the flight crew taken from the cockpit voice recorder indicated that they had little familiarity with the operation of the plane; they failed on several occasions to enter the proper information into the cockpit instruments, did not follow a proper flight plan, and had navigational difficulties prior to the crash.

Figure 3. Wake Turbulence

Figure 4. Learjet 45

Suggested PS factors associated to this accident: Culture, Codes and Standards, Competency, Operational Discipline, Working with Contractors, Stakeholder Outreach, Operating Procedures, Asset Integrity, Workforce Involvement, Process Knowledge Management, Auditing.

3.4 The Death of John F. Kennedy Junior - July 1999

John F. Kennedy's Piper Saratoga crashed into the sea during the night in the proximity of Martha's Vineyard, Massachusetts, also killing his wife and sister-in-law.

Kennedy did not hold an instrument rating. He was certified to fly only under visual flight rules. He was not certified to fly by instruments at night or in poor visibility conditions.

Take-off was delayed, which forced him to fly during the night and in poor visibility conditions. Although one of his instructors offered to take him, he decided to fly the aircraft himself.

Kennedy selected the wrong radio frequency, so he was not getting correct weather updates. This may have distracted him and caused spatial disorientation.

Figure 5. Piper Saratoga

Suggested PS factors associated to this accident: MOC, Hazard Identification and Risk Analysis, Culture (false sense of vulnerability), Operational Discipline, Competency, Training, Safe Work Practices.

4 Factors for human error

As explained above, human error is unintentional and is derived from situational or circumstantial factors that lead individuals to failure. This section will navigate into some aspects that have the potential to create the conditions for a mistake.

4.1 Causal Factors

There are different reasons that can lead to human error. It can happen when a person doesn't have the skills or the knowledge to perform a specific activity. People can also make mistakes when tired, stressed or distracted. Their actions can end up in failure when they follow wrong indications (verbal or written) or when the device or equipment that they operate is damaged (poor inspection / maintenance) or is not appropriate for the intended task (wrong selection or design).

In this context, focusing on the aspects that can lead to error should help to avoid its occurrence. The following are recommendations to have better control over these aspects and minimize the chances for humans to make mistakes:

- Have the knowledge required to perform the assigned tasks
- Get the skills necessary to develop intended activities
- Understand consequences derived from failures when performing actions
- Keep focused and get proper tools to avoid distractions
- Recognize situations of high level of stress.
- Identify signs of fatigue and get help or support.
- Ensure that procedures to execute a task are correct and up to date
- Make sure equipment is adequate for the intended task and that it is in good shape
- Verify that design of the equipment is appropriate and meets conditions required to perform the job

If structured processes can be established to manage and strengthen these aspects, they surely would reduce mistakes significantly during the execution of human activities.

On the other hand, this analysis finds that "human error" doesn't constitute a root cause for the occurrence of an undesired event. The "human error" itself is consequence of anomalies and weaknesses that reside in aspects that determine the conditions that lead a person to make a mistake **[4]**.

"Human error" then is a product in the sequence of events that leads to an undesired event, and certainly no more than perhaps an immediate cause. As a conclusion, under no circumstances could "human error" be determined as a root cause during the investigation of an incident.

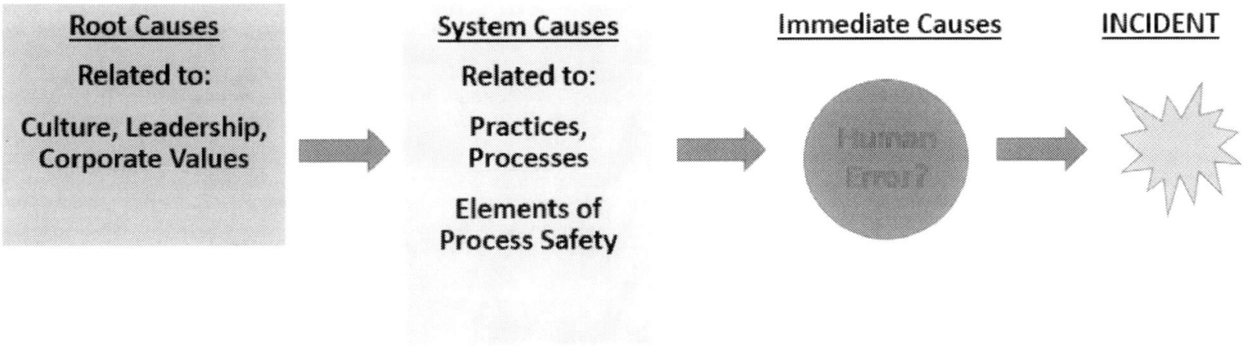

Figure 6. Human error in the sequence of causality

4.2 Elements of Process Safety after the Control of Human Factors

The elements of Risk Based Process Safety **[2]** can be used as a framework to build the processes required to have better control on the aspects that lead to the error.

The following table links the aspects that minimize chances of error with Risk Based Process Safety elements as a source of control mechanisms:

CONDITION	CORRESPONDING RBPS ELEMENT
• Knowledge required to perform assigned tasks	Training and performance assurance
• Skills necessary to develop intended activities	Process Safety Competency, Codes and Standards
• Understanding of consequences derived from failures when performing actions	Risk identification and hazard analysis, Process Safety Knowledge management, Emergency Management
• Focus and use of proper tools to avoid distractions	Conduct of Operations, Safe Work Practices
• Recognition of high level of stress situations	Process Safety Culture, Workforce Involvement,
• Identification of signs of fatigue	Conduct of Operations, Safe Work Practices

• Equipment adequate and in good shape for intended task	Conduct of Operations, Asset Integrity and Reliability, Operational Readiness
• Correct and up to date procedures to execute a task	Operational Procedures, Safe Work Practices
• Equipment design appropriate and meeting conditions required to perform the job	Codes and Standards, Risk identification and hazard analysis, Operational Readiness

Table 1. Tackling the aspects that allow human error with RBPS elements

5 Human Error: The Battle Horse of the Power Cultures

Historically human error has been identified as a common cause in most of the events, accidents and major catastrophes of the industry. The traditional approach of personal safety that tends to focus the management of risks on the attention and behavioral aspects of workers has been extended by companies to the evaluation of Process Safety incidents.

In organizations with a poor Safety Culture, human error is frequently identified as a main cause of the majority of incidents and anomalies. This practice tends to put responsibility for failures on the individuals and misses a great opportunity to go deeper and find existing weaknesses and gaps in other key aspects, such as leadership, corporate values, organizational culture, operational discipline and other Process Safety elements in which root causes can be found, as it has been indicated in this paper.

Some organizations that are managed by a "power culture" take advantage of this situation to hide weaknesses and bad practices of their leaders by repeatedly placing the blame on the workforce, which may open a bigger window to disaster.

6 Conclusions

5.1 There is no such root cause as human error. Behind the mistakes of the individuals are aspects that determine the conditions that induce poor human performance.

5.2 The framework provided by Risk Based Process Safety can be used to build on processes that allow a better control of critical aspects that can lead to human error.

5.3 The aviation crews and the operators of the petrochemical plants face similar hazardous environments. However, the aviation industry increasingly relies more on the technology of the aircraft and its instrumentation, limiting actions of the crew during emergencies.

5.4 When causes associated to failures are placed on individuals, organizations miss a great opportunity to identify the root causes and introduce effective actions that improve culture, safety and human performance.

7 References

Web resources:

[1] Federal Aviation Administration. Lessons Learned. Common Accident Common Themes.

Available at web link http://lessonslearned.faa.gov/ll_main.cfm?TabID=4.

Accessed 2017-02-04

Written material

[2] CCPS. *Guidelines for Risk Based Process Safety*. Center for Chemical Process Safety, American Institute of Chemical Engineers, New York, NY, 1996.

[3] CCPS. *Guidelines for Preventing Human Error in Process Safety*. ISBN: 978-0-8169-0461-7. Published August, 2004

[4] CCPS. *How to Design Human Factors into a PSM System to Ensure Effectiveness and Deliver Risk Reduction Performance*. Global Congress Presentation. April 11, 2016

Additional references not cited:

[5] Definitions. Dictionary.com. Available at http://www.dictionary.com/browse

Best Practices for Writing Operating Procedures and Trouble-Shooting Guides

William G. Bridges, President
PROCESS IMPROVEMENT INSTITUTE, INC. (PII)
1321 Waterside Lane, Knoxville, TN 37922
Phone: (865) 675-3458
Fax: (865) 622-6800
e-mail: wbridges@piii.com

Revonda Tew, Principal Engineer & Instructor
PROCESS IMPROVEMENT INSTITUTE, INC. (PII)
1321 Waterside Lane, Knoxville, TN 37922
e-mail: rtew@piii.com

Prepared for Presentation at
13th Global Congress on Process Safety
San Antonio, TX
March 27-29, 2017

UNPUBLISHED

Best Practices for Writing Operating Procedures and Trouble-Shooting Guides

William G. Bridges, President
PROCESS IMPROVEMENT INSTITUTE, INC. (PII)

Revonda Tew, Principal Engineer & Instructor
PROCESS IMPROVEMENT INSTITUTE, INC. (PII)

Keywords: Procedures, Human Factors, Risk control, process safety management, PSM

Abstract

There is no complete, best practice guideline or textbook for writing operating procedures and trouble-shooting guides. This paper presents the proven, best approach for developing accurate operating procedures and for ensuring the page formatting and step writing are optimized to reduce human error rates. This approach and the 32 rules established in this paper are based on the foundations set by Swain and others (in 1970) for control of human error rates, but uses experiences from more than 100 sites where this approach has been successfully followed. The approach and rules for developing operational troubleshooting guides (procedures for responding to process deviations such as those needed for Human IPL) are again the best approaches found and have been applied successfully since the early 1990s. Several case studies are provided that show the gains from following this approach. The guidelines in this paper build upon ones presented in 1999 at CCPS and 2016 at GCPS.

Background

Operating procedures have always been crucial to the safety, quality, and productivity of process systems. With the advent of new safety and quality standards such as OSHA's process safety management (PSM) regulation, OSHA's personal protective equipment (PPE) regulation, EPA's risk management program (RMP) regulation, and the ISO 9000 quality standard, many companies are facing the daunting task of developing or upgrading their procedures to satisfy varied and sometimes complex and conflicting requirements.

Although there are overlapping characteristics that the various regulations and standards share, each approaches the procedure-writing process from a somewhat different perspective (e.g., quality or safety, protection of workers or protection of the public/environment), and the required level of detail for documenting the procedures differs greatly between these regulations and standards. For instance, the ISO 9000 quality standard advocates documenting procedures and work instructions that impact quality and suggests keeping the procedures as simple as possible, while the OSHA PSM regulation requires detailed procedures that address: all operating modes (startup, shutdown, etc.), operating limits, consequences of deviations, means to avoid hazards, safety and health considerations, and safety systems and their functions. How can companies develop and maintain procedures that ensure productivity and simultaneously satisfy the different regulations and standards? The key is to remember the ultimate goal of the regulations and standards: to reduce human errors that can impact quality, productivity, and/or safety.

Procedure-related errors are errors that occur because some characteristic of the procedure caused task performance to fail. ***This is currently the most critical human factor at most sites since 90% of accidents have at least one root cause related to mistakes within procedures.*** Reducing these procedure deficiencies can reduce human error rates by a factor of 2 to 10, normally. [1, 2]

The concepts and rules presented in this paper are applicable to all industrial settings. Most importantly, this approach **is effective in satisfying both safety and quality concerns.** With this approach, procedures can and have been developed that:

- Provide a sound basis for training
- Provide the users and supervisors with an understanding of what can go wrong and how it can go wrong
- Provides management with an understanding of the inherent risk (to quality, productivity, safety, etc.) of an activity (i.e., the activity addressed by the procedure) and an appreciation of the safeguards that are being relied upon to mitigate this risk.

Three (3) case studies are presented (with data from other sites as well), which illustrate this approach, along with supporting the benefits of the accurate and clear procedures. These case studies' results are representative of the benefits seen at more than 60 sites where the same approach has been followed.

Fundamentals of Writing Effective Procedures

When developing procedures, or any written material, always remember that the burden of written communication is on the writer, not on the reader. It should never be up to the reader to decipher the meaning of a message.

- Engineers who write procedure manuals "tend to include information that is more relevant to them than it is to the operators actually running the unit" [Sutton, 1992].[3]
- When writing procedures, your guiding principle should always be: if the written procedures are difficult to read, difficult to locate (e.g., if titles do not have a clear meaning), or inconvenient to use, they will seldom be used [Swain, 1983].[1]

Deficient Procedures are the most prevalent problem in process industries since procedures have not traditionally been developed from the perspective of optimizing human factors; instead procedures have been traditionally developed to meet a compliance requirement. Examples of procedure deficiencies (inaccuracies) include:

- Incorrect/incomplete/nonexistent (most procedures we have audited have been only 70-75% accurate – the inaccuracies include missing critical steps, steps as written are not what needs to be done, or the steps are out of sequence)
- No/misplaced/incorrect information in warnings (for example, a warning should **never contain** the action to take; it should instead **emphasize** the action to take) and warnings should be located BEFORE the step applicable to the warning
- Poor format and presentation rules

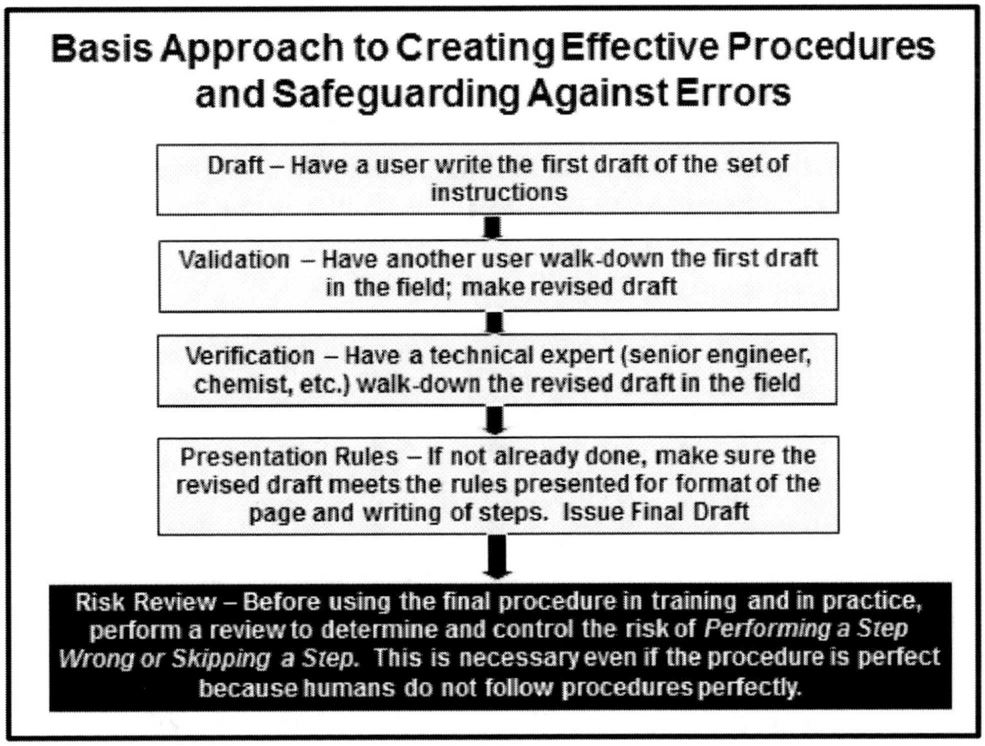

Figure 1: Best Approach to Developing written Procedures – courtesy of Process Improvement Institute, Inc. (PII), copyright 2003-2017[4]

The flowchart above describes the basic sequence of steps for writing effective procedures. Note that some basic rules are contained in the flowchart, such as "operators write operating procedures," "maintenance craft-persons write maintenance procedures," and "Lab technicians write lab procedures." Procedures are NOT written by engineers or department superintendents. This is a critical to ensure procedures are written in the user's common language and are written at the right level of detail.

Universal Purposes of a Procedure[4]

1. A written guide to accomplish initial training of employees
2. A quick reference manual for experienced (trained) employees

The Right Information to Write – Content is KEY[4]

The procedure procedural content controls the human error related to "procedure-based" error – if content is wrong, the trainee is very likely to learn the step wrong (trainer uses the procedure as a basis for training or the trainee refers to the procedure step that is wrong when executing the procedure).

Even if the content is correct, not following page formatting best practices can still increase human error rates.

If the content of the step is wrong it does not matter if the presentation is clear. If the content is perfect, you can further reduce error rates by 50% to 70% by presenting the information properly.

Ask experienced mechanics or operators to describe a simple task that they perform. You will quickly see that even simple tasks can involve many steps. You should consider the following questions when writing each procedure step:

- How is the step performed?
    - Does the action involve interacting with a computer terminal, an automatic controller, or simple devices (gauges, valves, micrometers)?
- Can the actions be performed as written and in the sequence written?
    - Can the equipment be operated as specified?
    - Can the steps be physically performed?
    - Does the user have the training or experience to understand and carry out the action using the information available, or is additional information needed?
- Does the user need to be alerted of any potential hazards (Cautions or Warnings) or need any supporting information (Notes) before performing the action?
- Does the user need to know specific operating ranges or limits to:

  - Perform this action
  - Recognize the successful completion of the action
  - Recognize an actual or potential problem
  - Make a decision

  For example, does the user have to verify temperature or pressure visually? Is the user waiting for a computer command or verification? How should the process react? Will exceeding a limit result in an accidental chemical release, personal injury, or quality concern? What is the chance of not operating within safe operating limits, and what would be the consequences? If operation of the component is critical and a deviation will probably cause a failure, state the corrective actions. If failure is unlikely, keep going.

- Is needed information found on an instrument, panel, or monitor or is it in the procedure or another source such as a graph, table, drawing, or specification sheet? Should this information be included in the procedure or be referenced?
- What is the next logical step? How is the next step affected by what is performed in the current step?
- What are the results of improper task performance? If an action is critical, spell it out in detail. For example, when starting a special pump that may bind or cavitate if not vented and primed properly, the procedure may require more information than simply stating "start the pump."
- Is the action frequently performed? Is it an action that is easily overlooked? Is this a complex piece of critical equipment that has a low frequency of use? These questions can help determine the level of detail in your procedure.
- Is the action performed so infrequently or is it so complicated that the user is not sure how to do it? Is the action so complicated that nobody is ever certain it's done right the first time?
- Is the decision point clearly defined if a decision is required? Unclear decision points can cause arguments and delays in performing actions.

To address the concerns listed above, a few simple rules are needed for developing content:

- The procedures should be written at the level of someone who has just completed the basic training for that task. Do not write the procedure for someone just hired or for the 10+ veteran.
- Clearly identify ahead of time what activities need procedures to help ensure error rates are controlled low enough, and identify which activities are considered common "skills" of all of the staff.

  Example: Starting a pump is typically considered a "skill." This means a procedure step can simply be "Start the benzene recycle pump (P-119)." The procedure does not need to explain how to start the pump. But, starting a pump is a skill that nonetheless must be learned; it requires doing several sub-steps related to positioning of intake and discharge valves, how the pump is throttled at startup, checking local pressure gauge, etc. So, a training module is needed for starting pumps of various types, and the operator needs to learn this necessary skill. Once the skill is learned, it will be applied to so many various cases and so often, that a procedure is no longer needed as a "refresher" on how to "start the pump."

- The first draft of the procedure must be walked-down in the field by another user. Simply reading through the procedure does not catch enough of the mistakes.
  The revised draft should be reviewed in the field by a technical staff person, such as a process engineer (for operating procedures).
- The procedure needs to be checked to ensure it follow the page format and step writing rules described in the next section. This ensures the accurate steps are clearly presented.

The target is to reach an accuracy of 95% or better (so, no more than one wrong or missing step out of 20 steps). The following have been observed in the field by PII staff by direct data collection at more than 110 sites/plants around the world, based on walk-down review of several to dozens of procedures at each unit/plant:

- At 95% accuracy or better and when the same procedures follow 80% of the rules for procedure clarity (presented next), then most users will follow the written procedures and will try to keep the procedures up-to-date
- At about 85% accuracy or less, about half of the users stop using the procedures
- At about 70% accuracy or less, less than 10% of the users will refer to the procedure or will try to keep it up-to-date. ***So, the written procedures are useless***
- Unfortunately, the typical operating procedure walked down by PII staff (accompanied by senior operators, a process engineer, and a shift supervisor) is about 75% accurate (so one step in four is missing or wrong). Usually these inaccurate procedures also follow less than half of the best practices for procedure format, presented next.

PHA/HAZOP of Procedures

Just as critical as getting the content accurate and presenting the steps clearly, is performing a PHA/HAZOP of the written steps to determine if you have enough independent protection layers (IPLs) against errors that the "imperfect" operators will eventually make when using the "near perfect" written procedures. This analysis is described in detail in Chapter 9 of the *Guidelines for Hazard Evaluation Procedures*, 3rd edition, 2008[5]. **For an update on the methods for such analysis and for a discussion of the extra emphasis on PHA of procedures by US OSHA, refer to the more recent paper (Bridges, Marshall; 2015).**[6, 7, 8]

Procedure Clarity

The importance of the accuracy of the procedure steps were discussed above. The clarity of the procedure steps (how they are written and how the page is formatted) is also important. Following best practices for step and page format reduces human errors by a multiplying factor of 3 to 5.[1, 9]

The best practice rules for writing and validating procedures have been published for many years *(see Bridges & Williams, 1997; Madden & Bridges).*[11, 12] These rules include best practices for formatting of the pages and steps. These have been gradually improved over the past decades and now are incorporated into PII's training materials.[4]

Table 1 below is a checklist (in auditing format) based on the current set of best practice rules for developing accurate and clear operating, maintenance, and other work-instructions/procedures.

TABLE 1 PROCEDURE QUALITY CHECKLIST *(courtesy PII, copyright 2003-2017)*[4]

#	Issue	Response
	Procedure Content Checklist	
1	Is the procedure drafted by a future user of the written procedure? (Engineers should not author procedures to be used by operators or maintenance staff.)	
2	Is the procedure validated by a walk-down in the field by another future user of the procedures?	
3	Is the procedure reviewed and commented on by technical staff (engineers or vendors)?	
4	Is the procedure checked versus the Page and Step format rules below?	
5	Is a hazard review of step-by-step procedures performed to make sure there are sufficient safeguards (IPLs) against the errors that will occur eventually (when a step is skipped or performed wrong)?	
6	Is the content measured using "newly trained operators" to judge the % of steps that are missing, steps that are confusing or wrong, and steps that are out-of-sequence? (A score of 95% accuracy of content is good; 98% should be the targeted average.)	
	Page Format Checklist	
1	Is the title of the procedure the largest item on the page?	
2	Is the procedure title clear and consistent with other titles, and does it uniquely describes the topic?	
3	Are the document control features the smallest items on the page?	
4	Are temporary procedures clearly identified?	
5	Is white space used effectively? • Is there one blank line between each step? • Does indentation help the user keep their place? • Are the margins large enough to reduce page congestion?	
6	Is type size is 12 pt font or larger?	
7	Is mixed case used for words of steps, with ALL CAPS used only for special cases (such as IF, THEN, AUTO, and WARNING)?	
8	Is the step number very simple (single level of number used)? Only an integer?	
9	Have sections or information not necessary to performing the steps been moved to the back or to another part of the manual or training guide?	
10	Are section titles bold or larger than the text font? Do sections have clear endings?	
11	Is the decision on electronic presentation versus hard copy correct? Are electronic linkages to procedures clear and accurate and easy to use? If paper is chosen, is the color of the paper appropriate?	
12	Is the overall page format (such as Outline format or Two-Column [T-Bar] format) appropriate to the use of the procedure?	
13	Are play script features added for tasks that must be coordinated between two or more users? • Play script is normally used when there are two or more hand-offs of responsibility for steps.	
14	Are rules followed for formatting of Warnings, Cautions, and Notes? (See annotated rules, such as Warnings are for worker safety and Warnings must clearly standout from rest of page.)	
	Step Writing Checklist	
1	Is each step written as a command?	
2	Is the proper level of detail used throughout? This is judged based on:	

	• *Who* will use the procedures • Same level of detail used in similar procedure steps	
3	On average, is there only one implied action per instruction? Best practice is to average 1.2.	
4	Does the procedure indicate when sequence is important? • If sequence matters, each step should be numbered (with an integer or letter) • If sequence does not matter, bullet lists should be used	
5	Are only common words used? Apply "education" level test (5 grade reading level is best)	
6	Do all acronyms, abbreviations, and jargon aid understanding? • Develop a list of such terms for use in procedures *and* communication. • Use terms that users use (within reason)	
7	Is each step specific enough? No room left to guess/interpret: • The meaning of a word or phrase (Check vs. Make sure) • The intent of a step or series of steps • A desired quantity or value • To what equipment the step applies	
8	Is the procedure free of steps that require in-your-head calculations? • Values expressed as ranges rather than targets with error bands • Conversion tables, worksheets, or graphs provided where needed	
9	Are graphics to the user's advantage? • No explanatory paragraphs or lengthy instructions that could be replaced by a picture • No impressive graphics that provide no real advantage	
10	Are references to the user's advantage? • No lengthy explanations or instructions that could be replaced by branching to a reference • No references to a procedure that references still another • No gaps or overlaps between this procedure and a referenced document • If branching, must branch to a procedure, not to a specific step in a procedure	
11	Are rules followed for writing warnings, cautions, and conditional steps? • No more than 2 per page • No actions within a warning or caution (actions must always be numbered steps) • Warnings and Cautions contain descriptions of potential consequences	

Each of these rules is explained in detail in the course *Writing Effective Operating and Maintenance Procedures*, by PII[4] (visit www.piii.com for more details). Each rule is backed by experimental or observed data that indicate there is a clear advantage in error reduction by using the rule. Below are a couple of examples from others on procedure writing rules.

> ***Proper Page Format:*** *In one study, Haney took a procedure presented in narrative format and converted it to column format. He then asked experienced technicians who had been completing the procedure using the narrative format to complete it as presented in the column format. He found that when the technicians used the column format, their errors were reduced by a factor of 3 compared with the narrative format [Haney, 1969].*[10]

> ***Including Supplemental Checklists:*** *Checklists can be effective aids to performance and often should be used to supplement procedures presented in any format. Checklists can be particularly effective in terms of preventing errors of omission. Some of the literature supports using checklists only for simple procedures that do not need to be performed under time or other constraints [Sutton, 1992].*[3] *However, there is also support for using checklists for emergency procedures in which a core set of steps is used to respond to emergency situations, particularly when the actions follow a simple sequence of steps for shutting down a process [Connelly, 1992].*[13] *Overall, experimental data have shown that the probability of errors of omission and commission can be reduced by a factor of 3 to 10 when checklists are*

used as reminders of key steps in a work instruction [CMA's A Managers Guide to Reducing Human Error, 1990].[9]

For more data on the history and development of these rules, see Bridges & Williams (1997)[11] and Madden & Bridges (2016)[12] especially refer to originating data and discussion for most of the rules (Swain, 1983).[1] **The procedure on the next page follows the rules** and is an example of the T-Bar format. Several other page formats are also valid.

Trouble-Shooting Guides (Procedures for Responses to Critical Deviation Alarms)

Trouble-shooting guides are a special form of operating procedure. They are written for the actions we want the operators to take to recover from a process deviation, *before an emergency situation occurs.* They are called guides since rarely can we predict the process conditions at the time the action is required. Trouble-shooting guides (and necessary training and drills) are required for any action that is considered a Human IPL. The Action Limit is what we show as the Min or Max in a Trouble-Shooting Guide. The action limit triggers the demand to use the trouble-shooting guide.

The starting criteria should be true before going to the effort of developing a procedure for response (before developing a trouble-shooting guide):

- The response is typically still possible, but it is time dependent. The time available is called the process safety time (PST). The operator must complete the diagnosis, make the necessary change(s), and make sure they are out of harm's way by the end of the Maximum Allowable Response Time (MART). [14]
- We usually set an alarm or a pre-alarm to trigger this action. This is usually before the shutdown triggers (ESD occur automatically) or release points (PSV set points) are reached
- The Min and Max shown in the Trouble-shooting Guide table are not the absolute safety limits for a system, but are instead some values that leave us some time to take action to prevent from reaching the absolute limits.
- There is still time to prevent or avoid the final consequence that could occur if we reach the ultimate limits of the process. Usually, we want the MART to be ½ or less of the PST, and we want MART > 10 minutes for trouble shooting in the field/plant and MART > 5 min for trouble-shooting only from the control room. [14]

Figure 2: Example of Procedure that Follows Most of the Rules in Table 1

PII CHEMICAL COMPANY
Unloading Monomer from Tanker to Storage

Page 1 of 2
Revised 11/4/02, Printed 2/21/16

Unloading Monomer from Tanker to Storage

STEPS	DETAILS
1. Wear standard PPE, plus rubber gloves and full-face organic respirator.	Standard PPE includes hard hat, safety glasses with side shields, and steel-toe shoes.
2. Weigh in the tanker.	Record weight on GROSS line of form.
3. Check bill of lading.	... to verify correct type of material is in tanker.
4. Sign in tanker driver.	Driver must sign in as a visitor and be escorted at all times.
5. Take Certificate of Analysis to QC.	***CAUTION: To avoid contamination, DO NOT unload until you receive approval from QC.***
6. Spot tanker.	A) Direct driver to location. B) Verify brake is set. C) Chock the wheels on at least one side between the two rear axles. D) Ground the tanker (attach grounding strap to an unpainted metal surface).
7. Have the storage area operator make sure the storage tank can hold contents of tanker.	***WARNING:*** *Failure to perform this step can result in an overflow and spill of hazardous Monomer.* Perform **Tank Gauging Procedure (SOP-01-804)** or check the storage tank load cell readout. Storage tank operating limits are stenciled on the control panel.
8. Place 5-psig nitrogen pad on tanker.	A) Connect from local 5-psig nitrogen drop to aft bottom nitrogen connection on tanker. B) Open nitrogen supply valve. C) Then open valve at tanker.
9. Connect Monomer unloading hoses.	A) Verify hoses and gaskets are in good condition. (Hoses are normally stored on hose rack at unloading spot.) B) Connect from tanker to air pump. (Use aft bottom connection on tanker. Remove cap.) C) Connect from air pump to storage tank. (Connect to storage tank line labeled "from TW.")

If the unit has a good PHA/HAZOP, then it is best to extract information from the HAZOP (or What-If) analysis tables to start the development of each guide. The guide is then finished with input from the process experts.[4]

Table 2 Examples of Creating a Trouble-Shooting Guide from a HAZOP Table

HAZOP Table Entry	**Trouble-Shooting Guide Entry**
Cause: *Bypass valve is open or passing*	Make sure the bypass is tightly closed
Safeguard: *Isolation valves for the vessel*	Isolate the vessel, if necessary
Safeguard: *Relief valve*	Make sure the relief valve block valves/relief path are open

The key categories of information needed in a trouble-shooting guide are:[4]

- IMMEDIATE ACTION (by system or by operator)
- DECIDE IF ALARM is REAL
- FINDING and FIXING the CAUSE
- FIX or BYPASS PROBLEM

The example below is of a trouble shooting guide for one critical alarm/action.

Figure 3: Example of Trouble-Shooting Guide that Follows Best Practices

Trouble-Shooting Guide

Alarm or Indicator:	**PAL 4446 – Low Pressure Alarm for Suction of Organic Feed Pump 40-PM-18.445**

Action Limit:	5 kPa		
Consequence:	Possible pump seal failure, releasing or spraying organic waste into the berm.		
Process Area:	FB&D Incinerator; Liquid Organic Liquid Feed	**Oper. Mode:**	Normal
Drawing #s:	D-400-PI-013		

IMMEDIATE ACTION (by system or by operator)

- DCS should shut down the organic feed pump (40-PM-18.445).
- From the DCS display, MAKE SURE the organic feed pump is shutdown.
- HAVE the field operator check for leaks near the organic feed pump.
- IF there is a large leak/release, THEN use the ESD switch to shutdown the unit and then follow/complete the shutdown and isolation procedure, OPS-ESD-117.
- IF there is a minor leak or no leak, THEN:
    - COMPLETE the rest of the trouble-shooting,
    - and DECIDE how to contain the leak for now,

DECIDE IF ALARM is REAL

- From the DCS, CHECK the pressure and feed tank level trends. IF the trends indicate the alarm if valid, THEN continue with finding the cause or fixing or bypassing the problem.

FINDING and FIXING the CAUSE

- CHECK valves upstream of the organic feed pump to see if any are closed too far, including checking ESD valves.
- CHECK, by feel with hand, if the heat tracing is on; IF Not, then TURN ON or open heat trace valves
- MAKE SURE nitrogen to the pump seal is at the normal operating pressure.
- CHECK if the line if plugged or frozen (skill)

ABC Chemical Company Prepared by: *Printed copy of this procedure is good for one job task duration.*	OPS-76-TSG-233	Incinerator Unit Revised 8/24/2015 Printed 2/21/16 Page 1 of 2

Figure 3: Example of Trouble-Shooting Guide that Follows Best Practices (continued)

Trouble-Shooting Guide

Alarm or Indicator:	**PAL 4446 – Low Pressure Alarm for Suction of Organic Feed Pump 40-PM-18.445**

- CHECK if the line if plugged or frozen (skill)
- CHECK if the level is actually low, use the Organic Feed Tank.
- IF the cause is low level in the feed tank, THEN resolve the problem if necessary based on cause that is found (skill)

FIX or BYPASS PROBLEM

- IF necessary, SHUT DOWN the Unit to allow fixing of the problem.
- FOLLOW the proper procedure to resolve problems (repair procedure, line clearing procedure, etc.)
- IF the decision is made to continuing operation without all equipment in normal condition, THEN:
    - FOLLOW Temporary Operating mode, if there is a temporary procedure already written for this possible problem/condition
    - FOLLOW MOC procedures to obtain approval for any non-standard temporary operating procedure or mode

END

Optimal Presentation of Trouble-Shooting Information

Ideally trouble-shooting information should be imbedded in the basic process control system (DCS) so that the operators can access the information on demand, with the click of mouse or key. Using the DCS for display of the steps for response to alarms (trouble-shooting), to be displayed "on demand" is becoming more of the norm each year.

Comments on Current Industry Consensus Guides for Writing Procedures

None of the consensus textbooks available for procedure writing cover all of the major points of this paper. The best "guidelines" for writing procedures currently exist in the course notebooks used by PII (and likely in the course notebook used by ABS Consulting, as they had similar origins).

Notably, the AIChE/CCPS textbook, *Guidelines for Writing Effective Operating and Maintenance Procedures*, 1996[15]:

- Contains NO description at all on the development, format and use of Trouble-Shooting Guides
- Mentions only about half of the rules listed in this paper (and the related PII course notebook) for formatting of pages and steps
- Emphasizes technical staff or professional procedure writers writing procedures, rather than the Users themselves writing the procedures (it violates the first rule of procedure writing)
- Fails to cover the competency needed by procedure writers
- Recommends and emphasizes having subsections in each work instruction/procedure (between the Title of the task and Step 1) rather than relegating such information either to other sections of the operating manual or rather than turning this information into steps
- Fails to mention the need for a PHA of procedures that focuses on hazard review of deviations from the steps; this is required by best practices and by US OSHA for PHAs under PSM

A better example, though still missing key best practices, is the ASM Consortium *Guidelines for Effective Procedural Practices,* 2010[16]:

- Includes a good discussion on the need and thought processes related to handling abnormal situations (trouble-shooting or response to critical deviations), but there is little description on the format and detailed content of the trouble-shooting guides. The guideline fails to differentiate between routine procedures for startup, shutdown, and online maintenance, and those used for response to critical alarms (trouble-shooting)
- Mentions only about half of the rules listed in this paper (and the related PII course notebook) for formatting of pages and steps
- Mentions that experience users themselves should write their own procedures, but it does not emphasize this rule and leaves the authorship open to engineers or a committee. It specifically mentions using retired senior operators for writing operating procedures. While this is one good way to retain institutional knowledge, we have found that the current users do a better job overall, once they have achieved competency in how to write procedures
- Mentions the competency needed for procedure writers in general
- Mentions the need for a PHA of procedures, but does not reference the method documented in Chapter 9 of *Guidelines for Hazard Evaluation Procedures*, CCPS/AIChE, 3rd edition, 2008,[5] or other related papers on the topic. Also, the description of the PHA is misleading because it focuses on how PHAs should be used in writing procedure content (which is fine for trouble-shooting guides) but fails to emphasize that a step-by-step analysis of deviations from the steps are needed to ensure that the process has enough IPLs against mistakes that can occur when imperfect humans use the procedure.

Though this guideline has deficiencies, it is worth the investment and provides useful insight not found elsewhere. Just be sure to address the deficiencies noted above.

Implementation of This Approach

PII staff has helped more than 60 sites implement the approach above. The typical approach for successful implementation is:

1. Perform a gap analysis to determine the average accuracy of procedure content, average clarity (as judged versus the procedure format rules), and identify any examples of excellent procedure writing.
2. Convince management of the problem and get their buy-in on the solution.
3. Train (1.5 to 2 days for operators; 1 day for lab or maintenance staff) and coach the users of procedures (normally for 3 to 8 days, to have them achieve full competency) so they can be effective authors. Normally, 20 to 25% of the users (hands-on operators, hands-on maintenance and inspection staff, lab staff) need to be trained as authors. Not all of those trained will make good writers. The organization needs to judge who does the best jobs. About 15 to 20% of the staff should end up being authors.
4. Get organized using project management principles for tracking and scheduling the effort. Use master spreadsheets such as those used at REC Silicon (case study presented later).
5. Use a Traveler form (described later) to make sure each new procedure gets a proper amount of review by other users and by technical staff, such as engineers.
6. Reward procedure writers for their efforts. Especially ensure there are no disincentives for taking the extra effort to write procedures.
7. Track the progress on the effort; fix problems as they arise; enforce schedules (to show management cares).
8. Perhaps use specialized software to make it easy to output the content of procedures into different user-friendly formats such as training tools, training competency records, and field checklists. WordPerfect (formerly from Corel) made this relatively easy with the use of Primary and Secondary merge files; MS Word makes this more difficult. As a result, some specialized software have been developed to help keep formats consistent and to help reformat the content, on demand.
9. Issue procedures through a document control system that does not add extra workload. This will of course make Management of Change (MOC) to procedures easier to track through the steps of change request, to risk review and technical review, and finally to issuing (and replacement of prior versions).
10. Of course, ensure there is JSA performed for procedures that have "personal injury" potential and perform a PHA (or mini-PHA for an MOC) of the steps in the procedures whenever there is "process safety" potential.
11. Train all affected workers in the new procedures.
12. Identify the best writers to become the trainers of new authors, over the long run.

CASE STUDY 1 – REC Silicon (note, the general approach used for the REC Silicon effort is the same used in nearly all of PII's procedure writing projects, including those described in CASE STUDY 2 and 3; the forms and general approach described in detailed only in CASE STUDY 1 to allowing presenting the approach in the context of actual plant work).

REC Silicon is one of the world's largest manufacturer and suppliers of Silane Gas and specialty silicon gases, Granular Polysilicon from Fluid Bed Reactor technology for solar photovoltaic applications and ultra-pure float zone Polysilicon for the electronic industry.

SOP Improvement Initiative

REC production facilities in the USA have been actively implementing process safety for many years, with a renewed interest in 2013 to exceed OSHA PSM requirements by meeting industry best practices. REC commissioned gap analyses at both the Butte and Moses Lake production sites to identify areas for PSM best practice implementation. One common finding was that the operating procedures and maintenance procedures needed improvement with respect to accuracy of the procedural steps and presentation clarity (page format and step format). The following are audit results pertaining to procedures:

- Procedure Inaccuracy
- Technical Staff were viewed as owners, not shop floor
- Annual review process ineffective
- Procedure formats needed improvement for reducing human error
- Inconsistent use of Warning, Caution, and Note
- Procedure Naming – not consistent
- Modes of operation contained in one procedure title
- Procedure accessibility was difficult in the legacy database; multiple searches needed to find tool for job
- Procedure Safety Reviews – did not utilize recognized methodologies for procedure safety reviews; i.e., there were no PHA of procedures with process hazards or JSA of tasks with personal hazards

Step 1 - Procedure Author Training

In early 2014, REC Silicon adopted Process Improvement Institute's (PII's) procedure author training module (*Writing Effective Operating and Maintenance Procedures*[4]) with minor customizations. The training consists of a 2-day class room lecture and about 6 exercises, followed by practical procedure development and field validation steps in the 2-day coaching phase. The classroom course covered all aspects of procedure writing with the focus on techniques that will lower human error rates when using the procedures. Topics included:

- human factors overview
- getting the content right

- page format rules and selection
- step format rules
- addressing operating limits (including creating trouble-shooting guides
- nuances of emergency shutdown and emergency operating procedures
- implementation guides and high level procedures for the task of getting all of the work instructions written
- examples/exercises on each topic

Step 2 - SOP Management System Revisions

REC Silicon revised its management systems to require SOP writing competency for all personnel developing or changing procedures and to reach full operator certification. The unit trainers have taken over leading the procedure author training. Operators took over procedural ownership from technical staff since these are the people performing the work in the field. Additionally:

- The quality and accuracy of procedures improved and subsequently the quality of operator training and certification program due to developing procedure writing competency
- The quality of the annual reviews and audits improved tremendously with the rewritten procedural format and content changes.

Step 3 - Developing Troubleshooting Guides Engineering Standard

REC Silicon developed an engineering standard governing the content and format for troubleshooting guides:

- *CCPS Guidelines for Engineering Design for Process Safety*[17] is the basis for determining the safe operating limits.
- Production engineers are the owners of these procedures.
- Engineering competency is required to develop the safe operating limits table based on the available process safety information.
- A consequence of deviation is then developed by the Process Hazard Analysis team.
- Involvement of certified operators is required to develop steps to correct or avoid exceeding the safe limit.
- REC silicon risk ranked its process areas and focused on developing safety troubleshooting guides for high risk process first.
- Over thirty six (36) guides were developed for the Moses Lake Facility.
- These have been developed in conjunction with PHA updates to the safe operating limits.
- Formal training and competency testing of these safety procedures.

Step 4 – Beginning and Managing The Re-Write Effort:

- Each operating department listed all existing operating procedures into the procedure development tracking form (also called a Traveler, because the form travels with the draft procedure until the procedure is final; see Figure 4 on the next page).

- Each department also developed a list of "skills" that are used across multiple procedures and that operators or maintenance staff are required to attain before using complex procedures.
- Procedures were screened for safety, equipment damage, or quality risk to determine whether 2, 3 or 4 peer field walk-downs were required (see validation traveler workflow in Figure 4).
- The procedure re-write assignments were divided up between all four rotating shifts and the superintendent assigned certified authors (e.g., operators trained/coached as procedure writers) from their pool.
- Each department manages their pool of qualified authors. Cross shift validation was utilized to ensure all operating approaches were considered. The assigned author was responsible for consolidating red line comments.
- A **Validation Traveler** was developed and used to ensure all steps in the SOP re-writing workflow were completed.

Figure 4 - A **Traveler form** shown to the right accompanies each new or revised procedure through the writing and editing/review process. When complete, the traveler form is then kept in the QA file for the procedure to validate long-term how the initial procedure was developed.

Procedure Development Traveler - Moses Lake			
New Document Number:		Area:	
Procedure Title:			
Procedure Type:			
Author:			
Initial Draft Completion Date:			
Author Walkdown Date:			
Author Revision 1 Completion Date:			
Peer 1:		Shift:	
Peer 1 Walkdown Date:			
Optional Peer 2:		Shift:	
Peer 2 Walkdown Date:			
Optional Peer 3:		Shift:	
Peer 3 Walkdown Date:			
Optional Peer 4:		Shift:	
Peer 4 Walkdown Date:			
Author Revision 2 Date:			
Engineer/Verifier:			
Verification Completion Date:			
Author Revision 3 Completion Date:			
QA Person:			
Author Revision 4 Completion Date:			

- A color coding process was used to flag SOP content that would be deleted, or moved into the unit training manual.
- A total of 184 Moses Lake procedures were tracked through the development, review, and revision process.
- **The overall tracking form below was used to develop KPIs to show accomplishments and to meet set completion dates.**

Figure 5 Procedure Development Project Status Table

***Images and layouts copyrighted by PII, 2008-2017*[4]**

Procedure Development Tracking Form - Moses Lake

							Initial Development			Validation				
Item	New Doc #	Old Doc #	Area	Procedure Title	Proc Type	Author(s)	Initial Draft	Author Walkdown	Author Rev 1	Peer 1 Walkdown	Opt. Peer 2 Walkdown	Opt. Peer 3 Walkdown	Opt. Peer 4 Walkdown	Author Rev 2
1	FBP.3.xxx	FBP.3.160	FBR	Hydrogen Booster Compressor Cleanup	Cleanup	LT/FG	3/19/14	3/19/14	3/20/14	3/21/14	3/21/14			
2	FBP.3.xxx	FBP.3.160	FBR	Hydrogen Booster Compressor Operation and Op Restrictions	Operation	LT/FG	3/20/14	3/20/14	3/20/14	3/21/14	3/21/14			
3	FBP.3.xxx	FBP.3.160	FBR	Hydrogen Booster Compressor LOTO	LOTO	BCA/OL	3/19/14	3/19/14	3/21/14					
4	FBP.3.xxx	FBP.3.160	FBR	Hydrogen Booster Compressor Standby	Standby	BCA/OL	3/20/14	3/20/14	3/21/14					
5			4.0	STC/TCS from Receiving Tank to STC/TCS Storage Transfer Through Metering Line	Transfer	ME/RR	3/19/14	3/19/14	3/20/14	3/20/14	3/21/14	3/21/14	3/21/14	
6			4.0	STC/TCS from Receiving Tank to STC/TCS Storage Transfer Through Bulk Line	Transfer	ME/RR	3/19/14	3/19/14	3/20/14	3/20/14	3/21/14	3/21/14	3/21/14	
7			4.0	H2 Recycle Compressor Switching to Standby	Standby	ME/RR	3/20/14	3/20/14	3/20/14	3/20/14	3/21/14	3/21/14	3/21/14	
8			4.0	H2 Recycle Compressor Cleanup	Cleanup	ME/RR	3/20/14	3/20/14	3/20/14	3/20/14	3/21/14	3/21/14	3/21/14	

- From PII's procedure author training,[4] the Quality and Content checklist was modified and used as quick reference guide to make sure that all the content development and formatting rules are being followed during the re-write process. We used Microsoft Word and made templates for the T-bar, Flat and Consequence of Deviations.

<u>Step 5 - Procedure Management of Change</u>

REC Silicon adopted formal safety review methods for procedures and were included in a new procedure management of change procedure. The REC Silicon IT department developed a custom tool to execute these requirements in an electronic workflow. Benefits included improving consistency and quality of MOC procedural safety and health reviews and improving the overall quality of the SOPs with respect to clarity and content; and helped identify human factor deficiencies such as with human machine interface (equipment accessibility, operator feedback, labeling, missing or deficient instruments/controls).

Overall Summary of the procedure writing initiative:

For the Moses Lake facility, 398 procedures (operating, maintenance, and trouble-shooting guides) have been developed through this initiative (new and rewrites combined.) REC Silicon has seen great improvements based on the following feedback from workers involved in the initiative:

- Operators are receiving better work instructions
- Sequence of steps is well understood
- Much easier to audit operator compliance to procedure
- **Standardizes how we do work – protects our equipment from damage, protects our people, protects the environment**
- No longer need 20 year operator tribal knowledge and interpretation to complete procedure
- Procedures are not cluttered anymore, they just have the necessary information
- Procedure element now requires involvement of many shop floor workers

- JSA technique is alive on the shop floor
- Procedure can be trusted
- Technical staff no longer doing all procedure work. Freed up to optimize processes and realize other cost saving benefits.

REC Silicon's procedural initiative's success was due to forming a cross-functional project team of key decision makers from all the correct disciplines and having a project sponsor who could remove barriers. REC Silicon held weekly meetings, set short term goals, and tracked KPIs to meet those goals. Leadership was key to support the initiative and the related culture change.

CASE STUDY 2

As with REC Silicon, PII worked with ROQUETTE in Keokuk, IA (USA) to develop operating procedures for a process that a food additive. There were many procedures for this process and the detailed procedures for each task had never been written. Many of the older staff were being replaced with younger staff and the company decided to take this opportunity to develop detailed procedures to help ensure the consistency of training and coaching of the new workers.

Since there were limited experienced staff, the company trained the newly hired staff (many of whom had years of experience at other companies) in how to write and validate operating procedures. In addition, since ROQUETTE could not spare sufficient workers for the writing effort, PII provided 2 senior staff onsite for 2 weeks. Five (5) workers were trained and coached in procedure writing. These PII senior staff and the newly trained writers (who were also new operators) then worked with the experienced operators to outline the procedure for each task and then develop the best sequence of steps for each task. About 50 tasks were identified that needed procedures. 24 were drafted in the first week onsite and a similar number on the second week onsite. The forms shown in Figure 4 and 5 above were used extensively in this project.

In between the onsite visits by PII, then newly drafted procedures were walked down by the senior operator and by the other new operators. The revised procedures were finally issued for use. Even though the newly hired operators were inexperienced in the process operation, the ones that were trained as procedure writers developed several procedures (with the senior operator input) without help from the PII writers.

The company is considering expanding this effort to the many other process in Keokuk and at their other sites around the world.

CASE STUDY 3

As with REC Silicon and ROQUETTE in the USA, PII worked with Bahrain National Gas Company (BANAGAS) in Bahrain to develop operating procedures for their two gas process plants and for the gas compression stations upstream of the gas plants. Although the native language of most of the workers is Arabic or Hindi, the official language of the company is English.

This effort was part of a larger project to full implement PSM (and ongoing project that began in 2015 and is scheduled for completion in 2018). There were hundreds of operating procedures for these processes and the detailed procedures for each task had never been written. Similar more than 50 procedures were identified for maintenance activities related to repair, inspection, test, PM, calibration, etc.

The company has many very experience operators and maintenance technicians and many new operators. PII trained 12 operators and 7 maintenance technicians how to write and validate operating procedures. PII provided 2 senior staff onsite for 3 days of coaching following the 2-day course. These PII senior staff and the newly trained writers then worked with the other experienced operators to develop the best sequence of steps for each task and to get the procedures properly formatted. More than a dozen procedures (a combination of operating procedures and maintenance procedures) were written while PII was onsite; the main function of PII staff was to coach newly trained writers as they developed procedures for their departments. About 12 of the 18 BANAGAS writers that were trained and coached reached reasonable competency by the end of the first week (2 days of training and 3 days of coaching). During subsequent visits to the site by PII staff and between site visits via e-mail, PII continue to critique and coach the new minted procedure writers at BANAGAS.

As in the other two companies listed in these case studies, the forms shown in Figure 4 and 5 above were used extensively in this project, and continue to be used as the effort is ongoing. In addition, PII developed a flowchart for the procedure development process (see Figure 6) and developed a procedure writing Guide for use by the BANAGAS procedure writers.

The procedure writing effort is nearing completion, with very minimal input required from PII staff.

OTHER CASE STUDIES

PII has used the approach above since 2003 at more than 60 plants/sites in North America, South America, Australia, Europe, and the Middle East. The approach has consistently yielded great results; accurate procedures and clear formats. In these cases, the local operators and maintenance staff (the trained and now competent writers) put in 95% to 99% of the procedure writing effort.

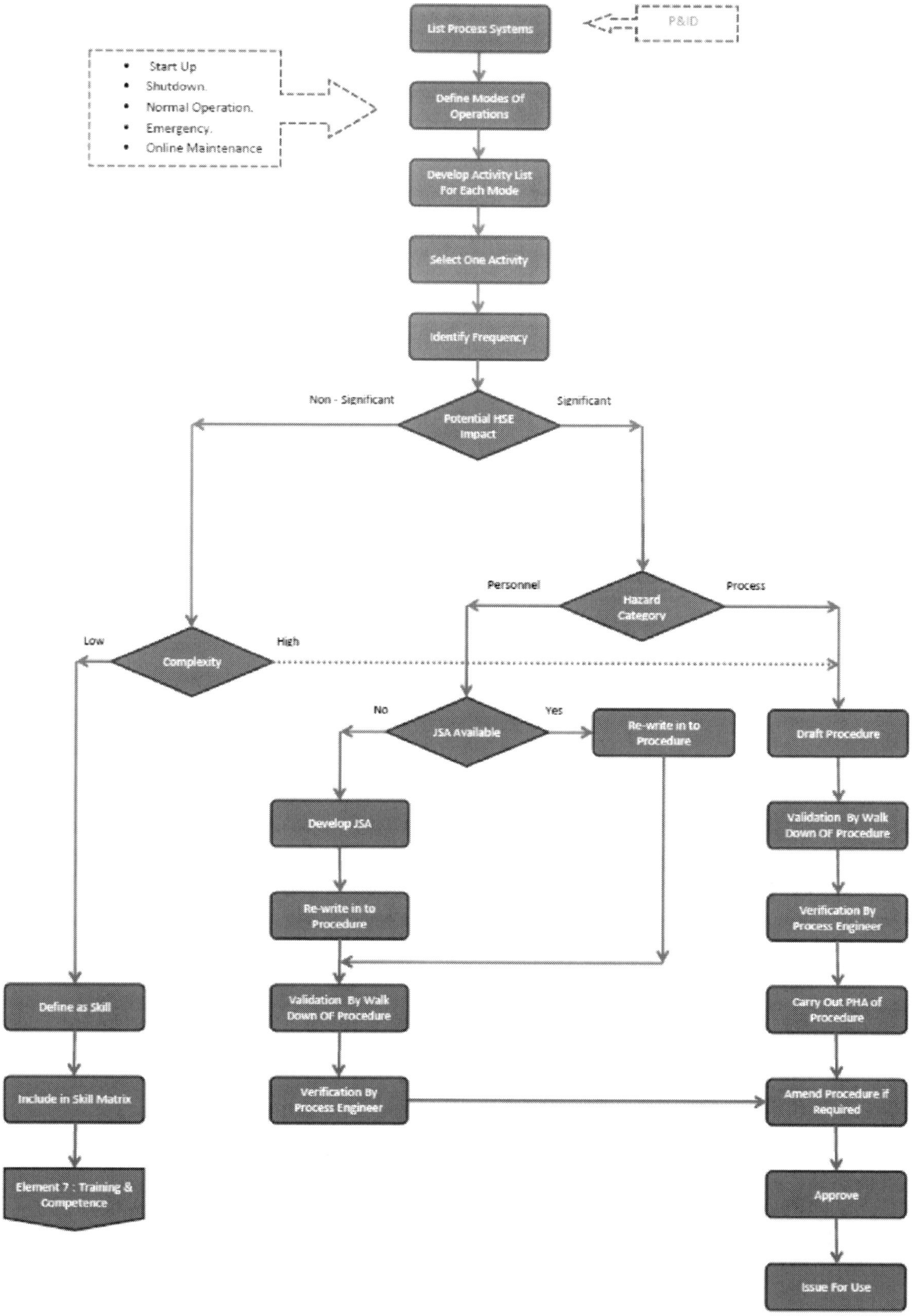

Figure 6 Procedure Development Flowchart (copyright PII, 2015-17)

Conclusions

Having clear and accurate procedures is a must. Poor procedures show up as one of the root causes in 90% of accidents. The approach described in this paper has worked for more than 35 years for one of the authors, and continues to work today. Controlling risk is mostly about controlling human error rates, and having accurate and clear procedures is one key to this control. One other gain from the approach described in this paper is that the users become the owners, and therefore they have a vested interest to keep the procedures up to date and clear.

No doubt, writing all of the procedures for a facility is a large task. But, it can be accomplished by site staff, with support from company process safety or quality control leadership, after enough users become competent authors. Attaining this competency takes less than 2 weeks of work, on average.

Acronyms Used

AIChE– American Institute of Chemical Engineers
ASM – Abnormal Situation Management consortium
BANAGAS – Bahrain National Gas Company
CCPS – Center for Chemical Process Safety (a division of AIChE)
CMA – Chemical Manufacturer 's Association, now American Chemical Council (ACC)
FBR – Fluidized Bed Reactor
HAZOP – Hazard and Operability Analysis
IPL - Independent Protection Layer
JSA – Job Safety Analysis
LOPA – Layer of Protection Analysis
MART – Maximum Allowable Response Time
MOC – Management of Change
NUREG – Nuclear Regulation, US Nuclear Regulatory Commission
OSHA – Occupational Safety and Health Administration, US Department of Labor
PHA – Process Hazard Analysis
PII – Process Improvement Institute, Inc.
P&ID – Piping & Instrumentation Diagram
PSI – Process Safety Information
PSM – Process Safety Management
PST – Process Safety Time
SOP – Standard Operating Procedure

References

1. Swain, A. D., Guttmann, H. E., *The Human Reliability Handbook - with Emphasis on Nuclear Power Plant Applications*; NUREG/CR-1278, U.S. Nuclear Regulatory Commission, Division of Facility Operations, 1983.

2. Gertman, D., Blackman, H., Marble, J., Byers, J. and Smith, C., *The SPAR-H Human Reliability Analysis Method,* NUREG/CR-6883, U.S. Nuclear Regulatory Commission, Office of Nuclear Regulatory Research, 2005.
3. Sutton, Ian S., "Writing Operating Procedures," *Proceedings of the International Conference on Hazard Identification and Risk Analysis, Human Factors and Human Reliability in Process Safety*, AIChE, Center for Chemical Process Safety, New York, NY, 1992.
4. *Writing Effective Operating and Maintenance Procedures*, Course Notebook, Process Improvement Institute, Inc., 2003-2017.
5. *Guidelines for Hazard Evaluation Procedures*, 3rd Edition, 2008, CCPS/AIChE.
6. "Necessity of Performing Hazard Evaluations (PHAs) of Non-normal Modes of Operation (Startup, Shutdown, & Online Maintenance)", W. Bridges and Mike Marshall (US OSHA), *18th Annual International Symposium, Mary Kay-O-Connor Process Safety Center, College Station, TX, October 2015. (also updated at 12th GCPS, 2016)*
7. "How to Efficiently Perform the Hazard Evaluation (PHA) Required for Non-Routine Modes of Operation (Startup, Shutdown, Online Maintenance)," W. Bridges and T. Clark, *7th Global Congress on Process Safety*, Chicago, AIChE, March 2011.
8. Bridges, W.G., et. al., "Addressing Human Error During Process Hazard Analyses," *Chemical Engineering Progress*, May 1994
9. *A Manager's Guide to Reducing Human Error*, 1990, Chemical Manufacturers Association (CMA)
10. Haney, R. W., "The Effects of Instructional Format on Functional Testing Performance," *Human Factors*, 11, 1969, pp. 181-187.
11. "Effective Procedure Writing – A Two-Part Series (Part 1: Reducing Risk of Worker Errors; Part 2: Addressing Process Safety Management Requirements for Operating Procedures)," *Chemical Engineering Progress*, 1997, W. Bridges with T. R. Williams (both of JBFA).
12. Madden, L., and Bridges, W., Best Practices for Writing Operating Procedures and Trouble-Shooting Guides, L. Madden and W. Bridges, *12th Global Congress on Process Safety*, Houston, TX, AIChE, March 2016.
13. Connelly, C. S., "Emergency Operating Procedure Writing for Oil Refinery Process Units," *Proceedings of the International Conference on Hazard Identification and Risk Analysis, Human Factors and Human Reliability in Process Safety*, AIChE, Center for Chemical Process Safety, New York, NY, 1992.
14. *Guidelines for Initiating Events and Independent Protection Layers*, 2015, CCPS/AIChE.
15. *Guidelines for Writing Effective Operating and Maintenance Procedures*, 1996, CCPS/AIChE.
16. *ASM Consortium Guidelines – Effective Procedural Practices*, 2010, ASM Joint R&D Consortium.
17. *Guidelines for Engineering Design and Process Safety*, 2nd Edition, 2012, CCPS/AIChE

478864 Process Safety Management: It's Not Just for Chemcal Plants!

Abstract:

Process Safety: It's Not Just for Chemical Plants & Refineries!

Presented By

Richard W. Sarnie, CSP, P.E., ARM-E

Director, CCPS Projects

Center for Chemical Process Safety

American Institute of Chemical Engineers

In this presentation Mr. Sarnie will discuss lessons learned and relearned over decades in manufacturing and cold storage facilities on preventing and mitigating process safety incidents (LOSS OF CONTAINMENT EVENTS) and how these lessons learned can and should be applied across many types of manufacturing and cold storage operations such as the food, beverage, pharmaceutical and nutraceutical industry. He will:

- Present a rationale as to why a **risk-based** rather than **regulatory-based** approach to managing process safety and other risk exposures is prudent
- Discuss the continuing movement by many organizations to more formal and comprehensive enterprise risk management systems
- A review of the typical steps or phases an organization usually takes in developing and improving management system processes
- Provide an overview of chemical industry best practices for managing process safety (with a focus on AIChE CCPS recommendations as described in the **CCPS Guidelines for Risk Based Process Safety** and many other CCPS books)

- Describe numerous process safety management system deficiencies (root causes) associated with serious chemical process safety incidents that have repeatedly occurred during the last 50+ years
- Discuss how proven process safety management practices and use of multiple layers of protection can help better manage risks in manufacturing and cold storage facilities.

Hit the Ground Running (Safely):

Process Safety for the Early Career Engineer

Jerry J. Forest, CCPSC
Celanese
225 E. John Carpenter Freeway Suite 1200, Dallas, TX 75234
Jerry.Forest@Celanese.com

Don Connolley, CSP
BP Corporation
Don.Connolley@bp.com

Amy Gray, Ph.D., PE, CFEI
ESI
aegray@engsys.com

Prepared for Presentation at
American Institute of Chemical Engineers
2017 Spring Meeting and 13th Global Congress on Process Safety
San Antonio, TX
March 26 – 29, 2017

Hit the Ground Running (Safely):

Process Safety for the Early Career Engineer

Jerry J. Forest, CCPSC
Celanese
225 E. John Carpenter Freeway Suite 1200, Dallas, TX 75234
Jerry.Forest@Celanese.com

Don Connolley, CSP
BP Corporation
Don.Connolley@bp.com

Amy Gray, Ph.D., PE, CFEI
ESI
aegray@engsys.com

Keywords: New Hire, Training, Competency, Mentoring

Abstract

Engineers entering the workforce are frequently expected to participate in process safety elements such as PHA, MOC, incident investigation, mechanical integrity programs and more. Although there is a recent ABET requirement to include process safety in engineering curricula, many schools do not help students define what to expect in their first year.

This paper will discuss common process safety duties, activities, and responsibilities expected of new engineers. The paper will follow the recently published CCPS book *Process Safety for Undergraduates and Engineers*. It will also introduce the AIChE's Health and Safety Division's Professional Outreach concept on mentoring. The mentoring project will provide a platform to connect experienced process safety engineers with student and local chapters, and individuals who are seeking assistance with specific process safety topics.

1. Introduction

When engineers enter industry for the first time they are often surprised that one of their first assignments includes an onboarding process that can include several years of training. The material in these professional development courses is usually not covered in school. Companies may use different onboarding tools and the early career engineer must learn the company specific methodologies for process safety elements.

This paper discusses an overview of how new engineers can hit the ground running with respect to process safety when first entering industry. It includes a brief overview of the Accreditation

Board for Engineering and Technology Inc. (ABET) requirements for university chemical engineering students, typical onboarding training in the first few years of employment, and the type of process safety duties a new hire engineer can expect. The paper also discusses a new pilot process safety mentoring program from the American Institute of Chemical Engineers (AIChE) Safety and Health Division designed to help new hire engineers hit the ground running safely by drawing on those who have years of industrial process safety experience.

2. T2 Laboratories and ABET Requirements

In 2007, an explosion occurred at T2 Laboratories in Florida, killing 4 employees and injuring 32 including members of the surrounding community. The U.S. Chemical Safety Board (CSB) investigated the incident and found that the primary cause of the incident was a lack of understanding of the reactive hazards involved with the process. A key recommendation from the incident is addressed to the AIChE:

"Work with the Accreditation Board for Engineering and Technology Inc. to add reactive hazard awareness to baccalaureate chemical engineering curricula requirements." [1]

AIChE responded to this recommendation by working with ABET to require not just reactive chemical hazards in chemical engineering curricula, but all chemical process hazards. [2] Given that ABET accredited schools must address all chemical process hazards in their undergraduate curricula, the first way early career engineers can hit the ground running is by understanding the process safety topics presented in their curricula.

The Center for Chemical Process Safety (CCPS) has gone beyond the original intent of the CSB recommendation with two additional initiatives. The first is publication of *Introduction to Process Safety for Undergraduates and Engineers* (hereafter referred to as *Introduction to Process Safety*) and the second is the undergraduate process safety education initiative.

Introduction to Process Safety supplements Crowl and Louvar's *Chemical Process Safety: Fundamentals with Applications* [3] by introducing the student to Risk Based Process Safety. The book is designed to be used as a stand-alone course book or as supplemental material for several chemical engineering courses. *Introduction to Process Safety* was published in 3rd quarter of 2016 and early adopters include Louisiana State University and Kansas State University. Celanese is presently evaluating making the book required reading for all new hire engineers and is considering an onboarding course from the material.

The second initiative in progress from CCPS that goes beyond the original intent of the T2 Laboratories CSB recommendation is an undergraduate process safety initiative. This initiative builds on AIChEs existing Safety and Chemical Engineering Education (SaChE) program and *Introduction to Process Safety* with four key elements:

- Curriculum – enhancing SaChE e-learning modules and translating to multiple languages
- Teaching – Educate professors about process safety with workshops at industrial sites
- Addressing gaps – with industry led "Boot-Camps" for students on campus and
- Maintaining focus – funding key research in process safety [4].

The early career chemical engineer has an opportunity to hit the ground running safely by taking advantage of the material presented by their undergraduate curriculum and by utilizing the resources available by the CCPS. These include the aforementioned SaChE modules, "Boot-Camps", and the resource books *Introduction to Process Safety for Undergraduates and Engineers* and *Chemical Process Safety: Fundamentals with Applications*.

3. Onboarding & Training for New Hires

Unfortunately for the new hire engineer, each company will address the various process safety elements in different ways and with differing methodologies. For example, while management of change (MOC) processes may have common elements of risk identification and mitigation, approval, and action item identification and closure, the methodologies used for these elements may differ among companies. While the fundamental principles will look familiar to what was presented in the undergraduate curriculum, the specifics particular to each company may be unfamiliar. Therefore, it is very common for an early career engineer to undergo an onboarding process that might include several years of professional development training.

The final onboarding content is a function of the hiring company and the job position of the early career engineer. The process safety courses given can be mandatory or optional and the material can be delivered by an internal training department or offered by third party vendors. Whatever the case, the onboarding process safety training represents an excellent opportunity for the early career engineer to hit the ground running.

Early career engineers typically are immediately immersed in the following risk based process safety elements: process safety information (PSI), process hazard analysis (PHA), management of change (MOC), pre-startup safety reviews (PSSR), conduct of operations, mechanical integrity (MI), standard operating procedures (SOP), training, metrics, and incident investigation. The extent to which early engineers contribute is usually based on training and experience. Therefore, it is in the engineer's best interest to pursue at a minimum, initial training on these topics.

4. What to Expect in the First Year

Although this is a paper on how to hit the ground running with regard to process safety, we would be remiss if we did not mention occupational safety expectations for the early career engineer. Note the majority of this section is taken from *Introduction to Process Safety* [5]. It is safe to say that most U.S. chemical manufacturers place an enormous emphasis on occupational safety – to the extent that they have been criticized by OSHA for placing an emphasis on occupational over process safety. Nevertheless, all engineers should understand their company's expectations for occupational safety and become a model leader by demonstrating behaviors consistent with standards and regulations and exceeding expectations set by the company leadership in this area. Activities that early career engineers can expect include:

- Leading and participating in safety meetings and shift safety meetings
- Unit inspections
- Life critical procedure audits
- 8 – 24 hours of occupational safety training per year

- And more, depending upon an individual company's requirements

4.1 Social Interactions

Engineers entering the manufacturing environment are thrust into social interactions for which many are ill prepared. Operators and maintenance crafts have often been part of the same team for years and have unique understanding of their units even if they do not have the vocabulary to express the physical processes in the same way as the new engineers. The main thing for an early career engineer to understand is to approach these groups with great respect. They have a deep understanding of process that is only gained by experience. To become a mentee of these experienced individuals, one must first respect their knowledge, abilities, and attitudes gained by experience.

This section comes with a warning. The early career engineer who adopts an attitude of knowing more than those hourly employees that have run the manufacturing units for years will most certainly fail. The reason simply is the early career engineer does not know as much as these experienced hands. The engineer may possess an understanding of first principals, but must also gain an understanding of the physical process in order to apply first principals.

Learning from operators and crafts will accelerate the learning curve for a new engineer and provide a way to apply technical skills learned at university in a practical way that helps the company achieve results. Interacting with the operators and crafts is also a way to eventually gain their respect and trust, and learn how to function well as a team.

4.2 New Skills

Many assignments in the university curriculum include group projects, formal reports, and working with teams. They are designed to prepare students in doing the same activities in their new engineering jobs. New skills in industry related to process safety include assembling process safety information, leading project teams, training operators, writing standard operating procedures, scribing or leading hazard reviews and management of change hazard reviews, delivering presentations and writing formal technical reports (just to name a few skills).

One of the most important skills to acquire is the ability to concisely document reasoning and thinking related to solution of items. Findings and action items are generated from process hazard analysis, management of change, incident investigation, audits, and other team meeting activities. The early career engineer is often assigned much of this important workload and must find solutions to the problems and document with proof that concern is addressed. In many cases, this documentation is regulatory requirement and becomes part of the operating record for the life of the process.

4.3 Conduct of Operations & Operating Discipline

One process safety element that has been overlooked in university engineering curricula is conduct of operations. Conduct of operations is a term that "describes human factors and tools that are necessary to produce repeatability in performance and consistency in results" [5]. While much of process safety is concerned with maintaining the integrity of engineered controls, conduct of operations is concerned with maintaining the integrity of administrative and human controls.

The day in a life of an operator can be described by discrete activities performed. These include: reporting to work outfitted, making relief in the field, attending shift change meetings, informal and formal evaluation rounds, issuing permits, preparing equipment for maintenance or operation, operating equipment, and documenting each activity. Early career engineers are often asked to participate in several or all of these aspects of operation.

For example, operators usually document their activities in a log book. Engineers should monitor the operator logs to gain a deeper understanding of unit activities. When problems are documented, engineers should take action. Failure to do so leads to an acceptance of problems and a normalization of deviance. Furthermore, engineers might be asked to create logs and periodically train operators on how to use them.

Other examples of an engineer's role in conduct of operation include participation in equipment evaluation. An engineer might create evaluation sheets by mapping out how an operator should travel through the unit, and define exactly what equipment should be evaluated and how it is to be evaluated. Then, by periodic monitoring of the evaluation rounds, engineers better understand normal operation and are better equipped to solve problems.

Early career engineers are often called to participate in all aspects of conduct of operations and should be familiar with the operating discipline tools to be effective. This familiarity is only gained by experience as it is rarely formally taught in training.

4.4 Engineering Discipline

Engineering discipline is closely related to operating discipline in that it describes activities that engineers and technical personnel do to monitor key operating parameters in the plant to keep the unit running at peak efficiency and normal safe operation. This new skill involves the selection and monitoring of key performance indicators and determination of, monitoring, and evaluating safe operating limits. Engineers monitor trends and take action to prevent process safety incidents based on their understanding of the hazards and risks gained from process safety information and hazard identification and risk assessment processes of their company.

4.5 Process Safety Duties

Early career engineers will be active in most of the 20 elements of risk based process safety. In addition to the company specific training described earlier, participation in these elements will be required. A few common areas are shown in table 1, summarized from *Introduction to Process Safety* [5].

Table 1: Process Safety Activities for New Engineers

Element	Activity
Process Knowledge Management	Ensure accuracy of PSI Be familiar with Safety Data Sheets Develop reactivity matrix Develop and update design basis calculations Develop and update P&IDs Verify electrical classification Maintain equipment files
Hazard Identification and Risk Assessment	Participate in HIRA, & PHA Assemble required PSI Review applicable RAGAGEPs Compile industry incidents Scribe, document reviews Resolve findings
Operating Procedures	Develop and update operating procedures
Safe Work Practices	Write and approve permits
Asset Integrity and Reliability	Identify equipment covered by inspection, testing, and preventative maintenance (ITPM) programs Write ITPM procedures Analyze inspection results & maintain records
Contractor Management	Approve and train contractors Write contractor procedures
Management of Change	Identify changes Participate in MOCs Assemble required PSI Resolve items
Operational Readiness	Participate in readiness reviews Resolve items
Emergency Management	Learn emergency response plans Participate in drills
Incident Management	Participate in investigations Resolve items
Metrics	Record and maintain data Analyze data
Audits	Assemble requested information for audits Become a certified auditor

The main point of this section is that there is much more to process safety for early career engineers than is generally presented in the university curricula even with the new ABET requirement. While many companies will have onboarding and training to prepare the engineers, many of the skills illustrated above are gained by experience. This is just a sampling of what early career engineers can expect in their first year.

To help engineers prepare for these activities in their first few years, the Safety and Health Division of AIChE is piloting a mentoring program.

5. AIChEs Safety and Health Division Pilot Mentoring Program

The American Institute of Chemical Engineers (AIChE) Safety and Health Division's mission is:

"To share, promote, and shepherd process and personal safety for individuals in the field of process safety."

In support of the mission the Division has formed several committees including Professional Outreach, Student Outreach, Communication, and Membership. The Professional Outreach Committee has identified a potential avenue to help the early career engineer hit the ground running safely by creation of a process safety mentoring program.

The objective of the mentoring program is to connect AIChE members experienced in process safety with members new to the area of process safety for the purpose of providing guidance and a resource point. It is envisioned that eligible mentors will be engineers who have at least 10 years of experience in process safety. This experience could have been gained through a variety of pathways. Reviews of the volunteer mentors' experience will be performed by the Safety and Health Division Mentoring Committee.

Eligible mentees will be early or transitional career professionals with less than 5 years of experience in the area of process safety. This career can be in a variety of industries. Reviews of potential mentees' eligibility will be performed by the Mentoring Committee.

The mentoring program is in a pilot stage with plans to focus on Louisiana to solicit mentors and mentees. It is expected that applications for mentors and mentees will be available on the Safety and Health Division website and will include information on where to return information. The mentor application will include name and company info, time commitment desires, areas of expertise, and a brief statement on why the applicant wants to be a part of the Mentoring Program. The mentor should attach an updated resume and provide 3 references. Mentee applications will require similar information and can be submitted by individuals or engineering supervisors interested in pairing mentors with groups of young professionals.

The Safety and Health Division Mentoring Committee will review all Mentor applications and approve or reject based on experience, reputation in industry, and referral comments. Mentors and Mentees will be paired by the Safety & Health Division Mentoring Committee. Decisions will be based on location, industry, and desired commitment level.

At the time of writing this paper, the pilot mentoring program is in its infancy. Those interested in being either a Mentor or Mentee should contact a member of AIChEs Safety and Health Division Executive Committee.

When it is fully established, we anticipate that this Safety and Health Division Mentoring Program will provide an excellent opportunity for the early career engineer to hit the ground running.

6. Conclusion

The early career Chemical Engineer quickly learns that education and competency are an integral part of job duties and in fact, will be an ongoing part of the job during the course of a career. A large portion of this continuous learning is gaining expertise in occupational and process safety; skills that often are not covered in depth at many universities.

It is in the early career engineers interest to gain competency in process safety as the elements of process safety are weaved throughout all aspects of an engineer's job. The CCPS has published many resources that provide excellent background in process safety. Most companies have an onboarding process that details several years of process safety training. Engineers will also gain "hand-on" experience through the early assignments given in common process safety elements such as MOC and PHA. And if successful with the pilot Mentoring Program, the AIChE Safety and Health Division may have an additional tool to help the engineer hit the ground running (safely).

7. References

[1] U.S. Chemical Safety and Hazard Investigation Board. *Investigation Report T2 Laboratories, Inc. Runaway Reaction (Four Killed, 32 Injured)*, CSB Report No. 2008-3-I-FL, 2009

[2] U.S. Chemical Safety Board, http://www.csb.gov/csb-applauds-aiche-s-response-to-including-reactive-hazard-awareness-in-college-chemical-engineering-curriculum/ (accessed January 29, 2017)

[3] Crowl, Daniel, A. and Louvar, J. F., *Chemical Process Safety: Fundamentals with Applications*, 3rd ed., Prentice Hall, Englewood Cliffs, NJ (2011)

[4] Center for Chemical Process Safety, http://www.aiche.org/ccps/community/undergraduate-process-safety-learning-initiative (accessed January 29, 2017)

[5] Center for Chemical Process Safety, Introduction to Process Safety for Undergraduates and Engineers, 1st ed., John Wiley & Sons, Inc., 2016

478550 Thermal Decomposition of Mono-Nitrated Toluene (MNT) with Additives

Abstract:

Abstract: On September 21st 1992, an explosion happened in Castleford, UK, at the distillation unit, owned by Hickson & Welch Ltd. The incident killed 5 people and caused nearly 200 injuries. The main reason is due to the decomposition of mono-nitrated toluene (MNT). Nitrated aromatic compounds present a potential run-away reaction threat to the chemical process industry, due to its unstable structure and extensive heat release during upset situations. This study employs the Advanced Reactive System Screening Tool (ARSST) to study the effects of various common contaminants on MNT's thermal decomposition, including sodium nitrate, sodium hydroxide, and sodium carbonate. Experimental parameters and results such as 'onset' temperature, pressure increase rate, temperature increase rate and maximum temperature are present. Sodium hydroxide and sodium nitrate can lower the decomposition onset temperature of MNT and make the consequence more severe. Thermokinetics of the decomposition reaction is shown and SADT prediction of UN25 kg package storage is calculated.

479379 Addressing Challenges of Controlling Change By Implementing Best MOC Practices

Abstract:

Effectively controlling change in a plant environment results in a cumbersome process with an inefficient paper-based Management of Change process or use of a traditional and complex electronic Management of Change (MOC) system. Recent research indicates that the majority of chemical incidents are the outcome of poor implementation of Process Safety Management (PSM) and the lack of efficient MOC processes. As a result, MOC workflow steps are often skipped or prolonged resulting in an ineffective process that does not capture the change fully or safely; approvals are delayed, Process Safety Information (PSI) is not being updated, Pre-Startup Safety Review (PSSR's) are not being performed, and MOCs are not being closed in a timely manner.

The goal of this paper is to address industry challenges with controlling change, and how Siemens implements best practices for an effective MOC process.

479243 Process Change Impact on Relief System Design

Abstract:

Process change impact on relief system design

(Understanding equipment impact and the evaluation warranted)

Various scenarios can cause the pressure within equipment to exceed design pressure. Appropriately sized relief devices are one of the most common methods of protecting equipment and personnel operating this equipment. A key challenge to being in compliance is the dynamic nature of processing plants. Numerous changes can be expected to happen over the lifecycle of a processing plant. Management of Change, one of the Process Safety Management (PSM) elements, requires equipment impact evaluation when these changes occur to ensure adequate pressure relief. This presentation talks about numerous types of changes which will impact equipment and warrant overpressure protection re-evaluation.

479152 Identification and Local Impact Analysis of Projectile Hazard in the LNG Industry

Abstract:

With a strong growth in natural gas production in the United States, a number of liquefied natural gas (LNG) facilities have been proposed and authorized. These LNG facilities have sufficient safety, security, risk assessment and mitigation plans developed in close coordination with local, state, and federal authorities. However, to continue to ensure safe operation, it is vital to continue to improve the scope of risk assessment and management strategies. Recent incidents have highlighted the need to consider the potential hazard and risk from projectiles. This paper studies the sequence of phenomena that occur as part of the local impact of a projectile. The main objectives of this study were to: 1) list projectile (missile/fragment) parameters such as source, type, and characteristics (mass, diameter, velocity) and 2) conduct a local impact analysis for identified potential projectile hazard scenarios at an LNG facility. The paper concludes that the five different calculated thicknesses for concrete penetration, scabbing, and perforation, and steel penetration and perforation, depend directly on the kinetic energy and diameter of the projectile. Furthermore, the authors recommend that operators conduct a projectile hazard analyses for LNG facility projects, which would be useful in making recommendations for the thickness and type of storage tanks and other equipment and for other appropriate mitigation measures.

478563 Thermal Hazard Analysis and Chemical Incompatibility Tests with Custom High-Pressure Crucibles Made from Commonly Used Metals and Alloys

Abstract:

The assessment of reactive hazards and chemical incompatibilities is necessary for safe operations in the chemical and pharmaceutical industries. Commercial ARC (Accelerating Rate Calorimeter), RSST (Reactive System Screening Tools) and high pressure crucibles for Differential Scanning Calorimetry (DSC) can be used for this purpose [1-2]. However, ARC and RSST require significant capital investment and require relatively large amounts of sample (at least 1 g).The alternative, commercial high pressure crucibles for DSC, can also be expensive (e.g., several hundred U.S. dollars for each pair of gold-plated high-pressure crucibles) when multiple samples need to be investigated. These requirements make thermal hazard analysis prohibitively expensive for research laboratories that do not focus on chemical process safety, and where materials under development may be available in limited quantities (e.g., milligrams).

To address these short-comings of current hazard screening methods, inexpensive custom DSC high-pressure crucibles made from different commonly used metals and alloys (stainless steel 304, stainless steel 316, titanium, brass and tantalum) have been constructed in-house using a micro-welding system containing a jewelry spot welder and a microscope. Chemical compounds are tested in different crucibles made from different metal/alloys, which simulates different real-world storage, transportation, and process operations in the chemical and pharmaceutical industry. The cost for each pair of crucibles is less than two U.S. dollars. In a typical experiment, only 1-5 mg of sample is required. Kinetic parameters can be extracted from only a few DSC runs with these crucibles, allowing for prediction of the chemical's thermal behavior at larger scales. This method provides an inexpensive and efficient way to perform thermal hazard analysis and incompatibility tests. The application of these crucibles is demonstrated with case studies of ammonium nitrate (AN) and ionic liquids. Predictions of thermal behavior at larger scales have been verified with ARC tests.

References

1. Oxley, Jimmie C. 2003. "Handbook of Thermal Analysis and Calorimetry. Vol. 2: Applications to Inorganic and Miscellaneous Materials M.E. Brown and P.K. Gallagher, Editors 9 2003 Elsevier B.V. All Right Reserved. 349" 2: 349–69.

2. Han, Zhe, Sonny Sachdeva, Maria I. Papadaki, and M. Sam Mannan. 2014. "Ammonium Nitrate Thermal Decomposition with Additives." Journal of Loss Prevention in the Process Industries. Elsevier Ltd, 1–9. doi:10.1016/j.jlp.2014.10.011.

13TH GLOBAL CONGRESS ON PROCESS SAFETY

Do Dust Hazards Analysis (DHA) Really Differ from a Process Hazards Analysis (PHA) or a Process Analysis (PA)?

Michael E. Robertson

4094 Roslindale Ln
Buford, GA 30519
processesmatter@gmail.com

Prepared for Presentation at
American Institute of Chemical Engineers
2017 Spring Meeting and 13th Global Congress on Process Safety
San Antonio, TX
March 26 – 29, 2017

Do Dust Hazards Analysis (DHA) Really Differ from a Process Hazards Analysis (PHA) or a Process Analysis (PA)?

Michael E. Robertson

4094 Roslindale Ln
Buford, GA 30519
processesmatter@gmail.com

Keywords: NFPA 652, Dust Hazards Analysis, DHA

Abstract

Do Dust Hazards Analysis (DHA) Really Differ From a Process Hazards Analysis (PHA) or a Process Analysis (PA)?

This author reviews the latest NFPA 652 requirement on Dust Hazards Analyses (DHA) and compares it to OSHA PSM's Process Hazards Analyses (PHA) requirement and other process analyses required by industry specific NFPA dust regulations. Though process analyses have been a part of NFPA regulations for several years and in the OSHA PSM regulation for 25 years, the 2016 version of NFPA 652 has clearly defined that a process analysis called Dust Hazards Analysis must be done where there is a combustible dust risk. The subject of this paper is the Dust Hazards Analysis.

The aim of the paper is to find commonality in dust hazards analysis, process hazards analysis and process analysis for tips that may limit the number of analyses a company may have to perform. These tips provided will assist a PHA leader in doing their DHA's. This will not be an exhaustive review of standards or regulations, but an overview and guide for those starting DHA's. It has become obvious that NFPA regulations are modeling a PSM framework and some discussion will be made on the subject in the area of risk assessment and management systems. Some comments will be made concerning the software that can be used, but no software will be sold or marketed. The author is not sharing examples of DHA's from his company due to confidentiality and timing of his DHA's.

Some of the Dust Hazards Analysis differences detailed are the emphasis on dust inside and outside of equipment, the use of NFPA dust thickness calculations, building reviews, and frequency of re-validating the analyses. How does a Dust Hazards Analyses affect the process safety information required, team make-up or report after the review? What are the real requirements for a DHA leader?

1 Introduction

There have been several dust related incidents that resulted in loss of life and severe damage to industrial properties. Imperial Sugar in Savannah, GA (2008) and West Pharmaceutical in North Carolina (2003) are two that top the list. Since the early 2000's, the Chemical Safety Board (CSB) has been recommending the Occupational Safety and Health Administration (OSHA) develop standards surrounding combustible dust. The National Fire Protection Association (NFPA) has had guidelines in several of its standards concerning dust for years.

Table 1. Industry Specific NFPA Guidelines Pertaining to Dust

61 – Agriculture
484 - Metal
654 - General
664 - Wood

NFPA 654, 664, 61, 484, 86 and 499 have dealt with combustible dust for years.

It is 2017 and while OSHA still has not set forth comprehensive dust standards, NFPA has published a comprehensive guideline on the management of combustible dust as 2016 NFPA 652 Standards on the Fundamentals of Combustible Dust. This guideline requires dust hazards analysis (DHA) to be complete by September 7, 2018. long with renewed regulations, risk assessments are now becoming part of every safety performance system. NFPA has required process analysis and hazards analysis in the industry specific guidelines..

A DHA is similar to a Process Hazards Analysis (PHA), but there are some key differences that need to be understood by study leaders. PHA methodology is well- defined especially from the OSHA 1910 PSM regulation, the CCPS book *Guidelines for Hazards Evaluations* **[2]**,and other information from new and vintage PHA courses like Primatech's Process Hazards Analysis Leaders' Training and Process Improvement Institute's PHA Leader's Training.

The similarities between PHA's and DHA's are in the methodology. The methodologies used to conduct a PHA, such as HAZOP, Checklist, What-If, are applicable for identifying and assessing hazards in dust handling processes. Expertise with PHA's can assist with the completion of DHA's. The differences begin when the building has to be evaluated for combustible dust hazards. Process Analysis is defined in NFPA 664-12 – Standard for the Prevention of Fires and Explosions in Wood Processing and Woodworking Facilities. [3]. However, NFPA did not offer much instruction on how to do a Process Analysis (PA).

There are other tools that have to be employed for the building evaluation. Pictures, audits, root cause analysis and cause and effect diagrams can be used for the out of the process review.

Detailed building structure evaluations may require a code inspection or some other engineering tools.

This paper evaluates the elements of dust hazards analysis. Comparisons are made between dust hazards analysis, a process hazards analysis methodology called HAZOP (HAZard and OPerability Study), and a process analysis as described in older NFPA guidelines. These comparisons are made to assist PSM companies in aligning PHA's and DHA's and to assist non-PSM companies on resources to assist them in conducting DHA's.

2 What Is a Dust Hazards Analysis (DHA)

NFPA 652 requires that all facilities handling or creating combustible dust must do a Dust Hazards Analysis. A dust hazards analysis is a systematic review of processes and buildings where dust is created or handled and could create a deflagration.

Per the CCPS process safety glossary: A combustion that propagates by heat and mass transfer through the un-reacted medium at a velocity less than the speed of sound.

This requirement has many businesses scrambling in a similar way to the release of the PSM standard in 1992, which left many oil and gas and chemical companies reeling. The dust hazards analysis is not a PSM requirement nor is it a type of process hazards analysis. While dust hazards analyses are new, analyzing your process for dust hazards has been a part of industrial specific NFPA guidelines for a few years now. Now with NPFA 652, OSHA through Recognized and Generally Accepted Good Engineering Practice(RAGAGEP) has the standards it needs for inspection and enforcement to address the potential for a dust deflagration in a facility.

The dust hazards analysis must be complete by September 2018. The DHA requirement is set forth in Element 7.0 of the 2016 version of NFPA 652. There is a good example of the DHA in Appendix B of NFPA 652. Operating areas are reviewed for the probability of deflagration. Classifications are 'Not A Hazard', 'Maybe a Hazard' and 'Deflagration Hazard'. If the probability exists, then the owner must document how they manage or mitigate that risk. If the owner does not have current controls to prevent the deflagration, steps must be taken to document a hazards management plan with steps to reduce the risk.

Within a process system or building, when there is a sufficient quantity of combustible dust, the following conditions must be considered and documented:

1) Oxidizing atmosphere
2) Credible ignition source
3) Credible suspension mechanism

These are items from the fire triangle used in most fire prevention training. These are key in the dust hazards analysis and must be treated with high importance. Housekeeping also must be properly reviewed and documented to prevent the third item mentioned above – credible suspension mechanism.

3 How Do Dust Hazards Analysis (DHA) Compared to Process Hazards Analysis (PHA)

A major difference between the DHA and a PHA is the consideration of dust risk in a building. PHA's do evaluate the risk to occupied buildings in relation to the potential for releases of highly hazardous chemicals covered by the PSM regulation. This risk is evaluated through facility siting studies and more detailed reviews like a Quantitative Risk Analysis (QRA) or Occupied Building Risk Assessment (OBRA).

Table 2. PHA and DHA Methodology Comparison

Process Hazards Analysis (HAZOP)	Dust Hazards Analysis	Process Analysis
Source: OSHA 1910.119 (e)	Source: NFPA 652 7.0 and Appendix B – Dust Hazards Analysis	Source: NFPA 664 4.0 and Explanatory Material A4.3
Covers the process system	Covers the process system and the building construction	A methodical review of the facility, each operation housed within, and the identification of where a hazard exists.
Gather Data – Drawings and Process Information	Gather Data – Lab Data, Drawings and Process Information	
Have a Qualified PHA Leader a Team with an Operator and a Process Expert, along with other Subject Matter Experts as needed	Have a Qualified DHA Review Leader and at least one other for a team.	
Set Objectives and Boundaries		
Determine the Study Nodes by marking up the Piping and Instrument Diagrams	Determine the nodes on the drawings and sketches	
Review HAZOP Guidewords	Review DHA questions from Appendix B of NFPA 652	
Describe Process Node	Describe Process Node	
Use all Guidewords in all Nodes	Use all questions in all Nodes	
Review Global Node, Human Factors, Management of Change and Incidents		

With the PSM regulation, PHA's are required in facilities with highly hazardous chemicals. These chemicals are listed in the PSM requirement. PHA's are a group of tools that are used to review processes handling highly hazardous chemicals. PHA's include HAZOP, Failure Mode and Effect Analysis (FMEA), What If Studies, and Checklists.

The DHA requires buildings to be reviewed for the possibility of a fire from dust build-up. Building construction must minimize the effects of a deflagration or explosion from a combustible dust.

From the table above, you can see some differences in PHA's and DHA's. DHA's do not require a review of management of change and incident review history. The team composition is not set in a DHA. P&ID's are the basis, but DHA's do not have a required standard drawing type although P&ID's are applicable for some aspects of dust handling.

4 DHA Toolbox

Similar to PHA's, multiple tools are needed to complete a DHA.A PHA method could be used to evaluate the risk associated with combustible dust in a process. However, beyond typical PHA methods, field audits, pictures and video recordings of the areas may be need of areas where dust can build up.Alternatively, a root cause analysis method like 5 Why method might also be applied in areas where a dust build-up to determine sources of dust leakage.. The key is to locate where dust gets out of a process or accumulates beyond its minimum explosive concentration (MEC) in a process, why the dust gets outside the confines of the process and what can be done to prevent the leakage.

Whereas chemical analysis and laboratory data are helpful in a PHA, parameters like Maximum Explosion Concentration (MEC), Minimum Ignition Temperature (MIT) and Kst are very key in determining dust combustibility and risk for a process and building. NFPA 652 has information in the use of these parameters. There is also good information in NFPA's Guide to Combustible Dust and the AIChE Book on Solids Handling Equipment. This paper will not cover in detail any laboratory or chemical analysis of dust. [4], [5]

5 Documentation for Review

Piping and instrument diagrams (P&ID's) have long been the basis of the PHA techniques, including HAZOP, Checklist, and What-If. Supporting documentation may include SDS's, process descriptions, management of change documents, incident history and standard operating procedures. Plant layouts and general arrangement drawings will help orient the team and provide information for facility siting and items in the global node.

P&ID's will help for the process analysis in the DHA. You may also need general arrangement drawings and assembly drawings where machines are involved. Plant layouts and 3D renderings may help with the building analysis portion of the DHA. Pictures and video of areas where dust buildup can occur will also be helpful.

Lastly, plant conveying, processing and finishing OEM information such as screen sizes, motor power, pneumatic flow rates, and duct sizing are key in understanding out the equipment affects the creating or lack of creation of combustible dust. Information on seals and design pressures must be available for comparison with operating parameters.

The bottom line is DHA documentation is as important to the process as Process Safety Information is for PHA. However, it might not be as standard or available as PSI is for a PHA. Some of the information may also be very old since plant layouts and renderings might come from the original start of the plant.

6 The Dust Hazards Analysis Process

The DHA process can start like a PHA with an overview, process description and agenda. The DHA must cover the fugitive combustible dust that gets out into the building and the building's ability to withstand deflagrations and explosions.

Documentation must be available to follow the flow of the materials that contribute to creation of combustible dust. Equipment like fans and screens are especially sensitive since those components can create dust suspensions and impact the Minimum Explosion Concentration of the system. Critical laboratory must be available for discussion since it will provide the data to determine combustibility.
As with a PHA, a tour of the area would be very critical to see the areas were combustible dust accumulates the most. A start to the process is chosen and each part is reviewed by asking a series of question. Below are the questions from Appendix B of NFPA 652 used in a DHA.

Table 3. DHA Process Questions ([1] – B.4.5.1.1 – B.4.5.1.5)

I.	Is the particulate deflagrable (explosible)?
II.	Is the particulate suspended in air?
III.	Is there sufficient concentration to propagate a flame front?
IV.	Are there competent igniters available?
V.	What hazard management is in place?
VI.	Is there metal detection, spark detection, bonding and grounding, or other hazard management means in place?

These questions are asked for each piece of equipment, piping and ducting. The answers will should lead the team or competent person to the hazard management system measures in place and the gaps that will need attention from site management. The process must continue until the entire process is analyzed and all the outcomes have been reached.

As mentioned previously, Appendix B of NFPA 652 (2016) provides a good example to do a DHA. For those in the wood industry, the example is actually of a wood processing plant.

7 Outcomes

NFPA 652 requires that results from the Dust Hazards Analysis be documented. This includes necessary action items requiring change to the process materials, process itself, operations or facilities. NFPA 652 has requirements for managing the hazards under the heading of Hazards Management Program.

The results from a Process Hazards Analysis are the hazard scenarios, a list of safeguards currently in place, and any action items required to ensure that risk is effectively managed within the process handling highly hazardous materials. This usually involves a risk ranking system as well.

The Dust Hazards Analysis does not require a risk ranking system. Where a hazard exists, identification and evaluation of the fire scenario shall include safe operating ranges, safeguard identification, recommendations for additional safeguards as needed to properly manage the fire event, and lastly a plan for implementation.

8 Leaders

A meeting has a leader and teams have leaders. DHA's and PHA's are no different. Each review must have a leader. Her qualifications are listed in Table 2. There are no certifications or extensive training requirements. Many believe that certification is coming for PHA leaders because the competency of those who lead PSM is being discussed in government and industry especially as we enter the next 25 years of PSM.

Many experts agree that an experienced PHA leader can also make an effective DHA leader. This is a good start, but the DHA leader must also be familiar with the hazards of combustible dust. Additionally, it would be prudent to have a leader who is experienced in the process being reviewed or ensure that a process expert is present as a team member. I would review the qualifications listed in CCPS's book on Hazards Evaluation and information.

Table 4. Comparison of Leader Qualifications

Dust Hazards Analysis Leader	Process Hazards Analysis Leader
Must be able to lead DHA's Must be familiar with the hazards of combustible dust	Must be trained in PHA techniques Must be re-qualified every 5 years

9 Frequency

PHA's are done initially to establish a base evaluation and then every five years after the initial PHA to address changes or revisions to covered processes. . DHA's are done once for the original process and for any major upgrades. There is no renewal period, but the initial reviews have to be done by September 7, 2018.

10 Management Systems

What is a dust hazards management system? How does a dust hazards management system compare to the standard elements of PSM? Chapter 9 of NFPA 652 mentions its elements as listed in Table 6. Both PSM and NFPA 652 have the requirement of risk assessment that has been the subject of this paper.

PSM is embedded in many of the NFPA guidelines. Though the thought of PSM to many firms is frightening, PSM offers a good framework for managing risk whether it is combustible dust or other system or process risks.

Table 5. Management Systems Comparison

Dust Hazards Analysis Management System (Ch. 9)	Elements of Process Safety Management
1. Operating procedures and practices	1. Standard Operating Procedures
2. Inspection testing and maintenance	2. Mechanical Integrity
3. Training and Hazard Awareness	3. Training
4. Emergency Planning and Response	4. Emergency Planning and Response
5. Incident investigations	5. Incident investigations
6. Management of change	6. Management of change
7. Management system review,	7. Compliance Auditing
8. Contractor training	8. Contractor Safety
9. Employee participation	9. Employee Participation
10. Document retention	10. Process Safety Information
11. Ignition Control (Covered in Chapter 8)	11. Hot Work Permits
12. DHA	12. Trade Secrets
	13. Pre-Start Up Safety Review
	14. PHA

11 Conclusions

DHA's must be completed. It is similar but different than a PHA. Expertise with PHA's can assist with the completion of DHA's. The similarity between PHA's and DHA's is in the methodology of reviewing processes for dust. The differences start when the building has to be evaluated for combustible dust. Here other tools have to be employed. These tools as previously discussed may vary from audits and inspections to root cause analysis tools. Detailed building structure evaluations may require a code inspection or some other engineering tools.

12 References

[1] NFPA 652-15 – Standards on the Fundamentals of Combustible Dust, National Fire Protection Association, September 7, 2015.

[2] CCPS. *Guidelines for Hazards Evaluations.* Center for Chemical Process Safety, American Institute of Chemical Engineers, New York, NY, 2007.

[3] NFPA 664-12 – Standards on the Fundamentals of Combustible Dust, National Fire Protection Association, 2011.

[4] NFPA– Guide to Combustible Dusts, National Fire Protection Association, 2012.

[5] CCPS. ***Guidelines for Safe Handling of Powders and Bulk Solids.*** **Center for Chemical Process Safety, American Institute of Chemical Engineers, New York, NY, 2005.**

478758 A Leading Indicators Based Decision Support Tool to Predict Blowout Events

Abstract:

Predicting a potential blowout scenario timely and efficiently is often a challenging task due to the complexity of drilling or other well intervention activities. Blowouts are usually preceded by gas kicks and early prediction of gas kick events can provide some precious time window to take control of the well in concern. A leading risk indicators based approach for predicting gas kick or blowout events has been introduced in this work. A causal factors based step-by-step approach for identifying comprehensive sets of leading indicators are presented and discussed. Leading indicators were divided into two broad sections - real-time indicators and long-term organizational safety performance indicators. With the real-time indicators, different decision support algorithms are developed which can be used as training tools. A real-time indicator may not always successfully lead to a well control event but at the same time the situation needs to be assessed carefully considering other factors and indicators. Such uncertainties are taken into account while developing the algorithms. For evaluating the relative importance of different leading indicators Bayesian Network models are developed. The key causal factors for well control barriers failure are identified by conducting fault tree analysis and appropriate indicators are linked with relevant causal factors in a Bayesian model. At the end, a list of efficient leading risk indicators are presented for predicting primary well control barrier failure which can initiate gas kicks.

Consequence Modeling at the Eastman Kingsport Site

David Skelton
Principal Engineer
Eastman Chemical Company
P.O. Box 511, Building 18
Kingsport, TN 37662
davidskelton@eastman.com

Michael James
Senior Chemical Engineer
Eastman Chemical Company
P.O. Box 511, Building 18
Kingsport, TN 37662
mjames@eastman.com

Prepared for Presentation at
13th Global Congress on Process Safety
San Antonio, TX
March 26-29, 2017

UNPUBLISHED

Consequence Modeling at the Eastman Kingsport Site

Michael James
Senior Chemical Engineer
Eastman Chemical Company
P.O. Box 511, Building18
Kingsport, TN 37662
mjames@eastman.com

David Skelton
Principal Engineer
Eastman Chemical Company
P.O. Box 511, Building 18
Kingsport, TN 37662
davidskelton@eastman.com

Keywords: Dispersion modeling, consequence analysis, risk assessment, dosage, shelter-in-place, temporary haven, evacuation, toxicity, toxic chemical release, toxic load, downwind concentration, blast overpressure, vapor cloud explosion, pressure vessel burst

Abstract

The intent of this paper will be to review the methods utilized at the Eastman Kingsport site in determining magnitude of consequences of toxic, thermal and blast overpressure scenarios. This will include discussion on dispersion modeling, dose-response methods for toxic and thermal probits as well as assessing occupant vulnerability through the use of BakerRisk's Explosion Risk Cooperative tools.

1. Potential Consequences of Loss of Containment

Figure 1 is a typical event tree for potential losses of containment [1].

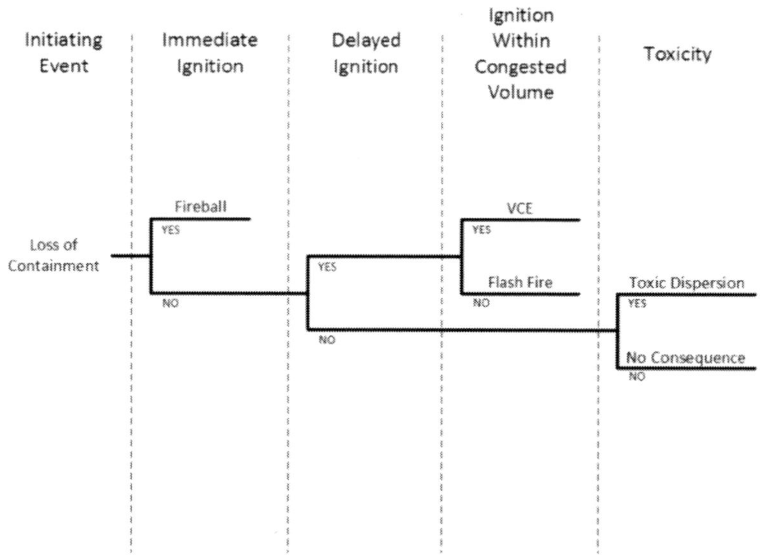

Figure 1. Potential Losses of Containment

The consequences of loss of containment in Figure 1 detail the potential release scenarios that are chiefly of concern. The method for consequence modeling for each type of scenario is described in this paper.

2. Source Terms

Figure 2 details the potential source terms that will need to be developed for each type of scenario.

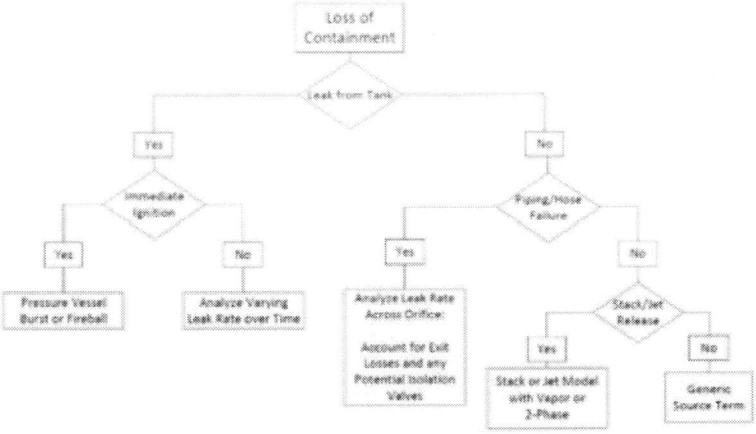

Figure 2. Potential Source Terms.

The source term may have any of the following characteristics:

- Initial momentum
    - Based on the linear velocity across the orifice and other physical properties of the release, the source term may need to account for significant initial momentum of the cloud.
    - In addition, if the linear velocity is sufficiently high, the source of the release may ingest air, resulting in a jetting condition. As a rule of thumb, 10 m/s is a lower limit for potentially jetting releases [2]. There are specific dispersion models to account for jetting conditions during a release.

- Pool Formation
    - Most dispersion models have an integrated pool evaporation model to determine the dynamic evolution of vapor from a liquid spill.
    - If the preferred evaporation model is outside of the dispersion model, the evaporation rate results will need to be analyzed for integration into the dispersion modeling tool.
- Flashing
    - Leaks across an orifice may partially or completely flash from liquid to vapor based on the pressure differential.
    - Modeling tools may report choke conditions at the outlet.
    - Further flash analysis may be required to determine the temperature and liquid/vapor composition from choke to atmospheric conditions.

- Piping Layout between Pressure Source and Leak Point
  - Ensure that all pressure losses are accounted for. This includes entrances, exits, elevation changes, elbows, pipe diameter changes, and other fittings.

- Indoor Release
  - If the release occurs indoors, analyze the concentration changes over time inside the structure based on the source term calculations of the release.
  - Concentration gradients throughout the structure may need to be ignored and perfect mixing assumed.
  - If the vapor portion of the leak will cause overpressuring of the indoor area, determine the amount of vapor required to be vented in order to balance pressure in the room.
  - A flash block can be employed to assess the amount of vapor generated from a leak that will need to be relieved from an indoor area.
  - Flash blocks can also be used to account for vapor suppression systems inside of buildings.
  - This level of modeling is beyond the scope of this paper. However, methods for assessing indoor releases can be referenced from "Consequence Modeling of Dynamic Source Terms" [3].

Presence of the above factors may result in models that require expert judgment and the development of conservative estimates. Most dispersion modeling tools do not have the levels of sophistication required to accurately model these types of source terms. However, the results of a dispersion model employed with conservative estimates may be suitable for site risk assessment.

The following section will discuss the potential leak scenarios and how source terms can be developed, accounting for each of the above factors.

3. Blast Modeling

Simple blast models can be used to predict potential consequences of explosions in lieu of more accurate but complicated methods such as computational fluid dynamic (CFD) models. The TNT equivalency model developed by the U. S. Navy and Air Force was largely used for years, even in industrial settings. Over time it became clear that the TNT model did not predict the behavior of all explosions accurately enough, particularly in regard to vapor cloud explosions more commonly occurring in refining and chemical producing industries. Today several simplified are available to estimate the potential of consequences of explosions, each method with its own strengths and weaknesses.

3.1 Vapor Cloud Explosion (VCE) Modeling

Baker Strehlow Tang (BST) is the simplified method predominantly used at Eastman Kingsport site to model vapor cloud explosions [4][5][7][8]. The BST model is based on the premise that only the portion of a flammable vapor cloud that is confined or contained within a congested area will significantly contribute energy to an explosion. Unconfined or uncongested gas clouds are more likely to result in flash fires or deflagrations. The first step to perform a BST analysis is to identify the confined/congested volume of the plant that has the potential to harbor a flammable cloud of vapor. Examples might include piping and equipment, closely spaced railcars or cars, solid walls, ceilings and floors. A flammable cloud is most often generated from a release of material as a result of an equipment failure, a leak, or a spill. Flammable clouds can drift a considerable distance under the right conditions, so it's important to account for all the equipment within a reasonable radius of the congestion. Once the confined/congested volume is calculated, two parameters contribute the greatest influence on explosion strength using BST methodology; the energy contained within the flammable vapor cloud and flame speed as it burns through the available fuel in the congested region. The first parameter is calculated using the spreadsheet shown in Figure 3. It is a simple tool developed by Eastman to calculate volume based on dimensions of the confined cloud, and then estimate the energy contained with the calculated volume.

ESTIMATION OF VAPOR CLOUD EXPLOSION ENERGY

The user enters dimensions (in ft.) of vapor cloud subject to confinement & obstructions, ambient temperature (F), and the chemical involved in the designated boxes. Thermodynamic data listed in the lookup table to the right is used, or if not available, an approximation is made based on typical hydrocarbons.

The spreadsheet will output the participating energy to be used in V-CLOUD for prediction of pressure and impulse by the Baker-Strehlow method. The methodology is based on the paper "Vapor Cloud Explosion Analysis" by Baker, Tang, and Scheier; Prepared for the AIChE Loss Prevention Symposium, April 1994.

Width (ft)	200	
Height (ft)	50	
Depth (ft)	100	
Volume (ft3)	1000000	
Volume (m3)	2.83E+04	
Ambient Temp (F)	90	
Ambient Temp (C)	32	
Chemical	Acetic Acid	
MW	60.05	
Hc(kJ/mole)	919.73	(heat of combustion in KJ/mole)
Hc(J/m3)	3.67E+07	(heat of combustion in J/m3)
Hc (BTU/lb)	6.59E+03	(heat of combustion BTU/lb for ref.)
Hcs(J/M3)	3.48E+06	(heat of combustion of a stoichiometric mixture with air)
Energy (J)	9.86E+10	
Energy (in-lb.)	8.72E+11	(total available energy in confined / obstructed volume)
% Participation	100	
Equiv Energy in-lb.	8.72E+11	(net energy of explosion in-lb.)

Figure 3. Spreadsheet used to determine energy content of vapor cloud

For a single congested area or multiple areas with little separation distance so they can be effectively considered a single area of congestion, the results from the spreadsheet are used as an input in VCloud®. VCloud® is a simple software tool developed by the BakerRisk Engineering Consultant to produce Pressure/Impulse (P-I) diagrams (this is not to be confused with PI Diagrams representing the response curve for structures and might better be referred to as the side-on blast pressure and impulse as a function of standoff distance). Notice the check box denoted "Plot Using Co-op Energy Factor." Membership in the BakerRisk Engineering Consultant Explosion Research Cooperative (ERC) includes access to experimentally verified energy factors that can reduce conservatism and provide a more accurate prediction of explosion effects. The first page of inputs for VCloud® is illustrated in Figure 4.

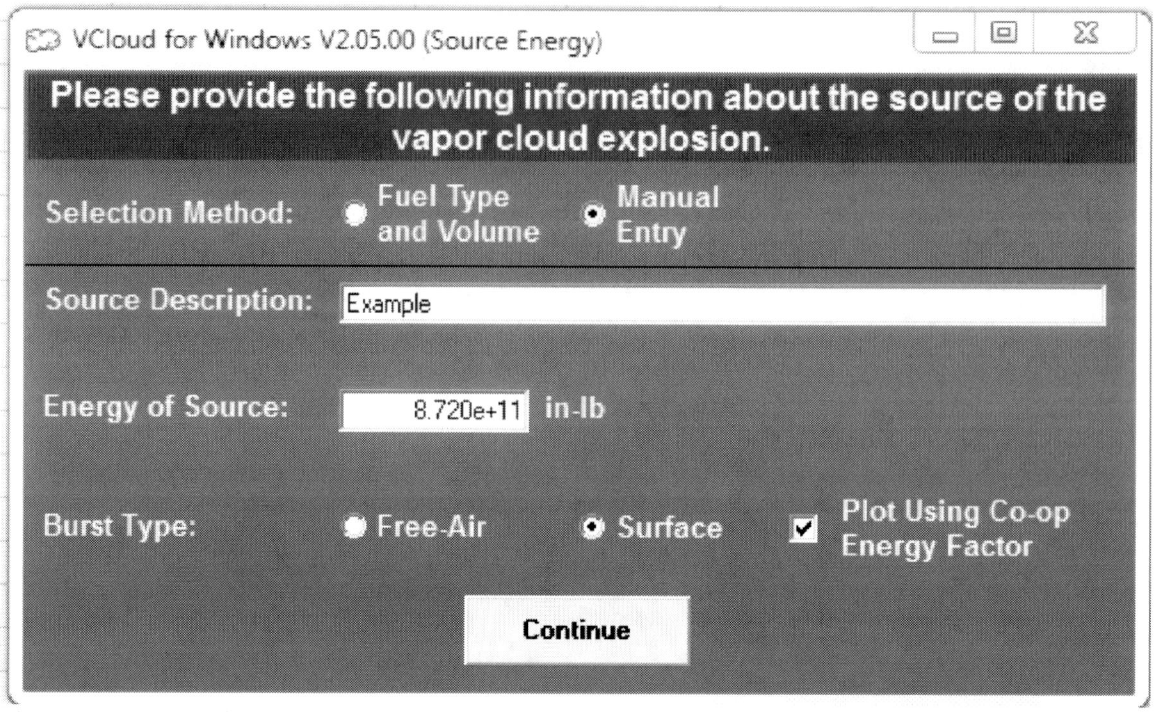

Figure 4. VCloud® Input to Determine Energy Content of Vapor Cloud. Image Courtesy of BakerRisk Engineering Consultant Explosion Research Cooperative.

The second parameter greatly influencing explosive strength when using BST methodology is flame speed, measured as a ratio to the speed of sound or Mach number. Four criterion most influence how fast a flame front burns through fuel: congestion, confinement, fuel reactivity, and the distance to free venting.

- Congestion
  - Burning rate is greatly dependent on the surface area of the flame as it travels through the fuel source. Turbulence causes the flame front to furrow and can greatly increase the surface area which contributes to increasing the speed of the advancing flame. In an industrial setting, turbulence can be generated by piping, equipment, gravel surfaces, vehicles, or any obstacle(s) that inhibits simple linear progression of the burning flame front.
  - The greater the induced turbulence, the faster a flame front will progress increasing the likelihood of detonation. BST method calculations qualify congestion into three categories; high, medium or low congestion. Congestion categorization is the most subjective parameter in BST methodology. Experience is probably the most important tool used to determine degree of congestion.
  - A good rule of thumb is if process equipment occupies an area but can be easily walked through, it is low congestion. If you can manage to slip through an area with difficulty, it's medium congestion. If it is impossible to physically pass through an area, it's high congestion.

- Confinement
    - Any solid surface that prevents the propagation of the flame in a given direction is considered confinement, with the exception of the ground. The BST method models a spherical, free-air vapor cloud explosion, but multiplies the energy of the explosion by a factor of 2 to account for ground reflection of the blast wave from its origin surface.
    - BST calculations categorize confinement in four ways; 3D, 2D, and 2½ D, and 1D.
        - 3D Confinement: A flame front burning through fuel is capable of expanding in all three dimensions with the exception of the ground. It applies to all free-field processing areas not bound by solid walls or ceilings.
        - *2D Confinement:* The propagation of the flame front is free to expand in only two axes because one dimension is blocked by a solid obstacle. A common example is a processing unit located between a solid concrete ceiling with open walls. It also applies if a processing unit is located between two solid walls with no roof. It is important to consider the aspect ratio of 2D confinement, which is the ratio of total width to height of the confined volume. Higher aspect ratios constrict venting capacity and allow additional pressure accumulation over those with a shorter distance to free venting. Typical values are available in a drop down menu in VCloud®, ranging as high as a 25:1 aspect ratio. Anything higher than 25:1 will not provide significantly more pressure accumulation, nor are aspect ratios greater than this commonly found in the field. Any aspect ratio greater than 4:1 should be represented in VCloud as 25:1 to maintain a conservative estimate.
        - 2½ D Confinement: This is a hybrid condition where neither 3D nor 2D accurately describe confinement. For instance, processing equipment contained within a pre-engineered building with lightweight metal walls and roof that will confine the flame front for a time but release at low pressures, before the impending explosion is fully realized. The roof and walls provide some containment, but in short order the blast wave is capable of 3D expansion. Tightly packed equipment on the boundaries of the congested area generating an explosion can also be considered 2½ D confinement, since it provides a small amount of venting, but not enough to consider it 3D expansion.
        - 1 D Confinement: Flame burn in a linear direction, such as inside a pipe.

- Fuel Reactivity
    - The BST method categorizes fuel reactivity in only three ways; low, medium, and high reactivity. Fuel reactivity is based on the laminar burning velocity at a specific concentration, determined by a standard testing procedure in stagnant conditions.
    - Fuels with a laminar burning velocity less than 45 cm/s are considered to have a low reactivity. Fuels with laminar burning velocities above 75 cm/s are considered highly reactive. Everything between is considered to be of medium reactivity.
    - Table 1 illustrates reactivity of a few common chemicals for reference here. NFPA 68 is a good reference for finding laminar burning velocities of many more chemicals so the reactivity category can be determined.

Table 1. Relative Reactivity

Reactivity	Fuel	Group	Relative Reactivity	Formula
Low	Methane	Alkanes	1	CH4
Low	Formic Acid	Acids	1	HCOOH
Low	Acetic Acid	Acids	1	CH3COOH
Medium	Ethane	Alkanes	3	C2H6
Medium	Propane	Alkanes	3	C3H8
Medium	Butane	Alkanes	3	C4H10
High	Ethylene	Alkenes	6	C2H4
High	Acetylene	Alkynes	7	C2H2
High	Hydrogen	Inorganics	7	H2

(Courtesy Eastman Chemical Company)

- Distance to Free Venting
  - Distance to free venting is simply the shortest distance a flame must travel to exit the area of congestion/confinement to find free space.
  - After free venting, turbulence is less likely to accelerate the flame front and under most circumstances it will begin to slow. This reduces the possibility of a transition from deflagration to detonation (DDT).

These four criterion are now input into VCloud® in the format shown in Figure 5. From the provided data, a Mach number (M_f) is assigned to the flame front.

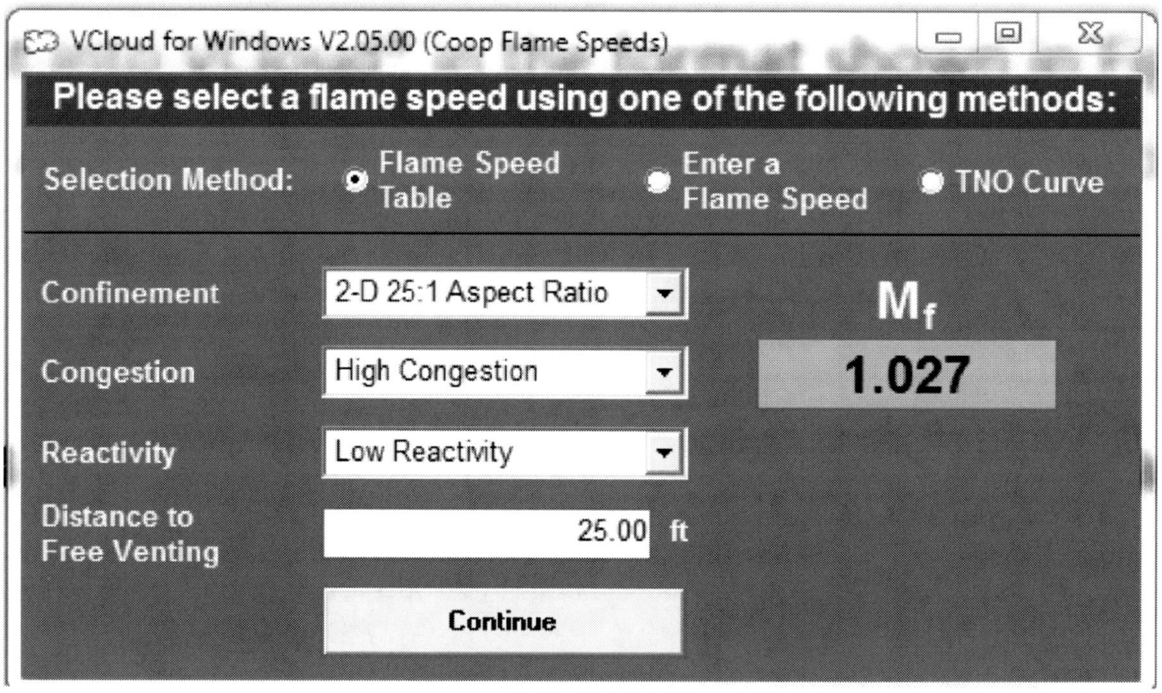

Figure 5. Vcloud® Input to Determine Mach Number. Image Courtesy of BakerRisk Engineering Consultant Explosion Research Cooperative.

A Mach number greater than one indicates the flame front accelerated faster than the speed of sound and beyond, meaning a detonation is more likely to occur. Some typical flame speeds in Mach numbers are provided in Table 2 for reference, with respect to fuel reactivity and degree of congestion.

Table 2. Flame Speeds for Soft Ignition Sources [9]. Provided as a reference but not relevant to the example calculation.

Flame Speed in Mach Numbers (M_w) for Soft Ignition Sources

Flame Expansion	Fuel Reactivity	Obstacle Density		
		High	Medium	Low
ID	High	5.2	5.2	5.2
	Medium	2.27	1.77	1.03
	Low	2.27	1.03	0.294
2D	High	1.77	1.03	0.588
	Medium	1.24	0.662	0.118
	Low	0.662	0.471	0.079
3D	High	0.588	0.153	0.071
	Medium	0.206	0.100	0.037
	Low	0.147	0.100	0.037

VCloud® uses the provided inputs to produce a Pressure/Impulse (PI) diagram. The PI diagram is used to determine the overpressure and impulse experienced by a target at a given distance. Figure 6 is an example of a PI diagram produced by VCloud®. The distance between the target and the explosion is manually input to determine the blast effects at that specified distance. The target at 150 feet from the explosion will experience a 10.12 psig side-on overpressure with a 316.05 psi-ms impulse. This is more than enough to completely demolish an unreinforced masonry building. A key attribute of vapor cloud explosions that differs from high explosives is a much higher impulse. Impulse is more threatening to rigid construction than overpressure, because it tends to distort supporting structural members beyond their plastic limits.

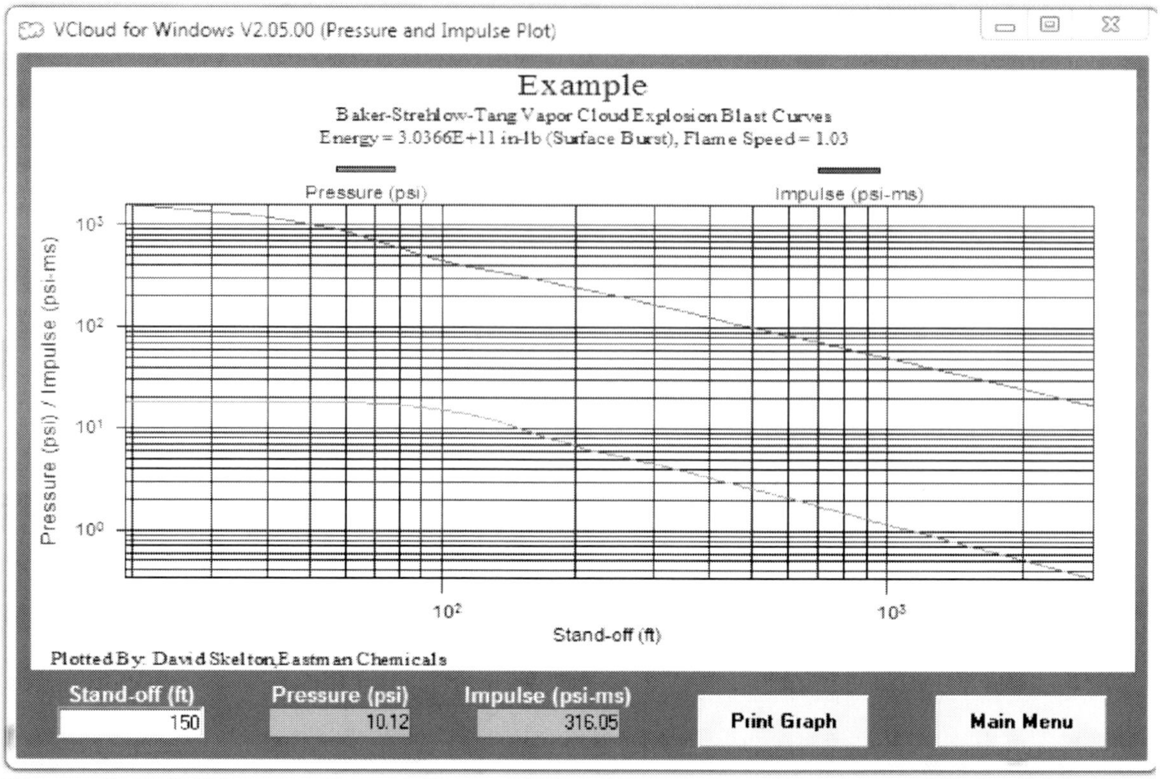

Figure 6. P-I Diagram Output From VCloud®. Image Courtesy of BakerRisk Engineering Consultant Explosion Research Cooperative.

Once a PI diagram is prepared, it can be used to predict the potential damage to conventional industrial buildings from the modeled blast event. Another software tool developed by the ERC can be used for this purpose, the Blast Evaluation and Screening Tool, affectionately known as BEAST®. BEAST® uses empirically based correlations from accidental explosions, shock tube tests, and government sponsored high explosive tests to predict the level of damage imparted by an explosion to a given building or structural component. Originally developed in 1995, BEAST® has been updated over the years to include more building and component damage curves, improve the user interface, and add more building types. The most recent version of BEAST® can evaluate 19 unique building types and many types of individual structural components, such as windows and doors. The first step is to define the building for evaluation. An example using the current BEAST® interface is included in Figures 7a and 7b. Required inputs are type of construction (drop down menu), building dimensions, blast overpressure and impulse (from PI Diagram), and whether wall and roof loading are side-on or reflected.

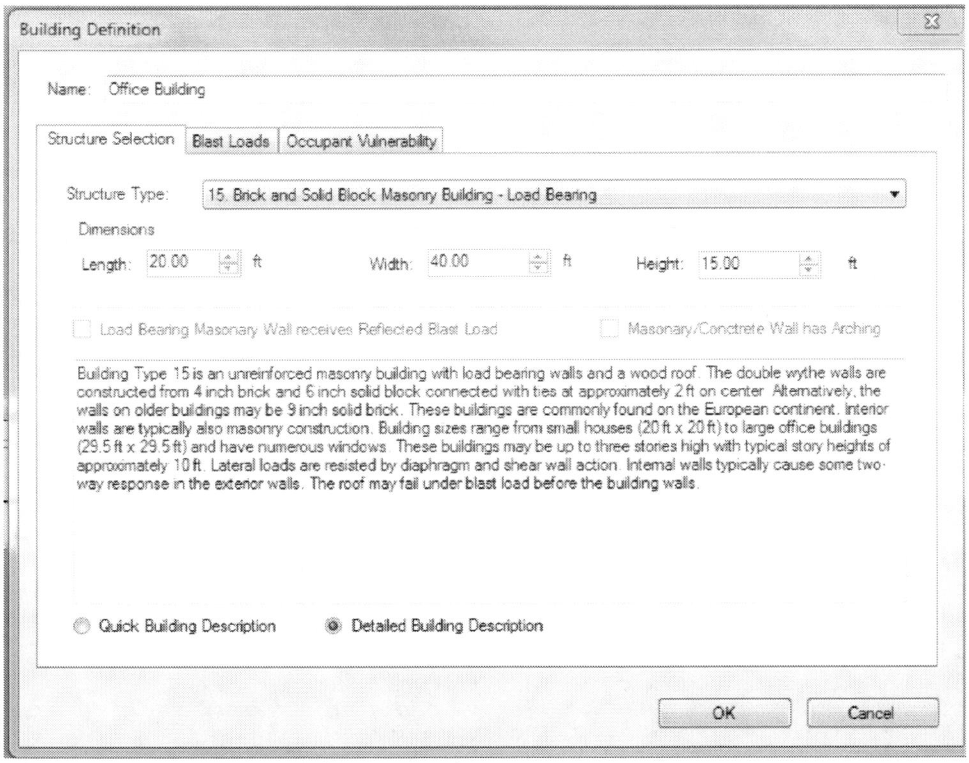

Figure 7a. Beast® Building Selection.

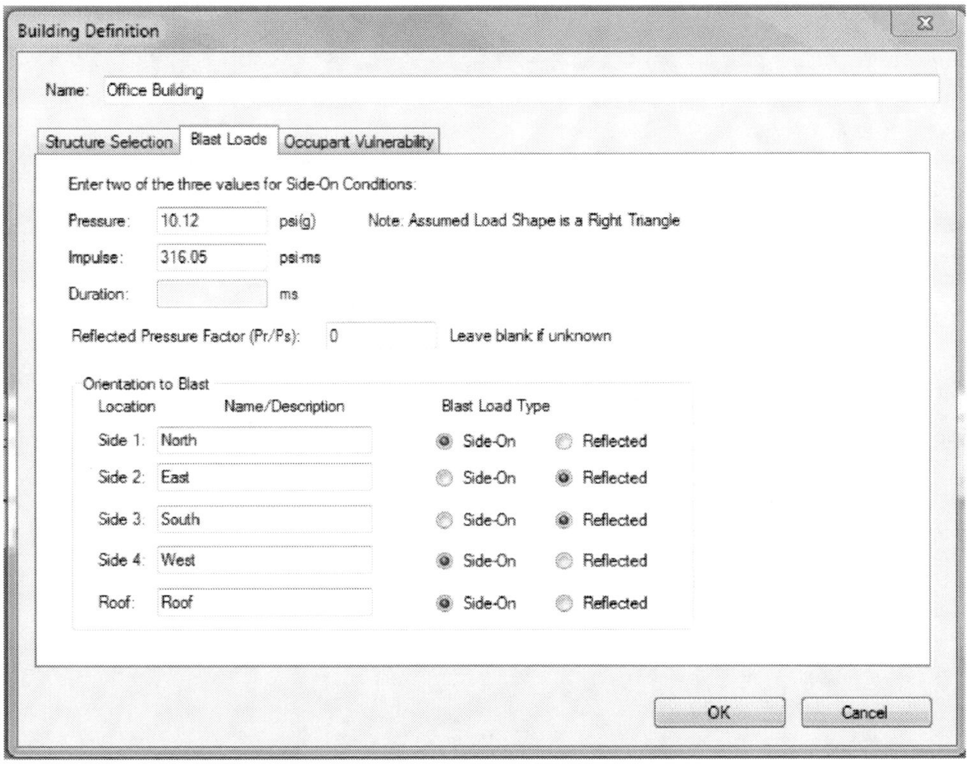

Figure 7b. Beast® Building Selection (cont'd).

Images Courtesy of BakerRisk Engineering Consultant Explosion Research Cooperative.

Given the appropriate inputs, BEAST® predicts a damage level to the building and an occupant vulnerability. Figure 8 illustrates the output provided from BEAST®.

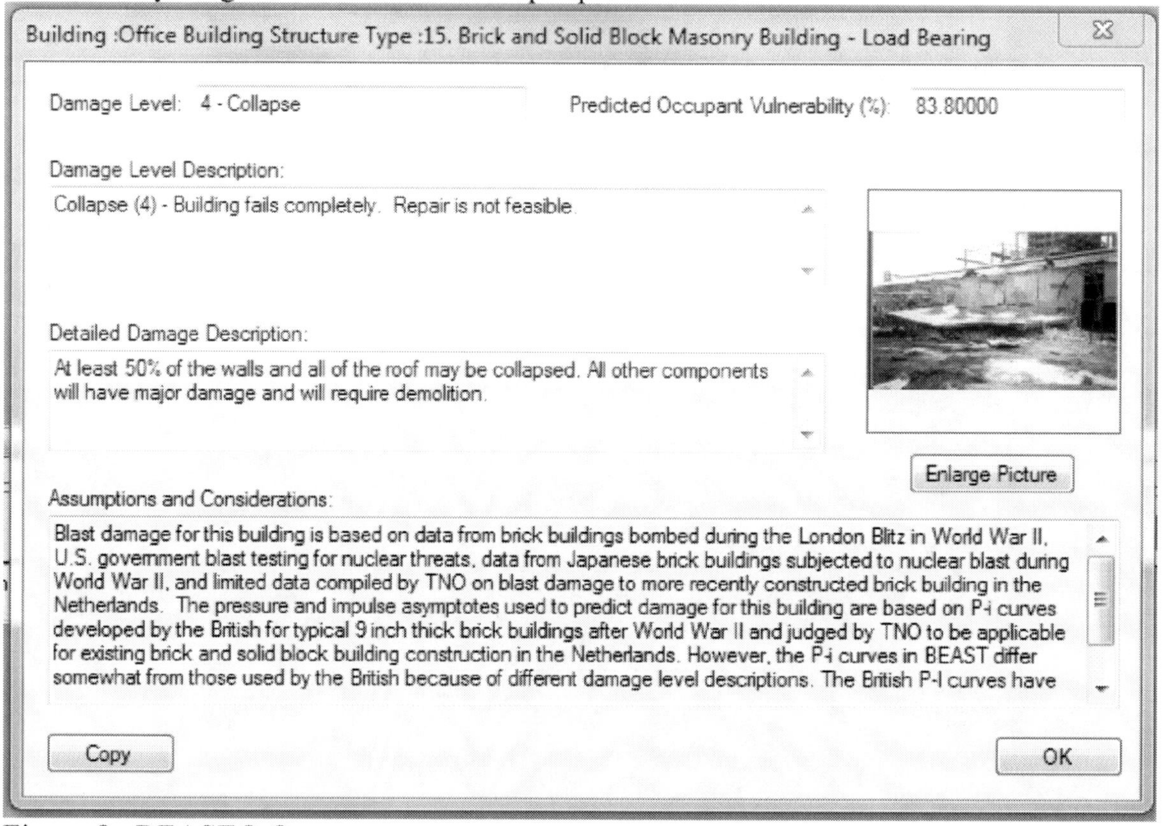

Figure 8. BEAST® Output

Of course, not all explosions result in the complete destruction of a building like the extreme example provided. BEAST® categorizes building damage using the criteria found in Table 3 [6]. A Building Damage Level up to 2B is still survivable by most occupants.

Table 3. BEAST® Building Damage Levels

Building Damage Level	Damage Description
1	Onset of visible damage to building walls facing explosive source. Repairs only needed for cosmetic reasons. Building is reusable following an explosion.
2A	Localized building damage, primarily to components on building walls facing the explosive source. Building performs function and can be used; however, major repairs are required to restore integrity of structural envelope. Total cost of repairs is moderate.
2B	Widespread building damage. Building cannot be used until major repairs are completed. Total cost of repairs is significant, approaching replacement cost of building.
3	Building has lost structural integrity and may collapse due to environmental conditions (i.e. snow or wind). Total repair cost exceeds replacement cost of building.
4	Building fails completely. Repair is not feasible.

Courtesy of BakerRisk Engineering Consultant Explosion Research Cooperative

4. Development of Source Terms for Potential Vapor Dispersion Release Scenarios

Methods for determining the source term for releases from tanks, liquid pools and piping systems are discussed here. All examples provided here do not reflect actual potential release conditions. For all examples, any relation to true conditions is incidental.

Prior to the modeling methods discussion, it is important to review a tool for determining losses due to piping layout.

IoMosaic SuperChems® v 8.21 (2016) allows for development of piping layouts to detail the piping diameters, elbows, and other potential connections. In simple models, as will be discussed in this paper, this can be used as the outlet piping for a pressure source (e.g., tanks) to a release point.

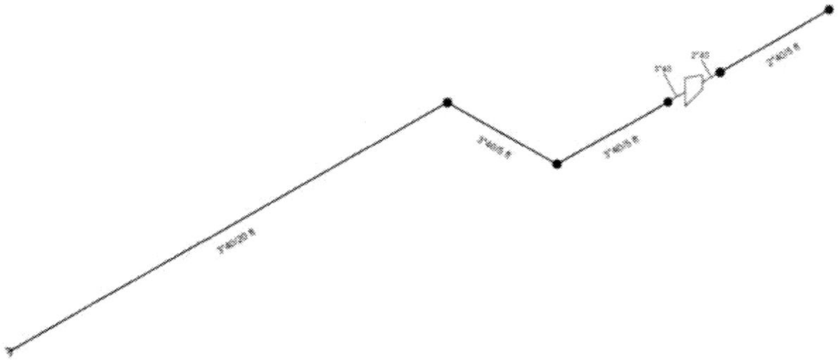

Figure 9. Example Piping Layout.

Figure 3 details a layout from the source at 3 inches with 2 90-degree elbows and a reduction to the 2-inch leak point. The release will be from a guillotine break on the 2-inch line.

4.1 Release from Tank

A reliable model for release from vessels is contained in ioMosaic SuperChems® v8.21 (2016).

This example will model the release of a liquid blanketed by nitrogen. The "Vessels Containing 2-Phases Model" allows for estimation of the vapor and liquid contents of a vessel, accounting for nitrogen blanketing or air in the head space. This same model can be utilized for relief from the vapor phase. There are alternative models for gas-filled vessels.

The overall process for using SuperChems® to model this is as follows:

1. Development of objects for mixture composition, vessel dimensions, piping layouts and other objects. Ensure piping layouts are properly applied.

   The following steps are within the "Vessels Containing 2-Phases" Model.

2. Input the vessel pressure and an initial estimate for overall composition inside of the vessel. The overall composition includes both the vapor and liquid phases. In this example, the modeled vessel is filled to 80% with methyl isocyanate and has a nitrogen blanket.

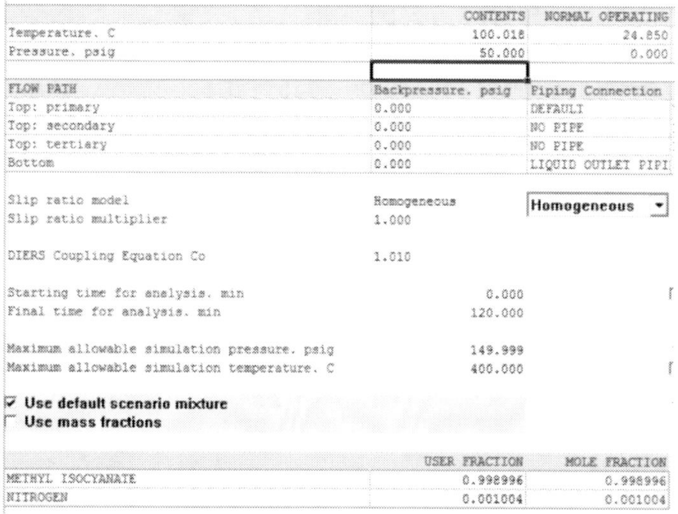

Figure 10. Vessels Containing 2-Phase Inputs

   Uncheck the "Use default scenario mixture" as defining this model is an iterative process.

3. Use "Specify Fill Fraction" to input the approximate vessel fill level. This example uses an estimate 80% fill level.

   The model it calculates the vessel temperature at the specified overall composition.

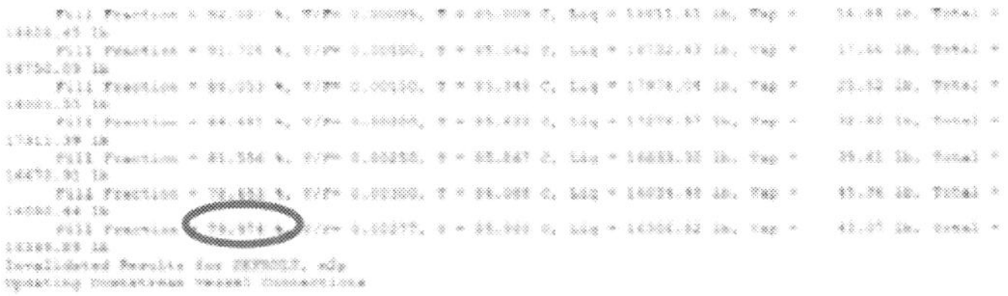

Figure 11. Model Iterations to determine vessel temperature at 80% liquid fill level.

4. Iterate over steps 2 and 3, varying the composition and re-entering the fill fraction until the modeled vessel temperature matches the true conditions. Increasing content of lower boilers will decrease the modeled vessel temperature.

   In the example, the composition was iterated until the storage temperature was approximately 25 deg C.

	CONTENTS	NORMAL OPERATING
Temperature, C	24.918	24.850
Pressure, psig	50.000	0.000

FLOW PATH	Backpressure, psig	Piping Connection
Top: primary	0.000	DEFAULT
Top: secondary	0.000	NO PIPE
Top: tertiary	0.000	NO PIPE
Bottom	0.000	LIQUID OUTLET PIPI

Slip ratio model	Fauske	Fauske
Slip ratio multiplier	1.000	
DIERS Coupling Equation Co	1.010	
Starting time for analysis, min	0.000	
Final time for analysis, min	120.000	
Maximum allowable simulation pressure, psig	149.999	
Maximum allowable simulation temperature, C	400.000	

☐ Use default scenario mixture
☐ Use mass fractions

	USER FRACTION	MOLE FRACTION
METHYL ISOCYANATE	0.991100	0.991100
NITROGEN	0.008900	0.008900

Figure 12. Final composition.

5. Running the model will determine the flow rate through the piping layout and guillotine break.

6. When the run is complete, a report will appear. The "Profiles-Bottom" tab is the most useful for this source term model.

TIME, min	VESSEL CONTENTS TEMPERATURE, C	VESSEL CONTENTS PRESSURE, psig	PRESSURE AT VESSEL BOTTOM, psig	BOTTOM PIPING LIQUID MASS FLOW RATE, lb/min	BOTTOM PIPING GAS/VAPOR MASS FLOW RATE, lb/min	Bottom Flow reduction factor (Actual flux with pipe/Ideal Flux)	BOTTOM FLOW NOZZLE PRESSURE, psig	BOTTOM FLOW DISCHARGE PRESSURE, psig	BOTTOM FLOW DISCHARGE TEMPERATURE, C
0.0000	24.92	50.000	56.394	2.005E+03	0.000	0.405	56.392	-0.001	25.094
0.0246	24.88	47.142	53.517	1.966E+03	0.000	0.394	53.516	0.000	25.051
0.0506	24.85	44.305	50.662	1.871E+03	0.000	0.383	50.660	-0.001	25.006
0.0782	24.81	41.472	47.809	1.775E+03	0.000	0.372	47.808	-0.001	24.960
0.1077	24.77	38.642	44.961	1.679E+03	0.000	0.360	44.959	-0.001	24.913
0.1394	24.73	35.817	42.116	1.583E+03	0.000	0.347	42.115	-0.001	24.864
0.1735	24.69	32.996	39.275	1.485E+03	0.000	0.334	39.274	-0.001	24.814
0.2104	24.64	30.181	36.440	1.387E+03	0.000	0.321	36.439	0.000	24.761
0.2506	24.60	27.371	33.610	1.289E+03	0.000	0.306	33.609	0.000	24.706
0.2946	24.55	24.568	30.787	1.190E+03	0.000	0.291	30.786	0.000	24.649
0.3433	24.50	21.773	27.971	1.090E+03	0.000	0.276	27.970	0.000	24.588

Figure 13. Vessels Containing 2-Phases Output.

The liquid and vapor release rates out of the piping as well as the discharge temperature and pressure are reported. The discharge pressure will be above atmospheric if there is choked flow inside the piping.

If the choke pressure is significantly above atmospheric, running some select points through the "2-Phase Expansion and Aerosol Model" to determine the final vapor/liquid composition and temperature is recommended.

7. As the release is liquid in this case, evaporation from a liquid forming pool will next be modeled. As stated previously, several dispersion models have integrated pool models. The SuperChems® pool evaporation will be employed in this case, and the output from the model will be input to an external dispersion model.

4.2 Liquid Pools

As with tank leaks, SuperChems® will be utilized for determining the source term of a liquid pool formation.

An important consideration is the variety of weather conditions over which a release could take place. As solar incidence is a major factor in pool evaporation, ensure that the most probable conditions for both daytime and nighttime are employed. Also, to account for solar incidence, ensure that the appropriate time of day and site coordinates are input for the scenario in SuperChems®. In addition, diked area dimensions and substrate composition can be specified.

SuperChems® pool evaporation model allows for a static release rate over a set period of time. As most releases will be dynamic in nature, this may involve expert judgment.

Once a characteristic release rate, duration and temperature have been determined, this information can be entered into the pool evaporation model.

Liquid discharge temperature. C	24.209
Liquid discharge rate. lb/min	583.903
Liquid discharge duration. min	1.949

Use default scenario mixture
Use mass fractions

COMPONENT	USER FRACTION
METHYL ISOCYANATE	1.0000
NITROGEN	0.0000

Average molecular weight	57.0520

Figure 14. Pool Evaporation Model Inputs

The model can perform evaporation on multi-component mixtures. In this case, the evolution of any dissolved nitrogen is not of concern. The liquid rate has been imported from the Vessels Containing 2-Phases model, and the source concentration has been set to pure methyl isocyanate.

Based on the inputs, the model will provide a conservative estimate for the evaporation of methyl isocyanate from the formed pool.

Time. min	OVERALL POOL EVAPORATION RATE. lb/min	Pool diameter. ft	Pool surface temperature. C
0.017	0.000	3.146	4.209
0.017	0.258	3.146	-5.182
0.017	0.257	3.146	-5.235
0.017	0.257	3.147	-5.287
0.017	0.256	3.147	-5.339
0.018	0.257	3.147	-5.389
0.028	0.249	3.171	-5.439
0.038	0.242	3.241	-7.525
0.049	0.252	3.371	-8.514
0.059	0.273	3.553	-8.996
0.070	0.302	3.765	-9.216
0.080	0.335	3.988	-9.306
0.090	0.371	4.212	-9.334
0.101	0.409	4.431	-9.335
0.111	0.447	4.644	-9.322
0.122	0.485	4.849	-9.305

Figure 15. Pool Evaporation Model Output.

Figure 15 details the most critical data points in the pool evaporation model. Understanding these will allow for straightforward input to a dispersion model.

While SuperChems® does have an integrated dispersion model, the author's preference is SAFER Trace® v10.2 (2012). Implementation of pool models into dispersion models is discussed later in this paper.

4.3 Leaks from Piping

For lengths of piping with isolation, it may be appropriate to treat the isolated volume as a vessel and analyze the varying leak rate over time. This section considers the situation where the pressure differential across the orifice may be assumed constant. In addition to the previous source terms, SuperChems® is utilized in determining the source term.

The "Source Term Wizard" can be employed to specify an upstream pressure and temperature to determine the flow rate across a specified piping layout or idealized nozzle. The example here will determine the release rate of methyl isocyanate pressurized to 50 psig and released through a guillotine break in the 2" segment of piping from Figure 9.

OPTIONS	
Liquid Density	Databank
PVT Behavior	Equation of state
VLE Search Direction	Dew point

☐ Use mass basis data to compute fractions
☐ Mixture is soluble in water
☐ Use an existing mixture

COMPOUND	FORMULA
METHYL ISOCYANATE	C2H3NO

Inlet temperature. C	25.00000
Inlet pressure. psig	50.00000
Back pressure. psig	0.00000

☑ Initial state is subcooled liquid

Figure 16. Liquid Source Term Wizard Inputs

The release is treated as a subcooled liquid in order to allow the model to estimate if any portion of the release will flash across the opening. It is not expected to have any flashing due to the release conditions in comparison to the chemical's vapor pressure at 50 psig. However, this model is employed for consistency.

** INLET CONDITIONS	
Inlet temperature. C	25.0
Inlet pressure. psig	50.0
Surroundings/back pressure. psig	-3.974E-06
Slip ratio exponent eta	0.001
Vapor quality	1.E-04

** EXIT CONDITIONS	
Exit temperature. C	25.160
Exit pressure. psig	-1.7331E-04
Barometric pressure at site elevation. psig	-3.974E-06
Mixture exit velocity. ft/s	21.213
Exit Mach Number	0.006
Maximum Mach Number	0.006
Exit rho*u*u in kg/m/s2 or in Pascals	38695.90
Maximum rho*u*u in kg/m/s2 or in Pascals	38695.90
Exit steady state impulse. lbf	68.146
Maximum steady state impulse difference. lbf	418.675
Pipe length travelled / total piping length	0.998
Flow rate. lb/min	1714.327

COMPONENT	MW	TOTAL FLOW. lb/min	LIQUID FLOW IN. lb/min	LIQUID FLOW OUT. lb/min	VAPOR FLOW IN. lb/min	VAPOR FLOW OUT. lb/min
METHYL ISOCYANATE	57.052	1713.817	1713.817	1713.817	0	0

Figure 17. Source Term Wizard Outputs.

The results from this model include liquid and vapor release rates as well as exit pressure and temperature. Had flow been choked in the piping, the exit pressure would be above atmospheric. As with previous models, it is recommended to utilize a flash block to determine conditions at atmospheric pressure for releases from systems with choked flow.

For releases from holes or cracks in piping, the piping layout can be developed with an orifice at the end of the layout. The orifice area would be equal to the area of the hole, and the orifice length would be equal to the width of the piping. In addition, due to constraints with the software, a short pipe segment of with the same diameter as the leak should be placed at the exit of the segment.

The results for the above cases can directly be input to a dispersion model, as described in the next section.

5. Application of Source Terms to Dispersion Models

This section will describe the inputs to dispersion models based on the results from the two-phase leak from the sour stream.

Section 5 will review how to utilize the data from the dispersion models in order to assess the hazard posed to a hypothetical shelter 500 ft downwind from the release.

5.1 Test for Jetting

In two-phase or vapor release cases, the linear velocity across an opening may be sufficient to cause jetting. As described previously, there are specific dispersion models for jetting conditions. The rule of thumb for jetting regime is 10 m/s [1]. A simple ideal gas calculation can provide an estimate for linear velocity, given a known release rate, hole size and temperature. This can be expanded to a real gas calculation for highly non-ideal cases. The temperature used in this calculation should be the flashed temperature to atmospheric pressure. The rate employed should be that of the vapor phase for two-phase releases as this will be the largest volumetric component once flashed to atmosphere.

In this example, a 200 psig stream of sour hydrocarbons at 150 deg C is released across a 1-inch hole in 6-inch piping. While several VLE models are available for hydrocarbon mixtures, ideality is assumed in this example.

Piping Isometric / Layout	6-INCH LINE WITH 1-INCH CRACK
Site Name	DEFAULT
Liquid Density Method	Databank
VLE/PVT Method	Equation of state
** INLET CONDITIONS	
Inlet temperature. C	150.691
Inlet pressure. psig	200.0
Surroundings/back pressure. psig	-3.974E-06
Slip ratio exponent eta	0.001
** EXIT CONDITIONS	
Exit temperature. C	133.870
Exit pressure. psig	138.432
Barometric pressure at site elevation. psig	-3.974E-06
Mixture exit velocity. ft/s	190.056
Exit Mach Number	1.0
Maximum Mach Number	1.0
Exit rho*u*u in kg/m/s2 or in Pascals	393239.785
Maximum rho*u*u in kg/m/s2 or in Pascals	393239.785
Exit steady state impulse. lbf	165.061
Maximum steady state impulse difference. lbf	6037.864
Pipe length travelled / total piping length	0.998
Flow rate. lb/min	455.119

COMPONENT	MW	TOTAL FLOW. lb/min	LIQUID FLOW IN. lb/min	LIQUID FLOW OUT. lb/min	VAPOR FLOW IN. lb/min	VAPOR FLOW OUT. lb/min
HYDROGEN SULFIDE	34.082	2.052	2.052	0.866	0.001	1.186
n-PENTANE	72.150	347.584	347.547	268.991	0.037	78.593
HEXANE	86.175	77.840	77.835	66.892	0.005	10.949
n-OCTANE	114.231	27.515	27.515	26.078	0.001	1.438

Figure 18. Flow of stream across 1-inch crack.

The exit conditions show that the flow has choked. As previously discussed, the "2-Phase Expansion and Aerosol Model" will be employed to determine the composition and temperature at atmospheric pressure. In addition, this model will report the linear velocity and cloud diameter when flashed to atmospheric pressure. The linear velocity can be used to determine if jetting is possible.

**TWO-PHASE EXPANSION / DEFAULT

	INITIAL	FINAL
Flow diameter. ft	8.3333E-02	2.3754E-01
Velocity. ft/s	190.42	651.59
Vapor density. lb/gal	0.24	0.02
Liquid density. lb/gal	4.20	5.19
Mixture density. lb/gal	0.98	0.04
Temperature. C	133.87	45.56
Pressure. psig	138.43	0.00
Vapor/Feed (V/F) mole fraction	0.21	0.71
Mass flow rate. lb/min	455.12	455.12
Available energy. BTU/lb	26.45	
Modified Weber number for thermal breakup	392.30	
Critical Weber number for mechanical breakup	12.01	
Viscosity group number	0.01	
Stable droplet diameter. Microns	39.50	
Thermal breakup droplet diameter. Microns	39.50	
Mechanical breakup droplet diameter. Microns	44.18	
Number of droplets	4.9421E+10	
Droplet mass. lb	4.4271E-11	

COMPONENT	TOTAL FLOW	INITIAL LIQ	FINAL LIQ	INITIAL VAP	FINAL VAP
HYDROGEN SULFIDE. lb/min	2.053	0.866	0.038	1.186	2.015
n-PENTANE. lb/min	347.681	269.076	80.158	78.605	267.523
HEXANE. lb/min	77.862	66.912	35.926	10.950	41.936
n-OCTANE. lb/min	27.523	26.085	23.812	1.438	3.711

Figure 19. Sour Stream Contents after Expansion.

Per Figure 19, the temperature of the release cloud will be 46 deg C at atmospheric pressure and will have a linear velocity of 652 ft/s (199 m/s). This is an order of magnitude above the 10 m/s guidance, and should result in significant jetting.

5.2 Input to Jet Model

The dispersion modeling tool utilized for these examples is SAFER Trace® 10.2 (2012).

Figure 20 details the inputs to the jet model based on our example. This is focusing on the vapor portion of the model, ignoring liquid entrainment.

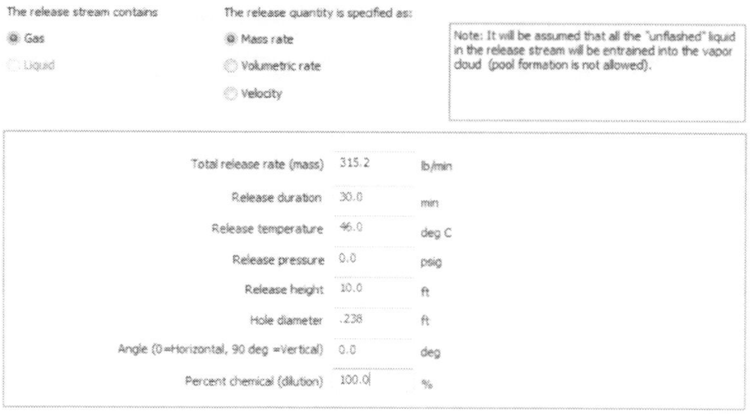

Figure 20. Jet Model Inputs.

The dispersion outputs will report concentrations and plots related to concentration of H2S in air.

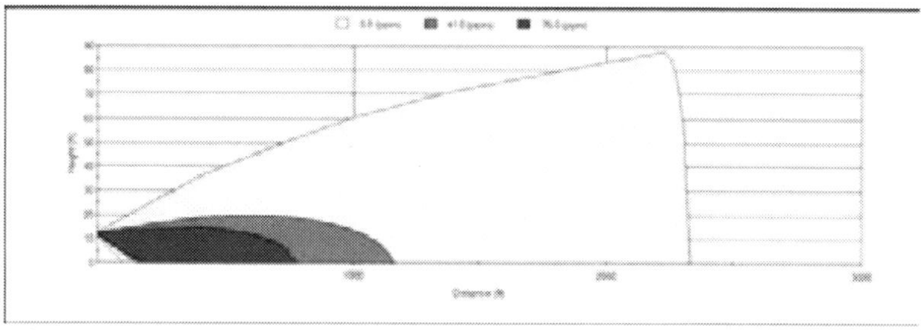

Figure 21. Jet Model Dispersion Output. AEGL concentration limits based on 10-minute exposure to H2S [10].

The H2S concentration profile at the outside of a shelter 500 ft downwind from the release point is shown in Figure 22.

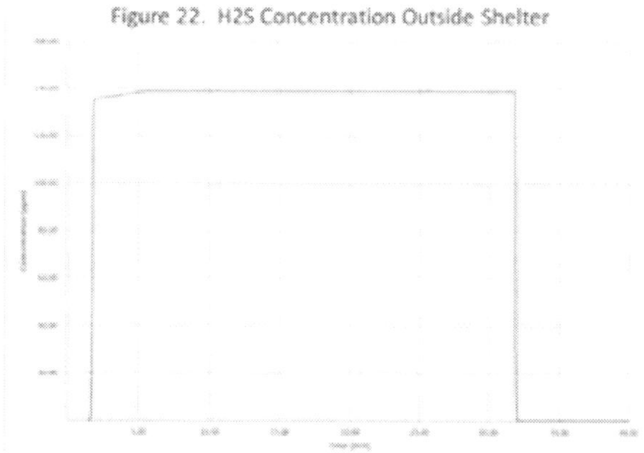

Consequence to personnel in this shelter will be analyzed using a probit function, detailed in Section 5.

5.3 Modeling Pool Evaporation

The liquid portion of the example two-phase release will be treated as the source of a liquid pool.

Spill Surface	DEFAULT
Liquid discharge temperature. C	45.560 <<<
Liquid discharge rate. lb/min	139.934
Liquid discharge duration. min	15.000

☑ **Use default scenario mixture**
☐ **Use mass fractions**

COMPONENT	USER FRACTION
HYDROGEN SULFIDE	0.0006
n-PENTANE	0.6394
HEXANE	0.2399
n-OCTANE	0.1200
Average molecular weight	80.5399

RUN OPTIONS

Figure 23. Liquid Pool Model Inputs

The concentration profile of evaporate is highly dynamic over the studied period of the release.

Figure 24. Pool Evaporate Concentration Profile

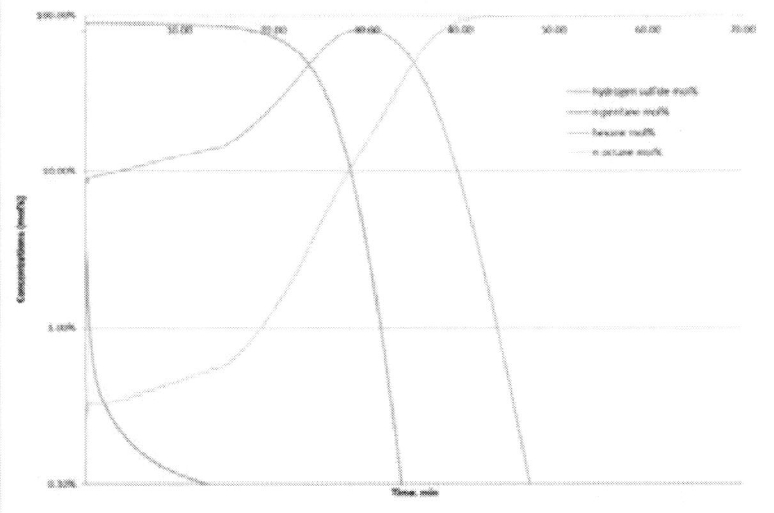

While the concentration is highly dynamic, the main toxicity concern is with H2S. The majority of H2S appears to evaporate in the first 15 minutes. An average rate will be developed based on the total evaporated mass of each component over that time. The effective concentrations and dispersion parameters are shown in Table 4.

Table 4. Molar Concentrations and Dispersion Parameters

Chemical	Mol%
Hydrogen Sulfide	0.1%
Pentane	87.0%
Hexane	12.4%
Octane	0.5%
Average Evaporation Rate	57.0 lb/min
Temperature	2°C
Duration	15 min

The data from this model is input to the SAFER Trace® "Single Phase" steady state model, per Figure 25.

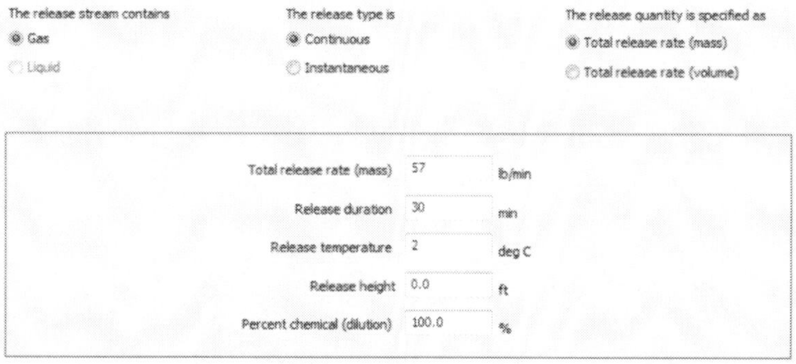

Figure 25. Steady State Model Inputs.

The downwind dispersion of the cloud, focusing on H2S concentrations in air is shown in Figure 26. The concentration is negligible at the 500-ft shelter, so the contribution from evaporation to toxic impact can be ignored.

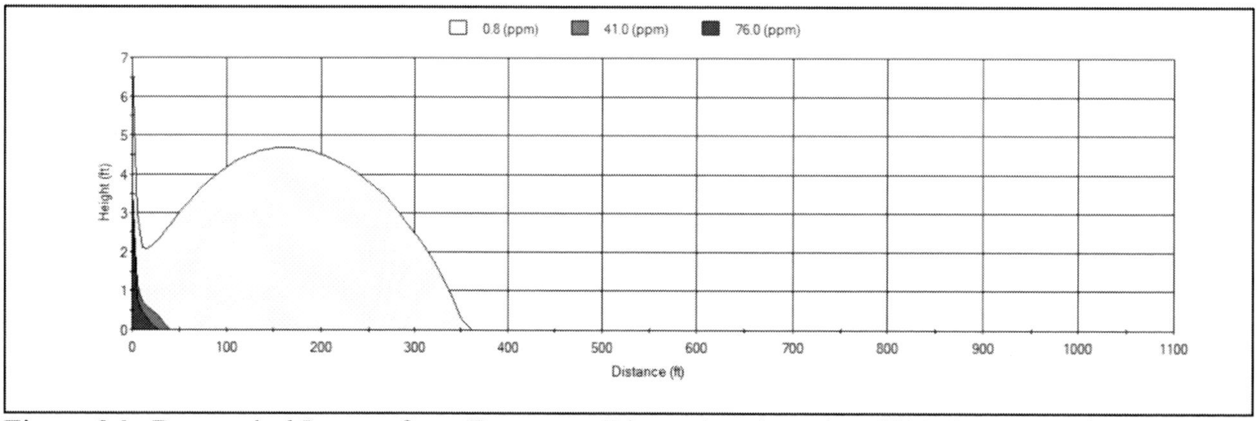

Figure 26. Downwind Impact from Evaporate Dispersion, based on H2S Concentration in Air.

6. Magnitude of Consequence Determination Using Probits

The methods utilized here are described in detail in "Simplified Methods of Using Probit Analysis in Consequence Analysis" [1].

To summarize, the impact to individuals may calculated using dose [5].

$$Dose = \int C^n dt \qquad \text{[Eq. 1]}$$

Dose may be transformed to probability of fatality using a probit transformation [5].

$$Y = A + B \log(Dose) \qquad \text{[Eq. 2]}$$

The constants "A", "B", and "n" are specific to the chemical. The Y value, or probit score is referenced on a probit table to determine probability of studied consequence.

Table 5. Probit table.

%	0	1	2	3	4	5	6	7	8	9
0	—	2.67	2.95	3.12	3.25	3.36	3.45	3.52	3.59	3.66
10	3.72	3.77	3.82	3.87	3.92	3.96	4.01	4.05	4.08	4.12
20	4.16	4.19	4.23	4.26	4.29	4.33	4.36	4.39	4.42	4.45
30	4.48	4.50	4.53	4.56	4.59	4.61	4.64	4.67	4.69	4.72
40	4.75	4.77	4.80	4.82	4.85	4.87	4.90	4.92	4.95	4.97
50	5.00	5.03	5.05	5.08	5.10	5.13	5.15	5.18	5.20	5.23
60	5.25	5.28	5.31	5.33	5.36	5.39	5.41	5.44	5.47	5.50
70	5.52	5.55	5.58	5.61	5.64	5.67	5.71	5.74	5.77	5.81
80	5.84	5.88	5.92	5.95	5.99	6.04	6.08	6.13	6.18	6.23
90	6.28	6.34	6.41	6.48	6.55	6.64	6.75	6.88	7.05	7.33
—	0.0	0.1	0.2	0.3	0.4	0.5	0.6	0.7	0.8	0.9
99	7.33	7.37	7.41	7.46	7.51	7.58	7.65	7.75	7.88	8.09

Source: Finney, 1952 [11]

The Microsoft Excel (2016) function "NORMSDIST" will convert the probit score to a probability.

The Equation 1 is combined with the "NORMSDIST" function to determine impact to occupants of the shelter, per Table 6. This is assuming that the air handling provides 10 air changes per hour to the shelter.

Table 6. Probability of Consequence to Sheltered Occupants.

Time (min)	Dosage ($ppm^{1.43}$-min)	% Probability
1.45		0%
1.46	-	0%
1.47	-	0%
1.48	-	0%
1.49	-	0%
1.50	-	0%
...	...	...
57.00	44,794	0%
58.00	45,354	0%
59.00	45,902	0%
60.00	46,436	0%
61.00	46,957	0%
62.00	47,466	0%
63.00	47,963	0%
64.00	48,448	0%
65.00	48,922	0%

In this example, the occupants can shelter for over an hour with 0% probability of consequence. Probit transformations can be utilized to study probability of fatal exposure from toxicity, thermal radiation, and blast overpressure. In addition, they can analyze probability of burn severity.

7. Conclusion

This paper has summarized the methods utilized in determining impact to personnel from toxic exposure and vapor cloud explosions. The covered models include pipe breach, tank release, and jetting models. These are highly utilized models for consequence analysis. In addition, the paper summarized methods to utilize dispersion modeling data to fully quantify impact to occupants of shelters downwind from releases.

For blast modeling, this paper provided an overview of one simplified method to predict explosion overpressure, impulse, and damage to buildings within the radius affected by the explosion. The method utilized is Baker-Strehlow-Tang (BST), one of several methods engaged by industry for this purpose. Most of the software tools used to perform the analysis were developed by the BakerRisk Engineering Consultant Explosion Research Cooperative, including VCloud® and BEAST®.

For all consequence models, more rigorous methods such as computational fluid dynamic (CFD) modeling will produce more accurate results. However, the methods discussed in this paper are excellent screening tools whose value lies in their ease of use and are suitable for risk analysis.

References

[1] Baybutt, Paul, "Using Layers of Protection Analysis to Evaluate Fire and Gas Systems", Process Safety Progress, Volume 31, Issue 3, 2011

[2] "SAFER Trace® Online Help", SAFER Systems, 2003

[3] James, Michael D., "Consequence Modeling of Dynamic Source Terms", Process Safety Progress, Volume 35, Issue 4, 2016

[4] Baker, Q. A., Tang, M. J., E. A. Scheier, and G. J. Silva, "Vapor Cloud Explosion Analysis," Process Safety Progress, Vol. 15, No. 2, 1996, pp. 106-109

[5] A. Sari, "Comparison of TNO Multienergy and Baker-Strehlow-Tang Models," Process Safety Progress 30 (2011), pp. 23-26

[6] BEAST® Help File, 2000 BEAST® Report, 2000

[7] "VCE Evaluation Method Comparison," The Quest Quarterly, Volume 4, Issue 1, Spring 1999

[8] Baker-Strehlow-Tang Methodology, Guidelines for Evaluating the Characteristics of Vapor Cloud Explosions, Flash Fires, and BLEVEs, CCPS, New York, 1994

[9] Pierorazio, A.J., J.K. Thomas, Q.A. Baker and D.E. Ketchum (2005) "An Update to the Baker-Strehlow-Tang Vapor Cloud Explosion Prediction Methodology Flame Speed Table," Process Safety Progress, 24(1): 59-65

[10] "About Acute Exposure Guideline Levels (AEGLs)", 2016, https://www.epa.gov/aegl/about-acute-exposure-guideline-levels-aegls (Accessed Mar 2, 2017)

[11] D. J. Finney, *Probit Analysis*, Cambridge University Press, New York 1952

Relief Systems Design: The Importance of Accounting for Static Head

Nicholas N. Cristea
Smith & Burgess, LLC
7600 W Tidwell Rd, Ste 600, Houston, TX 77040
Nicholas.cristea@smithburgess.com

Zubin Kumana
Smith & Burgess, LLC
7600 W Tidwell Rd, Ste 600, Houston, TX 77040
Zubin.kumana@smithburgess.com

Guomin Wang
Smith & Burgess, LLC
7600 W Tidwell Rd, Ste 600, Houston, TX 77040
Guomin.wang@smithburgess.com

Prepared for Presentation at
American Institute of Chemical Engineers
2017 Spring Meeting
13th Global Congress on Process Safety
San Antonio, Texas
March 26-29, 2017

Relief Systems Design: The Importance of Accounting for Static Head

Nicholas N. Cristea
Smith & Burgess, LLC
7600 W Tidwell Rd, Ste 600, Houston, TX 77040
Nicholas.cristea@smithburgess.com

Zubin Kumana
Smith & Burgess, LLC

Guomin Wang
Smith & Burgess, LLC

Keywords: Relief Systems Design, Static Head, Upstream, Downstream

Abstract

The process of evaluating an entire facility's relief and flare system design can be time consuming and costly. During the process, there will be systems to be evaluated that have liquid scenarios that may be protecting other assets within the facility. If the static head for the liquid scenarios is not accounted for, then the analysis can become even more costly due to the analysis not being performed correctly the first time. The result could be not meeting code by not adequately protecting system. This paper will look at two different cases where the static head was not accounted for, and how the issue was mitigated in the most cost-effective way. First, when dealing with the liquid heading on an offshore platform, there are many different levels on the platform and ensuring that the static head is accounted for is important to protect the different assets in the liquid handling train. Second, when evaluating a column system, it is important to ensure that the relief device provides protection for every ASME Section VIII Div 1 stamped equipment within the system. These two case studies highlight the necessity of accounting for static head in relief systems design.

1 Introduction

In the process safety industry, relief devices are set such that at least one relief device set pressure is equal to the lowest system MAWP within the protected system, to meet the requirement of ASME Section VIII Div 1 to provide adequate protection against overpressure. ASME Section VIII Div 1 UG-125 states "All pressure vessels within the scope of this Division, irrespective of size or pressure, shall be provided with overpressure protection in accordance with the requirements of UG-125 through UG-138, or with overpressure protection by system design in accordance with requirements of UG-140, or a combination of the two". ASME Section VIII Division VIII UG-125(c) and (1) also state that "when a pressure relief device is provided, it shall prevent the pressure from rising more than 10% or 3 psi (20 kpa), whichever is greater, above maximum allowable working pressure" and "when multiple pressure relief devices are provided and set in accordance with UG-134(a), they shall prevent the pressure from rising more than 16% or 4 psi (30 kpa), whichever is greater, above the maximum allowable working pressure". The vessel is normally hydrotested without accounting for the potential impact of inlet piping to the relief devices. Therefore, when the elevation difference between the top of the vessel and the relief device becomes significant [2], the vessel may experience a pressure higher than its allowable accumulation due to the static head, especially during liquid relief scenarios. The equation below is a conversion from the liquid height to the equivalent liquid head (psig).

$$P = 0.433\ x\ SG\ x\ H$$

P: liquid head in psi

SG: Specific Gravity

H: liquid head in feet

Equation 1: Liquid height to pressure conversion

This paper will look at two different cases where the static head was not accounted for and how the issue was mitigated for each case. The first case will look at a blocked system liquid outlet scenario on a platform where the vessel of interest has the same MAWP as the protecting vessel on an upper level. The second case will look at the liquid relief case (i.e. overfilling) and the mitigation for a column system that contains several vessels. These two cases will show the importance of accounting for static head and the potential mitigation options for these types of concerns.

2 Case 1 – Produced Water Desanders and Accumulators on an Offshore Platform

2.1 System overview

Figure 1 below is an illustration of case study 1. A produced water desander on the upper deck of a platform is fed by a booster pump. The desander then feeds the desander accumulator. The desander and the accumulator both have the same MAWP of 350 psig. The maximum discharge pressure of the booster pump is 415 psig. In the event of a blocked outlet, the relief device PSV-1 (4" P 6") on the desander is intended to provide adequate protection for the desander and the downstream accumulator. The relief device PSV-2 (1" E 2") on the accumulator is intended only for the fire scenario, based on the assumption that the upstream relief device (PSV-1) would limit the maximum deliverable pressure to the accumulator to 350 psig for the blocked outlet scenario. However, the vessels are located on different levels of the platform. The elevation difference was never accounted for in the choice of relief device set pressure. When PSV-1 is relieving at 385 psig, the pressure inside the accumulator would be the relief pressure (385 psig) plus the static head due to the difference in elevation. If the difference in elevation is large, the downstream accumulator may no longer be adequately protected by PSV-1, because the vessel would experience a pressure higher than its allowable accumulation. The elevation difference between the two relief devices illustrated in Figure 1 is roughly 27 feet. The static head from the elevation difference is about 12 psig. This means that when the desander is at the relief pressure, the pressure in the accumulator would be 397 psig, which exceeds the 10% allowable accumulation [2] (385 psig) of the accumulator.

2.2 *Case 1 mitigation*

The common mitigation option to account for the static head is to lower the relief device set pressure on the desander to account for the static head, so that the downstream accumulator would be adequately protected during the liquid relief event. However, for this particular case study, a more detailed engineering analysis revealed that the installation was acceptable as-is.

However, it was determined that there are technically two relief devices protecting the system in the event a blocked outlet on the accumulator. The allowable accumulation for the accumulator can go up to 16% of the system (406 psig) because of the multiple relief devices. Since the manual valve between the two systems will remain open (the initiating event is the closure of the downstream block valve), the accumulator is still protected. If the manual valve between the two systems were to be closed, then the blocked outlet case on the accumulator would no longer be applicable as the source of pressure would be lost as well.

If the static head is greater than 21 psig (~49 feet), the accumulator may experience overpressure in excess of the allowable (a pressure of 406 psig). The potential mitigation would be to lower the set pressure of the desander relief device to account for the static head such that the maximum pressure did not exceed this limit. For this case study, PSV-1 set pressure is not required to be lowered because the accumulator is still within the 16% allowable accumulation after accounting for the static head. Additionally, if the relief device set pressure were lowered, the potential exists that the relief device may be undersized, and additional mitigations may be required to resolve the concerns.

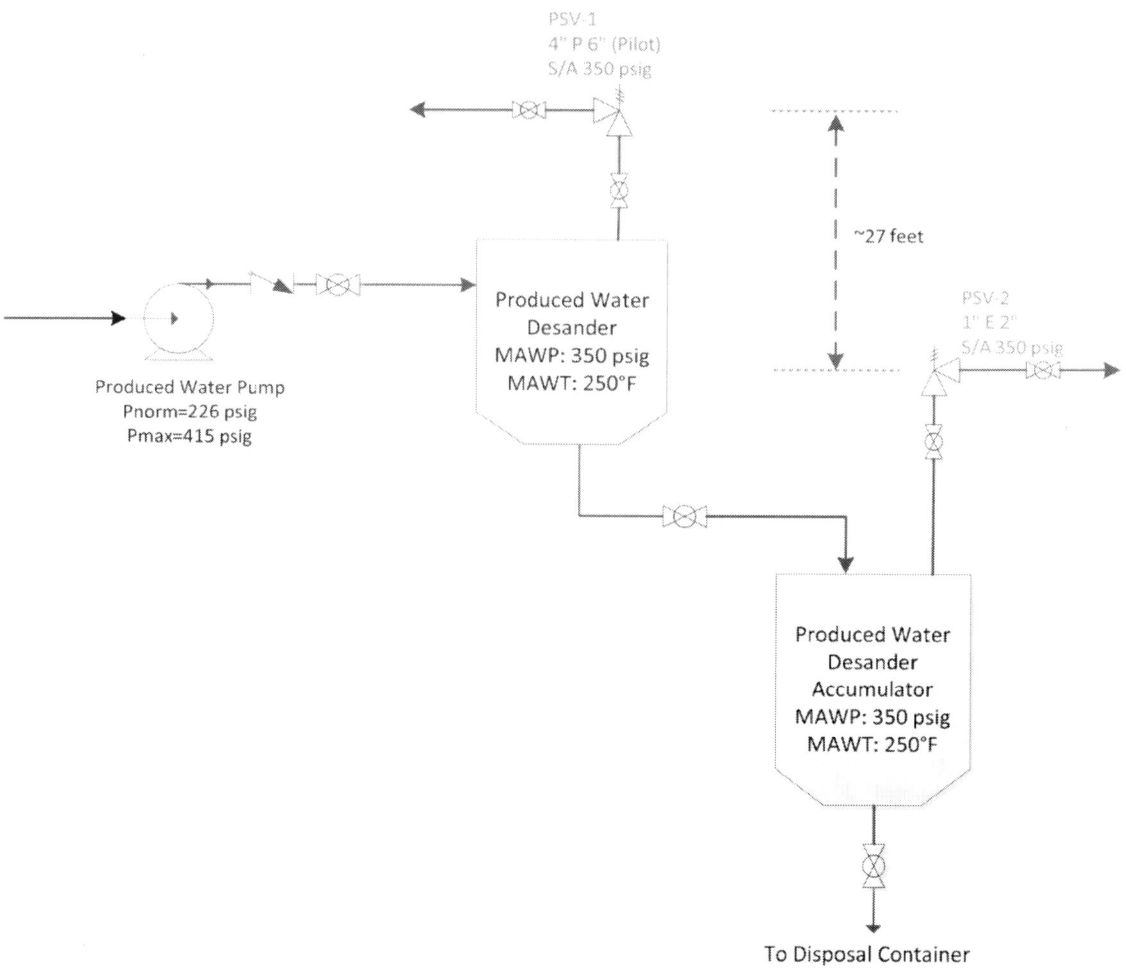

Figure 1, Case 1 – Produced Water System Overview

3 Case 2 – Common Column System

3.1 System Overview

Figure 2 below is a simplified column system that contains several pieces of equipment. The top of column is 50 feet from grade. The reboiler is located at grade. The relief device (PSV-3) is 5 feet above the top of the column and set at 330 psig. The lowest system MAWP is 330 psig based on the MAWPs of the column and the reboiler (shell side). This case study only looks at the "overfilling" scenario for the column. As the lowest system MAWP and the set pressure are the same, it may appear at first glance that the relief device is set correctly to protect the system. However, during the overfilling scenario, there are 55 feet of static head on top of the reboiler (assuming the reboiler is flooded). Therefore, the pressure that the reboiler (shell side) experiences during the liquid relieving event is higher than the allowable accumulation. For this case study, the specific gravity of the fluid is assumed to be 0.7, which would result in roughly 17 psig of static head. This means that during the relief event, the pressure in the reboiler (shell side) is 380 psig (~15% accumulation), which exceeds the 10% allowable accumulation. Therefore, mitigations may be required as the reboiler is not protected due to the static head.

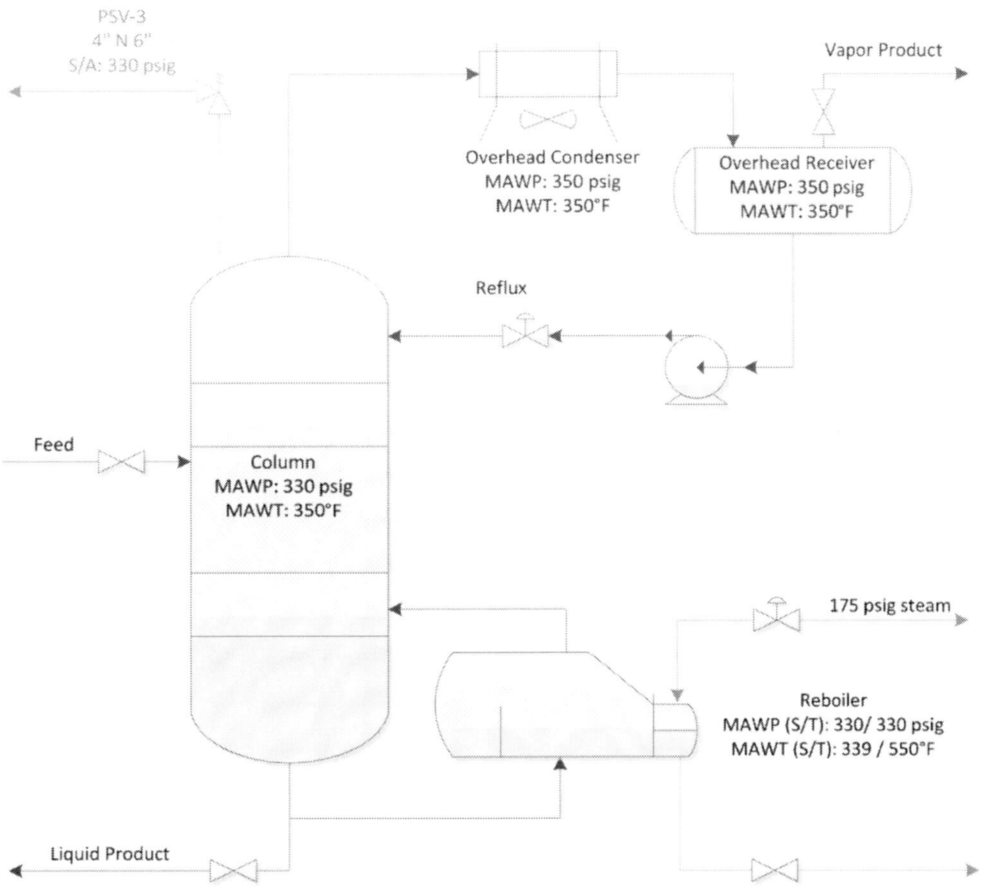

Figure 2, Case 2 - Common Column System Overview

3.2 Case 2 mitigation

There are multiple mitigation options for this type of concern. One mitigation option is to re-rate the reboiler to a higher MAWP to account for the static head. Another mitigation is to lower the set pressure of the relief device to account for the static head. Therefore, the new set pressure for this case study can be 314.5 psig or lower if the reboiler MAWP plus allowable accumulation is not exceeded by the relief pressure plus static head. When the relief device set pressure is lowered to account for the static head during the liquid relief events, engineers must confirm the relief device can still provide adequate capacity at the reduced pressure for all applicable overpressure scenarios. Also, the operating pressure of the column needs to be considered to ensure that the lower set pressure would not result in simmering of the relief device during normal operation. In practice, many plants would also have high independent liquid level alarms and fill time criteria to remove the potential liquid relief scenarios by taking credit for operator responses. Another option is to look at risk assessing the concern as the MAWP is not exceeded by 16% accumulation.

There are many ways to mitigate this concern; this case shows one of the most common mitigation option.

4 Conclusion

The two case studies discussed in this paper show the need to consider the static head when designing the relief system. The issue of static head should be considered during the design phase to ensure the system is code compliant. The cases studies show that the system may not be protected by code, but may be able to be risk accepted based on what the resulting pressure at the vessel is question becomes. It is important during the design phase to be cognizant on the elevation of the systems and the protecting relief devices to ensure the static head has been considered. This can be especially true when designing an offshore platform with multiple levels. Also, when commissioning columns or other vessels, it is important to consider the MAWP of all the vessels in the system to ensure all aspects of the system are protected by the relief device. The case studies give a brief overview of some of the possible mitigation options for static head concerns. It is the responsibility of the engineer reviewing the information to select the option that best fits the facility's needs.

5 References

[1] ASME Section VIII, "Boiler and Pressure Vessel Code," American Society of Mechanical Engineers, 2013.

[2] ANSI/API Standard 520 Part 2, "Sizing, Selection, and Installation of Pressure-relieving Devices," American Petroleum Institute, 6th ed., 2015.

[3] ANSI/API Standard 521, "Pressure-relieving and Depressuring Systems," American Petroleum Institute, 6th ed., 2014.

478548 Study of Factors That Affect Dust Explosion Characteristics of Fibrous Dust Materials

Abstract:

Abstract for 13th Global Congress on Process Safety

Track: Annual Loss Prevention Symposium (LPS)

Study of factors that affect dust explosion characteristics of fibrous dust materials

Bharatvaaj Ravi, Yan-Ru Lin, Chad Mashuga, M. Sam Mannan

Mary Kay O'Connor Process Safety Center

Artie McFerrin Department of Chemical Engineering

Texas A&M University

College Station, TX-77843-3122, U.S.A

+ 1 (774)-262-5215, vaaj26@tamu.edu

Combustible dust hazard is a serious threat to the process industries. Out of the 281 combustible dust incidents reported by the U.S. Chemical Safety and Health Investigation Board (CSB), 66 incidents have been caused by wood dust. In spite of the National Fire Protection Agency's (NFPA) efforts in specifically catering to the wood working and wood-based products manufacturing industries and Occupational Safety and Health Administration's (OSHA) National Emphasis Program on combustible dust hazards, to create awareness and minimize wood dust hazards, the wood dust incidents continue to occur. The fundamental difference between traditional dusts and wood dusts is that the latter are flocculent and fibrous. They are characterized effectively by an aspect ratio (length to diameter ratio) rather than a single dimension, such as diameter, as in the case of traditional - spherical or near spherical - dusts. Various studies in the literature have revealed that the standard dust explosion testing procedures and the categorization of these non-traditional dust based on particle diameter (mesh sizes) might not be valid. This research is concerned with similar line of thought and exploring the effect of aspect ratio on the time of suspension of these fibrous wood dusts. The likelihood of ignition increases with the time of suspension given other conditions to form a combustible dust hazard. The study also deals with characterizing the participating and non-participating portions when a dust explosion occurs through analysis of samples before and after explosion testing. Does the dust concentration give the entire picture of explosion characteristics such as maximum explosion pressure and explosion pressure rise rate? Or is there a better indicator of these characteristics when it comes to flocculent dust? These are the questions that the current research is trying to answer.

Key words: Combustible dust, Fibrous dust, Dust explosion testing, Dust concentration, Dust explosion characteristics, Non-traditional dust.

High Integrity Pressure Protection System (HIPPS) and Implementation Challenges

Janardhanan Kallambettu P.Eng.
Principal Engineer-Control Systems
Bechtel Corporation
3000 Post Oak Blvd, Houston, TX 77459
jkallamb@bechtel.com

Prepared for Presentation at
American Institute of Chemical Engineers
2017 Spring Meeting and 13th Global Congress on Process Safety
San Antonio, TX
March 26 – 29, 2017

High Integrity Pressure Protection System (HIPPS) and Implementation Challenges

Janardhanan Kallambettu P.Eng.
Principal Engineer-Control Systems
Bechtel Corporation
3000 Post Oak Blvd, Houston, TX 77459
jkallamb@bechtel.com

Keywords: PHA, HIPPS, SIS, SIF, SIL, PLC

Abstract

Over the years, the process industries have heavily utilized pressure relief devices (PRDs) to protect the mechanical integrity of system components. While PRDs offer a straightforward and practical way of controlling overpressure in process plants, they do have some inherent limitations such as impacts to flare system design, environmental concerns, and some physical limitations in PRD sizing. In these cases, where the use of PRDs is limited, High Integrity Pressure Protection System (HIPPS) can be utilized as a substitute for PRDs.

While HIPPS offers a solution to address the PRD limitations, there are some challenges associated with a successful HIPPS implementation. The HIPPS are classified as Safety Instrumented Systems (SIS) and require compliance with the functional safety standard IEC 61511[**1**]. The aim of this paper is to guide the reader with the necessary tools to successfully perform all the safety lifecycle activities of HIPPS, per functional safety standards and codes. Challenges in the various phases of the safety lifecycle, from process hazard analysis (PHA) and Safety Integrity Level (SIL) selection to operation and maintenance of HIPPS, are discussed in this paper. Additionally, this article describes many types of HIPPS, their pros and cons, to choose the right system for a given application. The paper concludes with an example HIPPS to achieve high-level risk reductions and demonstrates the appropriate use of quantitative tools, such as fault tree analysis (FTA), during the HIPPS design process.

1 Introduction

HIPPS is an acceptable solution instead of PRDs as a means of overpressure protection for vessels and pipelines. Codes such as ASME Section VIII, Division 1[**2**] and recommended practices and standards such as API Standard 521[**3**], API 17O [**4**] and HSE, U.K [**5**] provide criteria for the usage of HIPPS. The reader is advised to refer to the cited references to understand in detail the selection criteria for HIPPS.

The HIPPS is a safety instrumented function (SIF) implemented in a SIS and, therefore, the codes and the standards referenced above require compliance with the functional safety standards IEC 61508[**6**] and IEC 61511.

The functional safety standards require adherence to a safety lifecycle model. The various safety lifecycle activities are:

- process hazard analysis;
- SIL selection;
- safety requirements specification (SRS);
- SIS design per SRS;
- design verification meeting the target SIL including systematic capability, and architectural constraints (redundancy);
- installation, commissioning, validation, operation, and maintenance per SRS; and
- change management.

The intent of this paper is to expose particularly process engineers and other non-control systems engineers, about specific aspects of the HIPPS in a straightforward language.

2 PHA and SIL Selection

Whenever a HIPPS is indicated instead of PRDs, process and process safety engineers assign an SIL 3 target to the HIPPS as supported by Lawley and Kletz's [7] requirement that the reliability is ten times better than that of a relief device. For clean services, the probability of failure on demand (PFD) of a relief device is typically two orders of magnitude, hence the SIL 3 target for a HIPPS. API Standard 521, while emphasizing the need for an appropriate SIL selection process per established risk tolerance criteria, also suggests SIL 3 target for HIPPS in the absence of a documented risk tolerance criteria.

An assumption, however, that the HIPPS's SIL target is always SIL 3 may not be correct in some circumstances. Applicable codes and standards mandate a requirement for a documented PHA to identify and examine all potential overpressure scenarios. The severity of consequences typically ranges from moderate to catastrophic and, based on the owner's risk tolerance criteria, the HIPPS target can vary from SIL 2 to SIL 4. For example, if the likelihood of an initiating event causing the overpressure is 1 in 10 years and the owner's risk criteria for catastrophic severity is 1in 100,000 years, and in the absence of any other independent protection layers or conditional modifiers, the required risk reduction is 10,000, a SIL 4 target.

Designing a SIL 4 HIPPS is a significant challenge and not recommended. The functional safety standards discourage such high-level risk reduction measures by a single SIF. Every effort should be made to remove or reduce the hazard at the source, introduce other risk reduction measures which are not based on the SIS or minimize the likelihood of the causes of the hazardous event. If the above steps do not help, then use of multiple independent HIPPS, each with a lower integrity, to meet the overall risk reduction requirements should be considered. See section 6 for a hybrid HIPPS design solution meeting an overall risk reduction of 10,000 or more.

3 Types of HIPPS

Many in the process industry believe that HIPPS always includes 2oo3-voted pressure transmitters (sensors), Triple Modular Redundant (TMR) safety PLC (logic solver), and 1oo2-voted ball valves (final elements). In reality, there are many types of HIPPS available to choose from based on the specific application.

HIPPS can be broadly classified as **Electronic**, **Mechanical** or **Hybrid** based on the technology used for the logic solver as shown in **Figure 1**.

A brief description, pros and cons, and applications for each of the above types are given in the following subsections.

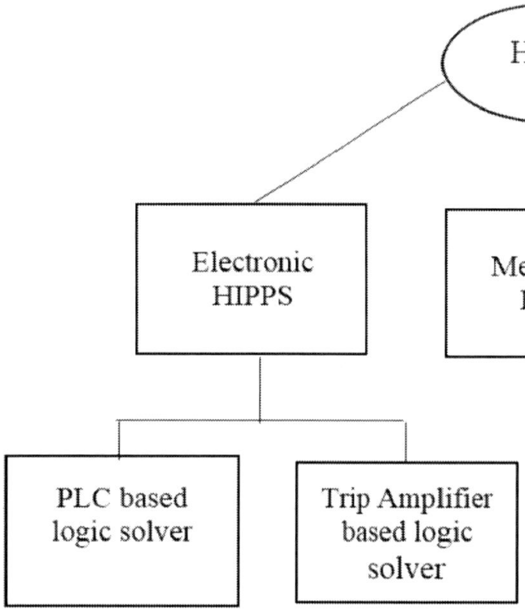

Figure1. Types of HIPPS

3.1 Electronic-HIPPS

An electronic-HIPPS uses either a programmable logic controller (PLC) or a trip amplifier system as a logic solver.

3.1.1 PLC based HIPPS

There are multiple varieties of PLCs equipped with microprocessor or solid state technology in a multitude of architectures such 1oo2, 1oo2D, 2oo3, 2oo4, and 2oo4D. Typically, all these PLCs will meet SIL 3 reliability. These PLCs are further categorized as complex devices and use application programming languages such as ladder logic or functional block diagrams. Refer to **Figure 2**.

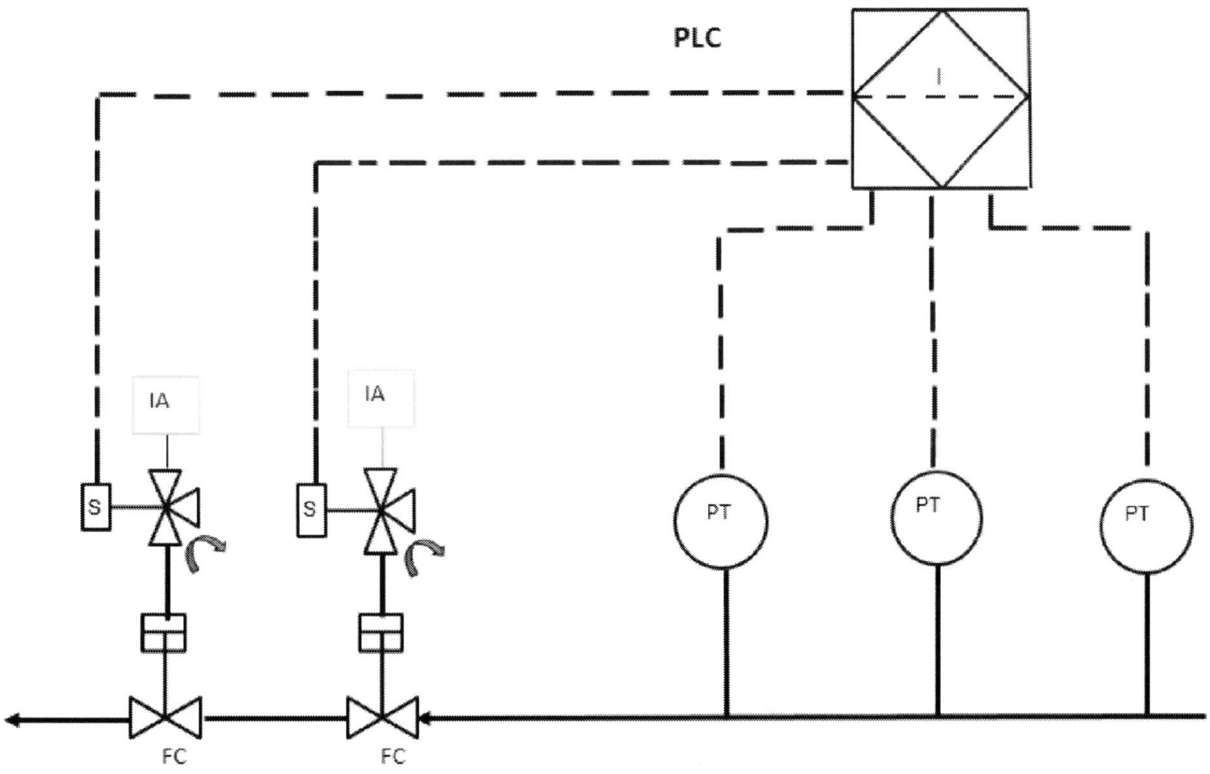

Figure 2. PLC based HIPPS

Pros:

- Flexible in developing the logic program and accommodates higher number of field inputs and outputs (I/Os);
- Equipped with special features for application programming such as variable valve closure speeds while closing the valve to mitigate consequences associated with slamming shut.
- Accommodate multiple HIPPS in a single logic solver within a plant facility;
- Facilitates partial stroke testing to meet target SILs or extend the testing interval, when practicable;
- Easily interfaces with the plant operator stations (human machine interfaces, HMIs).

Cons:

- Often needs specialized application programming expertise, in addition to vendor-specific training for selection of hardware and complex control strategies;
- Validation of the vendor claimed SIL against the safety manual is not a trivial exercise;
- Requires protection against cyber security threats;
- Potential for introducing systematic errors in application program;
- Application program using a low variability language (LVL) calls for special considerations per IEC 61508/IEC 61511 standards including application program safety requirements specification;
- Requires rigorous Factory Acceptance Test (FAT);
- Large footprint is needed -typically 800mmX600mmX1000mm panel;

- Extensive installation requirements such as controlled environment and special grounding;
- Requires complex, and redundant large capacity power supply;
- Voluminous documentation is involved;
- Requires a high level of competence in commissioning, validation, maintenance and operation; and
- Expensive for simple HIPPS applications ($50,000 for a logic solver alone).

Applications:

In a large facility, where many HIPPS are to be implemented for different hazardous event scenarios requiring complex functions or features, a PLC based HIPPS is recommended. For example, over pressure protection of multiple vessels, various piping segments, and process equipment within a facility to avoid loading of the flare system.

3.1.2 Trip-Amplifier based HIPPS

Trip-amplifiers are discrete logic solvers of low complexity. A trip amplifier accepts analog 4-20mA input from a field device such as a pressure transmitter and de-energizes contact output(s) when the process value exceeds the trip set point. Each sensor needs a dedicated trip amplifier. Refer to **Figure 3**.

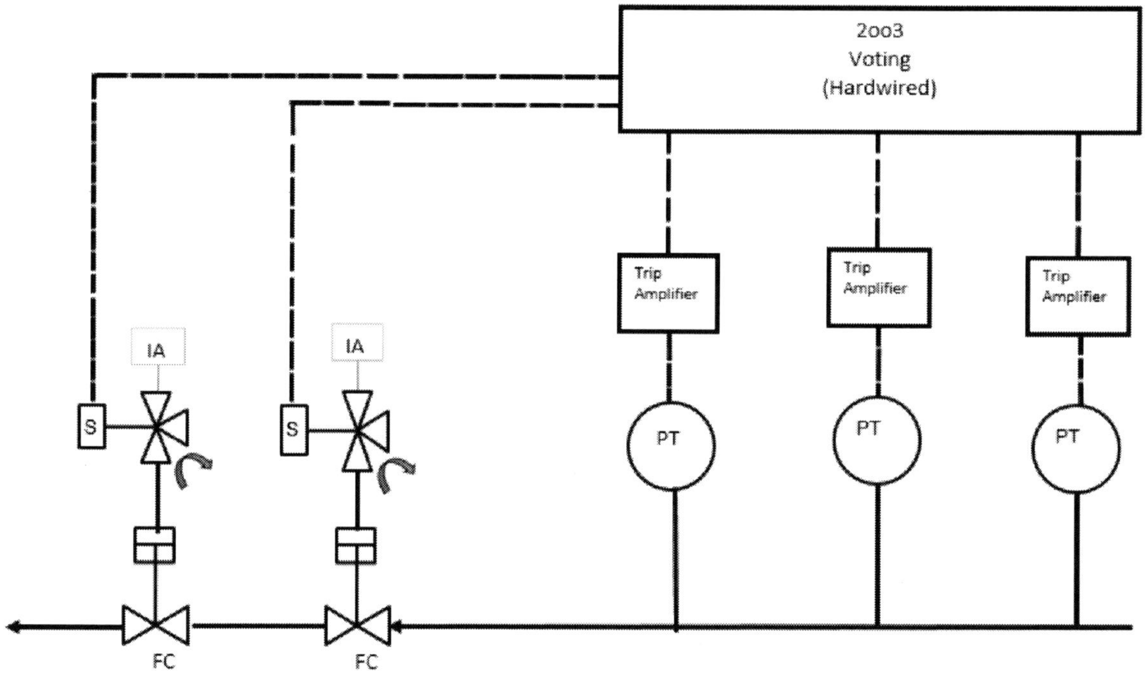

Figure 3. Trip Amplifier based HIPPS

Pros:

- Simple, reliable, with low design complexity;
- Meets SIL 2 reliability without redundancy for Trip Amplifiers and SIL 3 reliability with redundant architecture;
- Does not require application programming; classified as a device with fixed programming language (FPL) per IEC 61508/IEC 61511and, therefore, eliminates the need for application program safety requirements specification;
- No cybersecurity related issues;
- Less potential for systematic errors;
- Well suited for simple logic such as HIPPS logic;
- Low space requirement-typically DIN rail mounted;
- Low power device with no special requirements;
- No need for a controlled environment;
- Being a simple device FAT is not required for the logic solver;
- Easy to install and maintain; and
- Low cost; (logic solver range less than $10000)

Cons:

- Requires external devices for complex logic such as variable speed closing times for valves to avoid slam shut scenarios;
- Interfacing with operator stations requires additional I/Os in the plant DCS;
- Difficult to interface and facilitate partial stroke testing;
- Voting among the trip-amplifiers are external to the trip-amplifiers using the trip contacts and physical wiring which is less straightforward; and
- Implementation of maintenance bypasses is not practical;

Applications:

Well suited for applications involving straightforward, simple logic and for facilities which do not use PLCs. Also, appropriate for applications requiring a small footprint, low lifecycle cost and facilities not having multiple HIPPS. For example, remote pipeline compressor stations or storage tanks.

3.2 Mechanical-HIPPS

A mechanical HIPPS uses mechanical initiators. The mechanical initiator not only senses the over pressure but also serves as a logic solver to detect high pressure and actuate the valve. Mechanical HIPPS do not use any electrical or electronic components and, therefore, are inherently safer. Standalone versions of mechanical HIPPS are available which do not require any energy sources, such as instrument air supplies. Refer to **Figure 4**.

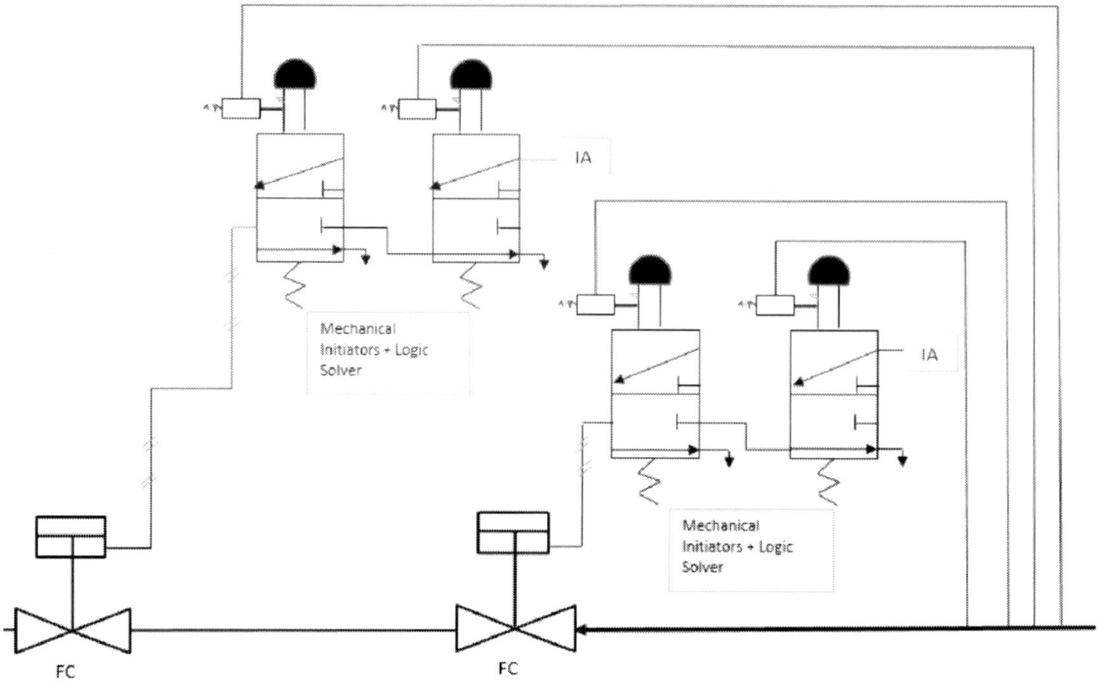

Figure 4. Mechanical HIPPS

Pro:

- Simple, reliable and minimally complex;
- Does not involve wiring and programming;
- Less susceptible to systematic errors;
- Cables, cable trays, conduits, etc. are not required;
- Can be made standalone without external energy sources;
- Single valve system can meet SIL 2 targets, redundant valve configuration can even meet SIL 3;
- Third party agencies certify an entire system in contrast to individual elements/subsystems of electronic HIPPS;
- Fewer components to fail;
- Easy to install, commission, validate, maintain and operate;
- Minimal documentation: no wiring diagrams or, installation drawings for logic solvers and initiators, application program SRS required;
- Competence in programming or maintaining a complex electronic system is not required;
- There are no EMC interference issues;
- There are no impacts due to cyber security threats;
- Suitable for harsh environments-no environmental climate control is needed;
- Easy to comprehend by operators and maintenance mechanics;
- Lowest life cycle cost among all the types of HIPPS;

Cons:

- Limited distance between the pressure sensing points, mechanical initiators and the valve- all components must be in the close proximities to avoid transportation time lags impacting the response times;
- Mechanical initiators need to be carefully selected as they are sensitive to setpoints and their accuracies;
- Interfacing with operator stations requires additional transmitters in the field and Inputs/outputs (I/Os) in the plant DCS;
- Requires external devices to facilitate partial stroke testing;
- Implementing maintenance bypasses consistent with the overall facility philosophy is not practical;

Applications:

Mechanical HIPPS are attractive for standalone applications in remote plant locations, for areas where electrical cabling is impractical and in facilities, like pipeline pump/compressor stations, where PLCs are not commonly used.

3.3 Hybrid-HIPPS

Hybrid-HIPPS are a combination of electronic and mechanical HIPPS.

Pros:

- High level of risk reduction- 10,000 (SIL 4) or higher can be obtained;
- Significant reduction of common cause and common mode failures due to diversity (mechanical and electronic);
- Low-level of systematic errors;

Cons:

- Complex system;
- Requires high level of competence in the installation, commissioning, validation, maintenance and operation;
- Requires a sizeable spare parts inventory;
- Many of the Cons of electronic and mechanical HIPPSs described above are applicable for the hybrid HIPPS as well.

Application:

High-pressure feed gas to a facility typically requires a high-risk reduction in the absence of non-instrumented IPLs. When a high level of risk reduction (greater than 10,000) is required, hybrid systems are well-suited as they are diverse systems with minimum common cause and common mode failures.

4 Safety requirements specification (SRS)

Functional safety standard IEC 61511 requires the development of a SIS safety requirements specification (SRS), including application program safety requirements, when applicable. HIPPS as a functional SIS, requires an SRS. There are a minimum of 29 SRS requirements specified by the standard. While every element is important, a few present unique challenges that can significantly impact the selection, design, and quantification of random hardware failures of the HIPPS.

4.1 Process safety time (PST)

Process safety time, per clause 3.2.52.1, IEC 61511, is the period between a failure occurring in the process or the basic process control system and the occurrence of a hazardous event, if the SIF is not performed. The SIF has to detect the failure and complete its action soon enough to prevent the hazardous event taking into account any lag in the process.

HIPPS that implements the overpressure protection function must complete the action to bring the process to a safe state within the process safety time. Quite often, the process safety times for HIPPS are less than few seconds. The valve subsystems may need quick exhaust valves, larger size air supply tubing, or solenoid valves to meet the response times. Therefore, the PST determination is an important specification which in turn determines various components within the HIPPS to achieve the response time.

4.2 HIPPS output action success criteria

Output actions of a SIF will prevent a particular consequence of concern only when the action is completed with certain performance criteria. For example, stating an action as to close a valve is inadequate. The criteria for the successful operation should be specified. Many HIPPS applications do not require tight shut-off requirements for valves. However, there are situations where larger leakage rates may impact the process safe states. In such scenarios, the process engineer must specify an acceptable leakage rate for the valves, and ensure the design shall have means to achieve such low leakage rates. The design may require unique valve types and trims. Also, additional pressure transmitters and isolation/bypass valves are required for measuring the extent of leakages.

Low leakage rate specification may impact the proof test coverages for the final elements in the quantification of random hardware failures (PFD_{avg}). Achieving a SIL 3 target with tight shut-off requirements is a significant challenge.

4.3 Proof test interval and overhaul interval (mission time)

IEC 61511 defines the proof test as a periodic test performed to detect dangerous (hidden) failures in a SIS so that, if necessary, a repair can restore the system to an "as new" condition or as close as practical to this condition.

The frequency of testing plays a significant role in determining the HIPPS's availability to perform intended safety function on demand. However, frequent testing may affect the production, introduce spurious shutdowns due to human error or affect the life of a device. If only a full stroke test is performed for a valve, then the test is imperfect since it will not detect all the covert failures. The undetected, hidden failures are detected and repaired during overhaul periods (mission time). The longer the mission time, the higher the PFDavg; therefore, it will be difficult to achieve SIL 3 targets with longer mission times.

Bypass valves and isolation valves are required for proof testing intervals shorter than the overhaul periods to avoid process interruption. Provision of bypass and isolation valves increases the complexity of the design, footprint, capital and the life cycle costs of the facilities.

4.4 Survivability

As per IEC 61511, the SRS shall specify any requirements for a SIF to survive a major accident event, e.g. the time required for the HIPPS valve to remain in operation in the event of fire shall be specified. Survivability specification impacts the selection of valve trims and additional fireproofing requirements.

4.5 Common cause failures

The SRS must specify the measures to identify and minimize common cause failures. Common cause failures will significantly impact PFD_{avg} for high SILs, such as SIL 3 and SIL 4. Common cause failures can be due to process conditions or similar redundant equipment configurations or systematic failures.

Physical and electrical separation and use of diverse devices which respond differently to the common stresses will minimize common cause failures. An effective functional safety management plan and verification processes will reduce systematic failures.

5 Quantification of random failure of HIPPS

The HIPPS design must be evaluated to ensure that the calculated failure measure (PFDavg) is less than the target related to the SIL as specified in the SRS.

The functional safety standard mandates assessment of the following as part of the SIL verification:

- Systematic capability of all the elements associated with a HIPPS meets the specified SIL target
- Calculated PFD_{avg} is less than the target PFD_{avg} of the selected SIL
- Hardware fault tolerance (HFT) complies with the required SIL target

5.1 Systematic capability

IEC 61511 defines the "systematic capability" as a measure (expressed on a scale of SC 1 to SC 4) of the confidence that the systematic safety integrity of a device meets the requirements of the specified SIL, in respect of the specified safety function, when the device is applied in accordance with the instructions specified in the device safety manual.

Systematic capability assessment is for addressing systematic failures which are not quantified and hence not included in the SIL verification calculations. IEC 61508 and IEC 61511explicitly identify a systematic capability parameter (SC). A systematic capability of "SC N" for a device means that the systematic safety integrity of "SC N" has been met when the device is applied following the instructions specified in the instrument safety manual. If a device is to be used in an SIL 3 application, then it must be assessed for the systematic capability of SC 3.

The systematic capability of a device depends on the nature of the device. If a device has no software, then only hardware systematic failure mechanisms are considered. If a device has both hardware and software, then both hardware and software failure mechanisms shall be considered.

There are agencies, such as TUV, Exida, and SIRA that perform systematic capability assessment of a product as per IEC 61508 and provide a certificate with a report specifying an SC parameter. Following a well-established functional safety management plan during detailed engineering, installation, operation and maintenance phases ensures that the systematic safety integrity of a device meets the requirements of the specified SIL.

5.2 PFD_{avg}

Many factors influence the PFD_{avg} of HIPPS function. A few of the factors that pose unique challenges are considered in this section.

5.2.1 Failure rates

There is always an uncertainty in the quantification of random failures. The failure rate is, probably, the least precise parameter used in performing SIL verifications [**8**]. Therefore, one must acknowledge that the results of the quantification are not real values, but rather a basis for comparing different design options and for monitoring the reliability performance during the operational phase of the safety lifecycle. The analysis is only a prediction. Historical performance is not the same as future performance.

If the failure rates are conservative, then achieving target SIL will be demanding. However, optimistic failure rates will result in inadequate protection integrity. Therefore, the failure rates of components used in the analysis should be sufficiently realistic.

There are many sources of failure rate data, such as application-specific, site-specific, company-specific, manufacturer-specific, industry-specific, and generic data. These failure rates can vary significantly by several orders of magnitude. While application specific or site specific data is the best option, but it can often be difficult to obtain. A detailed discussion on the selection of failure rates to use in the analysis is beyond the scope of this paper, and the reader is advised to refer to other sources [**9**].

5.2.2 Proof test interval, test coverage and mission time

As explained in section 4.3, a proof test may be imperfect based on the following:

- The extent of tests that could be performed to reveal the hidden failures. For example, if the valve is only full stroke tested without examining the internals then the proof test is imperfect;
- The conditions under which the proof test is carried out may differ from a real demand;
- Lack of competence of testing crew;
- Inadequate procedures.

Perfect testing is not a realistic assumption. IEC 61511/IEC 61508 requires consideration of the impact of imperfect testing in the analysis. In the SIL verification calculations, a proof test coverage (PTC) is considered for incomplete testing. The incomplete testing will increase the PFDavg. The lower the proof test coverage, the higher the PFD_{avg}. Undetected dangerous failures due to imperfect proof testing will be detected either during the complete overhaul period (when repaired or refurbished) or when the component is replaced or during an actual demand. Mission time is the terminology used by many reliability specialists for the above overhaul period.

For example, if an overhaul period or demand interval for HIPPS is ten years and the imperfect proof test interval for a valve is one year with a proof test coverage of 90%, then achieving an SIL 3 target for HIPPS is an arduous task. If tight shut-off requirement is specified, then the proof test coverage could be even less than 30%.

From the above, it is evident, that the significance of proof test interval, the effectiveness of proof test, and the mission time cannot be ignored and must be taken into account in the analysis.

5.2.3 Partial stroke testing

Partial stroke test will help to improve the PFD_{avg}, extend the proof test interval, or increase the safe failure fraction. A single valve with partial stroke test may achieve an SIL 2 integrity. However, for many HIPPS with short process safety times, partial stroke test may not be useful. Moreover, partial stroke tests may result in spurious trips and cause potential damage/wear to the valve seat due to the frequent testing.

A perusal of literature reveals that the partial stroke testing coverage is in the range of 40-80% [10]. However, the above-stated coverage is widely debated.

If the partial stroke testing is not automated, then it is not considered as the diagnostic testing per functional safety standards. Also, the analyst shall take into consideration the partial stroke testing interval with the repair time in the PFD calculations.

5.2.4 Common cause failures

Common cause failures impact the redundant components such as valves and sensors. The common cause failures occur simultaneously or, at least, within a short interval as to be perceived as simultaneous.

The following are the defenses against the common cause failures:

- Diversity (different types of valves and systems);
- Separation (physically and electrically);
- Staggered testing;
- Use of low complexity devices- simpler equipment less prone to common cause failures;
- Well administered and complied procedures;
- Environmental controls;

When diverse components are used, the common cause factor (Beta factor) is minimal. Using IEC 61508, Part 6 [5], Annex D, Tables, D.1, D.2 and D.3, the Beta factor can be estimated to be as low as 2% with diverse components. However, the diverse components must be suitable for the application with comparable performance without any degradation.

Also, staggered testing reduces the overall PFD_{avg} by a factor of 30 to 50% depending upon the voting configurations [**10**].

5.2.5 Modeling of components

Modeling of HIPPS elements in the analysis must include all the components that are involved in bringing the process to a safe state within the process safety time.

Many HIPPS valves are equipped with quick exhaust valves (QEVs) to reduce the valve closing time to meet the process safety time. Sometimes, redundant QEVs are provided in various voting configurations such 1oo2 or 3oo4, etc. The contribution of QEV failures to the PFD_{avg} is significant and cannot be ignored. Therefore, the reliability analyst must consider the QEVs with their appropriate voting architecture in the PFD_{avg} calculations. Similarly, the likelihood of plugging of sensing lines must be taken into account in freezing applications.

Any interfacing components such as barriers, isolators, interposing relays, and poppet valves, which are part of the safety instrumented function to achieve or maintain the process to a safe state must be modeled and analyzed.

5.3 Hardware fault tolerance (HFT)

Irrespective of the hardware reliability calculated (PFD_{avg}) for the design, the functional safety standard IEC 61511 specifies minimum levels of redundancy for each safety integrity level.

There are two approaches:

- Route 1_H: Determination of HFT based on the calculated safe failure fraction (SFF);
- Route 2_H: Determination of HFT based on field experience and field data (proven-in-use)

5.3.1 Route1$_H$

The Route1$_H$ approach uses SFF and the SIL target to determine the hardware fault tolerance. The SFF is defined as the sum of the potentially dangerous failures revealed by auto-test and those failures which result in a safe state, as a fraction of a total number of failures.

$$\text{SFF} = \frac{[dangerous\ detected\ failures(self\ test)+safe\ failures]}{Total\ failures}$$

The above is calculated for each element (sensor, complete valve assembly, logic solver). Two tables, Table 2 and Table 3, are provided in IEC 61508- Part 2. Table 2 is for "Type A" devices which are simple devices such as valves (no microprocessors and programming are involved), and Table 3 for "Type B" devices which are complex devices such as electronic transmitters, trip amplifiers, and PLCs. Based on the SFF, type of the device, and the SIL target, the hardware fault tolerance is determined using Table 2 and Table 3.

5.3.2 Route2$_H$

In Route 2$_H$, the hardware fault tolerance is specified by the functional safety standard for each SIL target if the device is selected based on well documented and verified field based failure rate data with 90% statistical confidence. For low demand application with SIL 1 or SIL 2 target, there is no need for any redundancies (HFT=0), and SIL 3 requires one hardware fault tolerance.

Unless the field failure data with 90% confidence levels are available, the Route 2$_H$ cannot be justified. There are third party agencies who provide Route 2$_H$ certifications and should be evaluated carefully before adopting the Route 2$_H$ approach based on those certifications.

6 High level risk reductions (RRF> 10,000)

As discussed in section 2, sometimes, PHA/SIL selection may indicate a required risk reduction greater than 10,000 in the absence of any non-instrumented protection layer or, human intervention is impractical due to process safety time considerations or safety reasons.

If there are no options other than providing HIPPS with a risk reduction greater than 10,000, then the functional safety standard IEC 61511 mandates certain requirements. The following are a few of the clauses and their requirements:

Clause 9.2.6: *If further consideration of application and confirmation that a risk reduction requirement >10,000....is still required, then consideration should be given to achieving the safety integrity requirement using a <u>number of protection layers with lower risk reduction requirements</u>. If the risk reduction is allocated to multiple protection layers, then such protection layers shall be independent from each other or the lack of independence shall be assessed and shown to be sufficiently low compared to the risk reduction requirements.*

Clause 9.2.7: *If a risk reduction greater than 10,000....is to be implemented, whether allocated to a single SIS or multiple SIS or SIS in conjunction with a BPCS protection layer, then a further risk assessment shall be carried out using a quantitative methodology to confirm that the safety integrity requirements are achieved.*

Clause 9.2.8: *If the risk reduction required for a hazardous event is allocated to multiple SIFs in a single SIS, then the SIS shall meet the overall risk reduction requirement.*

A hybrid HIPPS using the combination of electronic and mechanical HIPPS is a solution to achieve risk reductions greater than 10,000. Each HIPPS can be designed to be better than SIL 2, and the overall risk reduction will be better than 10,000. The performance of electronic and mechanical HIPPS are comparable.

High level of risk reduction is achieved due to the diversity and very low common cause failures between the electronic and mechanical HIPPS. All the subsystems of the HIPPSs are diverse with different technologies. Also, systematic failures are minimal. The integrity of such multiple HIPPS is best analyzed using a fault tree analysis (FTA).

An example fault tree analysis for a hybrid HIPPS using ISOGRAPH's Reliability Workbench™ is shown in **Figure 5**.

The example hybrid HIPPS achieved a risk reduction of 12,815(PFD_{avg} = 7.80E-05).

The FTA shows an approach for modeling of various components including the common cause failures. The failure rates assumed are generic for the purpose of demonstration. The actual failure rates and other factors are application specific and cannot be generalized.

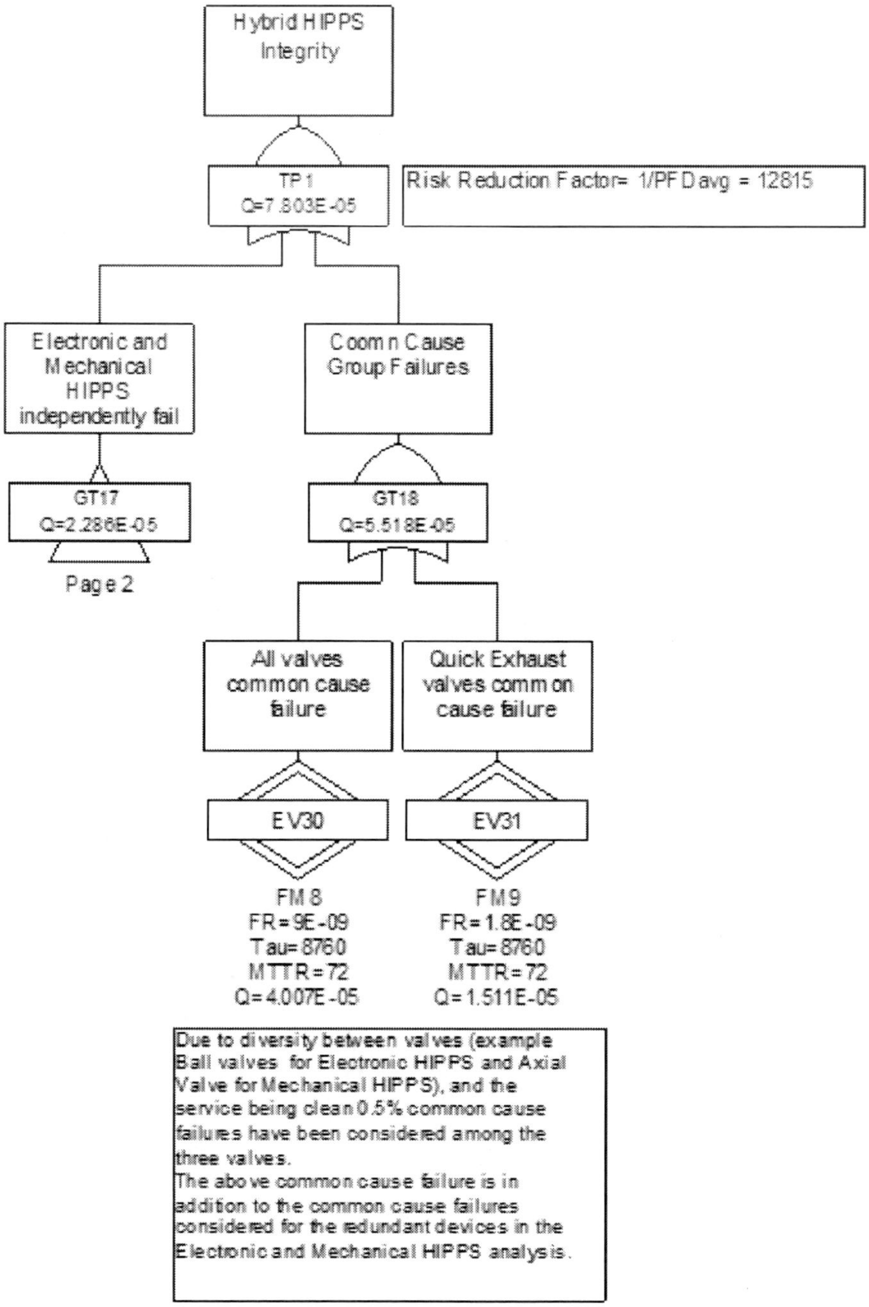

Figure 5. FTA for multiple HIPPS to achieve high risk reduction

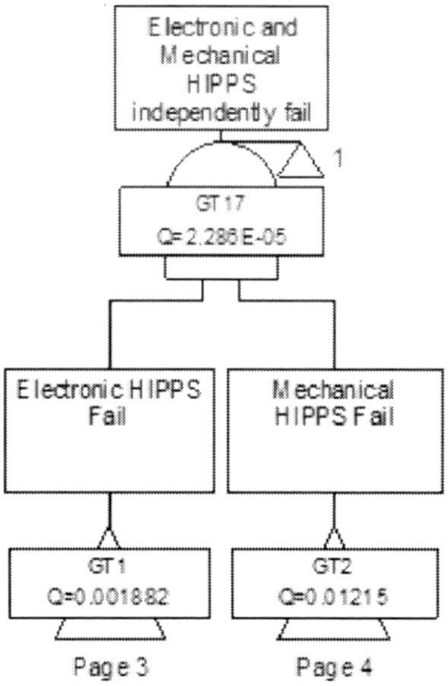

Figure 5. FTA for multiple HIPPS to achieve high risk reduction (contd..)

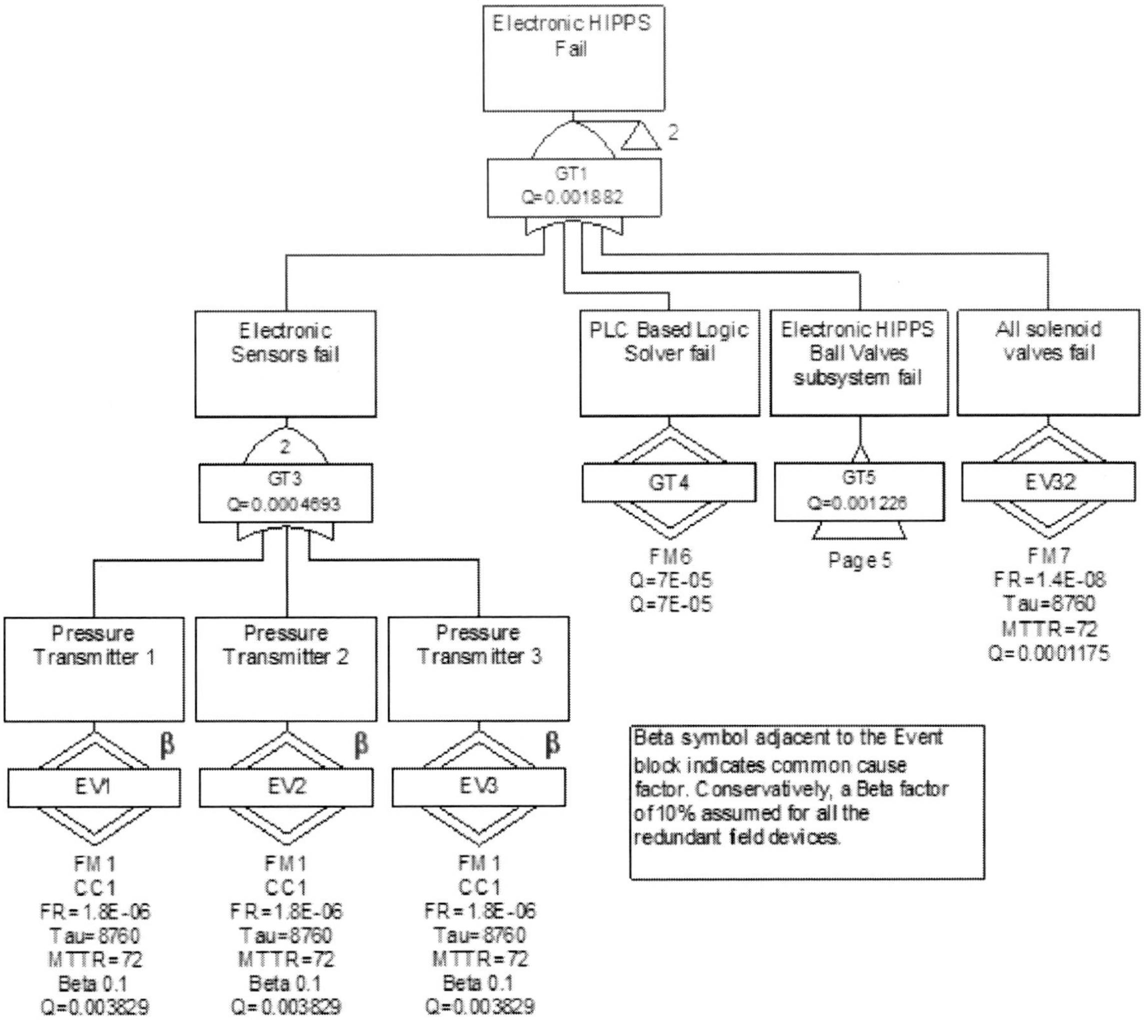

Figure 5. FTA for multiple HIPPS to achieve high risk reduction (contd..)

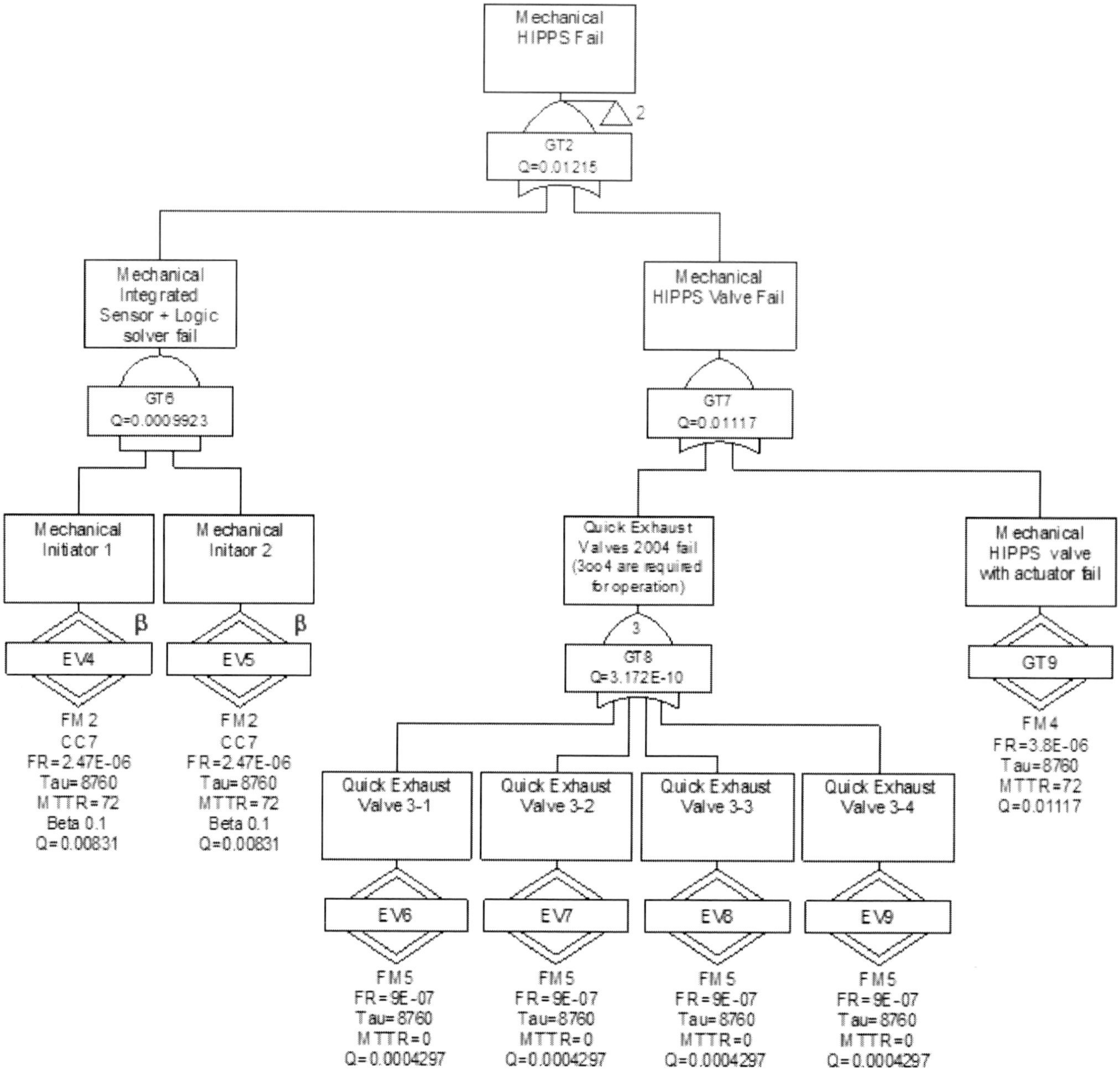

Figure 5. FTA for multiple HIPPS to achieve high risk reduction (contd..)

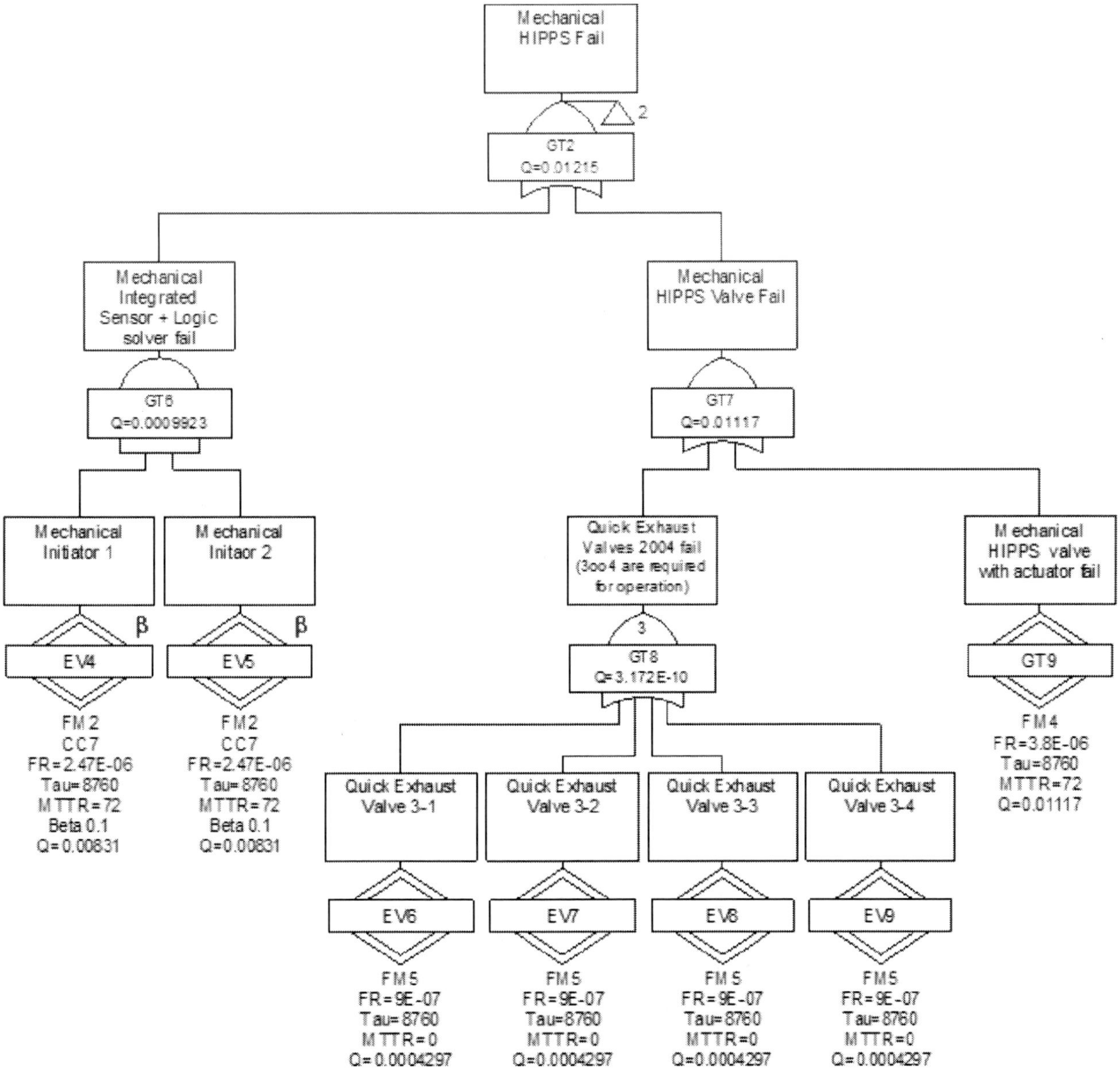

Figure 5. FTA for multiple HIPPS to achieve high risk reduction (contd..)

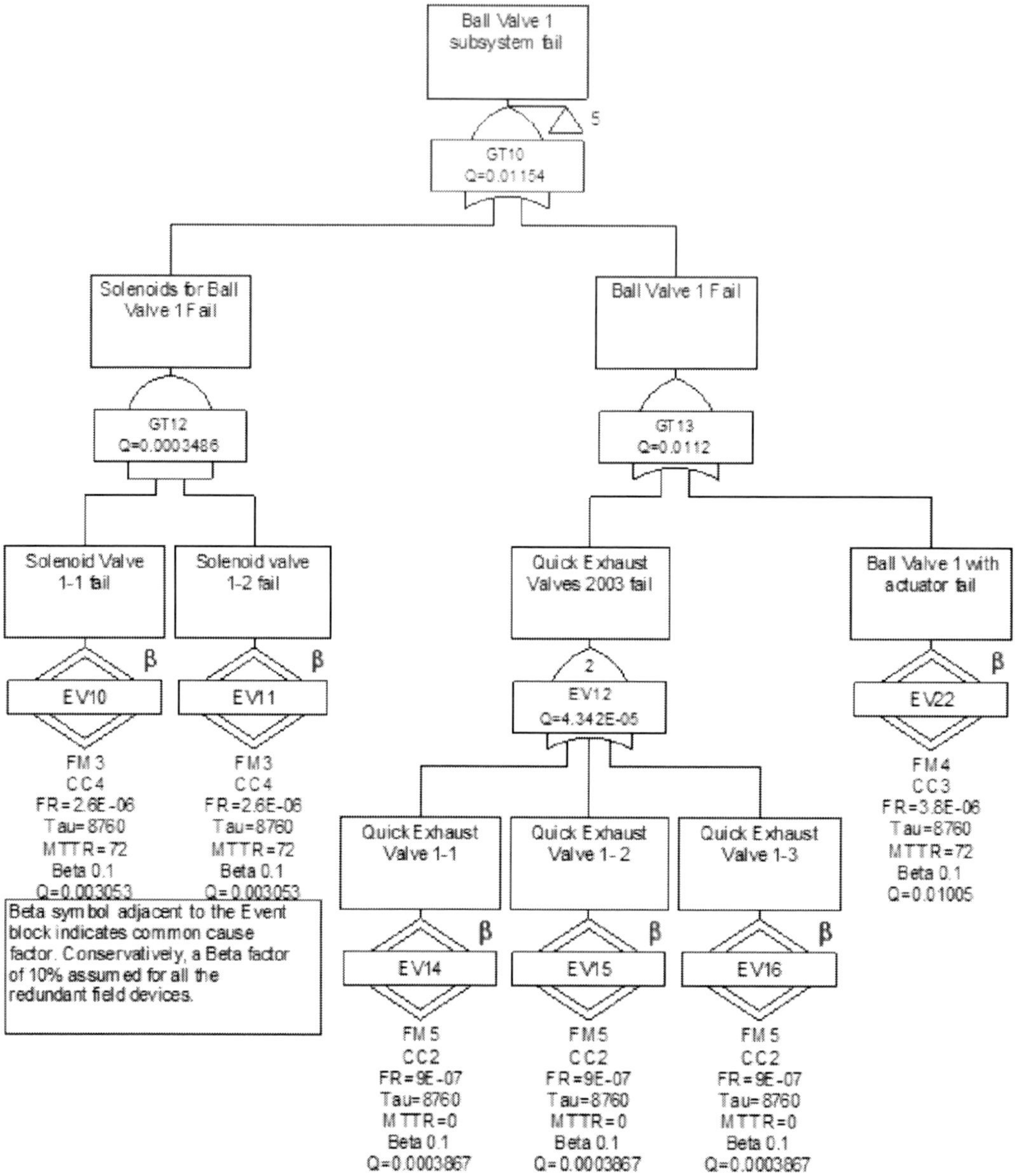

Figure 5. FTA for multiple HIPPS to achieve high risk reduction (contd..)

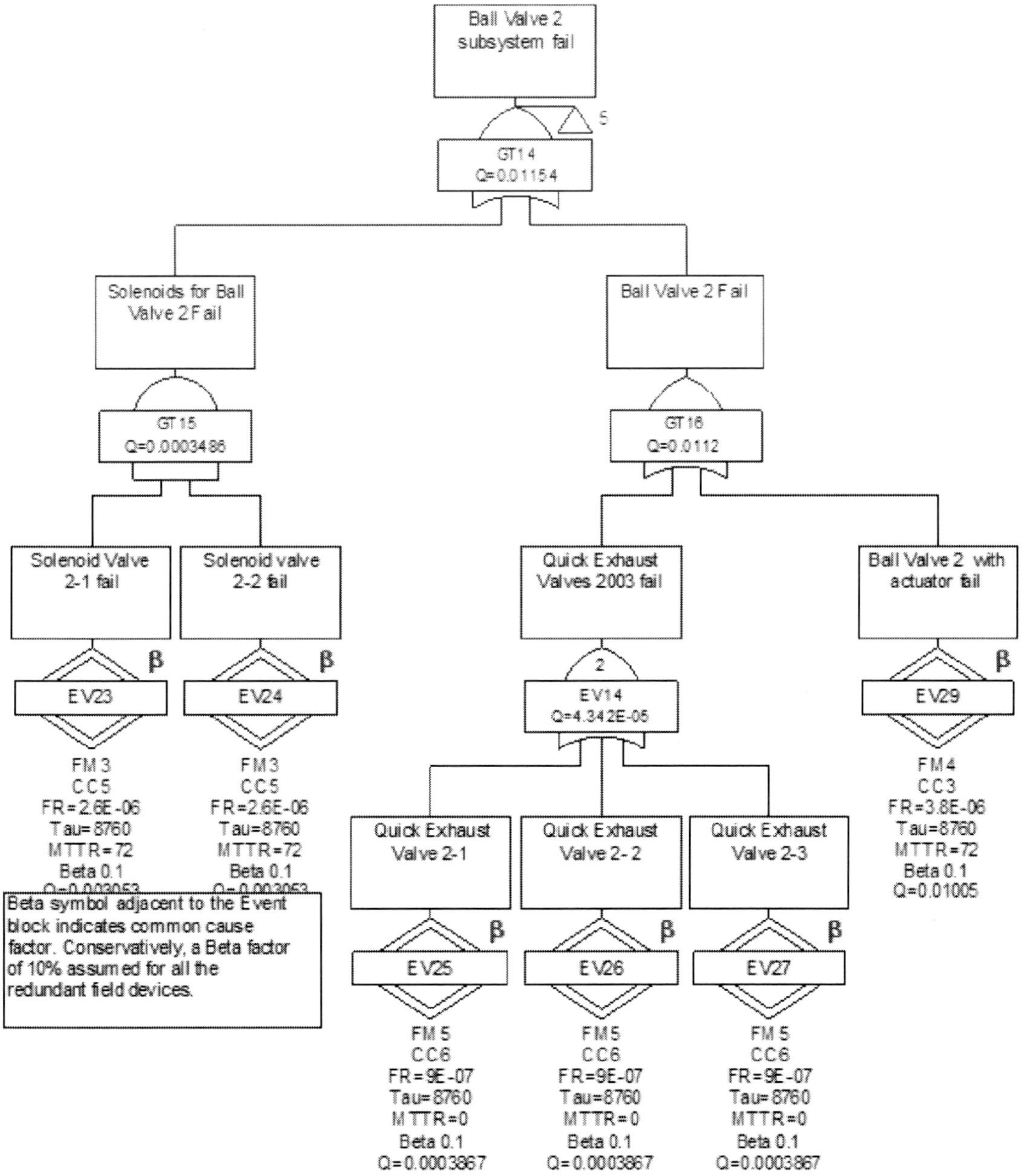

Figure 5. FTA for multiple HIPPS to achieve high risk reduction (contd..)

7 Installation, commissioning, operation and maintenance

Installation and commissioning of the HIPPS must be completed in accordance with the installation drawings, safety manuals, and the SRS.

Operation and maintenance is the longest period in the safety life cycle. The functional safety standards mandate continuous performance monitoring, including demand rates, failure modes and failure rates of various components. The actual performance must be compared with the predicted performance to identify any gaps so that remedial measures can be provided with appropriate safety review.

A robust change management process must be in place to ensure HIPPS integrity throughout the operation and maintenance phase.

8 Conclusion

Assignment of an SIL target and choosing an appropriate technology for a HIPPS is not trivial. In this paper, the significance of PHA and SIL selection for HIPPS has been re-emphasized and a brief account of available options for HIPPS to choose from, their pros and cons, and the selection criteria with examples are provided.

The design, installation, operation, and maintenance of HIPPS shall comply with the requirements of functional safety standards IEC 61508 and IEC 61511. In this article, the challenges in each of the above activities are discussed and appropriate solutions to mitigate the above challenges are provided.

In applications where a high risk reduction greater than 10,000 is required a hybrid HIPPS design approach is proposed. Using FTA demonstrated the modeling of complex hybrid HIPPS to achieve high level risk reduction.

9 References

[1] IEC. IEC 61511 Functional Safety: Safety Instrumented Systems for the Process Industry Sector- Part 1-3, Edition 2, Geneva, 2016.

[2] American Society of Mechanical Engineers (ASME), Boiler and Pressure Vessel Code, Section VIII-Pressure Vessels, New York, 2010.

[3] API, Standard 521, Pressure-relieving and Depressuring Systems, 6th Edition, Washington, DC, 2014.

[4] API, RP17O, Recommended Practice for Subsea High Integrity Pressure Protection Systems (HIPPS), 1 Edition, Washington, DC 2009.

[5] HSE, UK. SPC/TECH/OSD/31, Safety instrumented systems for the overpressure protection of pipeline risers, Available at

http://www.hse.gov.uk/foi/internalops/hid_circs/technical_osd/spc_tech_osd_31.htm, Accessed on Jan 29, 2017.

[6] IEC. IEC 61508 Functional safety of electrical/electronic/programmable electronic safety related systems- Part 1-7, Edition 2, Geneva, 2010.

[7] H.G. Lawley and T.A. Kletz. High-pressure-trip systems for vessel protection, Chemical Engineering, Vol 82, (1975).

[8] Dr. D. J. Smith, Reliability, Maintainability and Risk, 8th Edition, Butterworth-Heinemann, Waltham, MA, 2011.

[9] S. Hauge, S. Habrekke, M.A. Lundeteigen. Reliability Prediction Method for Safety Instrumented Systems- PDS Example Collection, SINTEF Technology and Society, Safety Research, 2010.

[10] Dr. D.J. Smith, and K.G.L. Simpson, The Safety Critical Systems Handbook, 4th Edition, Butterworth-Heinemann, Cambridge, MA, 2016.

475693 Managing Risk Associated w/ Electrical Energy Ignition Sources

Abstract:

In facilities that handle flammable hydrocarbons it is extremely important that all forms of potential ignition sources be carefully managed. This presentation will focus on key points associated with process electrical safety, open flame process heaters, and mechanical forms of ignition sources. OSHA 1910.119 has for many years required that hazardous area classification documentation remain in an evergreen state that requires re-certification at least once every five years. For most process facilities this OSHA mandate presents many problems. In most cases documentation has not been maintained in a current state, the facility perhaps has changed owners if not once many times over the years. The presentation attendee will be exposed to a common sense methodology that focuses primarily on minimizing electrical energy from becoming a source of ignition. A process for updating hazardous area classification documentation will be presented with tips on providing the end-user with an audit proof set of documentation. Once the hazardous area map has been developed then methods of properly documenting and conducting a site compliance assessment will be explained. This compliance assessment is a gap analysis that compares the installed electrical systems against the requirements of the National Electrical Code (NFPA 70 Article 500). Methods of presenting this documentation will be explained so the end-user has a clear understanding of the proper steps for remediation.

The presentation attendee will be exposed to the following:

- Managing risk associated with roadways that are impinged by hazardous classified boundaries
- Open Flame Process heaters encroachment by hazardous classified locations
- Methodology for area classification assessment *(it's determined by the process conditions)*
- Hazard's compliance for instrumentation and electrical system installations in hazardous process locations
- How to interpret documentation in support of the area classification assessments which include:
  - Plan Drawings (maps), Elevations, Site Specific Details. Legends, and notes
  - Assessment report documentation that addresses General Considerations, Identification of Point Sources of Emissions, Physical properties of Materials being Handled, Processed, or Stored, and Gas Dispersion Modeling Graphs and plots
- Methods for mathematical determination of hazard radius thru the concept of commercially available computer assisted gas/vapor dispersion modeling software.
- Application of common standards, codes and installation practices i.e. (NFPA, ISA, API and OSHA)

Detailed handouts as well as a complete audio-visual presentation are used to supplement the presentation.

479320 Review and Analysis of Recent Cyber Security Attacks

Abstract:

Process plants need to ensure that adequate safeguards are in place for a variety of upset conditions, including cyber security attacks, which can result in loss of product, damage to equipment, downtime, and potential safety and environmental disasters. Select cyber security incidents across a variety of industries will be reviewed to determine the modes of cyber security attack and their subsequent impacts. Specific areas of vulnerability will be identified with corresponding mitigation strategies. Possible solutions will be presented for incorporating cyber security into an existing process safety program to ensure a tolerable level of protection for high-consequence events.

478316 Resilience-Based Survival Analysis in the Event of a Process Upset: A Case Study of Utility Plant

Abstract:

Abstract: In a process plant system, safe and reliable operations are highly sensitive to the power or steam failure, loss of cooling water or nitrogen or instrument air. Most of these products and services from utilities plant play an important role as safety barriers. The main focus of this paper is to propose a model for survival evaluation of a process system under an upset condition. This model is based on the holistic resilience-based method that integrates both technical (process parameters variations) and social (policy/regulations, human and organizational) factors. The resilience methodology emphasizes non-linear dynamics, new types of threats, uncertainty, survival through upset conditions and recovery from catastrophic situations [1], [2].

The purpose of this paper is to maximize the performance of safety barriers that will be utilized in case of a process upset condition to prevent the escalation of the event into a catastrophic incident. In this work, a methodology is presented to select the best maintenance strategy [3] while considering the statutory safety regulations for maintenance of equipment like boilers, generators and more. Three different types of maintenance alternatives considered as part of this analysis are: predictive, preventative and condition-based. The key ideas of survival evaluation are presented using a utility plant as the case study.

References

1. Dinh, L. T., Pasman, H., Gao, X., & Mannan, M. S. (2012). Resilience engineering of industrial processes: principles and contributing factors. Journal of Loss Prevention in the Process Industries, 25(2), 233-241.
2. Jackson, S. (2009). Architecting resilient systems: Accident avoidance and survival and recovery from disruptions (Vol. 66). John Wiley & Sons.
3. Vassiliadis, C. G., & Pistikopoulos, E. N. (2001). Maintenance scheduling and process optimization under uncertainty. *Computers & Chemical Engineering*, *25*(2), 217-236.

478816 Model-Based Pressure Relief Design for Reactive Systems

Abstract:

Abstract

Reactive systems behave unpredictably under overpressure conditions. Varying composition impacts the physical properties and response of the mixture to temperature and time. Pressure generation in a reactive system is often a function of both vapor pressure and gas generation rate. Standard methodologies applied in non-reactive pressure relief system design cannot be relied upon to give appropriate sizing for reactive systems. Consistent with DIERS methodologies, model-based design provides accurate, reliable vent sizing. A comprehensive approach to modeling is presented which incorporates vapor-liquid equilibrium, reaction stoichiometry, and reaction rate.

Dynamic Modeling of Inbreathing Requirements for Low-Pressure Storage Tanks

Dona Abou-Chakra
ioMosaic Corporation
93 Stiles Road | Salem, NH 03079
chakra.d.nh@iomosaic.com

Georges A. Melhem, PhD, FAIChE
ioMosaic Corporation
93 Stiles Road | Salem, NH 03079
melhem@iomosaic.com

Ronald J. Willey, PhD
Northeastern University
360 Huntington Avenue | Boston, MA 02115
r.willey@neu.edu

Prepared for Presentation at
American Institute of Chemical Engineers
2017 Spring Meeting and 13th Global Congress on Process Safety
San Antonio, TX
March 26 – 29, 2017

Dynamic Modeling of Inbreathing Requirements for Low-Pressure Storage Tanks

Dona Abou-Chakra
ioMosaic Corporation
93 Stiles Road | Salem, NH 03079
chakra.d.nh@iomosaic.com

Georges A. Melhem, PhD, FAIChE
ioMosaic Corporation
93 Stiles Road | Salem, NH 03079
melhem@iomosaic.com

Ronald J. Willey, PhD
Northeastern University
360 Huntington Avenue | Boston, MA 02115
r.willey@neu.edu

Keywords: Thermal Inbreathing, API 2000, Storage Tanks, Venting.

Abstract

A Process Safety Office™ - SuperChems™ component based model is proposed to study the dynamic behavior of the thermal inbreathing of low-pressure storage tanks upon the occurrence of a sudden cooling rainstorm. The objective here is to establish a model that is capable of properly calculating the design relief capacity of the relief device to be installed on such tanks. The impact of phase-change, particularly vapor condensation, on the thermal inbreathing of the tank assessed. The model was validated against two large-scale experiment datasets and some other published theoretical models. The results were then discussed and compared to the relief requirements obtained using the API 2000 7th Edition method, which is the current standard for venting atmospheric tanks.

1 Introduction

Low-pressure storage tanks are protected against extreme weather changes following the API 2000 (7th edition) standard. The threat of creating a vacuum occurs when the contents of the tank vapor space are suddenly cooled by an intense rainfall causing the vapor space to contract and condense, hence, rapidly decreasing the internal pressure. This pressure drop will pull outside air into the tank's vapor space and acts as the driving force for thermal inbreathing.

The current limitation with the API standard is that it only addresses tanks storing non-condensable gases and have no liquid residue, where in fact most storage tanks contain condensable vapors such as water vapor which will have a big impact on the thermal inbreathing of the tank. This paper assesses the API 2000 relief requirement calculations for low-pressure storage tanks by comparing them to (1) the results of the dynamic simulations from a proposed SuperChems™ based model and (2) to several methods found in the literature (PTB W-22, Salatino, etc.) for both condensable and non-condensable vapor services.

2 API 2000 Methods

The tank contents are affected by the ambient conditions especially when the tank is uninsulated, which is the case for most storage tanks. This method uses air at normal conditions (0°C and 101.325 kPa) as the inbreathed gas. The maximum inbreathing rate, $\dot{V}_{IT}$, is primarily a function of tank volume, insulation, and a C-factor (found in Table 1 below) as shown in Equation (1) below:

$$\dot{V}_{IT} = C\,V_{tk}^{0.7} R_i \tag{1}$$

where $\dot{V}_{IT}$ is the maximum inbreathing volumetric flow rate of air at normal conditions (Nm^3/h), C is the dimensionless factor that depends on vapor pressure, storage temperature and latitude, V_{tk} is the tank volume (m^3), and R_i is the reduction factor for insulation.

Table 1. API 2000 *C*-factor for Thermal Inbreathing

Latitude, φ	C-factor			
	Vapor pressure similar or below hexane		Vapor pressure higher than hexane or unknown	
	Average storage temperature (°C)			
	< 25	≥ 25	< 25	≥ 25
φ < 42°	4	6.5	6.5	6.5
42° ≤ φ ≤ 58°	3	5	5	5
φ > 58°	2.5	4	4	4

The assumptions used for the scenario of "tank cooling by rain" are found in Annex E of API 2000 [1] and are listed below:

- Tank is filled with air only.
- Initial vapor space and shell wall temperature are in thermal equilibrium at 55 °C.
- Rain temperature and intensity are constant at 15°C and 225 kg/m^2h (This rainstorm has a probability of 1/100 year) respectively; rain angle is at 30° from the horizontal.
- Inside free convection heat-transfer coefficient is 5 W/m^2K.
- Outside film cooling heat-transfer coefficient is 5,000 W/m^2K.

2.1 *API 2000 Annex A Method*

The method found in Annex A was used prior to 7th Edition (2008) and is quite different than the current sizing method for venting devices. In order to quantify the thermal inbreathing for storage tanks volumes smaller than 3180 m^3, Annex A [1] assumes that the tank starts cooling from an initial temperature of 48.9°C and at a constant temperature rate of -56 K/h. The thermal inbreathing rate, $\dot{V}_{IT}$ (Nm³/h) is as follows:

$$\dot{V}_{IT} = 0.169\ V_{tk} \qquad (2)$$

where V_{tk} is the tank volume (m³). For tank volumes equal or greater than 3180 m³, the thermal inbreathing is assumed to be limited by a constant heat flux, $h\Delta T$, equal to 63 W/m². The thermal inbreathing rate in this case becomes:

$$\dot{V}_{IT} = 0.577\ A_{exp} \qquad (3)$$

where A_{exp} is the external tank area exposed to rain (m²). Table 2 below summarizes the thermal normal venting requirements as found in Annex A of API 2000. Volumes greater than 30,000 m³ are outside the scope of this method. However, interpolation is allowed for Table 2 as long as the tank volumes lie between the specified volume limits of 10 m³ to 30,000 m³. [1]

Table 2. API 2000 Annex-A Normal Venting Requirements Due to Thermal Effects [1]

Tank Capacity	Inbreathing	Out-breathing	
Column 1 [a]	Column 2 [b]	Column 3 [c]	Column 4 [d]
		Flash Point ≥ 37.8 °C or Normal Boiling Point ≥ 149 °C	Flash Point < 37.8 °C or Normal Boiling Point < 149 °C
m³	Nm³/h of air	Nm³/h of air	Nm³/h of air
10	1.69	1.01	1.69
20	3.38	2.02	3.38
100	16.9	10.1	16.9
200	33.8	20.3	33.8
300	50.4	30.4	50.4
500	84.5	50.7	84.5
700	118	71.0	118
1,000	169	101	169
1,500	254	152	254
2,000	338	203	338
3,000	507	304	507
3,180	537	322	537
4,000	647	388	647
5,000	787	472	787
6,000	896	538	896
7,000	1,003	602	1,003
8,000	1,077	646	1,077
9,000	1,136	682	1,136
10,000	1,210	726	1,210
12,000	1,345	807	1,345
14,000	1,480	888	1,480
16,000	1,615	969	1,615
18,000	1,750	1,047	1,750
20,000	1,877	1,126	1,877
25,000	2,179	1,307	2,179
30,000	2,495	1,497	2,495

3 Thermal Inbreathing of Non-Condensable Vapors

One of the major assumptions of API 2000 is that the tank experiencing inbreathing is empty (i.e. only filled with dry air) and the vapor space is considered non-condensable. This section will review recent work on inbreathing dynamics for non-condensable dry air services only.

3.1 Salatino's Model

Salatino et al. [2] compare their tank inbreathing model to the older version of API 2000 (1992) that utilized the method of Annex A: a method based on the maximum estimated heat loss rate of 63 W/m^2 upon sudden cooling by a rainstorm of 225 kg/m^{2}h rain density. They recognized that all existing thermal inbreathing models assume a uniform vapor-space temperature, thus his team's contribution was accounting for the temperature non-uniformity inside the tank by dividing the tank into two individual volumes: the shell and the roof.

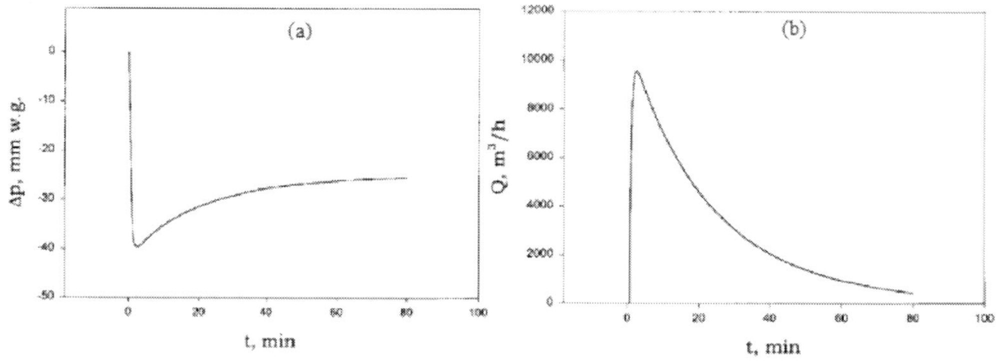

Figure 1. Pressure drop (a) and thermal inbreathing (b) profiles of Salatino's Model [2]

Salatino derived an equation for the maximum thermal inbreathing (Equation 4), stating that the maximum occurs at time t=0 (Figure 1) when the sudden weather change causes (1) vanishing of the radiative flux absorbed by the tank, (2) abrupt drop in ambient temperature and finally (3) rainfall.

$$\dot{Q}_{max} = -\frac{V}{T_G}\frac{dT_G}{dt}\bigg|_{t=0} = \frac{V}{T_{G,1}}\frac{1}{\tau}\left(T_{G,1} - T_{G,2}\right) \tag{4}$$

$$\tau = \frac{pV}{RT_G}\frac{c}{\sum A_i h_{iG}} \tag{4a}$$

where $\dot{Q}_{max}$ is the maximum thermal inbreathing rate (m^3/h), V is the tank volume (m^3), $T_{G,1}$ and $T_{G,2}$ are the gas temperatures (K) before and after the rainstorm, respectively. τ is the time constant that sums the effects of pressure, volume, heat transfer parameters (Equation 4a).

Salatino et al. further simplified the model by assuming an average value for the heat transfer coefficient of 4 W/m^{2}K and a gas temperature drop (after rainfall) of T_{G1}-T_{G2} ≅ 25°C claiming that these values will result in a conservative approximation of the inbreathing rate. Replacing these values in Equation (4) yields the simplified maximum inbreathing rate, $\dot{Q}_{max}$, which is a function of the tank volume, *V*, and the height-to-diameter ratio, *f*:

$$\dot{Q}_{max} = 2.6\, V^{2/3} \left(\frac{1 + 2f}{\sqrt[3]{f}} \right) \tag{5}$$

For a 63,000 m^3 cone-roof tank, Salatino predicted a maximum thermal inbreathing value of 9,900 m^3/h, using Equation 5. However, using the API 2000 Annex A method, the maximum inbreathing for the same tank was found to be 4,100 m^3/h. Salatino proved that his method is more conservative than the API 2000 equations and advised that an oversized thermal relief device is considered a safer design in this case.

3.2 Lloyd's Register Energy Americas, Inc. (LR) Model

Lloyd's Register (LR) Energy Americas, Inc. analytically solved the heat balances of a storage tank that is suddenly cooled by rain [3] and compared their results against ISO 28300, which is an equivalent of API 2000 6th edition (2009). The model results are summarized in Figure 2 below.

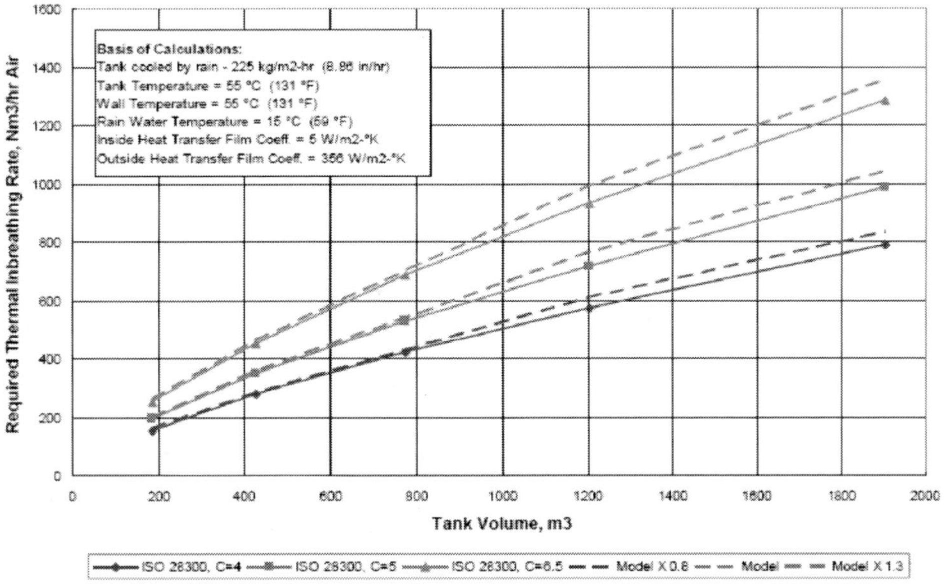

Figure 2. LR Model versus API 2000 6th ed. (ISO 28300)

The LR model was slightly more conservative than the API 2000 thermal inbreathing predictions by factors ranging between 0.5 and 1.3. Note that the C value in Figure 2 refers to the C-factor in the API 2000 thermal inbreathing in Equation (1) of this paper.

The model was then used to study the impact of several variables, some of which had minimal impact on the predicted thermal inbreathing (less than 5% error) such as the tank weight and initial gas temperature. Other variables, like the outside heat-transfer coefficient (Table 3) and rain intensity, had substantial impact on the calculated thermal inbreathing rates.

Table 3. LR Model - Impact of Outside Heat-Transfer Coefficient, h_o [3]

Case No.	Tank Volume m3	Predicted Thermal Inbreathing Rate, Nm3/hr		
		ho = 200 W/m2- K	ho = 300 W/m2- K	ho = 400 W/m2- K
1	186	169.6	191.2	205.1
2	427	308.3	341.4	362.7
3	775	475.1	520.7	548.6
4	1202	678.9	740.5	778.6
5	1903	935.2	1011.6	1058.5

The LR model was used to derive a relationship between the rain intensity (mm/h) and the heat transfer coefficient between the outer tank wall and rain (W/m^2K) as shown in Figure 3 below. This linear dependence relationship was implemented in SuperChems™ for this study.

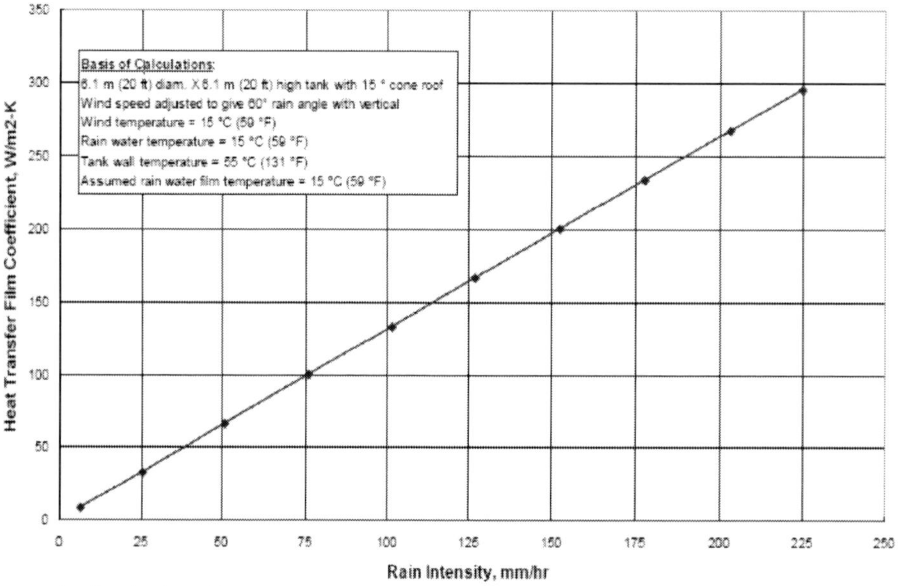

Figure 3. Rain/wall heat transfer coefficient versus of rain intensity

3.3 PROTEGO© Model

Scholz and Weber [4] from PROTEGO© derived the API 2000 7th Edition thermal inbreathing formula (Equation 1 of this paper) and determined that the maximum thermal inbreathing was a function of primarily tank volume with a coefficient that is less than or equal to 9.99:

$$\dot{V}_{max} = Cofficient \cdot V^{2/3} \tag{6}$$

The above equation is applicable to both flat-roof and conical-roof storage tanks. The derivation of Equation (6) was obtained by performing energy balances on the stored gas, the tank walls, and on the rainwater film. The differential equation and its solution are not included in this document, but can be found in Scholz and Weber's publication [4].

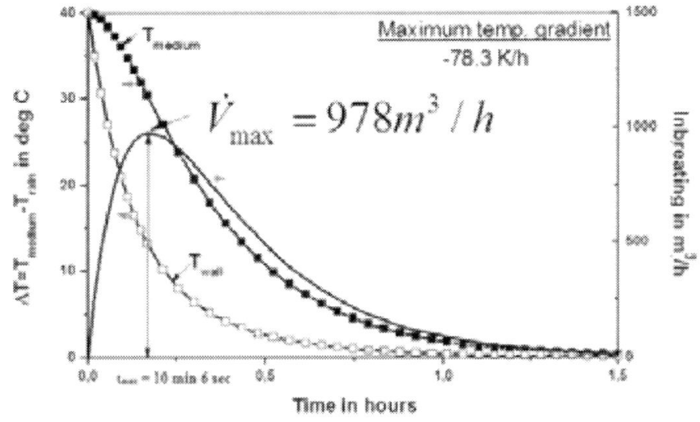

Figure 4. PROTEGO© Model Results

PROTEGO© then compared their results with the Foerster's model [5], whose main assumption was a fully immersed tank, which proved to be in good agreement with the published data. Figure 4 contains the results of a simulation using a conical-roof storage tank of 3,978 m^3 volume and an aspect ratio f = 0.6. Light rain was assumed (75 kg/m^2h) at 15°C. The inside and outside convective heat-transfer coefficients were set to 5 and 37 W/m^2K respectively.

PROTEGO© used their model to study 5 different types of condensable vapors and their effect on the maximum thermal inbreathing rate, however, the results will be discussed in Section 4 of this paper.

4 Thermal Inbreathing of Condensable Vapors

API 2000 neglects the possibility of vapor condensation for the thermal inbreathing/out-breathing relief calculation since only dry air is present in the tank i.e. the tank is empty. One reason to consider the effects of having condensable vapors is the steam cleaning that storage tanks undergo. Hot steam will occupy the tank after the cleaning process is complete and if a rainstorm occurs, the water vapor will condense/contract and pull a vacuum inside the tank.

4.1 Fullarton Model – Water Vapor

Fullarton [6] introduced two dimensionless factors in his model: (1) The reduction factor "η" that accounts for the tank wall thickness by factoring in the metal heat capacity of the walls and (2) the condensation factor "χ" that accounts for the stored-vapor condensation i.e. the additional condensation mass flux and the additional heat of condensation. The formulas of each "η" and "χ" can be found in Fullarton's paper [6].

It is expected that the tank wall temperature drops much faster than the temperature of the vapor stored inside the storage tank because of the metal's higher heat capacity. However, a thicker metal wall acts as a resistance to heat transfer. This means that the tank walls cool at a slower rate than when the wall thickness is neglected. The maximum inbreathing in this case will be lower, hence the reduction factor "η" which "reduces" the maximum thermal inbreathing of a storage tank with thicker walls.

Table 4. Range of χ-factor for Different Condensable Vapors

Condensable Vapor	χ-factor
Water vapor Heptane Hexane	1.5-2
Ethanol	2-2.5
Methanol	3-4

The condensation factor "χ", on the other hand, is primarily a function of the volatility of the condensable vapor being studied. For that reason, Fullarton established a range for the "χ" factor for different vapors as seen in Table 4 above [6].

Therefore, the maximum thermal inbreathing for a condensable vapor space per Fullarton can be obtained by using each of the Naumann approximation, $\dot{V}_N$, the reduction factor, η, and the condensation factor, χ.

$$\dot{V}_{K,max} = \dot{V}_N \eta \chi \quad (7)$$

One must keep in mind that properties of air at 288K were used to derive this equation in addition to the assumption of the initial temperature gradient ($T_{G,0} - T_A$) equal to 40°C. If other values are to be used, the above equation must be re-derived to account for the appropriate conditions.

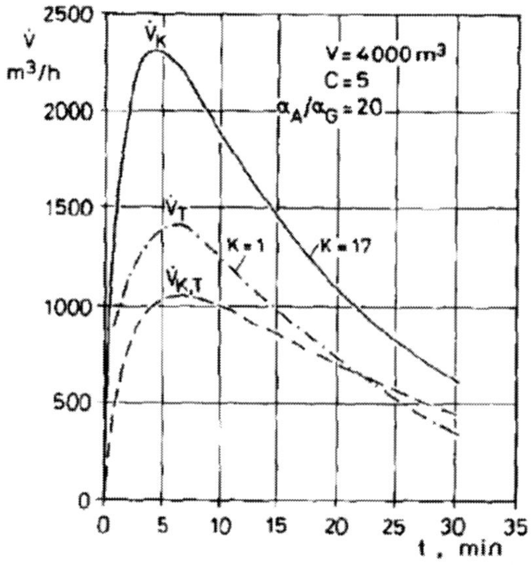

Figure 5. $\dot{V}_{inbreathing}$ profile – Fullarton Model [6]

Fullarton performed a dynamic simulation on a 4,000 m³ empty tank intended for the storage of ethanol. The initial gas temperature was 55°C and was suddenly cooled by rain at 15°C. The rain fall was assumed to be an extremely heavy downpour (probability of 1 every 100 years) with $\alpha_A/\alpha_G = 20$ and $\alpha_G = 5$ W/m²K. Figure 5 above shows the resultant inbreathing profile for each of dry air, K=1, and saturated water vapor, K=17. The maximum thermal inbreathing occurs at time t = 6 min where the inbreathing of the condensable vapor space ($\dot{V}_{K=17}$) is *1.6 times larger* than that of dry air ($\dot{V}_{K=1}$).

4.2 PROTEGO© Model – Condensable Vapors

PROTEGO© used their model [4] to obtain the inbreathing curves of five different vapors: methanol, ethanol, saturated water vapor, heptane and decane (Figure 6) and comparing them to the inbreathing curve of dry air in the same tank.

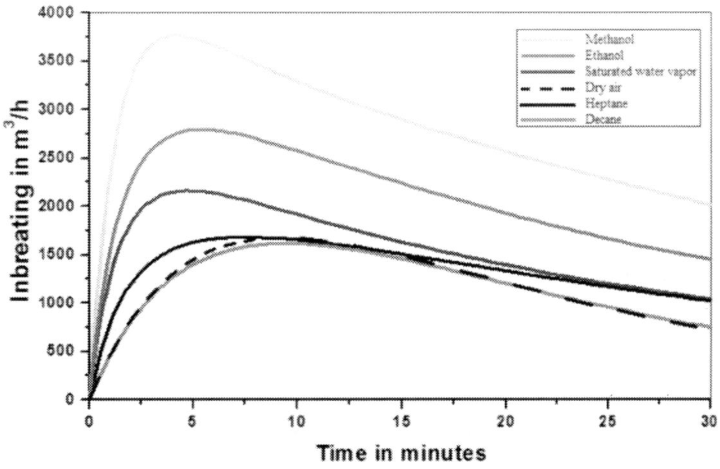

Figure 6. $\dot{V}_{inbreathing}$ profile of different vapors – PROTEGO© Model [4]

It can be seen that the condensation of volatile vapors has a significant impact on the inbreathing rate causing it to have a twofold increase. On the other hand, non-volatiles such as decane do not affect the inbreathing rate and yields the same inbreathing profile as with dry air.

Table 5. Condensation Factor "χ" Results

Vapor	$\chi_{PROTEGO}$	$\chi_{Fullarton}$
Methanol	4.76	3-4
Ethanol	2.23	2-2.5
Heptane	1.61	1.5-2
Water	1.29	1.5-2
Decane	1.01	N/A

PROTEGO© calculated the condensation factor "χ" for each of the simulated vapors and compared their values to that of Fullarton's (Table 4) and summarized the results in Table 5 above. The condensation factor values from this model are all in agreement with the published values [6] except the χ_{methanol} which is greater than the range specified by Fullarton (1986). A possible explanation for the latter is that PROTEGO© may have been more conservative (higher safety margin) when dealing with extremely volatile vapors like methanol.

5 Large-Scale Experimental Data

A thorough literature review yielded two sets of large scale data from actual experiments to study the thermal inbreathing of low-pressure storage tanks. Large scale datasets are preferred for model validation. The first dataset is for non-condensable gases (air) by *Sigel et al.* [7] and the second dataset is for condensable gases by *Holtkoetter et al.* [8].

5.1 Sigel's Experiment

Sigel performed actual data collection for the thermal inbreathing rate from a large tank of 617.5 m^3 capacity exposed to high solar radiation. The tank was empty without any trace of product vapor or liquid residue.

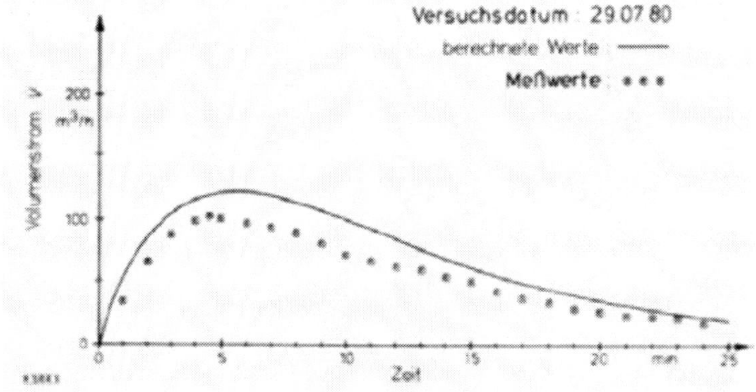

Figure 7. Sigel Experimental Inbreathing Profile

Sigel et al. [7] developed their own thermal inbreathing model to compare it with their experimental results. Figure 7 above displays the measured values of inbreathing (markers on the graph) versus the model (solid line). The inbreathing rates predicted by his model proved to be more conservative than the measured rates.

5.2 Holkoetter's Experiment

Holtkoetter performed an experiment on a 1.18 m^3 tank to measure the actual inbreathing rates and compare the results to his proposed model. This experimental investigation aimed at studying the effects of condensation on thermal inbreathing for three different vapors: water, isopropanol, and methanol.

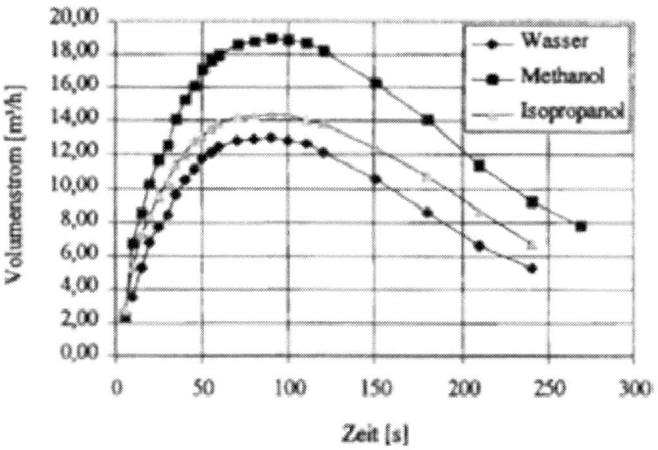

Figure 8. Holtkoetter Experimental Inbreathing Profile

The maximum thermal inbreathing is recorded after almost 95 seconds for all three gases. The maximum thermal inbreathing of methanol $\dot{V}_{max} \approx 19$ m^3/h is significantly higher than that for both water ($\dot{V}_{max} \approx 13$ m^3/h) and isopropanol ($\dot{V}_{max} \approx 14$ m^3/h) due to the fact that methanol is the most volatile vapor.

6 SuperChems™ proposed Model

The inbreathing profiles were generated using SuperChems™ dynamic simulation of storage tanks experiencing a sudden weather change. Dynamic models continually generate results that are used as the initial conditions for the next time interval calculation which perpetually drives the model to completion.

The proposed model first starts by defining a variety of objects to represent components of the system such as the storage tank vessel, the mixture that is stored, the piping segment and relief device, and the water sprays that are used to simulate a rainstorm. A built-in thermodynamics package is selected (modified Peng-Robinson EOS [9]) and scenarios are then created to collect the objects. Once the scenarios are created, single and multi-phase flow dynamics models are associated with the individual scenarios and the dynamic solutions for the inbreathing profiles are produced.

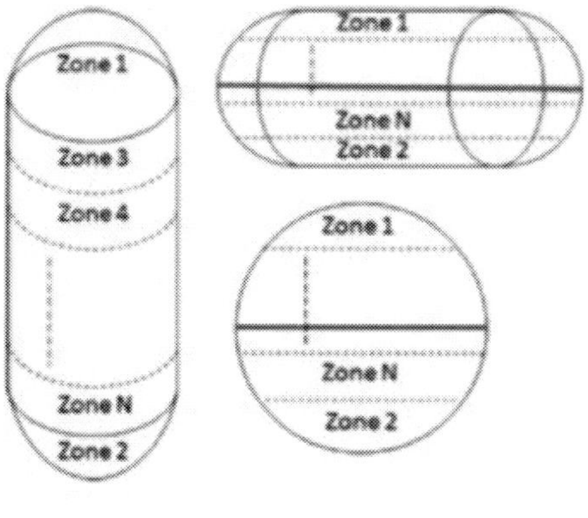

Figure 9. SuperChems™ Segmentation Scheme

An important feature of the software is wall dynamics which segments the vessel into multiple zones (Figure 9) allowing the user to account for heat conduction across the wall and between the zones using the effective conduction length method.

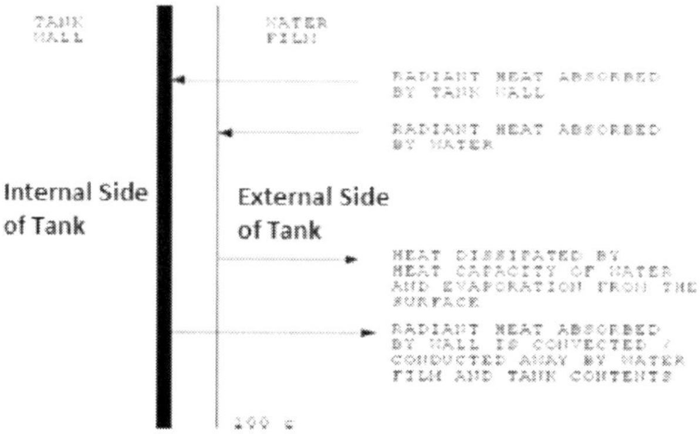

Figure 10: Heat Transfer Mechanisms for a Tank during a Rainstorm

Once segmentation is defined, heat transfer analysis is applied per wall segment (Figure 10) where the impact of rain, solar radiation, wind, and insulation is considered. Inside the tank, the wall-to-fluid heat transfer includes radiation and natural convection for the vapor phase; and nucleate, film and pool boiling for the liquid phase.

It should be noted that many parameters from the existing models and large-scale datasets were missing (vessel, dimensions, wall thickness, mixture composition, heat transfer coefficients, rain intensity, etc.) and had to be scientifically estimated in our dynamic simulations.

6.1 SuperChems™ Model Prediction for Non-Condensable Vapors (Large-scale Data)

The dimensions of the 617.5 m^3 empty tank used in Sigel's experiment [7] were introduced as input parameters to the SuperChems™ vessel object. The rainstorm used in this simulation had an intensity of 0.38 inch/h and a duration of 30 minutes with a temperature of 15°C. The inside heat transfer coefficient value was set to 5 W/m^2K as specified by API 2000 [1].

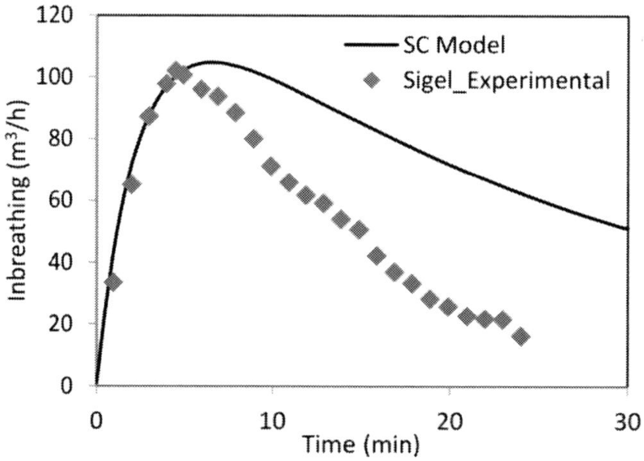

Figure 11. Simulation of Sigel's Tank using SuperChems™

The maximum value for inbreathing predicted by SuperChems was 104 m^3/h at time $t = 6.6$ min, as seen in Figure 11 above. This value is at a 2% deviation from the actual recorded maximum. The model's slope matches the first part of the inbreathing curve up until the maximum. Although the slopes of the decay of the inbreathing are different, they do not contribute to the design of the relief device because the most important value extracted from such studies is the maximum inbreathing that will act as the design relief capacity of the venting device.

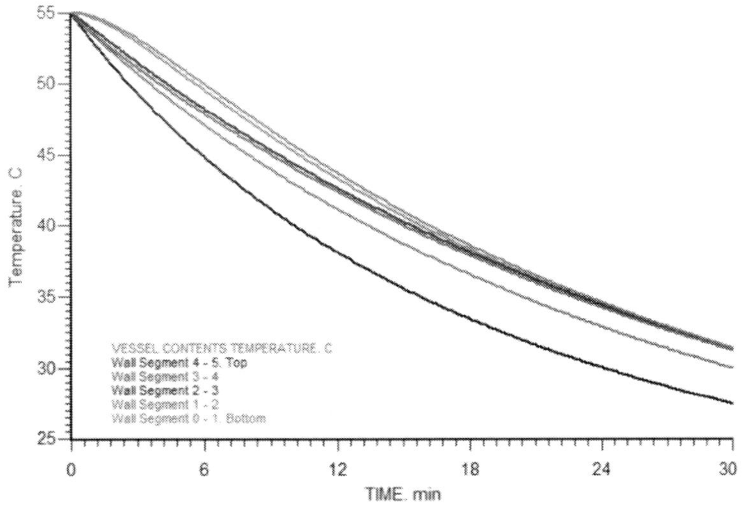

Figure 12: Temperature Profiles of Sigel's Tank using SuperChems™

Figure 12 above shows the temperature profiles of the wall segments and air inside the tank from the SuperChems™ simulation. Since Sigel [7] did not disclose the actual temperature profiles, therefore, we could not compare the computed temperature profile from SuperChems™ to the actual readings from Sigel's experiment.

6.2 SuperChems™ Model Prediction for Condensable Vapors (Large-scale Data)

Holtkoetter's experimental setup [8] consisted of a 1.18m^3 tank with walls that are 1.2 mm wall thick. The three vapors Holtkoetter studied were: methanol, isopropanol and water. For each type of vapor, the tank was filled with a small amount of liquid which was then vaporized using three infrared radiators (2.5-4.2 kW). After vaporization step was complete, water sprays were activated to cool the tank and the thermal inbreathing was recorded (Figure 8).

Using SuperChems™, the vessel was defined, then the rain-wall heat transfer coefficient, h_{rain}, was fixed at 600W/m^2K that corresponds to a rain intensity of 0.1262 kg/m^2s as specified by Holtkoetter. The duration of the rain for this simulation was 300 seconds. The h_{gas} was initially set to the API 2000 value of 5 W/m^2K, but was then adjusted to compensate for the vapor condensation inside the tank.

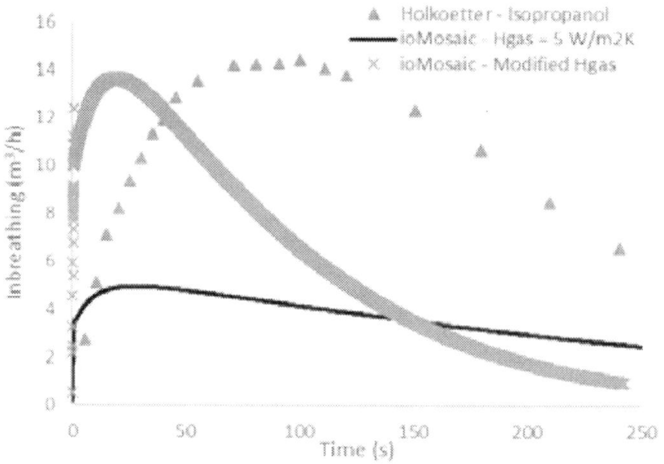

Figure 13. SuperChems™ Simulation of Holtkoetter's Tank with Isopropanol

Figure 13 illustrates the inbreathing profile for two different values of the free-convection heat transfer coefficient of the isopropanol vapor, h_{gas}, inside the tank. It can be clearly seen that using the API 2000 value $h_{gas} = 5 W/m^2K$ underpredicts the thermal inbreathing by almost 62%. After numerous iterations, a modified value for h_{gas} was found to be 30 W/m^2K. This higher value makes physical sense since the condensation of isopropanol occurring on the inner walls of the tank enhances the heat transfer between the vapor and the wall.

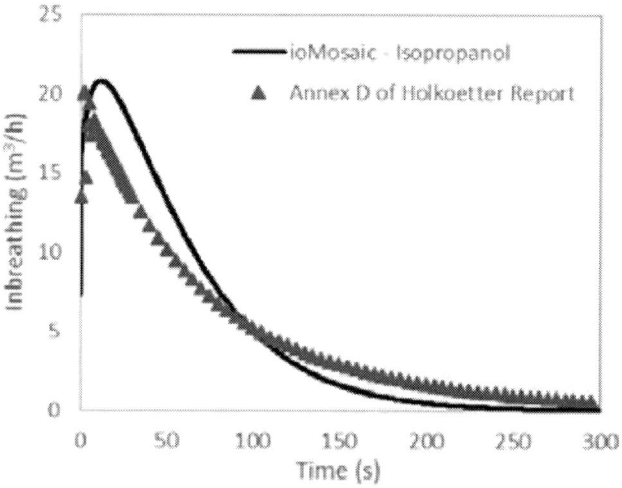

Figure 14. SuperChems™ Reproduction of Holtkoetter's Inbreathing Model (Isopropanol)

Although *Holtkoetter et al.* [8] performed experimental work to measure the maximum inbreathing, they did not, however, compare their experimental results to their proposed model. Hence, the SuperChems™ results were compared to Holtkoetter's model found in "Annex D" of his publication [8].

Figure 14 above proves that the SuperChems™ based model shows good agreement with the Holtkoetter model that considers spontaneous condensation. The predicted value for the $\dot{V}_{max}$ using SuperChems™ was 20.7 m^3/h and is at a +3% error from the value reported by Holtkoetter [8].

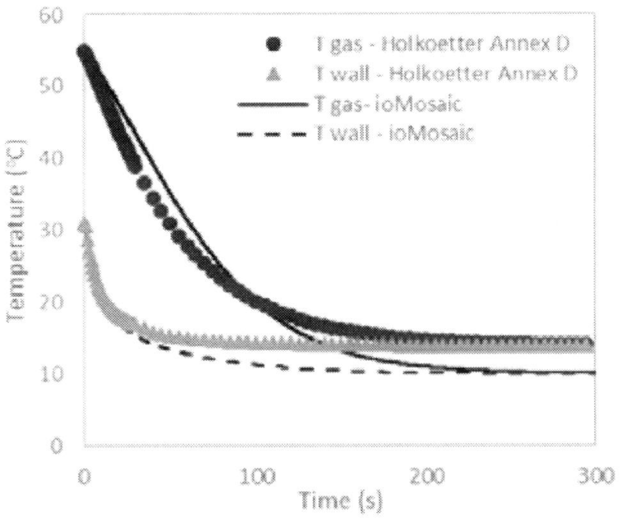

Figure 15. SuperChems™ Reproduction of Holtkoetter's Temperature Model (Isopropanol)

The gas and wall temperature profiles were almost accurately reproduced using the SuperChems™ model and are shown in Figure 15. Holtkoetter attributes the discrepancy between his model and his experimental result to the fact that the experiment was done in the month of December where the available water had a temperature of 8°C instead of 15°C. Colder rain will eventually lead to higher inbreathing rate values, therefore, the inbreathing should be inspected as a function of temperature to account for this 7°C difference in the water temperature. For more information about this method, please refer to the actual publication [8].

6.3 SuperChems™ Model vs. Published Models

In addition to validating our proposed model against large-scale experimental data, the SuperChems™ based model was also compared to the published models listed in Section 3 and 4 of this paper [10].

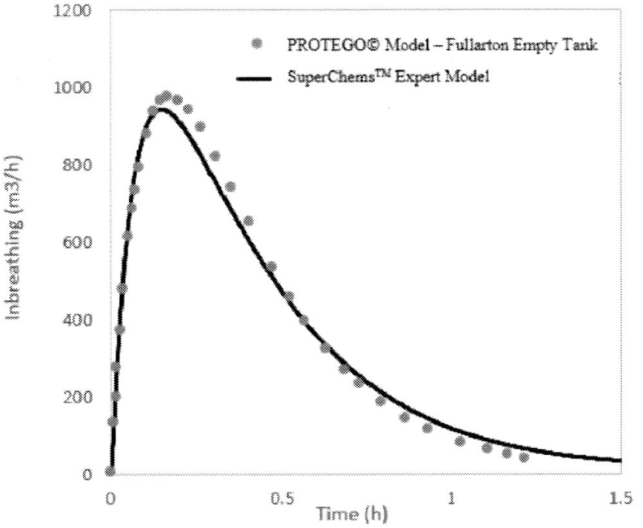

Figure 16. SuperChems™ Inbreathing Profile for Air (Fullarton's tank)

Figure 16 shows the predicted inbreathing profile of dry air using the SuperChems™ model and compared against the PROTEGO© model. The maximum recorded inbreathing from the SuperChems™ Expert simulation is around 940 Nm^3/h which has a -4% deviation from $\dot{V}_{max}$ predicted by PROTEGO©. The temperature and temperature-rate profiles were also calculated and can be found in the actual publication along with the fits for the different models and vapors using SuperChems™ Expert [10].

6.4 Modified Heat Transfer Coefficient

The heat transfer coefficient between the stored vapor and the inner-tank walls, h_{gas}, was modified for each type of volatile condensable vapor [10]. Table 6 below summarizes the values of h_{gas} used in the two-phase dynamic model for a tank-wall thickness of 5 mm. These results show that condensation "enhances" heat transfer thus providing evidence that use of the API 2000 value heat transfer coefficients h_{gas} will underpredict the maximum thermal inbreathing (Figure 13). Also, note that h_{gas} is directly proportional with the vapor's volatility.

Table 6. Modified h_{gas} Values for SuperChems™ Two-Phase Model

h_{gas} (W/m^2K)	Wall Thickness = 5 mm
Methanol	40
Ethanol	30
Water	20
Heptane	5
Decane	7

7 API 2000 Conservativeness

The maximum thermal inbreathing rates, V_{max}, obtained from the SuperChems™ model were compared to those obtained using the API 2000 thermal inbreathing equation, Equation (1), to assess the performance of the API 2000 inbreathing method.

In order to determine the appropriate value for the C-factor in Equation (1), the vapor pressure and storage temperature of the stored gas are needed.

- Since all the investigated vapors (methanol, ethanol, water vapor, etc.) do not have vapor pressures higher than hexane, the vapor pressure was classified as "Similar to Hexane".
- The storage temperature of the gas is 55°C, which is greater than 25°C, so according to Table 1, the C-factor has a *value of 5.*

The storage tanks in this study were all uninsulated, hence, the insulation reduction factor R_i has a value of one. The API 2000 inbreathing equation used for this case now becomes:

$$\dot{V}_{IT} = 5\, V_{tk}^{0.7} \tag{8}$$

To determine whether the API 2000 inbreathing relief requirement is adequate, a deviation percentage of the simulated (or experimental) results the API 2000 $\dot{V}_{max}$ method is computed as follows:

$$\% \, Deviation = \frac{\dot{V}_{max,SC} - \dot{V}_{max,\ API\,2000}}{\dot{V}_{max,\ API\,2000}} \times 100 \tag{9}$$

where $\dot{V}_{max,\ API\,2000}$ is the maximum thermal inbreathing obtained from Equation (8) (m^3/h) and $\dot{V}_{max,SC}$ is the maximum inbreathing from SuperChems™ model simulation (m^3/h).

7.1 API 2000 performance – Large-scale Experiments

Sigel et al. [7] used a tank volume of 600 m^3, which when replaced in Equation (8), results in a maximum inbreathing value of 440 m^3/h. Figure 17 compares $\dot{V}_{max}$ using the API 2000 method to the actual recorded $\dot{V}_{max}$ values during Sigel's experiment. The API 2000 inbreathing prediction is conservative for Sigel's empty tank where the relief device is designed to handle an inbreathing rate 4.2 times larger than the recorded $\dot{V}_{max}$.

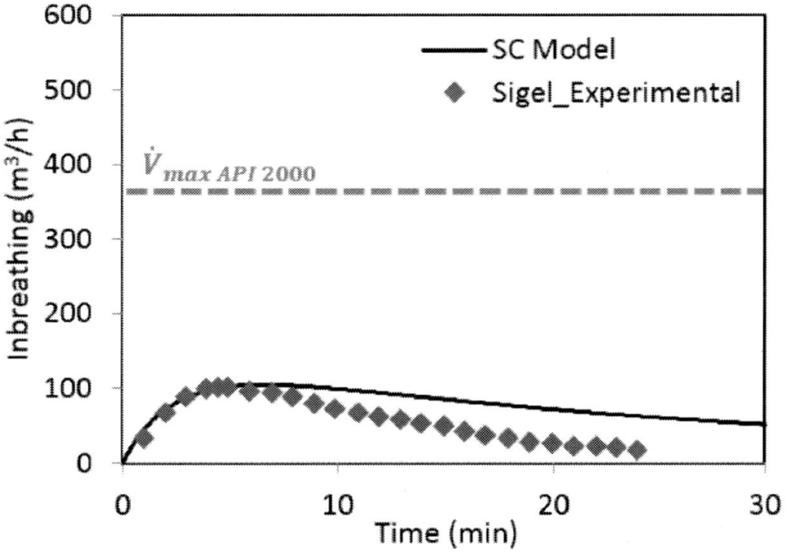

Figure 17. API 2000 Inbreathing Requirements – Sigel's Tank with Dry Air

On the other hand, *Holtkoetter et al.* [8] used a smaller tank with a capacity equal to 1.18 m^3 which results in a maximum inbreathing rate of 5.6 m^3/h according to the API 2000 method – Equation (8). Figure 18 shows that the API 2000 inbreathing relief requirement is insufficient for water, isopropanol, and methanol vapors in Holtkoetter's tank.

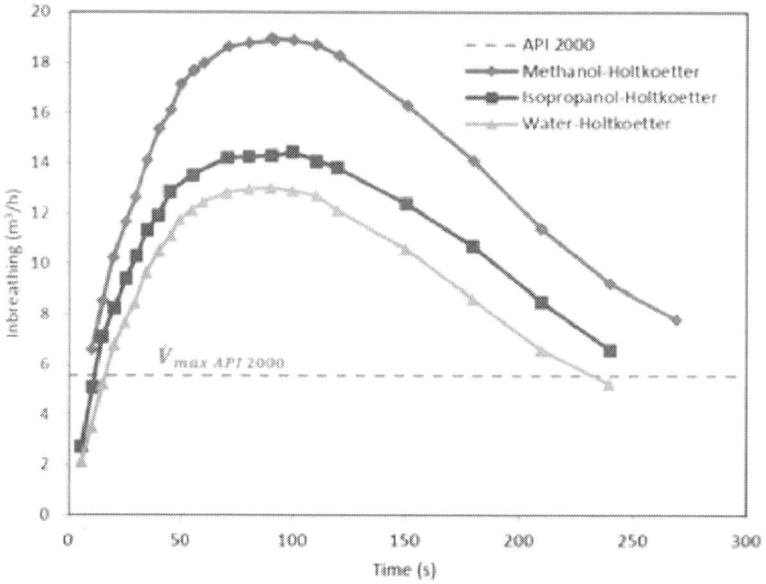

Figure 18. API 2000 Inbreathing Requirements – Holtkoetter's Tank

The recorded inbreathing rates are 2.3 to 3.7 times larger than the rate calculated by API 2000; this leads to an undersized relief device that puts the storage tank at a high risk of failure if a sudden change of weather occurs.

7.2 API 2000 performance – Computer Model

The tank size used in Scholz and Weber's model [4] was 6,283 m^3 to study the condensable vapors. Replacing the tank volumes in Equation (8), results in a maximum inbreathing, $\dot{V}_{max}$ = 2,278 m^3/h. Figure 19 graphically shows the API 2000 $\dot{V}_{max}$ in comparison to the dynamically simulated inbreathing profiles for the different types of vapors, including dry air.

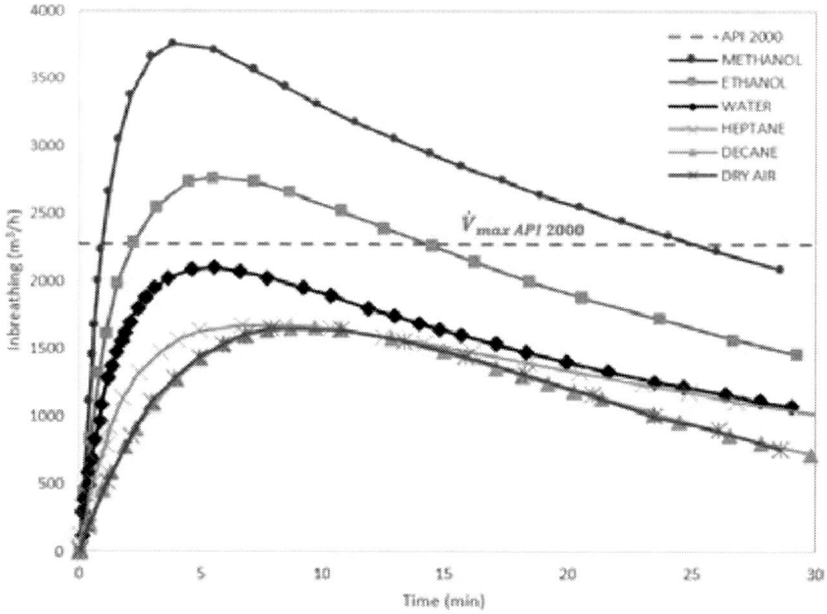

Figure 19. API 2000 Inbreathing Requirements – Different Types of Vapors

It can be seen that API 2000 method has a conservative prediction for the maximum thermal inbreathing of: (1) dry air, (2) heptane, (3) decane, and (4) water services. However, the API 2000 method underpredicts the relief requirement for the thermal inbreathing of a tank saturated with *methanol* and *ethanol*. The relieving capacity, for this case, has to be 1.7 times higher for methanol and 1.2 times higher for ethanol.

8 Conclusion

A dynamic simulation of the thermal inbreathing of low pressure storage tanks experiencing sudden change of weather conditions was performed using a SuperChems™ model. This model was validated by successfully reproducing the thermal inbreathing and temperature profiles of large scale experimental datasets and other available inbreathing models.

The thermodynamic variables specified in the API 2000 7th Edition standard have been tested for adequacy. It was found that the gas/wall free-convection heat transfer coefficient, $h_{gas} = 5W/m^2K$, specified by API 2000 was adequate when the tank was filled with non-condensable vapors like air. However, higher values were suggested for condensing vapors with a range of 10 W/m^2K to 45 W/m^2K depending on the type and volatility of the vapor. As for the rain/wall heat transfer coefficient, the API 2000 value of 5,000 W/m^2K proved to be too conservative, and even unrealistic, therefore leading to an oversized relief device. Suggested values for the rain/wall heat transfer coefficient were found to be in the range of 12 to 600 W/m^2K depending on the intensity and angle of the rainfall.

Finally, the API 2000 7th Edition method for calculating thermal inbreathing was compared with several tanks of different vapor services. The method proved to calculate an adequate relief requirement for tanks storing air, heptane, decane, and even water vapor in one case. However, the application of the API 2000 thermal inbreathing method was not conservative enough when the vapor services became more volatile, like in the case of isopropanol, ethanol, and methanol, that resulted in high risk of failure for these types of storage tanks.

9 References

[1] API Standard 2000, Venting Atmospheric and Low-Pressure Storage Tanks, 7th Edition, March 2014

[2] P. Salatino, G. Volpicelli and P. Volpe. "On the design of thermal breathing devices for liquid storage tanks", Process Safety and Environmental Protection, 77.6, pp. 354-359, 1999.

[3] D. Brooks, "Thermal Inbreathing Requirements of Low Pressure Storage Tanks at Elevated Temperatures". Lloyd's Register Energy Americas, Inc. Spring DIERS Users Group Meeting, Presentation, May 2012.

[4] R. Scholz and R. Weber, "Cooling Down Industrial Tanks". Prepared for PROTEGO Braunschweiger Flammenfilter at API 2000 Meeting Presentation, Fall 2010.

[5] H. Foerster, K. Schampel, and H. Steen, "Atmungsvorgänge infolge von Witterungseinflüssen an Lagertanks für brennbare Flüssigkeiten", Physikalisch Technische Bundesanstalt (PTB) W-22. Braunschweig Univ. Phys.-Tech. Bundesanst, 1984.

[6] D. Fullarton, J. Evripidis and E.U. Shclünder, "*Influence of Product Vapour Condensation on Venting of Storage Tanks*", Institut für Thermische Verfahrenstechnik, Universität Karlsruhe (TH), 1986.

[7] R. Sigel, B. Kuxdorf, R. Meiss and H. Schwarz, "Praktische und theoretische Untersuchung der Beatmung von Festdachtanks". Chemie Ingenieur Technik, 55.5, pp. 377-379, 1983.

[8] T. Holtkötter, J. Shang and H.G. Scheckerm "Behälteratmung - Experimentelle Untersuchungen und Entwicklung eines Vorhersagemodells" (Master Thesis), 1995.

[9] G.A. Melhem, R. Saini, and B. M. Goodwin, "A Modified Peng-Robinson Equation of State," *Fluid Phase Equilibria,* 47, pp. 189-237, 1991.

[10] D.D. Abou-Chakra, G.A. Melhem, and R.J. Willey, "Dynamic Modeling of Inbreathing Requirements for Low-Pressure Storage Tanks" (Master's thesis), 2016. Retrieved from **http://hdl.handle.net/2047/D20209806**

478223 Studying the Relationship Between Inherently Safer Design and Equipment Reliability

Abstract:

Risk associated with a process depends on both likelihood and consequence of an incident. Inherently safer design is based on the use of technologies and chemicals with intrinsic properties that reduce or eliminate hazards. This philosophy is mainly focused on reducing the consequences associated with an incident. However, recent research has revealed that applying principles of inherent safety can have an effect on the likelihood element of risk as well. One of the major drawbacks of implementation of an inherently safer design (ISD) is that it can lead to risk migration. Thus, attempting risk reduction through implementation of ISD by mitigating the consequence element of risk can undesirably lead to an increase in risk by affecting the likelihood element of risk. This likelihood element of risk is the mathematical complement of system reliability. This study mainly focuses on assessing the effect of implementation of ISD principles on the reliability of a chemical process as it progresses along different stages of process design and manipulating this relationship to achieve maximum economic benefits while simultaneously minimizing risk associated with the process.

In this study, safety indices are used for quantification of inherent safety of a process. The choice of safety index depends on the stage of chemical project under consideration. Reliability of an equipment involved in a process is determined through the use of offshore and onshore reliability data (OREDA) and center for chemical process safety (CCPS) reliability databases for process equipment. The reliability of process system is then determined from the reliability of the individual equipment involved in the process. Lastly knowing values of safety indices and reliability of systems for competing designs during different stages of a chemical project (such as process selection stage, conceptual stage etc.), the required relationship between ISD and reliability is established for that particular stage.

Applying this methodology to process selection stage reveals that selecting a process that's inherently safer in terms of the involved reactions, chemicals and equipment might lead to risk migration, since the selected process might have lesser system reliability as compared to other competing process. When this methodology is applied to conceptual stage, where-in the competing processes are similar in terms of reactions and chemicals and vary only in terms of the equipment involved, it is observed that a process that has lesser number of equipment might be more reliable and inherently safer as compared to other competing processes. Lastly, when this methodology is applied for a specific equipment such as a reactor where the reliability is modeled as a function of design parameters of the reactor, it has been observed that within a certain range of design parameters, the reactor becomes sufficiently reliable and inherently safer.

Thus, this study provides a framework for designing of process systems and equipment that are reliable, economically profitable and inherently safer during different stages of project.

479377 Pressure Relief Analysis - It's about More Than Just Compliance

Abstract:

Pressure relief valves and disposal systems are the last layer of defense in petrochemical plants and refineries. Therefore, there are many codes and standards that must be followed to design those systems. Failure to ensure safe plant operations can result in fatalities, injuries, and environmental releases which can result in significant cost and poor public relations.

OSHA's CFR 1910.119 - Process Safety Management of Highly Hazardous Chemicals is designed to manage the hazards associated with facilities which handle highly hazardous chemicals. Many of the requirements in PSM are also valuable for facilities which are not governed by the requirement and application of these principals may improve safety for all facilities.

This paper will present the proposal that Process Hazard Analysis (PHA) be the heart of process for ensuring safe operations to prevent negative impact on companies' image and goodwill with the community.

480249 How to Leverage Limted Design Data Along with Historical Incident Data to Assess Inherent Process Safety During Early Stage of Design?

Abstract:

Inherent safety level quantification remains as one of the challenges for the inherent safety concept to gain industrial acceptance comprehensively. In current practice, most of the time safety assessment is conducted once the process design has been completed. At this stage of design implemented strategies look to control the hazard level. Any design modification necessary as a result of unfavorable safety performance at this stage of design, would be costly when compared to alteration during conceptual design stage [1]. It is beneficial to design industrial processes that are user friendly, inherently safer and have a level of tolerance for the deviations from ideal performance, which will eventually reduce the number of hazardous accidents. Therefore, both industrial experts and researchers are interested in safety indices that can be applied during the early design stage of a process, where the extent of knowledge regarding the process is very limited. Again, the available indices to quantify inherent safety still leave room for enhancement to represent process stream condition more accurately as a mixture rather than pure component(s) and to account for the interaction of all process parameters in determining the level of inherent safety [2].

In this work, a property based methodology to quantify inherent safety level has been proposed by assessing chemical and process unit risk at earlier stage of design. This will assist process designers in producing safer designs by the application of inherent safety principles in a more efficient and cost effective manner. It would offer process/design engineers an alternative to include safety without waiting until detailed design stage, where safety costs increase. Degrees of freedom at this stage would include alternative route selection, equipment sizing, and selection of operating conditions. The additional benefit of safety consideration in the early design stages is that it decreases the level of dependence on safety systems. The proposed safety assessment method here considers the perspective of five main safety factors associated with chemicals such as flammability, explosiveness, corrosiveness, toxicity & reactivity. These factors were linked to the process operating conditions (temperature and pressure). The process unit risk are calculated by classifying them into six main categories (i.e., pump, fired heat exchanger, pressure vessels, etc.) based on API 14C standard and identifying the undesirable events associated with each category. Major incidents database and the probability of unwanted incidents were used to link the undesirable events to operating conditions; this allows for the calculation of unit safety factors using the limited available information. Lastly, an enhancement technique to reflect the contribution of the individual components in the mixture is introduced based on the mixture properties, where the severity score was calculated for the mixture of chemicals in each unit at the actual operating conditions (whenever applicable) and multiplied by the flow rate of the mixture.

This proposed methodology of safety assessment gives a more intuitive, easy-to-understand way to represent the chemical and equipment safety factors posed by each individual unit. The steps in the methodology calculate, compare and prioritize the level of inherent safety of process streams during simulation work that influences the abovementioned safety factors in order to offer the user an insight into the risk associated with alternative design decisions. By prioritizing the streams based on the potential for chemical and equipment safety factors, the design engineers can easily identify the critical streams/units to be considered for improvement in order to avoid or minimize the risk. This methodology of quantifying inherent safety level will provide users another dimension to consider when selecting optimal design with multi objectives (e.g. techno, economic, and environmental).

The assessment of inherent safety level using the proposed method is demonstrated through a case study. HYSYS simulation is used to design two alternative designs for ammonia production.

1. Thiruvenkataswamy, P., et al., *Safety and techno-economic analysis of ethylene technologies*. Journal of Loss Prevention in the Process Industries, 2016. **39**: p. 74-84.

2. Leong, C.T. and A.M. Shariff, *Process route index (PRI) to assess level of explosiveness for inherent safety quantification*. Journal of Loss Prevention in the Process Industries, 2009. **22**(2): p. 216-221.

Inherently Safer Design from Overpressure Protection Viewpoint

Chris Ng
Sachin Kanade
Nancy Faulk
Siemens Energy, Inc.
4615 Southwest Freeway, Suite 900
Houston, TX 77027
chris.ng@siemens.com

Prepared for Presentation at
American Institute of Chemical Engineers
2017 Spring Meeting and 13th Global Congress on Process Safety
San Antonio, TX
March 26 – 29, 2017

Inherently Safer Design from Overpressure Protection Viewpoint

Chris Ng, Sachin Kanade, and Nancy Faulk
Siemens Energy, Inc.
4615 Southwest Freeway, Suite 900
Houston, TX 77027
chris.ng@siemens.com

Keywords: pressure relief, process safety, inherent safety

Abstract

Pressure relief and flare system design is mostly evolving or iterative throughout the various engineering process design stages. This iterative nature stems from the need to capture the impact of engineering design changes on this critical layer of protection, which often is designed as the last line of defense for overpressure events.

The systematic review of pressure relief and flare system design basis at various engineering design and project execution phases will not only help to ensure adequate overpressure protection, but can also highlight opportunities for inherently safer processes and project cost savings. Often, evaluating trade-offs between process safety design improvements against project economics may be challenging.

In this paper, several case studies are presented to demonstrate how inherently safer design options and cost-effective solutions could be implemented across the various engineering design and project execution phases to minimize gaps in terms of overpressure protection between early and as-built designs.

1 Introduction

Figure 1 depicts the stages of a typical engineering project lifecycle, excluding decommissioning. Each stage has associated activities in terms of selection and refinement of design details which contribute to the overall integrity of the facility and economic feasibility of the project. The general approach involving timely review of pressure relief and flare system design could be applied to both large projects involving major Capital Expenditures (CAPEX) as well as smaller projects with limited scope.

Figure 1: Typical CAPEX Project Life Cycle.

During the early phases of design such as conceptual design and Front End Engineering Design (FEED) phases, the scope, objectives, and requirements are first defined, followed by development of a process basis (i.e., heat and material balance), Process Flow Diagrams (PFDs), and preliminary Piping and Instrumentation Diagrams (P&IDs). Equipment, line and instrument lists are then prepared, and preliminary equipment sizing and specifications for process and utility systems are generated. Design options are selected, and design drawings such as equipment layout plans are completed. A budget is developed and costs are estimated.

During later phases such as Detailed Engineering and Procurement, the remaining design information is added, such as vendor data, drains and vents, and piping details. The actual piping layout is verified and updates are incorporated. In addition, electrical, instrumentation, structural and building requirements to support the project are specified. Often Hazard and Operability (HAZOP) reviews are carried out. Typically procurement of the main equipment and auxiliaries takes place followed by construction, installation, commissioning, and Pre-Startup Safety Reviews (PSSRs), in preparation for start-up, operation and maintenance.

However, what is often overlooked during the early phases is input from pressure relief and effluent handling system design engineers. This paper will describe how early involvement by these engineers could help validate relief system design and reduce overall costs and delays later. The case studies described in Section 3 below offer examples of the potential savings.

Relief systems engineering analysis should be performed during various process design and project implemented phases, which allows identification of potential gaps as well as opportunities for improvement. Typical activities include:

- Overpressure scenario analysis
- Determination of relieving loads
- Selection of relief technology
- Sizing of relief systems
- Evaluation of relief device inlet loss and backpressure effects
- Depressuring analysis
- Hydraulic evaluation of common relief headers
- Flare radiation calculations
- Dispersion analysis, including suitability of atmospheric release
- Acoustic induced vibration (AIV) analysis
- Presenting findings and developing mitigation
- Developing comprehensive documentation
- Reviewing impacts of design changes and updating documentation to capture changes are they are implemented.

It is recommended that the reviewer(s) be well-versed in the appropriate industry standards, e.g., American Petroleum Institute (API) Standards 520 Parts I and II, 521, and 2000 [1][2] [3][4]; American Society of Mechanical Engineers Boiler and Pressure Vessel Code (ASME BPVC) Sections I and VIII [5][6]; as well as corporate guidelines and site-specific requirements. In addition, it is recommended that the reviewer establish and utilize a consistent approach for review of all designs, irrespective of the Engineering, Procurement, and Construction (EPC) companies working on the project.

2 Benefits of Early Review of Relief Systems Requirements and Design

2.1 Early Involvement Allows Critical Input

Early involvement of relief system engineers can allow critical input to be incorporated which influences selected design options from an overpressure protection point of view, thus promoting an inherently safer design. By careful consideration of a number of factors, the design team has the potential to reduce overpressure relief and flaring requirements which in turn can reduce sizing of relief devices and associated piping, as well as flare components such as headers, knockout drums, flare stack, and flare tip. Potential reduction of environmental emissions is also realized. The examples below demonstrate some ways inherently safer design options and cost-effective solutions were implemented to minimize gaps in terms of overpressure protection between early and as-built design. In addition, early review can help to reduce or minimize the need for later design changes or mitigating measures, which could add significant costs and delay completion schedule for the project.

2.2 Identify Opportunities to Increase Inherent Safety

2.2.1 Centrifugal Pump Design

Consider a centrifugal pump as a pressure source for downstream equipment, as in Figure 2. As it stands, P-1 is capable of overpressuring the downstream vessel, V-1, in the event of a blocked outlet or an inlet control valve failure.

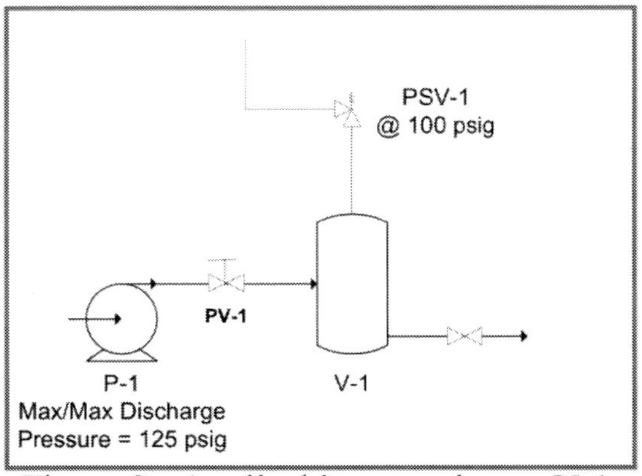

Figure 2: Applicable scenarios on V-1.

By properly sizing this pump during the design phase, however, it may be possible to minimize the relief rates for blocked outlet and control valve failure-open overpressure scenarios of the downstream equipment or even to eliminate the applicability of such scenarios altogether. Working with the process and mechanical engineers as well as rotating equipment designers, the maximum pump impeller size or speed and/ or maximum pump suction pressure could be carefully specified within process operation limits, which in turn limits the maximum pump discharge pressure. By reducing the maximum discharge pressure of the pump to a value less than the downstream equipment Maximum Allowable Working Pressure (MAWP), the above-noted scenarios may not be applicable, as in Figure 3a. An alternative to this, as shown in Figure 3b, is to ensure that the downstream equipment MAWP would be higher than the maximum discharge pressure of the upstream pump. These engineering design choices, which would result in minimizing relief loads or eliminating applicable scenarios, helps make the process inherently safer.

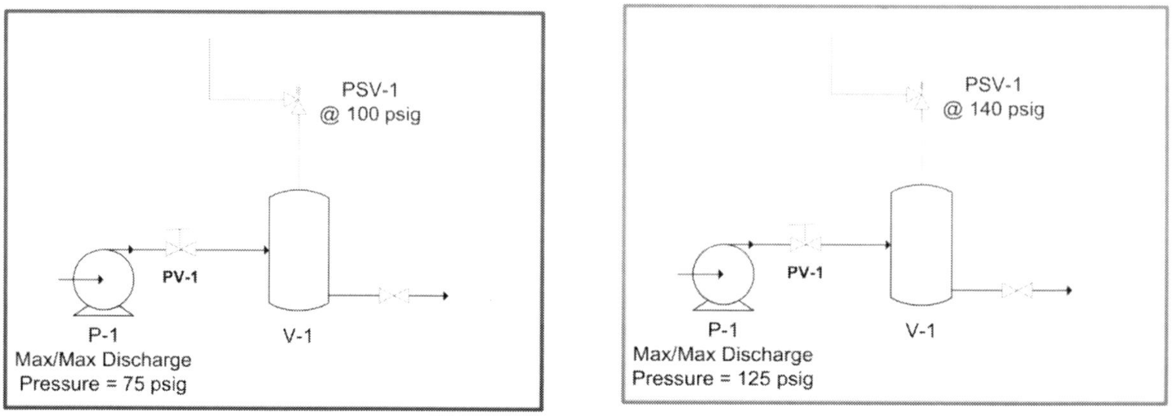

Figure 3: a) Limiting pump discharge pressure. b) Increasing vessel design pressure.

2.2.2 Pump and Compressor Discharge Check Valve Design

A related example involving pumps (and compressors, for that matter) is to ensure appropriate discharge check valve(s) are allocated to mitigate potential for reverse flow, impacting upstream equipment. In some instances, such as in Figure 4, feed could continue into an upstream vessel such as a feed surge drum if the liquid outlet pump is stopped with a discharge check valve stuck open. If this relief scenario is not identified by the relief designer, the relief valve on the feed surge drum may not provide adequate relief capacity.

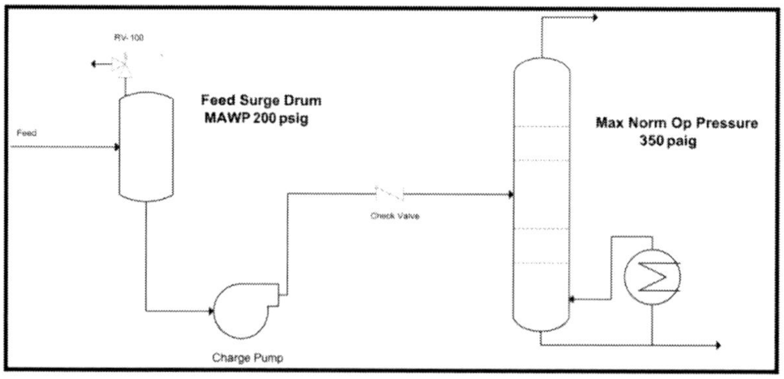

Figure 4: Applicable scenario on Feed Surge Drum

Allocating dual check valves in series of similar (or diverse) design may be required to minimize the backflow rate in the event of a check valve failure; if they meet the requirements set forth in API STD 521, the backflow rate may be limited to check valve leakage instead of check valve failure [7]. Careful specification of the MAWP of the upstream vessel may also help mitigate the overall relief requirement. Consideration of such issues early on is critical to avoid potentially costly design changes later as well as to increase the safety of the process; however, this is commonly missed due to factors such as improper management of change or poor communication between different groups designing upstream and downstream systems.

In a similar fashion, an early design review of several capital projects was performed, focusing on inherently safer design. One recommendation was proposed and accepted to raise the design pressure of the feed surge drum to mitigate the scenario in which a shutdown of a downstream charge pump results in reverse flow to the surge drum, while normal feed to the surge drum continues. By doing so, the feed and reverse flow sources would no longer be capable to overpressure the drum. Early review made the design inherently safer as the scenario was no longer applicable. Without this recommendation, installation of either a safety-instrumented system (SIS) or three 8T10 relief devices would have been necessary.

2.2.3 Control and Bypass Valve Design

Early input is critical in the selection and sizing of a control valve and its bypass valve, since the relief designer must account for the scenario in which *both* the control valve and its bypass valve fail open, as discussed in API STD 521 on "Pressure-relieving and Depressuring Systems". Early engagement of a pressure relief systems expert will help optimize the control valve/bypass and pressure relief requirements to provide adequate overpressure protection within the process operation envelope and requirements, while maintaining industry design and safety best practices. In one instance, a 2" gate valve was specified as the bypass for a 1" control valve, as shown in Figure 5. During a later design stage, it was discovered that this installation resulted in a very undersized relief device for the scenario described above, and thus the need to replace the bypass valve with a different and smaller-capacity type of valve, such as a globe valve. As in this instance, if the nominal size of the installed bypass valve is bigger than the control valve size, it necessitates careful review and possible re-design.

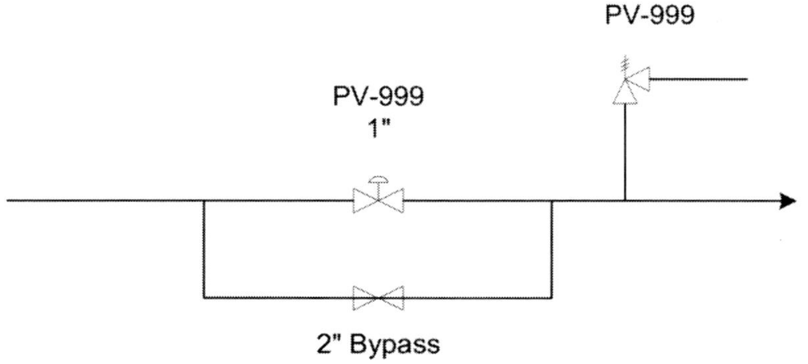

Figure 5: Bypass valve initially specified results in undersized relief device

2.2.4 Static Head

Static head is often overlooked in the preliminary design phases as equipment layouts may not yet be available. Relief designers should encourage vessel designers to specify vessels to be designed to be liquid full at the requested MAWP. Alternatively, relief device set pressures can be reduced to compensate for static head effects provided that the vessels can be filled above the top weld seam (into the top head). In another instance, the effects of static head on the set pressure of multiple relief valves had not been accounted for by an EPC company during detailed engineering. The potential for the internal pressure of affected equipment to exceed MAWP plus allowable accumulation per code was identified, allowing the relief valve set pressures to be adjusted prior to field installation.

2.2.5 Detailed Design Reviews

An independent pressure relief design audit was conducted for a large operating company. Relief design analysis and calculations had been performed by an EPC company as part of an expansion project. A review was conducted during the detailed engineering phase, during which it was found that most of the relief valve and associated piping upgrades which were being proposed by the EPC company were not required. This review included one application for which the EPC company had recommended upgrades to 6 relief valves (including recommendation to install two additional 8T10 relief valves), based on a recirculation flow rate which had been incorrectly included in a blocked outlet scenario relief rate estimate. Upon review, it was concluded that the existing relief valves were adequate and did not need to be upgraded. This reduced the potential for oversized relief capacity.

2.2.6 Design Gaps

In another example of early design review, engineers were able to identify a proposed design gap which would have resulted in costly and unnecessary modifications to a flare system. The proposed design project involved tie-ins of multiple large atmospheric relief valves to the associated flare for a process unit, resulting in inadequate flare capacity and the necessity of making expensive upgrades to this flare system. By performing detailed flare hydraulic modeling, relief engineers were able to show that re-routing the additional relief loads to an alternate adjacent flare having some spare capacity would be more practical.

2.2.7 Design of Feed Surge Drum

During a design review of a feed surge drum, additional examples of recommendations were developed to further increase the inherent safety, such as:

- By limiting the maximum discharge pressure of a feed pump to the proposed design pressure of the downstream equipment, a blocked outlet overpressure scenario was no longer credible for that equipment.
- Rerating the feed/bottom exchangers to account for static head in the event of overfilling in the associated distillation column.
- Car-sealing open (CSO) appropriate valves to ensure that various overpressure scenarios would no longer be considered as design contingencies and relief paths would be available for impacted equipment.

2.2.8 Car-Sealing Strategic Valves

In a final example, several recommendations were developed and adopted prior to start-up for a Coker Vapor Recovery Unit (VRU) revamp project. These recommendations include the following:

- Re-rating equipment
- Car-sealing closed (CSC) the control valve bypass valves
- Reducing the control valve flow coefficient (C_v) to limit maximum flow through the valve in the event of failure open
- Increasing the vessel design pressure to account for static head
- Adding dual check valves to mitigate reverse flow

These changes helped make the overall design inherently safer, so fewer safeguards were required to reduce the risks associated with overpressure scenarios for multiple systems.

2.3 Relief Systems Engineer's Role

Numerous other examples of inadequate design issues as well as missed opportunities to increase the inherent safety of the design are routinely identified by the relief systems engineers after the design has already progressed to a late stage or even after field implementation. Design review can begin early in the process, when preliminary P&IDs are issued for HAZOP or other types of Process Hazard Analysis (PHA). The relief systems engineer should also validate the design once piping isometrics are issued for approval and construction or after other design changes are proposed later. Taking these steps can help to ensure that design shortfalls can be resolved at the earliest opportunity. Making design changes early on is easier and less expensive than during later design stages and even more so than after commissioning and startup, thereby preventing subsequent shutdown, delay or major re-design work.

3 Critical Factors for Success

All pressure relief and effluent handling design documentation and supporting data should be stored in a central repository, for which access can easily be provided to all who have a need for such information. This access can be controlled, so that most viewers will have "read-only" access, and only authorized personnel can make required changes associated with Management of Change (MOC) items, thus keeping overall content up-to-date (evergreen) throughout the life of each facility as MOC activities, major expansion projects, or process modifications are implemented. These activities should trigger design review by relief systems engineers to ensure that relief design continues to be designed to provide adequate relief for all credible overpressure scenarios and that associated documentation is kept up-to-date.

In order for the review process to achieve the desired results, the three parties which are commonly involved (relief systems team, operating company, and EPC company) should work together to understand the goals of each party and how the design inputs from each party should be integrated in the various stages of engineering design and project execution. The following factors have been identified as critical for success:

- Conducting independent relief system design review at the right time in the project scope

- Communication among all parties
- Sharing of critical design data
- Effective management of change
- Discussing and resolving identified issues promptly
- Integrating relief system design review team with project design and execution teams

4 Conclusion

As shown in several examples in this paper, the systematic review of pressure relief and flare system calculations at various process design phases will not only help to ensure adequate overpressure protection, but can also highlight opportunities for inherently safer processes and promote overall project lifecycle cost savings. In addition, such a review process can easily be extended to resolve similar design issues commonly addressed during HAZOP reviews and integrated with the MOC workflow.

5 References

[1] American Petroleum Institute, *API Standard* 520, *Sizing, Selection, and Installation of Pressure-Relieving Devices, Part I-Sizing and Selection*, 9th Edition, 2014.

[2] American Petroleum Institute, *API Standard* 520, *Sizing, Selection, and Installation of Pressure-Relieving Devices, Part II-Installation*, 6th Edition, 2015.

[3] American Petroleum Institute, *API Standard 521, Pressure-relieving and Depressuring Systems,* 6th Edition, 2014.

[4] American Petroleum Institute, *API Standard 2000, Venting Atmospheric and Low-pressure Tanks*, 7th Edition, 2014.

[5] American Society of Mechanical Engineers, *ASME BPVC Sec I, Rules for Construction of Power Boilers*, 2015 Edition.

[6] American Society of Mechanical Engineers, *ASME BPVC Sec VIII-1, Rules for Construction of Pressure Vessels*, 2015 Edition.

[7] API STD 521, §4.4.9.3

Disclaimer:

The information contained in this document represents the current view of the authors at the time of publication. Process safety management is complex and this document cannot embody all possible scenarios or solutions related to compliance. This document is for informational purposes only. Siemens makes no warranties, express or implied, in this paper.

exida®

Three Reasons Major Event Risk Management Fails on Capital Projects

Denise Chastain-Knight

P.E., CFSE

exida®

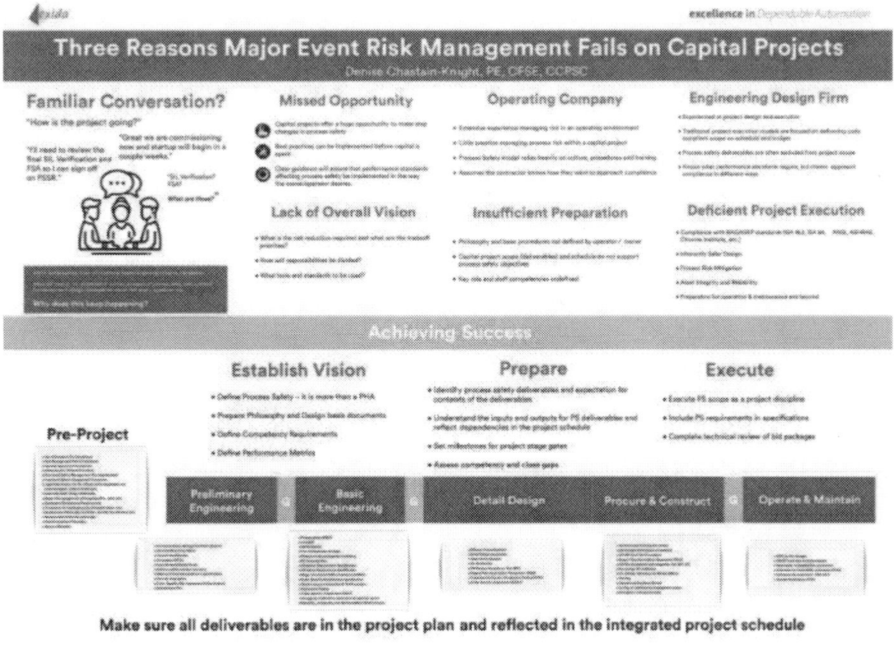

exida

About the Presenter

- **Denise Chastain-Knight, PE, CFSE, CCPSP**
  - BS ChE, Georgia Institute of Technology
  - Senior Functional Safety Engineer
  - More than 31 years experience in engineering design, process improvements, process risk analysis and safety instrumented system implementation
  - Experienced in the chemicals, resins, coatings, monomers, polymers, pharmaceutical, silicate, bio-fuels, petroleum, power…
- **Affiliations**
  - CCPS Technical Steering Committee
  - Past Chair NCEES PE (ChE) exam development committee
  - Fellow AIChE

 3

exida

Familiar Conversation?

How is the project going?

Great we are commissioning now and startup will begin in a couple weeks

I'll need to review the final SIL Verification and FSA so I can sign off on PSSR

SIL Verification? FSA? What are those?

 4

Familiar Conversation?

Similar conversations occur all to often as capital projects are nearing completion.

Often the topic of the conversation involves an aspect of process safety that is critical for management of high consequence events that pose a significant risk.

Why does this keep happening?

 5

Missed Opportunity

- Capital projects offer a huge opportunity to make step changes in process safety
- Best practices can be implemented before capital is spent
- Clear guidance will assure that performance standards effecting process safety be implemented in the way the owner/operator desires.

 6

Operating Company

- Extensive experience managing risk in an operating environment
- Little practice managing process risk within a capital project
- Process Safety model relies heavily on culture, procedures and training
- Assumes the contractor knows how they want to approach complaince

 7

Engineering Design Firm

- Experienced at project design and execution
- Traditional project execution models are focused on delivering code compliant scope on schedule and budget
- Process safety deliverables are often excluded from project scope
- Knows what performance standards require, but clients approach compliance in different ways

 8

PS Activities that Usually Happen

- Hazards Identification and Risk Analysis
- Relief system design
- Compliance with code mandated standards (ASME VIII, NFPA 30, NFPA 70)

 9

PS Activities that Often Fail

- Compliance with RAGAGEP standards (ISA 18.2, ISA 84, ANSI, ASHRAE, Chlorine Institute)
- Inherently Safer Design
- Process Risk Mitigation
- Asset Integrity and Reliability
- Preparation for operation & maintenance and beyond

 10

exida

Why We Fail

- No or inadequate overall vision
- Insufficient preparation
- Deficient project execution

Communication

exida

Lack of Overall Vision

- What is the risk reduction required and what are the tradeoff priorities?
- How will responsibilities be divided?
- What tools and standards to be used?

exida

Insufficient Preparation

- Philosophy and base procedures not defined by operator owner
- Capital project scope (deliverables) and schedule do not support process safety objectives
- Key role and staff competencies undefined

exida

Deficient Project Execution

- Underestimate of the workload required and dependencies with other scope objectives
- Process Safety /Risk Management not treated as a discipline
- Reactive not proactive

exida

How Project work

- Stage gate process
- Scope identified, schedule and budget developed to support scope
- Detail built in layers. Early definition important to minimize rework
- Often involve many contractor teams working on a subset of the scope

 15

exida

Establish Vision

- Define Process Safety – it is more than a PHA
- Prepare Philosophy and Design basis documents
- Define Competency Requirements
- Define Performance Metrics

 16

Prepare

- Identify process safety deliverables and expectation for contents of the deliverables
- Set milestones for project stage gates
- Understand the inputs and outputs for PS deliverables and reflect dependencies in the project schedule
- Assess competency and close gaps

 17

Execute

- Execute PS scope as a project discipline
- Include PS requirements in specifications
- Complete technical review of bid packages

 18

exida

Pre-Project Documents

- List of Standards for Compliance
- Risk Management Plan & Procedures
- Hazards Assessment Procedures
- Cybersecurity Policy & Procedure
- Functional Safety Management Plan (organization)
- Functional Safety Management Procedures
- Approved vendors list for critical/safety equipment and instrumentation
- Alarm Philosophy
- Inherently Safer Design Philosophy
- Major risk management philosophies (fire, relief, etc)
- Contractor Competency Requirements
- Procedures for required studies (dropped object, etc)
- Maintenance Philosophy (isolation, clearing requirements etc)
- Measurement & Metrics philosophy
- Communications Philosophy
- Spares philosophy

 19

exida

Preliminary Engineering

- Functional Safety Management Plan (project)
- Chemical Reactivity Matrix
- Hazards identification
- Conceptual SIF list
- Hazards Identification Study
- Define credible accident scenarios
- Major and long lead equipment specifications
- Controls Description
- Cyber Security Risk Assessment & Gap Analysis
- Maintenance Plan

 20

exida

Basic Engineering

- Project phase HAZOP
- CHAZOP
- HIPPS Studies
- Fire and Explosion Analysis
- Chemical release dispersion modeling
- SIL Determination
- SIS Safety Requirements Specification
- SIF Safety Requirements Specification
- Stage 1 Functional Safety Assessment (FSA1)
- Cyber Security Requirements Specification
- Countermeasure Inspection & Test Procedure
- Penetration Testing
- Cyber Security Assessment (CSA-1)
- Emergency notification, evacuation & response review
- Reliability, Availability, and Maintainability (RAM) analysis

21

exida

Detailed Design

- Balance of specifications
- Detail Design documents
- Alarm Rationalization
- SIL Verification
- SIS Factory Acceptance Test (FAT)
- Stage 2 Functional Safety Assessment (FSA2)
- Cybersecurity Factory Acceptance Testing (CFAT)
- Cyber Security Assessment (CSA-1)

22

Procure and Construct

- Technical and financial bid reviews
- Operating & Maintenance Procedures
- SIS/SIF Proof Test Procedures
- Stage 3 Functional Safety Assessment (FSA3)
- SIS Site Acceptance and Integration Test (SAT, SIT)
- Pre-startup SIF Validation
- Pre-startup Cybersecurity Review (CSA 3)
- Training
- Operational Readiness Review
- Configure maintenance management system
- Emergency management plan

 23

Operate & Maintain

- MOC for ALL changes
- SIS/SIF Proof tests & results analysis
- Revalidation of Detailed Risk Assessment
- Cybersecurity Vulnerability Assessment (CVA)
- Cybersecurity Assessment - CSA 4 & 5
- Periodic Revalidation of PHA

 24

A Comparative Analysis of Safety Rating System

Linlin Wang[a]

[a]Mechanical and electrical engineering college, China University of Petroleum (East China)

Abstract: With the development of petrochemical enterprises, the scale and petrochemical enterprises frequent accidents, the high risk of petrochemical enterprises has been more and more extensive attention. It's particularly important to establish a sound safety rating system for the timely detection and elimination of risk factors, reduce the probability of accidents appears. However, in China there is less research and application on safety rating system. This paper summarizes the similarities and differences between the various systems and puts forward the development of China's safety rating system.

Keyword: Safety Rating System; Petrochemical enterprises

1 Introduction

Petrochemical enterprises have a lot of flammable and explosive, toxic substances, and the installation of large-scale and continuous production, can easily cause major accidents. In 2013 Sinopec Huangdao Oil pipeline explosion, 62 people were killed, 136 people were injured. With the occurrence of petrochemical enterprises accidents, the safety production of petrochemical enterprises has become one of the issues that government and national focus on. Evaluation and rating of the status quo of enterprise security can promote the enterprises to find their own problems, develop improvement goals and implement continuous improvement. At the same time, it can also improve the government safety supervision departments to find out the status quo of enterprises, determine the key supervision enterprises, and optimize the allocation of resources, enhance the safe production level of enterprises in China, it is also an important means to improve business efficiency. The Chinese government's security supervision department attaches great importance to the assessment of enterprise safety, since 2006, has issued a "Basic norms of enterprise safety production standards", " General specification for safety standardization of hazardous chemicals", "Guidelines for oil industry safety production standardization "etc., The government promotes the implementation of the safety production standardization assessment throughout the country, take the "enterprise self-evaluation, government review" approach to the enterprise rating, and achieved important work results. But in the implementation a lot of problems were also found. Based on the status quo of safety rating of petrochemical enterprises in China, this paper analyzes and compares the implementation process of enterprise safety rating system such as ISRS (International Safety Rating System) ,NOSA five-star comprehensive evaluation system and VPP project implemented by OSHA. It is expected to provide support for the selection and development of our country's petrochemical enterprise safety rating tool.

2 Status Quo of Safety Rating of Petrochemical Enterprises in China

Safety standardization rating is developed, implemented and reviewed by China's relevant laws and regulations, industry standards. It is led by the State Administration of Safety Supervision and implemented on a national scale. It is the most extensive and highest commitment safety rating system in China. In addition, some petrochemical enterprises received the support of government, and independently introduced tools such as ISRS and NOSA. For example, 2014 China Petroleum West Pipeline Company reached the rate 7 of ISRS, and became first enterprise which reached this level in China.

Based on the safety standardization assessment of petrochemical enterprises in China, it is found that there are three main problems in the process of implementing safety standardization assessment in our country:

(1) Safety standardization system and some other implementation systems, such as HSE system, have incompatible problems, resulting in enterprises to repeat engaging in several sets of systems, repeat records, repeat systems. Companies not only feel cumbersome, but also waste human and material resources;

(2) Enterprises have low motivation and a bad attitude to practice safety standardization. Standardization of safety production is a measure imposed by the government to improve the safety level of the enterprise, and enterprises do not really realize its importance. So that the work is easy to become formalize and surface, and did not achieve the purpose of improving the level of safety, but only to let the enterprises to meet the rating and inspection;

(3) China's safety standardization assessment consists of three grades. The grade of the score range is quite large. And the flexibility of the assessment is not sufficient. Besides, in the absence of experimental circumstances, the assessment has a low recognition for some enterprises which take a new process or a new management approach.

3 System Introduction

3.1 ISRS（International Safety Rating System）

ISRS（International Safety Rating System）is a rating system of international security measure, and improving and showing the level of health, safety, environmental and business performance and other aspects of the organization, designed to help organizations manage risk and promote continuous improvement of the evaluation system. Det Norske Veritas (DNV) international safety rating system (ISRS) eighth edition published in 2009 after decades of practicing and improving.

ISRS [7] international safety rating system has 15 elements, each element is divided into several sub-elements of a total of 119 sub-elements, and list the number of points for each sub-element of the assessment, the total number is 661. 15 elements are: leadership, planning and administration, risk assessment, human resources, compliance assurance, project management, training and skills, communication and marketing, risk control, asset management, procurement and contractor management, emergency preparedness , the event learning, risk monitoring, and review the results, in which risk management (including 3 risk assessment,9 risk control, 14 risk monitoring) system for the trunk. During the assessment process, the customer selects the subroutine according to the grade to be evaluated, as follows::

Grade	1	2	3	4	5	6	7	8	9	10
Level range	1-4				5-6		7-8		9-10	
Required number of subroutines	30				60		90		119	
The number of customer chooses	20				15		10		0	
The total number of subroutines	50				75		100		119	
The lowest score of each program%	10	15	20	25	30	35	40	50	60	70
Average minimum score%	20	30	40	60	60	70	70	80	80	90
Live tour lowest score %	65	65	65	65	70	75	80	85	90	95

3.2 NOSA integrated five-star management system

In 1951, the International Occupational Safety Association (NOSA) created a five-star safety management system by conducting a comprehensive hazardous and harmful factors identification and risk assessment to identify all possible hazard identification and develop appropriate management for reduce hazard and risk. NOSA management system regards hazard identification and risk management as the core, Heinrich's "iceberg theory" as the basis and PDCA manner as its operation mode. System consists of five main elements, 72 elements, more than 300 minimum standards, the implementation of 1200 rules and standard instructions (against NOSA Integrated Five Star system revision and reorganization, the system consists of 75 elements). System evaluation content mainly in the following two aspects:

(1) Performance evaluation

Evaluation content of NOSA five-star rating system mainly consists of five areas, particularly in the following table:

NOSA integrated five-star rating system

project	Unit content	The number of elements	score	Share of the share
1	Building and housekeeping (factory) management	11	450	15
2	Mechanical electrical and personal protection	17	600	20
3	Fire and other emergency risk	8	400	13.3

	management			
4	HSE accident recording and investigation	5	400	13.3
5	HSE management organization	31	1150	38.4
total		72	3000	100

Through five major categories of 72 elements of qualitative evaluation, the project almost involved all activities of production and operation. It evaluates each element from the "four levels, three aspects, five grades" to carry out a comprehensive calculation to measure whether the risk of the element has been effectively controlled. "Four levels" means: risk awareness, the file system, compliance, implementation effect. "Three areas" means: safety, health, environment. "Five grades" means: 0, 25, 50, 75, 100 five scoring levels。

（2） Experience assessment

Introducing DIIR (Disabling Injury Incidence Rate) rate of accidents concept of enterprise, and performance assessment combined score, drawn 1 to 5-star level. DIIR shows annual percentage of the total number of employees were injured, according to NOSA claims, statistics should include employees of the enterprise, the tertiary industry and outsourcing construction team personnel. DIIR calculated as follows：

$$\text{DIIR} = \frac{\textit{The number of accidents caused by work} \times 200000}{\textit{Number of employees working hours per year}}$$

Where：DIIR——Disabling injury incidence rate；

The number of accidents caused by work - refers to the work caused by or caused by the injury, so that the injured cannot or cannot fully perform the daily designated job work up to one or more working days；

200000-- base by 100 people work for a year (50 weeks), 40 hours a week calculated；

Number of employees working hours per year - Total number of employees × Actual hours per person per year

The final combination of the above two aspects of the results given the final star rating results, the corresponding assessment of the star method as shown in the following table：

NOSA integrated five - star rating system assessment standard table

Star rating	**score （%）**	**DIIR （%）**
NOSCAR	≥95	≤0.8
Five star	≥91	≤1.0
Four star	≥75	≤2.0
Three star	≥61	≤3.0
Two star	≥51	≤4.0
One star	≥40	≤5.0

NOTE: DIIR = the number of industrial accidents × 200000 (base) ÷ annual working hours of employees

3.3 VPP（The Voluntary Protection Programs）

The Voluntary Protection Programs (VPP) is a safety and health management system to prevent workplace accidents and death, by allowing enterprises to implement additional higher than the US Occupational Safety and Health Administration (OSHA) Voluntary standards measures to achieve excellent safety and health management goals. OSHA is through a detailed site assessment and comprehensive site accident rate / illness rate to assess whether the corporate site has a good protection system。

The major elements of the program: management leadership and employee involvement; worksite analysis; hazard prevention and control and safety and health training. To assess something like this, there are three main forms of evaluation methods, including the inspection of documents, on-site inspection and interviews。

Leadership and employee participation	Written safety and health management system, management commitment and leadership, Authority and Line Accountability (authority and responsibility), contractor employees, employees' participation
Workplace analysis	Baseline Hazard Analysis baseline risk analysis, hazard analysis regular activities, regular inspection, hazard reporting systems, accident / incident investigation, trend analysis
Hazard Prevention and Control	Whether there are effective measures to combat control and control, as well as written contingency plans safety and health training

In addition, team members also need to The Total Case Incidence Rate (TCIR)以及The Days Away from Work, Restricted Activity and Job Transfer (DART) Case Incidence Rate evaluated to confirm the company level must be below the national average. After a series of reviews, OSHA decides that the applicant's qualifications are Star, Merit, Demonstration program, if the applicant fails to meet the requirements, return the application, and VPP level is relatively flexible, for those who do not meet the Star designing 1-Year Conditional Star Participants and Star Demonstration Programs.

3.4 Safety standardization assessment

Safety production standardization refers to the establishment of safety production responsibility system, the development of safety management systems and operating procedures, troubleshooting and monitoring of major hazards, the establishment of preventive mechanisms to regulate production behavior, so that the production links in line with relevant safety laws and regulations and standards Standard requirements, people, machines, objects, ring in a good state of production, and continuous improvement, and constantly strengthen the standardization of enterprise safety construction.

Safety standardization evaluation mainly taken independent assessment, external review of the way, self-assessed score of 60 points or more companies can apply for formal standardization rating, the final assessment is divided into one, two, three, one is the highest, from the security Production supervision and management departments to review the classification of supervision and management. During the review process, the following aspects are mainly reviewed: objectives, organization and responsibilities, safety production, laws and regulations and safety management system, education and training, production equipment and facilities, operation safety, hidden trouble investigation and management, major hazard monitoring , Occupational health, emergency rescue, accident learning, performance assessment and continuous improvement of the thirteen major aspects of compliance assessment, the final score to score how much. The enterprises that need to be graded need to carry out self-evaluation on a yearly basis to determine the safety standardization status of the enterprises and to re-order them every three years to urge enterprises to continuously improve the standardization work level.

4 System comparison results

Through the understanding of the information, compared the above systems, found the following the same and different:

1. Purpose

In essence, the above mentioned rating methods are all aimed at checking and enhancing the enterprise to achieve a certain level of security. However, the basic purpose of the rating system is different because of the different backgrounds of the development period. For example, the NOSA five-star rating system was developed in the 20th century, the rapid development of industrial and mining in South Africa and the quality of personnel and other issues caused the high rate of injury accidents, so its basic purpose is to reduce the high incidence of work accidents. In the final analysis, the rating is only a means and security is the purpose.

2. Applicable scope

ISRS is available in 11 languages and has been implemented on thousands of sites worldwide. The industries involved include marine, oil and gas, energy, nuclear, etc., and ISRS is introduced into the oil industry in China, such as China Petroleum West Pipeline Company reached the international security rating of 7.

NOSA five-star rating also can evaluate several kinds of industries including mining, forestry, railways, health care and natural gas, but its vast application in China is applied to the mining industry.

The VPP assessment program is applicable to all companies that use OSHA standards. VPP's participation companies include construction, agriculture, marine, oil and gas, health care and other industries.

Since the implementation of the Safety Production Standardization, under the guidance of the government, Safety Production Standardization has been applied in most enterprises in China, including industry and trade, power industry, transportation, hazardous industries, non-coal mines, construction industry, fireworks, coal industry, Its application covers most of the production enterprises.

3. Evaluators

Safety Production Standardization rating system take the way of "enterprise independent assessment then external assessment " to carry out the rating, Safety Supervision and Management Departments supervises the classification. The external review is conducted by the relevant experts. Similarly, the NOSA five-star rating system is evaluated by the company's team of experts. In the composition of the assessment team, VPP projects and ISRS systems have made innovations.

The VPP site assessment team is composed of experts from all areas of OSHA and SGE (The Special Government Employee), which has performed on-site evaluation. The implementation of the SGE project is intended for enterprises employees to participate in the process of VPP's on-site assessment. The implementation of this innovative project is more conducive to government and corporate information and ideas sharing.

ISRS needs to form an evaluation team consisting of two or more persons prior to evaluation, and to ensure the depth of the HSEQ field in evaluation, the team members contain relevant experts and do not restrict internal / external members, and All evaluation members will receive DNV training courses.

4. Rating method

Whether it is ISRS system, NOSA, VPP or safety standardization rating system are all required to compare the relevant standards and scoring rules with the actual situation of enterprises, through the document inspection, on-site visits and consultation to decide the company's compliance to standards.

However, NOSA and VPP projects need to combine other indicators before the final results. VPP not only requires companies to meet or exceed OSHA standards, but also requires companies to maintain below the average rate of disease released by National Bureau of Labor Statistics. And NOSA combines the evaluation rate of qualified with DIIR ratio to determine the stars should be given.

5. Follow up later

The "PDCA" model has been adequately applied in several rating systems, "plan, implement, check, action", adopt the infinite loop method of " find problems then solve problems, and then find the problem, and then solve the problem" to promote the continuous improvement of enterprise safety management system and continuously improve the level of enterprise security.

But for the rating implementation unit, the follow-up degree is different from others. Among them, the VPP project's follow-up work is very good. VPP participants need to submit an annual self-assessment report, OSHA's site assessment team will confirm whether the enterprise has made a corresponding improvement, and re-examination time interval varies by the enterprise level, such as star business, every 3-5 years conduct a reassessment. For companies with a one-year correction, if they do not meet the requirements after one year, they will withdraw the VPP application.

System comparison table

	ISRS	**NOSA**	**VPP**	**Safety Standardiza tion**
Basic Purpose	get management	Reduce the	prevent	Promote

	system to work more effectively	incidence of industrial accidents	workplace accidents and death	enterprises to achieve standardization and standardization
Applicable scope	ISRS is available in 11 languages. The industries involved include marine, oil and gas, energy, nuclear, etc.,	Including mining, forestry, railways, health care and natural gas, but its vast application in China is applied to the mining industry	all companies that use OSHA standards	Covers most of the production enterprises, such as industry and trade, power industry, transportation, hazardous industries, non-coal mines, construction industry, fireworks, coal industry
Evaluators	Two or more people (DNV training)	Company experts team	OSHA experts in various fields and SGEs who have conducted on-site evaluations	Independent assessment, external review The external review is conducted by the relevant experts
Rating method	Performance evaluation	Performance evaluation and DIIR	companies to meet or exceed OSHA standards, low disease rate	Performance evaluation
Assessment Frequency		Self-evaluation Twice a year Re-rating depend on clients	Self-evaluation once a year re-examination time interval varies by the enterprise level	Self-evaluation once a year Every three years have re-examination
Grades	10	6	Have Merit、Star、Demonstration program etc., grade flexibility is high	3
Voluntary	voluntary	voluntary	voluntary	Forced

5 Conclusion

Combine comparison and thinking above, for the Chinese petrochemical enterprise safety rating system selection, provide the following suggestions:

1. For small private enterprises, Safety Standardization has been fully compatible with security requirements, for other state-owned enterprises, large-scale Sino-foreign joint ventures can choose to introduce foreign security rating system.
2. ISRS system covers the ISO standard and OSHA standard requirements, and also developed a special version of ISRS-Pipe, the rating system also covers the majority of indicators in China oil safety management system, ISRS has been applied in a lot of petrochemical enterprises in China But the ISRS system uses the lowest score of the total score in the first-level elements as the criterion for determining the final grade, so it is not easy to reflect the advantages of the enterprise in the final level judgment.
3. NOSA five-star rating system considers the performance evaluation and DIIR value together in the final evaluation to get the overall level, but when the enterprise finally decided to use the NOSA system, NOSA was originally used for mining assessment should be considered; For the company implementing OSHA standards can apply for the VPP program because VPP determines the enterprise level by checking whether the company is performing a higher-than-OSHA standard.

Acknowledgement

This work is supported by the National Natural Science Foundation of China (21406271) and the China Scholarship Council "Innovative Talents International Cooperation Program" (Liu Jin Mei [2016] 7988). Thank you very much.

Reference

1. https://www.dnvgl.com/services/isrs-for-the-health-of-your-business-2458
2. https://www.nosa.co.za/

478942 The Difficulty and Prospect of the Development of Quantitative Assessment Technology in the Field of Chemical Safety in China

Abstract:

Abstract: With the increasingly high demand for safety of chemical industry quantitative assessment technology has been widely recognized in chemical industry safety field because of its more accurate assessment results. Nowadays many existing quantitative assessment techniques,(such as quantitative risk assessment (QRA), reliability assessment, safety integrity level assessment (SIL) .etc)have been gradually introduced into the filed of China chemical safety to provide more assistance for the risk management of chinese chemical enterprises. However, some basic data of the quantitative assessment on chemical industry in China are lost due to the lack of systematic statistical analysis of data(such as equipment failure data, accident event data .etc), which leads to the situation that our country often adopt foreign related data when conducting a quantitative assessment. In result, results of quantitative assessment are out of their authenticity. In addition, because there is no more specific and more operational assessment guidelines for the part of quantitative assessment techniques issued in our country, it also affects the accuracy and scientific of the results. According to analyzing some misunderstandings and difficulties in chinese chemical safety field when carrying out the quantitative assessment technology, the paper puts forward some suggestions on the change and improvement of related industries to make quantitative assessment technology to provide greater help for the development of chemical safety field in China.

PROCESS
SAFETY MATTERS

Recipe for a *Sizzling* HAZOP

Poster Presentation to 13th GCPS

Angus Keddie, Process Safety Matters

PROCESS
SAFETY MATTERS

Benefits of Running an Effective HAZOP

- Commercial: Meeting schedule; Risk optimisation of facility
- Safety: Pick up all significant potential hazards
- Psychological: Limiting stress on participants
- Chairman: High probability that the chairman will be invited back

PROCESS
SAFETY MATTERS

Recipe Success? – Key Components

- Recipe Disasters......
- Provenance
- Preparation
- Supporting Team
- Kitchen Knives
- Ingredients
- Spice and Flavour
- Oven Conditions
- ....and How to Avoid Them
- Ta Dah!

PROCESS
SAFETY MATTERS

Recipe Disasters

PROCESS SAFETY MATTERS

Recipe Disasters

- Behaviour to avoid if Effective HAZOP to be achieved:
  - Falling asleep
  - Fighting with team members
  - Allowing fighting between team members
  - Silence of participants
  - Permitting interminable discussions on minor issues
  - Scribing yourself
  - Working well in excess of 7 hours per day

PROCESS SAFETY MATTERS

HAZOP Chair Provenance

To paraphrase the Bard "Some or born to HAZOPness, some achieve HAZOPness and some have HAZOPness thrust upon them"

There are around 5000 HAZOP chairmen worldwide all of them being equal, but some are more equal than others.

In order to optimise process safety for existing and new facilities worldwide, excellent PHSERs need to be carried out. A key ingredient of this is the performance of the facilitator. What personal characteristics combine to create this performance?

HAZOP Chair Provenance

- Undergraduate degree in Chemical Engineering
- Professional exposure to relevant aspects of industry:
  - Process Design
  - Operations/Commissioning
  - Team Leadership/Management
  - Process Safety
- Top up with Masters in Process Safety
- Modal age range: 40 to 60
- CPD of IChemE Training Course
- Observation of Other respected HAZOP chairs to evolve best practice
- But like all great recipes, a little variety doesn't do any harm and may enhance the final product.

PROCESS SAFETY MATTERS

HAZOP Chair Characteristics

Desirable attributes for the leader include:

- wide experience of all the stages of process hazard studies, including quantitative risk assessment;
- extensive experience as a HAZOP study member and, preferably, some as a HAZOP study scribe;
- training in the leadership of HAZOP studies;
- technical competence and the ability to quickly understand the system and its operation;
- meticulous attention to relevant detail;
- good analytical thinking;
- motivational skills including the encouragement of creativity and open speaking;
- independence from the project itself with no other direct responsibility to the project manager other than completing the scope of the HAZOP study.

PROCESS SAFETY MATTERS

Fail to Prepare……

- Hold a Pre-meeting with key project stakeholders to establish significant draft tenets:
  - Project Background and Scope
  - Stakeholder Drivers
  - Optimal Methodology
  - Key Hazards
  - Putative Assumptions
  - Principal Personalities
- Outcomes and Terms of Reference inform Intro presentation and initial approach

The Scribe - Sous chef or trusty side kick?

LinkedIn HAZOP Discussion:

- **Does a PHA scribe need to have a technical background? – Ron Brachten, Nov 2014**
- **AK contribution: scribal attributes -**
  - Fast and accurate typing skills
  - Native or excellent English language skills
  - Sufficient *confidence* to capture the consensus and to overlook the inevitable spelling and grammar gremlins
  - *Stamina*
  - Good humour
  - Interest in the process
  - Threshold technical knowledge

  Technical knowledge is handy but….

HAZOP Tools – Sharpen those knives

- Climate controlled room, good natural light
- A1 PIDs on walls or second screen
- Availability of Flip charts or White boards
- Wifi/LAN access to upload project tech info and wider references
- Good food during breaks and lunch
- Effective HAZOP recording software

PROCESS
SAFETY MATTERS

Ingredients

- Meat and 2 Veg:
  - Process Description
  - Noded P&IDs
  - Intro Presentatiion
- Side Dishes
  - PFDs
  - Cause and Effect Diagram
  - Report from any previous HAZOP(s), HAZID
  - HAZOP Terms of Reference
  - Facility Design Standards
  - Plot Plan
  - Line List
  - Piping Classification Overview
  - Toxicity/flammability data for materials being handled
  - Relevant National/International Codes/Laws/Standards

PROCESS
SAFETY MATTERS

The texture and flavour – the local element!

The experience of HAZOP can be vary significantly depending on technical, cultural and stakeholder perspectives:

Stakeholders

- Operators focus on operability and consistency of design. (gooey and sweet)
- The EPC contractor's focus is on schedule and no changes (hard and dry)
- The operator's safety team focus on strict code compliance **(tart and bitter)**

Technical

- Upstream – no steady state; little standardisation
- Downstream – Large through puts/inventories; replication

Cultural

- "Many nations separated by a common language – HAZOP!"

PROCESS SAFETY MATTERS

Oven Conditions – HAZOP Ambience

The team works better if relaxed and motivated – try to make this happen early on:

- Introduce yourself to team members
- Get everyone to introduce themselves round the table, including what contribution they expect to make
- Directly engage all team members in the first 30 mins
- Allocate responsibilities based on member skills and attributes
- If team is flagging, call a break
- It two team members become adversarial, take a time out
- Smile

PROCESS SAFETY MATTERS

Recipe Disasters – and their remedies

- Behaviour to avoid if Effective HAZOP to be achieved:
  - Falling asleep – stand up and walk around when facilitating
  - Fighting with team members – develop excellent rapport with team from the off;
  - Allowing fighting between team members – call time out; verbal warnings; firing culprit(s)
  - Silence of participants – engage in meetings (and in breaks).
  - Permitting interminable discussions on minor issues – set time limits for discussion; "is there anything we need to record at this point?"
  - Self Scribing – refuse!
  - Working well in excess of 7 hours per day – planning; negotiate schedule; identify key risks up front; prioritise nodes (Shell approach – 6 working hours per day, 4 days per week, 4 weeks out of 5)

PROCESS
SAFETY MATTERS

Ta Dah!

Angus Keddie
Director
Process Safety Matters
+44 (0) 7515897869
+44 (0) 1344 455800
info@processsafetymatters.com

477894 Bridging the Divide - OHS and Process Safety

Abstract:

The occupational health and safety field is well established, with a number of professional bodies, undergraduate programs and a comprehensive body of knowledge. Conversely, the process safety field has evolved largely over the past 40 years as an engineering discipline, with a focus on engineering solutions. Both of these fields utilise a risk management basis for decisions and actions. But both fields still largely operate independently of each other. There is much to be learned from sharing knowledge and responsible collaboration to deliver safer plants and operations from an OHS and a process safety perspective. This paper will explore how we can bridge the divide between OHS and process safety to achieve better safety outcomes. This includes better education and information flow between both professions as well as leveraging off existing systems to advance process safety.

20 Years Of Personal Process Safety Lessons

Steve Streblow
ACS Engineering
10914 Milano Ct.
Richmond, TX 77406
streblowsr@yahoo.com

Prepared for Presentation at
American Institute of Chemical Engineers
2017 Spring Meeting and 13th Global Congress on Process Safety
San Antonio, TX
March 26 – 29, 2017

20 Years of Personal Process Safety Lessons

Steve Streblow
ACS Engineering
10914 Milano Ct.
Richmond, TX 77406
streblowsr@yahoo.com

Keywords:

Amine tower; carbon dioxide; communication and restriction; competency; compliance; culture; documentation; failure analysis; intimidation; investigation; lessons learned; material of construction; mechanical integrity; pressure; stress; risk assessment; vessel repair.

Abstract

This paper presents in detail one of the many process safety issues encountered by the author in his 20 years of industrial experience. This issue deals with an amine tower, but touches upon the following topics: carbon dioxide; communication and restriction; competency; compliance; culture; documentation; failure analysis; intimidation; investigation; lessons learned; material of construction; mechanical integrity; pressure; stress, risk assessment and vessel repair.

It describes better ways to address the situation and applicable RAGAGEP. It is hoped that sharing this information will enable others to learn from the past and prevent similar issues in the future.

1 Introduction

This paper presents in detail one of the many process safety issues encountered by the author in his 20 years of industrial experience. This issue deals with an amine tower, but touches upon the following topics: carbon dioxide; communication and restriction; competency; compliance; culture; documentation; drawings; failure analysis; intimidation; investigation; lessons learned; material of construction; mechanical integrity; pressure; stress, risk assessment and vessel repair.

It describes better ways to address the situation and applicable RAGAGEP. It is hoped that sharing this information will enable others to learn from the past and prevent similar issues in the future.

2 Amine Tower Leak

In one of the operating companies I worked for, I was supposed to be the technical support for many gas processing facilities, which included amine systems for removal of carbon dioxide. I was verbally informed of a small leak in one of the facilities amine regenerator towers which was stopped by placing a screw in the side of the tower, as shown in the photo (taken later) immediately below.

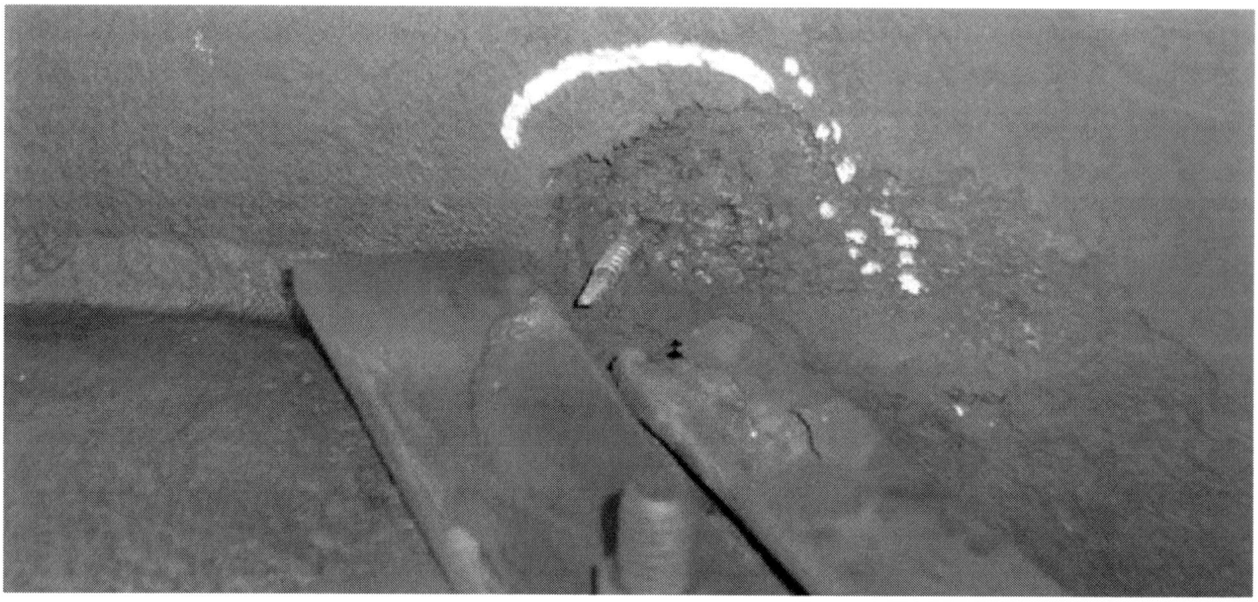

I was shocked for several reasons. First, I was not consulted for my technical opinion on how to proceed. Second, because after I communicated with the plant, they stated they were told by a service provider that they normally use that this type of repair was proper.

So much of this situation seemed wrong to me, so I provided the plant suggestions to restrict access to the area nearby the tower, while I determined how to proceed. The plant did put this area restriction in place.

I started to research what to do. First, I verified the pressure vessel law, as one can do by checking NB-370 at:
https://www.nationalboard.org/PrintAllSynopsis.aspx?Jurisdiction=Select
The location was in the state of Texas, which did not have a pressure vessel law. I next tried to find RAGAGEP for this. I called some pressure vessel manufacturers and had some brief conversations, which helped me focus my search. I found out that to keep the ASME code status, the repair should have been performed through a company which has an R stamp and filled out the appropriate paperwork. The R stamp status of a company can be confirmed at:
http://www.nationalboard.org/ManufacturerDirectory.aspx
The RAGAGEP to be used to maintain ASME code compliance is NBBI NB 23 (NBIC), National Board Inspection Code, Parts 1-3, reference 1.

If the vessel is a non-code vessel or if ASME code status does not need to be maintained, then one can refer to RAGAGEP API 510, Pressure Vessel Inspection Code: In-service Inspection, Rating, Repair, and Alteration, Tenth Edition, May 2014, reference 2.

After seeing that there was much technical knowledge needed for this activity that I did not have, I checked with management to see if there was someone better qualified than I inside the company, and I was disappointed to learn that there was not. Management did not seem concerned about the lack of expertise or the fact that I was being asked to work outside my area of expertise. It would be good to have a dedicated mechanical integrity expert in the company, since there is a great breadth and depth of technical knowledge needed.

I then tried to get the results of the previous mechanical integrity inspections for the tower, which I was told existed, but I was not provided. This withholding of information without any explanation seemed suspicious to me. This is a weakness in the culture. So, I tried to determine what the inspections should be for this item.

I tried to get the paperwork on the tower screw "repair", but none was provided, again showing a cultural problem.

I informed the plant that the company they were using did not have an R stamp and the screw action was not a proper repair.

I suggested contacting an inspection company to perform some external vessel wall thickness inspections, to see if additional thin spots were present. The external inspection was performed, which showed some additional thin areas.

I was verbally informed of a second small leak in the subject amine regenerator tower which was stopped by placing a screw in the side of the tower, as shown in the photo (taken later) immediately below.

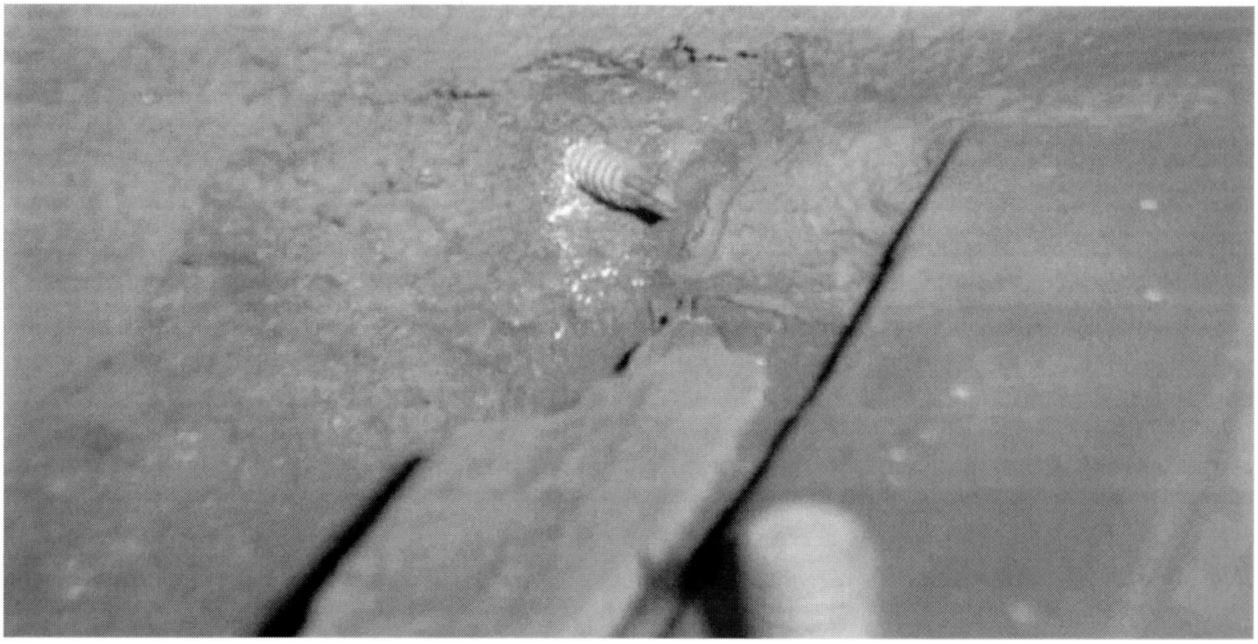

It was disappointing that the facility took the same actions as last time, when they had been advised otherwise. I notified my management, and they did not seem concerned. I again reminded the plant that the company they were using did not have an R stamp and the screw action was not a proper repair.

I was verbally informed of a third small leak in the subject amine regenerator tower which was stopped by placing a screw in the side of the tower, as shown in the photo (taken later) immediately below.

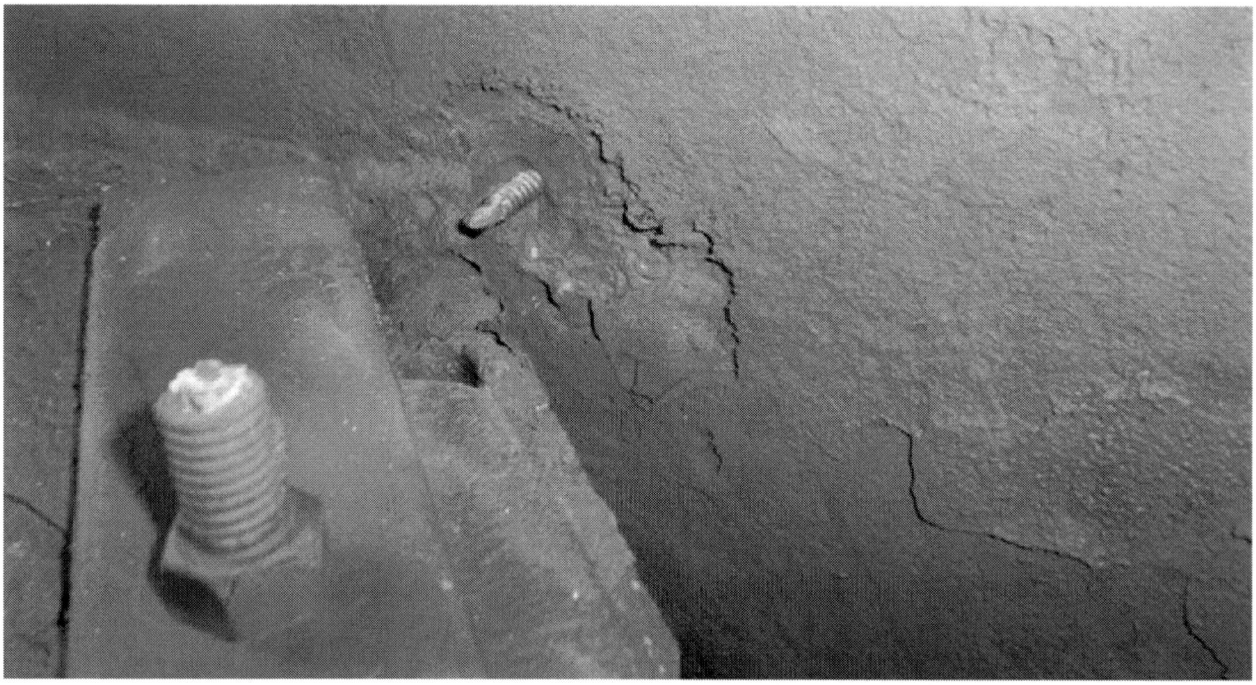

It was disappointing that the facility took the same actions as the last two times, when they had been advised otherwise. I again reminded the plant that the company they were using did not have an R stamp and the screw action was not a proper repair.

This time when I notified my management, they may have over reacted, since they now changed their direction drastically in stating that the right thing to do was shut down the facility, again without trying to get input from me or the plant manager. I was concerned about this drastic change in direction. When I spoke with management about this, they said this was a great success, since the facility was shut down in response to the latest leak. I informed management that I did not view it as a success to take this action in response to the third instance, but would have viewed it more positively if more action was taken on the first issue. A consequence analysis could have been conducted to determine the extent of the consequences. Then the likelihood could have been determined. Combining the consequence and the likelihood results in the risk, which could be used to help make a quantitative risk based determination, rather than a subjective decision.

I then asked management if we could share the learnings from this one site with the other 30+ sites in the company, and they said no, they would not do so until the incident happened at 3 or more facilities. It greatly disturbed me that management would wait for more issues to happen rather than sharing knowledge to prevent future incidents. This shows another cultural weakness.

Since the facility was now shutdown, it presented an opportunity to do some internal inspection, which was done. This effort resulted in many pictures, including the ones in this paper, as well as many wall thickness readings. The wall thickness readings were as thin as 0.110".

The original vessel wall thickness was supposed to be 0.375", with 0.125" corrosion allowance, which shows that the end of life wall thickness should be 0.250", well above the 0.110" measured. The vessel had a 50 psig design pressure, was 78" ID, with an 18 Ft skirt, 64 feet shell height and was made from SA-516-70.

I put together a spreadsheet, based on ASME Boiler and Pressure Vessel Code, Section VIII, Division I, 1980, reference 9, to evaluate what wall thickness was needed to handle different scenarios. Joint efficiencies were taken from reference 9, UW-12. Allowable stresses and overstress for combined loads were from reference 9, UG-23.

Seismic and wind speed were considered by calculating the moment at the bottom head to shell weld, and then converting to metal thickness.

I got the original calculations for the tower, and found that the original seismic calculations were in error. Seismic tension resulted in a required wall thickness of 0.048", which is less than the 0.110" measured wall thickness, so the wall thickness was acceptable. Seismic compression resulted in a required wall thickness of 0.113", which is just above the 0.110" measured, meaning the wall is not thick enough. A useful reference for seismic design is
Guidelines for Seismic Evaluation and Design of Petrochemical Facilities, Second Edition, ASCE, Reston, Virginia, 2011, reference 10.

For the wind loading, the tower was divided into 4 different zones. Wind tension resulted in a required wall thickness of 0.072", which is less than the 0.110" measured wall thickness, so the wall thickness is acceptable. Seismic compression resulted in a required wall thickness of 0.169", which is above the 0.110 measured, meaning the wall is not thick enough. A useful reference for wind loading is ASCE 7-05, Chapter 6, Figure 1A Basic Wind Speed – Western Gulf of Mexico Hurricane Coastline, reference 11. A second useful reference is Wind Loads for Petrochemical Facilities and Other Industrial Facilities, ASCE, Reston, Virginia, 2011, reference 12.

Reference 9, UG-27 was used to determine the minimum thickness for internal pressure and liquid height. The thickness needed for circumferential stress is 0.176", which is above the 0.110 measured, meaning the wall is not thick enough. The thickness needed for longitudinal stress is 0.052", which is less than the 0.110" measured wall thickness, so the wall thickness is acceptable. The thickness needed for compressive stress is 0.085", which is less than the 0.110" measured wall thickness, so the wall thickness is acceptable.

For the combined load temperature, pressure, level, wind and seismic, the thickness needed for longitudinal stress is 0.144", which is above the 0.110 measured, meaning the wall is not thick enough. The thickness needed for compressive stress is 0.306", which is above the 0.110 measured, meaning the wall is not thick enough.

Management tried to direct me to just solve the issue by reducing design pressure until the wall was thick enough. However, this was not a reasonable path forward, for several reasons, such as that if the same corrosion allowance was used, the remaining life (0.110" measurement minus 0.125" corrosion allowance) was 0. Also, there was no mention of the other aspects that would need to be considered, such as increased relief valve sizing at lower pressure. Also, the wall was not thick enough for several other cases listed above, such as seismic and wind, which are independent of the design pressure. Reducing the design pressure from 50 to 15 psig, meant that the wall thickness was still not adequate for the combined loading.

I also did some process simulation to determine if there could be changes made to the process and/or equipment to reduce the corrosion, such as:

- Changing type of amine
- Changing flowrate of amine
- Changing tower diameter
- Changing tower materials of construction
- Changing heat input to the reboiler

I was told to stop on this effort, which meant no better design would be in place when the next tower has problems. Production again won out over process safety when I was told to get a new tower the same as the last one.

A fitness for service assessment could have been conducted, however, production again won out over process safety when I was told to get a new tower the same as the last one.

A risk based mechanical integrity program could have been developed and put into place, but it wasn't. The emphasis was on production over process safety.

A root cause analysis could have been conducted, but the emphasis was on production and appearance over process safety. This is demonstrated by the operations direction to rapidly take the tower offsite to keep the plant appearance attractive. There was not an effort to engage in dialogue to discuss the issue. A better approach is to engage in dialogue with employees and seek their input and comments on suggestions. When management eventually realized the value of the investigation process, the tower was already destroyed. This was unfortunate, because the same type of failures starting appearing at other sites.

I told management that they needed to institute a mechanical integrity group. They eventually did, however, when I was informed of the head of the mechanical integrity group and the recent hire for our area, I could see that they were not qualified. I informed management of this, and I was told that was not my concern. This again shows a poor process safety culture. No attempt was made to resolve my valid concern, just to dismiss it.

I finally put many items in writing to my management, and was provided an individual visit by HR from headquarters. In this closed-door meeting, I was told many unacceptable things, including that I should not write about process safety, but management is glad to discuss orally. This shows a poor process safety culture. I resorted to writing when it was clear that oral was not effective. I did not agree with HR. I was then sent to an individual training session with the head of compliance, which was titled "Write Now, Pay Later." It is clear that the company tried to restrict internal written communication on safety and health, when it is an employee right to write written complaints to external agency, OSHA. This does not show a safety culture.

3 Additional RAGAGEP

One good resource is Guidelines for Mechanical Integrity Systems, AIChE CCPS, 2006, Wiley, reference 3. For pressure vessels, a useful item is "Table 9-1 RAGAGEPs for Pressure Vessels".

Another useful item in reference 3 is "Table 9-13 Mechanical Integrity Activities for Pressure Vessels". It has example activities and typical frequencies for:

- New Equipment Design, Fabrication, and Installation
- Inspection and Testing
- Preventive Maintenance
- Repair

Reference 3 also has Technical Basis for Activity and Frequency, Sources of Acceptance Criteria, Typical Failures of Interest, Personnel Qualifications, Procedure Requirements, Documentation Requirements for the following categories:

- New Equipment Design, Fabrication, and Installation
- Inspection and Testing
- Preventive Maintenance
- Repair

Reference 3, Chapter 11 contains an introduction to a risk based approach for mechanical integrity.

Reference 3, Section "12.4 Equipment Failure and Root Cause Analyses" provides useful information such as:

- "Step 1 - Evaluate Conditions at the Site."
- "Step 2 - Perform Preliminary Component Assessment."
- "Step 3 - Preserve "Fragile" Data Sources."
- "Step 4 - Perform Macroscopic Examination."
- "Step 5 - Perform Microscopic Examination (if warranted)."
- "Step 6 - Perform More Detailed Testing (if warranted)."
- "Step 7 - Test Under Simulated Conditions (if warranted)."
- "Step 8 - Make the Failure Mechanism Determination and Develop Corrective Actions."

Listed below are several additional API documents that contain excellent information related to mechanical integrity, with specific applicable sections listed.

API Recommended Practice 571, Damage Mechanisms Affecting Fixed Equipment in the Refining Industry, First Edition, December 2003, is reference 4. See sections:

- "4.2.14 Erosion / Erosion-Corrosion"
- "4.3.6 CO2 Corrosion"
- "5.1.1.1 Amine Corrosion"
- "5.1.2.2 Amine Stress Corrosion Cracking"

API Recommended Practice 572, Inspection of Pressure Vessels (Towers, Drums, Reactors, Heat Exchangers, and Condensers), Second Edition, February 2001, is reference 5. See sections:

- 7 for reasons for inspection
- 8 for causes of deterioration
- 9 for frequency and time of inspection
- 10 for inspection methods and limitations
- 11 for methods of repair

API Recommended Practice 579, Fitness for Service, First Edition, March 2000 is reference 6. Reference 6 states "The ASME and API design codes and standards for pressurized equipment provide rules for the design, fabrication, inspection and testing of new pressure vessels, piping systems, and storage tanks. These codes do not address the fact that equipment degrades while in-service and that deficiencies due to degradation or from original fabrication may be found during subsequent inspections. Fitness-For-Service (FFS) assessments are quantitative engineering evaluations which are performed to demonstrate the structural integrity of an in-service component containing a flaw or damage. This Recommended Practice provides guidance for conducting FFS assessments using methodologies specifically prepared for equipment in the refining and petrochemical industry. The guidelines provided in this recommended practice can be used to make run-repair-replace decisions to help ensure that pressurized equipment containing flaws which have been identified by inspection can continue to operate safely."

API Recommended Practice 580, Risk-based Inspection, First Edition, May 2002 is reference 7. Reference 7 states "1.1 PURPOSE -The purpose of this document is to provide users with the basic elements for developing and implementing a risk-based inspection (RBI) program. The methodology is presented in a step-by-step manner to the maximum extent practicable."

API Recommended Practice 581, Risk Based Inspection Technology, Second Edition, September 2008 is reference 8. In reference 8, "PART 1- INSPECTION PLANNING USING API RBI TECHNOLOGY", see sections:

- 4 for RBI Concepts
- 5 for Pressure Vessels

In reference 8, "PART 2 - DETERMINATION OF PROBABILITY OF FAILURE IN AN API RBI ASSESSMENT", see section 8 for "Damage Factor – Amine Cracking".

Reference 8, "PART 3 - CONSEQUENCE ANALYSIS IN AN API RBI ASSESSMENT" covers hole size, release rate calculation, release inventory, impact of detection systems, flammable and explosive consequences, toxic consequences, non-flammable non-toxic consequences, component damage and personnel injury consequence areas, financial consequences.

4 Conclusions

This paper presented in detail one of the many process safety issues encountered by the author in his 20 years of industrial experience. This issue deals with an amine tower, but touches upon the following topics: carbon dioxide; communication and restriction; competency; compliance; culture; documentation; failure analysis; intimidation; investigation; lessons learned; material of construction; mechanical integrity; pressure; stress, risk assessment and vessel repair.

It describes better ways to address the situation and applicable RAGAGEP. It is hoped that sharing this information will enable others to learn from the past and prevent similar issues in the future.

5 References

[1] NBBI, NB 23 (NBIC), National Board Inspection Code, Parts 1-3.

[2] API 510, Pressure Vessel Inspection Code: In-service Inspection, Rating, Repair, and Alteration, Tenth Edition, May 2014.

[3] Guidelines for Mechanical Integrity Systems, AIChE CCPS, 2006, Wiley.

[4] API Recommended Practice 571, Damage Mechanisms Affecting Fixed Equipment in the Refining Industry, First Edition, December 2003.

[5] API Recommended Practice 572, Inspection of Pressure Vessels (Towers, Drums, Reactors, Heat Exchangers, and Condensers), Second Edition, February 2001.

[6] API Recommended Practice 579, Fitness for Service, First Edition, March 2000.

[7] API Recommended Practice 580, Risk-based Inspection, First Edition, May 2002.

[8] API Recommended Practice 581, Risk Based Inspection Technology, Second Edition, September 2008.

[9] ASME Boiler and Pressure Vessel Code, Section VIII, Division I, 1980.

[10] Guidelines for Seismic Evaluation and Design of Petrochemical Facilities, Second Edition, ASCE, Reston, Virginia, 2011.

[11] ASCE 7-05, Chapter 6, Figure 1A Basic Wind Speed – Western Gulf of Mexico Hurricane Coastline.

[12] Wind Loads for Petrochemical Facilities and Other Industrial Facilities, ASCE, Reston, Virginia, 2011.

Identifying Opportunities of Enhancing Safeguard Stewardship through IPL Rationalization

SreeRaj R Nair
Technical Safety, Chevron U.S.A Inc.
Houston, USA
snair@chevron.com

Bernadette Thornton
Safety Systems, Chevron
ETC, Houston, USA
bernadette.thornton@chevron.com

[Rev2]

Prepared for Presentation at
American Institute of Chemical Engineers
2017 Spring Meeting and 13th Global Congress on Process Safety
San Antonio, TX
March 26 – 29, 2017

Identifying Opportunities of Enhancing Safeguard Stewardship through IPL Rationalization

SreeRaj R Nair
Technical Safety, Chevron U.S.A Inc.
Houston, USA
snair@chevron.com

Bernadette Thornton
Safety Systems, Chevron
bernadette.thornton@chevron.com

Keywords: Process Hazard Analysis, Layer of Protection Analysis, Independent Protection Layers, Rationalization

Abstract

Ensuring effective safeguard stewardship is an integral part of the journey to address the challenge of moving from 'I think" to "I know" that a major accident hazard is managed. One of the steps in achieving effectiveness is to ensure that the process design is robust and that the safeguards provide the necessary protection to reduce the process risk.

Process Hazard Analysis and Layer of Protection Analysis are carried out to identify hazards, determine the potential consequence and evaluate the adequacy of the safeguards. Opportunities for further risk reduction are identified and risks are eliminated or reduced through established risk management programs. Though the study methodologies have been used for decades, there are opportunities for improvement.

This paper specifically addresses the opportunities in the evaluation of safeguards during the design of major capital projects. It explores some of the key concerns including lack of independence and variations in modifier application that could lead to over or under engineering; this could in turn may result in increased risk exposure. Additionally, this paper explains how the use of a defined methodology for the rationalization of Independent Protection Layers can assist in overcoming these challenges. The significance of performing IPL rationalization at appropriate stage of engineering progress is key in ensuring effective safeguards through design. A comparison of the advantages of rationalization exercise performed at various project phases is also discussed in the paper.

1 Introduction

A critical starting point in process safety management lies with the identification of facilities with potential for high impact (Safety and Environment) events and an understanding of the high consequence scenarios. For the high impact facilities, a detailed Process Hazard Analysis is conducted to determine the scenarios of concern. The potential impact from realizing those scenarios and the likelihood considering the safeguards (prevention and mitigation) is evaluated. A process safety management system is established with a system of "controls" (including critical safeguards verifications, and validation) to prevent the occurrence of these incidents or reduce their severity.

Layers of Protection Analysis (LOPA) was introduced as a detailed frequency evaluation method compared to Hazard and Operability Studies and as a simpler version compared to Quantitative Risk Assessment. This method is used to assess the adequacy of independent protection layers (IPL) and to ensure that required risk reduction is achievable (i.e., there will be no gaps post implementation of the risk reduction plan). LOPA is a systematic approach for applying IPL which follows guided principles and rulesets as prescribed in the industrial standards (AIChE 2001). CCPS publication (AIChE, 2007) provides guidance on choosing and evaluating initiating event frequency and independent layers of protection.

An independent protection layer (IPL) is a device, system, or action that is capable of preventing a scenario from proceeding to its undesired consequence independent of the initiating event or the action of any other layer of protection associated with the scenario (AIChE, 2015).

An effective IPL should be independent of the initiating causes, survive the initial event, and available to perform at the expected level for the intended duration. In order to determine the effectiveness, each safeguard which has been identified as a potential IPL needs to have the entire core attributes; the attributes are independence, specificity, integrity, reliability, auditability, access security, management of change. IPL can be either 'Active' or 'Passive' with Active IPL being those with the ability to detect (sense a condition or problem), decide (whether an action or intervention is necessary or not) and deflect (action capable of preventing consequence in a timely manner) and Passive IPL being those which primarily deflect to reduce the impact.

The following sections review some of the challenges in determining the effectiveness of active IPL and how to overcome the challenges through a structured approach known as IPL rationalization. It is followed by discussion on the inherently safer approach and its significance in improving the performance of the LOPA results. The potential economic benefits of IPL rationalization (Capital and Operational) are also evaluated with relevant project case studies.

2 LOPA Process Concerns

After evaluation of multiple projects, we have determined that the major LOPA concerns tend to fall into three categories: lack of independence, variations in modifier application, and appropriate documentation. The challenges in IPL related documentation can be addressed effectively by implementation of Process Safety Information management plan.

2.1 Lack of Independence

The first issue, lack of independence, relatively an easy one to eliminate, yet it continues to appear in reviews. Most of this is due to a failure to understand packaged equipment controls and a failure to identify the critical actions required by a safety function. Packaged equipment controls typically are not well defined in early Process Hazard Analysis (PHA) / Hazard and Operability (HAZOP) studies. Assumptions concerning the independence of systems and poorly defined functionality tend to lead to IPL credit being given to these systems in error. For example, vibration monitoring systems tend to send requests to other systems, such as the Safety Instrumented System (SIS), to complete critical actions such as the closure of valves or motor stops. If a Safety Instrumented Function (SIF) exists with the same critical actions, then credit cannot be taken for both. The critical actions of each would have to be defined to remove the conflict and the system common cause would need to be reviewed to ensure that the combination will provide the appropriate layers of protection. Failure to validate packaged equipment independence as the project progresses leads to discovery late in the project that the required number of IPL is unavailable.

2.2 Modifier Application

Differences in the team make-up and the facilitation of the LOPA studies leading to differences in the application of enabling events and conditional modifiers and therefore poor application of IPLs. Enabling events and conditional modifiers are applied to change the likelihood of a consequence or severity of a consequence, so these take into account factors such as the likelihood of personnel being in the area and the phase of operation. Failing to apply enabling events and conditional modifiers properly can lead to major operational issues, including alarm floods and excessive instrumentation. For example, if one facilitator applies a conditional modifier to a scenario and another facilitator does not, the required risk reduction is greater for the second facilitator and will require the application of additional IPL. This inconsistency in the application of enabling events and conditional modifiers leads to under-design (with gaps) and over-design (most-often) of instrumented systems and, potentially requiring additional maintenance and testing requirements.

Effective IPL rationalization can reduce the concerns of lack of independence and modifier application by providing an opportunity to look at the application of IPL on a facility basis rather than a scenario by scenario basis.

3 IPL Implementation Challenges

Poor application of IPL in the processes could result in facilities operated at elevated risk levels. The key challenges in the implementation of LOPA findings include the following:

- Risk reduction during design primarily relies on engineered systems, leaning to the route of 'Add-on' safety;
- Lack of credits for non-engineered / soft barriers as additional layers of protection;
- Operational inefficiency due to overly engineered protection system (multiple interlocks, trips for dedicated systems designed without considering overall facility reliability requirements);

- Regulatory requirements – prescriptive versus performance based approach;

Kletz and Amyotte [Ref. 3] have said, "Intensification, substitution, attenuation, and limitation of effects produce inherently safer design because they avoid hazards instead of controlling them by adding protective equipment. The term 'inherently safer' implies that the process is safer because of its very nature and not because equipment has been added to make it safer". However, this very point is often overlooked during design and often adding additional layers (mainly instrumented and hardware) are considered as improvement to safety. Industries' traditional solution to increased levels of hazard has been through more controls. For example, an offshore facility designed to be operated as 'normally unmanned' could become 'normally manned' just because of the manning distribution required to ensure the effectiveness (through inspection and maintenance) of the 'added on' engineered safeguards.

Projects have become more complex and personnel involved in projects rarely see the effect of their designs, so design complexity is less of a consideration and the more cost effective solution, typically an instrumented one, is implemented to ensure enough safeguards are present to manage the risk. Unfortunately, this leads to the massive problem of facilities having to manage large numbers of safeguards and IPL.

While industry is moving towards performance based approaches to define risk reduction, prescriptive regulatory environments still exist. Prescriptive environments ensure a minimal level of risk reduction is implemented, but they also can add challenges to the implementation of a performance based approach. Prescribed testing and testing procedures may provide some reduction in the overall risk of a facility, the tests rarely are sufficient to provide assurance of the effectiveness of the prescribed functions as IPL; therefore, additional testing must be performed on top of the regulatory required testing potentially requiring more personnel to complete ensure the functions meet their risk reduction target.

4 How to Overcome the Challenge

Eliminating hazards using the inherently safer design route is the most effective way to overcome the challenge of safeguard stewardship simply because there will be only fewer safeguards as IPL (Kletz and Amyotte, 2010). Time and effort can be spent on managing the asset integrity of pipe, vessels, and major equipment rather than managing hundreds of safeguards. In such cases, the projects compare options for managing risk. If elimination is not a viable option considering safety and operability factors, decisions are made to proceed with acceptable levels of risk using adequate safeguards/IPL. The following words of wisdom shall be considered as part of ISD approach (Nair 2009, Nair 2017, Stuart, et. al 2003, Kletz and Anyotte, 2010):

- More is not always safer
- Identify all hazards (you can't eliminate or manage a hazard which have not been identified)
- Design against accidental loads
- Design plants for easy and safe maintenance (*reliable equipment which requires less maintenance*)
- Simplify the process and minimizing the requirement to rely on human intervention

While companies acknowledge that management of IPL is necessary, the number of IPL being identified prioritizes finding ways to reduce the load on maintenance and testing personnel. One way to reduce the load is by using IPL rationalization. IPL rationalization can be defined as a process of defining the most effective set of IPL to provide the appropriate risk reduction while also reducing the number of management activities and operator tasks and the maintenance and testing load a facility must endure. So the question of how to define the most effective set of IPL must be answered.

IPL rationalization uses the results of a LOPA to identify IPL which are unnecessary to provide the facility with the risk reduction required. The process requires a functional safety Subject Matter Expert (SME), a process engineer and an operations representative to review the application of enabling events and conditional modifiers, identify SIF with multiple Safety Integrity Level (SIL) and ensure that SIF, critical alarms and basic process control system (BPCS) loops identified as IPL are, in fact, effective. During a LOPA, many of these items may be addressed, but as projects require more and more facilitators to make large HAZOP more manageable, more and more issues creep into the design. The smaller group allows for a more in-depth review of independence, consistency in identification of the SIF and its actions, and a way to compare the results of one study with another study to ensure IPL chosen between similar units and similar scenarios provide the appropriate risk reduction while applying the same enabling events and conditional modifiers. Most importantly, IPL rationalization provides a way for the project to review the IPL from a bird's eye view as well as scenario by scenario to identify whether IPL are appropriate to mitigate the risk for the overall facility as well as the individual scenarios. The most surprising element of the process is that it requires very little time to complete, especially with the correct pre-work completed.

The process of IPL rationalization is simple. First, the team must first identify the order of preference for implementation of IPL. This preference will take into account the facility's maintenance staff size, the competency of the staff and the operating basis of the facility. For example, some owners may prefer to take more SIFs as they have a good management program for them and no management program for other engineering IPL. Other owner may prefer other instrumented IPL over SIF as they do not expect to have the competency they would prefer to manage the IPL.

Secondly, the team reviews the application of enabling events and conditional modifiers to guarantee their effectiveness or applicability. Next, the IPL are reviewed in order of preference and where different IPL have been provided for scenarios with the same consequence, the use of different IPL is either rationalized or the IPL are revised to provide the most effective set. Unused IPL are either removed or removed from the IPL list.

Finally, the team reviews any recommendations to identify if there are LOPA gaps that may be addressed using existing IPL or whether new IPL must be created.

Experience has demonstrated that the process can be effectively used for brownfield facilities as well. However, it need to be noted that there could be differences in the IPL preferences and that unused IPL are less commonly found as facilities choose to use existing IPL rather than add new ones, even if the final result may be less efficient.

5 A Representation of the Benefits

The benefits of completing IPL rationalization depends on three things:

1. The objective of the project
2. The timing of the study
3. The regulatory environment of the facility

The objective of the project may affect IPL rationalization results as looking for ways to eliminate or reduce the numbers of a particular IPL type or may require the addition of IPL in other categories. This activity may not be a reduction of IPL, but rather a reallocation of IPL. Alternatively, the project may be looking to reduce the number of IPL overall by ensuring the application of IPL is streamlined. Illustrations of IPL rationalization benefits in terms of IPL and potential economic saving are given in Figure 1 and Figure 2. A recent project looking to reduce the number of alarms was able to achieve a 90% reduction in alarms with little effect to the number in other IPL categories due to the proper application of conditional modifiers. If one assumes that maintenance of these alarms requires approximately $100 per year and that the number of alarms was reduced from 1000 to 100, then it can be found that the facility saves approximately $90000 per year in alarm maintenance alone (*representative figures only*).

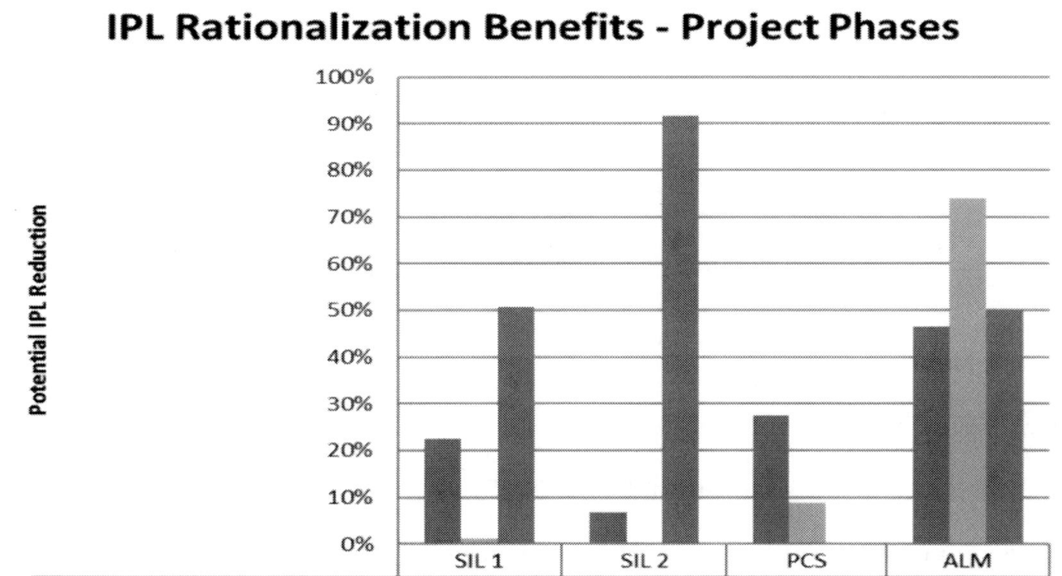

	SIL 1	SIL 2	PCS	ALM
Project 1 - Basic engineering	23%	7%	27%	46%
Project 2 - Detailed design	1%	0%	9%	74%
Project 3 - Post commissioning	51%	92%	0%	50%

Figure 1: IPL Rationalization Benefits – At Different Phases of Project

The timing of the study has a significant effect on the outcome of the IPL rationalization. Performing rationalization before the design is complete (i.e., before construction) provides opportunity to eliminate IPL, such as SIF, that are found to be unnecessary (Figure 1). A project may choose to keep the functions or alarms for operational or reasonable risk reduction justification/demonstration; but take the opportunity to reduce the testing or maintenance associated with these. In addition, the team may be able to identify designs which could be modified to provide a more inherently safer facility. A study performed late in the project will not achieve the same benefits. They may be able to make reductions in some areas, but if the design or installation is almost complete, there is only limited opportunity to remove unnecessary IPL (Figure 1). Alarm priorities may be reduced and maintenance and testing requirements may be relaxed, but the facility will likely be forced to retain most IPL. Performance of IPL rationalization post-commissioning provides even less opportunity for savings. IPL have not only been designed and installed, but they have already been commissioned and, in many cases, changing maintenance and testing requirements will require changes to maintenance managements systems that likely have already been completed, potentially adding cost short-term. A review of projects showed potential IPL reduction of 25-30% for projects in Basic Design / Front End Engineering Design (FEED) and 10% reduction for those in Detailed Design or later.

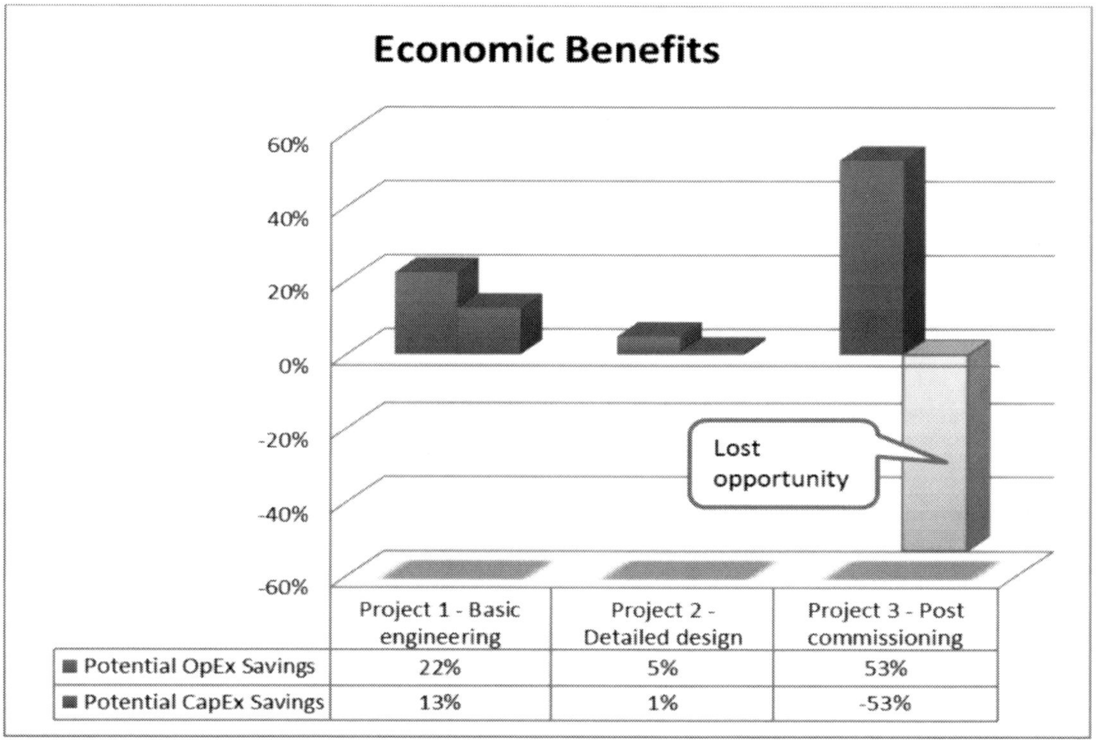

Figure 2: IPL Rationalization and Potential Economic Savings

When taking into account both Capital Expenditure (engineering design, construction, installation) and Operational Expenditure (inspection, testing and maintenance), the savings to a facility can be significant if the appropriate evaluation is carried out in early project stages; even in later stages of the project, the suggested approach can reduce IPL requirement and remove some of the maintenance and testing cost of the so-called "gold plating" that occurs on projects (Figure 2).

Finally, the regulatory environment can have an impact on the study. Where regulators follow a prescriptive regimen for safety instrumented systems (e.g., U.S. Gulf of Mexico), then the benefits of IPL rationalization are realized through the removal of extensive testing and maintenance related to IPL, but not through removal of IPL or through the removal of regulatory required testing. The potential to gain savings still exists though, so the exercise provides benefit by removing the "gold plating" of the facility.

6 Conclusion

Critical safeguards identified, performance requirements established and an evergreen process to ensure safeguards remain in place are key to keep high consequence scenarios from occurring. Through safety studies and by following industrial codes and standards, multiple layers of protection are factored in to prevent or mitigate potential events. However, it is experienced that often the designs and facility become complicated due to 'add-on' safety. Such facilities could experience operational inefficiency, high cost of maintenance and potentially higher risk exposure to personnel.

IPL rationalization should be part of the design engineering process to overcome the challenges and a systematic approach will help in:

- establishing 'effective' safeguard
- optimizing IPLs
- lifecycle benefits

The paper reviewed some of the challenges in determining the effectiveness of IPL and how to overcome the challenges through a IPL rationalization exercise. The paper discussed the inherently safer approach and its significance in improving the performance of the LOPA results. The timing of the exercise has a significant effect on the outcome of the IPL rationalization. The potential economic benefits (Capital and Operational) are also evaluated with relevant project case studies.

7 References

1. CCPS. "Guidelines for Independent Protection Layers and Initiating Events", American Institute of Chemical Engineers, New York, NY, 2007
2. CCPS, "Guidelines for Initiating Events and Independent Protection Layers in Layer of Protection Analysis", American Institute of Chemical Engineers, New York, NY, 2015
3. CCPS, "Layer of Protection Analysis – Simplified Process Risk Assessment", American Institute of Chemical Engineers, New York, NY, 2001
4. Nair SreeRaj, "Determining the criteria for evaluation of toxic hazards, Journal of HSE & Fire Engineering", ASFE, Issue 2 March 2009
5. Nair SreeRaj, "'Inherently Safer' – the Concept and its Application in Oil & Gas Sector," International Conference on Safety and Fire Engineering, SAFE-'17, 2017
6. Stuart Chia, Kieran Walshe and Ed Corpuz, "Application of inherent safety challenge to an offshore platform design for a new gas field development – approaches and experiences," Symposium Series No. 149, IChemE, 2003
7. Trevor Kletz and Paul Amyotte, "Process Plants: A Handbook for Inherently Safer Design," Second Edition, 2010

Process Safety Site Assessment Program (PSSAP) – 5 years of Assessments

Andrew P. Broadbent
American Petroleum Institute (API)
1800 West Loop South, Suite 475
Houston, TX 77027
broadbenta@api.org

Prepared for Presentation at
American Institute of Chemical Engineers
2017 Spring Meeting and 13th Global Congress on Process Safety
San Antonio, TX
March 26 – 29, 2017

Process Safety Site Assessment Program (PSSAP) – 5 years of Assessments

Andrew P. Broadbent
American Petroleum Institute (API)
1800 West Loop South, Suite 475
Houston, TX 77027
broadbenta@api.org

Keywords: Process safety, assessments, benchmarking, protocols, scoring, rating, weighting

Abstract

The API Process Safety Site Assessment Program was developed in 2011 as part of the AFPM and API Advancing Process Safety Programs. PSSAP began conducting assessments in 2012. By the end of 2016, the API Site Assessment program will have conducted 52 General (7 protocols) and 27 HF Alkylation/RP 751 assessments. These assessments have been conducted 59 different refineries and petrochemical facilities. The areas assessed are: Process Safety Leadership, Management of Change, Mechanical Integrity, Safe Work Practices, Operating Practices, Facility Siting, Process Hazards Analysis, and HF Alkylation/RP 751.

This presentation will provide a brief overview of the program and present some the learnings, trends and benchmarking developed from the assessments based on the five-year history of the program.

In 2010, the American Petroleum Institute (API), American Fuel & Petrochemicals Manufacturers (AFPM) and industry joined together to address improving process safety performance with the goal of preventing process safety events at refineries and petrochemical plants. API, AFPM and the industry wanted this effort, known as Advancing Process Safety, to be:

- Focused on higher risk process safety activities;
- Refining & petrochemical focused;
- Voluntary; and
- An avenue to communicate and share experiences and knowledge.

Advancing Process Safety (APS) consists of six different programs which are:

- Process Safety Metrics and Analysis (RP 754)
- Process Safety Hazard Identification
- Event Sharing
- Focused Improvement
- Regional Meetings
- Process Safety Site Assessments Program (PSSAP).

The intent of all the programs is to raise the level of process safety performance at the sites to prevent process safety events.

The management of these programs is split between API and AFPM. API manages the Process Safety Site Assessment Program, Process Safety Metrics and Analysis, and the Focused Improvements effort. AFPM manages Process Safety Hazard Identification, Event Sharing and Regional Networks.

Benefits/Opportunities

The Process Safety Site Assessment Program is perhaps the most hands-on of the six programs since it involves a team of assessors being onsite at a refinery or petrochemical plant for an entire week. The program uses credible, third-party assessors who have specific expertise in one or more of the eight areas being assessed. The team utilizes a set of protocols that are applicable to both refineries and chemical plants. These protocols were developed through a collaborative effort of subject matter experts from a wide range of companies.

Through these site assessments, the program provides the sites with feedback regarding what they are doing well and what improvements they can make to their process safety programs. Additionally, the program provides the sites with benchmarking data so the sites can compare how they are doing versus their industry peers.

The assessments are not process safety management compliance audits, and so we refer to them as assessments and not audits. We speak to process safety rather than process safety management so people will understand that these assessments are not the three-year process safety management (i.e., 29 CFR 1910.119) compliance audits, rather they are assessments of how sites are performing relative to 'what good looks like" based on the experience of industry and the expert assessors. As you will see, the assessments only cover seven areas of process safety (compared with the 14 areas in OSHA 1910.119) as well as HF alkylation/API RP 751.

Protocols

The eight protocols address:

- Process Safety Leadership
- Operating Practices
    - Procedures
    - Shift Handover
    - Operator Rounds
    - Alarm Management, and
    - Workplace Conditions
- Mechanical Integrity – Fixed Equipment
- Safe Work Practices
    - Procedures
    - Work Permit Procedures
    - Utilities, etc.
- Management Of Change
- Process Hazards Analysis
- Facility Siting
    - Permanent Buildings
    - Portable Buildings, and
    - Tents
- HF Alkylation/API Recommended Practice (RP) 751

The first seven elements are areas that could prevent a process safety event from occurring or reduce the severity of an event (i.e., Facility Siting). The eighth area – HF alkylation – deals with process safety in this specialized operation. Assessments involving the first 7 protocols are referred to as "General" Assessments and the HF Alkylation/RP 751 ones are known as "HF In-Depth RP 751" Assessments.

Note: Since the HF Alkylation protocol specifically follows API RP 751, the HF-In-Depth RP 751 Assessment does meet the RP 751 requirement to conduct an audit once every three years.

Sites may also choose to do just some of the protocols (for example, MOC and Facility Siting) rather than all seven at a time. These are known as "Partial" assessments.

These protocols were developed by industry for industry utilizing 54 subject matter experts from a wide range of companies. The protocols were developed, balloted using the API standards development consensus process, and then piloted at three sites. The protocols were then reviewed and modified based on learnings from the pilot assessments, including input from the assessors.

Keys to the Program

There are three very important components the PSSAP program, which are:

- Benchmarking
- Mentorship
- Expertise of the Assessors

Benchmarking

Refineries and petrochemical plants always like to see how they compare to their peers in the industry. In the PSSAP program this is done by providing a score for each protocol and each protocol question.

The 7 General Assessment protocols have a total of 577 questions and the HF In-Depth protocol has 394 questions.

When developed, each protocol question was given a weighting indicating the question's importance to preventing a process safety event from occurring. The question's weighting is a fixed number between 1 and 4 in half increments (i.e., 0, 0.5, 1, 1.5, etc.) with 4 indicating the most importance. During the assessment the assessor will rate each question based on the information and evidence the site provides. This is done on a scale of 0 to 4, also in half increments. The score is a simple calculation of the weighting times the rating.

Keeping the scoring objective is done through ensuring the protocols are well written so that different assessors will understand and interpret the question the same way. Consistency is also addressed through the training provided and calibration sessions with the assessors at the start of the assessments. Additionally, the PSSAP Program Manager provides oversight so that a consistent approach is used by the assessors.

Mentorship

One of the most important pieces of the program is the information exchange the sites get with the assessors. As mentioned, the assessors have considerable experience and expertise based on their years of service in the industry and through performing PSSAP assessments.

We inform the sites (we typically visit the site six months prior to the assessment to set up all the logistics), to get the most value out of the assessment, that their employees should be interviewing the assessors as much as the assessor is interviewing the employee.

As an example, on one assessment the assessor sat down with three site employees for an hour and discussed various ways to manage car seals.

We encourage sites to take advantage of the time with the assessors throughout the week, including the daily debriefs, lunch time discussions, and the closing conference of Friday.

Assessors

As mentioned, the assessors are major piece of the PSSAP program. They are handpicked based on their expertise and industry background/knowledge. Typically they are retired from industry companies, though some work for consulting firms (or have their own firm) but worked in the industry at some point in time.

An assessor candidate with appropriate qualifications and references is selected to attend a 1-1/2 day training course which explains the API Process Safety Site Assessment Program to them. The candidate must undergo an interview with a three-person panel – the PSSAP Manager and two industry representatives before being selected to be an assessor. Not all candidates that attend the training get selected. Also, assessors are picked for specific protocols (i.e., we do not look for Mechanical Integrity assessors to be Operating Practices experts and vice versa). Currently, over 70 people have inquired about being an assessor and we have around 20 in our assessor pool.

To date, a typical assessment team is averaging 39 years of industry experience. Typically the team has 7 assessors.

An in-depth HF alkylation unit assessment follows the API RP 751 Recommended Practice in detail. This assessment can be done at the same time as a "basic" assessment or completely separate and consists of 5 assessors. As noted, an In-Depth RP 751 assessment can be used by the site as their 3-year RP 751 audit.

Assessments

An assessment gets started when the site contacts the PSSAP Program Manager to schedule a week for the assessment. It is important that the site assessment date, be selected well in advance – at least 6 months – so appropriate time is available for planning, including scheduling the assessors.

A pre-assessment visit is then scheduled and conducted at the site by the PSSAP Program Manager. This meeting serves two purposes. First, it allows the site management team to learn more about the program and ask any questions. The second purpose is to meet with the site contact (the person who will work with the PSSAP Program Manager to ensure all arrangements are made prior to the start of the assessment) to go over the logistical needs for the assessment - including conference rooms, scheduling interviews, documents needed prior to the assessment, etc.

Conducting the Assessment

On the week of the assessment the team arrives on Sunday afternoon/evening ready to go. The assessors will have already spent about 8 hours reviewing site documents, provided by the site through an FTP file share site, so they will be well versed on the site's programs relative to the protocols they are assessing.

Monday begins with an opening conference to review the ground rules and usually a quick tour of the site. After the tour, each assessor interviews key people to learn more about the site's management systems, programs and procedures. These interviews have been pre-scheduled by the site using a template provided to them by the PSSAP Program Manager.

The assessors also work to spend as much time in the field as possible. The intent is to not only review whether the site has good management systems, but to provide them with insight on how well their systems are being implemented in the field.

On Tuesday, Wednesday, and Thursday, the assessment team meets with site staff, spends time in the field, and holds a debrief with the site personnel, typically around 4:00 pm. These daily debriefs have been extremely beneficial as they are excellent opportunities for site personnel to continue to interact with the assessors. Often when the debrief has been completed, there are follow-up conversations with the assessors so site staff can gain more insights.

Thursday night and Friday morning is when the assessors compile all the information and the protocol ratings. The assessors enter observations and ratings for each line item in each protocol, thus the facility is provided with a considerable amount of information.

On Friday, an hour to an hour and a half closing conference is held with the site personnel, including the site management team. The site receives a PowerPoint overview, the completed protocol spreadsheets, and some preliminary benchmarking data. The draft report is provided to the site the following week. Distribution of the reports is as designated by the site.

When the assessment team leaves the site on Friday, the assessors return all hard copy documents and delete any electronic documents that were provided to them by the site.

Handling Report & Data

The site will typically be given two to three weeks to comment on the draft final report. This review period is not a discussion of the scores, but whether the wording of the Observations and Good Practices is clear. A final report is then issued by the program manager to the site.

Once the final report is issued, the data and information including the good practices and observations, with the permission of the site, are blinded and entered into the database for trending and benchmarking purposes. API does not ask for corrective action plans or updates.

Program Status, Experience & Keys

As of year-end 2016, we have conducted 53 General site assessments, 27 HF alkylation/API RP 751 in-depth assessments and 5 Partial assessments. The program has been to 28 different companies at 60 different locations (note: several companies have had a site assessment at more than one site.)

Item	Total	Average per Assessment
Protocol Questions	> 41,200	577 Basic 394 HF
Interviews	> 8,000	>100
Observations	> 2,800	~ 35
Good Practices	> 2,400	~ 30

Below are a couple of comments received from sites that have been assessed:

- "You cannot fool these Assessors — they have

done *your* job for 30 years"

- *"In one week the assessment team found issues that our internal teams could not find in 3+ weeks, and it was stuff we needed to know and address. Well focused, deep dive process."*

Benchmarking Data

The benchmarking data is provided in several formats. For example, the data is analyzed and presented by:

- Protocol
- Sub-protocol (some protocols have multiple areas of review – Shift Handover, Operator Rounds, etc.)
- Specific protocol sections (groupings of questions for specific areas)
- Management Systems vs. Implementation (having the procedure vs. how is it being done in the field)
- Size of site
- Size of company
- Quartile

Benchmarking data updates are provided generally every six months to sites that have had an assessment.

Data Review – Good Practices and Observations

Below are some examples of the types of Good Practices and Observations that the assessors have been identifying through the site assessments. If you note the last bullet on both slides this is an example of exactly what we are after. At some sites it's an item that needs to be addressed while at other sites we have seen them doing it well.

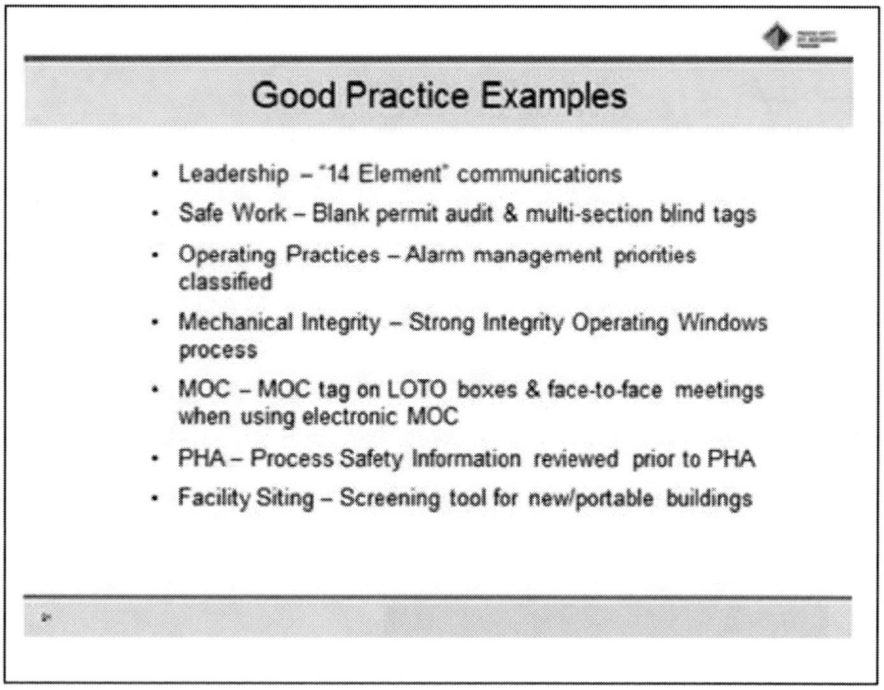

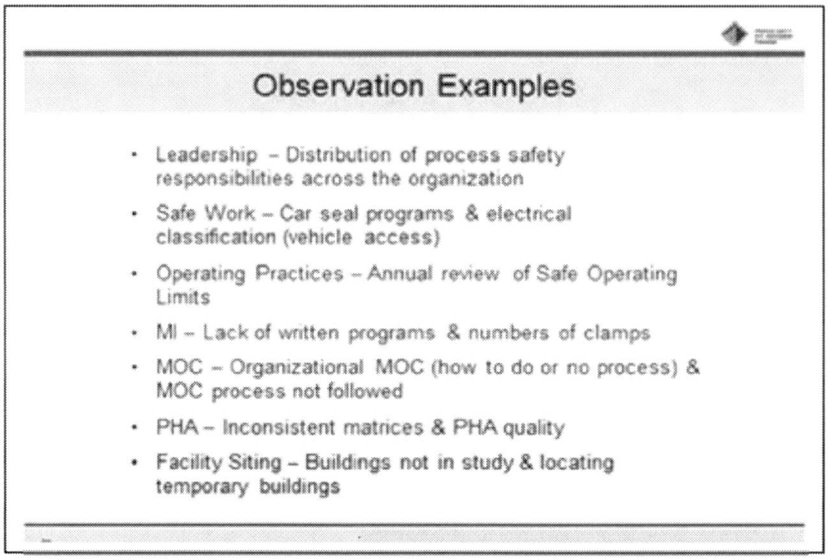

This points to a key element of the site assessment program - sharing the information from sites doing it well with others that are working to put systems/programs/procedures in place at their site. A variety of methods are used for sharing the practices learned through the site assessment program including:

- The Advancing Process Safety Program
    - Regional networks
    - Hazard Identification program
- Industry Conferences and Seminars
- API standards

The data, information, and learnings from the site assessments have been used to focus resources to develop informational and guidance documents regarding areas identified as improvement opportunities. For example, an informational document for mechanical integrity regarding small bore piping was developed based on observations during PSSAP site assessments and data from the 754 metrics and Event Sharing database..

Conclusion

In conclusion, the “boots on the ground” performance of site assessments began April of 2012. Feedback from these assessments has told us that the process of preparing for an assessment (going through the protocols prior to the assessment) and the mentorship/interaction with the assessors has been very beneficial to the sites. And now that we have done enough sites to allow the sharing of benchmarking data the sites are getting even more value from the program.

Alarm Management Meets SIS

SIS Alarm Requirements

How many alarms must be configured per the SIS Standard (ISA-84 Part 1) for an SIF?
Zero

How many alarms must be configured per the AM Standard (ISA-18.2) for an SIF?
Zero

How many alarms are necessary for an SIF to correctly perform its function?
Zero

Pre-alarm – What makes sense?

Pre-alarm before an SIS impending trip?
True – This gives the operator an opportunity to prevent the trip, keeping the plant running and reducing the demand frequency on the SIF.

Pre-alarm should always be on the BPCS control or indicator tag rather than the SIS transmitters?
False – Choice of where to put the pre-alarm is up to the facility, and it's a good idea to document this in the Alarm Philosophy.

Each transmitter in a voting arrangement should alarm individually?
False – Are we trying to keep our plants operating safely or do we just throw multiple alarms at the operator?

SIS and BPCS transmitter disagreements among redundant SIS transmitters should be alarmed?
True – With this alarm in place, there is no need for a pre-alarm on more than one transmitter.

Introduction

Most plants focus on meeting the alarm management requirements laid out in ISA 18.2. Even if the proper approach is taken in configuring the BPCS, there is one area that is sorely neglected, the SIS system. SIS is addressed based on the requirements laid out in ISA 84. Unfortunately, this standard does not address alarming related to SIS events. ProSys contends that a successfully implemented alarm management philosophy is only effective when taken as a holistic view of the BPCS and SIS systems.

Unified Alarming Scheme

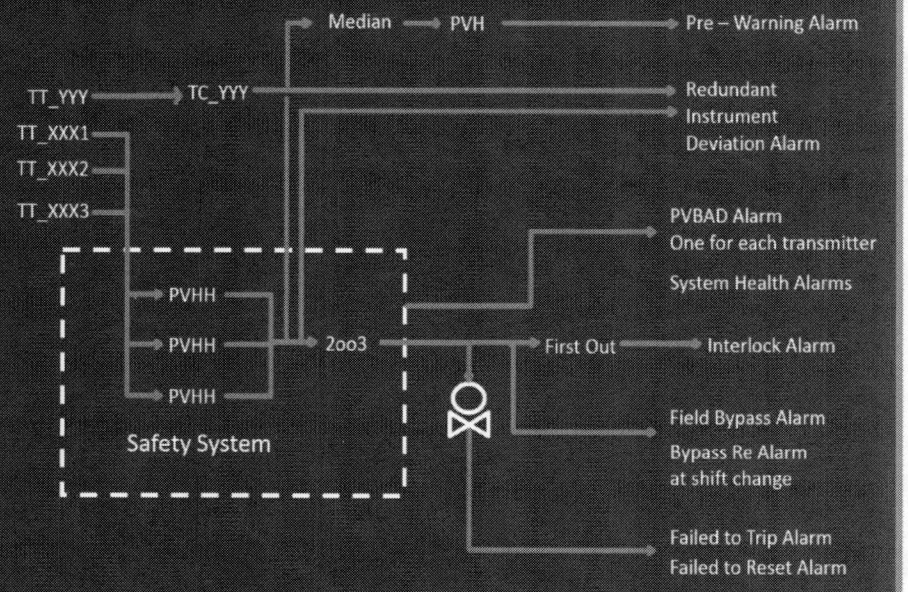

Trip – What makes sense?

First out alarm indicating the reason for a trip?
True – There is no better way to announce to the operator that a trip has been initiated.

Alarm SIS valve moving to trip position?
False – Going to the trip position is the expected response to an SIS trip – better, provide a failed to trip alarm, which makes noise if the valve doesn't work as expected.

Bypass – What makes sense?

Alarm board operator initiated SIS bypass?
False – Although a common practice, this alarm adds little or nothing to the integrity of the system: "Hey Joe/Mary/Sam, you just pressed a button!"

Alarm field operator initiated SIS bypass?
True – This may come as a surprise to the console operator, so it's not a bad thing to annunciate.

If a bypass is in place, trip point alarm on the SIS transmitter?
False – Why give the operator an alarm that isn't reliable or won't be believed? Find an alternative monitoring plan.

Re-alarm on SIS Bypass?
Neutral – Bypass and shift changeover procedures exist, but do we follow them 100%? This is a facility decision. If used, re-alarm shortly after shift change.

PROSYS® Maximize Your Plant™

www.ProSys.com

ENHANCING PROCESS SAFETY MANAGEMENT INTEGRATION TO IMPROVE PROCESS SAFETY AND PROFITABILITY IN THE FUELS AND PETROCHEMICALS INDUSTRY

Thomas D. Watson, P.E.
TDWatson Consulting LLC
Pittsburgh, PA USA
twatson@tdwatson.com

Ronald D. Fuchs, P.E.
RD Fuchs Engineering LLC
Lake Jackson, TX USA
rdfuchs13@aol.com

Harold Phillippi, P.E.
Independent Process Safety Consultant
Charleston, SC USA
butch.phillippi@gmail.com

Keywords: PSM, Process Safety Management, Relief Systems, Incident Investigation, Mechanical Integrity.

Abstract

Process Safety Management (PSM) in the Fuels and Petrochemicals industry has the paradox of being well established and yet constantly challenged to improve its effectiveness and impact on process safety and profitability. Such a paradox is a natural consequence of an industry exhibiting a positive process safety culture that maintains a sense of vulnerability and sees the imperative to improve process safety. This paper addresses the frequently asked question, "How do we continue to improve our organization's process safety culture?" and presents opportunities and examples for organizations to enhance their existing PSM systems and culture by further integrating PSM elements both within an organization's established PSM systems and its other systems, processes, and procedures. The paper illustrates integration and enhancement opportunities using Relief System Design Basis (RSDB) documentation, a component of the PSM element - Process Safety Information (PSI). Illustrations show RSDB integration with the PSM elements of Mechanical Integrity (MI), Process Hazard Analysis (PHA), and Incident Investigation (II). The paper then illustrates opportunities to increase profitability by highlighting RSDB integration with partial downtime throughput opportunities and improved project definition. The illustrations challenge organizations that truly "own" a process safety culture to look for additional opportunities to further enhance and integrate established PSM systems into their organization to further improve safety and profitability.

1. Introduction

The Fuels and Petrochemicals Industry in partnership with governing regulatory agencies have established effective Process Safety Management (PSM) systems to handle hazardous materials and energies over the past several decades. Examples include the U.S. Occupational Health and Safety Administration (OSHA) PSM Standard and the European Commission Seveso II Directive. The Center for Chemical Process Safety (CCPS) offers an extensive reference list of regulatory/industry/system requirements.[1] The catalyst for these industry systems have often been the tragic circumstances around industry incidents and catastrophes. Today, PSM in the Fuels and Petrochemicals Industry has the paradox of being well established and yet constantly challenged to improve its effectiveness and impact on process safety and profitability. Such a paradox is a natural consequence of an industry exhibiting a positive process safety culture that maintains a sense of vulnerability and sees the imperative to improve process safety.[2] The vulnerability and sense of urgency are well founded as the industry continues to experience preventable incidents. The Marsh Report states the 100 largest property damage losses in the hydrocarbon industry from 1974 to 2013 had a combined property damage cost of approximately US$34 billion.[3] Organizations demonstrating a committed process safety culture continually challenge and address the question, "How do we improve our organization's process safety culture?"

Organizations that have successfully managed the PSM implementation and sustainment phases have already experienced and moved past the common misconception that process safety and profitability have conflicting objectives. These organizations further realize they must move beyond a "separate but equal" PSM system established among their current business management systems, and fully integrate PSM into their systems to further improve their organization's impact on process safety and profitability. Recent efforts in the Fuels and Petrochemicals Industry to improve process safety through management system integration include industry efforts such as:

- The American Fuel and Petrochemical Manufacturers (AFPM) "Advancing Process Safety Programs," which promotes sharing PSM learnings across member organizations of AFPM
- The CCPS publication containing guidelines for integrating management systems and metrics to improve process safety performance.[4]

This paper identifies and presents specific examples and opportunities for organizations to continue to enhance their existing PSM systems and process safety culture by further integrating PSM elements both within an organization's established PSM systems and its other business systems, processes, and procedures. The paper illustrates integration and enhancement opportunities using Relief System Design Basis (RSDB) documentation, a component of the PSM Process Safety Information (PSI) element. Specifically, the examples illustrate RSDB integration with the PSM elements of:

- Mechanical Integrity (MI)
- Process Hazard Analysis (PHA)
- Incident Investigation (II)

RSDB integration with other business processes also present opportunities to directly improve profitability. Two examples presented here include RSDB integration with:

- Production planning (partial downtime throughput opportunities)
- Improved project definition

The following sections provide a general introduction to RSDB and illustrate the five integration opportunities in more detail.

2. An Introduction to RSDB and its Relationship to PSM

Within the Fuels and Petrochemicals Industry, many processes involve highly hazardous chemicals. To adequately assess and address a process' hazards an organization must adequately define the Process Safety Information of the process. Process Safety Information (PSI), includes information regarding the process' hazards, technology, and equipment.[5] Examples of hazard information include toxicity and permissible exposure limits. Examples of technology information include a process' chemistry and operating limits. Finally, examples of equipment information include materials of construction and the Relief Systems Design Basis (RSDB). There are many references for what constitutes RSDB and how to effectively manage RSDB documentation.[6] The American Petroleum Institute states RSDB documentation should demonstrate that all pressure-containing equipment has been identified and the relief requirements adequately analyzed. Documentation should include specific engineering guidance, methodology, analysis, and reference information that collectively reflects Recognized and Generally Accepted Good Engineering Practices (RAGAGEP). [7]

RSDB is a key component of PSI, but RSDB is also an important component in maintaining other Process Safety Management (PSM) elements such as Mechanical Integrity (MI), Process Hazard Analysis (PHA) and an Incident Investigation (II). Furthermore, RSDB can be a significant aspect of process production constraints and project opportunities associated with processes. The next sections discuss integrating RSDB into the aforementioned PSM elements, production planning, and project development processes.

3. RSDB Integration with Mechanical Integrity (MI)

Integrating Relief System Design Basis (RSDB) analysis with Mechanical Integrity's (MI's) Risk Based Inspection (RBI) through a combined risk assessment can identify acute process equipment vulnerabilities. Mechanical Integrity (MI) activities ensure the proper design, installation, operation, and maintenance of process equipment. Risk Based Inspection (RBI) is one component of an effective MI program that links risk assessment and mitigation activities with appropriate inspection and process control activities. RSDB analysis, on the other hand, identifies the applicable relieving scenarios and required relieving rates affecting pressure-containing process equipment to ensure the relieving scenarios do not expose the equipment to operating conditions outside of design limits. The American Petroleum Institute's Recommended Practice (RP) 581 identifies the interdependency of the Pressure Relief Device (PRD) with the fixed equipment the PRD is protecting.[8] RP 581 evaluates PRD criticality to set mechanical integrity activities such as inspection frequency. The RBI methodology combines the PRD's anticipated demand rate with the probability of failure on demand of the PRD to determine an overall risk for the PRD and the process equipment the PRD protects. While the API RP 581 methodology is sound, when the RSDB documentation is not available or the RSDB analysis has identified deficiencies in the PRD design, the RBI methodology develops shortcomings that should be addressed.

As an example, consider a process vessel containing a flammable hydrocarbon liquid. An RBI of the vessel defines a risk associated with the vessel, the vessel's PRD and the anticipated overpressure scenarios. The RBI analysis is predicated in part, however, on the PRD adequately protecting the vessel in the anticipated overpressure scenarios should the PRD open. The RSDB documentation identifies fire as the only applicable overpressure scenario but concludes the PRD is undersized. The RSDB can provide additional insight into the level of overpressure that would occur (and the anticipated consequence). The additional insights would significantly impact the RBI risk assessment and potentially warrant a reassessment of inspection and mitigation actions for the PRD and the protected vessel.

The example illustrates how integrating RSDB with MI's RBI can identify acute equipment vulnerabilities and aid in further prioritizing MI efforts and RSDB deficiency resolution, including risk mitigation efforts. Consequently, integrating both RSDB and MI can further enhance PSM integration and improve process safety and profitability.

4. RSDB Integration with Process Hazard Analysis (PHA)

Integrating RSDB with PHA can reduce costs and improve the effectiveness of both the PHA and the RSDB. As most readers are aware, a Process Hazard Analysis (PHA) in the United States and many other countries, is a requirement of a Process Safety Management (PSM) program, (for example, 29 CFR 1910.119 requires PHAs and periodic revalidations). No matter which methodology a PHA employs [Hazard and Operability study (HAZOP), "what-if," Failure Modes and Effects Analysis (FMEA), etc.], capable teams must perform the analysis. The teams typically range from 4 to 8 (or more) persons, consisting of owner's personnel and often outside consultants/specialists. No matter who is present, the participants represent a necessary and significant staff resourcing cost. Furthermore, each PHA finding should be well documented, especially those requiring additional evaluation and stewardship before the findings are declared either a confirmed issue or not. The PHA finding documentation and corresponding follow-up/stewardship also represent a significant staff resourcing cost. Finally, actions required to address the confirmed issues, whether facilities changes, work process changes, or administrative changes represent significant costs.

With the PHA costs sufficiently identified, now consider how integrating RSDB with PHA reduces costs and improves PHA and RSDB effectiveness. PHAs always identify questions and concerns about the adequacy of the over (or under) pressure protection systems as well as items such as Maximum Allowable Working Pressure (MAWP), hydrotest pressure, materials of construction, etc. Furthermore, PHAs require the most recent drawings, heat and material balances, and other Process Safety Information (PSI). Similarly, the most recent RSDB should have many (if not all) of these items assembled and documented, with any changes since the most recent RSDB tracked and provided as a part of the facility's Management of Change (MOC) program. The overlap of PSI and resources between RSDB and PHA is substantial. Consequently, completing RSDB analysis prior to and integrating it with the PHA process and follow-up activities introduces cost savings. Common savings occur relating to over (or under) pressure findings. If properly completed and documented, the RSDB can provide the PHA participants with immediate feedback regarding the adequacy of pressure protection devices, which in most cases are code items and must be addressed. A benefit of having a quality RSDB done prior to the PHA is that the over or under pressure concerns can be quickly resolved. On many occasions, no action item follow-up work is necessary. This reduces costs both during the PHA and the PHA follow-up. Conversely, the PHA serves as an audit check of the RSDB as a capable team reviews the RSDB basis as needed.

As a quick example, a PHA team questioned the adequacy of an exchanger shell's design pressure during a blocked-in condition. The maximum pressure generated from a blocked-in condition from a manual valve closure was greater than the design pressure by approximately 10 percent. The PHA team referenced the RSDB. The RSDB previously identified the same issue and investigated the situation in more detail. The RSDB identified the exchanger's MAWP to be approximately 20 percent more than the noted design pressure, and the hydrotest pressure was based on 1.5 times the MAWP. The U-1 form that was part of the RSDB package clearly indicated this. The RSDB package also noted the exchanger shell was adequate based on ASME VIII, Appendix M, M-5.7, in this situation, because the pressure was below 116% of MAWP and it would be immediately noted by operators, the valve closure was not identified as an issue and the PHA follow-up item was closed. Because the PHA utilized the RSDB, the PHA saved costs on discussion and documentation of an action item and the necessary follow-up and tracking of a code related item. Conversely, the PHA team validated the RSDB, enhancing PSM integration and improving process safety and profitability.

5. RSDB Integration with Incident Investigation (II)

Incident Investigation (II) is also a mainstay of any effective personnel or Process Safety Management (PSM) program. II is an opportunity to learn from mistakes, system failures, equipment failures, and/or overlooked vulnerabilities. II is also an opportunity to evaluate the effectiveness of preventive and control measures installed to prevent or mitigate an incident. In the CCPS Vision 20/20 Characteristics of Companies with Great Process Safety Performance the authors write: "To reduce incidents, employers and employees must enthusiastically support a culture that is driven to learn from many sources including benchmarking, near misses and incidents, and jobs done well."[9] For incidents involving release from a relief device and/or a flaring event, II intersects Relief System Design Basis (RSDB) and together they provide a learning opportunity that enhances PSM and improves process safety.

Yogi Berra said; "When you come to a fork in the road, take it." When an incident occurs investigate it. Definitions may vary for what constitutes an incident. Indeed, there are legal requirements for investigation of safety and environmental incidents. These form the baseline for what organizations must do but do not limit organizations from doing more. For instance, some organizations with very effective PSM programs investigate all flaring events. The organization calling the event an incident, a near miss, or flaring event is least important, while the organization's learning opportunity from the event is most important.

By integrating RSDB with II organizations can do real case study evaluation of the RSDB's assumptions and technical predictions. Integration can help answer such questions as:

- Did the scenarios of the RSDB represent the incident that occurred?
- Did the relief valve effectively handle a release rate lower than the controlling case without issues such as chattering?
- Were the release rate, pressure limitation, and system temperature change what the calculations predicted they would be?
- Did the flare handle the relieving event without an environmental release or safety issue? For example: flare smoking, toxic release, or high temperature exposure of personnel and adjacent equipment.

RSDB integration with II may facilitate getting to the root cause of the incident. The RSDB contains valuable information about the relief system response and may help identify the original process design intent. If the relief system response was not as intended, the relief system may be the cause of the incident. Additionally, if there are layers of protection designed to make flaring or release valve opening an infrequent event but flaring is occurring often, the preventing layers of protection may need further evaluation.

The RSDB and II integration may also help identify additional inspections and corrective actions. In this case II also integrates Mechanical Integrity (MI). For example, damage to the relief system may have occurred from too-high or too-low temperature or pressure. Mechanical damage resulting from corrosive chemicals or relief valve chattering may have occurred. Additionally, the RSDB may shed light on damage that the incident caused to other process equipment. Finally, if the RSBD does not accurately predict what will occur during such an incident there should be a corrective action to change the RSDB and possibly modify the relief system. Consequently, the II serves to further validate the RSDB and the RSDB enhances the II. An important point to note: incorporating RSDB into II requires someone knowledgeable in RSDB be on the II team.

In conclusion, a relief device release or flaring event II that does not consider the RSDB leaves a valuable tool unused. Integrating RSDB with II enhances PSM integration and improves process safety and profitability by gaining additional benefit from both RSDB and II.

6. RSDB Integration with Production Planning

Opportunities often present themselves in complex facilities to ensure process safety and increase profitability by integrating Relief System Design Basis (RSDB) with production planning. Many facilities integrate processing units in a complex manner. Such facilities often have multiple production constraints preventing individual processing units from running maximum capacities simultaneously. Production constraints may include downstream processing capacity, feed qualities, processing unit downtimes, permitting limits, or even facility RSDB constraints. When a facility RSDB constraint limits production, understanding the relationship between other production constraints and the facility RSDB constraint becomes critical to ensure process safety and optimal profitability. Hence the need and opportunity to integrate RSDB with production planning.

Before presenting RSDB and production planning integration examples, there are a few points to note. First, RSDB analysis for an individual processing unit typically establishes the unit's safe operating window at a maximum capacity. A facility RSDB, however, reflects multiple processing units operating simultaneously. Second, due to the myriad of production constraints a complex facility may have, the facility RSDB often reflects a limited set of facility production constraints (often with a very conservative basis). Third, in some instances, the facility RSDB may constrain various processing units to a fraction of their maximum production capacity. Finally, when production constraints setting the facility RSDB change, production planning needs to evaluate alternative production options that the facility RSDB does not adequately reflect.

In one example, a complex facility relief system constraint of excessive relief system backpressure for low set point relief devices exists. The excessive backpressures result from a potential utility failure scenario impacting multiple processing units. This facility RSDB constraint limits the facility's production. When the processing unit with the low set point relief devices is down for maintenance, however, the facility RSDB constraint relaxes. Thus, opportunities present themselves for other processing units to increase production. In another example, a complex facility schedules a portion of the facility relief system maintenance downtime with processing units having a substantial impact on the facility relief system constraint. During the downtime, however, the startup timing for the relief system is delayed due to discovery work and opportunities arise to increase profitability by minimizing production cuts within a newly defined facility RSDB constraint.

7. RSDB Integration with Project Development

Integrating Relief System Design Basis (RSDB) with projects presents opportunities to improve project definition, increase project profitability (through project cost savings), and enhance Process Safety Management (PSM). Having a preliminary well documented RSDB at the on-set of a project — often called the Front End Engineering Design (FEED) stage — will identify the various scenarios that should be a part of the sizing of not only the individual relief devices, but also the relief device collection system functional requirements. The potential costs of the project's impact on the relief devices and systems can then be reviewed and optimized in coordination with the project's functional equipment. The following examples best illustrate these points.

In a larger project a liquid feed drum was required. The feed drum's initial design was 50 psig, and the feed sources ranged from 35 psig to 100 psig. Because the feeds to the drum were greater than the drum's proposed design pressure, the overpressure device and corresponding system had to be designed for liquid overfill. All the sources that were greater than 50 psig represented an overfill potential of around 70,000 barrels a day, and consequently the relief device and the corresponding collection system size and costs were unattractive. Upon further investigation and review of the options, the project team increased the design pressure of the feed drum to 110 psig, which had minimal cost differential when compared to the necessary relief system design. Thus, the 110 psig design was chosen and the relief device was then sized for fire and some potential back-flow on stoppage of the feed pumps. Cost savings were in the millions of dollars.

A second project involved design and installation of equipment that had to be tied into the flare system. A preliminary RSDB was done early in the project. Upon analysis of some typical global scenarios such as loss of power, loss of cooling water, etc., the new project's primary relief device(s) would require increasing the collection system's piping and flare size. Furthermore, the sizing of the relief device and piping alone had a large estimated cost.

The relief device's size was driven by the main tower's selected design pressure and steam reboiler pressure available. The project team, in coordination with the RSDB personnel, were tasked with optimizing the reboiler and the relief system. During the optimization phase, the team identified 2 key items: 1) the reboiler steam had only one pressure available, but it could be let down to a lower pressure, and 2) at a slightly lower pressure the steam could easily provide the available heat necessary for the process.

Additional investigation by the team revealed that the tower material and thickness had not initially taken advantage of the potential MAWP. The material thickness needed was ~0.82 inches, and 0.875 inches was to be selected. This allowed for a MAWP higher than design. However, because the steam pressure was reduced, good engineering practice would require that the design consider failure of the steam pressure control valve.

Failure of the steam pressure control valve would provide sufficient heat to overpressure the tower based on the new design pressure and the relief device to the flare was still needed. To address this failure, the steam system (downstream of the pressure let-down valve) was provided with its own pressure relief valve which could be vented to atmosphere.

With the team's recommended changes noted above, the only scenario for the main tower relief device was fire, which had a lower cost of installation and did not require an update of the flare system. Having the RSDB done early in the project allowed the designers and project team to make some easy (paper) cost effective changes, and did not require any modifications to the existing collection and flare systems.

The examples above demonstrate organizations integrating RSDB during project development to improve project definition, reduce project cost, and even make processes inherently safer. Organizations realize further PSM enhancements by integrating a process unit's existing RSDB documentation with project development activities. In addition to the benefits illustrated above, organizations incorporate project impacts on a process unit's RSDB with minimal duplication of effort (minimal cost), and enhance the process unit's RSDB integrity. From an overall PSM perspective, the project development integration enhances both the PSM and project development system.

8. Conclusion

This paper presents opportunities for organizations to enhance their PSM systems and process safety culture by integrating Relief System Design Basis (RSDB) with the following PSM elements and business processes:

- Mechanical Integrity (MI)
- Process Hazard Analysis (PHA)
- Incident Investigation (II)
- Production planning
- Project development.

Organizations that truly "own" a process safety culture should use each successful opportunity implementation as a spring board to further enhance PSM integration and improve process safety and profitability.

9. References

1. CCPS, "Guidelines for Integrating Management Systems and Metrics to Improve Process Safety Performance", Appendix A, Center for Chemical Process Safety, American Institute of Chemical Engineers, New York, NY 2016.

2. CCPS, "Vision 20/20 Booklet – Committed Culture", Center for Chemical Process Safety, American Institute of Chemical Engineers, New York, NY 2016.

3. Marsh and McLennan Companies, "The 100 Largest Losses", 23rd edition (2014).

4. CCPS, "Guidelines for Integrating Management Systems and Metrics to Improve Process Safety Performance", Center for Chemical Process Safety, American Institute of Chemical Engineers, New York, NY 2016.

5. Occupational Health and Safety (OSHA) "Process Safety management of Highly Hazards Chemicals" 29CFR1910.119(d).

6. Thomas Watson, "AM-13-24 Considerations for Effectively Managing Emergency Relief Systems Design," American Fuels & Petrochemicals Manufacturers Annual Meeting, 2013.

7. American Petroleum Institute, "API Standard 521 – Pressure-relieving and Depressuring Systems", section 4.7.3, sixth edition (2014).

8. American Petroleum Institute, "API RP 581 – Risk-Based Inspection Technology, Part 1", Section 7, second edition (2008).

9. Center for Chemical Process Safety's Vision 20/20: "Characteristics of Companies with Great Process Safety Performance" (2013, McCavit, Berger, and Nara).

479098 Reactive Hazards Management: Lessons Learnt from Major Incidents

Abstract:

Chemical process industry has long suffered from major incidents due to lack of proper understanding of reactive hazards. The methyl-isocyanate leak in Bhopal, India, has been by far the worst industrial incident in history which led to the development of many process safety regulations. However, the formation and implementation of regulations has not prevented the occurrence of these incidents. Many incidents such as Concept Sciences Inc. where a hydroxylamine manufacturing unit was destroyed, the explosion at the Sterigenics International Inc., ethylene oxide sterilization facility, Bayer CropScience Pesticide Chemical Runaway Reaction Explosion and West Fertilizer Explosion have highlighted the importance of having a better understanding of the reactive hazards. The focus of this presentation is to identify different types of reactive hazards through examples of chemicals incidents and analyze how a better understanding of these hazards could have avoided these incidents.

479245 The Study of Safety Evaluation Technique for High Temperature Process

Abstract:

There are many accidents of a firing furnace and kneading machine by overheating and the abnormal injection of the product. When the gas which suddenly expanded by thermal decomposition destroys plant equipment and enters the explosion concentration range, a gas explosion is caused and damage spreads. The material which it is easy to decompose is invested by mistake by a firing furnace and has the trouble that the pressure of the furnace abnormally rises to in our factory. Also, troubles such as an explosion of the plumbing by the thermal decomposition of the polymer by the overheat, vent explosion by the thermal decomposition by the false injection of an additive having low heat stability has broken out with kneading machine

These troubles are caused by the thermal decomposition by a sudden temperature rise. The temperature program of the analytical instrument which is generally used by a thermal analysis is 10-20 /min. The conventional thermal analysis equipment cannot evaluate the temperature rise rate at the time of the trouble exactly to become than 20 /min. It becomes the problem that the slow temperature rise may be different from rapid heating up in a decomposing reaction.

We produced a device capable of quickly raising temperature to solve these problems.

This device is comprised of the vertical plumbing which it sealed and heating furnace, and it is very simple structure. The sample holder is fixed to the upper part in the plumbing by a magnet. And the holder drops in the heating zone by keeping a magnet away from the plumbing and is heated rapidly. The materials of the plumbing are SUS316L, and, as for the use upper limit temperature, 500 , the use upper limit pressure are 1MPa. Because the deterioration of SUS materials by the heat is intense when it is raised the temperature of the device to 750 , the plumbing is disposable.

The inside of the plumbing is replaced with inert gases such as nitrogen, argon. The maximum temperature rising rate is 150 /min (furnace temperature: 750).

We introduce an example of the trouble analysis of a firing furnace and kneading machine using this device.

[Case example 1] Study of firing furnace trouble

After burning with a new additive at 720 , abnormal pressure rise happened in a furnace, and a plant stopped. After measuring quantity of outbreak gas using the rapid heating device, it was recognized that gas of 7 times was generated and a gasification rate tripled in comparison with a conventional additive.

[Case example 2] Study of kneading machine trouble by overheating

The additive which had low heat stability was spent mistakenly in large quantities to a kneading machine, and a vent board exploded. After performing a rapid temperature rise examination under a kneading temperature condition, as for the pressure raising rate when an additive disintegrated, 1.8MPa/min, the quantity of gasification became 190L/kg. The additive concentration became (mixing ratio with the resin) more than at least 75%, and it was recognized that a large amount of additives were invested mistakenly.

Fortunately, this example does not become the big accident with resulting in injury or death. In this explosion example, a vent board penetrated a checker plate. It might become the casualty when there was a person near a kneading machine.

In comparison with the plant of the petrochemical and fine chemistry that heat risk analytical technique establishes, the firing furnace and kneading machine is often run without detailed risk assessment. The rapid heating device becomes the powerful tool to do the heat risk evaluation of these high temperature processes.

In future, it is going to be improved the rapid heating device changes the plumbing materials, and to use it to the temperature ranges more than 1,000 .

Handling uncertainty in LOPA and SIF design

Akshat Khirwal
IRESC
Unit 2801, Enterprise Square 3
39 Wang Chiu Road
Kowloon Bay
Hong Kong
Akshat.khirwal@irescglobal.com

Arjan Abeynaike, Venkatesh Sourirajan
IRESC
Arjan.abeynaike@irescglobal.com, Venkatesh.s@irescglobal.com

Keywords: Uncertainties, LOPA, SIL, SIF design, reliability

Abstract

Layer of protection analysis (LOPA) is a widely used semi-quantitative method for determining the reliability requirement for safety instrumented functions (SIFs). An advantage of LOPA is that it specifies a numerical reliability target - the average probability of failure on demand (PFD_{avg}). However, uncertainties in the input parameters and simplifications within the LOPA method lead to a margin of error in the target PFD_{avg} that is often not accounted for in the design of the SIF. Further uncertainties are introduced when carrying out a reliability analysis of the SIF design to ensure that the target PFD_{avg} is met. The result of overlooking such uncertainties is that the target PFD_{avg} may be set low and the calculation of the actual PFD_{avg} achieved by the design may be optimistic, leading to a SIF that is under-designed. The need to account for uncertainty is emphasized in the latest version of IEC 61511, the international standard for SIF design.

This paper investigates and highlights the sources of such uncertainties in LOPA and SIF reliability analysis. In particular, the quality of the data used as input parameters to LOPA and reliability models has a large effect on the accuracy of the results. These parameters include failure rates, conditional modifiers, beta factors for common cause failures, and proof test coverage factors. Available data for these parameters can be sparse and may not be representative of the conditions in which the SIF is applied. Where the confidence intervals of the data are available, these can be incorporated into the calculations to reach a more conservative result. Alternatively, a safety factor can be applied to try and capture all the uncertainties. In this approach, the target PFD_{avg} obtained from LOPA is reduced by the safety factor to give a more stringent target PFD_{avg} for the SIF design to meet. The methods for handling uncertainty are reviewed in this paper.

1 Introduction

SIL Determination is the process of assigning a Safety Integrity Level (SIL) to Safety Instrumented Functions (SIFs) following any of the methods i.e. Risk Graph, Risk Matrix or Layer of Protection Analysis (LOPA); as laid down in IEC 61511 and IEC 61508 [1] [2]. Of these methods, the semi-quantitative LOPA approach is increasingly being adopted for SIL Determination as compared to other qualitative approaches such as Risk Graph or Risk Matrix. An advantage of LOPA is that it specifies a numerical reliability target for a safety instrumented functions (SIF), which is expressed in terms of the target average probability of failure on demand (target PFD_{avg}), or target risk reduction factor (RRF, where RRF = 1/ PFD_{avg}). This target PFD_{avg} then forms the basis for design of the SIF wherein a reliability analysis is carried out to ensure that the target PFD_{avg} is met. However, uncertainties in the input parameters and simplifications within the LOPA method lead to a margin of error in the target PFD_{avg} that is often not accounted for in the design of the SIF. Further uncertainties are introduced when carrying out a reliability analysis of the SIF design. Without accounting for these uncertainties in the assessment, the SIF may be under-designed since LOPA may result in under prediction of the target PFD_{avg} or the reliability analysis may show that the target PFD_{avg} has been achieved although this may not be the case. This paper has reviewed the sources of uncertainties in the LOPA method and design of the SIF (SIF reliability analysis) and provides a way forward for handling such uncertainties.

2 Uncertainties in LOPA

Initiating event frequency, consequence severity, and likelihood of failure of Independent Protection Layers (IPLs) are the key parameters used in LOPA to approximate the risk of a scenario in case the SIF fails to act when the demand arises. In addition to the above, conditional modifiers [3] may also be used to factor in specific parameters pertaining to the demand on the SIF. The simplicity of LOPA allows it to be easily understood by engineers of all disciplines, leading to its widespread use. The drawback of this simplicity is the greater uncertainty in the results, as described in the following sub-sections. The governing equations and the LOPA worksheet used for assessing the scenarios discussed in this paper are referenced from IEC 61511-3 [1] and CCPS, Simplified Process Risk Assessment [4].

2.1 Initiating event frequency

The demand on any SIF may arise due to one or more events, each with a frequency in the range of once per year to once in plant life (for example, 30 years). There is some approximation involved in the selected initiating event frequency values especially when using generic databases. For instance, the source data in the generic database may not be appropriate for application to the facility under review or there may not be enough confidence in the dataset. Power failure or failure of other utilities such as cooling water or steam is a good example. The frequency of these failures from generic databases is typically $1\text{x } 10^{-1}$ per year (once in 10 years) [4]. However, in the author's experience, based on feedback from many operators, these values are likely to be somewhere between once per year to once in ten years. Similarly, the frequency of pump trip based on generic databases is $1\text{x } 10^{-1}$ per year, however operational experience suggests that a value in the range of once in 2 years to once in 3 years may be more appropriate. An analysis of the OREDA [5] data also provides a value close to once in 3 years. Additionally, when there are multiple demand scenarios for a particular SIF, failure to consider the cumulative frequency of all the demand scenarios can also lead to under-estimation of the overall initiating frequency. A simple approach is to consider a suitable approximation for the cumulative demand frequency [6] [7]. In addition, common cause events specific to the facility is also a contributor which needs to be identified. Without such considerations, the required target PFD_{avg} may be under-predicted. All of the above introduce uncertainties in the LOPA assessment noting that the initiating event frequency is the starting point of a LOPA assessment. Therefore, care must be taken in selecting an appropriate value for this parameter.

To better understand the dependence on the initiating event frequency, consider a typical distillation column as shown in Figure 1. There are many scenarios such as overhead condenser failure, reflux pump failure, reboiler control failure, etc. which may lead to a demand on the high high pressure SIF provided on the column. For the sake of brevity, only the case of Reflux Pump trip is used further in this example.

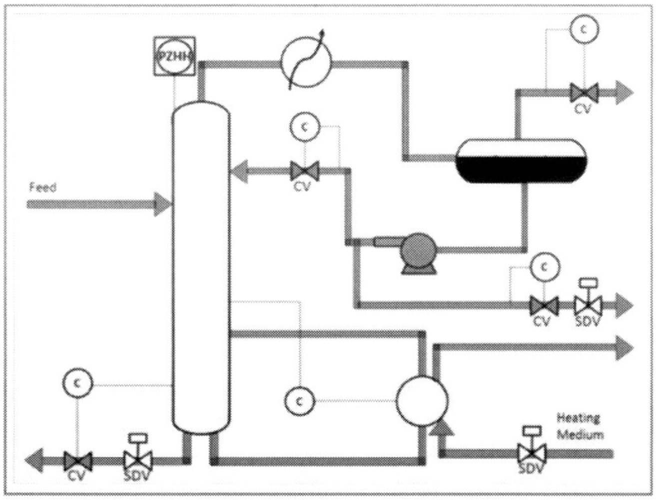

Figure 1. Typical Distillation Column

Assuming an initiating event frequency of 1 x 10^{-1} per year for pump trip and considering a Target Mitigated Event Likelihood (TMEL) of 1 x 10^{-5} per year for multiple fatalities along with an ignition probability of 0.1 and taking credit for pressure safety valve (PSV) on the column (PFD_{avg} of 1 x 10^{-2}), this would result in a SIL 1 with Risk Reduction Factor (RRF) 10 (target PFD_{avg} of 1 x 10^{-1}). However, if the initiating event frequency for pump trip is changed to 3.33 x 10^{-1} per year (once in 3 years), this translates to SIL 1 with RRF of 33 (target PFD_{avg} of 3 x 10^{-2}). This example demonstrates the dependence of initiating event frequency selection on the target PFD_{avg}.

2.2 Consequence Severity and Conditional modifiers

The TMEL assigned to the SIF is directly associated with the degree or the severity of the consequence assumed (e.g. single fatality: 1 x 10^{-4} per year, multiple fatalities: 1 x 10^{-5} per year). This in turn has a direct impact on the SIL assigned to the SIF as well as its target PFD_{avg}. Therefore, careful consideration must be given to the selection of the consequence severity during a LOPA review. For instance, the impact area or hazard zone due to a loss of containment event can be determined using consequence modelling [7], the result from which can be used to assign the severity (e.g. single fatality vs multiple fatalities). Additionally, this can be used to determine the relevant conditional modifiers such as occupancy factor within the impact zone. This approach is more reliable than making coarse assumptions which may introduce uncertainties in the analysis. Such refinement can also be applied to other modifiers such as ignition probability where using a more specific value representative of the hazardous fluid and/or the facility features may be appropriate as compared to use of generic values.

If we revisit the example of typically distillation column presented in Section 2.1; considering an initiating event frequency of 3.33 x 10^{-1} per year and assuming a TMEL of 1 x 10^{-4} per year (single fatality), an ignition probability of 0.1 and taking credit for PSV on the column (PFD_{avg}

of 1 x 10^{-2}), this would result in a RRF of 3 (target PFD_{avg} of 3 x 10^{-1}), which is below SIL 1 requirement. However, if a higher severity (TMEL: 1 x 10^{-5} per year corresponding to multiple fatalities) is selected based on consequence modelling results and an ignition probability of 0.3 is selected based on refinement, the assigned SIL will increase significantly to SIL 2 with RRF 100 (target PFD_{avg} of 1 x 10^{-2}). Although this example shows an increase in the assigned SIL, the application of detailed assessment for determining severity could also show a reduction.

2.3 *Independent Protection Layers*

Independent Protection Layers (IPLs) are devices, systems or actions that prevent the realization of an undesired consequence (e.g. the consequence the SIF is safeguarding against). Each IPL has an associated PFD_{avg} which is assigned and used in the LOPA methodology for SIL Determination. For example, process deviation alarm with sufficient time for operator intervention has a PFD_{avg} of 1 x 10^{-1}, an independent BPCS control has a PFD_{avg} of 1 x 10^{-1} while a PSV has a PFD_{avg} of 1 x 10^{-2}.

2.3.1 Effect of Common Cause Failures

One of the fundamental assumptions in the LOPA approach is that these IPLs are completely independent and do not share any common cause failures (CCF). In reality, multiple protection layers may be impacted or their performance may be affected due to similar environmental conditions or deviations from desired process parameters. Corrosion or deposition leading to simultaneous degradation of a pressure safety valve and a shutdown valve is one such example. Other examples of CCF include scenarios where multiple PSVs are required to meet total relief requirements, or where an alarm and control, although independent safeguards, both reside on the BPCS. Additionally, dual check valves for preventing backflow could also be impacted by CCF. Although such scenarios may take a period of time to develop and the likelihood may be low, such CCF can have a direct impact on the PFD_{avg}. For example, consider a case where the safety system comprises a SIF and an IPL and the target PFD_{avg} for this system is 1 x 10^{-3}. The PFD_{avg} of the IPL is 5 x 10^{-2}. If the SIF and IPL fail independently only, then the target PFD_{avg} required of the SIF is (1 x 10^{-3}) / (5 x 10^{-2}) = 2 x 10^{-2}. However on inclusion of CCF, even if a low probability of CCF is assumed (1% of the PFD_{avg} of the IPL), the target PFD_{avg} for the SIF then becomes 1 x 10^{-2}.

The authors do acknowledge that for the purpose of time-effective LOPA review, it may not be practical to dwell deeper into CCF for IPLs as it may require an in-depth analysis (for which sufficient details may also not be available at the time of LOPA review). However, it is important to be aware of the inherent uncertainties introduced in the analysis by not accounting for the CCF associated with IPLs.

2.3.2 Effect of Multiple Protection Layers

There is another inherent uncertainty associated with IPLs apart from CCF. The PFD_{avg} for any protection layer (whether mechanical protection, alarm with operator response, BPCS or SIF) is not static but increases over time until a required test or maintenance activity is conducted to check for faults [4]. This relationship between PFD and time is shown graphically in Figure 2. In LOPA, the PFD_{avg} values of the IPLs are multiplied together to estimate the PFD_{avg} of the overall safety system [4]. This calculation, while simple to perform, may tend to underestimate the overall PFD_{avg}. In a more quantitative analysis of reliability, the time-dependent PFDs of each protection layer are multiplied together to yield a result similar to Figure 3, which shows the PFD of the safety system. The PFD_{avg} for the system can then be calculated.

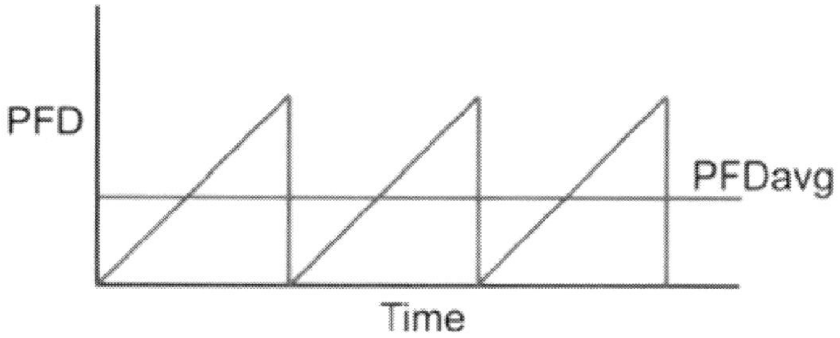

Figure 2. PFD as a function of time for a single protection layer. Periodic testing / maintenance was assumed to detect all faults in this example.

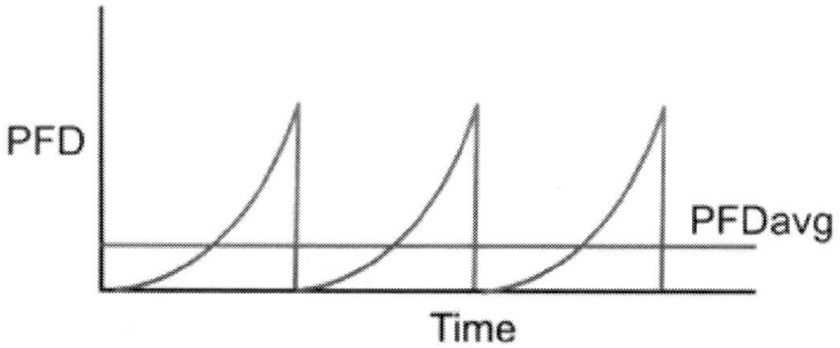

Figure 3. PFD as a function of time for a safety system with two protection layers (one SIF and one other IPL). Proof testing was assumed to detect all faults in this example.

To illustrate the difference in the results provided by LOPA and a more rigorous analysis, the same example from Section 2.3.1 is used. LOPA estimates the PFD_{avg} of the SIF to be 2×10^{-2} (excluding CCF), considering one IPL with a PFD_{avg} of 5×10^{-2} per year and a target overall PFD for the system of 1×10^{-3} (see calculation in Section 2.3.1). Using the PDS method of reliability analysis [8] gives a SIF PFD_{avg} of 1.5×10^{-2}, i.e. LOPA overestimates the target PFD_{avg} by 33%. As the number of protection layers increases, this error becomes more significant, as shown in Table 1. For two or more IPLs with a SIF, the error in the target PFD_{avg} for the SIF is at least 100% using the PDS method; i.e. the LOPA value for target PFD_{avg} is double that given by the PDS method. This gross overestimation in the target PFD_{avg} for the SIF suggests that the LOPA method may not be sufficiently accurate where two or more IPLs are considered. In such cases a more rigorous analysis may be conducted outside of the LOPA session (refer Section 3). Such rigorous approach, if adopted, may also help reduce the error for cases where only one IPL is applicable.

Table 1: Effect of the number of IPLs on the error in PFD_{avg} for a safety system as calculated by LOPA

Number of IPLs	Error in target PFD_{avg} for SIF
0	0%
1	33%
2	100%
3	220%
n	$\frac{2^{n+1}}{n+2} - 1$

2.4 Assessing the cumulative uncertainty for multiple parameters in LOPA

The above sections have discussed how the uncertainty of each individual parameter affects the target PFD_{avg} calculation. In practice, however, the cumulative effect of the uncertainties for multiple parameters also needs to be considered. As a simple illustration and keeping in line with the discussion above, the same example of the distillation column high high pressure SIF is used. As shown in the table below, the changes in various parameters results in a significant difference in the target PFD_{avg} requirements. One additional parameter presented in the table below is the use of PFD_{avg} of 3×10^{-2} for IPLs considering multiple PSVs are required to meet relief requirements. As expected, the required RRF is much more stringent when the uncertainties are modelled. However, as highlighted in earlier sections, depending on the SIF under review, in some cases, the requirements may also be less stringent upon further refinement.

Table 2: Effect of cumulative uncertainty of multiple parameters in LOPA

Description*	Initiating Event Frequency	IPL	Ign. Prob.	Total Mitigated Event Frequency	TMEL	Target PFD_{avg}	RRF
Case 1	1.00E-01	1.00E-02	1.00E-01	1.00E-04	1.00E-04	1.00E+00	1
Case 2	3.33E-01	1.00E-02	3.00E-01	1.00E-03	1.00E-04	1.00E-01	10
Case 3	1.00E-01	1.00E-02	1.00E-01	1.00E-04	1.00E-05	1.00E-01	10
Case 4	3.33E-01	1.00E-02	3.00E-01	1.00E-03	1.00E-05	1.00E-02	100
Case 5	1.00E-01	3.00E-02	1.00E-01	3.00E-04	1.00E-04	3.33E-01	3
Case 6	3.33E-01	3.00E-02	3.00E-01	9.00E-04	1.00E-04	1.11E-01	9

3 Approaches for handling uncertainty in LOPA

Based on the various modes of introduction of uncertainties in the LOPA analysis as discussed in the preceding sections, the question that arises is whether the direct use of target PFD_{avg} from LOPA as input to SIF design is optimal or is it desirable to investigate possible ways to refine the target PFD_{avg} from LOPA to ensure the SIF is not under-designed. Some ways to refine the target PFD_{avg} are discussed in the following paragraphs.

One proposed approach for dealing with uncertainty without compromising the simplicity that makes LOPA an attractive method is to conduct a more rigorous risk assessment outside of the LOPA session for high consequence / high risk scenarios to confirm that the SIF reliability target PFD_{avg} obtained from LOPA is sufficient. Such an assessment may include a quantitative reliability analysis such as a Fault Tree Analysis or Markov Model to obtain a more accurate estimate of the required target PFD_{avg}. These methods can also take into account common cause failures and the time dependency of PFD. In fact the most recent version of IEC 61511 [1] requires that a quantitative methodology be used for cases where the required risk reduction is greater than 10,000 (target PFD_{avg} from LOPA is equal to or lower than 1×10^{-4} per year). By using LOPA as a screening tool, this strategy can ensure that high consequence / high risk scenarios receive the more rigorous assessment they deserve, while for lower risk scenarios the more time-efficient LOPA can be considered sufficient.

Alternatively, a number of different cases can be considered for each LOPA scenario, such as in Table 2. Each case considers different values for the various parameters used in the LOPA calculation, resulting in a range of RRFs, from the more conservative to the more liberal. Taking Table 2 as an example, the results range from an RRF of 1 to 100. A final RRF at some point in between these two extreme cases may be selected for the SIF and carried forward into the SIF design. In this way, a level of uncertainty can be accounted for without being overly conservative.

Another approach is to introduce a "Safety Factor" to the reliability target set during LOPA. In this approach, it is recommended that the target Risk Reduction Factor obtained from LOPA should be increased by a safety factor to give a more stringent target for the SIF design to meet. For instance, if the target based on results from LOPA is RRF 100, the SIF design can conservatively consider a target PFD_{avg} associated with RRF 300. Similarly, if the target is RRF 500, then SIF design may consider RRF 800. However, when the RRF is at or close to borderline values, the SIF design may consider the next interval, e.g. for RRF 800 and above, the SIF design may consider RRF 1300 or alternately the RRF used can be justified by conducting a more rigorous quantitative analysis as mentioned above. Some engineering judgement should be applied when selecting this safety factor depending on the level of conservatism adopted in the initial LOPA assessment. As can be inferred from Table 2 above, when uncertainties are accounted for, the target PFD_{avg} can actually vary by one order of magnitude. Applying a safety factor to the results will therefore account for the uncertainties.

4 What PFD_{avg} to assume when Risk Graph Method is adopted for SIL Determination?

Risk Graph Method [1] is the alternate method commonly used for SIL Determination. Unlike the semi-quantitative LOPA approach, which specifies a numerical reliability target (PFD_{avg}) for the SIF in addition to the assigned SIL, the Risk Graph approach relies on the qualitative definitions for the parameters involved to deduce the assigned SIL without setting an actual numerical target (PFD_{avg}). Furthermore, this target PFD_{avg} value from Risk Graph is provided as a range. Hence, where Risk Graph method is used, it is not an uncommon practice to use the upper bound of the PFD_{avg} associated with the assigned SIL. For example, the LOPA method may provide an RRF of 40 (PFD_{avg} of 2.5×10^{-2} per year) for a SIL 1 function, while the same SIL 1 function following the Risk Graph method may adopt the upper bound RRF of 10 (PFD_{avg} of 1×10^{-1} per year).

Consider the same example of a distillation column as discussed in Section 2. If the column high high pressure SIF is assessed using the Risk Graph approach assuming that a trip of Reflux Pump could be expected once a year, taking credit for the pressure safety valve on the column as an IPL (credit of 100) and taking no credit for occupancy and avoidance; the resulting analysis show that Risk Graph method also determines a SIL 2 which is similar to the results from LOPA are accounting for uncertainties. The key difference being the overall PFD_{avg} of 3.33×10^{-3} per year (RRF 300) obtained from the LOPA which will be more stringent than the PFD_{avg} of 1×10^{-2} per year (RRF 100) in case the upper bound value of SIL 2 is used.

The above example shows that using the output from the Risk Graph directly may lead to under-prediction of the target PFD_{avg} and thereby leading to under design of the SIF. The concept of "Safety Factor" as discussed in Section 3 may therefore also be applied where Risk Graph is used. Based on this, it is recommended that the SIF design can use the mid-point of the upper bound and lower bound values based on PFD_{avg} range applicable to the SIF corresponding to the assigned SIL. The safety factor may also be selected on a case by case basis depending on the severity of the associated consequences (for example, using a value above the mid-point of upper and lower bound may be justified for the most severe consequences).

5 Uncertainties in SIL reliability analysis

The uncertainties in terms of obtaining an appropriate target PFD_{avg} for a SIF using LOPA methodology are discussed in Section 2. It is however, worth noting that potential for introduction of uncertainties into the analysis is not limited to this step alone. Moving on in the safety life cycle, the next step after LOPA involves a reliability analysis to ensure that the target PFD_{avg} is met for each SIF. This activity is commonly termed as "SIL Verification" in the industry, but strictly with reference to IEC 61511, it is considered within the domain of "SIS Design and Engineering". The target PFD_{avg} that is considered for the reliability analysis should be the same as the target PFD_{avg} that was calculated in LOPA (a safety factor may be applied as described in Section 3). Where Risk Graph methodology is applied, the target PFD_{avg} can be a mid-point value in the assigned SIL band as suggested in Section 4.

This reliability analysis also involves a number of input parameters, some of which may introduce uncertainties into the analysis. It is important to understand and evaluate the nature of these uncertainties associated with input parameters such as failure rate data for SIF components, beta factor for common causes and proof test coverage factors in SIF reliability analysis as discussed in the subsequent sub-sections. In this section, PFD_{avg} results that were calculated using exSILentia software are presented to illustrate the effects of the input parameters on the PFD_{avg}. The numerical PFD_{avg} results are only applicable for the specific systems and assumptions used in the calculations.

5.1 Failure Rate Data

Failure rate data for the individual SIF components is the basis of the reliability analysis. It is therefore of utmost importance that the data used is based on sufficient sample size and is representative of the conditions in which the SIF is applied. The quality of the data used as input parameters to reliability models has a large effect on the accuracy of the results.

Ideally, an appropriate method for determining the failure rate is to collect field data from a large, randomized sample of units. The operational time until failure should be recorded for each unit. Such data mining however; is a long term project, requiring the co-operation of multiple operators, ideally from a wide range of different facilities and operating environments. For this reason, such studies are rare and normally "generic data", i.e. data applicable for a broad type of component (such as pressure transmitter or shut-down valve) is used rather than data for a specific model or a specific environment. The OREDA database [5] is one example, which provides such generic data. Alternately, some vendors may commission a field study to collect failure data specific to their components or build their own failure data based on reported failures. However, given the smaller sample size as well as the gaps in ensuring that each failure is reported appropriately by the operating facilities, there will be uncertainties in the reported data and these values may tend to be on the optimistic side. For example, while the failure rate for a ball valve from a generic data source is 9.8×10^{-7} per hour [9], several vendors of ball valves report failure rates of less than 1.0×10^{-7} per hour. Another issue is that such data may only be sourced from a limited number of facilities operating in a specific environment and hence, this data may not be applied to a range of facility types or operating environments.

One approach to iron out these uncertainties is to conduct a Failure Mode and Effect Analysis which is a workshop activity bringing together a group of experienced personnel (operations, maintenance, instrumentation, design, etc.) for brainstorming the possible failure modes associated with each component and its associated failure rate using operational experience and field data. The quality of such an exercise is highly dependent on the experience of the participants and their engineering judgement. Hence, even though conducting such an exercise will be fruitful, some margin of error is still expected.

5.2 Beta Factor

Typically, the overall reliability of a SIF is estimated based on both independent and dependent failures. For functions with redundant components, multiple components may fail simultaneously due to a common failure mode (e.g. stress, maintenance errors, etc.). These dependent failures are the percentage of the failures attributed to the common causes and are denoted as Beta Factor in the reliability analysis.

Taking note of the uncertainties in failure data for components as highlighted in earlier sections, it is worth highlighting that the chances of availability of detailed field data for beta factors covering a large sample size is even lower, thereby leading to a greater degree of uncertainty. IEC 61508 [2] advocates a method for estimation of the beta factor using a questionnaire and a scoring system. Based on the scoring system, typically the beta factor value ranges from 1 - 10%. This is a somewhat subjective approach. In reality, beta factors higher than 10% have also been reported in the industry [10]. This may however represent an extreme operating environment pertaining to a particular facility. Therefore, some engineering judgement is required.

It is to be noted that beta factor has a direct impact on the PFD_{avg} achieved for systems with redundant components. Since common cause failures are often the dominating contributor to PFD_{avg}, an increase in beta factor leads to a significant increase in PFD_{avg}. This relationship can be seen in the PFD_{avg} results given in Figure 4, for SIFs with various configurations (1oo2, 2oo3, etc.). Hence, selection of a low beta factor may lead to underestimating the PFD_{avg} and potentially also under design of a SIF. For configurations without redundancy (e.g. 2oo2), the beta factor has little impact on the PFD_{avg}.

In the author's experience, some references and guidelines from operating companies provide recommended beta factor values. However, taking note of the findings presented in this section, it should be ensured that such recommended values have a sound basis and are as accurate as possible. Over-estimating beta factor value may lead to over-design of the SIF which in turn also has an associated cost impact. Conversely, under-estimating the beta factor may lead to SIF being under-designed.

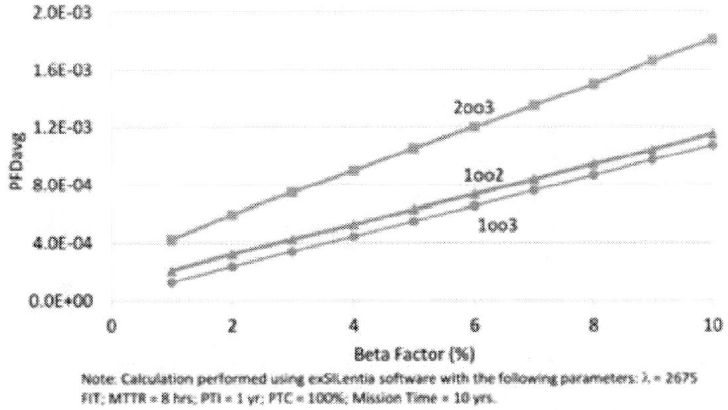

Figure 4. Impact of Beta Factor on PFD_{avg} calculation

5.3 Proof Test Interval

A proof test is performed periodically to detect failures and restore the system / component to an "as good as new" condition. The actual proof test interval may be governed by factors associated with SIF design and operating philosophy specific to a facility. However, the selected interval forms the basis of SIF design since the reliability of the SIF can be compromised with regard to the desired target PFD_{avg} in case the proof test interval used for SIF design is not followed. The uncertainty in this regard lies in whether the recommended proof test interval is rigidly followed or is only conducted on as-need basis in the actual operational phase. Operators may tend to postpone a shutdown for testing purposes if it may result in a loss of production, thereby compromising the SIF reliability. The emphasis to rigidly adopt the recommended proof test intervals therefore rests with the Operating Company and the kinds of safety culture they advocate or follow.

5.4 Proof Test Coverage

One of the important parameters associated with proof tests is the coverage factor which represents the fraction of the potential failures that will be identified during a full proof test. A perfect proof test (i.e. coverage factor of 100%) is unlikely. As far as the authors are aware, there is also no field data available to determine the effectiveness of proof tests. Vendors may recommend a coverage factor based on a detailed FMEA exercise for the associated components. Such an FMEA can identify which failure modes will be successfully captured by a proof test, and thus estimate the coverage factor. Proof test coverage factors can sometimes be lower for components with good diagnostic coverage because the remaining failure modes that are undetected by diagnostics are more difficult for a proof test to detect. In the author's experience, typical coverage factors range from 80 - 99% for sensors and logic solvers and 70 - 80% for final elements (e.g. shutdown valves). While this is currently the best approach for determining the coverage factor, the uncertainty in the result is usually not estimated.

The proof test coverage factor has a significant impact on the calculated PFD_{avg} of a SIF element, as can be seen in Figure 5. For example, for an element with a 1oo1 configuration, decreasing the coverage factor from 90% to 80% resulted in a 46% increase in the PFD_{avg}. Given this effect, caution should be exercised when selecting coverage factors for use in SIF design since they will impact the risk reduction that can be achieved. The actual test performance should ideally be verified to match the assumptions (this becomes especially critical when higher factors e.g. 90% or above are used).

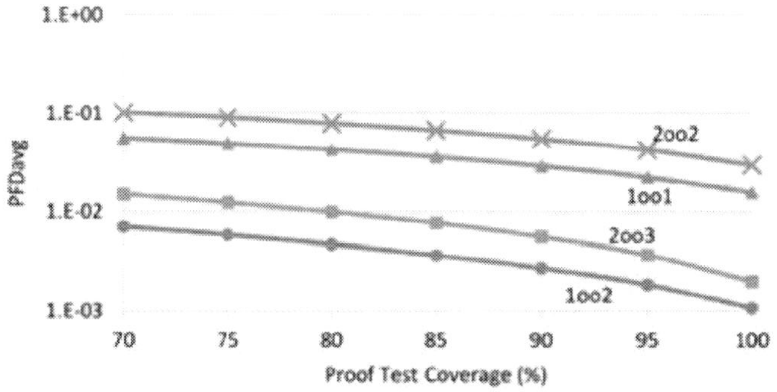

Note: Calculation performed using exSILentia software with the following parameters: λ = 1810 FIT; MTTR = 8 hrs; PTI = 2 yrs; Mission Time = 20 yrs; beta factor = 5%

Figure 5. Effect of Proof Test Coverage Factor on PFD_{avg} for various voting configurations

An alternative method for considering failures that are not detected by a proof test is to include a "probability of test independent failures", or P_{TIF}. This approach, which is advocated in the

PDS method [8], differs from the proof test coverage approach in that P_{TIF} is a constant contributor to the PFD_{avg}, whereas the effect of proof test coverage on the PFD_{avg} increases over time. Even when using P_{TIF}, however, there remains uncertainty in the value used for this parameter, with implications on the accuracy of the calculated PFD_{avg}.

5.5 SIF Operational Life

Operational life of a SIF refers to the life time of the various components of the SIF, sometimes also denoted as mission time. This also serves as a key input in SIF design. In the author's experience, typically a mission time in the range of 20 to 30 years is assumed. Lower mission times such as 5 or 10 years may be impractical in terms of actual implementation on site since this implies that the corresponding components will be replaced or completely overhauled after a 5 or 10 year period, which also has a significant cost impact. This assumption on operational life of the SIF has a direct impact on the risk reduction achieved by a SIF, provided that all other parameters are kept constant. Figures 6 and 7 show that a higher mission time is likely to result in a higher PFD_{avg} (or in other terms, a lower risk reduction). Figure 7 shows that the relationship between mission time and PFD_{avg} is almost linear, with a steeper gradient at a lower proof test coverage. This shows that the mission time has a greater impact on PFD_{avg} when the proof test coverage is lower. If the proof test coverage is taken as 100%, the mission time has no effect on PFD_{avg}. However, as discussed in Section 5.4, 100% proof test coverage is not likely to be realistic.

While operators may typically consider the cost of replacing equipment when deciding upon mission time, consideration should also be given to the effect that mission time will have on the reliability of SIFs. A shorter mission time can help SIFs to meet the target PFD_{avg} and may even allow a lower level of redundancy to be used for the SIF components. This is particularly relevant for SIL 3 rated SIFs, which may require a high level of redundancy as well as high reliability of components, resulting in a high capital cost. Reducing the mission time may help optimize the capital cost of the SIF. A cost benefit analysis may be conducted to justify optimization of the mission time. However, some engineering judgement must be exercised in order to keep the mission time within practical limits. In this regard, 15 years may be considered as a practical lower limit for mission time but this may vary on a case by case basis also depending on the owner's operating philosophy. In such cases, it is important to be aware that in case the capital replacement is not performed after 15 years, this will potentially leave the system exposed to higher probability of failures.

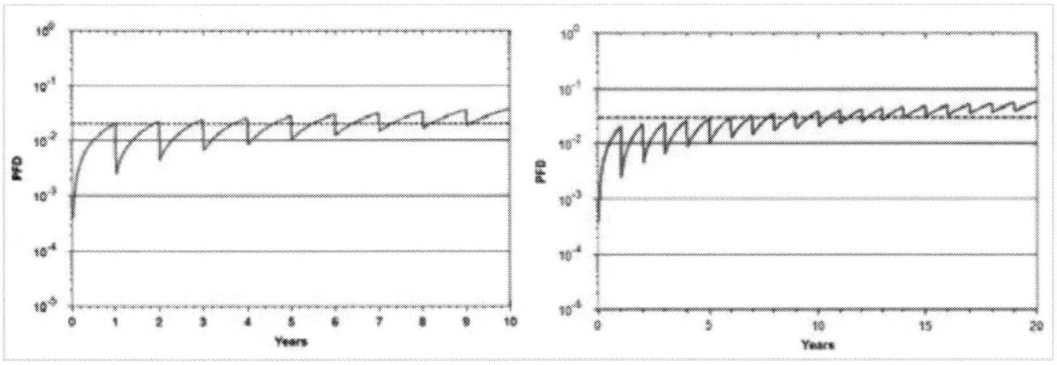

Figure 6. PFD_{avg} vs SIF Operational Life / Mission Time (Years)

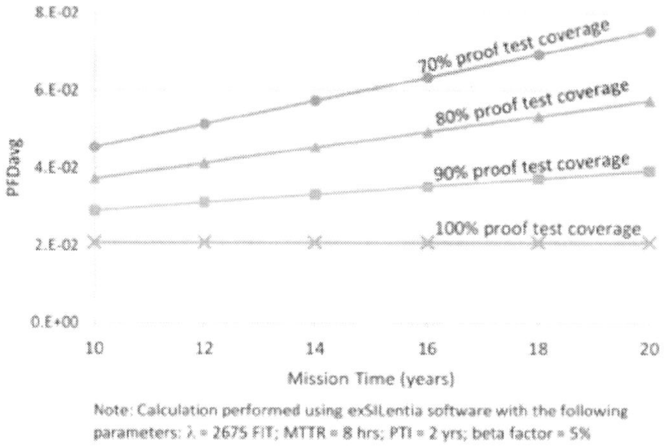

Figure 7. Effect of Mission Time on PFD_{avg} for a SIF with 1oo1 configuration

5.6 Assessing the cumulative uncertainty for multiple parameters in SIF Reliability Analysis

The above sections discuss how the uncertainty of each individual parameter affects the PFD_{avg} calculation. In practice, however, the cumulative effect of the uncertainties for multiple parameters also needs to be considered. As a simple illustration, Table 3 presents PFD_{avg} values calculated for a high pressure SIF with 1oo2 component configuration, using three different combinations of beta factor and proof test coverage, representing a best case, base case and worst case scenario. The beta factor was varied between 1% and 10%, while the proof test coverage was varied between 70% and 90%, which are ranges that can be reasonably expected for these parameters. The PFD_{avg} for the worst case scenario was more than double that of the base case and more than five times that of the best case. In these scenarios, beta factor was found to be the main contributor to the variation in PFD_{avg}, as the uncertainty in the beta factor may be +100% / -60%, whereas the proof test coverage may typically have an uncertainty of only ±13%.

Table 3: PFD_{avg} for a high pressure SIF in three scenarios

Scenario	Beta Factor	Proof Test Coverage	PFD_{avg}	Variation from base case
Best Case	1%	90%	2.93E-04	-70%
Base Case	5%	80%	9.88E-04	0
Worst Case	10%	70%	2.10E-03	+113%

Note: Calculation performed using exSILentia software with the following parameters: λ = 600 FIT; MTTR = 8 hrs; PTI =3 yrs; Mission Time = 15 yrs.

6 Conclusion

In this paper, the authors have highlighted multiple modes through which uncertainties are introduced into the LOPA review and subsequently in the reliability analysis for SIF design. Through working examples, this paper demonstrated that without accounting for such uncertainties, the SIF can easily be under-designed or over-engineered. The emphasis on accounting for such uncertainties is also highlighted in the most recent edition of IEC 61511. This paper investigated further into the nature of uncertainties associated with various input parameters associated with LOPA and SIF reliability analysis including approaches using which such uncertainties can be accounted appropriately.

7 Acknowledgement

The authors wish to acknowledge the support received from IRESC including some of the information included in this paper, as well as the contribution from Ms. Nicole Yeung.

8 References

[1] The International Electrotechnical Commission (IEC), Functional Safety – Safety Instrumented Systems for the Process Industry Sector, IEC-61511, Second Edition

[2] The International Electrotechnical Commission (IEC), Functional Safety of Electrical/ Electronic/ Programmable Electronic Safety Related Systems, IEC-61508, First Edition

[3] Guidelines for Enabling Conditions and Conditional Modifiers in Layer of Protection Analysis, CCPS

[4] Layer of Protection Analysis, Simplified Process Risk Assessment, CCPS

[5] Offshore and Onshore Reliability Data (OREDA), Volume 1 – Topside Equipment, 6th Edition, 2015

[6] Akshat Khirwal, Venkatesh S., Consideration of Cumulative Frequency of Demand Scenarios in SIL/LOPA Review, Hazards Asia Pacific, Kuala Lumpur, Malaysia, 2015

[7] Akshat Khirwal, Venkatesh S., Inherently Safer Design & Robust Protection Layers for High Consequence Hazards, Mary Kay O'Connor Process Safety Center, 19th Annual International Symposium, College Station, Texas, USA, 2016

[8] Reliability Prediction Method for Safety Instrumented Systems, PDS Method Handbook, SINTEF, 2010 edition

[9] Safety Equipment Reliability Handbook, Exida, Third Edition, 2007

[10] S. Hauge et al., Common cause failures in safety-instrumented systems: Using field experience from the petroleum industry. Reliab. Eng. Sys. Saf. 151 (2016), 34-45.

479264 Fundamental Research at Purdue Process Safety Research Center

Abstract:

Fundamental Research at Purdue Process Safety Research Center

Significant industrial incidents continue to occur such as fires, explosions and gas releases resulting in serious injuries, loss of life, and damage to the environment and facilities. Such incidents are not solely germane to the oil & gas industry, but occur in a wide variety of industries including agriculture, pharmaceuticals, and manufacturing. Two years ago the 'Purdue Process Safety and Assurance Center' (P2SAC) was established within the School of Chemical Engineering to conduct fundamental research to prevent such incidents. The presentation will address the four research projects the Center is currently pursuing with funding from its initial Sponsors: BP, Chevron, ExxonMobil, Honeywell, Eli Lilly and Shell.

The initial projects include:

- **optimal placement of gas detectors to prevent fires and explosions**: the research uses CFD-based dispersion simulations and rigorous mathematics to perform optimal placement over hundreds of thousands of leak scenarios and potential detector locations.

- **design & synthesis of catalysts to prevent undesired hazardous byproducts**: zeolite catalysts are widely used in the refining industry and certain structures can promote undesired byproducts leading to potential over-pressures and explosions. Zeolite synthesis and characterization methods are being developed.

- **real time process optimization to reduce risk**: Novel continuous manufacturing processes can reduce the risk related to powder handling. Continuous manufacturing, as with pharmaceuticals, requires efficient process monitoring and control.

- **application of inherently safer separation technology** to significantly reduce processing volume, solvent use, and extent of unit operations. Advanced mixed solvent extraction & simulated moving bed techniques are used to recover polycarbonates and flame retardants from plastic waste.

Indirect benefits from the Center are the depth of the core Chemical Process Safety course required of all undergraduates, as well as undergraduate process safety research initiatives in such as areas as dusts, reactive chemicals and nano-materials.

Current focus is on growing the Center and encouraging participation by a wide variety of industries such as pharmaceuticals, agriculture and manufacturing, in addition to oil & gas, and chemicals.

10-13-16

479277 Bringing New Life to Process Safety

Abstract:

<>Abstract:

Whilst risk cannot be eliminated, it can be better managed. The world of Process Safety Management continues to evolve as organizations find better ways of managing risks. In the simplest form, managing the risk is to reducing inherent process risk to the tolerable level.

First step in managing risks is to identify source of Hazards and quantifying risk associated to Hazards. Traditionally organizations heavily rely on expert judgment and brainstorming methodologies like HAZOP for finding operational and design risks. In addition, quality of such analysis are directly related to the precision of assumptions and information from operation and maintenance of the equipment under control.

In this session we will explore how organizations are using new techniques such as dynamic risk visualization, cumulative risk techniques, operational risk assessments and cyber Hazop's to advance the boundaries of process safety management.

This paper will explore how existing systems and data sources, risk visualization and intelligence techniques are being used to bring static process safety data to life. Flat "1 dimensional" information such as lists, forms and databases are now being transformed into "2 dimensional" dynamic information such as plot plans, barrier models, risk matrix, bow ties etc. that provide greater context of relationships between information.

This information will resonate with engineers and technician who rather working with meaningful information of operation and maintenance systems instead of some abstract numbers and concepts that need extensive expertise in the reliability engineering.

We also will look at how advances in Operational Risk Assessments and Cumulative Risk allow operators to determine in real time if the overall cumulative risk is tolerable before actions are taken, devices are removed from service, maintenance activities are deferred etc. In other words, leading indicators and metrics will be in disposal of operators to proactively facilitate decision making for managing risk. This could avoid incidents that developing in the operation due to degradation of control and safety systems or deviations from initial operation and maintenance procedures and may lead to severe incidents.

And finally, since new technologies and architectures for instrumentation, control and safety systems introduced new risks of cyber intrusions and attacks that need to be considered in the risk management and the world of security standards (ISA 62443-3-2 Security Risk Assessment and System Design) and safety standards (ISA-TR84.00.09-2013 Security countermeasures related to Safety Instrumented Systems and Associated IACS) advanced in the last few years, we will look at how Cyber Process Hazard Analysis (PHA) techniques are being applied to provide the same systematic approach to security as traditionally applied to safety.

"Good process safety management does not happen by chance and require constant active engagement", in this paper we suggested novel approaches that can revive and improve the process safety.

479367 Developing an Effective System for Action Item Resolution

Abstract:

PSM action items are are regularly identified through many different means in a typical chemical processing facility. Knowing what to do with these action items and how best to proceed is no easy task. Not all action items are created equally, or even with the same goal in mind. An action item identified in a PHA may present a direct hazard to an employee, while an action item generated from a pressure relief systems study may present a documentation concern, a deviation from current practices, a direct hazard to an employee, or any combination thereof. Many facilities have a system in place for shepherding action items to completion, but these systems typically assume all action items are created equally. In this paper, we will demonstrate the importance of having a system that handles each action item in the most effective way; from evaluating the nature of the concern, considering the risks associated with the action item, developing the action plan, to completing the action plan. We will also cover whose input should be considered at each stage in the process to help ensure the best solutions possible are developed.

13TH GLOBAL CONGRESS ON PROCESS SAFETY

Where Does Management of Change (MOC) Begin?

Juliana Schmitz
Praxair, Inc.
Av. Pastor Martin Luther King Jr., 126, bloco 10/ala A,
Corporate Shopping Nova América, Rio de Janeiro, RJ, Brazil
juliana_schmitz@praxair.com

Patti Jones
Praxair, Inc.
175 E Park Dr, Tonawanda, NY, 14150, USA
patti_jones@praxair.com

Gabriel Drummond
Praxair, Inc.
Av. Pastor Martin Luther King Jr., 126, bloco 10/ala A,
Corporate Shopping Nova América, Rio de Janeiro, RJ, Brazil
gabriel_drummond@praxair.com

Prepared for Presentation at
American Institute of Chemical Engineers
2017 Spring Meeting and 13th Global Congress on Process Safety
San Antonio, TX
March 26 – 29, 2017

Where Does Management of Change (MOC) Begin?

Juliana Schmitz
Praxair, Inc.
Av. Pastor Martin Luther King Jr., 126, bloco 10/ala A, Corporate Shopping Nova América, Rio de Janeiro, RJ, Brazil
juliana_schmitz@praxair.com

Patti Jones
Praxair, Inc.
175 E Park Dr, Tonawanda, NY, 14150, USA
patti_jones@praxair.com

Gabriel Drummond
Praxair, Inc.
Av. Pastor Martin Luther King Jr., 126, bloco 10/ala A, Corporate Shopping Nova América, Rio de Janeiro, RJ, Brazil
gabriel_drummond@praxair.com

Keywords: MOC, engineering, process safety

Abstract

Management of Change is often affiliated with an operating facility. However, several projects have the common challenge of being developed within the constraints of an aggressive schedule, considering the time from project kick-off until unit start-up. In order to comply with the deadlines, early releases of engineering design documentation become subject to changes. During project development, these changes may come as scope changes, supplier updates, etc. that may affect more than one engineering discipline and any on-going construction/assembly activities. Another change source may be non-conformances identified during the construction/assembly phase. In this case, considering project milestones and costs, an option taken into account is to leave the system or portion the way it is instead of re-working to match the original design. For all of the given cases, a robust Engineering Management of Change (EMOC) process must be in place throughout the project life cycle to guarantee modifications are properly identified, communicated, and evaluated by technical experts for possible effects on process and design safety prior to implementation, or at a minimum, prior to startup of the system/subsystem. In addition, an EMOC process gives the opportunity to analyze the reasons for the modifications in order to avoid increased costs and design errors on future projects, leading to improved quality control. This paper presents some previous accidents related to management of changes, and addresses EMOC implementation, work process and elements of a successful EMOC program.

1 Introduction

Preliminary design activities begin early in the process to define scope, cost and product delivery requirements. Changes are a natural part of these efforts.

As the project progresses through the engineering detail design phase, a process hazard analysis (PHA) is conducted. At this moment, a multidisciplinary team evaluates if the process is safe and if any modifications are required. The project team is informed of the revised documentation and proceeds with each skill activity.

Subsequent changes may occur as, for example: updated supplier documentation, the team that will effectively operate the plant in the future may raise a new need for the project, some may be identified during detail design requiring changes and change requests during the construction phase. Any modification after the PHA may not be properly communicated to the adequate professionals and analyzed, resulting in a possible unsafe condition.

This paper focuses on project execution phases until plant startup and turnover to operations where the Management of Change (MOC) work process begins. It also explores the challenges identified when developing an Engineering Management of Change (EMOC) work process and shares an approach taken within Praxair, Inc.

2 Why is the Engineering Management of Change Important?

Process Safety is a relatively new engineering area, being developed from the early 60's and mostly based on catastrophic incidents, which negatively affected the industry. Kerin [5] confirms this point of view and lists a series of landmark events, their key learnings and influences in process safety. These events had different causes, mostly related to poor management of various safety areas, such as hazard identification, training, maintenance, management of change (MOC) and others.

The numerous and notable MOC related accidents that happened in the industry over the last century showed the world the need of a well-developed and maintained MOC procedure.

2.1 Titanic, 1912

The sinking of the Titanic, in 1912, is a well-known catastrophic incident. "The design of the ship included 44 water tight compartments giving rise to the perception that the ship was unsinkable and in the event of an accident would be its own lifeboat." [2] An accelerated construction schedule was the catalyst in making changes that proved to be fatal. A robust management of change process and a cultivated safety culture may have prevented the incident.

2.2 Flixborough, 1974

Venart [3] describes the Flixborough accident, which occurred on June 1st, 1974, causing 28 fatalities and several property damages near the Nypro plant in North Lincolnshire, UK. In this plant, there was a process area responsible to produce caprolactam, an intermediary for the Nylon production. Caprolactam production was conducted through the oxidation of the cyclohexane, a highly flammable chemical. There were six reactors in series where this oxidation reaction was performed.

Two months before the explosion, a crack was observed in the 5th reactor of the chain. With the objective of reducing the impact in production, the responsible plant personnel decided to connect directly the 4th and 6th reactors while the 5th was removed for maintenance. To proceed with the change, a "dog-leg" pipe was used to connect the reactors 4 and 6, while the original installation considered a straight pipe between these reactors. This is a modification that clearly changed the characteristics of fluid flow between reactors 4 and 6. The dog-leg line support was a temporary installation, made of scaffolding bars. These changes were never evaluated in detail and during start-up. On June 1st, the bellows connected to the reactors flanges failed due to an overstress and tons of cyclohexane was released into the atmosphere. With the vapor dispersion, an ignition source was found in the plant, and a vapor cloud explosion occurred resulting in fatalities, as well as significant property damage. [4]

2.3 CONDEA Vista Company, 1998

U.S. Chemical Safety and Hazard Investigation Board [5] promotes a discussion about the reaction vessel explosion in CONDEA Vista Company, on October 13th, 1998. In this unit, linear alkyl benzene was produced by mixing aluminum chloride and hydrocarbons. Around 3 months before the incident the process was modified, involving a mixing of powdered aluminum with hydrogen chloride to form the aluminum chloride. After this change was implemented, the bottom nozzle of the reactor got plugged due to the formation of a sludge residue, which the operators were not able to reach and remove during maintenance. Considering this scenario, the chemist of the site and the process engineer decided to use water to dissolve the sludge and to absorb the heat generated due to an exothermic reaction between the water and the sludge. This process was started and once it was apparently under control, the process engineer and chemist left the plant.

The shift supervisor decided to make a new change and inject steam at the nozzle. The process supervisor who was not in the plant agreed with the new procedure. At the shift turnover, the day shift supervisor talked with the night shift supervisor about the steam injection. The night shift supervisor took responsibility to inject steam and oriented an operator to execute this procedure.

After a few minutes of continuous steam injection, the reactor exploded, leaving three workers injured and property damage estimated in $13 million USD. Among other shortcomings related to the catalyst change that resulted in sludge formation, U.S. Chemical Safety and Hazard Investigation Board **[5]** affirms that this accident highlights the value of having an authorization or approval step as part of an MOC system for abnormal situations. If such a procedure had been in place, a technical manager would have reviewed the proposed procedure involving steam and may have detected its deficiencies.

2.4 Prevention

Among others, these accidents demonstrate the importance of Management of Change in process safety. This was emphasized when MOC was included as one of the main pillars of the Process Safety Management (PSM) standard, issued by the U.S. Occupational Safety and Health

Administration and of the Risk Management Plan of the Environmental Protection Agency (EPA).

However, not only changes performed after the plant startup will result in safety accidents. Engineering Management of Change (EMOC) is also a key area to prevent accidents and should be addressed in a proper manner by industrial companies.

One example of a poor EMOC related event can be an updated drawing sent by an equipment supplier during the project execution phase indicating a decrease of a vessel Maximum Allowable Working Pressure (MAWP). If this modification is not communicated to the proper professionals, the pressure relief device set point designed for the equipment may not be adjusted as required, which can lead to lack of vessel overpressure protection and possible vessel rupture.

Another example can be a change of a pipe routing during project phase, without revision of the flexibility study. This could result in pipe over stress, leading to a line rupture and energy release.

Several other examples of poor EMOC related accidents could be discussed showing proper management of changes during engineering is indispensable for the safety of the plant, workers and environment. Figure 1 indicates some sources of modifications that could impact the safe startup/operation of the system.

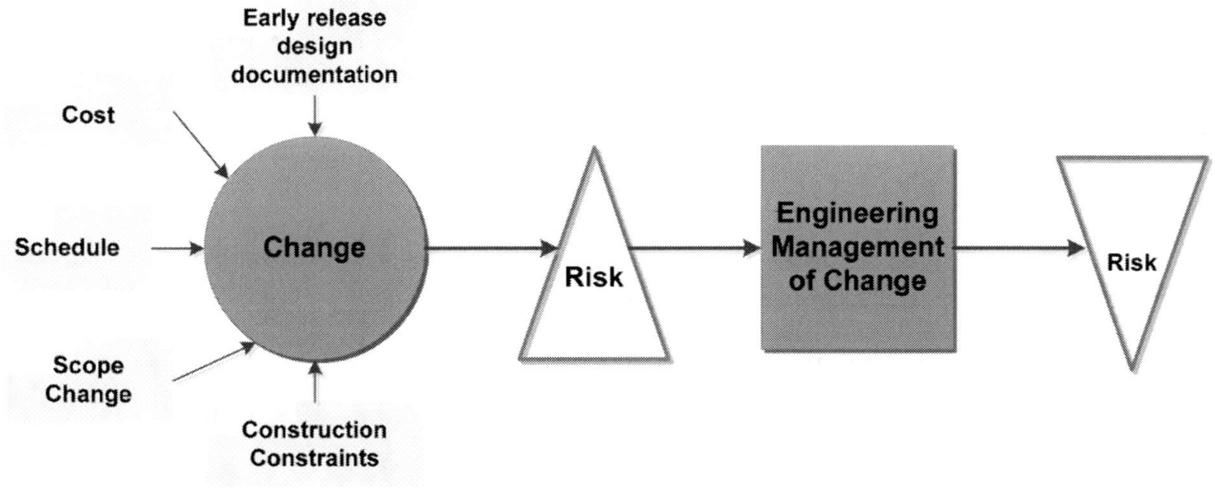

Figure 1 Integrated management of change system

Unfortunately, EMOC use is not as widespread as MOC, which still gives space for industrial accidents to happen. The main purpose of defining a work process for managing changes, even during the project development phases, is to guarantee all changes are adequately communicated, reviewed, new hazards are identified and analyzed and a mitigation plan is defined to ensure that a safe installation and operation is delivered to the customer.

3 EMOC Implementation

A change is known as a deviation from the original design. Modifications may come as scope changes, supplier updates, or other sources, and they may affect more than one engineering discipline and any on-going construction/assembly activities.

"Changes in construction projects are very common and likely to occur from different sources, by various causes, at any stage of a project, and may have considerable negative impacts on item such as costs and schedule delays." [6] If there is no formalized change management process, impacts not only on cost and schedule may be observed, but also process safety events, as mentioned on the previous section of this paper.

According to CCPS [7], the level of detail covered by the management of change system should be based on

- The hazards or risks of the process
- The existing process safety culture at the location

The company should develop a system for the control of modifications implemented on the project under design, construction, commissioning or startup phases. This work process should include approval of the change by a qualified professional or team, depending on the complexity of the hazards associated.

Based on the hazards involved in the specific process, the type of changes that need evaluation by an engineering expert from the appropriate discipline, as well as changes that do not, shall be defined when designing the specific EMOC work process for a certain company. For example; changes that can affect safety systems, critical operating parameters, shutdown set points, relief valve settings, materials of construction, etc. should undergo EMOC approval. Minor corrections to documentation or clarifications may not.

The management of change work process may not be the same in every industry and this does not mean one is better than another. The culture of the company and the location may require different applications of the MOC activities. [7]

3.1 Critical Phase

Changes during construction inherently introduce unexpected risk, especially under schedule and cost pressure. Some changes that can affect process safety may not be obvious to construction personnel. A simple yet important example is relocating a support on the discharge of a relief valve due to interference. Another may be substituting flange bolt material. Regardless of the origin of the change, evaluation and approval should take place prior to starting up the system/subsystem. Identification of all sources which can generate an engineering change must be considered as a natural part of the overall work process.

4 Proposed EMOC Work Process

Typically, during the engineering phase, several design safety meetings are conducted, such as P&ID Reviews, Layout Reviews, PHAs and FSSs. These reviews shall be conducted with the participation of all engineering disciplines related to the plant design. After these project landmark meetings, several changes may occur and their evaluation is a critical part of risk management throughout the project life cycle. At this moment, after the main safety reviews are performed, the EMOC work process is expected to begin.

As shown on Figure 2, all changes related to the project design are identified and communicated by the professional who initiated the modification, to a single point of contact within the project team.

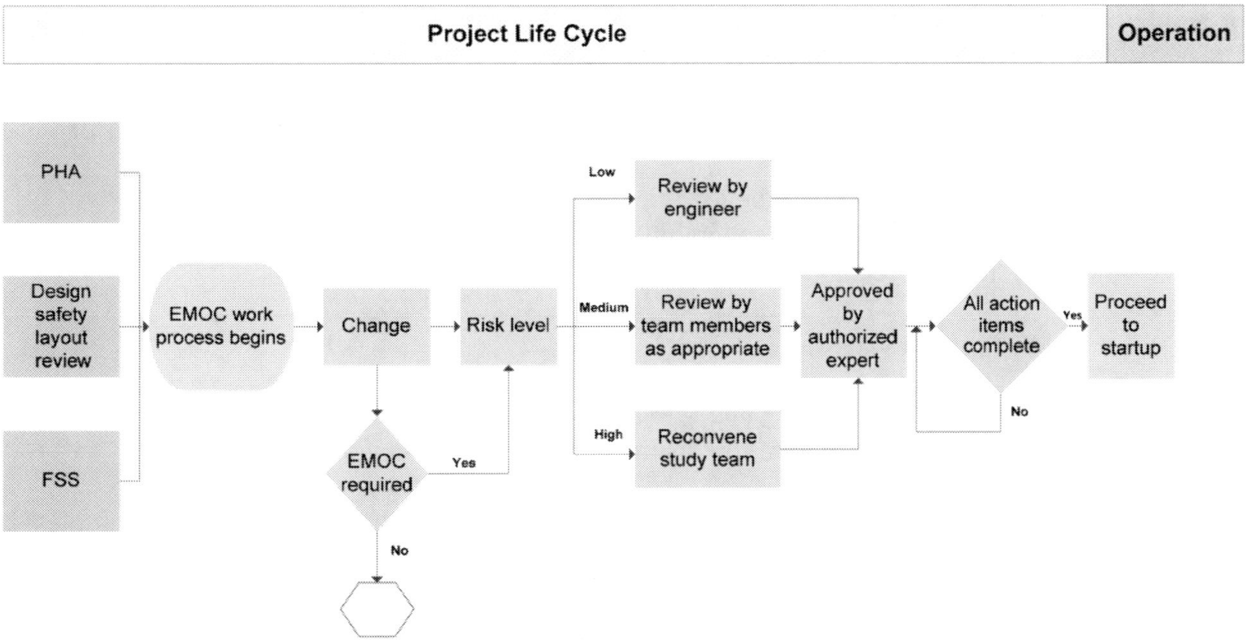

Figure 2 High-level EMOC Workflow Diagram

In cases where a modification is only a correction, clarification or replacement in a kind, there is no need to go through the full EMOC work process. The change can be recorded in a simple spreadsheet containing the type of change, a brief explanation and a mitigation plan to make sure nothing has been missed.

The modification should then be classified as low, medium or high risk per the EMOC owner and the level of review is defined based on the associated risk. The risk ranking criteria shall be described on the company's EMOC program; e.g.:

- Low: addition of a local pressure indicator on the design for maintenance purposes only.
- Medium: manual isolation valve removal, while valve not indicated at PHA as part of a hazard scenario.

- High: customer addition of a tie in point near the pipeline battery limit, which could lead to misdirected or reverse flow, with potential for process safety impact.

After reviewing the change, an approval should be required by a Subject Matter Expert (SME) before proceeding to change implementation and mitigation plan. The plant will be allowed to start-up only after all EMOC items were reviewed and approved.

It will be responsibility of the company implementing EMOC to determine which kind of changes will require a complete review, the risk categories that will be available to be assigned, both qualitatively and/or quantitatively and the required level of review according to these risks. In the example presented in Figure 2, a low risk results in a review by an engineer, a medium risk results in a review by the team members and a high-risk change will require the engineering team involved in the HazOp, Design Safety Layout review or the FSS evaluations to reconvene.

Several different approaches are possible, as long as they are simple, easy to understand by those responsible for implementation and provide a robust EMOC work process.

5 Elements of a Successful Engineering Management of Change Program

Successful EMOC programs define the key elements applicable to their industry. Some will be common such as commitment from leadership, roles and responsibilities, etc. Included below are some elements to take into consideration.

- Commitment from Leadership;
- Roles and responsibilities as they apply to specific skill sets (i.e. process engineering, controls system engineering, manufacturing, construction management, etc.);
- Clear definition of changes that must undergo engineering evaluation, including examples;
- Definition of changes that are not considered EMOC related;
- Risk level associated with the change;
- Approvers limited to subject matter experts;
- A streamlined work process focused on function over form, aligned with the various phases of the project life cycle;
- Identification of change sources;
- In process verification at predetermined milestones;
- Robust training that includes internal and external examples accompanied by a test with feasible scenarios;
- Associated PSI updated accordingly;

- Auditable;
- Evergreen
  - Periodic calibration of the work process;
  - Input from those responsible to implement the process;
  - Captures internal data to enhance understanding of what constitutes a change that requires EMOC review.
- Lessons learned captured and implemented on future projects;
- Leverage key learnings from external events in where MOC was identified as the cause or a contributing factor.

6 Conclusion

An aggressive schedule may lead to early release of design documentation that is still subject to change, which could result in the execution of PHA studies earlier than the optimal period. Although the early study may provide an opportunity to identify key risks earlier in the process, a significant number of action items and changes will probably occur, which may affect the integrity of the PHA.

A well-defined and implemented EMOC program is an essential part of risk management. The methodology applied allows systematic control of project changes which helps to ensure process, personnel and environmental safety. An additional benefit is facilitating a proactive means to analyze the underlying cause for unplanned changes. These lessons learned can assist on future projects to manage cost and improved quality control.

The main purpose of the management of change is to avoid loss of primary containment (LOPC), especially of hazardous chemicals, which could lead to catastrophic consequence. Once the EMOC process is successfully implemented, the reduction of quality issues will become a natural part of the process, resulting in a safer and higher quality facility/assembly/operation. An effective Engineering Management of Change work process is critical to the safe commissioning and startup of a chemical plant.

7 References

[1] T. Kerin. The evolution of process safety standards and legislation following landmark events – What have we Learnt?, Process Safety Progress 35 (2016), 165-170.

[2] CCPS. *Incidents that Define Process Safety*, Center for Chemical Process Safety, American Institute of Chemical Engineers, New Jersey, 2008.

[3] J.E.S Venart, Flixborough: The explosion and its aftermath. Process Safety and Environmental Protection 82 (B2): 105 – 127 (2004).

[4] C. Sadee, D.E. Samuels and T.P. O'Brien, The Characteristics of the explosion of cyclohexane at the Nypro (UK) Flixborough plant on June 1974, Journal of Occupational Accidents, 1 (1976/1977) 203 – 205.

[5] U.S. Chemical Safety and Hazard Investigation Board, Management of Change (No. 2001-04-SB) (2001).

[6] Q. Hao et all, Change Management in Construction Process, International Conference on Information Technology in Construction, Santiago, Chile, 2008

[7] CCPS. *Guidelines for Management of Change*, Center for Chemical Process Safety, American Institute of Chemical Engineers, New Jersey, 2008.

479400 Effectively Utilizing Calarp/RMP/PSM Contractors to Assist in Optimizing Process Safety Culture, Compliance with Standards, and Workforce Involvement

Abstract:

Since the inception of Occupational Safety and Health Administration (OSHA) Process Safety Management (PSM) in 1992, facilities have frequently contracted process safety consultants to satisfy regulatory needs, as well as assist in complying with industry best practices.

The primary objective of this paper is to provide tangible suggestions for effectively utilizing process safety contractors for assisting in optimizing process safety culture, compliance with regulatory requirements and industry best practices, as well as tips to improve workforce involvement and coordination. The core of this paper will detail various management styles of the prevention program and an evaluation of influences on process safety culture based on varying degrees of involvement from the facility and usage of contractors. This paper aims to provide an in-depth analysis of the pros and cons of contractor involvement and its efficacy relative to the scale of the facility and intricacy of corporate structure based on anecdotal evidence and industry experience.

Key topics include:

- Overview of different types of process safety culture in relation to prevention program management and contractor usage,
- The Challenge - Striking an optimal balance between contractor use and facility involvement whilst maintaining economic viability,
- The Challenge for Multi-Facility Corporations - Ensuring changes in corporate standards are successfully communicated to all children facilities ranging from documentation updates and organizational changes,
- Suggestions and Improvements - Implementation of effective prevention program management systems to ensure workforce involvement and accountability, as well as practical allocation of contractor resources to accomplish critical tasks whilst balancing facility resource commitment.

A Streamlined Approach for Full Compliance with SIF Implementation Standards

William G. Bridges, President
PROCESS IMPROVEMENT INSTITUTE, INC. (PII)
1321 Waterside Lane, Knoxville, TN 37922
Phone: (865) 675-3458
Fax: (865) 622-6800
e-mail: wbridges@piii.com

Art Dowell III, Principal Engineer
PROCESS IMPROVEMENT INSTITUTE, INC. (PII)
2437 Bay Area Blvd PMB 260
Houston, TX 77058
Phone: 713-865-6135
e-mail: adowell@piii.com

Prepared for Presentation at
13th Global Congress on Process Safety
San Antonio, TX
March 27-29, 2017

UNPUBLISHED

13TH GLOBAL CONGRESS ON PROCESS SAFETY

A Streamlined Approach for Full Compliance with SIL Implementation Standards

William G. Bridges, President
PROCESS IMPROVEMENT INSTITUTE, INC. (PII)
e-mail: wbridges@piii.com

Art Dowell III, Principal Engineer
PROCESS IMPROVEMENT INSTITUTE, INC. (PII)
e-mail: adowell@piii.com

Keywords: Safety Instrumented System, SIS, Safety Integrity Level, SIL, LOPA, Layer of Protection Analysis, Independent Protection Layers, IPLs, human factors

Abstract

Many companies put FAR too much redundant effort into determining what SIL (safety integrity level) is needed and then verifying the SIF (safety instrumented function) design will give the SIL targeted. This paper shows how to apply the qualitative definition of independent protection layers (IPLs) within the setting of a process hazard analysis (PHA) to get most of the gain from Layer of Protection Analysis (*LOPA)* without doing a LOPA (without using numerical values). We show how we use a PHA team to identify when a SIF is needed and to select the proper target SIL. This portion of the SIL evaluation and the identification and labeling of the IPLs during the PHA/HAZOP does not take any longer than a normal PHA/HAZOP, once the right habits are established. Note that this approach eliminates the need for a separate SIL Evaluation Study to identify the SIFs and select the target SIL. Then, this paper describes how to perform the SIL Verification and Safety Requirements Specification (SRS) remotely, again without the need for a redundant team meeting. This approach has been used at many sites and for thousands of SIFs.

Background

Typically, hazardous consequences, their causes, and safeguards are identified in a PHA/HAZOP (process hazard analysis / hazard and operability analysis)[1]. The PHA may determine if additional protection layers are required or if additional analysis is required.[2]

In many organizations, a separate LOPA (layer of protection analysis)[3] meeting is held to determine which HAZOP safeguards are independent protection layers (IPLs) and to determine if additional IPLs are needed to meet the risk tolerance criteria. If SIFs are needed, the required risk reduction (PFD, probability of failure on demand) is determined in LOPA, which then sets the SIL.

Some organizations may even have a separate SIL evaluation study, which is certainly redundant to a well-run PHA and is wasteful if the PHA is properly staffed and managed and/or if any subsequent LOPA are performed well.

Fundamentals of SIL Assessment

The lifecycle requirements for safety instrumented functions are specified in industry standards IEC 61511 (ANSI/ISA 84.00.01-2004, Part 1)[4,5]. The required steps to specify the SIS (safety instrumented system) are:

1. Safety Instrumented Function (SIF) identification.
2. Determining the Safety Integrity Level (SIL) for each SIF.
3. Designing the SIF to meet the required SIL.
   SIL Verification Calculation (actually, this calculation is iterative with step 3, but the end calculation is a deliverable that proves 3 is correct). The SIL standards allow the following methods to use to determine the required SIL (ANSI/ISA 84.00.01-2004, Part 3, Section 3.8)[4]:
   - Fault Tree Analysis (FTA), Full Quantitative Risk Analysis (QRA)
   - LOPA – simplified quantitative analysis
   - Qualitatively within PHA (HAZOP, etc.)

   > "A qualitative method may be used as a first pass to determine the required SIL of all SIFs. Those which are assigned a SIL 3 or 4 by this method should then be considered in greater detail using a quantitative method to gain a more rigorous understanding of their required safety integrity." – *ANSI/ISA 84.00.01-2004, Part 3, Section 3.8*[4]

Table 1 compares risk analysis approaches for Steps 1 and 2.

FTA and QRA are very detailed and labor-intensive and would be required for complicated systems, particularly where there is not complete independence between safeguards.

LOPA is a simplified analysis tool that requires fewer resources, but is heavily dependent on the requirement that all the safeguards be completely independent.

However, for many processes, there are a number of scenarios (cause leading to a consequence) where the PHA team can readily determine sufficient independent protection layers are already in place.

Table 1: Comparison of Risk Analysis Approaches

Approach	Assessment Methods	Risk Judgment Method	Risk Judgment Method	Estimated Range of the Results
Qualitative Only	HAZOP, FEMA	Expert Voting, focusing on site data	Capable >95% of time	Plus or minus 1/2 order of magnitude
Simplified Quantitative	LOPA, Risk Graph	Multiplication of statistical averages of general failure rate data; with broad assumptions on management systems	Needed on about 5% of the scenarios	Plus or minus 1 order of magnitude
Full Quantitative	FTA, ETA, QRA, HRA		Needed for less than 0.01% of the scenarios	Plus or minus 1 order of magnitude

Figure 1 contrasts the typical approach to SIL assessments with the optimized streamlined approach.

Typical SIL Determination versus Streamlined Approach

In the typical approach,

- **Step 1**: The PHA/HAZOP is performed using a multidisciplinary team of engineers, operators, instrumentation staff, maintenance staff, and perhaps vendors. Then in a separate SIL study, a multidisciplinary team of engineers, instrumentation staff, maintenance staff, and perhaps vendors brainstorm the hazard scenarios (again) and decide which are candidates for SIFs. A semi quantitative approach, such as RiskGraph, or layer of protection analysis, determines if an SIF is required.
- **Step 2:** The RiskGraph dictates the required SIL, or the LOPA compares the mitigated risk to the risk tolerance criteria and determines the SIL from the order-of-magnitude risk reduction needed.

In contrast, in the streamlined approach,

- **Step 1:** The PHA/HAZOP is performed using a multidisciplinary team of engineers, operators, instrumentation staff, maintenance staff, and perhaps vendors (to this point the same as the typical approach). Within the PHA/HAZOP, the team estimates the risk of each scenario using qualitative judgment (based on experience, knowledge, and memory of site-specific data). The PHA/HAZOP team then qualitatively determines if the best risk reduction method is an SIF.
- **Step 2:** The PHA/HAZOP team qualitatively determines the SIL needed (typically by judging the level of redundancy needed based on the value they place on each IPL and the new or modified SIF needed)

Figure 1: Comparison of the Typical and the Optimized Streamlined Approach from PII for SIL Assessments (Steps 1 and 2)

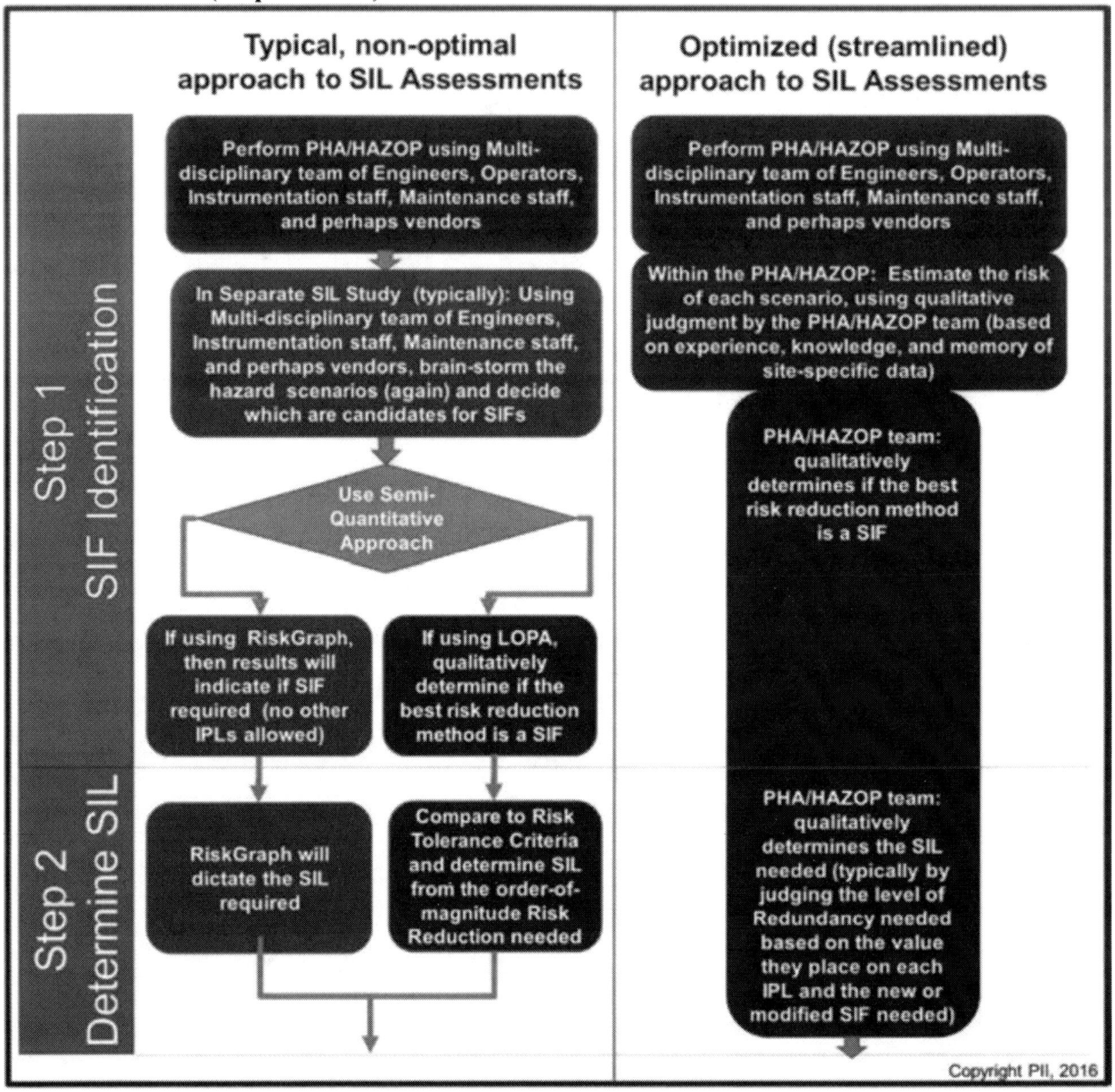

Figure 2 also contrasts the typical approach to SIF design and verification with the streamlined approach. Steps 3 and 4 are iterative.

In the typical approach,

- **Steps 3 and 4**:
  - Decide on basic architecture of SIF, from experience with other SIFs.
  - Decide on vendors of components
  - Provide full design of SIFs from experience

- Perform trial SIL verification calculations using software such as exSILentia™ using actual component data
- If the target SIL is achieved, issue the SIL verification calculation report and the safety requirement specification (SRS)
- If the target SIL is not achieved, revise the design and recalculate.

Figure 2: Comparison of the Typical and the Optimized Streamlined Approach from PII for SIL Assessments (Steps 3 and 4)

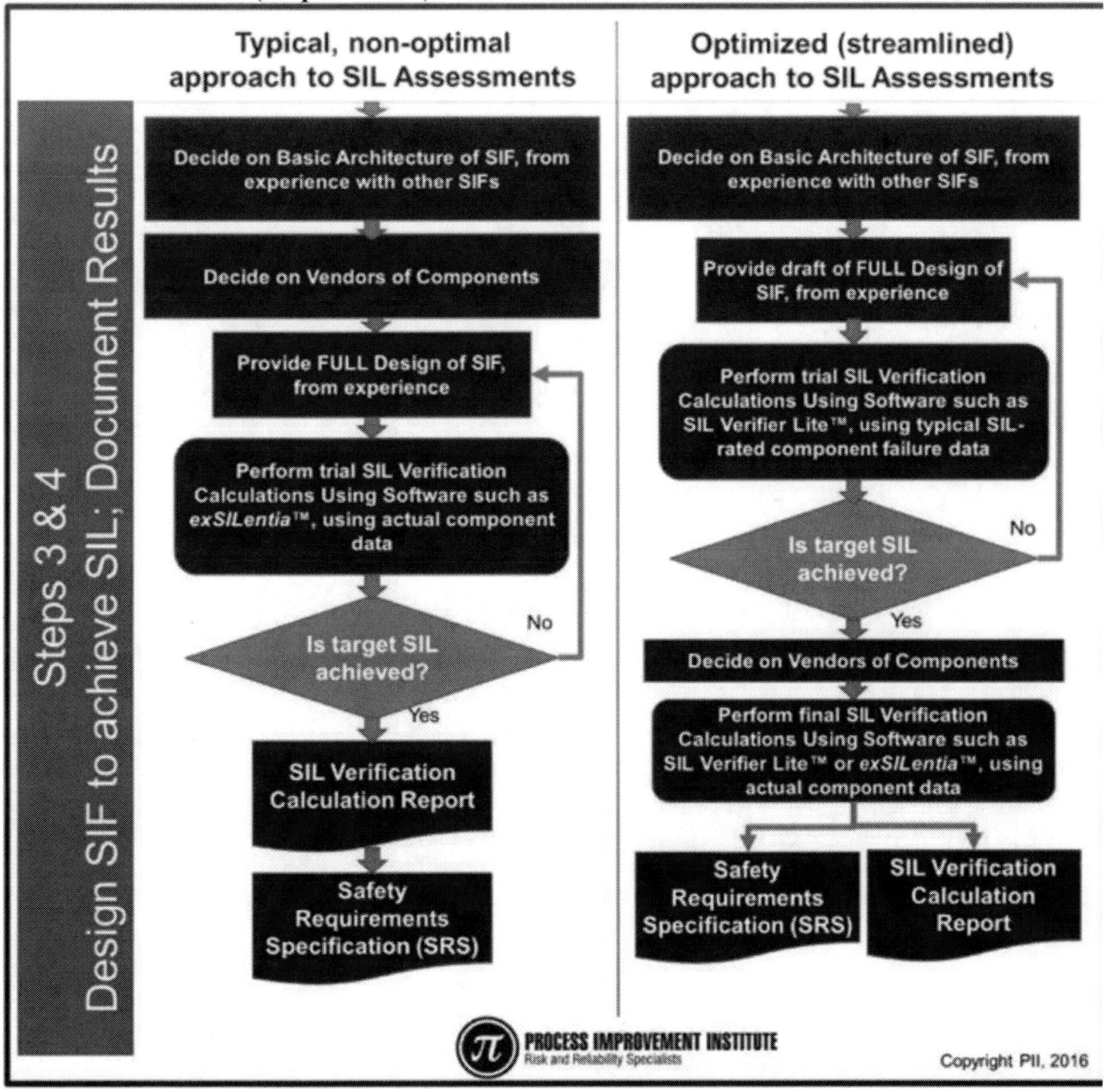

In the streamlined approach,

- **Steps 3 and 4**:
  - Decide on basic architecture of SIF, from experience with other SIFs.

- Provide draft of full design of SIF, from experience.
- Perform trial SIL verification calculations using software such as SIL Verifier Lite™ using typical SIL-rated component failure data.
- If the target SIL is not achieved, revise the design and recalculate.
- If the target SIL is achieved,
    - Issue the SIL verification calculation report and the safety requirement specification (SRS).
    - Decide on vendors of components.
    - Perform final SIL Verification Calculations Using Software such as *SIL Verifier Lite*™ (from PII) or *exSILentia*™ (from exida) using actual component data.
    - Issue the SIL verification calculation report and the safety requirement specification (SRS).

The advantage of the streamlined approach is determination of the need for an SIF and of its SIL quicker and with fewer resources than the typical approach.

Requirements for the Qualitative Streamlined Approach for SIL Determination

For the qualitative streamlined approach to be used effectively, there are important requirements for the PHA/HAZOP team.

- PHA/HAZOP team leader is knowledgeable and proficient in LOPA.
- PHA/HAZOP team members have the following information in memory:
    - Consequence severity.
    - Initiating cause frequency.
    - Independent protection layer (IPL capability).
    - Risk tolerance criteria.

An experienced PHA team, who is knowledgeable in the process technology, can effectively and efficiently analyze a scenario and mentally perform a layer of protection analysis. The team determines the consequence severity and the initiating cause frequency. A highly competent team understands which safeguards are truly independent protection layers and can judge the risk reduction provided by each. Then the team can determine if additional IPLs are required to meet the risk tolerance criteria. For some scenarios, an SIF may be required and the team can determine the required probability of failure on demand for the SIF and can thus determine the required SIL.

A best practice is that the PHA team recommends the functional description and the SIL for the proposed SIF. The functional description does not specify the details of the SIF design, rather it specifies what process variables should be considered and what action should be taken to detect the upset and to prevent the consequence. For example, the functional description may say "monitor the level of a vessel and stop the flow into the vessel to prevent overflow".

If the PHA team is uncertain, or if the scenario is complex, then the PHA can recommend a detailed LOPA be done for the scenario.

These conclusions are based on more than 8000 PHA/HAZOPs, 3000 LOPAs, and about 100 QRAs and HRAs. We have observed PHA participants becoming knowledgeable about LOPA and IPL requirements. The participants then began to apply the principles of LOPA spontaneously in the PHA. With good leadership, the PHA team could then easily move to the full streamlined approach.

The experience and competency of the PHA leader and the PHA team is extremely important. If the PHA team is not comfortable with the principles of LOPA, and especially the principles of IPLs, the streamlined approach is not appropriate. Likewise, the PHA team must have the required information and understanding in its memory. Otherwise, the qualitative approach can degenerate into the "arm wrestling" arguments that PHA teams used qualitatively to determine SIL requirements before the introduction of LOPA.

Suggestions for the Streamlined Approach for SIF Design

Likewise, there is a streamlined approach for the SIF design. Organizations have developed a recipe book of typical SIF designs for SIL 1, SIL 2, and SIL 3. Depending on the required SIL, the designer can choose a design from the recipe book (as described in Figure 2, "from experience"). The recipe book provides a starting point for the detailed SIF design. Figure 3 shows a small sample of SIF designs that could be included in a recipe book.

Figure 3: Typical SIF Architectures for Information

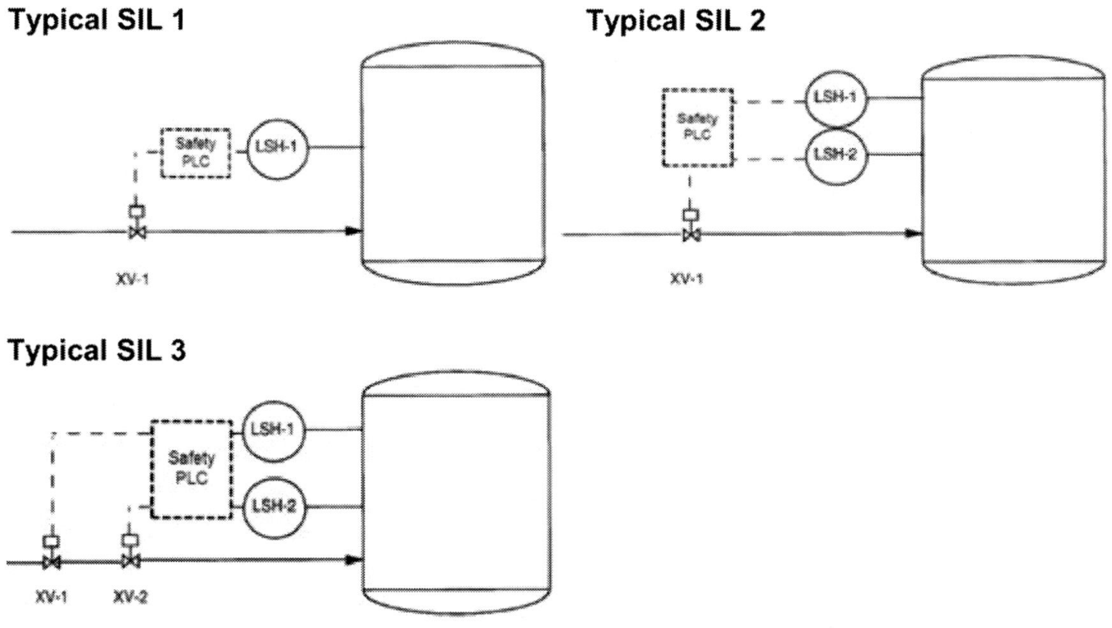

Case Study 1

An example of the streamlined approach is summarized in Table 2 for a facility with storage spheres (in China; the PHA and other analyses were performed in 2014). The table shows two deviations from the PHA. IPLs are identified and existing SIFs are assigned an SIL, based on the architecture. (Note that the achieved SIL for each of the SIFs (existing or recommended) must be confirmed by a SIL verification calculation that includes the failure rate of the components, the voting architecture, and the proof test interval.)

For the high-level deviation, the high-level SIF was assigned SIL 1. The overflow through the pressure equalization line to other spheres was determined to be an IPL.

For the low level deviation, the low level indication and low level alarm were determined not to be an IPL because they are part of the initiating cause. The PHA team concluded that feeding from two spheres at all times is an IPL for this facility because of the unlikelihood for both spheres to have low level at the same time. The PHA team also thought that an additional SIF level alarm with operator response to switch tanks within 60 minutes could be a possible IPL, if the action of the operator is quick enough. The team wrote a recommendation to ensure that the operator response could be an IPL.

In this case study, the PHA team was able to determine in the PHA meeting which safeguards were IPLs. They also determined the nominal SIL for existing SIFs (to be confirmed in a later SIL verification). And the team determined that operator response could be an IPL if there was a documented procedure and an annual drill. For these two deviations, the PHA team did the work that would be done in a later LOPA meeting in the typical approach.

Next, the SIL Verification was performed for the SIL 1 SIF (and all other SIFs in the unit). This SIL Verification was accomplished using exSILentia™ software and the analysis and documentation took less than 2 hours.

Thus, the entire requirements for a SIL identification, SIL target level determination, and SIL Verification were accomplished very quickly compared to typical approaches.

Table 2: Storage Sphere PHA

No.: 2 XXXX storage spheres xxx-T-XX A/B/C/D/E/F/G/H/I/J/K/L (1 of 12)					
#	**Dev.**	**Causes**	**Consequences**	**Safeguards**	**Recommendations**
2.1	High level	Too much flow to one sphere from XX Plant (through their pump; about 40 bar MDH)	High pressure (see 2.5)	High level SIF with level sensors voted 2oo2, to close inlet valve - **SIL 1** Overflow thru pressure equalization line to other spheres (through normally open [NO] valve) - **IPL**	
2.2	Low level	Failing to switch from the sphere with low level in time (based on level indication)	Low/no flow - Liquid from spheres through high pressure product pumps to the vaporizer (see 4.2) Low/no flow - Unqualified liquid from spheres back to Plant (see 6.2)	Level indication and low level alarm, inspected each year, per government regulation (**not IPL; part of the cause**) 9 other spheres with possibly enough level to switch to Feeding from two spheres at all times, so unlikely for BOTH spheres to have low level at the same time - **IPL** Two level indication from SIS level transmitter, with low level alarm, with more than 60 min available to switch tanks (SIF driven alarm and response) - **possible IPL, if action of the operator is quick enough**	Rec 4. Make sure the Human IPL of response to low level in all spheres and tanks is described in a trouble-shooting guide (like an SOP) and practiced once per year per unit operator. This will make this response a valid IPL.

Case Study 2

In 2014 and 2015, an identical approach to that of Case Study 1 (the streamlined approach of this paper) was performed for a large gas producer in South America. In that work, the PHAs for more than 500 nodes plus all procedures of all operating modes was completed in about 10 weeks of PHA meetings. The PHA team identified more than 800 SIFs, along with their target SILs, using the approach in Steps 1 and 2.

Next, staff within PII but separate from the PHA team performed about 60 LOPA to clarify certain issues, some related to SIFs. But less than 5% of the SIFs were established with LOPA; 95% were established with the PHA alone.

Finally, staff separate from the PHA performed SIL Verification of the 800+ SIFs, using exSILentia™. The verification was for a combination of SIL 1 and SIL 2 SIFs; a couple of SIL 3 SIFs were also analyzed and debated. The relatively few LOPAs took about 2-3 hours each (due to the complexity of the 5% of scenarios that went to LOPA) and each SIL Verification took about 2-3 hours to perform and document.

Caution – the issue of human factors for SIL 2 and SIL 3 SIFs[6]

There is an important aspect of important aspect of SIFs and the possible achieved SIL that is not addressed in this paper. Relying on a high integrity SIF, without accounting for human error during interventions, is also a waste of resources. You may believe that you have reduced the risk by a factor of 100 for a SIL 2 SIF, or a factor of 1000 for a SIL 3 SIF, when in fact the human factors during commissioning and proof testing may degrade the overall risk reduction factor to about 10. If the human factor is accounted for in the SIF design phase and the SIL verification phase, then the human error can also be prevented to allow the higher reduction in risk that is targeted.

Conclusions

Based on experience in a multitude of PHAs, LOPAs, QRAs, and HRAs, we have seen that experienced, competent PHA teams with an experienced, competent leader can effectively apply the principles of LOPA to many scenarios during the PHA meeting. The team can determine the approximate SIL of existing SIFs, the team can determine which safeguards are truly IPLs, and the team can recommend additional IPLs, including SIFs with specific SILs. If the scenario is too complex, or the PHA team is uncertain, the team can recommend a detailed LOPA. The streamlined qualitative makes better use of resources and time than the typical approach with separate meetings for PHA and LOPA.

It is critical that the PHA leader is experienced and competent in LOPA. It is critical that the PHA team has in its memory the data for consequence severity, initiating cause frequency, IPL requirements, and the risk tolerance criteria.

Acronyms Used

AIChE– American Institute of Chemical Engineers
CCPS – Center for Chemical Process Safety (an AIChE technology alliance)
HAZOP – Hazard and Operability Analysis
HRA – Human Reliability Analysis
IEC – International Electrotechnical Commission
IEF – Initiating Event Frequency
ETA – Event Tree Analysis
FMEA – Failure Mode Effect Analysis
FTA – Fault Tree Analysis
HRA – Human Reliability Analysis
IPL - Independent Protection Layer
LOPA – Layer of Protection Analysis
PHA – Process Hazard Analysis
PII – Process Improvement Institute, Inc.
QRA – Quantitative Risk Analysis
SIF – Safety Instrumented Function
SIL – Safety Integrity Level
SIS – Safety Instrumented System
SRS – Safety Requirements Specification

References

1. *Guidelines for Hazard Evaluation Procedures*, 2008, CCPS/AIChE.
2. Bridges, W., & Dowell, A., "Identify SIF and Specify Necessary SIL, and other IPLs, as part of PHA/HAZOP," *12th Global Congress on Process Safety*, AIChE, 2015.
3. *Layer of Protection Analysis: Simplified Process Risk Assessment,* 2001, CCPS/AIChE.
4. *Functional Safety: Safety Instrumented Systems for the Process Industry Sector - Part 3: Guidance for the Determination of the Required Safety Integrity Levels*; ANSI/ISA 84.00.01 Part 3, 2004.
5. IEC 61511, *Functional Safety: Safety Instrumented Systems for the Process Industry Sector - Part 1: Framework, Definitions, System, Hardware and Software Requirements*, International Electrotechnical Commission
6. Bridges, W., and Thomas, H., "Accounting for Human Error Probability in SIL Verification Calculations," *8th Global Congress on Process Safety,* Houston, AIChE, April 2012.

479509 The Art of the Audit: Being Audit Ready, Everyday

Abstract:

Have you ever made an appointment for an annual physical and suddenly join a gym? We all know we should be exercising every day, but it becomes more important before going in for that checkup. In reality, not exercising has consequences to your health and not your doctors, so why do we do this? To impress them, instead of doing it to take care of ourselves? This phenomena of doing things you do not normally do in preparation for an event is unfortunately how too many of us treat audits. We hear an auditor is coming and end up running around trying to organize paperwork, convince the operators to clean up the plant, and everyone in management becomes much more attentive. Does this behavior really set the proper tone to develop a safety culture? Auditors are there to check up on our compliance of rules that benefit only us to follow, particularly in the field of safety.

A successful operation will be 'Audit Ready' every day. 'Audit ready' means that operations are in compliance with rules, personnel are safe and exhibit safe behaviors, and all of the team members understand why they perform the functions of their job. Compliance tools, personnel awareness and routine self-audits are three key topics in becoming ready for an audit every day.

Personnel awareness is among the most vital areas to ensuring safe and effective operations, as well as successful audits. Audits involve people from across almost all departments, and as such mirror the requirements of multifunctional teams in everyday operations.

Compliance tools are important to track work that is being completed, and create convenient locations to store data which proves compliance. This can include tracking the progress of MOCs through weekly notifications, using databases to track when procedures need to be reviewed, and more. Visibility of what needs to be accomplished is important.

Routine self-audits help us to find gaps in compliance, and reduce complacency. These can include a simple spot check of one area of compliance, or be more comprehensive, and there are pros and cons to both. A spot check keeps everyone on their toes and takes a short enough time such that they can be done more frequently.

Ensuring sustainable records and checking everyday habits will go a long way to creating a safe operating environment, and with that a workforce that is audit ready, every day.

479546 Economic Justification for Better Maintenance in a Safety System

Abstract:

It is not difficult to think of numerous ways in which human factors might impact functional safety. For example, imperfect repair, improper calibration, faulty installation, etc. can all contribute to decreases in functional safety performance. However, if actions are to be taken to mitigate these impacts then it is essential to have some quantitative measure of how this impacts the PFDavg, so that the actions' effectiveness can be assessed. We will show how a simple measure system called the Site Safety Index™ (SSI) is used to adjust safety metrics, including the PFDavg and the MTTFS, on a site specific basis. The lifecycle economic impact of these metrics is then used to show how improved operations and maintenance can be justified. Economic justification for better maintenance can be easily verified through this approach being further mindful of the safety culture.

479547 How Important Is Realistic Failure Rates for Safety Instrumented Functions?

Abstract:

A given Safety Instrumented Function design must be verified to meet the required SIL level by calculating probability of failure on demand. However, this does require realistic failure rate data for each device. Fortunately, many device manufacturers are getting their products certified per IEC 61508 and many of the certificates now have failure rate data. Other manufacturers supply analysis reports with failure rate data. But how good is this data supplied by the manufacturers? All such data must be checked to make certain it is applicable for the process industries as some data may have been derived based on machine safety or automotive applications.

This paper will show some examples of manufacturer provided failure rate data and a methodology created to validate device type failure rate data for the process industries.

479917 Enabling Sustainable Relief System Management Programs through Independent Reviews

Abstract:

Overpressure protection and flare systems usually represent the last line of protection for personnel and the environment in industrial facilities operating under high temperatures and pressures. During the design phase of either Greenfield or Brownfield projects, assessing the basis for major relief systems and making clear recommendations to reduce the risks associated with each system has helped companies realize substantial safety and economic benefits. Independent and safety focused reviews, with an emphasis on value added solutions has enabled organizations to set the right foundation for a sustainable relief system management program. This, in turn, has helped enable the facilities in meeting the ever-changing regulatory needs in a timely manner.

This paper will give a typical breakdown of the workflow process for each of the entities involved in the project - the client, the contractor, and the independent reviewer. Four distinct work activities are typically identified for such a review - development of preliminary overpressure protection, engineering design of relief systems, validation of the design, and validation of as-built documentation. After each stage, the output from the independent analysis is compared with the contractor's work and updates are performed before continuing to the next stage. This paper will also give an overview of these stages and discuss the various steps and value added for each stage of the project.

Jack Rabbit II: Experimental Large-Scale Chlorine Release Trials

Mark Whitmire[1], Shannon Fox[2], Thomas Spicer[3], Leo Stockham[4], Steven Hanna[5], Joseph Chang[6], Michael D. Sohn[7], Thomas Mazzola[8], Ronald Meris[9], Damon Nicholson[10], Andy Byrnes[11]

[1]Noblis, Falls Church, VA, USA; [2]DHS S&T Chemical Security and Analysis Center, Aberdeen Proving Ground, MD, USA; [3]University of Arkansas, Fayetteville, AR, USA; [4] Leidos, Albuquerque, NM, USA; [5]Hanna Consultants, Kennebunkport, ME, USA; [6]ANSER, Falls Church VA, USA; [7]Lawrence Berkeley National Laboratory, Berkeley CA, USA; [8] Engility, Lorton, VA, USA; [9]DTRA, Fort Belvoir, VA, USA; [10]Dugway Proving Ground, UT, USA; [11]Utah Valley University, Provo, UT USA

INTRODUCTION

Chlorine is a potent toxic inhalation hazard (TIH) chemical, and each year millions of tons of it are transported in bulk via road, water, and rail. Consequently, an accidental or deliberate chlorine release near a populated area may cause a considerable number of fatalities, and would create other health, safety, and environmental problems. Hazard prediction models are inconsistent with the evidence and observations from chlorine disasters, and the understanding of such releases is otherwise insufficient to guide and inform an effective emergency response. So beginning in 2010, DHS and its Health, Safety and Environment (HSE) partners and stakeholders planned the Jack Rabbit (JR) Program. The program was led by the DHS Chemical Security Analysis Center (CSAC) located at Aberdeen Proving Grounds-Edgewood Area MD.

OVERVIEW

The Jack Rabbit (JR) program is comprised of a series of projects that includes JRI (2010), JRII Phase 1 (2015), and JRII Phase 2 (2016). The JRI (Fox and Storwold, 2011) project was completed to develop, test, and evaluate a mechanism for the controlled, rapid release of liquefied, pressurized gases (specifically, ammonia and chlorine), to observe the transport and dispersion of a disseminated vapor and aerosol cloud, and to investigate chemical deposition and reactions with water and soil. The project also evaluated instruments, test methods, and strategies that may be used to study more industrial scale tests. Using either chlorine or ammonia, two 'pilot' tests and eight 'record' tests of either 1 or 2 tons were completed. For chlorine, among other findings, the results demonstrated that downwind transport and turbulent mixing are initially resisted by a dense persistent gas and aerosol cloud, that source phenomena were non-linear with increasing release volumes, and that the reactivity with soil containing water and organic matter is an important removal mechanism.

JACK RABBIT II

JRII leveraged and extended the work of JRI with controlled chlorine field experiments on a scale corresponding to railcars and tank-trucks; that is, 5 to 20 tons each. The goals were to collect data on the release source, measure chemical reactions with the environment, observe and evaluate exposure effects on equipment and infrastructure, and assess urban impacts using a Mock Urban Environment (MUE) or an unobstructed dissemination and downwind dispersion environment, both located on the Urban Test Grid (UTG) at Dugway Proving Grounds UT. Data was collected in real time via ground based and unmanned aerial vehicle systems, video cameras, and spectroscopic instruments, in situ point detectors, path integrated detectors, and concentration and dosage determination instruments. The data was thoroughly and qualitatively reviewed and shared with all project participants in prescribed and common formats. The generated data and findings will improve the understanding of and operational hazard prediction modelling for TIH releases, and lead to more effective emergency response and training, national preparedness, and mitigation strategies.

URBAN TEST GRID (UTG)

The UTG (Nicholson, Hedrick, and Lian, N; 2017) was the designation for the JRII chlorine release site surveyed and built specifically for the JRII tests; it was at Dugway Proving Grounds west of Granite Mountain on Goodyear Road. It included the playa and adjacent support areas, concrete pad, gravel pad, and surveyed grid extending 11 km downrange and aligned roughly south to north.

The UTG included a circular, 0.5 ft. deep, 82 ft. diameter, rebar reinforced concrete disseminator vessel pad with 3 embedded temperature sensor stations, and a 1 inch steel lip to contain chlorine liquid. The center of the circular pad was the center of the UTG site, Northing 4445633.945, Easting 288109.182 (Latitude, 40.134347; Longitude, -113.487076). The concrete pad rested on a 24 inch deep, 400 ft. × 600 ft. gravel pad level with the concrete pad and symmetric about it.

Figure 1. Chlorine release from the disseminator on the concrete pad, with CONEX container array on the gravel pad of MUE, looking upwind (south) from atop the 2 x 3 CONEX container stack. The dense cool cloud remains low as it moves downwind and laterally around and through the Mock Urban Environment made from CONEX containers.

Winds predominate from approximately 165° from June to September, so the UTG was oriented about a centerline from that direction. The UTG included the playa within a 90° angle bisected by the UTG centerline with vertex at the disseminator vessel location, and extending to the 11 km arc. The "Near Field" was the UTG region from the disseminator up to and including the 2 km arc and inside the 90°

angle; the "Extended Field" was the UTG region past the 2 km arc to the 11 km arc and inside the 90° angle.

The Mock Urban Environment (MUE) was constructed on the gravel pad and surrounding the concrete disseminator pad using a staggered row array of 86 CONEX containers (CONtainer EXpress shipping containers, either 18 or 40 ft. long) to observe and study the transport and dispersion of dense gas around and above urban structures. The array included a two wide by three high (2 x 3) stack of six 20 ft.-CONEX containers located directly downwind of the source, and two 2-axle trailers nearby with insulation, dry wall, and trailer skirting for infiltration studies. Aside from the CONEX containers and trailers, up to seven vehicles were included in the array. Three fire trucks, one ambulance, and three cars were located on the north edge of the gravel pad during Trials 1-3. During Trial 4, four vehicles were placed on the playa past the 250 m arc; during trial 5, three vehicles were moved between the rows of CONEX containers but adjacent to the concrete disseminator pad, and three vehicles remained on the playa past the 250 m arc.

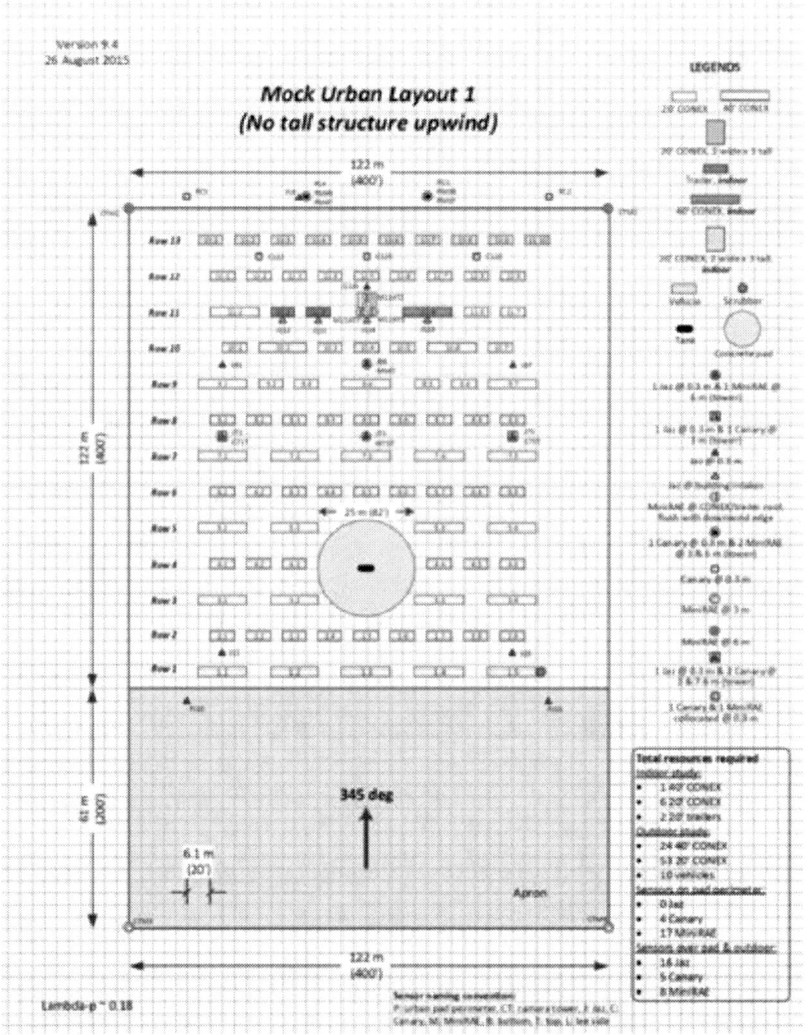

Figure 2 Mock Urban Environment (MUE) schematic showing gravel pad around circular pad for Chlorine disseminator; CONEX containers and instrument locations are also shown.

Instrument locations for the UTG "Near Field" and "Extended Field" were surveyed, recorded, and reported. Point sensors included about 170 MiniRAE 3000s equipped with volatile organic compound

(VOC) photoionization detectors; about 60 ToxiRAE Pro units; a Gasmet™ DX-4000 (FTIR) gas analyzer; 28 Ultraviolet (UV) Canary units for UTG monitoring and indoor studies; and 16 Ocean Optics Jaz instruments for UTG monitoring and indoor studies. Standoff Detectors included two Cerex Monitoring Solutions UV Sentry chlorine-specific integrated path detector, three Ultraviolet (UV) Light Detection and Ranging (LiDAR) systems for monitoring the moving chlorine gas cloud, and elastic backscatter LiDAR systems for monitoring the moving aerosol cloud. A GRIMM Aerosol portable laser aerosol spectrometer and dust monitor was also used for particle sizing model using laser scattering techniques. Other equipment included remotely operated high-definition (HD), high-speed (HS), and FLIR® IR Camera SC660 micro-bolometer cameras positioned throughout the test grid to visualize and record the chlorine release and resulting plume.

Meteorology equipment included 34 Portable Weather Information Display Systems (PWIDS) collecting data at 2 m; three 32 m towers each equipped with sonic anemometers collecting data at heights between 2 and 32 m; and 31 3D ultrasonic anemometer to gather wind and turbulence statistics, and fluxes of heat and momentum. Energy Balance Station (EBS) provided in situ measurements of the surface vertical fluxes of sensible heat, latent heat, net radiation, and soil surface heat flux. The MiniSODAR™ high-frequency Doppler acoustic sounding system designed to measure the atmospheric vertical wind profile from 15 to 200 m. Numerical gridded forecast meteorology using embedded observations and real-time meteorological data from the test grid were used for HPAC modelling leading up to each trial and to determine "GO/NO GO" test status.

DOWNWIND DISPERSION AND TRANSPORT ENVIRONMENT

For tests in 2016, and other than two vehicles, a CONEX container, and a modular office trailer used for follow-up infiltration experiments, the UTG was cleared of the simulated buildings and obstructions used in 2015. The project goals were to monitor and collect the cloud concentration data from the initial phase of a very dense, two-phase cloud near the source to dispersion of the vapor cloud farther downwind and up to 11 km away using point detectors and stand-off spectroscopic instruments.

DISSEMINATION

The disseminator vessel for use as the chlorine release source during the tests was located at the center of the concrete pad with its major axis perpendicular to the UTG centerline. The 2000 gal. tank was mounted in a 3-dimensional dynamic mass measurement load cell system. Other vessel mounted instrumentation included a redundant system for determining the remaining mass of chlorine in the vessel, differential pressure sensors on the vertical axis at various depths in and below the vapor space, guided wave radar liquid level measurement, and instrumentation for absolute pressure and direct contact temperature profiles inside the vessel. At 80-90% operating capacity, the vessel could contain from 8.5 to 10 tons chlorine.

Each release was accomplished using a blind flange with exploding bolts rated to safely hold the pressure of a vessel loaded with liquid chlorine at ambient temperature. To blow the flange and release the chlorine, the bolts were remotely detonated; when the bolts broke, the blind flange was blown off with the aid of the chlorine weight and head pressure, and chlorine flowed through the vessel's circular opening. A quick release valve was not used to simulate the release because of the likelihood of internal flashing flow in the piping and valves; that is, the mass release rate would be less than what would be observed from a breach of containment. In 2015, every release was directly down (180°) through a six inch hole. In 2016, each release was either directly down (Test 6); at 135° and downwind (Test 7), or directly up (Test 8). After Tests 7 and 8 any liquid chlorine remaining in the vessel was discharged via a remotely operated valve in the bottom of the tank. Before inert gas purging and chlorine refill operations, the ejected blind flange was replaced using standard bolts. The last release was from a tanker truck that had been removed from service, and containing up 19 t chlorine. The tanker truck's skin was breached with a remotely activated explosive shape charge.

Figure 3. Test Release 6; 10 Tons chlorine on August 31, 2016. A moment after the release the liquid chlorine may be seen splashing onto the concrete pad. As the chlorine aerosolizes and vaporizes it cools the surroundings and condenses water vapor.

OBSERVATIONS AND FINDINGS

From the JRII trials the program observed that:

- Upon breaching the tank, liquid chlorine exits the tank and rapidly flashes to a two phase aerosol and vapor cloud with a liquid layer deposited on the pad, ultimately all vaporizing and transporting downwind.
- The escaping chlorine was found to auto-refrigerate to its -34 degree Celsius boiling point and lower due to further evaporative cooling outside the tank.
- The force of the escaping chlorine jet generated strong initial turbulence and momentum to drive the expanding cloud over nearby Conex structures.
- Evacuation of the tank during 5-ton releases took approximately 35 seconds.
- Evacuation of the tank during 9-ton releases took approximately 50 seconds.
- Structures were observed to channel the escaping and expanding cloud laterally.
- Significant upwind movement of the chlorine clouds was observed due to the momentum of the escaping gas, expansion of the gas, and gravity-slumping.
- Visible clouds were observed to persist in UTG for approximately 20-30 minutes depending upon release amount and winds.
- Long visible clouds observed to stretch from UTG to approximately 2 km as downwind dispersion progressed.

- Data sets from Jack Rabbit II have been quality-assured, finalized, and recently released, enabling in-depth studies to begin on cloud concentrations, movement, and behavior to be quantified.

EMERGENCY RESPONSE EXPERIMENTS

For each test, researchers from Utah Valley University placed two "Witness Boards" on surveyed arcs adjacent to the disseminator in the urban environment to observe the effect of liquid chlorine and high concentrations of chlorine gas (approaching 25%) on common building components, urban surfaces, and fire equipment. Go-Pro® and HD color surveillance cameras captured the exposure on video. The selection of the test materials emphasized high organic content (creosote treated telephone pole and railroad tie, road asphalt, shingles, rubber, and bare wood), but copper, aluminum, galvanized steel, and carbon steel coupons were included. The copper and carbon steel surfaces began to be discolored 1-2 hours after the exposure, and the chlorine sometimes dissolved the petroleum based materials.

INDOOR INFILTRATION STUDIES

The Defense Threat Reduction Agency (DTRA) is evaluating various hazardous airborne contaminant applications for transition to DOD CBRN consequence models. Their current indoor application models presume that any contaminant is inert and neutrally buoyant, thus the models may not be useful for predicting the indoor effects of a dense, reactive gas like chlorine. Lawrence Berkeley National Laboratories (LBNL) proposed a series of experiments to DTRA and the JR program to assess the applicability of the models for predicting the indoor concentration of chlorine in the event of an outdoor chlorine release.

LBNL successfully completed a series of indoor infiltration experiments in 2015 in support of the DTRA effort. The LBNL experiments were completed to understand and characterize the dominant infiltration pathways for reactive dense gases into buildings, the uptake of such gases into structures, and the re-emission of them from the structures after the peak of the plume had passed. Three single-story CONEX structures, a single three-story CONEX structure, and 2 modular office trailers placed on the gravel pad were used to represent indoor office, school, or home environments. Follow-on studies were completed in 2016 using a single-story CONEX structure at the same, but unobstructed site. The CONEX structures were configured for controlled experiments of airflow and transport in single- or multi-story structures, and modified to include insulation and temperature-control. Chlorine penetration into and through the structure, and transfer of it across the building envelope was monitored via controlled artificial "leaks" and exhaust fans. The model offices were furnished with realistic insulation, surface materials, windows, and HVAC system characteristics, thus the simulated modular offices helped to relate the data from the controlled CONEX structures to real buildings. The data may be used for calibration, verification and/or validation of the indoor model, and to support DOD concepts of operation for sheltering in the event of an outdoor CBRN release.

PROGRAM STATUS

The program is entering a new phase after completing field testing. In 2017, the program will be processing and reporting data from 2015 and 2016, and transitioning it for permanent data storage, and to stakeholders. CSAC will also be presenting a JR overview, narrative, and plans in presentations to interested groups. The program will be proposing, designing, managing, and performing small and large scale experiments during chamber testing, and reducing and analyzing data to demonstrate or elucidate the reaction of chlorine with various substrates to suggest large-scale toxic inhalation hazard (TIH) chemical release mitigation strategies, inform emergency response strategies that may reduce the impact to affected populations and infrastructure, calculate the value of modeling parameters, and otherwise support stakeholders interests.

ACKNOWLEDGEMENTS

This research has been supported by the US Department of Homeland Security (DHS) and the US Defence Threat Reduction Agency (DTRA). Additional support (funding, personnel and instruments) is coming from numerous partners, with the list growing with time. We appreciate guidance from many scientists who are members of the planning groups and/or have provided comments on the scientific approach.

REFERENCES

1. Nicholson, D., Hedrick, A., and Lian, N., Final Data Report for Jack Rabbit (JRII), 2017

2. Fox, S. and D. Storwold, 2011: Project Jack Rabbit: Field Tests. Chemical Security Analysis Center, Science and Technology Directorate, U.S. Department of Homeland Security. CSAC 11-006, 162 pp.

479394 Functional Safety - Pitfalls and Good Practices

Abstract:

The implementation of effective functional safety is valuable in ensuring that safety critical control and protective systems are appropriately specified, designed, installed and maintained. Through the International Electro-technical Commission (IEC) 61508 and 61511 suite of functional safety standards, the concepts of the Safety Lifecycle and Safety Integrity Levels (SIL) are now well known. However, DNV GL's experience from numerous projects across different industry sectors suggests that there remain challenges to be overcome.

DNV GL has conducted a number of Functional Safety Assessments (FSAs) on various electrical/ electronic and programmable electronic safety systems in the process sector. The FSAs are a mandatory element of IEC 61511 which provide an independent judgment on the integrity of the safety system and should be performed by competent independent Functional Safety Practitioners. At least one FSA must be performed on a Safety Instrumented System and should be completed prior to the introduction of hazards. Through DNV GL's experiences in functional safety assessments this paper will discuss;

- How to avoid common mistakes in the management and implementation of functional safety;
- Examples of good practice in the application of functional safety;
- How good practice can help optimize test regimes, increase plant availability and reduce maintenance workload whilst ensuring the system continues to provide the required risk reduction.

General Method for Uncertainty Evaluation of Safety Integrity Level (SIL) Calculations – Part 2 Analytical Methods

Raymond "Randy" Freeman
S&PP Consulting

For Presentation at the American Institute of Chemical Engineers
2017 Spring Meeting in San Antonio, Texas -- March 26-30, 2017

Abstract

The IEC 61511 standard requires a verification calculation that a proposed design for a safety instrumented function (SIF) achieves the desired safety integrity level (SIL). The evaluation of the safety integrity level of a new or existing safety instrumented system requires detailed calculations based on the failure rates of the device and the planned maintenance/testing cycle for the system. The failure rates of the devices are often taken from standard failure rate tabulations of equipment. The maintenance and testing plans are developed based on plant experience. All of the data used in the SIL calculations are uncertain. This paper develops a general method for uncertainty analysis of the SIL calculations. The general method is based on the application of probability theory - variance contribution analysis (VCA) – to the equations presented in ISA TR 84.00.02-2115. An example is worked to demonstrate the methodology.

Background

The calculation of the probability of failure on demand (PFD) is a common engineering task when designing an interlock or safety system that is to be in compliance with IEC 61511 [Ref. 1]. The calculation of the PFD is often done using approximate equations defined in the ISA TR84.00.02 technical report [Ref. 2]. The simplified equation method in the ISA report is commonly used and is based on the use of reliability block diagrams where the field sensors, Safety Instrumented System (SIS) logic solver and final control elements are considered independent of each other in the sense of not sharing common devices or systems. The PFD is then calculated using the failure rates of the devices, planned test intervals, vendor supplied estimates on diagnostic coverage of the devices and an allowance for the potential for common cause failures. Almost all of these parameters are uncertain. The failure rate data is often taken from generic data sources which show wide ranges in the observed values.

Because of the uncertainty in the parameters, the design engineer makes allowances in the design by the use of safety factors or rules of thumb to improve the chances that the final interlock installation will work as intended. Since each engineer has a different set of safety factors and rules of thumb, two designs may differ significantly in the way a hazard is controlled.

A more formal method for handling the underlying uncertainty in the calculation of the PFD of an interlock is needed. Previously, Freeman and Summers [Ref. 3] published an uncertainty analysis of the PFD of an interlock. Two different methods were used in this analysis:

- Monte Carlo Simulation
- Variance Contribution Analysis (VCA)

Monte Carlo Simulation requires that the engineer build a model of the interlock using specialized computer software. The use of VCA requires that the sensitivity of the interlock model be determined either by numerical methods or by direct analytical calculations. The Freeman paper demonstrates that VCA can be used for the uncertainty analysis. However, the paper does not present a complete analysis method that can be applied to any system defined in the ISA technical report TR84.00.02. The goal of this paper is to develop a general set of analytical equations that will allow VCA to be used in uncertainty analysis of any interlock developed per the IEC 61511 standard.

Review of Interlock Design

The IEC standard looks at an interlock as a series of three major components (see Figure 1). First is the sensor set which sends an indication of an abnormal event to a logic solver. The logic solver determines is the signal from the sensor meets the conditions required to activate (trip) the interlock. If the interlock is to be activated, a signal is sent to the final control elements to take action (stop flow, close valve, etc) to prevent a process safety event from occurring.

Based on the devices selected the design engineer then evaluates whether the proposed interlock design meets the design criterion for probability of failure on demand. Typically, the design engineer is given the target PFD as a statement such as "provide a SIL-2 interlock to stop flow on high level in a vessel." A SIL-2 interlock requires a PFD of at most 0.01 or a risk reduction factor of at least 100. The needed risk reduction is often determined in a Layer of Protection Analysis (LOPA) [Ref. 4] and represents the managements decision on how risk is to be managed in a system. The resulting SIL-2 interlock design must have a high likelihood of achieving the target risk reduction as the LOPA team may be relying on it to be part of an overall risk reduction plan to prevent a process safety incident.

The ISA technical report (TR 84.00.02) [Ref. 2] provides shortcut methods for the evaluation of the PFD of an interlock design. Tables 1 and 2 are taken from the ISA technical report [Ref. 2]. Typical parameters need to compute the PFD include:

- DC is the diagnostic coverage;
- DI is the diagnostic interval;
- TI is the proof test interval,
- λ^D is the dangerous failure rate;
- MTTR is the mean time to restore the system to operation
- β is the common cause failure parameter that is always is between 0 and 1

In practice most if not all of these parameters are uncertain. Using parameter values obtained from plant records, vendor data or generic data bases the design engineer proceeds to develop the model for the interlock.

What is needed is a means to rapidly evaluate the impact of parameter uncertainty on the interlock PFD. The remainder of this paper applies the methods of variance contribution analysis (VCA) to determine the mean (expected value) and the standard deviation of the interlock PFD for a given design.

Table 1. Simplified PFD_{avg} formulas for Non-Repairable System without considering CCF, Diagnostics or MTTR.			
Configuration	**Function**	**PFD based on "Average before" failure rate**	**Appendix A Equation Number**
1oo1	F_1 =	$\left[\lambda^D \times \frac{TI}{2}\right]$	A-1
1oo2	F_2 =	$\left[(\lambda^D)^2 \times \frac{TI^2}{4}\right]$	A- 2
1oo3	F_3 =	$\left[(\lambda^D)^3 \times \frac{TI^3}{8}\right]$	A-3
2oo2	F_4 =	$\left[\lambda^D \times TI\right]$	A-4
2oo3	F5 =	$\frac{3}{4} \times \left[(\lambda^D)^2 \times TI^2\right]$	A-5
3oo3	F6 =	$3 \times \left[\lambda^D \times \frac{TI}{2}\right]$	A-6

Where TI is the proof test interval, λ^D is the dangerous failure rate.

Table 2. Simplified PFD_{avg} Formulas for Repairable System considering CCF, Diagnostics and MTTR			
Configuration	**Function**	**PFD based on "Average before" failure rate**	**Appendix A Equation Number**
1oo1	$F_7 =$	$\left[\frac{(1-DC)\times\lambda^D\times TI}{2}+\frac{DC\times\lambda^D\times DI}{2}+\lambda^D\times MTTR\right]$	A-24
1oo2	$F_8 =$	$\left[\frac{(1-DC)\times(1-\beta)\times\lambda^D\times TI}{2}+\frac{DC\times(1-\beta)\times\lambda^D\times DI}{2}+(1-\beta)\times\lambda^D\times MTTR\right]^2+\left[\frac{(1-DC)\times\beta\times\lambda^D\times TI}{2}+\frac{DC\times\beta\times\lambda^D\times DI}{2}+\beta\times\lambda^D\times MTTR\right]$	A- 25
1oo3	$F_9 =$	$\left[\frac{(1-DC)\times(1-\beta)\times\lambda^D\times TI}{2}+\frac{DC\times(1-\beta)\times\lambda^D\times DI}{2}+(1-\beta)\times\lambda^D\times MTTR\right]^3+\left[\frac{(1-DC)\times\beta\times\lambda^D\times TI}{2}+\frac{DC\times\beta\times\lambda^D\times DI}{2}+\beta\times\lambda^D\times MTTR\right]$	A-26
2oo2	$F_{10} =$	$2\times\left[\frac{(1-DC)\times\lambda^D\times TI}{2}+\frac{DC\times\lambda^D\times DI}{2}+\lambda^D\times MTTR\right]$	A-27

Table 2. Simplified PFD_{avg} Formulas for Repairable System considering CCF, Diagnostics and MTTR			
Configuration	**Function**	**PFD based on "Average before" failure rate**	**Appendix A Equation Number**
2oo3	F_{11} =	$3\times\left[\frac{(1-DC)\times(1-\beta)\times\lambda^{D}\times TI}{2}+\frac{DC\times(1-\beta)\times\lambda^{D}\times DI}{2}+(1-\beta)\times\lambda^{D}\times MTTR\right]^{2}+\left[\frac{(1-DC)\times\beta\times\lambda^{D}\times TI}{2}+\frac{DC\times\beta\times\lambda^{D}\times DI}{2}+\beta\times\lambda^{D}\times MTTR\right]$	A-28
3oo3	F_{12} =	$3\times\left[\frac{(1-DC)\times\lambda^{D}\times TI}{2}+\frac{DC\times\lambda^{D}\times DI}{2}+\lambda^{D}\times MTTR\right]$	A-29

Review of Variance Contribution Analysis (VCA) Methodology

The mean and variance of a function of random variables can be approximated using the method described by Haugen [Ref. 5] and applied by Freeman [Ref. 3, 6, 7]. Define an arbitrary function of a set of random variables, x_i, as:

Let

$$Y = F(x_i) \qquad \text{(Eq 1)}$$

The mean of Y can be estimated using the following approximation:

$$E(Y) = F[\, E(x_i)] \qquad \text{(Eq 2)}$$

Where:
E(Y) = expected value of random variable Y = mean of Y
$E(x_i)$ = expected value of random variable x_i = mean of x_i

The variance of Y can likewise be estimated as:

$$V(Y) = \sum_{i=1}^{n} \left[\frac{\partial Y}{\partial x_i}\right]^2 V(x_i) \qquad \text{(Eq 3)}$$

Where:

V(Y) = variance of random variable Y as defined above in Equation 1

$V(x_i)$ = variance of random variable x_i as defined above in Equation 1

Note that the variance is simply the square of the standard deviation. Using the variance will simplify the mathematics that is described below. The contribution of each independent random variable to the overall variance in the function is:

$$V(Y \text{ from } x_i) = \left[\frac{\partial Y}{\partial x_i}\right]^2 V(x_i) \qquad \text{(Eq 4)}$$

The relative contribution of each term to the overall variance V(Y) is a measure of the importance in the uncertainty in the particular random variable, x_i. In effect, this is a sensitivity analysis combined with a uncertainty evaluation. The variance contribution combines the sensitivity in the answer to changes in the uncertain random variable, x_i, with a measure of the uncertainty in the random variable, x_i. The overall variance in Y is found by summing the sensitivity weighted variances from each random variable.

Recommended Interlock PFD Uncertainty Analysis Method

Previously, Freeman and Summers [Ref. 3] suggested a framework for the inclusion of uncertainty analysis in the calculations completed for a new interlock per the ISA TR [Ref. 2]. The following uncertainty analysis method has been expanded to incorporate critical decisions that the process management must make in the design process.

1. Complete the interlock design using the methods outlined in IEC 61511 [Ref. 1].

2. Review with the process system management and determine if the proposed interlock should be considered repairable or non-repairable. A simple flow chart for this decision making is presented in Figure 2. The basic question is: "can the interlock be repaired safely while the process operates." This is a management question and management should be the one that decides the answer to this important question.

3. Create interlock performance equation as the mathematical model for the combination of sensor, logic solver and final control elements using the methods outlined in ISA technical report (TR84.00.02) [Ref. 2]. Equations needed to evaluate the uncertainty of the models specified in the ISS report are developed in Appendix A of this paper. For non-repairable systems, Table 3 can be used. For more complex systems such as non-repairable redundant systems with the potential for common cause failures, use the repairable equations of Table 4 and set DI, DC, and MTTR all equal to zero. For systems where common cause failures (CCF), Diagnostics and repair are to be considered, use the recommendations of Table 4. There are design situation where the equations developed in Appendix A will not exactly apply. Specifically, designs where the system contains redundancy subject to common cause failures (CCF) that should be considered non-repairable during operation. The failure models and the equations needed to evaluate the uncertainty in these systems are developed in Appendix B to this paper.

Table 3. Roadmap for Mean and Variance of Non-Repairable System Without Considering CCF, Diagnostics and MTTR			
Configuration	Function	Mean PFD Using Appendix A Equation	PFD Variance Using Appendix A Equation
1oo1	F_1 =	A-9	A-18
1oo2	F_2 =	A- 10	A- 19
1oo3	F_3 =	A-11	A-20
2oo2	F_4 =	A-12	A-21
2oo3	F_5 =	A-13	A-22
3oo3	F_6 =	A-14	A-23

Table 4. Roadmap for Mean and Variance of Repairable System Considering CCF, Diagnostics and MTTR			
Configuration	Function	Mean PFD Using Appendix A Equation	PFD Variance Using Appendix A Equation
1oo1	F_7 =	A-30	A-44
1oo2	F_8 =	A- 31	A- 86
1oo3	F_9 =	A-32	A-94
2oo2	F_{10} =	A-33	A-48
2oo3	F_{11} =	A-34	A-102
3oo3	F_{12} =	A-35	A-50

Note that in all of the calculations indicated by the equations referenced in Table 4, the expected value of any the random variables is used to complete the calculations.

4. Define the uncertainty in the parameters and variables of the interlock model specified in step 2. The uncertainty can be given as the upper and lower range of the possible values (uniform probability distribution), as the upper, lower and recommended values (triangular distribution), or as a mean and standard deviation (normal distribution). See the example above for guidance in the evaluation of safety instrumented system interlocks.

5. Compute the expected value of each variable in the interlock performance equation. The equations for the mean and variance for the uniform, triangular and normal probability distributions are presented for various probability distributions in Vose [Ref. 8]. The example interlock calculations used a triangular distribution to represent the uncertainty in the parameters.

6. Compute the expected value or mean of the interlock PFD using the mean value of each of the variables in Step 5. See Tables 4 and 5 for recommended equations.

7. Compute the sensitivity of the result from the interlock performance equation by use of the partial derivative of the basic interlock performance equation with respect to each of the variables as presented in Appendix A.

8. Compute the variance of the interlock performance equation PFD by use of the variance contribution using equations from Tables 3 or 4. This entails multiplying the variance of each of the uncertain variables in the basic interlock performance equation by the square of its sensitivity (obtained in step 7), as evaluated at the variable mean. Sum the resulting terms to obtain the overall variance of the PFD in the interlock performance equation. See Tables 4 and 5 for recommended equations.

9. Determine the level of risk that the owner/operator wishes to take that the final interlock will not work. Note that this is a management decision not an instrument engineer decision! In this paper, the 95% level of risk reduction has been used:
   - 5% chance of failure or a 95% chance of the interlock achieving the desired risk reduction

10. Assuming that the interlock owner operator wishes to take a low risk (5%) of the interlock failing to achieve its design target PFD, compute the 95% upper confidence limit on the computed PFD by use of the standard normal factor, Z, [Ref. 9] as:

$$Z = \left[\frac{xi - E(x)}{\sigma}\right] \qquad \text{(Eq 5)}$$

Where:

σ = standard deviation of the PFD of the interlock of interest from the interlock performance equation obtained from step 7. Note that the variance of a random variable is the square of the standard deviation of the random variable.

E(x) = the expected value of the PFD of the interlock of interest from the interlock performance equation obtained from step 5

For the 95% upper limit, Z = 1.645. Rearranging Eq. 5 allows for the direct calculation of the corresponding value of the 95% upper confidence limit on the PFD as:

$$X95\% = 1.645\,\sigma + E(x) \qquad \text{(Eq 6)}$$

Where:

X95% = the upper 95% limit on the computed PFD of the interlock of interest from the interlock performance equation.

Compare the 95% upper confidence limit on the PFD of the interlock of concern with that established as the desired PFD for risk reduction. If the 95% confidence of the RRF is greater than the desired RRF, the design is complete. If not, revise the design or change inspection test intervals to achieve the desired RRF. If it is not possible to achieve the desired target RRF economically, revisit the LOPA study accordingly to incorporate better information obtained in the uncertainty analysis. Improve the integrity of the LOPA IPLs or identify additional IPLs to drive the risk to a tolerable level. Continue this process until the computed RRFs are greater than the desired RRFs for risk reduction and risk management.

Example Interlock Problem

Previously, Freeman and Summers [Ref. 3] published an example in the use of VCA to determine the likelihood that the interlock would provide the risk reduction desired by the management and the LOPA team. The process system is shown in Figure 3. The process uses a compressor to increase the pressure of a process stream prior to additional processing. The gas being compressed is toxic and flammable. Of concern is a slug of liquid being sent to the compressor. If a liquid slug is sent to the compressor, significant damage to the compressor is expected with probable seal damage and a subsequent release of flammable and toxic material into the work area. A large fire and/or explosion could result if the release were to be ignited. If the release is not ignited, the nearby workers could be exposed to the toxic gas resulting in death or injury. A Layer of Protection Analysis (LOPA) review of this system has been completed. Among several recommendations, the LOPA team recommended the installation of a SIL-2 high level interlock in the Compressor Knock Out Drum to stop the compressor prior to liquids entering the system.

Figure 1 shows a simplified diagram for the interlock. Three level sensors are provided in the Compressor Knock Out Drum. The SIS logic solver will use 2oo3 voting to detect high level in the Compressor Knock Out Drum. The SIS logic solver will also monitor the difference in level signal from each of the three level sensors and will activate an alarm if the deviation is excessive. Two independent methods of stopping the compressor are provided. The SIS will directly signal the motor controller on the compressor to stop. In addition, the SIS logic solver will also signal two additional relays to open the power supply to the compressor motor. These two different shutdown will both be activated in the event of high level in the Compressor Knock Out Drum. Either one of the two shutdowns is capable of stopping the compressor by turning off the electric power supply to the motor.

Models

The first step in the calculation of the "goodness" of an interlock is to establish the model to be used in the calculations. Note that the sensors are 2oo3 voting and the final control elements are each 1oo1. Appendix A presents the equations for various models that can be used for describing this system.

The overall probability of failure on demand (PFD) of the interlock is given as:

PFD = PFDs + PFDsis + PFDfce (Eq 7)

Where:

PFD = Probability of failure on demand of the interlock as a whole

PFDs = Probability of failure on demand of the sensors (voting as 2oo3)

PFDsis = Probability of failure on demand of the SIS logic solver

PFDfce = Probability of failure on demand of the final control elements.

Since there are two final control elements arranged in series, the PFDfce = sum of the PFDs of the final control elements (Relay PFD and MCC PFD).

PFDce = PFDr + PFDmcc (Eq 8)

Where:

PFDr = Probability of failure on demand of the two relays voting as 1oo2 to shutdown gas compressor.

PFDmcc = Probability of failure on demand of the MCC to shutdown the gas compressor

For this example, the following selections are made to model the performance of the interlock.

Model for Sensors (2oo3)

The sensors are considered repairable as one sensor can be replaced while the other two function and provide continuous functionality of the interlock. There are three sensors that will be used in a 2oo3 voting system. From Appendix A, using Equation A-7, the model for the sensors becomes:

$$\text{PFDsavg} = 3\times\left[\frac{(1-DCs)\times(1-\beta s)\times\lambda^{Ds}\times TIs}{2}+\frac{DCs\times(1-\beta s)\times\lambda^{Ds}\times DIs}{2}+(1-\beta s)\times\lambda^{Ds}\times MTTRs\right]^2 +$$

$$\left[\frac{(1-DCs)\times\beta s\times\lambda^{Ds}\times TIs}{2}+\frac{DCs\times\beta s\times\lambda^{D}\times DIs}{2}+\beta s\times\lambda^{Ds}\times MTTRs\right] \qquad \text{(Eq 9)}$$

Where:

PFD_{savg} is the average probability of failure on demand of the sensors

DC_s is the diagnostic coverage for sensor failure

DI_s is the diagnostic interval for the sensors

$MTTR_s$ is the mean time to restore the sensors to functionality given a sensor failure

TI_s is the test interval for the sensors

β_s is the common cause failure parameter

λ^{Ds} is the failure rate to a dangerous condition for the sensors

MTTR is the mean time to restore the system from the time that failure occurs

Model for SIS Logic Solver

Use a fixed probability of failure on demand as specified by the vendor. Unless detailed design information is provided by the SIS logic solver vendor, this will be the normal default condition for most studies. For this problem a typical PFD of 1.30×10^{-4} was selected to represent the SIS Logic Solver.

Model for Final Control Elements

There are two separate paths to shutdown the gas compressor. First is by the SIS logic solver commanding the MCC to shutdown power to the the gas compressor motor. The second is for the SIS logic solver to issue a shutdown command to two interposing relays (R1 and R2) which will cause the power to the gas compressor motor to stop. Two different models are needed to the final control elements. These two subsystems are considered non-repairable in this analysis,

Model for Interposing Relays (1oo2)

There are two interposing relays (R1 and R2 in the interlock). From Appendix A, use Equation A-25 setting parameters DC=DI=MTTR=0 for the non-repairable interposing relays, the model becomes:

$$\text{PFDravg} = \left[\frac{(1-\beta r)\times\lambda^{Dr}\times TIr}{2}+\right]^2+\left[\frac{\beta r\times\lambda^{Dr}\times TIr}{2}+\right] \qquad \text{(Eq 10)}$$

Where:

PFD_{ravg} is the average probability of failure on demand of the relays voting 1oo2 to shutoff the gas compressor.

TI_r is the test interval for the relays

β_r is the common cause failure parameter

λ^{Dr} is the failure rate to a dangerous condition for the relays

Model for MCC (1oo1)

From Appendix A and using equation A-1 for the non-repairable MCC, the model becomes:

$$\boldsymbol{PFDmccavg} = \frac{\lambda^{Dmcc} * \mathbf{TImcc}}{2} \qquad \text{(Eq 11)}$$

Where:

PFD_{mccavg} is the average probability of failure on demand of the MCC to shutoff the gas compressor.

TI_{mcc} is the test interval for the MCC

λ^{Dmcc} is the failure rate to a dangerous condition for the MCC

Data

The calculation of the PFD of the interlock requires a set of data to be used to represent the system. Tables 5, 6, and 7 are taken from the Freeman-Summers paper [Ref. 3] and presents the data used to represent the interlock system. Note that these data were originally taken from generic data sources and do not represent any particular device or system.

Table 5. Uncertainty Data for Level Sensors Used in Example Interlock

(All variable probability distributions assumed to be triangularly distributed)

Variable	Min	Mode	Max	Mean	Variance
λ^{Ds}- Fail Dangerous Rate	2.84×10^{-3} Yr^{-1}	5×10^{-3} Yr^{-1}	8.5×10^{-3} Yr^{-1}	5.45×10^{-3} Yr^{-1}	1.36×10^{-6} Yr^{-1}
DC_s – Diagnostic Coverage *	0.8	0.9	0.99	0.897	1.51×10^{-3}
DI_s – Diagnostic Interval **	5.71×10^{-5}	5.71×10^{-5}	5.71×10^{-5}	5.71×10^{-5}	0
TI_s – Test Interval	4 Yr	5 Yr	6 Yr	5 Yr	5.56×10^{-2} Yr^2
β_s- Common Cause Failure Fraction *	0	0.02	0.1	0.04	4.67×10^{-4}
$MTTR_s$ - Mean Time to Restore ***	1.37×10^{-3} Yr	8.22×10^{-3} Yr	1.92×10^{-2} Yr	9.59×10^{-3} Yr	1.34×10^{-5} Yr^2

* Unit less

** 0.5 hours, assumed to be deterministic

***Min of 12 hr, Mode of 72 hours, Max of 168 hrs

Table 6. Uncertainty Data for Relays Used in Example Interlock (All variable probability distributions assumed to be triangularly distributed)					
Variable	**Min**	**Mode**	**Max**	**Mean**	**Variance**
λ^{Dr}- Fail Dangerous Rate	8.76×10^{-9} Yr^{-1}	2.00×10^{-3} Yr^{-1}	4.73×10^{-2} Yr^{-1}	1.64×10^{-2} Yr^{-1}	1.19×10^{-4} Yr^{-2}
DC_r – Diagnostic Coverage *	NA	NA	NA	NA	NA
DI_r – Diagnostic Interval*	NA	NA	NA	NA	NA
TI_r – Test Interval	1 Yr	1 Yr	2 Yr	1.33 Yr	0.056 Yr^2
β_r- Common Cause Failure Fraction	0	0.02	0.1	0.04	4.67×10^{-4}
$MTTR_r$ - Mean Time to Restore*	NA	NA	NA	NA	NA

* Not used in relay model

Table 7. Uncertainty Data for MCC Used in Example Interlock (All variable probability distributions assumed to be triangularly distributed)					
Variable	**Min**	**Mode**	**Max**	**Mean**	**Variance**
λ^{Dr}- Fail Dangerous Rate	1.74×10^{-4} Yr^{-1}	1.31×10^{-3} Yr^{-1}	3.00×10^{-2} Yr^{-1}	1.05×10^{-2} Yr^{-1}	4.76×10^{-5} Yr^{-2}
DC_r – Diagnostic Coverage*	NA	NA	NA	NA	NA
DI_r – Diagnostic Interval*	NA	NA	NA	NA	NA
TI_r – Test Interval	1 Yr	1 Yr	2 Yr	1.33 Yr	5.6×10^{-2} Yr^2
β_r- Common Cause Failure Fraction*	NA	NA	NA	NA	NA
$MTTR_r$ - Mean Time to Restore*	NA	NA	NA	NA	NA

* Not used in MCC model

Uncertainty Analysis Results

Using the VCA method to estimate the PFD of the interlock design yields the results shown in Table 8.

Table 8. Results of VCA on Example Interlock Design

Subsystem	Mean PFD	Variance of PFD
Sensors	6.41E-5	1.7657E-9
Logic Solver	1.34E-4	0
Relays	5.46E-4	2.0333E-7
MCC	6.98E-3	4.52E-5
Totals	**7.73E-3**	**4.54E-5**

Note: Calculations are carried to an adequate number of places to allow for the resulting PFD to be determined. Rounding to significant number of figures is done at the end.

The mean PFD of 0.00773 implies a mean risk reduction factor of:

RRF = 1/PFD = 1/0.00773 = 129 (Eq 11)

The variance of 4.54E-5 implies a standard deviation in the PFD of

Std Dev = $(\text{Variance})^{1/2}$ = 0.006737 (Eq 12)

The corresponding 95% level of confidence in the PFD:

X95% = 1.645 std Dev + Mean PFD (Eq 13)

X95% = 1.645 (0.006737) + 0.00773= 0.01881 (Eq 14)

The RRF for 95% certain PFD is

RRF95% = 1/0.01881 => 53 (Eq 15)

This is essentially the same result previously reported by Freeman and Summers [Ref. 3] using either Monte Carlo Simulation or numerical approximation methods for the VCA sensitivities. The calculated PFD at the 95% level of 0.01881 indicates that there is only a 5% chance that the interlock will provide an RRF worse that 53. Since this level is only SIL-1 capable and not SIL-2 capable as desired by management, a revised design will be needed to achieve the desired risk reduction or a different protective measure will be needed to control risk.

Acknowledgement

This paper is an expanded version of a paper entitled ''General Method for Uncertainty Evaluation of Safety Integrity Level (SIL) Calculations'' presented at the Mary Kay O'Connor 2016 Process Safety Symposium, October 26, 2016, College Station, TX. The author wishes to thank those individuals who gave comments and suggestions for improving this article.

References

1. International Society of Automation, Functional Safety: Safety Instrumented Systems for the Process Sector—Parts 1, 2, and 3, ANSI/ISA 84.00.01, 2004 [USA implementation of IEC 61511].

2. International Society of Automation, *Safety Integrity Level (SIL) Verification of Safety Instrumented Functions*, ISA TR84.00.02, Research Triangle Park, NC. (2015).

3. Freeman, R. And Summers, A, "Evaluation of Uncertainty in Safety Integrity Level Calculations," *Process Safety Progress*, Vol 35, No 4, pp 341-348, December 2016. Published Wiley Online Library (wileyonlinelibrary.com). DOI 10.1002/prs.11805, 23 January 2016

4. Center for Chemical Process Safety, *Layer of Protection Analysis – Simplified Process Risk Assessment*, American Institute of Chemical Engineers, New York, 2001

5. Edward B. Haugen, *Probabilistic Approaches to Design*, John Wiley, New York, 1968

6. R.A. Freeman, "Quantifying LOPA Uncertainty," *Process Safety Progress*, Vol 31, No. 3, pp 240-247, 2012

7. R. A. Freeman, "Simplified Uncertainty Analysis of Layer of Protection Analysis Results," *Process Safety Progress*, Vol. 32, No 4, pp 351-360, December 2013, Published online in Wiley Online Library (wileyonlinelibrary.com) DOI 10.1002/ prs11585.pdf, 3 June 2013

8. David Vose, *Risk Analysis – A Quantitative Guide*, 3rd Edition, John Wiley, 2008

9. Paul L. Meyer, *Introductory Probability and Statistical Applications*, John Wiley, New York, 1972

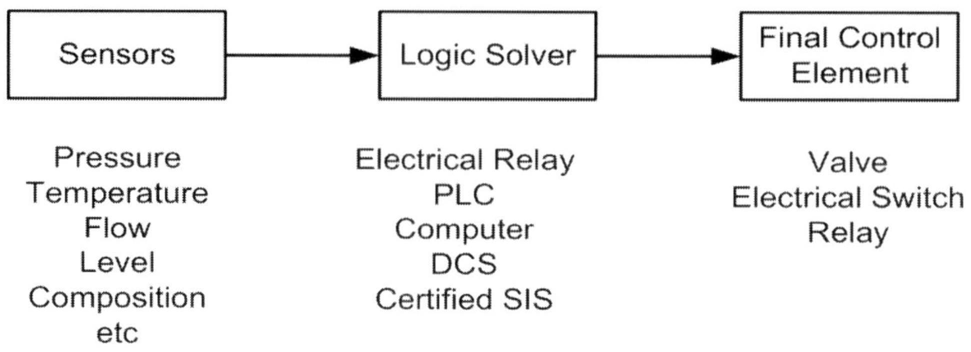

Figure 1. Typical Components of an SIS Interlock

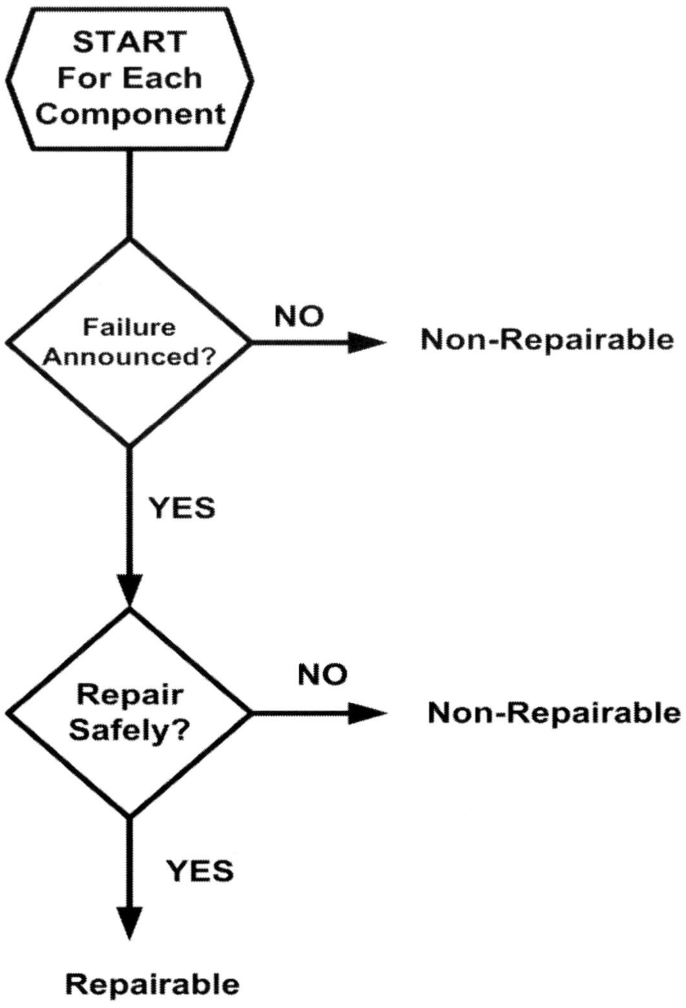

Figure 2. Decision Tree for Determination of Repairable or Non- Repairable

Figure 3. P&ID Sketch for Example Interlock

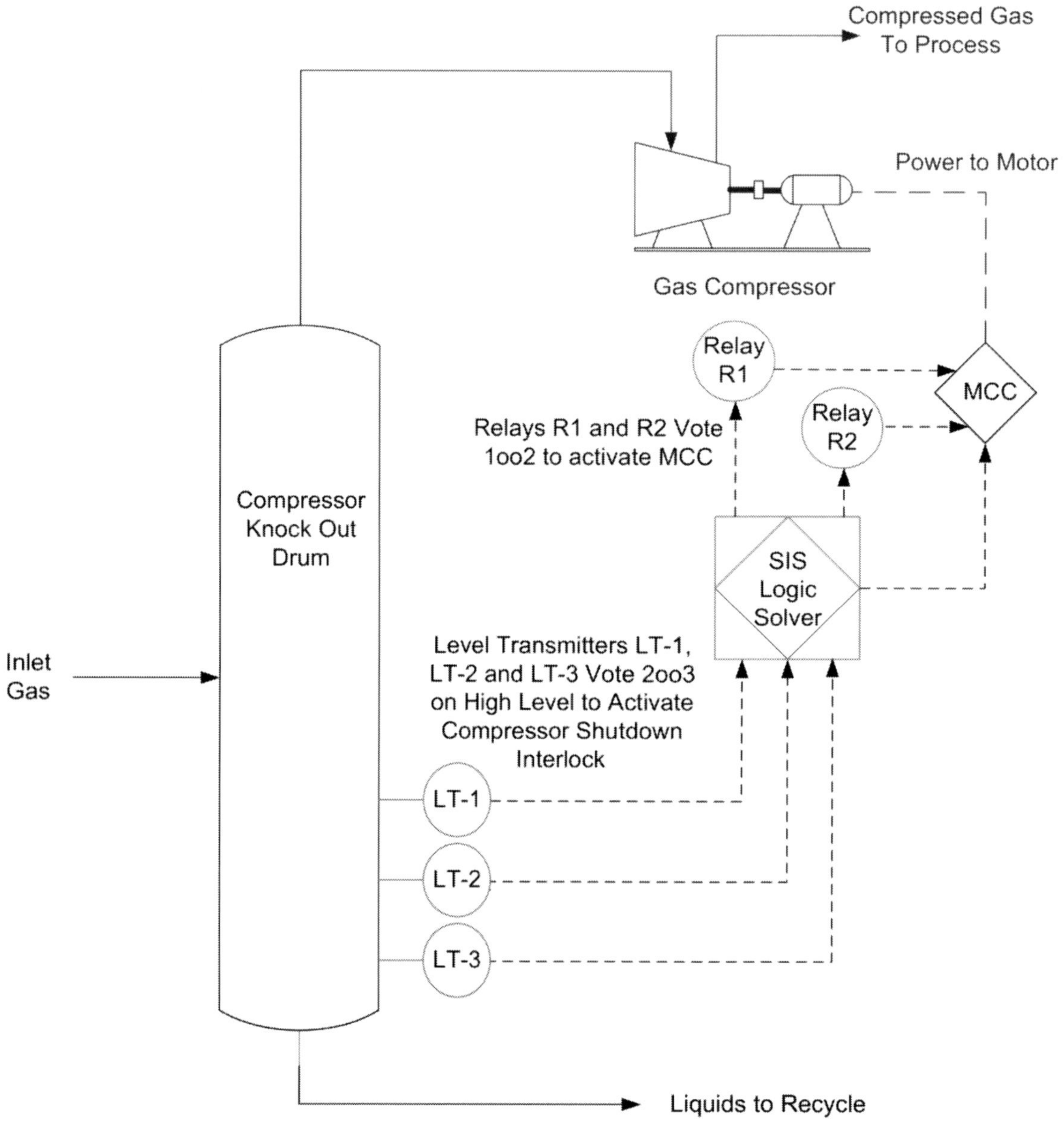

APPENDIX A -- DEVELOPMENT OF UNCERTAINTY EQUATIONS

This appendix develops the equations needed to evaluate the uncertainty in the SIL PFD as computed using the methods outlined in the ISA Technical Report [Ref. 2]. This appendix is divided into two sections:

- Non-Repairable Systems
- Repairable Systems

The uncertainty equations for other configurations not covered in the ISA report are developed in Appendix B.

NON-REPAIRABLE SYSTEMS

For systems that are considered non-repairable, the simplified equations of Table A.1 (taken from the ISA Technical Report [Ref. 2]) are used for the analysis.

Table A.1 Simplified PFD_{avg} formulas for Non-Repairable System without considering CCF, Diagnostics or MTTR.

Configuration	Function	PFD based on "Average before" failure rate	Equation Number
1oo1	$F_1 =$	$\left[\lambda^D \times \frac{TI}{2}\right]$	A-1
1oo2	$F_2 =$	$\left[(\lambda^D)^2 \times \frac{TI^2}{4}\right]$	A- 2
1oo3	$F_3 =$	$\left[(\lambda^D)^3 \times \frac{TI^3}{8}\right]$	A-3
2oo2	$F_4 =$	$\left[\lambda^D \times TI\right]$	A-4
2oo3	F5 =	$\frac{3}{4} \times \left[(\lambda^D)^2 \times TI^2\right]$	A-5
3oo3	F6 =	$3 \times \left[\lambda^D \times \frac{TI}{2}\right]$	A-6

Where TI is the proof test interval, λ^D is the dangerous failure rate.

The expected value of the probability of failure on demand is found by calculating the PFD using the expected value (mean) of the random variables.

Let y a function of some random variables, x_i, as:

$$y = F(x_i) \qquad \text{(Eq A-7)}$$

The mean of y can be estimated using the following approximation:

$$E(y) = F[E(x_i)] \qquad \text{(Eq A-8)}$$

Where:

$E(y)$ = expected value of random variable y = mean of y

$E(x_i)$ = expected value of random variable x_i = mean of x_i

For the configurations defined in Table A.1, the corresponding expected PFD are presented in Table A.2.

Table A.2 Expected Value of PFD_{avg}formulas for Non-Repairable System without considering CCF, Diagnostics or MTTR.

Configuration	Function	Expected PFD based on "Average before" failure rate	Equation Number
1oo1	$E(F_1)$ =	$\left[E(\lambda^D)\times\frac{E(\text{TI})}{2}\right]$	A-9
1oo2	$E(F_2)$ =	$\left[(E(\lambda^D))^2\times\frac{E(\text{TI})^2}{4}\right]$	A- 10
1oo3	$E(F_3)$ =	$\left[(E(\lambda^D))^3\times\frac{E(\text{TI})^3}{8}\right]$	A-11
2oo2	$E(F_4)$ =	$\left[E(\lambda^D)\times E(\text{TI})\right]$	A-12
2oo3	$E(F_5)$ =	$\frac{3}{4}\times\left[(E(\lambda^D))^2\times E(\text{TI})^2\right]$	A-13
3oo3	$E(F_6)$ =	$3\times\left[E(\lambda^D)\times\frac{E(\text{TI})}{2}\right]$	A-14

Where E(x) is the expected value of random variable x.

The variance of the PFD is found by taking the weighted sum of the variance of the random variables. The weighting function is the sensitivity of the PFD with respect to random variable. In mathematical terms the variance of a function may be found in terms of the variance of the random variables as:

Once again let

$$y = F(x_i) \qquad \text{(Eq A-15)}$$

The variance of Y can likewise be estimated as:

$$V(y) = \sum_{i=1}^{n} \left[\frac{\partial y}{\partial x_i}\right]^2 V(x_i) \qquad \text{(Eq A-16)}$$

Where:

$V(y)$ = variance of random variable y as defined above

$V(x_i)$ = variance of random variable x_i as defined above

The sensitivity of y with respect to a random variable x_i is:

$$\text{Sensitivity of y with respect to } x_i = \left[\frac{\partial y}{\partial x_i}\right] \qquad \text{(Eq A-17)}$$

For the configurations defined in Table A.1, the corresponding variance of the PFD are presented in Table A.3.

Table A.3 Variance of PFD_{avg} formulas for Non-Repairable System without considering CCF, Diagnostics or MTTR.			
Configuration	**Function**	**Variance in Function F_i based on "Average before" failure rate**	**Equation Number**
1oo1	$V(F_1) =$	$\left[\frac{E(TI)}{2}\right]^2 V(\lambda^D) + \left[\frac{E(\lambda^D)}{2}\right]^2 V(TI)$	A-18
1oo2	$V(F_2) =$	$\left[E(\lambda^D)^2 \frac{E(TI)^2}{2}\right]^2 V(\lambda^D) +$ $\left[E(\lambda^D)^2 \frac{E(TI)}{2}\right]^2 V(TI)$	A- 19
1oo3	$V(F_3) =$	$\left[3 \times E(\lambda^D)^2 \frac{E(TI)^3}{2}\right]^2 V(\lambda^D) +$ $\left[\frac{3}{8} E(\lambda^D)^3 E(TI)^2\right]^2 V(TI)$	A-20
2oo2	$V(F_4) =$	$E(TI)^2 V(\lambda^D) + E(\lambda^D)^2 \times V(TI)$	A-21
2oo3	$V(F_5) =$	$\left[\frac{3}{2} \times E(\lambda^D) \times E(TI)^2\right]^2 V(\lambda^D) +$ $\left[\frac{3}{2} \times E(\lambda^D)^2 \times E(TI)\right]^2 V(TI)$	A-22
3oo3	$V(F_6) =$	$\left[\frac{3}{2} \times E(TI)\right]^2 V(\lambda^D) +$ $\left[\frac{3}{2} \times E(\lambda^D)\right]^2 V(TI)$	A-23

Note that for the calculation of the variance functions, V(Fi), the random variables are evaluated at the expected value (mean).

REPAIRABLE SYSTEMS

For systems that are considered repairable, the simplified equations of Table A.4 are used for the analysis. The expected value of the probability of failure on demand (PFD) is found by substituting the expected value of each of the random variables into the corresponding equation. The expected value of the PFD for each of the system configurations is presented in Table A.5

Table A.4 Simplified PFD_{avg} Formulas for Repairable System considering CCF, Diagnostics and MTTR

Configuration	Function	PFD based on "Average before" failure rate	Equation Number
1oo1	$F_7 =$	$\left[\frac{(1-DC)\times\lambda^D\times TI}{2}+\frac{DC\times\lambda^D\times DI}{2}+\lambda^D\times MTTR\right]$	A-24
1oo2	$F_8 =$	$\left[\frac{(1-DC)\times(1-\beta)\times\lambda^D\times TI}{2}+\frac{DC\times(1-\beta)\times\lambda^D\times DI}{2}+(1-\beta)\times\lambda^D\times MTTR\right]^2+\left[\frac{(1-DC)\times\beta\times\lambda^D\times TI}{2}+\frac{DC\times\beta\times\lambda^D\times DI}{2}+\beta\times\lambda^D\times MTTR\right]$	A- 25
1oo3	$F_9 =$	$\left[\frac{(1-DC)\times(1-\beta)\times\lambda^D\times TI}{2}+\frac{DC\times(1-\beta)\times\lambda^D\times DI}{2}+(1-\beta)\times\lambda^D\times MTTR\right]^3+\left[\frac{(1-DC)\times\beta\times\lambda^D\times TI}{2}+\frac{DC\times\beta\times\lambda^D\times DI}{2}+\beta\times\lambda^D\times MTTR\right]$	A-26
2oo2	$F_{10} =$	$2\times\left[\frac{(1-DC)\times\lambda^D\times TI}{2}+\frac{DC\times\lambda^D\times DI}{2}+\lambda^D\times MTTR\right]$	A-27

Table A.4 Simplified PFD_{avg} Formulas for Repairable System considering CCF, Diagnostics and MTTR			
Configuration	**Function**	**PFD based on "Average before" failure rate**	**Equation Number**
2oo3	F_{11} =	$3\times\left[\frac{(1-DC)\times(1-\beta)\times\lambda^D\times TI}{2}+\frac{DC\times(1-\beta)\times\lambda^D\times DI}{2}+(1-\beta)\times\lambda^D\times MTTR\right]^2+\left[\frac{(1-DC)\times\beta\times\lambda^D\times TI}{2}+\frac{DC\times\beta\times\lambda^D\times DI}{2}+\beta\times\lambda^D\times MTTR\right]$	A-28
3oo3	F_{12} =	$3\times\left[\frac{(1-DC)\times\lambda^D\times TI}{2}+\frac{DC\times\lambda^D\times DI}{2}+\lambda^D\times MTTR\right]$	A-29

Where:

DC is the diagnostic coverage;

DI is the diagnostic interval;

TI is the proof test interval,

λ^D is the dangerous failure rate;

MTTR is the mean time to restore the system to operation

β is the common cause failure parameter that is always is between 0 and 1

Table A.5 Expected PFD Formulas for Repairable System considering CCF, Diagnostics and MTTR			
Configuration	**Function**	**PFD based on "Average before" failure rate**	**Equation Number**
1oo1	E(F_7) =	$\left[\frac{(1-E(DC))\times E(\lambda^D)\times E(TI))}{2}+\frac{E(DC)\times E(\lambda^D)\times E(DI)}{2} + E(\lambda^D)\times E(MTTR)\right]$	A-30
1oo2	E(F_8) =	$\left[\frac{(1-E(DC))\times(1-E(\beta))\times E(\lambda^D)\times E(TI)}{2}+\frac{E(DC)\times(1-E(\beta))\times E(\lambda^D)\times E(DI)}{2}+(1-E(\beta))\times E(\lambda^D)\times E(MTTR)\right]^2+\left[\frac{(1-E(DC))\times E(\beta)\times E(\lambda^D)\times E(TI)}{2}+\frac{E(DC)\times E(\beta)\times E(\lambda^D)\times E(DI}{2})+E(\beta)\times E(\lambda^D)\times E(MTTR)\right]$	A- 31
1oo3	E(F_9) =	$\left[\frac{(1-E(DC))\times(1-E(\beta))\times E(\lambda^D)\times E(TI)}{2}+\frac{E(DC)\times(1-E(\beta))\times E(\lambda^D)\times E(DI)}{2}+(1-E(\beta))\times E(\lambda^D)\times E(MTTR)\right]^3+\left[\frac{(1-E(DC))\times E(\beta)\times E(\lambda^D)\times E(TI)}{2}+\frac{E(DC)\times E(\beta)\times E(\lambda^D)\times E(DI)}{2}+E(\beta)\times E(\lambda^D)\times E(MTTR)\right]$	A-32

Table A.5 Expected PFD Formulas for Repairable System considering CCF, Diagnostics and MTTR			
Configuration	**Function**	**PFD based on "Average before" failure rate**	**Equation Number**
2oo2	$E(F_{10}) =$	$2\times\left[\frac{(1-E(DC))\times E(\lambda^D)\times E(TI))}{2}+\frac{E(DC)\times E(\lambda^D)\times E(DI)}{2}+E(\lambda^D)\times E(MTTR)\right]$	A-33
2oo3	$E(F_{11}) =$	$3\times\left[\frac{(1-E(DC))\times(1-E(\beta))\times E(\lambda^D)\times E(TI)}{2}+\frac{E(DC)\times(1-E(\beta))\times E(\lambda^D)\times E(DI)}{2}+(1-E(\beta))\times E(\lambda^D)\times E(MTTR)\right]^2+\left[\frac{(1-E(DC))\times E(\beta)\times E(\lambda^D)\times E(TI)}{2}+\frac{E(DC)\times E(\beta)\times E(\lambda^D)\times E(DI)}{2}+E(\beta)\times E(\lambda^D)\times E(MTTR)\right]$	A-34
3oo3	$E(F_{12}) =$	$3\times\left[\frac{(1-E(DC))\times E(\lambda^D)\times E(TI))}{2}+\frac{E(DC)\times E(\lambda^D)\times E(DI)}{2}+E(\lambda^D)\times E(MTTR)\right]$	A-35

The variance of the PFD of the repairable systems is found in the same manner as that in the non-repairable cases. Once again let:

$$y = F(x_i) \quad \text{(Eq A-36)}$$

The variance of Y can likewise be estimated as:

$$V(y) = \sum_{i=1}^{n} \left[\frac{\partial y}{\partial x_i}\right]^2 V(x_i) \quad \text{(Eq A-37)}$$

Where:

V(y) = variance of random variable y as defined above

$V(x_i)$ = variance of random variable x_i as defined above

The sensitivity of y with respect to a random variable x_i is:

In the case of the repairable systems, there are several more potentially uncertain or random variables:

- DC is the diagnostic coverage;
- DI is the diagnostic interval;
- TI is the proof test interval,
- λ^D is the dangerous failure rate;
- MTTR is the mean time to restore the system to operation
- β is the common cause failure parameter that is always is between 0 and 1

We must evaluate the partial derivative of the PFD function with respect to each and then combine them using equation A-37. The partial derivatives of function F7 with respect to each random variable are found as follows

$$\frac{\partial F_7}{\partial DC} = \frac{-\lambda^D \times TI}{2} + \frac{\lambda^D \times DI}{2} \quad \text{(Eq A-38)}$$

$$\frac{\partial F_7}{\partial \lambda^D} = \frac{(1-DC) \times TI}{2} + \frac{DC \times DI}{2} + MTTR \quad \text{(Eq A-39)}$$

$$\frac{\partial F_7}{\partial TI} = \frac{(1-DC) \times \lambda^D}{2} \quad \text{(Eq A-40)}$$

$$\frac{\partial F_7}{\partial DI} = \frac{DC \times \lambda^D}{2} \quad \text{(Eq A-41)}$$

$$\frac{\partial F_7}{\partial MTTR} = \lambda^D \qquad \text{(Eq A-42)}$$

We can now compute the variance of function F7 as:

$$V\left(F_7\right) = \sum_{i=1}^{n} \left[\frac{\partial F_7}{\partial x_i}\right]^2 V(x_i) \qquad \text{(Eq A-43)}$$

$$V\left(F_7\right) = \left[\frac{-\lambda^D \times TI}{2} + \frac{\lambda^D \times DI}{2}\right]^2 \text{V(DC)} + \left[\frac{(1-DC)\times TI}{2} + \frac{DC\times DI}{2} + MTTR\right]^2 \text{V}(\lambda^D) +$$

$$\left[\frac{(1-DC)\times \lambda^D}{2}\right]^2 \text{V(TI)} + \left[\frac{DC\times \lambda^D}{2}\right]^2 \text{(DI)} + \left[\lambda^D\right]^2 \text{V(MTTR)} \qquad \text{(Eq A-44)}$$

Equation 44 represents the variance of the probability of failure on demand of the 1001 configuration when the system is repairable. We note the following relationships between the 1001, 2oo2 and 3oo3 configurations:

$F_{10} = 2\ F_7$ (Eq A-45)

$F_{12} = 3\ F_7$ (Eq A-46)

We can directly determine the variance of functions F_{10} and F_{12} from the properties of the variance operator, V(X).

$V(F_{10}) = V(2\ F_7)$ (Eq A-47)

$V(F_{10}) = 4\ V(F_7)$ (Eq A-48)

$V(F_{12}) = V(3\ F_7)$ (Eq A-49)

$V(F_{12}) = 9\ V(F_7)$ (Eq A-50)

The determination of the variance of functions F_8, F_{10}, and F_{11} is done in a similar manner. To simplify the presentation of the derivation of variance of functions F_8, F_{10}, and F_{11}, we introduce two functions H and Q as:

$$\text{H} = \left[\begin{array}{l}\dfrac{(1-DC)\times(1-\beta)\times\lambda^D\times TI}{2} + \\ \dfrac{DC\times(1-\beta)\times\lambda^D\times DI}{2} + (1-\beta)\times\lambda^D\times MTTR\end{array}\right] \qquad \text{(Eq A-51)}$$

$$\text{Q} = \left[\frac{(1-DC)\times\beta\times\lambda^D\times TI}{2} + \frac{DC\times\beta\times\lambda^D\times DI}{2} + \beta\times\lambda^D\times MTTR\right] \qquad \text{(Eq A-52)}$$

We note the following relationships:

$F_8 = H^2 + Q$ (Eq A-53)

$F_9 = H^3 + Q$ (Eq A-54)

$F_{11} = 3\ H^2 + Q$ (Eq A-55)

The variance of F_8, F_9 and F_{11} are found as:

$V(F_8) = V(H^2 + Q)$ (Eq A-56)

$V(F_9) = V(H^3 + Q)$ (Eq A-57)

$V(F_{11}) = V(3\ H^2 + Q)$ (Eq A-58)

We will need the sensitivity of F_8, F_9 and F_{11} to each of the potential random variables.

$$\frac{\partial F_8}{\partial x} = \frac{\partial (H^2 + Q)}{\partial x} = \frac{\partial (H^2)}{\partial x} + \frac{\partial (Q)}{\partial x} \quad \text{(Eq A-59)}$$

Likewise

$$\frac{\partial F_9}{\partial x} = \frac{\partial (H^3 + Q)}{\partial x} = \frac{\partial (H^3)}{\partial x} + \frac{\partial (Q)}{\partial x} \quad \text{(Eq A-60)}$$

$$\frac{\partial F_{11}}{\partial x} = \frac{\partial 3(H^2 + Q)}{\partial x} = 3\frac{\partial (H^2)}{\partial x} + 3\frac{\partial (Q)}{\partial x} \quad \text{(Eq A-61)}$$

Using the chain rule for partial derivatives, we now write:

$$\frac{\partial F_8}{\partial x} = 2H\frac{\partial (H)}{\partial x} + \frac{\partial (Q)}{\partial x} \quad \text{(Eq A-62)}$$

$$\frac{\partial F_9}{\partial x} = 3H^2\ \frac{\partial (H)}{\partial x} + \frac{\partial (Q)}{\partial x} \quad \text{(Eq A-63)}$$

$$\frac{\partial F_{11}}{\partial x} = 6H\frac{\partial (H)}{\partial x} + \frac{\partial (Q)}{\partial x} \quad \text{(Eq A-64)}$$

We now evaluate the derivatives of function H with respect to each potential random variables:

- DC is the diagnostic coverage;
- DI is the diagnostic interval;

- TI is the proof test interval,
- λ^D is the dangerous failure rate;
- MTTR is the mean time to restore the system to operation
- β is the common cause failure parameter that is always is between 0 and 1

$$\frac{\partial(\mathrm{H})}{\partial DC} = \frac{-(1-\beta)\times\lambda^D\times TI}{2} + \frac{(1-\beta)\times\lambda^D\times DI}{2} \qquad \text{(Eq A-65)}$$

$$\frac{\partial(\mathrm{H})}{\partial TI} = \frac{(1-DC)\times(1-\beta)\times\lambda^D}{2} \qquad \text{(Eq A-66)}$$

$$\frac{\partial(\mathrm{H})}{\partial\lambda^D} = \left[\begin{array}{l}\frac{(1-DC)\times(1-\beta)\times TI}{2} + \\ \frac{DC\times(1-\beta)\times DI}{2} + (1-\beta)\times MTTR\end{array}\right] \qquad \text{(Eq A-67)}$$

$$\frac{\partial(\mathrm{H})}{\partial MTTR} = (1-\beta)\times\lambda^D \qquad \text{(Eq A-68)}$$

$$\frac{\partial(\mathrm{H})}{\partial\beta} = \left[\begin{array}{l}\frac{(1-DC)\times(-1)\times\lambda^D\times TI}{2} + \\ \frac{DC\times(-1)\times\lambda^D\times DI}{2} + (-1)\times\lambda^D\times MTTR\end{array}\right] \qquad \text{(Eq A-69)}$$

$$\frac{\partial(\mathrm{H})}{\partial DI} = \frac{DC\times(1-\beta)\times\lambda^D}{2} \qquad \text{(Eq A-70)}$$

In a similar manner we evaluate the partial derivatives of function Q with respect to each potential random variables.

$$Q = \left[\frac{(1-DC)\times\beta\times\lambda^D\times TI}{2} + \frac{DC\times\beta\times\lambda^D\times DI}{2} + \beta\times\lambda^D\times MTTR \right] \quad \text{(Eq A-71)}$$

- DC is the diagnostic coverage;
- DI is the diagnostic interval;
- TI is the proof test interval,
- λ^D is the dangerous failure rate;
- MTTR is the mean time to restore the system to operation
- β is the common cause failure parameter that is always is between 0 and 1

$$\frac{\partial(Q)}{\partial DC} = \frac{(-1)\times\beta\times\lambda^D\times TI}{2} + \frac{\beta\times\lambda^D\times DI}{2} \quad \text{(Eq A-72)}$$

$$\frac{\partial(Q)}{\partial TI} = \frac{(1-DC)\times\beta\times\lambda^D}{2} \quad \text{(Eq A-73)}$$

$$\frac{\partial(Q)}{\partial \lambda^D} = \left[\frac{(1-DC)\times\beta\times TI}{2} + \frac{DC\times\beta\times DI}{2} + \beta\times MTTR \right] \quad \text{(Eq A-74)}$$

$$\frac{\partial(Q)}{\partial MTTR} = \beta\times\lambda^D\times MTTR \quad \text{(Eq A-75)}$$

$$\frac{\partial(Q)}{\partial \beta} = \left[\frac{(1-DC)\times\lambda^D\times TI}{2} + \frac{DC\times\lambda^D\times DI}{2} + \lambda^D\times MTTR \right] \quad \text{(Eq A-76)}$$

$$\frac{\partial(Q)}{\partial DI} = \frac{DC\times\beta\times\lambda^D}{2} \quad \text{(Eq A-77)}$$

We can now determine the sensitivity of the functions F8, F9 , F11 to the random variables of interest. Restating equation A-62 for sensitivity of F8 with respect to the random variables.

$$\frac{\partial F_8}{\partial x} = 2H\frac{\partial(H)}{\partial x} + \frac{\partial(Q)}{\partial x} \quad \text{(Eq A-78)}$$

For F_8 we find

$$\frac{\partial F_8}{\partial x} = 2\mathrm{H}\frac{\partial(\mathrm{H})}{\partial x} + \frac{\partial(\mathrm{Q})}{\partial x} \qquad \text{(Eq A-80)}$$

$$\frac{\partial F_8}{\partial DC} = 2\left[\frac{(1-DC)\times(1-\beta)\times\lambda^D\times TI}{2} + \frac{DC\times(1-\beta)\times\lambda^D\times DI}{2} + (1-\beta)\times\lambda^D\times MTTR\right]\times$$
$$\left[\frac{-(1-\beta)\times\lambda^D\times TI}{2} + \frac{(1-\beta)\times\lambda^D\times DI}{2}\right] +$$
$$\frac{(-1)\times\beta\times\lambda^D\times TI}{2} + \frac{\beta\times\lambda^D\times DI}{2} \qquad \text{(Eq A-79)}$$

$$\frac{\partial F_8}{\partial TI} = 2\left[\frac{(1-DC)\times(1-\beta)\times\lambda^D\times TI}{2} + \frac{DC\times(1-\beta)\times\lambda^D\times DI}{2} + (1-\beta)\times\lambda^D\times MTTR\right]\times\left[\frac{(1-DC)\times(1-\beta)\times\lambda^D}{2}\right] +$$
$$\frac{(1-DC)\times\beta\times\lambda^D}{2} \qquad \text{(Eq A-81)}$$

$$\frac{\partial F_8}{\partial MTTR} = 2\left[\frac{(1-DC)\times(1-\beta)\times\lambda^D\times TI}{2} + \frac{DC\times(1-\beta)\times\lambda^D\times DI}{2} + (1-\beta)\times\lambda^D\times MTTR\right]\times$$
$$\left[(1-\beta)\times\lambda^D\right] + \beta\times\lambda^D\times MTTR \qquad \text{(Eq A-82)}$$

$$\frac{\partial F_8}{\partial \beta} = 2\left[\begin{array}{l}\frac{(1-DC)\times(1-\beta)\times\lambda^D\times TI}{2} + \\ \frac{DC\times(1-\beta)\times\lambda^D\times DI}{2} + (1-\beta)\times\lambda^D\times MTTR\end{array}\right] \times$$

$$\left[\left[\begin{array}{l}\frac{(1-DC)\times(-1)\times\lambda^D\times TI}{2} + \\ \frac{DC\times(-1)\times\lambda^D\times DI}{2} + (-1)\times\lambda^D\times MTTR\end{array}\right]\right] +$$

$$\left[\frac{(1-DC)\times\lambda^D\times TI}{2} + \frac{DC\times\lambda^D\times DI}{2} + \lambda^D\times MTTR\right] \qquad \text{(Eq A-83)}$$

$$\frac{\partial F_8}{\partial DI} = 2\left[\begin{array}{l}\frac{(1-DC)\times(1-\beta)\times\lambda^D\times TI}{2} + \\ \frac{DC\times(1-\beta)\times\lambda^D\times DI}{2} + (1-\beta)\times\lambda^D\times MTTR\end{array}\right] \times \left[\frac{DC\times(1-\beta)\times\lambda^D}{2}\right] +$$

$$\frac{DC\times\beta\times\lambda^D}{2} \qquad \text{(Eq A-84)}$$

$$\left[\frac{\partial F_8}{\partial \lambda^D}\right] = 2\left[\begin{array}{l}\frac{(1-DC)\times(1-\beta)\times\lambda^D\times TI}{2} + \\ \frac{DC\times(1-\beta)\times\lambda^D\times DI}{2} + (1-\beta)\times\lambda^D\times MTTR\end{array}\right] \times \left[\begin{array}{l}\frac{(1-DC)\times(1-\beta)\times TI}{2} + \\ \frac{DC\times(1-\beta)\times DI}{2} + (1-\beta)\times MTTR\end{array}\right] +$$

$$\left[\frac{(1-DC)\times\beta\times TI}{2} + \frac{DC\times\beta\times DI}{2} + \beta\times MTTR\right] \qquad \text{(Eq A-85)}$$

Using the sensitivities for F8 calculated above, the variance in F8 due to uncertain or random variables is found as:

$$V(F_8) = \left[\frac{\partial F_8}{\partial DC}\right]^2 V(DC) + \left[\frac{\partial F_8}{\partial TI}\right]^2 V(TI) + \left[\frac{\partial F_8}{\partial \lambda^D}\right]^2 V(\lambda^D) +$$

$$\left[\frac{\partial F_8}{\partial MTTR}\right]^2 V(MTTR) + \left[\frac{\partial F_8}{\partial \beta}\right]^2 V(\beta) + \left[\frac{\partial F_8}{\partial DI}\right]^2 V(DI) \qquad \text{(Eq A-86)}$$

Using the equation A-63, we now determine the variance of F_9.

$$\frac{\partial F_9}{\partial x} = 3H^2 \frac{\partial (H)}{\partial x} + \frac{\partial (Q)}{\partial x} \qquad \text{(Eq A-87)}$$

$$\frac{\partial F_9}{\partial DC} = 3\left[\frac{(1-DC)\times(1-\beta)\times\lambda^D \times TI}{2} + \frac{DC\times(1-\beta)\times\lambda^D \times DI}{2} + (1-\beta)\times\lambda^D \times MTTR\right]^2 \times \left[\frac{-(1-\beta)\times\lambda^D \times TI}{2} + \frac{(1-\beta)\times\lambda^D \times DI}{2}\right] + \frac{(-1)\times\beta\times\lambda^D \times TI}{2} + \frac{\beta\times\lambda^D \times DI}{2} \qquad \text{(Eq A-88)}$$

$$\frac{\partial F_9}{\partial TI} = 3\left[\frac{(1-DC)\times(1-\beta)\times\lambda^D\times TI}{2} + \frac{DC\times(1-\beta)\times\lambda^D\times DI}{2} + (1-\beta)\times\lambda^D\times MTTR\right]^2 \times \left[\frac{(1-DC)\times(1-\beta)\times\lambda^D}{2}\right] + \frac{(1-DC)\times\beta\times\lambda^D}{2} \quad \text{(Eq A-89)}$$

$$\frac{\partial F_9}{\partial MTTR} = 3\left[\frac{(1-DC)\times(1-\beta)\times\lambda^D\times TI}{2} + \frac{DC\times(1-\beta)\times\lambda^D\times DI}{2} + (1-\beta)\times\lambda^D\times MTTR\right]^2 \times \left[(1-\beta)\times\lambda^D\right] + \beta\times\lambda^D\times MTTR \quad \text{(Eq A-90)}$$

$$\frac{\partial F_9}{\partial \beta} = 3\left[\frac{(1-DC)\times(1-\beta)\times\lambda^D\times TI}{2} + \frac{DC\times(1-\beta)\times\lambda^D\times DI}{2} + (1-\beta)\times\lambda^D\times MTTR\right]^2 \times \left[\left[\frac{(1-DC)\times(-1)\times\lambda^D\times TI}{2} + \frac{DC\times(-1)\times\lambda^D\times DI}{2} + (-1)\times\lambda^D\times MTTR\right]\right] + \left[\frac{(1-DC)\times\lambda^D\times TI}{2} + \frac{DC\times\lambda^D\times DI}{2} + \lambda^D\times MTTR\right] \quad \text{(Eq A-91)}$$

$$\frac{\partial F_9}{\partial DI} = 3\left[\frac{(1-DC)\times(1-\beta)\times\lambda^D\times TI}{2} + \frac{DC\times(1-\beta)\times\lambda^D\times DI}{2} + (1-\beta)\times\lambda^D\times MTTR\right]^2 \times \left[\frac{DC\times(1-\beta)\times\lambda^D}{2}\right] + \frac{DC\times\beta\times\lambda^D}{2} \quad \text{(Eq A-92)}$$

$$\left[\frac{\partial F_9}{\partial \lambda^D}\right] = 3 \times \left[\frac{(1-DC)\times(1-\beta)\times\lambda^D\times TI}{2} + \frac{DC\times(1-\beta)\times\lambda^D\times DI}{2} + (1-\beta)\times\lambda^D\times MTTR\right]^2 \times \left[\frac{(1-DC)\times(1-\beta)\times TI}{2} + \frac{DC\times(1-\beta)\times DI}{2} + (1-\beta)\times MTTR\right] + \left[\frac{(1-DC)\times\beta\times TI}{2} + \frac{DC\times\beta\times DI}{2} + \beta\times MTTR\right] \quad \text{(Eq A-93)}$$

Using the sensitivities for F_9 calculated above, the variance in F_9 due to uncertain or random variables is found as:

$$V(F_9) = \left[\frac{\partial F_9}{\partial DC}\right]^2 V(DC) + \left[\frac{\partial F_9}{\partial TI}\right]^2 V(TI) + \left[\frac{\partial F_9}{\partial \lambda^D}\right]^2 V(\lambda^D) + \left[\frac{\partial F_9}{\partial MTTR}\right]^2 V(MTTR) + \left[\frac{\partial F_9}{\partial \beta}\right]^2 V(\beta) + \left[\frac{\partial F_9}{\partial DI}\right]^2 V(DI) \quad \text{(Eq A-94)}$$

Finally, in a similar manner the variance in F_{11} is computed as:

$$\frac{\partial F_{11}}{\partial x} = 6\mathrm{H}\frac{\partial(\mathrm{H})}{\partial x} + \frac{\partial(\mathrm{Q})}{\partial x} \qquad \text{(Eq A-95)}$$

$$\frac{\partial F_{11}}{\partial DC} = 6\left[\begin{array}{l}\frac{(1-DC)\times(1-\beta)\times\lambda^{D}\times TI}{2}+\\ \frac{DC\times(1-\beta)\times\lambda^{D}\times DI}{2}+(1-\beta)\times\lambda^{D}\times MTTR\end{array}\right]\times \left[\frac{-(1-\beta)\times\lambda^{D}\times TI}{2}+\frac{(1-\beta)\times\lambda^{D}\times DI}{2}\right]+ \frac{(-1)\times\beta\times\lambda^{D}\times TI}{2}+\frac{\beta\times\lambda^{D}\times DI}{2} \qquad \text{(Eq A-96)}$$

$$\frac{\partial F_{11}}{\partial TI} = 6\left[\begin{array}{l}\frac{(1-DC)\times(1-\beta)\times\lambda^{D}\times TI}{2}+\\ \frac{DC\times(1-\beta)\times\lambda^{D}\times DI}{2}+(1-\beta)\times\lambda^{D}\times MTTR\end{array}\right]\times \left[\frac{(1-DC)\times(1-\beta)\times\lambda^{D}}{2}\right]+ \frac{(1-DC)\times\beta\times\lambda^{D}}{2} \qquad \text{(Eq A-97)}$$

$$\frac{\partial F_{11}}{\partial MTTR} = 6\left[\begin{array}{l}\frac{(1-DC)\times(1-\beta)\times\lambda^{D}\times TI}{2}+\\ \frac{DC\times(1-\beta)\times\lambda^{D}\times DI}{2}+(1-\beta)\times\lambda^{D}\times MTTR\end{array}\right]\times \left[(1-\beta)\times\lambda^{D}\right]+ \beta\times\lambda^{D}\times MTTR \qquad \text{(Eq A-98)}$$

$$\frac{\partial F_{11}}{\partial \beta} = 6\left[\begin{array}{l}\frac{(1-DC)\times(1-\beta)\times\lambda^{D}\times TI}{2}+\\ \frac{DC\times(1-\beta)\times\lambda^{D}\times DI}{2}+(1-\beta)\times\lambda^{D}\times MTTR\end{array}\right]\times$$

$$\left[\left[\begin{array}{l}\frac{(1-DC)\times(-1)\times\lambda^{D}\times TI}{2}+\\ \frac{DC\times(-1)\times\lambda^{D}\times DI}{2}+(-1)\times\lambda^{D}\times MTTR\end{array}\right]\right]+$$

$$\left[\frac{(1-DC)\times\lambda^{D}\times TI}{2}+\frac{DC\times\lambda^{D}\times DI}{2}+\lambda^{D}\times MTTR\right] \quad \text{(Eq A-99)}$$

$$\frac{\partial F_{11}}{\partial DI} = 6\left[\begin{array}{l}\frac{(1-DC)\times(1-\beta)\times\lambda^{D}\times TI}{2}+\\ \frac{DC\times(1-\beta)\times\lambda^{D}\times DI}{2}+(1-\beta)\times\lambda^{D}\times MTTR\end{array}\right]\times\left[\frac{DC\times(1-\beta)\times\lambda^{D}}{2}\right]+$$

$$\frac{DC\times\beta\times\lambda^{D}}{2} \quad \text{(Eq A-100)}$$

$$\left[\frac{\partial F_{11}}{\partial \lambda^{D}}\right] = 6\left[\begin{array}{l}\frac{(1-DC)\times(1-\beta)\times\lambda^{D}\times TI}{2}+\\ \frac{DC\times(1-\beta)\times\lambda^{D}\times DI}{2}+(1-\beta)\times\lambda^{D}\times MTTR\end{array}\right]\times$$

$$\left[\begin{array}{l}\frac{(1-DC)\times(1-\beta)\times TI}{2}+\\ \frac{DC\times(1-\beta)\times DI}{2}+(1-\beta)\times MTTR\end{array}\right]+$$

$$\left[\frac{(1-DC)\times\beta\times TI}{2}+\frac{DC\times\beta\times DI}{2}+\beta\times MTTR\right] \quad \text{(Eq A-101)}$$

Using the sensitivities for F_{11} calculated above, the variance in F_{11} due to uncertain or random variables is found as:

$$V(F_{11}) = \left[\frac{\partial F_{11}}{\partial DC}\right]^2 V(DC) + \left[\frac{\partial F_{11}}{\partial TI}\right]^2 V(TI) + \left[\frac{\partial F_{11}}{\partial \lambda^D}\right]^2 V(\lambda^D) +$$

$$\left[\frac{\partial F_{11}}{\partial MTTR}\right]^2 V(MTTR) + \left[\frac{\partial F_{11}}{\partial \beta}\right]^2 V(\beta) + \left[\frac{\partial F_{11}}{\partial DI}\right]^2 V(DI) \qquad \text{(Eq A-102)}$$

PROCEDURE FOR DETERMINATION OF VARIANCE FOR REPAIRABLE 1oo2, 1oo3 AND 2oo3 SYSTEMS

Based on the above derivations we may now write the formulas for the variance of various configurations of repairable equipment.

Step 1. Determine the configuration of equipment to be studied.

Step 2. Evaluate Function H using the expected value of all random or uncertain variables, equation A-51

Step 3. Evaluate Function Q using the expected value of all random or uncertain variables, equation A-52

Step 4. Evaluate the sensitivity of function H with respect to the random variables using equations A-65 thru A-70.

Step 5. Evaluate the sensitivity of function Q with respect to the random variables using equations A-72 thru A-77.

Step 6. Evaluate the variance as:

a. For 1oo2 systems use equation A-86 for $V(F_8)$
b. For 1oo3 systems use equation A-94 for $V(F_9)$
c. For 2oo3 systems use equation A-102 for $V(F_{11})$

The end result of the above calculations is the determination of the mean and variance of the configuration used in the sensor, logic solver or final control element systems. Table 4 of the main body of this paper presents a cross reference roadmap for the determination of the mean and variance of various of the probability of failure on demand (PFD) of various hardware configurations. The mean and variance for the particular configuration is then returned to the overall calculations of the PFD of the interlock.

APPENDIX B -- UNCERTAINTY EQUATIONS FOR CONFIGURATIONS NOT COVERED IN ISA TR 84.00.02

The ISA Technical report [Ref. 2] does not define the calculation equations for the PFDavg for non-repairable systems where duplicated devices create the potential for common cause failures. This appendix derives the equations needed to evaluate the uncertainty for non-repairable systems with the potential for common cause failures.

To derive the needed functions for the non-repairable systems with common cause failure (CCF) potential, we start with the equations for the repairable system with common cause failure allowed (Table A.4). First we note that a system composed of only one set of components (1oo1) without duplication has no potential for a common cause failure. In addition, a review of the equations of Table 2 finds that systems 1oo1, 2oo2 and 3oo3 do not contain the common cause failure (CCF) parameter β. Therefore, systems 1oo1, 2oo2, and 3oo3 can be eliminated from consideration as common cause failures cannot lead to failure of the specified interlocks.

For the other situations [1oo2, 1oo3 and 2oo3] we note that no repair will be allowed during the mission of the non-repairable system. This observation implies that the diagnostic coverage (DC), the diagnostic interval (DI) and the mean time for restoration (MTTR) are all zero (0). We will set DC = DI = MTTR = 0

Substitution of the zero value for DC, DI and MTTR into the equations of Table A.4 yields the non-repairable system PFD equations where CCF is allowed. Table B.1 presents the results of these algebraic substitutions.

Table B.1 Simplified PFD_{avg} formulas for Non-Repairable System Allowing for CCF

Configuration	Function	PFD based on "Average before" failure rate	Equation Number
1oo2	F_{13} =	$\left[\dfrac{(1-\beta)\times\lambda^D\times TI}{2}\right]^2 + \left[\dfrac{\beta\times\lambda^D\times TI}{2}\right]$	B-1
1oo3	F_{14} =	$\left[\dfrac{(1-\beta)\times\lambda^D\times TI}{2}\right]^3 + \left[\dfrac{\beta\times\lambda^D\times TI}{2}\right]$	B-2
2oo3	F_{15} =	$3\times\left[\dfrac{(1-\beta)\times\lambda^D\times TI}{2}\right]^2 + \left[\dfrac{\beta\times\lambda^D\times TI}{2}\right]$	B-3

Where TI is the proof test interval, λ^D is the dangerous failure rate.

The expected value of the probability of failure on demand is found by calculating the PFD using the expected value (mean) of the random variables.

Let y a function of some random variables, x_i, as:

$$y = F(x_i) \qquad \text{(Eq B-4)}$$

The mean of y can be estimated using the following approximation:

$$E(y) = F[E(x_i)] \qquad \text{(Eq B-5)}$$

Where:

$E(y)$ = expected value of random variable y = mean of y

$E(x_i)$ = expected value of random variable x_i = mean of x_i

For the configurations defined in Table A.1, the corresponding expected PFD are presented in Table B.2.

Table B.2 Expected Value of PFD_{avg} formulas for Non-Repairable System Allowing CCF			
Configuration	**Function**	**Expected PFD based on "Average before" failure rate**	**Equation Number**
1oo2	$E(F_{13})$ =	$\left[\frac{(1-E(\beta))\times E(\lambda^D)\times E(TI)}{2}+\right]^2 + \left[\frac{E(\beta)\times E(\lambda^D)\times E(TI)}{2}\right]$	B-6
1oo3	$E(F_{14})$ =	$\left[\frac{(1-E(\beta))\times E(\lambda^D)\times E(TI)}{2}+\right]^3 + \left[\frac{E(\beta)\times E(\lambda^D)\times E(TI)}{2}\right]$	B-7
2oo3	$E(F_{15})$ =	$3\times\left[\frac{(1-E(\beta))\times E(\lambda^D)\times E(TI)}{2}+\right]^2 + \left[\frac{E(\beta)\times E(\lambda^D)\times E(TI)}{2}\right]$	B-8

Where E(x) is the expected value of random variable x.

The variance of the PFD is found by taking the weighted sum of the variance of the random variables. The weighting function is the sensitivity of the PFD with respect to random variable. In mathematical terms the variance of a function may be found in terms of the variance of the random variables as:

Once again let

$y = F(x_i)$ (Eq B-9)

The variance of Y can likewise be estimated as:

$$V(y) = \sum_{i=1}^{n} \left[\frac{\partial y}{\partial x_i}\right]^2 V(x_i) \qquad \text{(Eq B-10)}$$

Where:

V(y) = variance of random variable y as defined above

$V(x_i)$ = variance of random variable x_i as defined above

The sensitivity of y with respect to a random variable x_i is:

$$\text{Sensitivity of y with respect to } x_i = \left[\frac{\partial y}{\partial x_i}\right] \qquad \text{(Eq B-11)}$$

The variance of the PFD for the repairable cases 1oo2, 1oo3 and 2oo3 are presented in equations of Appendix A as:

- 1oo2, the variance is specified by equation A-86
- 1oo3, the variance is specified by equation A-94
- 2oo3, the variance is specified by equation A-102

These three equations are the basis upon which the equations for the non-repairable system with common cause failure (CCF) are developed.

Variance of PFD for Non-Repairable 1oo2 with CCF

To derive the variance for F_{13} we start with the general equation for the variance as:

$$V(F_{13}) = \left[\frac{\partial F_{13}}{\partial DC}\right]^2 V(DC) + \left[\frac{\partial F_{13}}{\partial TI}\right]^2 V(TI) + \left[\frac{\partial F_{13}}{\partial \lambda^D}\right]^2 V(\lambda^D) +$$

$$\left[\frac{\partial F_{13}}{\partial MTTR}\right]^2 V(MTTR) + \left[\frac{\partial F_{13}}{\partial \beta}\right]^2 V(\beta) + \left[\frac{\partial F_{13}}{\partial DI}\right]^2 V(DI) \qquad \text{(Eq B-12)}$$

The variance of a constant is zero. Thus, setting DC=DI=MTTR =0 results in the variance of these variables as becoming zero.

$$V(DC) = V(DI) = V(MTTR) = 0 \qquad \text{(Eq B-13)}$$

Thus, the variance for the 1oo2 case becomes:

$$V(F_{13}) = \left[\frac{\partial F_{13}}{\partial TI}\right]^2 V(TI) + \left[\frac{\partial F_{13}}{\partial \lambda^D}\right]^2 V(\lambda^D) + \left[\frac{\partial F_{13}}{\partial \beta}\right]^2 V(\beta) \quad \text{(Eq B-14)}$$

The evaluation of the sensitive's of the PFD may now be taken from appendix A equations as:

$\left[\frac{\partial F_{13}}{\partial TI}\right]$ from equation A-81

$\left[\frac{\partial F_{13}}{\partial \lambda^D}\right]$ from equation A-85

$\left[\frac{\partial F_{13}}{\partial \beta}\right]$ from equation A-83

Substituting DC=DI=MTTR=0 into these sensitivity equations yields:

$$\frac{\partial F_{13}}{\partial TI} = 2\left[\frac{(1-\beta)\times\lambda^D\times TI}{2} +\right]\times\left[\frac{(1-\beta)\times\lambda^D}{2}\right] + \frac{\beta\times\lambda^D}{2} \quad \text{(Eq B-15)}$$

$$\frac{\partial F_{13}}{\partial \lambda^D} = 2\left[\frac{(1-\beta)\times\lambda^D\times TI}{2} +\right]\times\left[\frac{(1-\beta)\times TI}{2}\right] + \left[\frac{\beta\times TI}{2}\right] \quad \text{(Eq B-16)}$$

$$\frac{\partial F_{13}}{\partial \beta} = 2\left[\frac{(1-\beta)\times\lambda^D\times TI}{2}\right]\times\left[\frac{(-1)\times\lambda^D\times TI}{2}\right] + \left[\frac{\lambda^D\times TI}{2}\right] \quad \text{(Eq B-17)}$$

Variance of PFD for Non-Repairable 1oo3 with CCF

To derive the variance for F_{14} we start with the general equation for the variance as:

$$V(F_{14}) = \left[\frac{\partial F_{14}}{\partial DC}\right]^2 V(DC) + \left[\frac{\partial F_{14}}{\partial TI}\right]^2 V(TI) + \left[\frac{\partial F_{14}}{\partial \lambda^D}\right]^2 V(\lambda^D) +$$

$$\left[\frac{\partial F_{14}}{\partial MTTR}\right]^2 \text{V(MTTR)} + \left[\frac{\partial F_{14}}{\partial \beta}\right]^2 \text{V(}\beta\text{)} + \left[\frac{\partial F_{14}}{\partial DI}\right]^2 \text{V(DI)} \qquad \text{(Eq B-18)}$$

The variance of a constant is zero. Thus, setting DC=DI=MTTR =0 results in the variance of these variables as becoming zero.

$$\text{V(DC)} = \text{V(DI)} = \text{V(MTTR)} = 0 \qquad \text{(Eq B-19)}$$

Thus, the variance for the 1oo3 case becomes:

$$\text{V(}F_{14}\text{)} = \left[\frac{\partial F_{14}}{\partial TI}\right]^2 \text{V(TI)} + \left[\frac{\partial F_{14}}{\partial \lambda^D}\right]^2 \text{V(}\lambda^D\text{)} + \left[\frac{\partial F_{14}}{\partial \beta}\right]^2 \text{V(}\beta\text{)} \qquad \text{(Eq B-20)}$$

The evaluation of the sensitive's of the PFD may now be taken from appendix A equations as:

$\left[\frac{\partial F_{14}}{\partial TI}\right]$ from equation A-89

$\left[\frac{\partial F_{14}}{\partial \lambda^D}\right]$ from equation A-93

$\left[\frac{\partial F_{14}}{\partial \beta}\right]$ from equation A-91

Substituting DC=DI=MTTR=0 into these sensitivity equations yields:

$$\frac{\partial F_{14}}{\partial TI} = 3\left[\frac{(1-\beta)\times\lambda^D\times TI}{2} + \right]^2 \times \left[\frac{(1-\beta)\times\lambda^D}{2}\right] + \frac{\beta\times\lambda^D}{2} \qquad \text{(Eq B-21)}$$

$$\frac{\partial F_{14}}{\partial \lambda^D} = 3\left[\frac{(1-\beta)\times\lambda^D\times TI}{2}\right]^2 \times \left[\frac{(1-\beta)\times TI}{2}\right] + \left[\frac{\beta\times TI}{2}\right] \qquad \text{(Eq B-22)}$$

$$\frac{\partial F_{14}}{\partial \beta} = 3\left[\frac{(1-\beta)\times \lambda^D \times TI}{2}\right]^2 \times \left[\frac{(-1)\times \lambda^D \times TI}{2}\right] + \left[\frac{\lambda^D \times TI}{2}\right] \quad \text{(Eq B-23)}$$

Variance of PFD for Non-Repairable 2oo3 with CCF

To derive the variance for F_{15} we start with the general equation for the variance as:

$$V(F_{15}) = \left[\frac{\partial F_{15}}{\partial DC}\right]^2 V(DC) + \left[\frac{\partial F_{15}}{\partial TI}\right]^2 V(TI) + \left[\frac{\partial F_{15}}{\partial \lambda^D}\right]^2 V(\lambda^D) +$$

$$\left[\frac{\partial F_{15}}{\partial MTTR}\right]^2 V(MTTR) + \left[\frac{\partial F_{15}}{\partial \beta}\right]^2 V(\beta) + \left[\frac{\partial F_{15}}{\partial DI}\right]^2 V(DI) \quad \text{(Eq B-24)}$$

The variance of a constant is zero. Thus, setting DC=DI=MTTR =0 results in the variance of these variables as becoming zero.

$$V(DC) = V(DI) = V(MTTR) = 0 \quad \text{(Eq B-25)}$$

Thus, the variance for the 2oo3 case becomes:

$$V(F_{15}) = \left[\frac{\partial F_{15}}{\partial TI}\right]^2 V(TI) + \left[\frac{\partial F_{15}}{\partial \lambda^D}\right]^2 V(\lambda^D) + \left[\frac{\partial F_{15}}{\partial \beta}\right]^2 V(\beta) \quad \text{(Eq B-26)}$$

The evaluation of the sensitive's of the PFD may now be taken from appendix A equations as:

$\left[\frac{\partial F_{15}}{\partial TI}\right]$ from equation A-97

$\left[\frac{\partial F_{15}}{\partial \lambda^D}\right]$ from equation A-101

$\left[\frac{\partial F_{15}}{\partial \beta}\right]$ from equation A-99

Substituting DC=DI=MTTR=0 into these sensitivity equations yields:

$$\frac{\partial F_{15}}{\partial TI} = 6\left[\frac{(1-\beta)\times\lambda^{D}\times TI}{2} + \right]\times\left[\frac{(1-\beta)\times\lambda^{D}}{2}\right] + \frac{\beta\times\lambda^{D}}{2} \qquad \text{(Eq B-27)}$$

$$\frac{\partial F_{15}}{\partial \lambda^{D}} = 6\left[\frac{(1-\beta)\times\lambda^{D}\times TI}{2}\right]\times\left[\frac{(1-\beta)\times TI}{2}\right] + \left[\frac{\beta\times TI}{2}\right] \qquad \text{(Eq B-28)}$$

$$\frac{\partial F_{15}}{\partial \beta} = 6\left[\frac{(1-\beta)\times\lambda^{D}\times TI}{2}\right]\times\left[\frac{(-1)\times\lambda^{D}\times TI}{2}\right] + \left[\frac{\lambda^{D}\times TI}{2}\right] \qquad \text{(Eq B-29)}$$

The evaluation of the functions for the variance are always done using the expected value of the variable. For example when the variable TI appears in an equation, the numerical evaluation of the function would be using E(TI) where E(TI) is the expected value of the random variable TI.

Comparison of Batch versus Continuous Process in Pharmaceutical Industry Based on Safety Consideration

Shiqi Chen
Mary Kay O'Connor Process Safety Center
Artie McFerrin Department of Chemical Engineering
Texas A&M University
College Station, Texas 77843-3122, USA Email address
shiqi_chen@tamu.edu

Yogesh Koirala, M. Sam Mannan
Mary Kay O'Connor Process Safety Center
Artie McFerrin Department of Chemical Engineering
Texas A&M University
College Station, Texas 77843-3122, USA Email address
ykoirala@tamu.edu, mannan@tamu.edu

Prepared for Presentation at
American Institute of Chemical Engineers
2017 Spring Meeting and 13th Global Congress on Process Safety
San Antonio, TX
March 26 – 29, 2017

Comparison of Batch versus Continuous Process in Pharmaceutical Industry Based on Safety Consideration

Shiqi Chen
Mary Kay O'Connor Process Safety Center
Artie McFerrin Department of Chemical Engineering
Texas A&M University
College Station, Texas 77843-3122, USA Email address
shiqi_chen@tamu.edu

Yogesh Koirala, M. Sam Mannan
Mary Kay O'Connor Process Safety Center
Artie McFerrin Department of Chemical Engineering

Keywords: Continuous Process, Hazard Identification, Pharmaceutical Manufacturing

Abstract

As opposed to the petro-chemical and bulk chemical industry, where continuous processes are widely applied, the pharmaceutical industry still primarily relies on traditional batch process due to the complexity of product, multi-step operation and low-volume production. Considering these conditions, the versatile batch process is more appropriate. Nowadays, driven by the contradiction between increasing demand for drugs and inefficient batch production mode, there is a trend in pharmaceutical industry, that is the transformation from traditional batch process to novel continuous process. Related projects and research that is aimed at analyzing this transition are conducted in worldwide, and the scale of these studies ranges from lab-scale reactions to overall arrangement of the factories.

Although continuous pharmaceutical process is thriving, the safety issue in this field is not promoted at the same time. Since the continuous pharmaceutical process is a novel technology, little information can be provided to evaluate its safety level. Moreover, process conditions are usually intensified for continuous process comparing to batch process. It also may bring potential risks and make continuous manufacturing inappropriate for some of pharmaceutical productions.

This research provides a comprehensive comparison for batch versus continuous pharmaceutical process by application of Dow's Fire and Explosion Index. In addition to this conventional safety evaluation, influences from production efficiency and specialties in pharmaceutical production are integrated into the comparison. Production of 2-methyl benzimidazole and peracetic acid via batch and continuous processes are conducted in this research. Via these integrative and systematic studies, safety levels of batch and continuous processes is analyzed and compared. The ways in which process conditions, production efficiency, and other requirements influence safety level for different production modes are illustrated.

1 Introduction

Continuous process is widely applied in petro-chemical and bulk chemical industry for its high production rate, automated operation and saving in cost[1]. In contrast, the pharmaceutical industry still primarily relies on traditional batch process. This aspect stems from intrinsic limitations in active pharmaceutical ingredients (APIs) production. Complexity of medical substances means multistep synthesis. An average of 8.1 steps is needed for one API synthesis[2]. Another limitation is production volume. Demand for drugs are not comparable with most of the bulk chemicals. These limitations make batch process dominate pharmaceutical industry for a long time.

However, due to the gradual expansion of the pharmaceutical market, batch process is behind the time because of its low production rate caused by low yield and inefficient operation mode. Driven by this contradiction between increasing demand for drugs and inefficient production mode, many pharmaceutical manufacturers are trying to shift from a traditional batch process to a novel continuous process. Experimental and industrial trials that are aimed at this transitioning have been done and exhibit the practicability of this new production mode. Several companies such as Lonza, Phoenix Chemical, and Corning make significant progress in this area[3-5]. Novel continuous process exhibits great potential for the pharmaceutical industry.

"Which process should be applied?" and "Is there enough benefit to move forward?" are the questions that are raised when an innovative process is introduced. To answer these questions, comprehensive comparison between batch process and continuous process is required.

This comparison should be specific in the early stages. For example, for a given substance, based on its intrinsic properties and demand in market, reasonable comparison is required to make a selection between batch and continuous process. After composing a conclusive summary that is based on number of different specific cases, the result of these comparisons can be modified and applied to generate a practical process screening methodology.

For the questions mentioned above, economic benefits, safety and environmental impacts should all be considered to get a reliable answer. In the batch versus continuous process comparisons, related economic analysis and environmental analysis have been done by many researchers[6-8]. However, there remains gaps in research and understanding of safety considerations in comparison of batch versus continuous process. Therefore, the need arises for related comparison within safety consideration and systematic methodology for safety evaluation.

In the following sessions, the development in continuous pharmaceutical process, the previous studies in process screening, selecting and comparing, and concept of inherent safety assessment are discussed. The statement of research objective, the description of methodology and results of case studies are illuminated as well.

2 Literature Survey

To decide whether to head to novel continuous process or not, evaluations of batch and continuous process are in demand for pharmaceutical companies.

In preliminary stage of the evaluation, fundamental information about reaction kinetics, heat transmission and mass transfer is collected and analyzed. The objective of this investigation is looking into the technical feasibility of chosen reaction. After these basic assessments for intrinsic properties of reaction, far-reaching analyze in economic, environmental and safety aspect can be achieved. In this more comprehensive comparison based on overall picture, more factors such as expenditure, wastewater treatment and loss control measure should be covered. Among these factors, which factor should be selected as criterion, how these factors interact, and what's the correlation between these factors and comparison results are questions require to be answered.

To answer these questions, some researchers put efforts in individual comparisons of manufacturing for specific active pharmaceutical ingredient (API). Most of these studies focus on economic advantage of continuous process. Spencer D. Schaber et al.[12] estimated the capital, operating and the overall cost for production of a given API from synthesis to tablet formulation by integrated continuous and batch pharmaceutical manufacturing respectively. Based upon the equal annual yield, the result shows that the capital expenditures for continuous production could be markedly lower than those of batch mode. Differences in operating and overall cost are not such significant but potential saving by material handling and labor in continuous process is explicitly illustrated in the study. Hessel, V. et al. compared continuous vs. batch pharmaceutical manufacturing for production of ibuprofen and artemisinin on the standpoint of cost and environmental impact. Capital expenditure and operating expenditure savings are up to 57.0% and 51.6% respectively for ibuprofen, while capital expenditure and operating expenditure savings are up to 19.6% and 29.3% respectively for artemisinin[6, 8]. These study shows the potential economic benefits in application of continuous pharmaceutical processes.

In addition to process comparison, some researchers developed general criterions or created systematic methodologies for process selection. Dominique M. Roberge et al.[13] classified reactions used in the fine and pharmaceutical industry into 4 categories according to kinetic properties. Based on this classification, the very fast reactions (reactions time less than 1s) and the fast reactions (reactions occurring from 1s to10min) would benefit the most from continuous process. These two types of reactions occupy 44% of all reactions investigated. Ryan L. Harman and Jonathan P. McMullen[14] developed a decision roadmap of flow process for use in lab-scale. Considering objectives of the study, the intrinsic properties of the chemistry and requirements for mixing, heat transferring, related questions are asked at the various nodes in this flow map to get the final choice. Gary S. Calabrese and Sergio Pissavini[15] established an conceptual methodology to evaluate applicability of continuous reactor in the form of reaction properties test. They measured applicability of flow reactor via series of questions. These questions covered issues in reaction properties and some requirements on process. Three "zones" that tagged with different level of applicability are defined. Different questions are included for each "zone". For any given reaction, the more answers with "yes" in the "zone", the higher possibility of this reaction to be at the level of applicability that the "zone" stands for.

3 Problem Statement

Although advantages of continuous process in pharmaceutical manufacturing are illustrated by many studies, it is still too early to claim that continuous process is the right direction to go forward. Several gaps in the literature survey are required to be focused on.

Firstly, most of process selection methodologies are confined to analyzing separate properties of reaction taken in process, rather than evaluation in a systematic way. For example, small inventory is usually identified as one of significant advantage of continuous process, however, it is possible that to achieve similar yield in the corresponding batch process, the reaction in continuous process requires more intensified conditions like higher temperature, higher pressure, and higher concentration of hazardous chemicals. Selection and comparison by independent and individual parameters without integration is not enough.

Secondly, existing comparisons between batch and continuous processes mainly focus on economic aspect, while lacking combination of safety issue and other important aspects from real production. Although the driving force for the application of this novel process is the economic interests, safety issues directly determine its feasibility. A choice between batch and continuous process needs to be made based on business performance within acceptable safety levels. Those comparisons that merely consider the profits are not enough.

Thirdly, the existing studies are limited to fixed production system. Potential problems in a different scale are not considered. In the pharmaceutical industry, gradually increasing production scales is required which is different from the bulk chemical industry. Hence, safety of process conditions needs to be guaranteed for all scale production. Comparison only based on fixed scale is not comprehensive.

Therefore, the primary objective of this study is to provide a comprehensive and systematic comparison between batch and continuous process used in pharmaceutical industry. Then on the basis of this analysis, this study will evaluate how main factors considered in Dow F&EI impact the safety level for these two different processes. Finally, this study will provide support to further improve existing methodology for process screening and selection for pharmaceutical industry.

It is worth mentioning that there are other risks such as chemical exposure for pharmaceutical manufacturing, however, this study mainly focuses on potential fire, explosion and reactivity incident. In addition, comparison in this study focuses on intrinsic properties of the reaction and process which are the major hazards of the reaction system. Therefore, other factors like loss control measures and emergency equipment are supposed to be effective for both batch and continuous systems.

4 Methodology

To measure the realistic fire, explosion and reactivity potential of batch and continuous process, Dow's Fire and Explosion Index (F&EI) methodology is applied as hazard index in this study, since it is the most widely used hazard index that recognized by the chemical industry[18].

From the first edition of Dow's Fire & Explosion Index Hazard Classification Guide in 1964, this index has been continuously improving though the past 50 years. The last version is the seventh edition that published in 1994. The purposes of this index include forecasting quantitative damage of potential explosion, reactivity incidents and fire; determining hazardous equipment which is possible to give rise to incidents in the system; conveying this F&EI risk potential to related managements and organizations. Since the aim of this study is to identify major hazard that affect selection, application of Dow F&EI will be mainly focus on its function of hazard identification.

In the Dow's Fire & Explosion Index Hazard Classification Guide, economic analysis is confined to financial loss after an incident. Correlation between safety and economic gain from production is not built. Actually, this kind of connection could be established via yield of reactions and geometry and inventory of units, since these factors take important role in both safety assessment and process gain. In the current study, this correlation between safety and profit is constructed on the foundation of Dow's Fire & Explosion Index.

5 Case Study I: Benzimidazole Derivative Synthesis

5.1 overview

Benzimidazole derivative is an important type of intermediates in pharmaceutical industry. It is widely exhibited in a variety of pharmaceutical substance, such as proton pump inhibitors, antimicrobials, antivirals, anticancer, anti-inflammatory, hormone modulators as well as depressants, antidiabetics and so on[22]. Famous drug with benzimidazole nucleus includes Omeprazole and Albendazole. Omeprazole is used to treat gastroesophageal reflux disease and peptic ulcer disease[23], and Albendazole is an effective drug in treatment of various parasitic worm infestations[24].

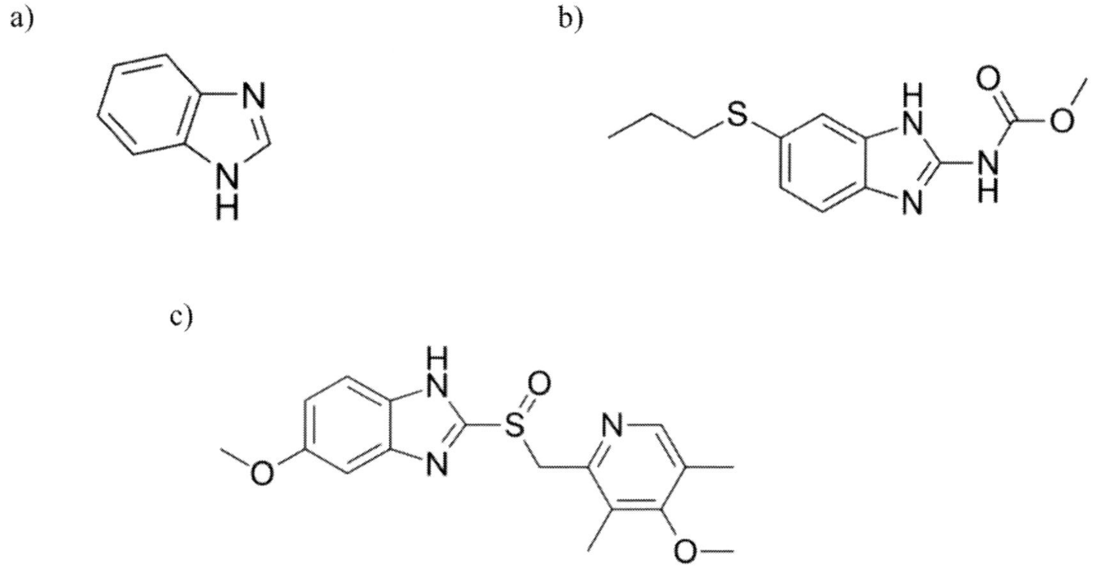

Figure 1. a) Molecular Structure of Benzimidazole, b) Molecular Structure of Albendazole, c) Molecular Structure of Omeprazole

2-methyl benzimidazole is a pharmaceutical-potential substance that belongs to this benzomidazole ring system. The condensation of o-phenylenediamines with carboxylic acids is one of the most common synthesis route for 2-methyl benzimidazole[25]. Conditions and production comparisons of o-phenylenediamines condensation by batch vs. continuous processes are showed in Table 1. These fundamental parameters are used for safety evaluation by Dow's F&EI in the following study.

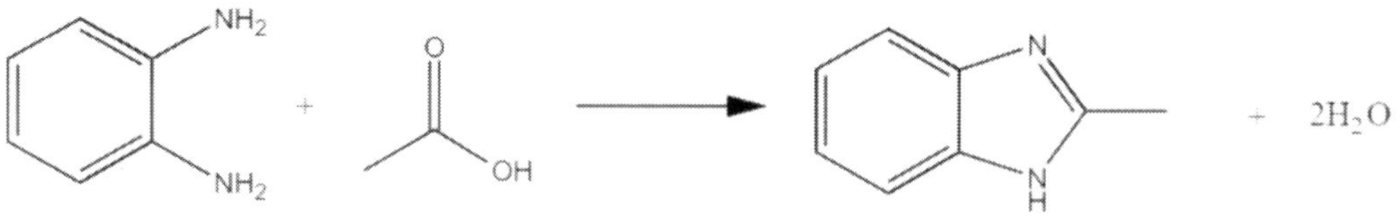

Figure 2. Synthesis of 2-methyl Benzimidazole by Condensation of o-phenylenediamine

Table 1. Production Conditions of Batch and Continuous Processes for Synthesis of 2-methyl benzimidazole

	Batch Process	Continuous Process
Temperature (℃)	200	270
Pressure (psi)	145	1885
Reaction Time (min)	5	0.5
Yield (%)	98	94
Substrate Concentration (M)	5	1
Reaction Volume (ml)	1000	4

In order to concentrate on study in intrinsic differences between two processes, and simplify the hazard identification in F&EI, several assumptions are adopted:

1. The reaction mass is perfect mixed in reactors;
2. Adjunct cooling systems in both processes are in perfect performances;
3. Emergency equipment is same and in perfect performances in both processes;
4. Scalabilities of both processes are good enough that reaction conditions remain same in all scales.

5.2. Safety Comparison Result

The comparison is conducted based on material properties and process features. Related explanation is stated below.

5.2.1. comparison on material

In this study, impacts from intrinsic properties of reactants are same for both processes. Material factors of all substances are listed in Table 2. According to Dow's Fire and Explosion Index Hazard Classification Guide, impact from the most hazardous material dominates the material factor of the whole process, therefore only impact from acetic acid is considered for both processes. In addition, although temperatures and composition are different in two processes and may lead to different chemical properties, these features are neglected in this part and contained in process comparison.

Table 2. Material Factor for Batch and Continuous Reactor

	Batch Reactor	Continuous Reactor
o-phenylenediamine	10	10
2-methylbenzimidazole	1	1
Acetic Acid	14	14

5.2.2. comparison on process

First of all, it's worth mentioning that albeit lots of distinctions exist between batch and continuous processes, comparisons in this study are confined to synthesis in reactor on account of its core role in the whole process: operation mode of reactor – continuously or intermittently – determines whether the process is continuous or not from the start. Therefore, the reactor is selected as the pertinent process unit without conventional unit screening procedure.

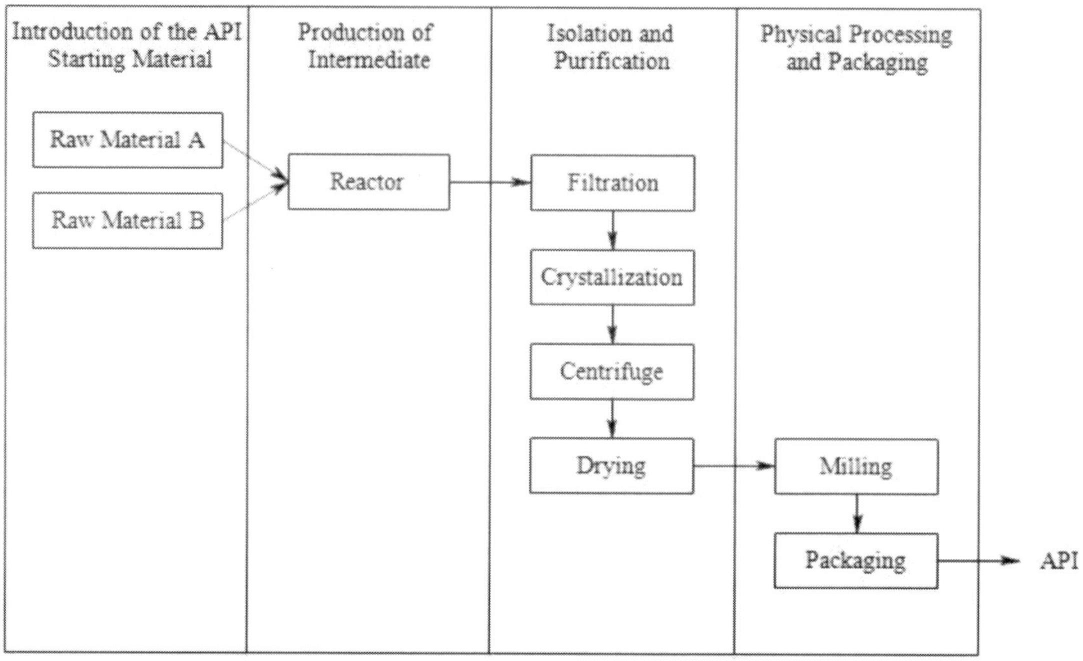

Figure 3. Procedure for Pharmaceutical Production

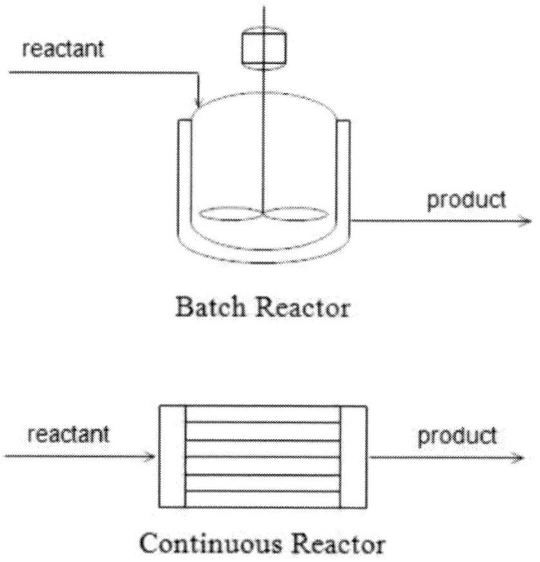

Figure 4. Differences in Structure between Batch and Continuous Reactors

On the foundation of safety evaluation by Dow's Fire & Explosion Index, quantitative judges between batch and continuous process are illustrated through comparison in sensitivity and productivity within safety consideration.

This sensitivity analysis focus on correlation between safety level and inventory of reactor. Inventory of unit plays an important role in the safety. In addition, this sensitivity analysis can also provide supportive information for scaling up procedure.

Inventory of unit plays an important part in the safety via quantity of flammable or unstable material. This potential is considered in both the General Process Hazards (F1) and the Special Process Hazards (F2) according to Dow's F&EI Hazard Classification Guide. In general, more the flammable or unstable material, higher the risk of fire and explosion within the process. For different unit (process unit or storage unit) and different phase (liquid, gas or solid), different correlation formula between inventory and penalty is given in the Special Process Hazards (F2) part of the Dow's F&EI Guide. Moreover, other issues such as whether these materials are processed indoor or outdoor, what's the temperature to handle these materials influences the safety level as well. These questions are included in the General Process Hazards (F1) part.

In this sensitivity analysis, F&EI that combines both the general and the special process hazards is expressed as a dependent variable that changes with the variation of total inventory of reactor. Other operation conditions remain the same for different inventory. Correlations between the F&EI and material inventory for batch and continuous process are shown in the Figure 5.

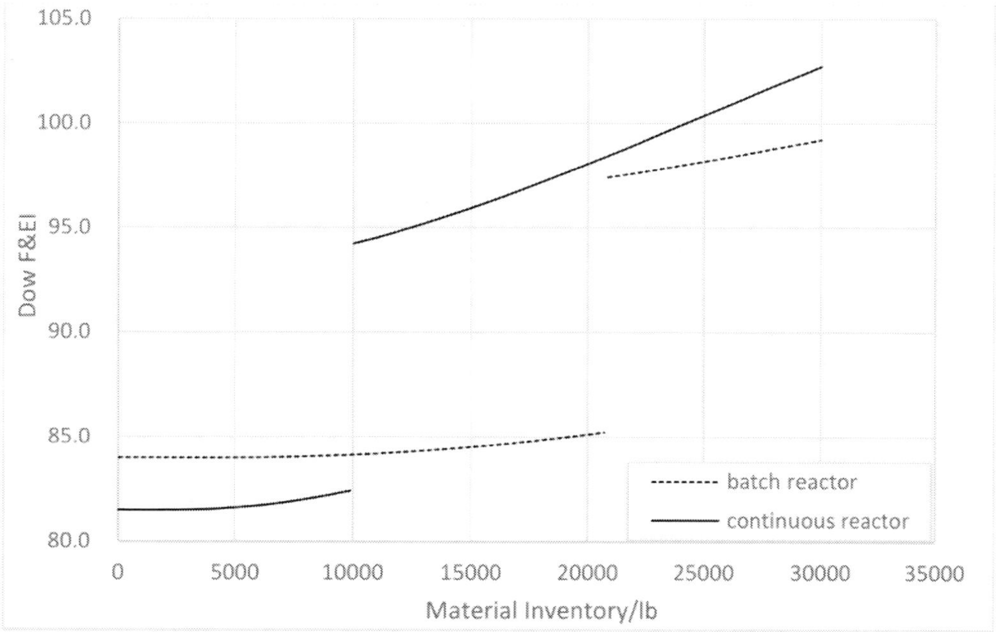

Figure 5. Analysis of Sensitivity to Inventory for Batch and Continuous Reactors Based on 2-methyl Benzimidazole Synthesis

for 0 lb < Inventory < 10000 lb:

F&EI of batch reactor is slightly greater than continuous reactor. This difference mainly comes from difference in usage of stirrer and reaction pressure. Due to requirement for homogeneous mixture, stirrer is necessary in batch reactor, while such a stirrer is not needed (or not allowed) in a continuous tubular reactor. Thus, stirrer, as a hazardous rotating equipment, adds extra penalty to batch process. As for difference in pressure, higher penalty is given to continuous process for its intensified condition. From the result, influence from rotating equipment is greater than that from pressure in this case.

Another observation is that, the magnitude of the difference in the F&EI between two reactors decreases with the growing inventory. This trend stems from the increasing impact from quantity of hazardous material. When inventory approaches zero, difference in F&EI is mainly caused by stirrer and distinct pressure aforementioned. As inventory increases, influence from unstable material extends. This influence depends on composition of mixture, or more specifically, concentration of acetic acid that defines the Material Factor (MF). Based on information of reactions, molar concentrations of acetic acid in batch and continuous reactor are 8.31M and 15.66M respectively. Thus, for similar total inventory, penalty from hazardous material in continuous reactor is greater than batch reactor.

when 10000 lb < Inventory < 20833 lb:

F&EI of continuous reactor is signally higher than batch reactor, and this difference grows along with the increasing inventory. In this range, part of influences come from rotating equipment, pressure and increasing inventory that mentioned above. More importantly, quantity of flammable and unstable liquids handled above boiling point exceeds threshold of 10,000lb for continuous reactor, therefore give a vertical rise to F&EI for continuous reactor, as shown in the Figure 5. This threshold is illustrated clearly in Dow's F&EI Hazard Classification Guide: when quantity of flammable and unstable liquids above its boiling point is less than 10,000 lb, a penalty of 0.60 is determined. A penalty of 0.90 is given when this liquids quantity above its boiling point is greater than 10,000 lb. In this case study, operation temperature of continuous reactor (270℃) is higher than boiling point of both o-phenylenediamine (257℃) and acetic acid (118℃), while operation temperature of batch reactor (200℃) merely higher than boiling point of acetic acid (118℃). Therefore, for inventory in the range of 10,000lb to 20833lb, total mass in continuous reactor is defined as unsafe, while merely acetic acid mass is defined as unsafe in the batch reactor. This divergence leads to different penalty in this range of inventory.

when 20833 lb < Inventory < 30000 lb:

F&EI of batch reactor drastically increases. However, F&EI of continuous reactor is still slightly greater than batch reactor, and the magnitude of this difference increases with the growing inventory.

This vertical rise in the F&EI of batch reactor is also caused by critical quantity of material above the boiling point. When total quantity in batch reactor goes beyond 20,833lb, quantity of flammable and unstable substance - acetic acid - also goes beyond threshold of 10,000 lb. Thus, penalty from material above boiling point is similar for both of processes.

The higher F&EI of continuous reactor originated from quantity of flammable and unstable material in this range of inventory. Since higher concentration of acetic acid is applied in continuous reactor, more proportion of substance is identified as unsafe for continuous reactor, thus higher penalty is given. This explanation is similar to that in range of low inventory (< 10,000 lb). However, unlike appearance in the low inventory range, influence from these flammable and unstable materials exceeds impact from pressure and stirrer. Therefore, continuous reactor becomes more dangerous for this range of total inventory.

It's worth mentioning that higher inventory is not considered here. Upper limit of inventory (30,000 lb) is set based on the general highest inventory of reactor used in pharmaceutical manufacturing.

Degree of hazard for batch versus continuous reactor used in this case study is shown in Table 3. In the range of inventory investigated, batch reactor is identified as moderate hazardous for a broader range of inventory, while continuous reactor is identified as intermediate hazardous for a broader range of inventory. Even so, It's worth mentioning that the F&EI values for both reactors are in a relative safe range (F&EI > 128 indicates heavy even severe hazard).

Table 3. Degree of Hazard for Batch and Continuous Reactors

Material Inventory X	Batch Reactor	Continuous Reactor
X < 15,152 lb	Moderate	Moderate
15,152 lb < X < 20,833 lb	Moderate	Intermediate
20,833 lb < X < 30,000 lb	Intermediate	Intermediate

In summary, F&EI for both batch and continuous reactors increase as material inventory increases. Moreover, F&EI of continuous reactor is more sensitive to change of inventory comparing to batch reactor. This higher sensitivity caused by comprehensive influence from higher concentration of unstable and flammable material, higher operation temperature and higher pressure it applied.

Evaluation in this part mainly focuses on safety of batch versus continuous process base on productivity considerations. Emphasis on productivity scale of this study primary comes from flexibility required for pharmaceutical production. Pharmaceutical industry is highly supply driven. Requirement on production rate is changed according to demand of market all the time. Therefore, better understanding on correlation between safety and productivity provides practicable support for research in process selection.

To approach this target, safety level is evaluated for the process at the same productivity capacity. In this part, F&EI is regarded as indicator for safety evaluation as above. Annual production is set as indicator for productivity. Related formula is shown below.

General calculations for productivity capacity are:

$$\text{Annual Production} = \text{Product Yield} \times \text{Annual Operation Time}$$

$$\text{Product Yield} = \text{Mass Flow of Reagent} \times \text{Conversion}$$

Mass flow is supposed to be proportional to reactor volume as set in the laboratory trial.

According to definition of fine chemical, annual production of fine chemical is limited to 1,000 tons per year[26]. Therefore, 1,000 tons per year is set as the upper limit in this evaluation.

As for product yield and annual operation time, the former is provided in Table 1, and expression for the latter is formed referring to the normal operation time[27].

Annual operation time for batch reactor is:

$$\text{Annual Operaton Time} = 300\ \text{day/yr} \times 24\ \text{hr/day}$$

Annual operation time for continuous reactor is:

$$\text{Annual Operation Time} = 8000\ \text{hr/yr}$$

Normally, in pharmaceutical industry, multi products are produced in one production line. Therefore, the same product cannot occupy the producing line for the whole year. In this study, 2-methyl benzimidazole is assumed to be produced in 2 months of one year.

In addition, circulating time in batch reactor need to be considered. According to the literature[11], the ratio of reaction time to total operation time for this reaction is shown below. Excepted reaction, other operations include filling, cleaning and cooling and other additional work.

$$\frac{\text{Reaction Time}}{\text{Total Operation Time}} = 0.185$$

Based on these formulas and assumptions, correlation between annual productions and F&EI is shown in the Figure 6.

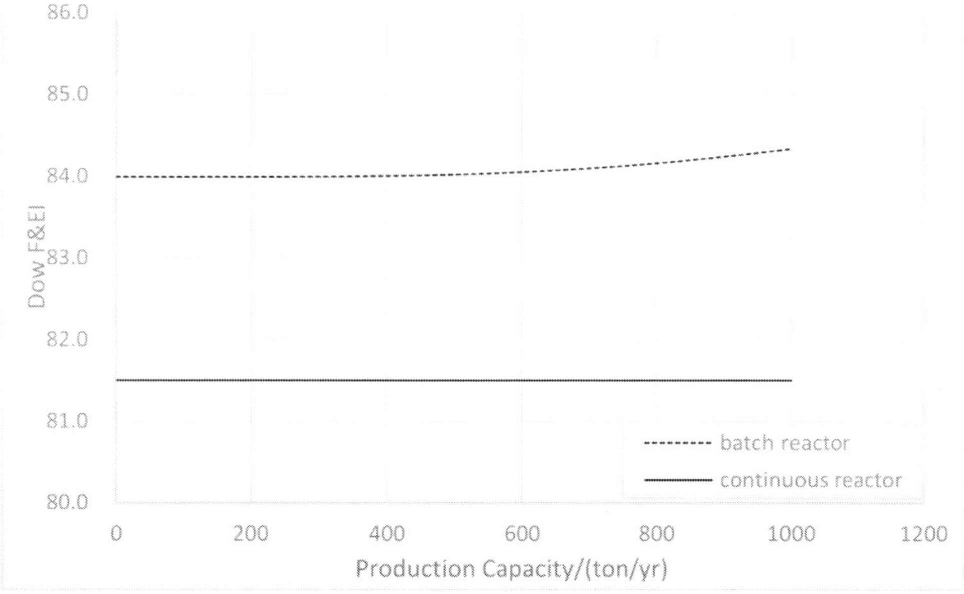

Figure 6. Analysis of Safety Based on Production for Batch and Continuous Reactors Based on 2-methyl Benzimidazole Synthesis

According to the Figure 6, F&EI value is constant in the range of practicable annual production (0 ~ 1000 tons per year). Plus, the F&EI for both reactors corresponds to a moderate level of hazard (61 < F&EI < 96).

This low F&EI resulted from relatively low demand for production. According to the sensitivity analysis, F&EI for both batch and continuous reactor increase as material inventory increases. Both processes could be intermediate hazardous once its inventory is higher than a specific value. However, these values are unapproachable in practical pharmaceutical manufacturing. Since demand for fine chemical, always less than 1,000 tons per year, confines scale of production.

Another observation is that, the F&EI of continuous reactor is always lower than that of batch reactor for same production. Plus, slight rise appears for batch reactor when production capacity exceeds 500 tons per year. These appearances predicate consistent superiority in safety for continuous reactor than batch reactor. Explanation is stated below:

Firstly, as stated in the sensitivity analysis, settled elements, such as operation pressure, temperature and use of rotating equipment, take major account for the difference at small-scale production. Advantages of continuous process in these aspects determine its superiority in small-scale pharmaceutical production.

Secondly, continuous reactor possesses higher space-time yield compared to batch reactor. This advantage stems from the intrinsic property of continuous operation mode - the low residence time and high flow throughput. That is to say, it costs fewer time and lower inventory for continuous reactor to produce same amount of product than batch reactor. These benefits in production efficiency influence safety level indirectly but significantly: Smaller inventory means fewer unstable materials in process, and shorter operation time means less chance to fail. This point is not considered in separate safety evaluation, but can be clearly illustrated by correlation between production capacity and F&EI. In this case, when demand on production growing, demand on inventory grows faster for batch reactor than continuous reactor. This faster rise in inventory lead to faster rise in amount of unstable material in process so as to the faster rise in F&EI value. It explains why slight rise appears for batch reactor when production capacity exceeds 500 tons per year.

In fact, the comparable batch and continuous reactor are rarely at the same level of volume. Referring to literature, for commercial production of fine chemical, the volumes of continuous reactor are usually in a range of 0.1 to 3 liter, while the volume of batch reactor is usually around the range of 2 to 10 cubic meters[6]. For this case, based on assumption of same annual production, ratio of batch reactor inventory and continuous reactor inventory is as much as one hundred.

6 Case Study II: Oxidation with Hydrogen Peroxide

6.1. overview

Hydrogen peroxide is one kind of widely used oxidation agent in pharmaceutical production. Typical reaction involving hydrogen peroxide includes Baeyer–Villiger oxidations for synthesis of lactone[9], N-oxidation of alkyl pyridine for synthesis of alkyl pyridine N-oxidant[28] and benzylic hydro-peroxide rearrangement for synthesis of Brivanib alaninate[29].

This common oxidation agent is also used in production of peracetic acid. Peracetic acid is produced via oxidation of acetic acid by hydrogen peroxide with sulfuric acid as the catalyst. Conditions for peracetic acid production by batch vs. continuous process are showed in Table 4[30].

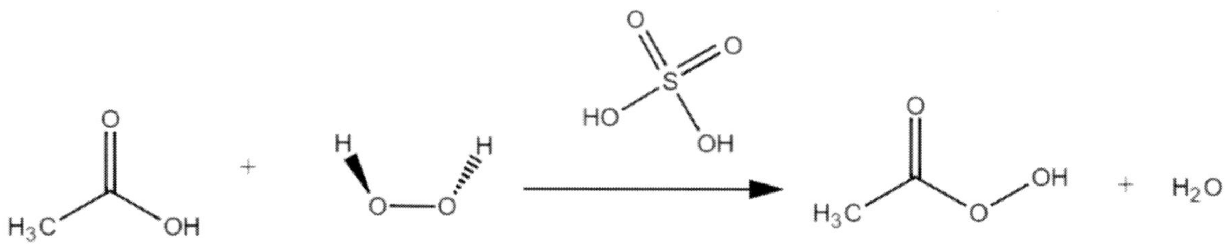

Figure 7. Synthesis of Peracetic Acid by Oxidation of Acetic Acid

Table 4. Production Conditions of Batch and Continuous Process for Synthesis of Peracetic Acid

	Batch Process	Continuous Process
Temperature (℃)	60	80
Pressure (psi)	14.5	14.5
Reaction Time (min)	30	15
Yield (kg/h)	170	20
Acetic Acid Concentration (M)	7.3	7.3
Reaction Volume (L)	4000	10

Assumption for this case study is listed below:

1. The reaction mass is perfect mixed in reactors.
2. Since the temperature for decomposition of hydrogen peroxide in peracetic acid synthesis reaction is higher than 120 ℃ [31], which is larger than operation temperature for both processes in the case study, decomposition of hydrogen peroxide is assumed to be inexistent.
3. Sulfuric acid is dissociated completely to H^+ and HSO_4^- , in addition, concentration of H^+ is assumed to be constant of 0.67 mol/L during the reactions for both processes.
4. Influence from sulfuric acid is not considered in safety evaluation.
5. Emergency equipment and cooling system are perfectly performed for both of processes.
6. Scalabilities of both processes are good enough that reaction conditions remain same in all scales.

6.2. safety comparison result

The comparison is conducted based on material property and process feature. Related explanations are stated in the following paragraph.

6.2.1. comparison on material

In this case, impact from material within the reactor is same for both of processes. Material factors for reactants involved are listed in Table 5. According to Dow's Fire and Explosion Index Hazard Classification Guide, influence from the most hazardous material dominates the material factor. Although material factor values are same for hydrogen peroxide and peracetic acid, since concentration of hydrogen peroxide is higher than peracetic acid, hydrogen peroxide is chosen to be the dominant material. Therefore, only impact from hydrogen peroxide is considered for both processes. Change on substances' property from temperatures are neglected in this part and reconsidered in process comparison.

Table 5. Material Factor for Batch and Continuous Reactor

	Batch Reactor	Continuous Reactor
Acetic Acid	14	14
Hydrogen Peroxide	24	24
Peracetic Acid	24	24

6.2.2. comparison on process

The major difference between these two production processes come from distinctions in reaction conditions and reactor constructions. These differences are shown in Table 4. Based on safety evaluation by Dow's Fire & Explosion Index, quantitative judges between batch and continuous process are illustrated through comparison in sensitivity and productivity within safety consideration.

Procedure for sensitivity comparison is the same as the first case study. Correlations between the F&EI and material inventory for batch and continuous process are shown in the Figure 8.

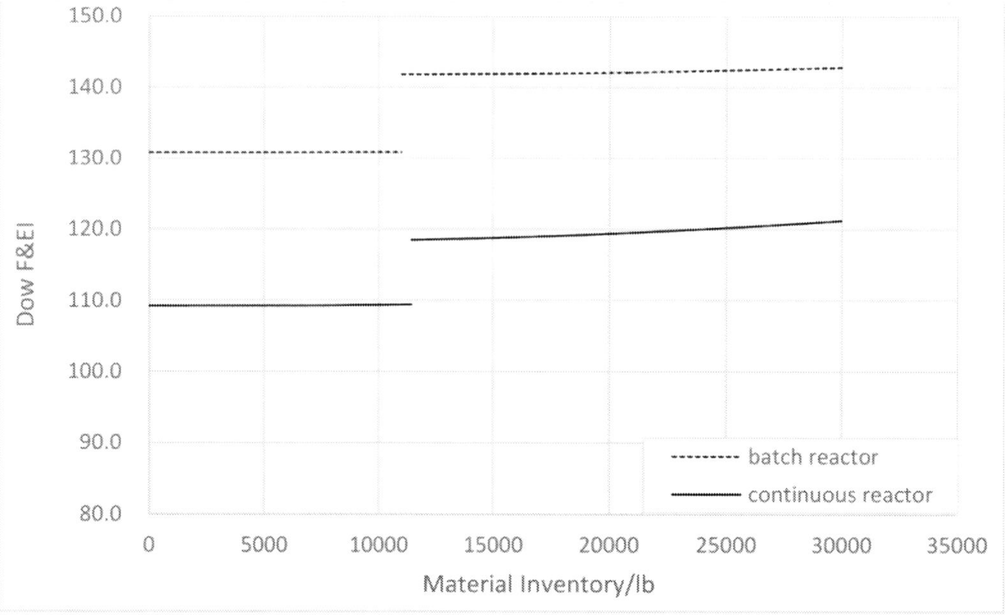

Figure 8. Analysis of Sensitivity to Inventory for Batch and Continuous Reactors Based on Peracetic Acid Synthesis

for 0 lb < Inventory < 11000 lb:

F&EI of batch reactor is larger than continuous reactor. This difference is almost caused by usage of stirrer in batch reactor. Application of such a rotating equipment increases penalty for batch process.

when 11000 lb < Inventory < 11430 lb:

F&EI of batch reactor is still higher than continuous reactor, but this difference grows drastically due to the sharp transition in F&EI of batch reactor. This transitions are caused by quantity of flammable liquids that handled above flashing point exceeds threshold of 10,000lb. This threshold is illustrated clearly in Dow's F&EI Hazard Classification Guide: when quantity of flammable liquids above their flashing point is less than 10,000 lb, a penalty of 0.30 is determined. A penalty of 0.45 will be given when this quantity of flammable liquids above their flashing point is larger than 10,000 lb. For peracetic acid production, operation temperature for batch reactor (60℃) and continuous reactor (80℃) is higher than flashing point of both acetic acid (39℃) and peracetic acid(40.5℃). Therefore, the value of inventory for higher penalty is determined by compositions of peracetic acid and acetic acid in reactors. Since concentrations of these two materials is slightly higher for batch comparing to continuous reactor, batch reactor approaches this transition in F&EI at a smaller total inventory.

when 11430 lb < Inventory < 30000 lb:

F&EI of continuous reactor exceed the transition caused by quantity of unsafe material above flashing point. However, it is still lower than F&EI of batch reactor.

Compared to the first case study, value of slope for both processes is similar low. It results in the consistent superiority of continuous reactor in safety level compared to batch reactor. This similarity in slopes comes from similarity in operation conditions. Differences only appear in operation temperature and hydrogen peroxide concentration. These differences are not such significant that lead to huge divergence between slopes. Thus, difference in F&EI value between two reactors does not change along with increasing inventory.

Degree of hazard of this case study is exhibited in Table 6. Although marked transitions appears for both reactor, safety levels for two reactors do not change based on scaling up of reactor size. The safety levels are majorly decided by settled parameters, such as operation pressure and usage of stirrer. These parameters do not change with inventory as variable. Hence, inventory does not play a significant role in judgement on degree of hazard.

Table 6. Degree of Hazard for Batch and Continuous Reactors

Reactor Type	Batch Reactor	Continuous Reactor
Degree of Hazard	Heavy	Intermediate

In sum, F&EI for both batch and continuous reactor increase as material inventory increases due to the increases of quantity above flashing point. However, since operation conditions are relatively mild and similar for these two reactors, both reactor do not exhibit high sensitivity when inventory is growing gradually.

For evaluation based upon productivity, assumption for practical production are set as following:

1. Confined by the conversion of this reaction route and volume of reactor, upper limit for annual production is settled to be 580 tons per year.
2. Same as the first case study, referring to normal operation time for this two operation mode[27], annual operation time for batch reactor is 8760 hour per day, and annual operation time for continuous reactor is 8000 hour per day.
3. Peracetic acid is assumed to be produced for 2 months in one year.
4. According to the literature[30, 32], the ratio of reaction time to total operation time for this reaction is 0.143

Based upon these formulas and assumptions, correlation between annual productions and F&EI is shown in the Figure 9.

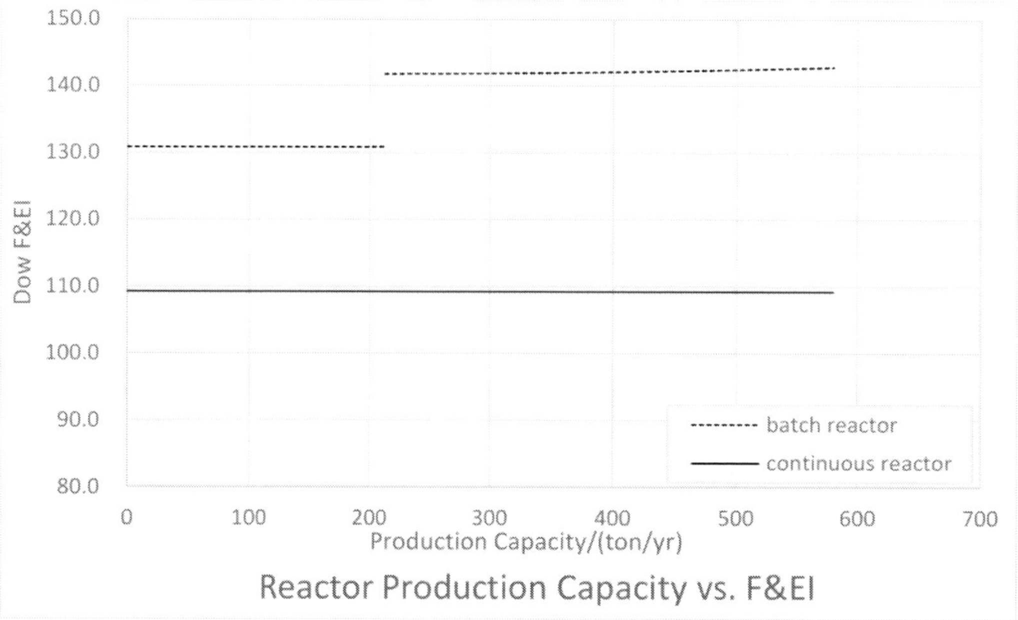

Figure 9. Analysis of Safety Based on Production for Batch and Continuous Reactors Based on Peracetic Acid Synthesis

According to the Figure 9, F&EI is stabilized at 109 in the range of practicable annual production for continuous reactor. For batch reactor, the F&EI value rise drastically when the annual production is higher than 220 tons per year.

For the batch production mode, relatively low production efficiency leads to substantial increase in reactor volume when demand on production grows. This increase in reactor volume is so significant that exceed the threshold for large quantity of flammable and unstable material above flashing point. Therefore, the F&EI value jump from 131 to 143 when the annual production is higher than 220 tons per year.

For continuous production mode, high production efficiency causes only a few change in volume hence only few change in F&EI in as well. This difference in F&EI between batch and continuous reactor means that, to produce the same amount of peracetic acid in the given period, the volume of continuous reactor is much smaller than batch reactor.

7 Result Summary

Figure 10 represents the comparisons of two case studies. The X-axis represents two different case studies, and the Y-axis represents the average Dow's F&EI value which can be used to define the degree of hazard.

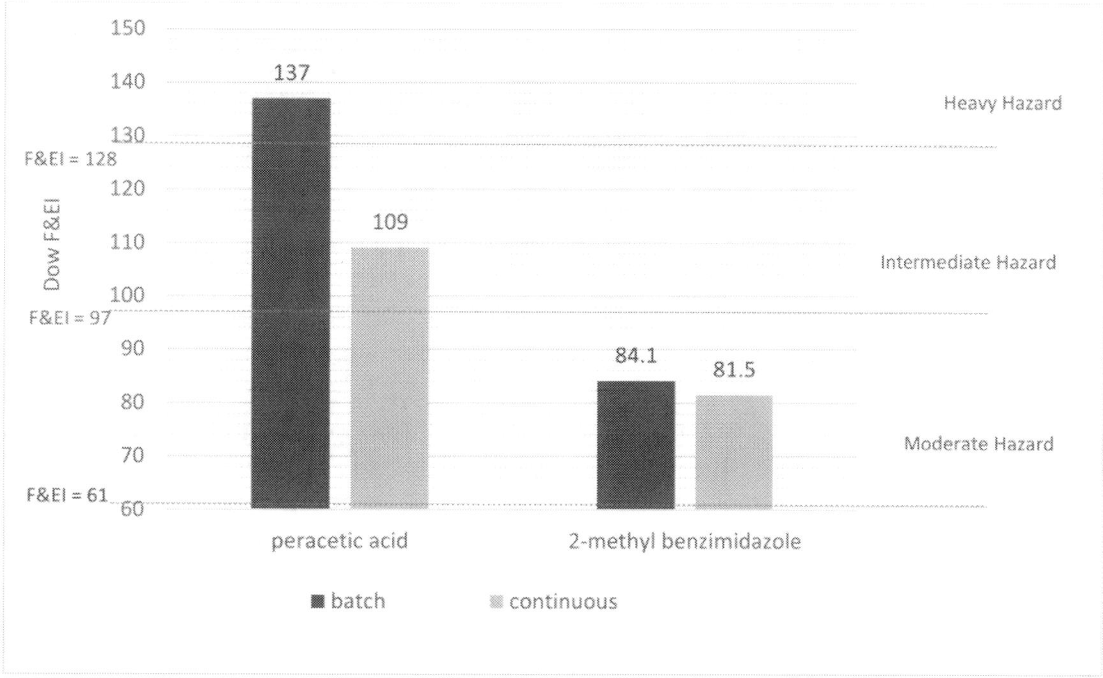

Figure 10. Comparison of Safety Level by Batch and Continuous Productions for Two Case Studies

Regardless of the reactor type, the safety level is predetermined by properties of the synthesis reaction. Although distinctions in F&EI appear when using different reactors for both cases, peracetic acid production is generally more hazardous than 2-methyl benzimidazole production, no matter which reactor it uses. This higher level of danger comes from the intrinsic properties, which consist of the flammability and reactivity of the chemicals involved in the synthesis of peracetic acid.

Taking batch and continuous reactors into consideration, distinction in F&EI comes from the differences in operation conditions and production efficiency. These varying factors such as temperature, pressure, and yield do not influence F&EI as individual factors. Instead, they are correlated and impact the F&EI as an ensemble. Difference in these ensemble decides the variance in F&EI value between reactors, then determines to which extent the selection of reactor can make a difference. For this study, the variance in F&EI value between batch and continuous reactor is greater for peracetic acid production (variance in F&EI = 28) than 2-methyl benzimidazole production (variance in F&EI = 2.6). Therefore, selecting a reactor is more important for peracetic acid comparing to 2-methyl benzimidazole production.

The importance of choosing a reactor is different for the two cases, the F&EI value for a continuous reactor is lower than a batch reactor for both of these cases. This consistency exhibits the superiority of a continuous reactor in safety.

8 Conclusion

A comprehensive comparison between the batch and continuous process used in the pharmaceutical industry based upon Dow's F&EI is made in this study. Two typical pharmaceutical productions, the production of 2-methyl benzimidazole and the production of peracetic acid are analyzed. Safety levels for batch and continuous reactors are compared based on both inventory and annual production being used as variables. The conclusions on the foundation of these two case studies are stated below:

1. The comparison of batch versus continuous processes in the pharmaceutical industry is complex. It's determined by reaction properties, operation conditions, structures of equipment, production efficiency, scale of production, and many other factors. To get a conclusive comparison between these two production modes, more systematic studies based on the integrated batch and continuous process need to be conducted for safety consideration.

2. Just focusing on the reactor in which part of API synthesis occurs, the continuous reactor is more sensitive to inventory in general. The magnitude of this sensitivity is determined by operation conditions such as operation temperature, operation pressure and concentration of hazardous material. This sensitivity to inventory might lead to problems for continuous production in a relatively large scale.

3. However, when considering practical pharmaceutical production, which is always in a small scale (no larger than 1000 tons per year), hazardous large-volume continuous reactor is not necessary. In addition, the space time yield of continuous production mode is usually higher than the batch production mode, application of a continuous reactor is relatively safer when compared to batch reactors.

9 References

1. Plumb, K., Continuous processing in the pharmaceutical industry: changing the mind set. *Chemical Engineering Research and Design* **2005,** *83* (6), 730-738.

2. Laird, T., What Type of Reactions Do Process Chemists Use on Scale? *Organic Process Research & Development* **2006,** *10* (5), 851-852.

3. Braune, S.; Poechlauer, P.; Reintjens, R.; Steinhofer, S.; Winter, M.; Lobet, O.; Guidat, R.; Woehl, P.; Guermeur, C., Selective nitration in a microreactor for pharmaceutical production under cGMP conditions. *Chimica oggi* **2009,** *27* (1), 26-29.

4. Proctor, L. D.; Warr, A. J., Development of a Continuous Process for the Industrial Generation of Diazomethane1. *Organic process research & development* **2002,** *6* (6), 884-892.

5. Roberge, D. M.; Gottsponer, M.; Eyholzer, M.; Kockmann, N., Industrial design, scale-up, and use of microreactors. *Chemistry Today* **2009,** *27* (4), 8-11.

6. Hessel, V.; Vural Gürsel, I.; Wang, Q.; Noel, T.; Lang, J., Potential analysis of smart flow processing and micro process technology for fastening process development: use of chemistry and process design as intensification fields. *Chemical Engineering & Technology* **2012,** *35* (7), 1184-1204.

7. Schaber, S. D.; Gerogiorgis, D. I.; Ramachandran, R.; Evans, J. M.; Barton, P. I.; Trout, B. L., Economic analysis of integrated continuous and batch pharmaceutical manufacturing: a case study. *Industrial & Engineering Chemistry Research* **2011,** *50* (17), 10083-10092.

8. Jolliffe, H. G.; Gerogiorgis, D. I., Plantwide design and economic evaluation of two continuous pharmaceutical manufacturing (CPM) cases: ibuprofen and artemisinin. *Computers & Chemical Engineering* **2016,** *91*, 269-288.

9. Gutmann, B.; Cantillo, D.; Kappe, C. O., Continuous - Flow Technology—A Tool for the Safe Manufacturing of Active Pharmaceutical Ingredients. *Angewandte Chemie International Edition* **2015,** *54* (23), 6688-6728.

10. Wakami, H.; Yoshida, J.-i., Grignard exchange reaction using a microflow system: from bench to pilot plant. *Organic process research & development* **2005,** *9* (6), 787-791.

11. Damm, M.; Glasnov, T. N.; Kappe, C. O., Translating high-temperature microwave chemistry to scalable continuous flow processes. *Organic Process Research & Development* **2009,** *14* (1), 215-224.

12. Schaber, S. D.; Gerogiorgis, D. I.; Ramachandran, R.; Evans, J. M. B.; Barton, P. I.; Trout, B. L., Economic Analysis of Integrated Continuous and Batch Pharmaceutical Manufacturing: A Case Study. *Ind Eng Chem Res* **2011,** *50* (17), 10083-10092.

13. Roberge, D. M.; Ducry, L.; Bieler, N.; Cretton, P.; Zimmermann, B., Microreactor technology: A revolution for the fine chemical and pharmaceutical industries? *Chem Eng Technol* **2005,** *28* (3), 318-323.

14. Hartman, R. L.; McMullen, J. P.; Jensen, K. F., Deciding Whether To Go with the Flow: Evaluating the Merits of Flow Reactors for Synthesis. *Angew Chem Int Edit* **2011,** *50* (33), 7502-7519.

15. Calabrese, G. S.; Pissavini, S., From Batch to Continuous Flow Processing in Chemicals Manufacturing. *Aiche J* **2011,** *57* (4), 828-834.

16. Kletz, T. A., What you don't have, can't leak. *Chemistry and Industry* **1978,** *6*, 287-292.

17. Kletz, T. A., *Plant design for safety: a user-friendly approach.* Taylor & Francis: 1991.

18. Fire, D., Explosion Index Hazard Classification Guide. *AICHE New York* **1994**.

19. Lawrence, D. Quantifying inherent safety of chemical process routes. © Duncan Lawrence, 1996.

20. Heikkilä, A.-M., *Inherent safety in process plant design: an index-based approach.* VTT Technical Research Centre of Finland: 1999.

21. Gentile, M.; Rogers, W.; Mannan, M., Development of a fuzzy logic-based inherent safety index. *Process Safety and Environmental Protection* **2003,** *81* (6), 444-456.

22. Bansal, Y.; Silakari, O., The therapeutic journey of benzimidazoles: a review. *Bioorganic & medicinal chemistry* **2012,** *20* (21), 6208-6236.

23. omeprazole. https://www.drugs.com/monograph/omeprazole.html.

24. albendazole. https://www.drugs.com/monograph/albendazole.html.

25. Patil, A.; Ganguly, S.; Surana, S., A systematic review of benzimidazole derivatives as an antiulcer agent. *Rasayan J Chem* **2008,** *1* (3), 447-460.

26. Patt, R.; Kordsachia, O.; Süttinger, R.; Ohtani, Y.; Hoesch, J.; Ehrler, P.; Eichinger, R.; Holik, H.; Hamm, U.; Rohmann, M., Ullmann's encyclopedia of industrial chemistry. Wiley: 2002.

27. Annual Production. http://www.eng-tips.com/viewthread.cfm?qid=389883.

28. Pineda-Solano, A.; Saenz-Noval, L.; Nayak, S.; Waldram, S.; Papadaki, M.; Mannan, M. S., Inherently safer reactors: Improved efficiency of 3-picoline N-oxidation in the temperature range 110–125° C. *Process Safety and Environmental Protection* **2012,** *90* (5), 404-410.

29. LaPorte, T. L.; Spangler, L.; Hamedi, M.; Lobben, P.; Chan, S. H.; Muslehiddinoglu, J.; Wang, S. S., Development of a Continuous Plug Flow Process for Preparation of a Key Intermediate for Brivanib Alaninate. *Organic Process Research & Development* **2014,** *18* (11), 1492-1502.

30. Ebrahimi, F.; Kolehmainen, E.; Turunen, I., Safety advantages of on-site microprocesses. *Organic Process Research & Development* **2009,** *13* (5), 965-969.

31. Kadla, J. F.; Chang, H., The reactions of peroxides with lignin and lignin model compounds. ACS Publications: 2001.

32. Ebrahimi, F.; Kolehmainen, E.; Oinas, P.; Hietapelto, V.; Turunen, I., Production of unstable percarboxylic acids in a microstructured reactor. *Chemical engineering journal* **2011,** *167* (2), 713-717.

478238 Risk Based Decision Making Methods for Evaluating Complex Technologies, a Comparative Analysis on Pipelines and Railroad for Crude Oil Transport

Abstract:

Decision making for long-term capital intensive projects involve large number of uncertainties. Traditional risk assessments are based on probabilities which are estimated from limited data sets and provide limited flexibility for complex technologies. Further, the environment and specific conditions under which the data is obtained may be unknown and yield to erroneous extrapolations. The method proposed in this work utilizes a novel method for estimation of the risk based on the net present value of technologies and provides the basis for a decision support tool based on the expected value theorem. This is demonstrated using the case of the safety and economic crisis in onshore crude oil transport in the continental United States. The exponential growth of crude oil production in the continental United States has led to crisis in the mid-stream industry with existing pipeline infrastructure under pressure to service the increased production. This fact is reflected in the 4000% increase in the carloads in a span of five years from 2009 to 2013. This has been compounded by the volatile fluctuation in crude oil price from $100 to $30 over the past few years. This has put enormous constraints on capital decision making. Since railroad infrastructure albeit aged, already exists in many regions, loading and unloading crude oil is relatively easy and their connection to the refining centers along the Gulf of Mexico apart from sites in the east and west coast make it an option to many producers. In this paper a novel decision making model is proposed using the concept of 'Net present value' of a technology to evaluate the option between railroad and as suitable mid-stream alternatives for the transportation of crude oil. The model is built on using a prescriptive decision making technique for the quantification of factors affecting the decision, incorporating also the utility of wealth of the decision maker. Monte-Carlo simulation is further used to obtain results and trends based on variation of initial parameters. This analysis can provide a basis for making decisions given the complex environment of crude oil transport operations and can be extended to a variety of geographies given that they satisfy the requirements of the model. The method can also be further extended to other technology evaluation given appropriate factors can be identified to build the model.

479492 Key Modeling Parameters and Process Safety Information for Effective Atmospheric Dispersion Modeling

Abstract:

Dispersion modeling is an important part of Quantitative risk assessment (QRA) documentation. Dispersion modeling can be performed using commercially available software. Variety of modeling parameters and Process Safety Information (PSI) data are needed for atmospheric dispersion modeling. Some of the parameters include weather, facility geographical location, flammability and toxicity data of the service fluid, radiation intensity, software tool, and model. Some PSI data needs include process and instrumentation diagrams (PIDs), Process Flow Diagrams (PFDs), relief system evaluation documentation, Heat and Material Balance (H&MB), and equipment data sheets. This paper describes the key parameters and their selection for effective dispersion modeling.

479373 Implementation of Hazard Mitigation Solutions: Consequence Vs. Risk-Based Building Specifications

Abstract:

Implementation of Hazard Mitigation Solutions: Consequence vs. Risk-Based Building Specifications

Travis Holland, Joshua Bruce-Black

Baker Engineering and Risk Consultants

3330 Oakwell Court, Suite 100

San Antonio, TX 78218

tholland@bakerrisk.com, jbblack@bakerrisk.com

Abstract

There are various strategies to mitigate occupant vulnerability and risk after completion of a consequence based facility siting study (FSS) or quantitative risk assessment (QRA). One potentially cost-effective mitigation strategy is to focus on reducing the occupant vulnerability and risk to personnel by upgrading existing buildings or designing new buildings. A building specification is commonly used to relay a small set of building design parameters based on the results of a consequence based FSS to engineers or contractors. However, with the industry shift towards risk based studies (QRAs), there has been a growing perspective that the information required to develop an appropriate building specification is more complicated given the increased number of variables involved in a QRA. Using the development of a new control room specification as an example, this paper will compare and contrast the steps and benefits/drawbacks involved in developing an appropriate building design basis employing a consequence and risk based approach. This paper will highlight how establishing tolerance criteria, emergency response strategy, personal protective equipment (PPE), and building occupancy early in the project definition stage promotes effective streamlined communication in the specification and reduces the differences between consequence and risk-based building specifications.

Generalization of Modeling Along-Wind Dispersion

Jessica M. Morris
Tom Spicer, Ph.D., P.E.
Ralph E. Martin Department of Chemical Engineering
University of Arkansas, Fayetteville, AR
Jmm065@uark.edu

Prepared for Presentation at
American Institute of Chemical Engineers
2017 Spring Meeting and 13th Global Congress on Process Safety
San Antonio, TX
March 26 – 29, 2017

Generalization of Modeling Along-Wind Dispersion

Jessica M. Morris
Tom Spicer, Ph.D., P.E.
Ralph E. Martin Department of Chemical Engineering
University of Arkansas, Fayetteville, AR
Jmm065@uark.edu

Keywords: along-wind, dispersion, finite duration, time-varying, atmosphere, releases, modeling

Abstract

The method to model along-wind dispersion is an important aspect of currently used commercial atmospheric dispersion models when considering time-limited or time-varying releases of toxic or flammable materials. Past wind tunnel experimental data has shown the validity of this approach. Inclusion of the effects of along-wind dispersion has proven to be critically important to modeling available field-scale data with current models for such time-varying releases. Many investigators have considered modeling along-wind dispersion from a theoretical point of view as well as basic parameterization of available field test data. This paper reviews past work and proposes a non-dimensionalized parameterization which shows improved agreement with available field-scale data.

1 Introduction

Continued world-wide industrial development has brought the wide spread use of industrial chemicals, many of which are hazardous. Whether released accidentally (or intentionally), air borne hazardous chemicals effect large populations quickly as evidenced by several industrial chemical accidents (e.g., Bhopal[1], Flixborough[2], Seveso[3]). Understanding and modeling hazardous chemical releases is vital to protecting the public. Many researchers have worked to assess these hazards using physical and computational models. Predicting the potential impact of hazardous gas releases in the atmosphere is imperative to properly determine where transport, storage, and use of such chemicals should be located. In the case of an actual loss of containment, understanding where air borne chemicals can travel is important to determine the proper response to an emergency.

While models exist to address these questions, there are opportunities to improve modeling methods particularly for (realistic) finite duration and time varying hazardous gas releases, which depend strongly on parameterizing the effects of along-wind dispersion (the elongation of the cloud in the along-wind direction). Modeling methods for along-wind dispersion are frequently based on established theoretical approaches which provide guidance for analyzing currently available field scale data. This paper considers experimental data and established theoretical methods to consider how the effects of along-wind dispersion can be improved.

2 Theoretical Description

2.1 Theoretical Models

Along-wind dispersion can be calculated under many different conditions and assumptions. There have been many established theories accounting for different conditions and assumptions. Scientists choose the along-wind dispersion coefficient as either a function of time or distance. Majority of scientists choose the along-wind dispersion coefficient as a function of the friction velocity. Some of the most prominent theories about along-wind dispersion are discussed below.

2.2 Batchelor

Batchelor (1964)[4] determined the Lagrangian Similarity Hypothesis. This hypothesis relates Eulerian and Lagrangian quantities when deriving theoretical solutions to problems. Applying known properties of fluid flow to the atmosphere, Batchelor was able to establish a generalized equations under neutral stability conditions, with no restraints on maximum height, for center line concentration at ground level and wind speed dependence with height. Batchelor derived many equations that are the foundation for modern theoretical equations commonly utilized. Batchelor's relationship between turbulence in a constant stress region on a boundary layer is a function of the friction velocity and time. This relationship and the Lagrangian Similarity Hypothesis serve as a template for along-wind dispersion equations to come.

Release height can be interpreted differently on its effect on dispersion. Restating Batchelor's similarity theory, "the statistical properties of the velocity of a marked fluid particle at time, t, after release at the ground depend only on friction velocity, u_* and t." This theory is then extended to the case of an elevated release: "The statistical properties of the velocity of a marked particle at time, t, after release at height, h, are the same as those of a particle released at the ground (z=0) at the instance $-t_1$, provided $t \gg t_1$ where t_1 is expected to be of the order of magnitude of the time-scale of the turbulence at height h, that is, of order h/u_*". Under Batchelor's similarity theory, after a sufficient amount of time, the height of release can be ignored.

2.3 Chatwin

Chatwin (1968)[5] was the first mathematician to theoretically derive a full equation with numerical values for the along-wind dispersion coefficient. He also derived the shear stress component of the along-wind dispersion equation. Chatwin used Batchelor's similarity theory along with the assumptions of the ability of turbulence to be described as an eddy diffusivity to solve for along-wind dispersion, σ_x. He makes further assumptions of neutral stability, ground release, and an unbounded boundary layer. The generalized equations for both overall along-wind dispersion, σ_x , and wind shear, σ_{xs} , by Chatwin are as followed

$$\sigma_x = 0.803\frac{u_* t}{\kappa} \quad (1)$$

$$\sigma_{xs} = 0.596\frac{u_* t}{\kappa} \quad (2)$$

where u_* is the friction velocity and κ is the van Karman's constant, approximately equal to 0.41.

2.4 *Van Ulden*

Van Ulden (1992)[6] derived an equation under slightly different dispersion definitions. Van Ulden made the distinction between absolute and relative dispersion. Wheatley and Chatwin were able to authenticate their equations with the assumption of a hypothetical atmosphere: where no large horizontal eddies exist. Taking into consideration this distinction of absolute and relative dispersion, van Ulden derived a new equation for σ_x, the total variance of the concentration distribution in the wind direction

$$\sigma_x{}^2 = \sigma_c{}^2 + \sigma_d{}^2 \quad (3)$$

where σ_c is the deterministic part (or centroid variance), and σ_d is the random part (or diffusive variance). The total longitudinal variance, σ_x, is dependent on three separate quantities: centroid variance, diffusive variance due to shear effects, and direct dispersion. These three variables were able to be solved for shear, centroid, and diffusive effects respectively as a function of friction velocity and time

$$\sigma_s = 0.5\frac{u_* t}{\kappa} \quad (4)$$

$$\sigma_c \simeq 0.54\frac{u_* t}{\kappa} \quad (5)$$

$$\sigma_d = 0.67\frac{u_* t}{\kappa}. \quad (6)$$

Under the assumption that $\sigma_c \simeq \sigma_s = 0.5\frac{u_* t}{\kappa}$, these variables can be substituted into equation (3) to get the total longitudinal variance. An exact equation for σ_x is obtained:

$$\sigma_x = 0.84\frac{u_* t}{\kappa} \quad (7)$$

Van Ulden not only derived an along-wind dispersion equation under neutral conditions from a ground release in an unbounded boundary layer, but also for an along-wind dispersion equation for very stable conditions. Under very stable conditions, the wind speed profile and eddy diffusivities change. When applying the new constraints for wind speed and eddy diffusivities in a different stability class, the total variance of concentration distribution in the wind direction is shown to be:

$$\sigma_x = \{(\frac{3}{4} - \frac{16}{9\pi})\frac{5u_*{}^3 t^3}{kL}\}^{1/2} \simeq 0.43(\frac{5u_*{}^3 t^3}{kL})^{1/2}. \quad (8)$$

where L is the Monin-Obukov length.

2.5 Summary

The relationship between advection cloud speed, time, and distance to determine the along-wind diffusion coefficient is complex. Changing conditions such as stability class, surface roughness, wind speed, function of distance or time, release height, and other parameters all have an effect on determining the along-wind diffusion coefficient. With the varying conditions, not all the equations are able to be integrated together into an overarching equation meeting all conditions. Different assumptions or criteria result in diverse equations for the same diffusion coefficient. An equation that takes into account all of these parameters is desired.

3 Correlation of Field Data by Hanna and Franzese

The 12 field and wind tunnel tests mentioned prior were analyzed by Hanna and Franzese to determine a suitable along-wind dispersion coefficient that resulted in an equation comparable to Chatwin's theoretical equation[5]Error! Bookmark not defined.. Results of the data and the predicted equation are shown below in Figure 1.

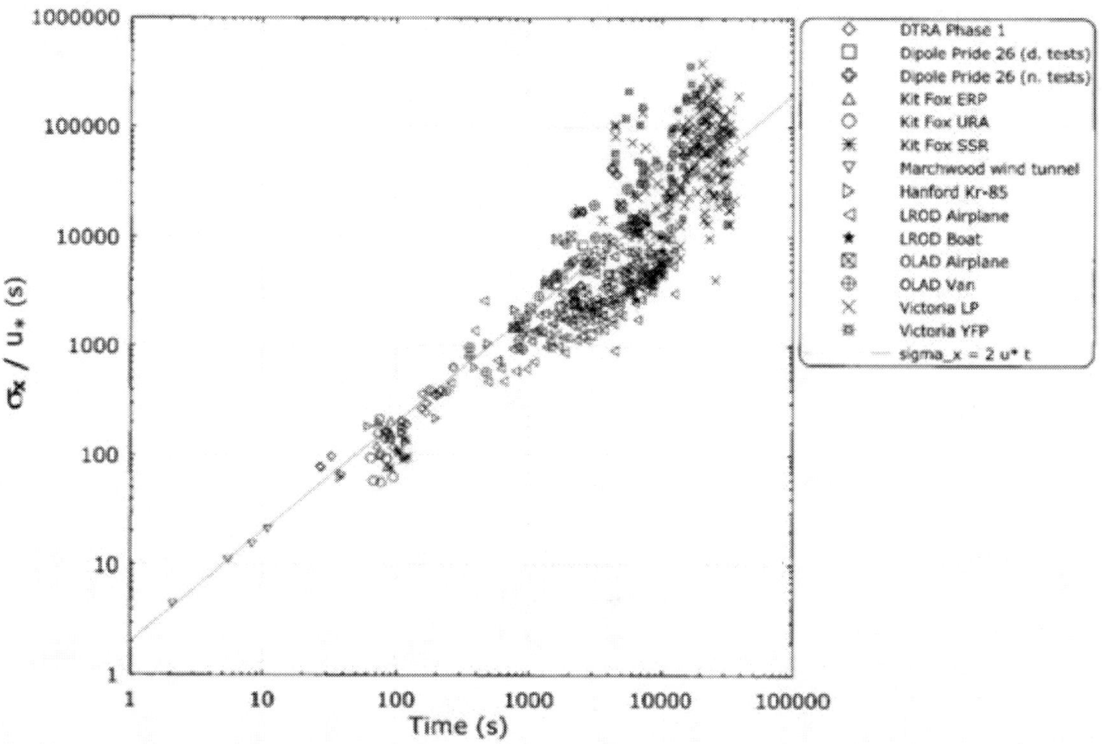

Figure 1. Hanna and Franzese results of along-wind dispersion coefficient divided by friction velocity and plotted vs time. Line represents $\sigma_x = 2u_*t$.[8]

4 Reconsideration of Field Data Using Dimensionless Groups

Figure 1 shows a correlation of the along-wind dispersion coefficient with the friction velocity as a function of time. While this is linear, their data can be up to a factor of 10 away from the predicted equation. There is also a noticeable trend of data beneath the predicted line when time is less than 10,000 seconds and above the predicted line when time is greater than 10,000 seconds. This is believed to be an effect of the near field vs. the far field. This equation for the along-wind dispersion coefficient does not take into effect the near field and far field which results in a very large portion data over predicted from 1,000-10,000 seconds and a large portion of data under predicted from 10,000 seconds and above. An improved equation to improve the correlation accounting for the error and trends is desired.

A non-dimensionalized equation was created to better predict the along-wind dispersion coefficient. The parameters of the non-dimensionalized equation were based off theory and assumptions regarding trends seen with atmospheric dispersion and along-wind dispersion. The non-dimensionalized equation allows for the along-wind dispersion coefficient to be predicted based on wind speed, distance, and stability class for each test in addition to the previous parameters of time and friction velocity.

The non-dimensionalized equation takes into account multiple theories and trends. Batchelor's Lagrangian Similarity Theory predicts σ_x to be a function of friction velocity, u_* , and time, t, while ignoring the release height[4]. Van Ulden predicts separate equations for both neutral and stable conditions[6]. Surface roughness has been shown to have an effect on wind speed[7]. The overall non-dimensionalized equation for along-wind dispersion is shown below

$$\frac{\sigma_x}{x} = \left(\frac{ut}{x}\right)^p \left(\frac{u_*}{u}\right)^q C_1 \tag{9}$$

where x is the distance in meters and p, q, and C_1 are all dimensionless constants. The $\left(\frac{ut}{x}\right)$ term accounts for the effect of wind speed, u, time, t, and distance, x. The $\left(\frac{u_*}{u}\right)$ term is to take into effect the wind speed, u, and friction velocity, u_*, which ultimately represents the stability class.

Data was accumulated and analyzed by Stephen Hanna and Pasqual Franzese in 1999 from 11 field tests and 1 wind tunnel test[8]. The accumulation of these 12 tests produced 446 data sets over varying conditions. Other tests were accumulated and stored in the Modeler's Data Archive[9] for future reference. The following tests were analyzed in hopes of a better fit for along-wind dispersion in atmospheric modeling. Of the 446 data sets, only six data points were not used in the analysis due to lack of recorded friction velocity and/or wind speed. Reference heights for recording the wind speed from each of the data sets vary. A constant reference height of 10m was implemented across all data sets for a more standardized data analysis. Data sets used in this analysis are described below.

4.1 *DTRA Phase* 1

The DTRA Phase 1 tests were carried out from the 9th -26th of September in 1996 at the Dugway Proving Ground on Pad 11 producing 23 ensemble averages[10]. Propylene was instantaneous released from air cannons at ground level. Each air cannon produced 0.36kg of propylene in a period of 0.2-0.4 seconds. DigiPIDs and Ultra-Violet Ion Collectors were both used at distances of 200 m, 300 m, 400 m, 800 m, and 1200 m along a fixed sampling line 1.5m above the ground. Increasing distances resulted in an increased number of air cannons to result in an increased mass of propylene for the increased distance. 1 air cannon was used for distances of 200 m and 300 m, 2 air cannons for 400 m, 4 air cannons for 800 m, and 8 air cannons for 1200 m. Wind speed is referenced at a height of 2m which ranged from 2-7 m/s. The Great Salt Lake Desert produced a surface roughness of less than 1mm.

4.2 Dipole Pride

Dipole Pride tests were grouped into day and night tests from the 4th -20th of November in 1996 at the Department of Energy Nevada Test Site – Yucca flat. SF_6 was released from 2 vertically mounted, pressurized cylinders on the back of a 5-ton flatbed truck. These point sources released 8-21.6 kg of mass in a duration of 1-3 seconds[11]. 90 whole air samplers, with 15 minute time averaging, and 6 continuous TGA-4000 SF_6 analyzers were placed in 3 lines at distances of 1 km, 10 km and 20 km[12]. Wind speeds were referenced at a height of 10 m resulting in wind speeds of 3-5 m/s during the day and 2-5 m/s during the night. Surface roughness at the Yucca flat was around 32 mm. The 23 total tests produced 10 adequate day releases and 5 adequate night releases.

4.3 Kit Fox (ERP, URA, SSR)

All Kit Fox releases took place at the National Spill Test Facility at the Nevada Test Site. The Equivalent Roughness Pattern (ERP) and Uniform Roughness Array (URA) tests took place from the 24th -31st of August in 1995 whereas the Smooth Surface Rough (SSR) tests took place the 11th -15th of September in 1995.[13] The tests were similar in all conditions other than surface roughness. The Kit Fox ERP tests had a surface roughness of 0.2 m, the Kit Fox URA tests had a surface roughness of 0.02 m, and the Kit Fox SSR tests had a surface roughness of 2 mm[14]. Carbon Dioxide was released from a ground level area source with dimensions of 1.5 m x 1.5 m x 1 m at a rate of 1 m/s resulting in 1.5 kg/s – 4 kg/s being released[15]. Concentrations were measured at distances of 25 m, 50 m, 100 m, and 225 m. 84 high and low infrared concentration sensors, solid state infrared CO_2 sensors, and sample bag Nova sensors, were all used to detect concentrations of CO_2 with an averaging time of around 20 seconds[16]. Wind speeds were referenced at a height of 2 m which resulted in wind speeds of 2-3 m/s for the ERP and SSR test and 3-4 m/s for the URA tests. All tests took place right before sunset which resulted in lower wind speeds with neutral to stable stability classes. The ERP and SSR tests resulted in 6 sufficient trials each and the URA tests resulted in 9 sufficient trials to be used in this analysis.

4.4 Hanford Kr-85

The Hanford Fr-85[17] tests took place from 18 August to 8 November 1967 at the Battelle Pacific Northwest Laboratory. A vial containing Krypton 85 at ambient pressure was crushed by a guillotine-like device at ground level, instantaneously releasing a 1 m puff of the radioactive gas. Concentration was detected by 64 halogen-quenched Geiger-Muller tubes at a distance of 200 m and an elevation of 1.5 m above the ground. 24 halogen-quenched Geiger-Muller tubes were positioned on 6 towers at a distance of 800 m at elevations of 0.8 m, 4.6 m, 10.7 m and 21.3 m. The time averaging associated with the Geiger-Muller tubes was 4.8 seconds. 8 trials were performed, but only 6 releases remained within the boundary of the array of sensors. The wind speed was taken at a reference height of 1.5 m which resulted in wind speeds of 1-8 m/s. The roughness was around 30 mm due to the sagebrush and steppe grass.

4.5 LROD

The Long Range Overwater Dispersion (LROD)[18] tests took place from 13th -26th of July in 1993 at the Pacific Missile Range Facility in Kauai, Hawaii. A crosswind line source of pressurized, liquid SF_6 was released from a high pressure hose from the back of a plane at an elevation of 90 m at a rate of 9.6 g/meter and a speed of 129 m/s over water. A total duration of the release of SF_6 was around 13 minutes for the 100 km distance. Concentration was measured at ground level from boats along with planes at various elevations using electron capture detectors (ECD) and TGA-4000 whole air samplers. The averaging time was around 30-120 seconds for the sequential bag samples. Noise associated with the continuous analyzer resulted in limited use. Concentration was detected from the plane by flying parallel to the dissemination line and 45 km south of the sampling line (5 km, 15 km, 30 km, 60 km, and 100 km). The plane, equipped with a continuous SF_6 analyzer, would fly to the dissemination line then back to 25 m. It would slowly ascend from 25 m to 2500 m for various heights. Testing over water resulted in a surface roughness of 0.02-0.86 mm. A reference height of 10 m was used for wind speed which resulted in speeds of 3-11 m/s. The LROD tests produced 11 ensemble averages.

4.6 OLAD

The Over Land Along-wind Dispersion (OLAD)[19] tests were similar to the LROD tests, except over land. These tests took place from the 8th -25th of September in 1997 at the Dugway Proving Ground, West Desert Test Center. SF_6 was released from both plane and truck line sources. The plane released 100 kg of SF_6 at an elevation of 100 m from a N_2 pressurized tank at a rate of 33.3 kg/min for a distance of 20 km. The truck released 15 kg of SF_6 at an elevation of 3 m from a mounted tank at a rate of 1.5 kg/min for a distance of 10 km. Concentrations were recorded with TGA-4000 whole air samplers, with time averaging of 15 minutes, and 2 continuous electron capture detector (ECD) analyzers at the end of each line at distances of 2 km, 5 km, and 10 km for the truck releases and 10 km, 15 km, and 20 km for the plane releases. Wind speed was measured at a height of 2 m which resulted in speeds of 0.5 – 7 m/s. The surface roughness was measured around 30 mm. These tests resulted in 3 ensemble averages for the aircraft release and 8 surface ensemble averages for the truck release.

4.7 Marchwood wind-tunnel

The Marchwood[20] wind tunnel tests are the only tests that took place in a wind tunnel. A mixture of propane and helium, resulting in neutral buoyancy, was released from a 15 mm ground level point source. The mixture was released for a duration of 0.5 – 12 seconds. 150-300 releases were summarized into 4 ensemble averages. A single FID measured the concentration with a 5 ms sampling interval at locations of 1 m, 3 m, 5 m, and 7 m. Wind speeds ranged from 1-4 m/s at neutral stability class. Surface roughness was measured at 0.3 mm for the wind tunnel.

4.8 Oceanside

The Oceanside[21] tests took place at the shoreline of Oceanside and Del Mar in Southern California. 20 μm particles of green and yellow fluorescent pigment (FP) were released by plane, boat, and land with standardized Dugway Proving Ground equipment. For the plane releases, the DPG model D-1 released 1500 g/min of FP for a duration of 10 minutes at an elevations of 61 m, 122 m, and 152 m over a 40 km distance. For the boat releases, a DPG Mark IV was used to release 110 g/min for a duration of 2 hours at sea level over 40 km. For instantaneous land releases, the DPG Mark IX was used at three different inland locations to release 10-50 g of FP instantaneously at ground level. Concentration of FP were recorded from rotorod samplers, sequential samplers, vertical samplers, 100 m square grids, dense sampling lines, and metronics drum impact samplers. Time averaging ranged from 5 minutes at coastal regions to 15 minutes at inland conditions. Surface roughness ranged from 30-50 m due to the coastal shoreline. The wind speed reference height varied from 8-120 m which resulted in wind speeds ranging from 1-5 m/s.

4.9 Fort Wayne

Fort Wayne[22] releases took place in the Summer and Fall of 1963 in Fort Wayne, Indiana. An aerosol of fluorescent pigment (FP) was released via 2 aircrafts in the form of an elevated, 91-244 m, line source. Dry green and yellow FP were released from the planes via FP dispensers supplied by the Dugway Proving Ground. The amount of release was measured by the difference in weight, before and after. The release was upwind of the 180,000 population urban setting with a river running through the city and agriculture surrounding. Rotorods, Gelman throw-away plastic filter holders, and Gelman paper tape samplers were used for concentration measurements of the FP. The tape provided an averaging time of 5 minutes where the filter holders had an averaging time of 30 minutes. Wind speed was measured at the release height and resulted in speeds ranging from 5-16 m/s.

4.10 Short Range Diffusion Experiment Series

The Short Range Diffusion Experiment Series[23] took place in 1979 in Japan. A 200 m x 200 m vacant lot covered in turf and surrounded by trees was used for the releases. NO_x was released instantaneously from a balloon at an elevation of 1.5 m. Chemiluminescence NO_x meters were used to detect concentration with a sampling time of <2 minutes. Sample lines were chosen at distances of 60 m and 100 m. Wind speeds were referenced at heights of 0.5 m, 3 m, and 9 m which resulted in speeds of 1-2 m/s.

4.11 Results and Discussion

A multivariable regression analysis, also known as multiple regression, is an extension of linear regression[24]. A regression analysis determines the relationships among variables.

With a multivariable regression analysis, the relationships between multiple independent variables can be determined. A regression analysis is a useful tool in understanding the changes of one variable with respect to others. The form of a typical multi-variable regression analysis is shown below

$$Y = a + b_1 x_1 + b_2 x_2 + \cdots + b_n x_n \qquad (10)$$

A multi-variable regression analysis can be applied to the non-dimensionalized equation. The non-dimensionalized equation is rearranged with the natural log of each variable to perform the multi-variable regression analysis. This equation is shown below

$$ln\left(\frac{\sigma_x}{x}\right) = p\, ln\left(\frac{ut}{x}\right) + q\, ln\left(\frac{u_*}{u}\right) + C_2 \qquad (11)$$

where $C_2 = \ln(C_1)$. A multivariable regression analysis with known data is used with this equation to determine the constants for p, q, and C_1. An analysis is done for the overall data and categorized into stability classes using macros in Excel[25].

Multiple parameters were analyzed for optimal results using the multivariable regression analysis. Data was quantified and analyzed based on individual tests, surface roughness, friction velocity, Monin-Obukhov length, distance, time, near field, and far field.

Data from all 12 experimental data sets referenced above were analyzed with a multi-variable regression analysis to determine appropriate exponents and a constant for an optimum non-dimensionalized equation. While many parameters were originally analyzed for the optimum non-dimensionalized equation, the quantification of near field and far field as a function of time was chosen as the most prominent parameter. The optimum non-dimensionalized equation was parameterized by near field with time less than 10,000 seconds and far field with time greater than 10,000 seconds. These equations are shown below in Table 1. The ratio, α, of the along-wind dispersion coefficient, σ_x, for the data vs σ_x for the predicted non-dimensionalized equation is used for comparison of the difference in equations. A non-parametric confidence interval was determined for the non-Gaussian distribution of α.[26] Hanna and Franzese's along-wind dispersion coefficient was analyzed in the same manner in Table 1.

Table 1. Comparison of along-wind dispersion equations with 90% confidence intervals

Model	Equation	90% Confidence Interval of $\alpha = \frac{\sigma_x\,(predicted)}{\sigma_x\,(data)}$	Range of Confidence Interval
Hanna and Franzese	$\sigma_x = 2u_* t$	$1.63 < \alpha < 1.88$	26%
Near Field Non-Dimensionalized	$\frac{\sigma_x}{x} = 1.72\left(\frac{ut}{x}\right)^{1.18}\left(\frac{u_*}{u}\right)^{1.10}$	$1.16 < \alpha < 1.31$	16%
Far Field Non-Dimensionalized	$\frac{\sigma_x}{x} = 0.05\left(\frac{ut}{x}\right)^{0.67}\left(\frac{u_*}{u}\right)^{-0.27}$	$0.85 < \alpha < 1.05$	20%

From Table 1, the range of the 90% confidence intervals for each of the equations is shown. The non-dimensionalized equations have a tighter correlation of the predicted values. The 90% confidence intervals shows that Hanna's predicted along-wind dispersion equation has a trend over predict the σ_x values. An interesting result from the multi-variable regression analysis is the determination of the coefficient for the $\left(\frac{u_*}{u}\right)$ term. The -0.27 exponential value essentially confirms that the stability class has no effect on the along-wind dispersion coefficient after a specific amount of time. A discontinuity is understood between the near field and far field non-dimensionalized equation and is currently being evaluated.

The results of the non-dimensionalized predicted σ_x values were divided by the friction velocity and plotted against the original data to determine their relationship. This relationship is shown below in Figure 2.

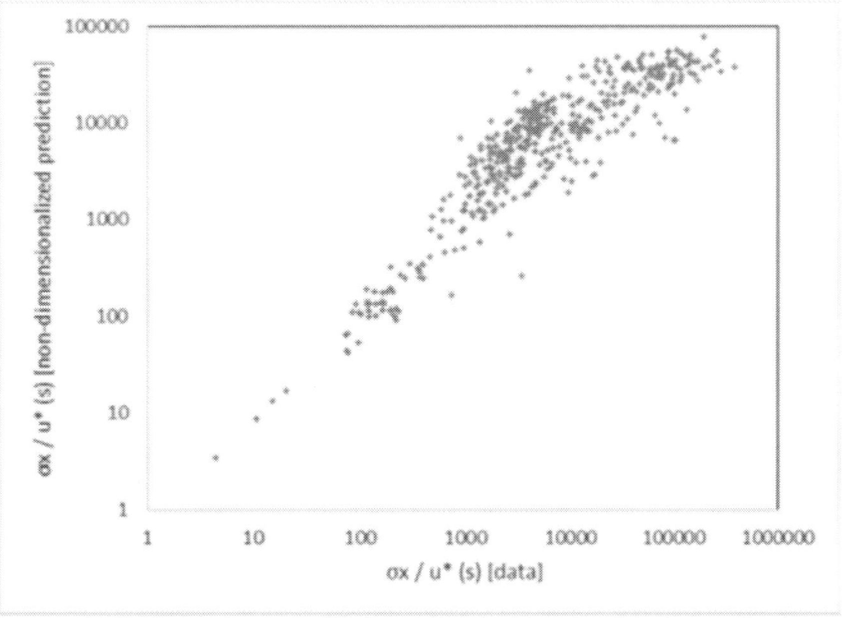

Figure 2. Along-wind dispersion coefficient divided by the friction velocity of the non-dimensionalized modeled values vs the original data.

As shown in Figure 2, the correlation between the data and non-dimensionalized predicted values shows a 45° line. This correlates a good fit for the predicted σ_x values when compared to the original σ_x values.

A visual representation of the non-dimensionalized predicted σ_x values, Hanna and Franzese's σ_x values, and the original σ_x values as a function of time are all shown below in Figure 3.

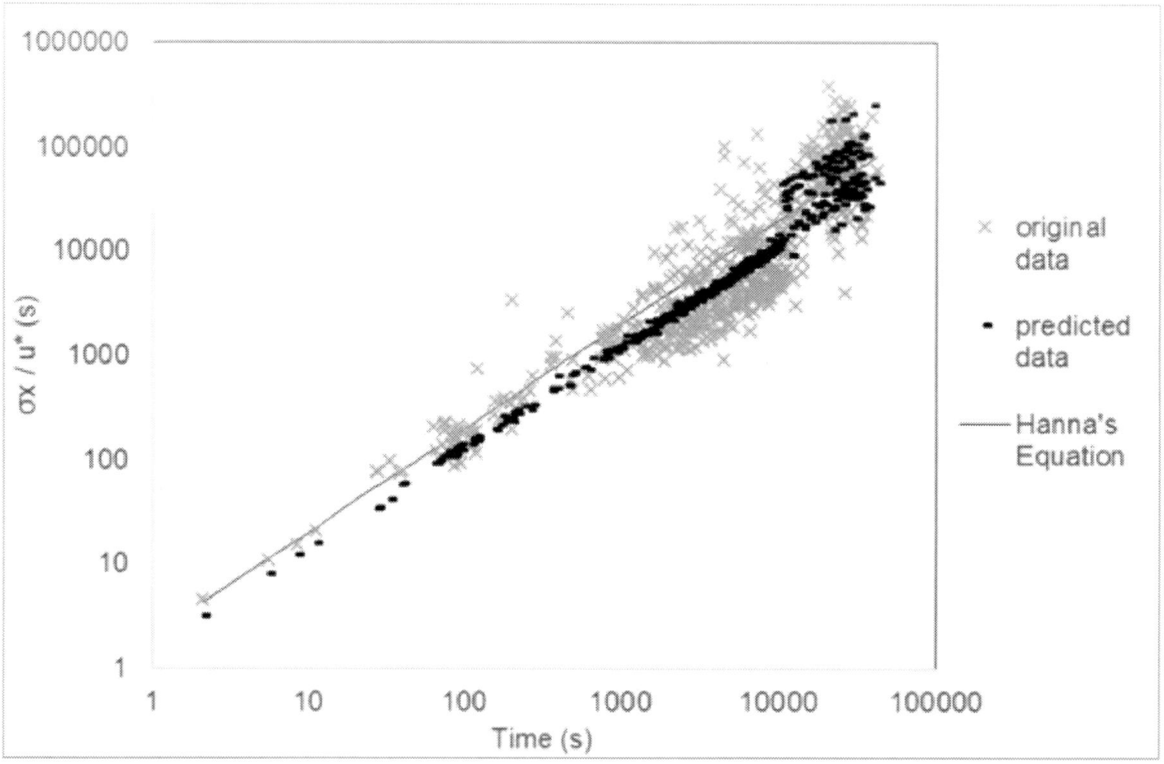

Figure 3. Time vs. the along-wind dispersion coefficient divided by the friction velocity as a function of the original data, non-dimensionalized predicted data, and Hanna's predicted data.

From this figure, the predicted values from both Hanna and Franzese's predicted data and the non-dimensionalized predicted data are shown in comparison with the original data. The non-dimensionalized equation takes into account additional parameters, other than time, which results in the spread of σ_x values as a function of time and additional parameters in comparison to the single line by the $\sigma_x = 2u_* t$ along-wind dispersion equation. The non-dimensionalized equation also takes into effect the near field and far field. This is clearly seen at times greater than 10,000 seconds where the σ_x values are spread over a wider range. Overall, the non-dimensionalized equation for along-wind dispersion proves better than the previous $\sigma_x = 2u_* t$ correlation.

5 Summary

Along-wind dispersion is critical when modelling toxic, flammable, or hazardous time-limited or finite-duration releases into the atmosphere. While many models exist from a theoretical standpoint or parameterization of experimental data, there is no overarching correlation to predict the along-wind dispersion coefficient. This paper took into effect established theoretical equations validated with experimental data to determine a better equation to model the effects of along-wind dispersion. A successful non-dimensionalized equation parameterized by the near field and far field was determined and validated against available field-scale data which showed improvement in comparison to previous equations. The non-dimensionalized equation for along-wind dispersion solved for in this paper will improve the existing atmospheric modeling with along-wind dispersion.

6 Acknowledgments

We gratefully thank our colleagues from DNV GL, Henk Witlox and Maria Fernandez, who provided insight and expertise that greatly assisted the research. We also thank Steven R. Hanna and Pasquale Franzese for sharing their accumulated data for further analysis.

7 References

[1] Boybeyi, Zafer, Sethu Raman, and Paolo Zannetti. "Numerical investigation of possible role of local meteorology in Bhopal gas accident." *Atmospheric Environment* 29.4 (1995): 479-496.

[2] Høiset, S., et al. "Flixborough revisited—an explosion simulation approach." *Journal of hazardous materials* 77.1 (2000): 1-9.

[3] De Marchi, Bruna, Silvio Funtowicz, and Jerome Ravetz. "Seveso: A paradoxical classic disaster." *The long road to recovery: Community responses to industrial disaster* (1996): 86-120.

[4] Batchelor, G. K. "Diffusion from sources in a turbulent boundary layer." *Arch. Mech. Stosow* 3.6 (1964): 661-670.

[5] Chatwin, P. C. "The dispersion of a puff of passive contaminant in the constant stress region." Quarterly Journal of the Royal Meteorological Society 94.401 (1968): 350-360.

[6] Van Ulden, A. P. "A surface-layer similarity model for the dispersion of a skewed passive puff near the ground." *Atmospheric Environment. Part A. General Topics* 26.4 (1992): 681-692.

[7] DNV GL. DNV GL Theory Manual for Unified Dispersion Model. London UK 2015.

[8] Hanna, Steven R., and Pasquale Franzese. "Alongwind dispersion—a simple similarity formula compared with observations at 11 field sites and in one wind tunnel." *Journal of Applied Meteorology* 39.10 (2000): 1700-1714.

[9] Chang, Joseph C., and Steven R. Hanna. "Modelers' Data Archive – A Collection of Atmospheric Transport and Dispersion Data Sets." (2010).

[10] Biltoft, C., Phase I of DSWA transport and dispersion model validation study. DPG-FR-97-058, (1997): 156.

[11] Biltoft, C., Dipole Pride 26: Phase II of DSWA transport and dispersion model validation study. DPG-FR-98-001, (1998): 130.
[12] Watson, T. B, R. E. Keislar, B. Reese, D. H. George, and C. A. Biltoft, The DSWA Dipole Pride 26 field experiment. NOAA Tech. Memo. ERL ARL-225, (1998): 99.
[13] King, S. Bruce. The Kit Fox Field Demonstration Project and Data Set. *International Conference and Workshop on Modeling the Consequences of Accidental Releases of Hazardous Materials*. (1999).
[14] Spicer, T. and Havens, J. Description and Analysis of Atmospheric Dispersion Tests Conducted by EPA. *International Conference and Workshop on Modeling the Consequences of Accidental Releases of Hazardous Materials*. (1999).
[15] Sheesley, D. et al.. Chemical Dispersion Modeling Project, 1995 Kit Fox Field Demonstration DOE Nevada Hazardous Spill Center. Volume I – Experiment Description and Data Processing. (1998).
[16] Hanna, S. R., and K. W. Steinberg, Studies of dense gas dispersion from short-duration transient releases over rough surfaces during stable conditions. *Air Pollution Modeling and Its Applications XI,* Plenum Press, (1996): 481–490.
[17] Nickola, P. W., J. D. Ludwick, and J. V. Ramsdell Jr., An inert gas tracer system for monitoring the real-time history of a diffusing plume or puff. *J. Appl. Meteor.,*9, (1970): 621–626.
[18] Bowers, J. F., G. E. Start, R. G. Carter, T. B. Watson, K. L. Clawson, and T. L. Crawford, Experimental design and results for the Long-Range Overwater Diffusion (LROD) Experiment. DPG/JCP-94/012, (1994): 531.
[19] Biltoft, C., S. D. Turley, T. B. Watson, G. H. Crescenti, and R. G. Carter, Over-Land Atmospheric Dispersion (OLAD) test summary and analysis. WDTC Doc. WDTC/JCP-99/048, (1999): 38.
[20] Robins, A. G., and J. E. Fackrell, An experimental study of the dispersion of short-duration emissions in a turbulent boundary layer. *Air Pollution VI,* C. A. Brebbia, C. F. Ratto, and H. Powers, Eds., WIT Press, (1998): 697–707.
[21] Smith, and B. L. Niemann, Shoreline diffusion program, Oceanside, CA. Vols. I and II. MRI-64-FR-860, (1969): 230 and 456.
[22] Hilst, G. R., and N. E. Bowne, A study of the diffusion of aerosols released from aerial line sources upwind of an urban complex. Vols. I and II. Rep. DA-42-007-AMC-38®, (1966): 231 and 583.
[23] Sato, J., F. Kimura, and T. Yoshikawa, The longitudinal spread of puff in the short range diffusion experiment. *Pap. Meteor. Geophys.,*32, (1981): 155–162.
[24] Draper, Norman Richard, Harry Smith, and Elizabeth Pownell. *Applied regression analysis*. Vol. 3. New York: Wiley, (1966).
[25] Davis, Richard A. "Practical Numerical Methods for Chemical Engineers: Using Excel with VBA". Lexington, KY, (2015).
[26] Conver, W. J. "Practical Nonparametric Statistics". Wiley, New York, (1971).

478500 Review and Analyze the Effects of Parameters for H_2s Gas Mixture Dispersion Models: Degadis and CFD Models

Abstract:

Due to the increasing demands of energy, natural gas explorers and developers are interested in exploring sour gas fields, which have a high concentration of hydrogen sulfide (H_2S). Because H_2S is highly corrosive and could reduce the heating values of natural gas, H_2S needs to be removed from natural gas before transfer. Then the removed H_2S gas will be converted to sulfide as the products in the Claus process. The H_2S mixture (*e.g.,* H_2S and CO_2) may cause corrosion to pipelines and thus, could lead to H_2S gas mixture release.

H_2S mixture behaves as a heavy gas, which travels down to the ground causing harm to the local population and the environment. In this paper, the DEnse GAs DISpersion (DEGADIS) model and the Computational Fluid Dynamics (CFD) model (*e.g.,* ANSYS Fluent) are implemented for the H_2S mixture simulation, and assumptions in the models are also studied. The effects of topography and meteorological parameters such as atmospheric temperature, wind speed, and humidity are investigated. Finally, the simulation results from the DEGADIS model and the CFD model for flat terrain with constant wind velocity are compared.

The simulation results show that the DEGADIS model, which is suitable for flat terrain and constant wind conditions, can be applied to simple terrain geography and stable atmospheric conditions. Whereas the CFD model, which considers unsteady wind speed and turbulence near the obstacles, can be utilized for complex topography. The series of simulation results can provide guidance for the risk assessment of H_2S mixture releases under different topographies and meteorological conditions.

478998 Considerations When Justifying the Purchase of an Operator Training Simulator

Abstract:

Highly skilled plant operators ensure higher levels of plant safety, reliability and profitability. An Operator Training Simulator accelerates both the time to competence and raises the competency level of an operator. The ARC Advisory Group recently surveyed companies who have invested in Operator Training Simulators (OTS). Two critical results from the survey are that over sixty five percent of those companies report fewer abnormal situations, fewer trips, and incidence avoidance attributable to the use of OTS. Over sixty percent of these companies report improved safety and reduced risk and lower liability attributable to the use of the OTS.

Many leading companies have embraced the use of Operator Training Simulators across their Enterprise however many facilities and companies have yet to embrace the idea. The most common reason given for not investing in such a solution is that the cost of an OTS cannot be justified. The nuclear industry mandates that a nuclear power plant must have an OTS whereas the American Petroleum Institute only recommends the use of Training Simulators. The Process industries overall will benefit by the greater adoption of training simulators and this paper discusses the different ways that can be used to justify the investment in an OTS.

479500 Consequence Analysis of Gas Explosion in Underground Common Utility Tunnels

Abstract:

The purpose of this study is to figure out the effects of blast overpressure in gas piping leakage in underground common utility tunnels through the explosion simulation tool FLACS (Flame Acceleration Simulator) and find and apply methods that can reduce the risk.

Common utility tunnels are facilities to jointly bury pipes in underground spaces. Placing gas piping in common utility tunnels has the advantages of reducing the risk of corrosion and preventing the physical damage due to civil engineering works and excavating work on the ground. However, when an explosion has occurred in a common utility tunnel, the consequences could be larger in that it is a closed space. Nevertheless, studies on underground common utility tunnels are currently limited to fire accidents. Therefore, risks due to explosions should be quantitatively studied.

Therefore, in this study, the effects of explosions were analyzed assuming cases where gases leaked from gas piping in underground common utility tunnels exploded. In addition, blast overpressures were evaluated for diverse scenarios regarding ignition sources, leakage locations, and discharge rates and the results were statistically derived. Blast overpressures high enough to cause dangerous situations of the underground common utility tunnels were observed. Methods that can reduce the risk, that is, methods that can reduce the frequency of explosions, and methods for designs of internal structures of common utility tunnels that can mitigate the consequences will be presented.

479158 Air Cooler Explosion Analysis Using CFD

Abstract:

Air coolers or fin-fans are located in most process plants. They cool process fluids using forced air flow over radiators containing the fluid. Air coolers initially appear to offer many features that would result in a significant explosion source should a flammable cloud reach the cooler. The radiator coils are closely spaced pipes which will accelerate flames. The metal box represents confinement and the turbulent air flow produced by the fans will accelerate flames similar to how congestion produces turbulence ahead of a flame front. Also, the flow of air can draw rich flammables near the ground up into the air cooler.

This study examines the severity of an air cooler explosion using computational fluid dynamics (CFD). The commercial software FLACS is used to examine the parameters mentioned above (congestion, confinement, and turbulence) and their effect on explosion severity. Comparisons between FLACS and the Multi-Energy Method are made and recommendations provided for assessing air coolers in non-CFD facility siting studies.

Revisions to the QMEFS Vapor Cloud Explosion Model

Jeffrey D. Marx and Benjamin R. Ishii
Quest Consultants Inc.®
908 26th Avenue N.W.
Norman, OK 73069
jdm@questconsult.com / bri@questconsult.com

Prepared for Presentation at
American Institute of Chemical Engineers
2017 Spring Meeting and 13th Global Congress on Process Safety
San Antonio, TX
March 26 – 29, 2017

Revisions to the QMEFS Vapor Cloud Explosion Model

Jeffrey D. Marx and Benjamin R. Ishii
Quest Consultants Inc.®
908 26th Avenue N.W.
Norman, OK 73069
jdm@questconsult.com / bri@questconsult.com

Keywords: vapor cloud explosion (VCE), Baker-Strehlow-Tang (BST) model, estimating flame speeds

Abstract

The Quest model for estimating flame speeds (QMEFS) was introduced in 2007, has since been published, and has been used by Quest and other companies to provide a detailed method for vapor cloud explosion (VCE) predictions. QMEFS provides a set of parameters that predict a flame speed, which then is used with the Baker-Strehlow-Tang (BST) blast curves to predict overpressure and impulse. The original intent of the QMEFS model was to provide better characterization of confined or congested areas, as well as fuel reactivities, which fit between the low, medium, and high categories offered by the BST model, so that the large differences in predictions between categories could be eliminated. In this way it provided a more refined system for predicting the consequences of a VCE, while remaining anchored within the flame speeds provided by the BST matrix.

In the years since QMEFS' publication, Baker has added to the BST flame speed matrix. In addition, use of the model has highlighted some of the areas where the model could be improved. This paper shows several ways in which the model has been enhanced to encompass a wider range of conditions than are covered by the BST model.

1 Introduction

Any release of a flammable fluid in a petrochemical facility has the potential to generate a flammable vapor cloud that, if ignited, could produce a vapor cloud explosion (VCE). Most VCEs that generate damaging levels of overpressure originate in what is called a potential explosion site (PES), where confinement and congestion contribute to flame acceleration and the generation of overpressure. When flammable vapors enter the PES and are ignited, the possibility of human injury/death, asset damage, or event escalation becomes a concern. The concern for human injury or death is most often addressed in the form of a building siting study. Because people are somewhat less likely to be injured or killed when outside, as compared to when inside a building, the siting study focuses on the potential impacts to buildings within and around petrochemical facilities. It then becomes the task of process safety professionals to estimate the potential for VCE events, and their resulting overpressure impacts on buildings.

The Baker-Strehlow-Tang (BST) model and the TNO Multi-energy (ME) model use blast curves to estimate the overpressure impacts generated by an exploding vapor cloud. Both approaches require the user to estimate the strength of the explosion as a function of the reactivity of the flammable material and the degree of confinement or congestion present in the cloud. This information is then used to determine which strength curve (in the case of the TNO model) or flame speed curve (in the case of the BST model) is used to calculate the overpressure of the explosion as a function of distance from the center of the explosion. It is not clear, however, how best to choose the correct curve when modeling an explosion.

There have been several papers published regarding blast curve selection for the ME model. These methods are generally presented as a qualitative ruleset which acts as an index into which of the 10 curves should be used. The BST model provides a simple set of guidelines that result in a selection of a flame speed, which corresponds to an overpressure vs. distance relationship. Three parameters are used to determine the flame speed: reactivity of the flammable gas, the degree of confinement of the flammable cloud, and the degree of obstruction due to obstacles within the flammable cloud. These parameters are represented by a set of what are effectively low, medium, and high choices for each parameter.

While the BST model provides reasonable guidance on choosing the explosion parameters, there are still areas that could be improved. One problem is that the BST flame speed matrix includes large jumps in flame speed between categories of confinement, reactivity, or congestion. This was first addressed by Baker in a 1998 paper [1] where a new confinement category, 2½-D, was added for those situations that were more confined than the 3-D case, but less so than the 2-D case. The flame speeds used for the 2½-D confinement were simply the arithmetic average of the flame speeds for the 2-D and 3-D cases for a given reactivity and congestion class. Even with this extension, large discontinuities remained among the categories for flame speed according to the prescribed methodology. These discontinuities drive the need for a new method to determine flame speed based upon quantifiable properties of the flammable gas and its surroundings that varies smoothly across the range of conditions that are found in actual process plants.

The creation of the QMEFS model [2] improved upon the basic BST scheme by combining it with the GAMES correlations [3] and providing a numerical resolution to the BST matrix. Integration of the GAMES model allows the strength of the explosion to be defined by numeric input parameters and avoids the low/medium/high categorizations.

2 Parameters Affecting Flame Speed

With the incorporation of the GAMES correlations, the resulting explosion flame speed is characterized by five parameters:

- Laminar burning velocity (LBV) is used as the characteristic property for fuel reactivity

- Volume blockage ratio (VBR) and average obstacle diameter define the level of congestion
- Confinement is characterized as the number of confining planes, or walls, that are available to confine the expanding gases.
- Flame run-up distance is used to characterize the flame acceleration path

The last part of explosion characterization is the explosive energy of the fuel/air mixture, which is defined by the gas concentration, the combustion energy of the fuel, and the volume occupied by the fuel/air mixture.

2.1 Reactivity

Fuel reactivity is a measure of the propensity of the flame front to accelerate and create overpressures or potentially undergo a deflagration-to-detonation transition (DDT). The BST model originally classified reactivity as high, medium, or low based on the fuel in the fuel-air mixture. The original rule set [1] defined propylene oxide, ethylene oxide, and hydrogen (or materials with a 0.8 m/s LBV or larger) as high reactivity; methane and carbon monoxide were classified as low reactivity, and everything else as medium reactivity. According to the current BST model, materials having a laminar flame speed greater than 0.75 m/s are considered high reactivity (to accommodate ethylene being high reactivity) while those having a laminar flame speed below 0.4 m/s are considered low reactivity [1], although some BST publications list the low boundary as 0.45 m/s. In 2016, the BST categories were expanded to include a very-low category for fuels like ammonia, which has an LBV in the range of 0.07 m/s to 0.15 m/s [4].

In the experimental studies that form the basis of the BST model, ammonia, methane, propane, and ethylene were used to define the very low, low, medium, and high reactivity categories. These categories produce confusion due to how they create boundaries between the BST categories. For example, the LBV of ethylene, depending on the cited reference, ranges between 0.64 and 0.83 m/s. With the current definition of 0.75 m/s as the high reactivity boundary, this makes ethylene a "borderline" fuel – one that may either be defined as medium or high reactivity. However, ethylene is explicitly categorized as a high reactivity material, as their experimental work used ethylene as the high reactivity material [5]. This then begs the question: what about ethylene oxide (LBV 1.0 m/s) or acetylene (LBV 1.6 m/s)? These two materials are clearly more reactive than ethylene, but are still categorized as high reactivity. The BST model effectively says that an ethylene explosion will be as severe as an acetylene explosion, when all other parameters are held constant.

2.2 Congestion

In the original BST model, low obstacle density was defined as having an area blockage ratio (ABR) of less than 10%, while high obstacle density was defined as an ABR of 40% or greater, and everything in between is considered medium [6]. Later BST guidelines have used the volume blockage ratio (VBR) as the parameter for classifying obstruction [5]. In practice both blockage ratios will affect how fast a flame accelerates, but, for a uniform obstacle field, the two are related simply by the pitch-to-diameter ratio of the obstacles. If the obstacles included in an ABR are not repeated quasi-uniformly throughout the obstacle field, then their effect would be more accurately portrayed by VBR.

The classifications of low, medium, and high are often too coarse for proper assignment of congestion to a particular PES. In the experimental programs that helped to define the BST model, VBRs of 1.5%, 4.3%, and 5.7% were used to represent the low, medium, and high categories. As with other VCE parameters, the differences between these categories leaves significant room for variation between the categories. In addition, no definition of 0% VBR is provided, nor is any relationship to VBRs higher than 5.7%.

2.3 Confinement

The effect of the confinement of the flammable cloud is taken into account by determining the number of dimensions in which the burning gas may expand. 3-D expansion allows the burning cloud to expand freely in all directions and results in the slowest flame acceleration and lowest overpressures. 2-D expansion, such as a flame between two flat plates, generates higher overpressures because the combustion gases have fewer directions in which to expand resulting is a higher flame acceleration.

Confinement is included in the BST model by identifying the number of dimensions that are available to the products of combustion for expansion. Expansion into free space (except for the plane of the ground) is considered 3-D expansion, expansion between two parallel planes is considered 2-D expansion, and expansion in a pipe is considered 1-D expansion. The 1-D case was removed in the more recent publications discussing the BST model [5]. To handle the case of a frangible or partially-confining plane, such as a very closely spaced pipe rack, the BST model added a 2.5-D classification which simply averaged the 2-D and 3-D flame speed results [5].

While the classification is fairly straightforward, especially after the 2.5-D category was added, some situations are still difficult to define. These include partial walls, pieces of equipment that may form more confinement than congestion, or surfaces that are porous such as grated decks.

2.4 Other Factors

Several researchers [5, 7, 8] have acknowledged that the overall dimensions of the flammable vapor cloud before ignition directly affect the final flame speed and consequent overpressures of the vapor cloud explosion. Since a flame will accelerate until it reaches a maximum sustainable value (or undergoes a deflagration to detonation transition, a DDT), the maximum dimension of a flammable cloud is expected to be an important variable in the creation of overpressure. As a point of reference, the length available for flame acceleration in the more recent BST tests was 15 meters [5].

In addition, research [7, 8] suggests that the scale of obstacles within the congested area also affects the maximum flame speed that may be achieved. As a flame front moves around and past obstacles, the turbulence that is generated can increase the flame surface area and accelerate combustion. The characteristic obstacle diameter for the BST tests was 2 inches (0.0508 meters); other test series (including the MERGE/EMERGE tests, varied the average obstacle diameter).

2.5 Parameters Comparison

The BST parameters discussed above are listed in Table 1, along with the corresponding numerical values that are equivalent to those applied in the QMEFS model. When the numerical values in Table 1 are applied in the QMEFS model, the results will match the predictions of the BST model.

Table 1. BST Flame Speed Parameters

BST Parameter	BST Category	Model Parameter	Corresponding Parameter Value
Reactivity	Very-Low (e.g., Ammonia)	S_l	0.08 m/s
	Low (e.g., Methane)		0.37 m/s
	Medium (e.g., Propane)		0.43 m/s
	High (e.g., Ethylene)		0.75 m/s
Obstacle Density	Low	*VBR*	0.015
	Medium		0.043
	High		0.057
Expansion	3-D	*nPlanes*	1
	2.5-D		1.5
	2-D		2

3 Model for Estimating Flame Speed

The QMEFS model was created to overcome some of the limitations described above. By incorporation of the GAMES correlations and the overpressure/impulse curves from the BST method, a model that provides the capability for more detailed descriptions of explosion scenarios has been created. Thus, QMEFS incorporates the modeling methodologies and experimental data sets that have been used in both the BST model and in the GAMES project.

3.1 Fundamental Correlations

The fundamental correlation is based on correlations developed from the GAMES project [3]. These correlations related overpressure to parameters based on the MERGE and EMERGE experiments [7,8]. These correlations can be combined with acoustic theory to relate flame speed to explosion parameters [2] to produce the equations below.

For no confining surfaces other than the ground (3D expansion):

$$2.4 * \frac{M_f^2}{1+M_f} * p_0 = 0.84 * \left(\frac{VBR*L_f}{D}\right)^{2.75} * S_l^{2.7} * D^{0.7} \tag{1}$$

For confinement between two flat planes (2D expansion):

$$2.4 * \frac{M_f^2}{1+M_f} * p_0 = 3.38 * \left(\frac{VBR*L_f}{D}\right)^{2.25} * S_l^{2.7} * D^{0.7} \tag{2}$$

where:

VBR = volume blockage ratio
L_f = maximum distance a flame may propagate in the obstructed region (i.e., the run-up distance), m
D = average obstacle diameter, m
S_l = laminar burning velocity (LBV) of the flammable gas, m/s
p_0 = the ambient pressure, bara
M_f = the flame speed relative to a fixed observer, expressed as a Mach number

To correct the flame speed to account for confinement in spaces other than one or two planes, a curve fit of the ratio of burned cloud to unburned cloud radii versus the number of confining planes was made. A curve fit was generated with an expansion ratio (α) equal to 7, a typical expansion ratio for common hydrocarbon fuels in air. This is also the value assumed in the derivation of portions of the BST model [9]. The equation resulting from this curve fit is:

$$NPF = \frac{1}{0.5035 - 0.0757 \cdot nPlanes} - 1.338 \tag{3}$$

where:

NPF = flame speed correction factor
$nPlanes$ = number of confining planes

For true 3-D spherical expansion, $nPlanes = 0$. For a typical explosion at grade level, $nPlanes = 1$. For 2-D expansion, such as a flame propagating between a floor and a ceiling, $nPlanes = 2$. For 1-D expansion, such as a flame front in a pipe, $nPlanes = 4$. Planes that are not completely solid or rigid may be accounted for using a fraction of a plane.

The flame speed correction factor, *NPF*, does not fully account for increases in flame speed as the number of confining planes increases. This is accomplished by applying the following factor to the flame speed when *nPlanes* is equal to 2:

$$\beta = \frac{M_{f,2D}}{M_{f,3D} \cdot NPF(nPlanes = 2)} \tag{4}$$

This can be implemented for nPlanes =1 to nPlanes = 4:

$nPlanes = 1$

$$M_f = M_{f\,3D} * NPF \tag{5}$$

$1 > nPlanes < 2$

$$M_f = M_{f\,3D} * NPF * \beta \tag{6}$$

$nPlanes >= 2$

$$M_f = M_{f\,3D} * NPF * [1 + (\beta - 1) * (nPlanes - 1)] \tag{7}$$

3.2 Flame Speed Limits

The equations above are implemented in the QMEFS model to define the flame speed (M_f) and then coupled with the latest BST flame speed matrix [4]. One limitation to the GAMES correlations is that they can predict unrealistic flame speeds. This is a direct result of the correlations fitting the data within the scope of the MERGE and EMERGE tests, and not being designed to extend beyond that scope. By coupling with the BST model, this problem is partially resolved because the BST matrix provides a set of upper limits to flame speeds (and by extension, the explosion strength).

Upon examination, it can be seen the GAMES correlations will predict higher and higher flame speed values as parameters such as L_f or S_l, increase. In reality, for a given obstacle configuration, geometry, and flammable gas mixture, there is a limit to the flame speed that can be achieved. This flame speed may be either subsonic or the cloud may undergo a deflagration-to-detonation transition (DDT) in which the flame front becomes a detonation propagating at roughly the Chapman-Jouguet (C-J) detonation velocity for the flammable mixture. For most common flammable hydrocarbons in air, this speed is roughly 1800 m/s or $M_f = 5.2$. For purposes of this model, any flame that accelerates to a calculated velocity greater than $M_f = 3.0$ is considered to have undergone DDT [10, 11] and the flame speed is set to the C-J velocity, $M_f = 5.2$.

The flame speeds suggested by the latest BST model are applied as the upper limit flame speeds to be used in this model, as they have been "scaled up" to account for industrial scale flammable clouds [5] and account for DDTs.

The result of these correlations is a model that estimates the flame speed based on VBR, D, S_l, L_f and the number of confining planes. The resulting flame speed is compared to the published BST model, which provides a matrix of flame speeds as a function of reactivity, obstacle density, and flame expansion. The BST matrix values are used as maximum flame speed values for the set of parameters given in Table 1. Values between the BST matrix elements are calculated by linear interpolation (as was done for Baker's 2½-D flame expansion category). This methodology provides an extended, systematic approach for estimating flame speeds resulting from the combustion of a flammable cloud in an obstructed and/or confined region. Predicted flame speeds are used with the existing BST blast curves to produce estimates of overpressure at a distance from the explosion source.

3.3 Expansion of the BST Matrix

The range of the QMEFS model is limited by the upper and lower bounds of the BST matrix elements for congestion, confinement, and fuel reactivity. Through application of this model to various petrochemical projects, the need to expand the matrix became obvious. As described above, the model is built to provide predictions for the *nPlanes* (confinement) parameter in the range of 1.0 to 4.0 (the model is not validated nor trusted above this value). Values of average obstacle diameter and flame run-up distance are built into the GAMES correlations, so naturally are accounted for within reasonable limits. This leaves the congestion (VBR) and fuel reactivity parameters as places to potentially expand the model.

The intent of model expansion is to widen the range of congestion and reactivity inputs so that it removes questions about its use in the "below low" and "above high" regions. With proper expansion of the model described above, the predicted flame speed following ignition of a flammable cloud may then be used with the BST blast curves to estimate the overpressure impacts in the area surrounding a vapor cloud explosion.

To accomplish this expansion, additional categories were added to the BST matrix for congestion and reactivity. This follows Baker's own expansion into the **very-low reactivity** category when ammonia explosion experimental results were published [4].

For a **very-high reactivity** category, hydrogen was assumed as the characteristic fuel, and represents a true upper bound for LBV. Expansion to the very-high category, for parts of the matrix with high flame speeds is simple because DDTs are expected in these conditions already. For other areas, published data was required for the basis of the expansion.

In the case of the **very-low congestion** category, open air (3-D flame expansion) explosion tests were performed by Harris and Wickens [12] as well as Sato and Iwabuchi [13] for a variety of materials. The **very-high congestion** category, several of the BFETS series of tests were used [14] due to their use of a VBR in the 10% range.

The expanded matrix applied in the new QMEFS model is presented in Table 2. The latest BST flame speed matrix values are outlined by bold lines in Table 2. Values outside these areas have been added as part of the QMEFS expansion.

While a 1-D confinement is not expected to be applied in large-scale industrial scenarios, the *nPlanes* = 5 (1-D) confinement category is included in this model to provide an interpolation point beyond *nPlanes* = 2 (2-D flame expansion).

Table 2. Expanded Flame Speed Matrix

Confinement	Reactivity	Congestion				
		Very-Low	Low	Medium	High	Very-High
1D	Very-High	DDT	DDT	DDT	DDT	DDT
	High	DDT	DDT	DDT	DDT	DDT
	Medium	1.029	1.029	1.765	2.265	DDT
	Low	0.5	0.5 *	1.029	2 *	DDT
	Very-Low	0.023	0.023	0.124	0.168	0.303
2D	Very-High	DDT	DDT	DDT	DDT	DDT
	High	0.588	0.588	DDT	DDT	DDT
	Medium	0.47	0.47	0.66	1.6	DDT
	Low	0.079	0.079	0.47	0.66	1.244
	Very-Low	0.008	0.008	0.049	0.069	0.130
3D	Very-High	0.5	DDT	DDT	DDT	DDT
	High	0.06	0.36	DDT	DDT	DDT
	Medium	0.03	0.11	0.44	0.5	1.084
	Low	0.025	0.026	0.23	0.34	0.678
	Very-Low	0.003	0.003	0.024	0.036	0.073

* Values adjusted from BST publication for model consistency

4 Application

It is helpful to test the QMEFS model outside of the published BST matrix, in the areas where the QMEFS matrix has been expanded. Large scale hydrogen experiments were performed in a hemispherical test rig with no congestion. The test volume was 300 m^3 with a base of 10.4 meters diameter [13] and center ignition. The tests were modeled with the latest BST model and the expanded BST model. The model parameters listed in Table 3 were applied to these calculations. Figure 1 presents the overpressure in relation to distance from center of the fuel-air mixture, for the test results and both BST and QMEFS predictions. The BST model does not have a category for a hydrogen vapor cloud with no congestion, thus the scenario falls into the high fuel reactivity and low congestion portion of the matrix. The result is an underprediction of the overpressure. The BST high fuel reactivity works well for fuel such as ethylene (LBV = 0.76 m/s), but will often underpredict the impacts of a hydrogen (LBV = 3 m/s) explosion. The expanded matrix allows a higher fuel reactivity category and a lower congestion category, and consequently a better prediction of the explosion impacts. As seen in Figure 1, the QMEFS result is a slight overprediction of the overpressure.

Table 3. Modeling Parameters for an Unconfined Hydrogen Explosion

Parameter	BST	Expanded QMEFS
Volume [m^3]	300	300
nPlanes	1	1
Reactivity	High	Very-High
VBR	N/A	0
Congestion	low	N/A
Run-up distance [m]	N/A	5.2

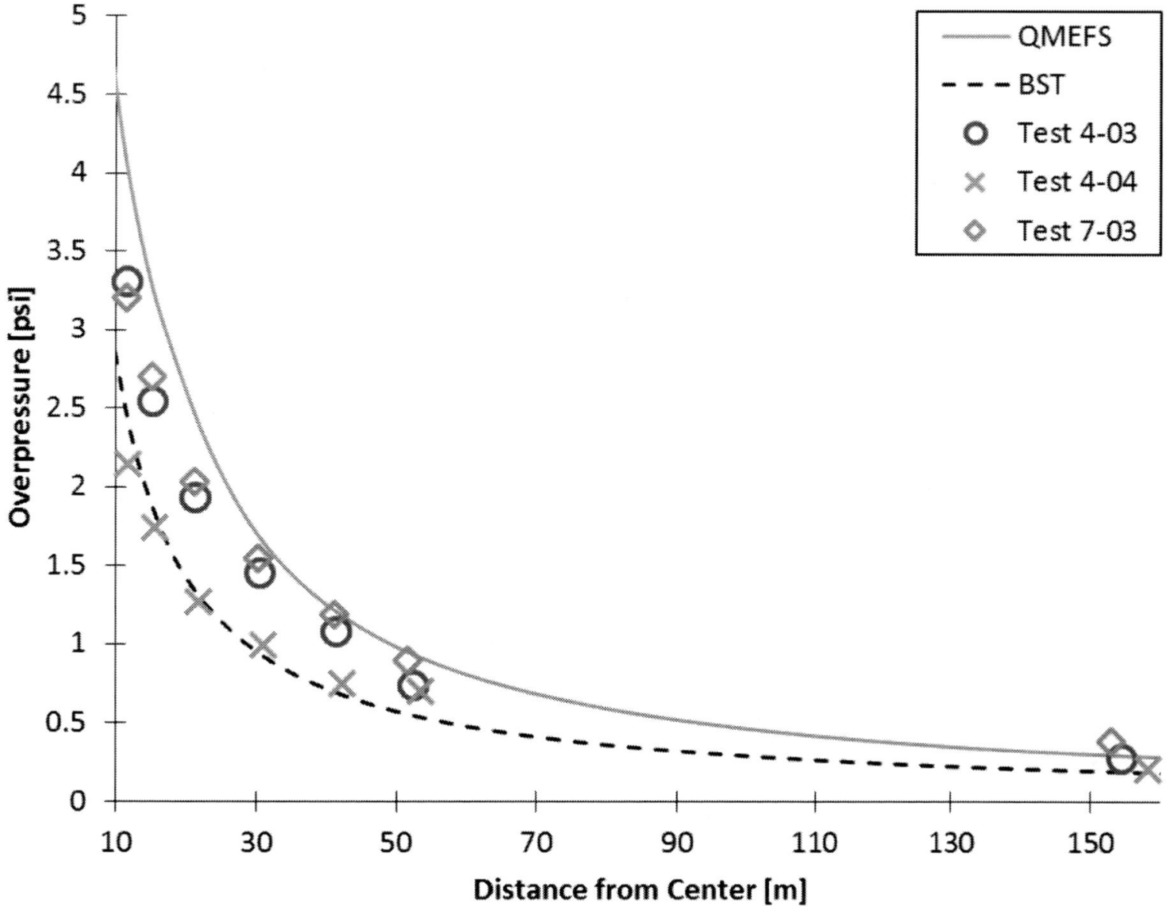

Figure 1. Overpressure vs. Distance for Open-Air Hydrogen Tests

As discussed in the paper, the distance that is available for the flame to accelerate, i.e. the run-up distance, can have a significant effect on the flame speed attained in a flammable gas cloud. The flame speeds presented in the original Baker-Strehlow (BS) model [6] were generated in a test rig whose largest dimension was less than 6 feet (1.8 meters) [5]. The flame speeds presented in the 2005 BST paper are based on tests where the largest dimension was 48 feet (14.6 meters) with the published flame speed results "scaled up" to account for the maximum size of a typical industrial plant [5]. One test of the QMEFS model is to determine the flame speed versus run-up distance for the test configurations used in the newest BST model when specific parameters are varied.

Figure 2 shows the results for the 3-D flame expansion (*nPlanes* = 1), medium reactivity (S_l = 0.43 m/s) case. The curves plotted in Figure 2 show six VBR values – the high (5.7%), medium (4.3%), and low (1.5%) congestion values used in the BST experiments, one intermediate value to show how the model behaves as the VBR is adjusted, a fifth value (1%) below the BST low category to demonstrate the behavior in the "less-than-low" congestion range, and the sixth value (10%) above the BST high category to demonstrate the behavior in the "higher-than-high" congestion range.

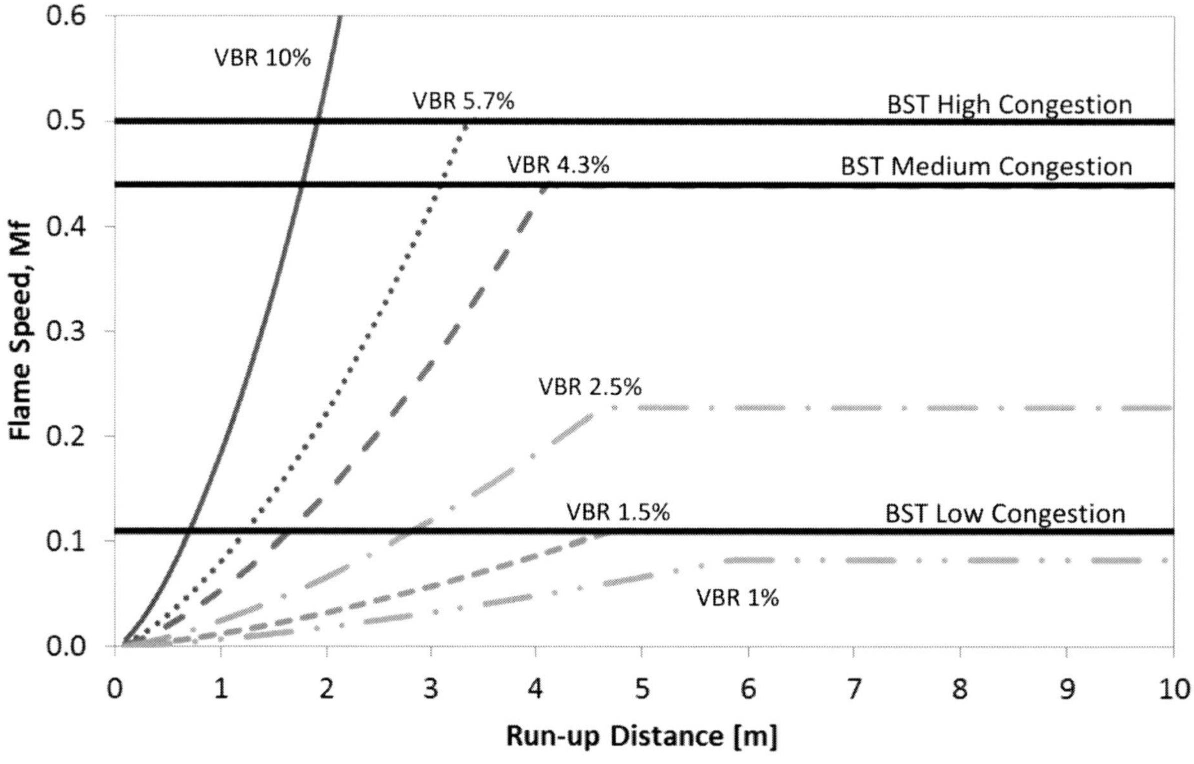

Figure 2. Flame Speed vs. Run-Up Distance; 3-D, Propane, D = 2"

Figure 3 demonstrates how flame speed varies with *nPlanes* and LBV. The basis of the figure is an explosion involving a medium VBR (0.043) and a 4-inch average obstacle diameter. The *nPlanes* parameter is varied between 1.0 (3-D flame expansion) and 4.0 (the maximum value recommend for use in the QMEFS model). The curves plotted in Figure 3 show five materials with various LBV values – ammonia (0.08 m/s), methane (0.37 m/s), propane (0.43 m/s), ethylene (0.76

m/s) and ethylene oxide (0.91 m/s). In the BST model ethylene oxide and ethylene are grouped together into the high category, but these two fuels have different burning properties as can be seen with their LBV values. Thus, by accounting for the LBV of the fuel, the new model provides a finer resolution for flame speed. In addition, as seen in Figure 3, the QMEFS model provides more resolution for the flame expansion parameter through the use of the *nPlanes* variable.

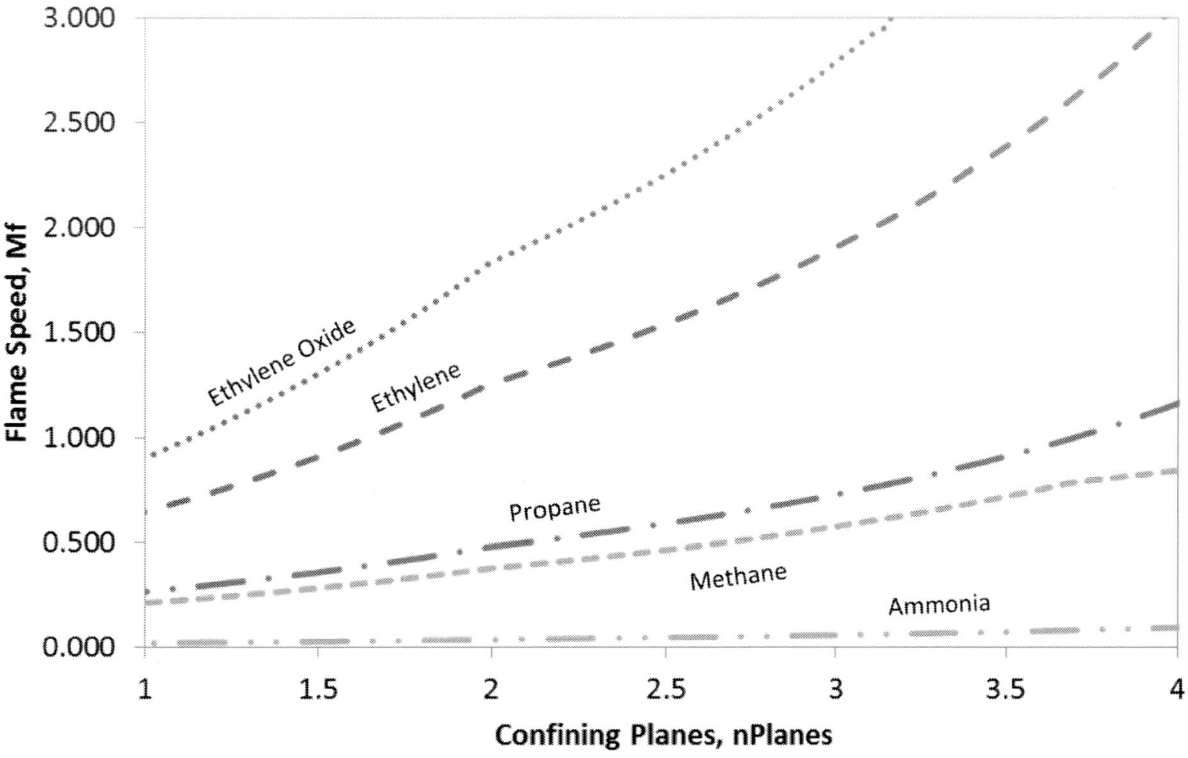

Figure 3. Flame Speed vs. Confining Planes; Medium VBR, D = 4"

5 Conclusions

The model described in this paper provides an improved method to describe the characteristics of a vapor cloud explosion over a wide range of conditions. It combines elements of the work conducted by Baker to develop the BST model, as well as the European MERGE/EMERGE projects; thus incorporating data from the two largest, modern vapor cloud explosion test projects.

Because the BST categories are resolved into numeric ranges, a higher resolution solution space and a clearer set of measureable inputs are achieved. With the ability to model the reactivity of a flammable gas cloud based on laminar flame speed instead of a low, medium, or high classification, the model is able to more accurately describe a wide range of flammable gases and mixtures. Expansion of the constraining flame speed categories to very-low and very-high then provides an even wider range of possibilities.

With the QMEFS matrix expanded to congestion (VBR) categories of very-low and very-high, the model provides a wider range of predictive capability for explosion scenarios. These expansions and enhancements are still confined by the flame speeds in the BST system, and are applied within the set of curves for overpressure and impulse. Thus, with the QMEFS model presented here, process safety experts have a more versatile tool to predict VCE impacts in petrochemical plants.

6 References

[1] Baker, Q.A., C.M. Doolittle, G.A., Fitzgerald, and M.J.Tang, "Recent Developments in the Baker-Strehlow VCE Analysis Methodology," *Process Safety Progress*, v.17, No.4 (1998).

[2] Melton. T.A. and Marx J.D., "Estimating Flame Speeds for Use with BST Blast Curves," Process Safety Progress, v. 28, March, (2009).

[3] Mercx, W.P.M., A.C. van den Berg, and D. van Leeuwen, A*pplication of correlations to quantify the source strength of vapour cloud explosions in realistic situations*. Final report for the project: 'GAMES', TNO report PML 1998-C53, Rijswijk, The Netherlands (1998)

[4] Thomas, K.J., Malik. D.R., Downes, A.M., Rodges, S., "Ammonia Vapor Cloud Explosion Testing and Implications for Facility Siting Studies, Consequence Analyses and Risk Assessments," *50th Loss Prevention Symposium*, AIChE, April, (2016).

[5] Pierorazio, A.J., J.K.Thomas, Q.A.Baker, and D.E.Ketchum, "An Update to the Baker-Strehlow-Tang Vapor Cloud Explosion Prediction Methodology Flame Speed Table," *Process Safety Progress*, v.24, no.1 (2005).

[6] Baker, Q.A., M.J. Tang, E. Scheier, and G.J. Silva, "Vapor Cloud Explosion Analysis," *28th Loss Prevention Symposium*, AIChE, April, (1994).

[7] Mercx, W.P.M., editor, *Extended Modelling and Experimental Research into Gas Explosions*, Final summary report of the project EMERGE, CEC contract EV5VCT930274, European Commission, Directorate General XII, Brussels, Belgium (1997).

[8] Mercx, W.P.M., A.C. van den Berg, C.J. Hayhurst, N.J. Robertson, and K.C. Moran (2000), *Developments in Vapour Cloud Explosion Blast Modeling*, Journal of Hazardous Materials 71 (2000) 301-319.

[9] Tang, M.J. and Q.A.Baker, "A New Set of Blast Curves from Vapor Cloud Explosion," *Process Safety Progress*, v.18, no.4 (1999).

[10] Oran, E.S. and V.M. Gamezo, "Origins of the deflagration-to-detonation transition in gas-phase combustion," *Combustion and Flame*, 148 (2007).

[11] Vasil'ev, A.A., "Estimation of Critical Conditions for the Detonation-to-Deflagration Transition," *Combustion, Explosion, and Shock Waves*, v.42, no. 2 (2006).

[12] Harris, R.J. Wickens, M.J., "Understanding Vapor Cloud Explosions – An Experimental Study," The Institute of Gas Engineers 55th Autumn Meeting, (1989).

[13] Sato, Y., Iwabuchi, H., Groethe, M., Merilo, E., Chiba, S., "Experiments on Hydrogen Deflagration," Journal of Power Sources v.159 (2006).

[14] Health & Safety Executive, "Explosions in Full Scale Offshore Module Geometries Main Report," Offshore Technology Report – OTO 1999 043. Prepared by BG Technology, April, (2000).

479444 Do We Need a Quantitative Risk Assessment (QRA) If My Consequence Analysis Facility Siting Presents a Problem?

Abstract:

There are three steps that a regular facility siting study (FSS) can go through. The first step is the basic separation table approach descrived in API 752. The second step is the consequence analysis facility siting and the third and final one is the quantitative risk assessment. It turns out, however, that the vast majority of the consequence analysis FSS studies quickly demonstrate that the impact to buildings can be more significant that what have been defined as acceptable. For such cases, the analysis normally turns into the QRA, exceedance curves, building structural assessment, etc. The idea of this paper is to cover a intermediate step in early stages to use computational fluid dynamics (CFD) technique in the early stage of the analysis without compromising the time schedule and the budget, this would include dispersion, radiation and explosion modeling. The more accurate results obtained with CFD modelling lead to cloud sizes which are mainly less conservative than traditional methods. The objective of this paper is to go a setp beyond a previous analysis for the dispersion using CFD, but rather make a full evaluation of a FSS using CFD.

13TH GLOBAL CONGRESS ON PROCESS SAFETY

Proposed Method for Investigating the Reactivity of Chlorine with Environmental Materials under Relevant and Controlled Conditions

Audrey Feuvrier, PhD.
Ralph E. Martin Department of Chemical Engineering
University of Arkansas, Fayetteville, AR
audrey.feuvrier@uark.edu

Tom Spicer, Ph.D., P.E.
Ralph E. Martin Department of Chemical Engineering
University of Arkansas, Fayetteville, AR
tos@uark.edu

Prepared for Poster Presentation at
American Institute of Chemical Engineers
2017 Spring Meeting and 13th Global Congress on Process Safety
San Antonio, TX
March 26 – 29, 2017

Proposed Method for Investigating the Reactivity of Chlorine with Environmental Materials under Relevant and Controlled Conditions

Audrey Feuvrier, PhD.
Ralph E. Martin Department of Chemical Engineering
University of Arkansas, Fayetteville, AR
audrey.feuvrier@uark.edu

Tom Spicer, Ph.D., P.E.
Ralph E. Martin Department of Chemical Engineering
University of Arkansas, Fayetteville, AR

Keywords: Chlorine reactivity, chlorine release, dry deposition, Jack Rabbit II

Abstract

Many common toxic industrial chemicals (TICs) such as chlorine and ammonia are very reactive with commonly encountered materials in the environment. The amount of TIC material reacted in the environment is an important factor in the impact of a release. Typically characterized with a deposition rate in atmospheric dispersion models, the reaction rate and total deposition quantities have predominately been studied in field scale experiments. Laboratory experiments have also been undertaken, but past investigations have inadequately quantified the available turbulence which can importantly impact surface reaction rates. The state of the art description of surface deposition represents the process as a series of resistances including the effect of atmospheric turbulence and surface reaction(s) depending on the surface material (substrate). Sponsored by the Chemical Security Analysis Center (CSAC) of the U.S. Department of Homeland Security and Transport Canada, an experimental program is underway to test potential substrates under known inlet gas concentrations with levels of turbulence that are comparable to those of the atmosphere. The objective is to develop a test methodology for subjecting a substrate to a reactive gas flow of known TIH concentration with turbulence levels consistent with those in the atmosphere. With data from repeat experiments, deposition of a reactive TIH chemical can be quantified with the proposed experimental protocol.

1 Introduction

Estimating the quantity of a Toxic Industrial Chemical (TIC) removed from the environment through chemical processes is critical to proper consequence assessment of accidental (or intentional) releases, particularly in urban areas [1]. TIC removal mechanisms play an important role in the rational assessment of risk, as these phenomena may significantly and quickly reduce the TIC concentration near a release [2,3], particularly for a reactive chemical such as chlorine. Typically characterized as dry deposition, reactions of chlorine with the environment was not included in the computational models used to predict the consequences of large releases of chlorine in the past [2], but new efforts are underway to characterize chlorine reactivity and incorporate such effects in consequence and risk assessments.

While species like sulfur dioxide and iodine have been the subject of extensive experimental studies [3], studies about deposition involving chlorine are scarce and show a great deal of variability. The high reactivity of chlorine as well as its toxicity make it difficult to study in a laboratory environment, and consequently, many of the studies are conducted as field studies [4]–[6], which restricts test conditions to a limited set of atmospheric conditions. In one set of laboratory experiments, Hill [7] studied the effect of deposition of several TICs on an alfalfa canopy in a chamber where humidity, light, and wind speed were controlled. He showed that chlorine deposition is rapid on alfalfa and that chamber air velocity has a significant effect on the chlorine uptake, but the level of turbulence in the chamber is unknown. Other experimental studies gave promising results such as [8]–[10], but these studies involved no other movement of the flow than the one produced by the mixing device or the circuit used to monitor the concentration level. The state of the art description of surface deposition, [11], represents the process as a series of resistances including the effect of atmospheric turbulence and surface reaction(s) depending on the surface material (substrate).

In this context, an experimental program, sponsored by the Chemical Security Analysis Center (CSAC) of the U.S. Department of Homeland Security and Transport Canada is underway to test potential substrates in a controlled environment. The project includes the design and characterization of a mixing chamber as well as development of a protocol for the measurement of chlorine removal rate for different samples of soil and vegetation. Concentration, velocity, and turbulence measurements will be made in the working section to confirm that the flow is uniform and consistent. It is also desired that the flow in the working section of the apparatus is consistent with the atmospheric flow with target turbulence intensities on the order of 10%. Working section flow speed will be variable, with a maximum in the section of 8 m/s, and expected velocities between 0.5 to 3 m/s. Passive flow conditioning will be used to eliminate non-uniformities in the flow. Recognizing that chlorine gas concentrations as high as 10% or 100,000 ppm may be used, apparatus materials of construction will be carefully chosen to resist damage and also to not remove significant amounts of chlorine during a test. Finally, to limit the hazards involved with the use of chlorine at high concentration level, the apparatus will need to fit under a 6 ft long (about 1.8 m) fume hood.

In the following, the proposed designs for the apparatus and its main elements are introduced. The choice of the mixer, fan, and materials will be discussed. In the second part of the article, the future protocol of measurement will be presented. Focus will be put on the means of measurement used and the characterization of the chamber. The steps to determine chlorine removal rates for several samples types under different environmental conditions will be considered.

2 Proposed design

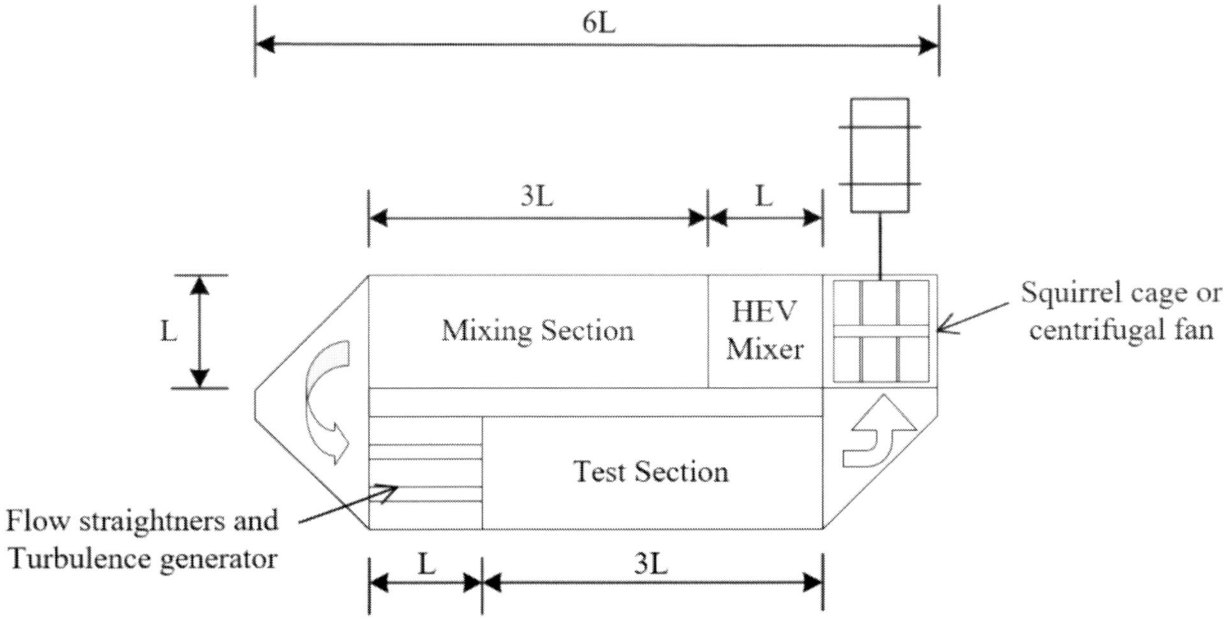

Figure 1. Preliminary apparatus design

Figure 2 shows the preliminary design of the apparatus following the criteria laid down previously using a simple closed return design. The test or working section is placed under the mixing section to facilitate access to the test section through a lab hood sash. The characteristic length, L, chosen to fit in a standard 6 ft hood, will be between 0.25 and 0.3 m, giving a total volume for the apparatus of between 0.16 and 0.28 m^3. A standard lecture bottle contains 454 g of chlorine, which would be sufficient to provide chlorine for 8 to 14 tests depending on the length L based on a maximum concentration of 100,000 ppm. The use of lecture bottles limits the quantity of chlorine required in the lab during the experiments and effectively eliminates the hazard associated with the storage of large quantities of chlorine on site. These volume values are approximate because it may be convenient to use a circular conduit for mixing and a square cross-section for the working section. A circular conduit reduces the required volume of the mixing section (with some minor reduction in power consumption), and a working test section with flat surfaces facilitates optical measurements during characterization and sample placement in the testing phase.

2.1 Mixer and flow conditioner

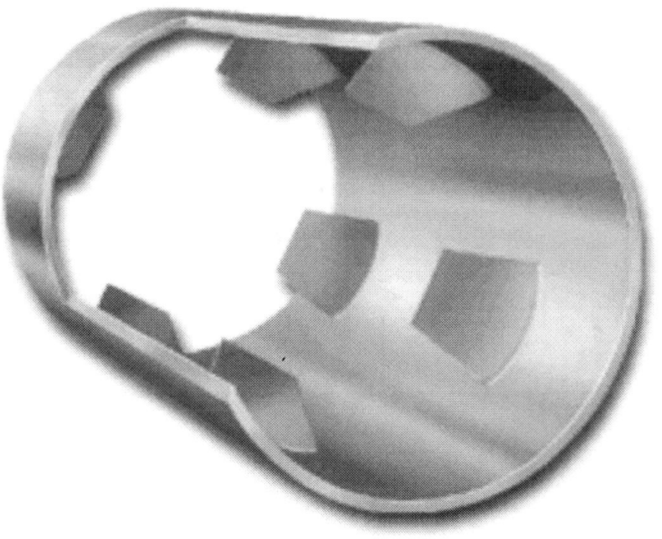

Figure 2. Typical High Efficiency Vortex (HEV) mixer [12].

Figure 3 shows a typical High Efficiency Vortex (HEV) mixer which is placed downstream of the fan. This type of mixer was selected because it provides efficient mixing with turbulent flow with minimal pressure drop. For a Kenics™ HEV mixer, Figure 4 shows the relative coefficient of variation, COV_R , defined as the quotient between the outlet and inlet coefficient of variation, as a function of the Reynolds number (defined as $U_{\infty} L/v$ where U_{∞} represent the characteristic velocity in the tunnel, L is the characteristic length and v is the kinematic viscosity of the gas). The coefficient of variation for the mixing of two fluid represents the quotient of the standard deviation of the concentration and the mean concentration. Therefore, the lower the COV_R, the better the mixture quality. The gray zone on Figure 4 represents the range of Reynolds number of this study. Based on Myers et al [13], the pressure drop for our (custom made) apparatus was estimated to be between 1.7 and 17.3 Pa, making it negligible. The HEV mixing process rely on the production of hairpin vortices downstream of the blades of the mixer combined with a pair of counter rotating longitudinal vortices [14]. A mixing section downstream of the device and equal to three times the length of the mixer is recommended to provide for sufficient time for the vortices to complete the mixing process and consequently obtain the predicted COV_R. Additional turbulence generated between the mixing section and the test section will only improve the COV_R before the flow enters the test section..

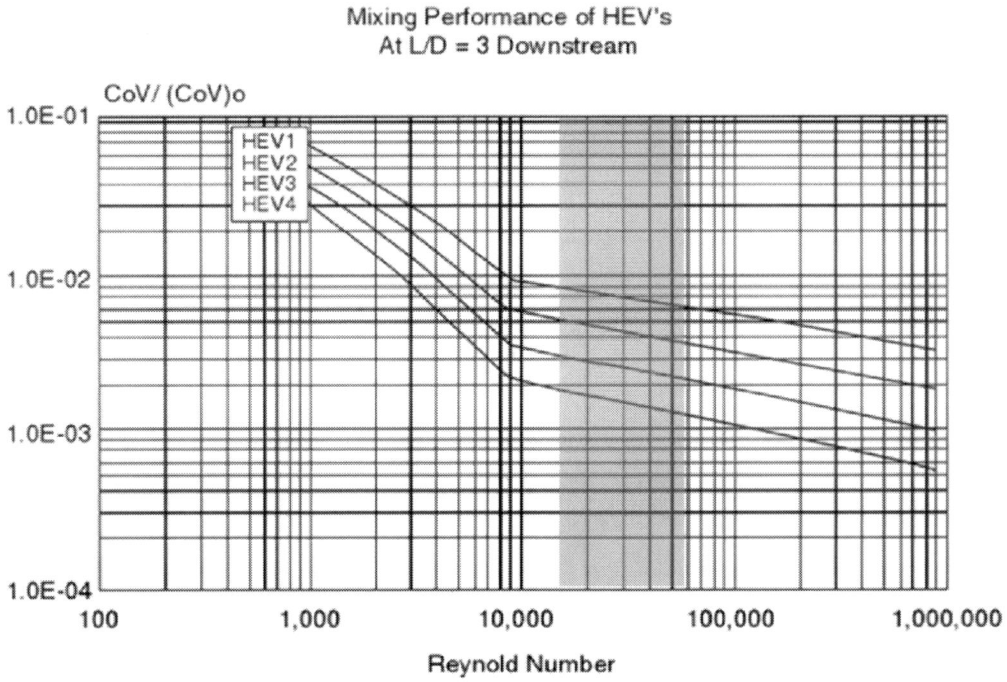

Figure 3. Relative coefficient of variation (COV_R) vs. Reynolds Number and number of tabs row for the Kenics HEV, [12].

The corners currently consist of two 45° bends with a short duct between them, and the 45° bends are preferred to 90° bends to prevent the accumulation of gases in the dead space. To help maintain good flow distribution in the channel, sets of vanes will be placed in each bend. Operating at low wind speed allows us to consider simple circular cross sectional shapes for the vanes [15].

After the two first corners, the flow conditioner section is planned to eliminate large eddies that might be present in the flow ensuring uniformity of the flow and simultaneously raising the turbulence intensity up to 10%. To achieve this, we plan to use a combination of screens or perforated plates. Samples will be placed in the test section through a removable side with provision for a tray in the floor of the test section to accommodate soil and/or plant roots. After experiments, ventilation of the apparatus will be achieved through two hatches on each bend between the mixing and flow conditioner sections.

2.2 The fan

On Figure 2, the motor of the fan is represented outside of the apparatus because chlorine would be expected to react and corrode any of its exposed surfaces. The motor shaft will be supported by Nylon or Teflon bearings located in the side of the apparatus to isolate the motor from chlorine exposure. A centrifugal (squirrel cage) fan is being considered to allow placement of the fan above the apparatus. The fan blades and shaft exposed to chlorine will be treated to limit damage due to chlorine exposure as discussed below.

2.3 Building materials

During the first phase of the experimentation, the apparatus will only use a mixture of air and an inert gases will be used for optic velocity and turbulence measurements and concentration measurements (including a propylene tracer). During this preliminary phase, the structure of the test section will be made with plywood, but two of its walls will be made with polycarbonate to allow visualization and optical measurements of the velocity in the section.

In the second phase of the experimentation involving chlorine, all elements in contact with the test mixture must be made with a material or coated with a material capable of resisting high chlorine concentrations. In addition to extending the workable lifetime of the apparatus, reducing the potential for chlorine to react with materials in the test chamber is also desired so that the reduction in chlorine gas concentration can be attributed to reaction with the sample tested as opposed to the chamber itself. In the case of the fan blades and shaft, a commercial coating is planned (such as Cerakote or titanium). Titanium is known to be highly reactive with dry chlorine, however 1% of water may be enough for passivation after mechanical damage to titanium under normal conditions [16]. For the mixing section, a metal or plastic (PVC) duct will be used since no optical access is required. The structure of the corners and the test section will be made with plywood. In the second phase, polycarbonate may be replaced by plywood. The flow straighteners will likely be made of a suitable metal. All surfaces except the fan blades and shaft are planned to be coated with Viton™, a coating material widely used in applications with chlorine in the chemical industry. We are currently evaluating whether the Viton™ coating under consideration will adhere sufficiently to plywood, and if it does not, then the entire apparatus may need to be fabricated from sheet metal with all internal surfaces painted with Viton™.

3 Protocol for future experimentations

3.1 Experimental resources and characterization of the mixing chamber

Several instruments will be used during the characterization of the chamber and concentration measurement. During both phases of the experimental measurements, the temperature, pressure and humidity of the ambient air will be monitored.

For the characterization of the chamber, the first step will be verifying that the flow entering the test section is uniform and fully mixed through visualization. The working section will be equipped with LED lights and a ROSCO 1500 fogger machine will provide the fog. The fog will be introduce downstream of the fan, at the same point planned for injection of the gases. Video recordings of the fog in the working section will be made with several velocities. After visualization experiments, velocity measurements will be made with a 2D-LDV (Laser Doppler Velocimetry) system using an Argon Ion 300 mW air-cooled laser. This single point technique is non-intrusive and allows simultaneous measurement of two components of the velocity with a good spatial and temporal resolution. Finally, the concentration measurements will be made using an Aurora Scientific mini-PID (photo ionization detectors) The PID measurement method relies on photoionization of a volatile organic compound (in this case a propylene tracer). Small access points will have to be made in the test section for the PID. In the characterization phase, a simulant gas mixture (using sulfur hexafluoride) with be used to reproduce the molecular weight of chlorine. Sulfur hexafluoride is non-toxic, non-flammable, non-reactive and can be used in the PID system.

In conclusion, the characterization of the mixing chamber will assess the quality of the mixing and the flow in the working section, the steps are listed below :

1. Visualization of the flow going in the test section.
2. Velocity profiles taken in the test section will ensure the uniformity of the flow and that the turbulence intensity level is respected.
3. Concentration profiles will provide information on the uniformity of the concentration of the simulant gases in the chamber.

3.2 Concentration measurements and evaluation of the reactivity of chlorine with different samples

After the characterization of the working section, the apparatus will be modified to undergo tests with high concentration of chlorine. The changes have been described in section 2.3. Furthermore, the tunnel will have to allow the access of the JAZ sensors in order to measure the concentration of chlorine. As with characterization of the chamber, the evaluation of the reactivity of chlorine with different samples will request several steps listed hereunder.

1. Characterization of the response of the apparatus to chlorine : Tests will be done without samples, so that the chlorine concentration decay due to the apparatus itself can be measured. A correction factor will be calculated and implemented in the following concentration measurements. Multiple tests will be done, in order to ensure of the repeatability of the measurements.
2. Removal rate estimation : Samples will be placed in the tray, allowing the removal rate of these samples to be estimated through concentration measurements.
3. Analysis of the results : The measurement of the concentration decay in the chamber and the study of the samples put in contact with chlorine will provide valid information in order to calculate the removal rate of chlorine for these different samples. Analytical techniques for determining the amount of chlorine uptake by the test material will depend on the material.

4 Conclusion

In this article, the design of an innovative apparatus or mixing chamber and its main elements have been presented in details. One of the main difficulties encountered was to scale the device to fit under a six feet long fume hood to limit the hazards involved with the use of chlorine at high concentration level in the laboratory. One other obstacle was to find materials and coating suitable and able to resist chlorine in the long run. The future means and procedures of characterization and measurement have been introduced as well. Once build and characterized, this apparatus will provide new data sets for chlorine deposition on vegetal samples and small objects in conditions similar to the atmospheric environment. In the end, this study aim at the improvement of the understanding and computational modeling of toxic inhalation hazards and the consequences to affected population and infrastructure.

5 Aknowledgment

This project is supported by the Chemical Security Analysis Center (CSAC) of the U.S. Department of Homeland Security and Transport Canada. The authors wishes to thank Katherine Elmore for her assistance with this project.

6 References

[1] M. B. Dillon, "The role of deposition in limiting the hazard extent of dense-gas plumes," *J. Hazard. Mater.*, vol. 164, no. 2–3, pp. 1293–1303, 2009.

[2] S. Hanna, S. Dharmavaram, I. Zhang, J., Sykes, H. Witlox, S. Khajehnajafi, and K. Koslan, "Comparison of Six Widely-Used Dense Gas Dispersion Models for Three Actual Railcar Accident," in *Process Safety Progress*, 2008, vol. 27, no. 3, p. 248–259.

[3] S. P. Arya, *Air pollution meteorology and dispersion*, 6th ed. 1999.

[4] J. P. Dimbour, A. Dandrieux, and G. Dusserre, "Reduction of chlorine concentrations by using a greenbelt," *J. Loss Prev. Process Ind.*, vol. 15, no. 5, pp. 329–334, 2002.

[5] G. Dusserre and J. Ollivier, "Small scale field experiments of chlorine dispersion," *J. Loss Prev. Process Ind.*, vol. 15, pp. 5–10, 2002.

[6] J. D. Hearn, R. Weber, R. Nichols, M. V. Henley, and S. Fox, "Deposition of Cl_2 on soils during outdoor releases," *J. Hazard. Mater.*, vol. 252, pp. 107–114, 2013.

[7] A. C. Hill, "Vegetation: A Sink for Atmospheric Pollutants," *J. Air Pollut. Control Assoc.*, vol. 21, no. 6, pp. 341–346, Jun. 1971.

[8] J. Hearn, J. Eichler, C. Hare, and M. Henley, "Cl_2 deposition on soil matrices," *J. Hazard. Mater.*, vol. 237–238, pp. 307–314, 2012.

[9] J. Hearn, J. Eichler, C. Hare, and M. Henley, "Effect of soil moisture on chlorine deposition," *J. Hazard. Mater.*, vol. 267, pp. 81–87, 2014.

[10] D. Brown, "The removal of toxic gases from the air by water, plants and soils.," 2006.

[11] M. L. Wesely and B. B. Hicks, "A review of the current status of knowledge on dry deposition," *Atmos. Environ.*, vol. 34, pp. 2261–2282, 2000.

[12] J. R. Couper, W. R. Penney, and J. R. Fair, *Chemical process equipment revised 2E: selection and design.* Gulf Professional Publishing, 2009.

[13] K. J. Myers, A. Bakker, and D. Ryan, "Avoid agitation by selecting static mixers," *Chem. Eng. Prog.*, vol. 93, no. 6, pp. 28–38, 1997.

[14] W. J. Gretta and C. R. Smith, "The Flow Structure and Statistics of a Passive Mixing Tab," *J. Fluids Eng.*, vol. 115, no. 2, p. 255, 1993.

[15] R. H. Perry, D. W. Green, and J. O. Maloney, *Perry's chemical engineers' handbook.* McGraw-Hill, 2008.

[16] Timet corporation, "Corrosion resistance of titanium," 1997.

Fire Proofing Cost Reduction Measures for LNG Plant

Raminaidu Girada
Senior Process Safety Engineer, Bechtel Oil, Gas & Chemicals, Inc.
3000 Post Oak Boulevard, Houston, TX 77056
rgirada@bechtel.com

Jegan Babu Arumugakan Thuraiswamy
Senior Process Safety Engineer, Bechtel Oil, Gas & Chemicals, Inc.
3000 Post Oak Boulevard, Houston, TX 77056
jbarumug@bechtel.com

Nino Nicotra
Process Safety Engineering Chief, Bechtel Oil, Gas & Chemicals, Inc.
3000 Post Oak Boulevard, Houston, TX 77056
anicotra@bechtel.com

Prepared for Presentation at
American Institute of Chemical Engineers
2017 Spring Meeting and 13th Global Congress on Process Safety
San Antonio, TX
March 26 – 29, 2017

Fire Proofing Cost Reduction Measures for LNG Plant

Raminaidu Girada
Senior Process Safety Engineer, Bechtel Oil, Gas & Chemicals, Inc.
3000 Post Oak Boulevard, Houston, TX 77056
rgirada@bechtel.com

Jegan Babu Arumugakan Thuraiswamy
Bechtel Oil, Gas & Chemicals, Inc.
3000 Post Oak Boulevard, Houston, TX 77056
jbarumug@bechtel.com

Nino Nicotra
Process Safety Engineering Chief, Bechtel Oil, Gas & Chemicals, Inc.
3000 Post Oak Boulevard, Houston, TX 77056
anicotra@bechtel.com

Keywords: Consequence Modeling, Liquefied Natural Gas (LNG) & Liquefied Petroleum Gas (LPG) and Safe Design

Abstract

The current industry practice in designing fireproofing requirements for LNG facilities follows the traditional approach of "Liquid Pool Fire". The common industry practice for fire proofing design criteria defines a fire potential equipment (FPE) as the equipment that handles flammable liquid above a specific threshold volume. The FPE generates fire scenario envelope whose dimension ranges are recommended by API 2218. Equipment and structural supports that falls inside the fire envelope are considered for fire proofing.

LNG plants are handling products with very low boiling point, and any leak to atmosphere results in an instant flashing into vapor phase or aerosols. The results from our recent works on pool fire consequence modelling has proved that most of the currently considered fire potential equipment seldom create pool fire. This consequence modeling results clearly demonstrates that the current fire proofing design approach is overdesign, that is not effective. The apparent saving in man-hours for not modeling the liquid release will result in a considerable amount of engineering efforts and material cost being spent on the design and installation of fire proofing for equipment and structure with potentially no real benefits to the inherent safety of the process plant. Considering the current economic trend, a change in approach using consequence modelling will effectively result in considerable cost savings in major LNG projects.

1 Introduction

Currently in LNG projects, fireproofing is considered for equipment supports, pipe rack support, structural supports and other auxiliary critical hydrocarbon carrying pipe supports within a fire proofing zone to improve its structural integrity during a fire. The principal value of fire proofing is realized during the early stages of a fire by providing additional time to fire fighters to control or extinguish a fire thus by delaying additional loss of containment which may escalate the incident. The fireproofing requirement is determined by identifying the fire potential equipment and the associated fireproofing zones. Fire potential equipment is defined as an equipment holding flammable liquid above a specific threshold volume, which will have the potential to create a liquid pool and subsequent pool fire if ignited. Per API 2218, the fireproofing zone is defined for every fire potential equipment, that extends 20 ft. to 40 ft. (6 m to 12 m) horizontally and 20 ft. to 40 ft. (6 m to 12 m) vertically from the source of liquid fuel. As a typical approach most projects consider fireproofing zones that extend 30 ft. (9m) horizontally and 30 ft. (9m) vertically from the source.

The above mentioned qualitative approach does not reflect the realistic possibility of pool fire scenario. Therefore, a quantitative analysis may be performed to analyze whether a pool fire is realistic for equipment handling multiple refrigerants and LNG. This analysis has been performed on some process equipment handling multiple refrigerants and LNG using Canary software by Quest Consultants.

2 General Technical Guidance

2.1 Scope of Fire Proofing

Fire scenario envelopes for an LNG plant are based on pool fire scenarios. Jet fires are typically not considered as credible for defining fire scenario envelops for fireproofing as described below:

- Jet fires are handled by Prevention, Intervention and Protection. Prevention is by designing the facility in compliance with general fire prevention activities such as those included in NFPA 59A and in the broad application of Process Safety Management(PSM). Once a pressurized jet fire occurs, the intervention and protection can be achieved by a high flow rate firewater stream directed on the point of impingement. This has been proved as an effective mitigation measure in preventing escalation with Jet fires. This will be achieved by strategically locating the fire water monitors around high risk process area, where jet fire is foreseen as a credible scenario.

- Emergency isolation and depressurization system will be provided to lower the pressure of specific isolated sections of the plant that may be involved in a localized jet fire scenario and to de-inventory of gas to flare.

- Efforts on similar LNG projects to extend the fire proofing systems for jet fire scenarios led to enormous impractical level of fireproofing requirement. This could also introduce other hazards such as corrosion.

- Jet fire represents a significant risk or concern for personnel life, working on offshore installations whereas property protection is the basis of pool fire for on-shore LNG facility or petroleum refinery or petrochemical facilities.

2.2 Pool Fire Formations

A pool fire may result via a number of scenarios and it begins typically with the release of flammable material from process equipment. If the material is liquid, stored at a temperature below its normal boiling point, the liquid will collect in a pool. The geometry of the pool is dictated by the surroundings (i.e., diking), but an unconstrained pool in an open, flat area is possible, particularly if the liquid quantity spilled is inadequate to completely fill the diked area. If the liquid is stored under pressure above its normal boiling point, then a fraction of the liquid will flash into vapor, with un-flashed liquid remaining to form a pool in the vicinity of the release.

The pool of flammable liquid may be ignited creating a pool fire. If the release of flammable material from a pressurized process equipment continues, then a jet fire is also likely. If the ignition source occurs at the very beginning of the release, then inadequate time is available for the liquid to form a pool and only a jet fire will result.

Pressurized liquefied gases such as LNG, methane, ethylene and propane will in general disperse as mists or vapor, but may possibly form a small pool if the source operates at low pressure conditions. An exception has to be made for butanes, pentanes and heavier hydrocarbons, which always create pool.

Refer to the following **Figure 1** for pictorial view of the pool fire formation.

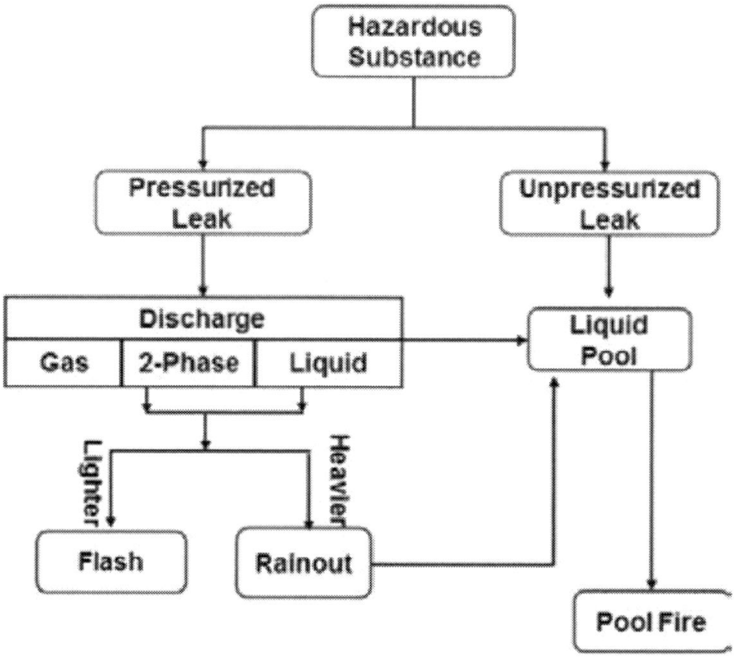

Figure 1: Pool Fire Formation

2.3 Determination of Pool Fire scenario

This section provides an overview of the methodology for determining pool fire existence.

In case of accidental release of flammable materials such as Pentanes and heavier hydrocarbons create liquid pool irrespective of their operating conditions. However, the lighter hydrocarbons such as butane and propane may create liquid pool only under specific operating conditions. The potential for creating liquid pool is based on certain parameters as described below:

- **Butanes**

  Butane releases potentially create liquid pool except for high pressure releases and slight sub cooling of release. Further consequence analysis is required to check if a liquid pool is created from the release.

- **Propane and Ethylene**

  Ethylene and propane liquids may or may not pool, depending on conditions. It has been found that a pool formation can be correlated on several parameters: (1) vessel operating pressure, (2) size of rupture hole for self-leakage vessel, (3) the amount of subcooling, which is also important but the assumption is the liquid will be slightly subcooled, and (4) confinement of *fire hazard area*. Together, these variables establish the size of the pool.

  Higher pressure release are more likely to vaporize and mist, and low pressure releases are more likely to pool. Based on a slightly subcooled liquid at pressure, the analysis is divided into three regimes (1) no liquid pools, (2) highly likely to form a significant pool fire, and (3) possible pool fire. Depending on the size of the release, slope, temperature and composition, a consequence model may be necessary to determine if a pool fire is warranted.

- **LNG**

  Similar to propane and ethylene liquids, LNG may or may not pool depending on several parameters: (1) vessel operating pressure, (2) size of rupture hole, (3) the amount of subcooling, and (4) confinement.

Therefore, a quantitative approach was performed in the following section.

3 Pool Fire Case Study

In this case study, the following light flammable fluids are modelled: methane, propane, ethylene and LNG. Based on the current fire proofing design approach a total of 28 equipment of the LNG plant that handles the above fluids are qualified as Fire Potential Equipment. All those leakage scenarios are modelled using Canary modelling tool, based on its operating condition.

3.1 Leak Scenarios

To provide a spill rate using CANARY, one has to postulate a leak scenario.

Typical hole size choices range from small (3-10mm) to full bore ruptures (diameter of pipe). Hole size selection can define the general level of risk (larger, high consequence releases are significantly less frequent than small, low consequence release).

CCPS book of "Guideline for Fire Protection in Chemical, Petrochemical, and Hydrocarbon Processing Facilities" defined hole sizes as small (3-10 mm), medium (10-50 mm), large (50-150 mm), and full-bore ruptures. Discussion around these hole sizes also points out that the large and rupture releases are not always the worst-case events due to release duration and dosage effects that should be discounted for.

The Canadian Standard Association (CSA) standards Z276 specifies design spills which define the potential extent of hazard zones that are not allowed to impact the public. Z276 defines, among other (smaller) release scenarios, that the design spill should be "The maximum flow of LNG that could be spilled, from the largest branch connection, 60 mm (2-inch) or smaller..."

Based on the above references, 2-inch is considered as an appropriate hole size for consequence modeling. Therefore, the pool fire modeling has been performed for 2-inch hole size. However, sensitivity analysis has been done considering 6-inch hole to check if larger hole size can result in pool formation. The modeling results for both 2-inch and 6-inch are presented below.

The release elevation was considered to be constant at 1 m height (the lowest height is deemed to result in higher consequences). The release directions in all scenarios were assumed to be horizontal. The consequence modeling input data shown below were used for the calculations. All scenarios were modeled using CANARY software developed by Quest. The software is widely used in industrial consequence modelling studies.

3.2 Atmospheric Conditions

The following more stringent atmospheric conditions that favors/sustains the pool scenario are considered:

Wind Speed: 2 m/s

Stability Class: F

Ambient temperature: 80 °F

3.3 Liquid Methane Release at High Operating Pressure

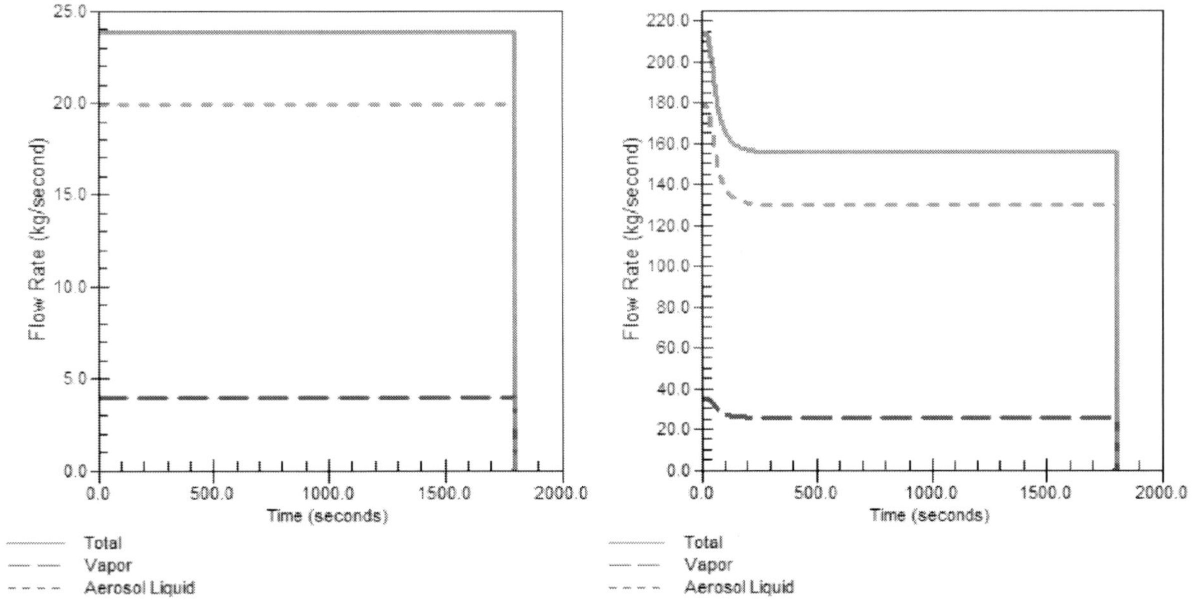

Figure 2: Liquid Methane Release – 2-inch (left figure) & 6-inch (right figure)

Consequence modeling for a liquid methane release was performed at specific process conditions for 2-inch and 6-inch hole size. As shown in the above figures, most of the release is flashing into aerosol liquid and remaining release is in vapor phase. No liquid pool is generated from the liquid methane release.

3.4 Liquid Propane Release

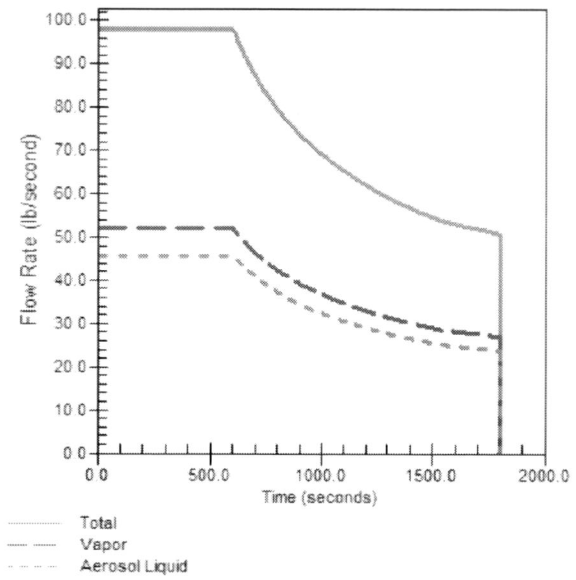

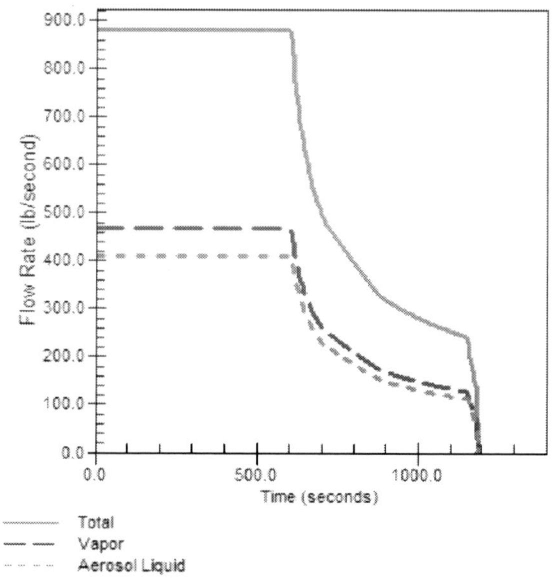

Figure 3: Propane Release – 2-inch (left figure) & 6-inch (right figure)

Consequence modeling for a propane release was performed at specific process conditions for 2-inch and 6-inch hole size. As shown in the above figures, the release is flashing into aerosol liquid and vapor phase. No liquid pool is generated from the propane release.

3.5 Liquid Ethylene Release

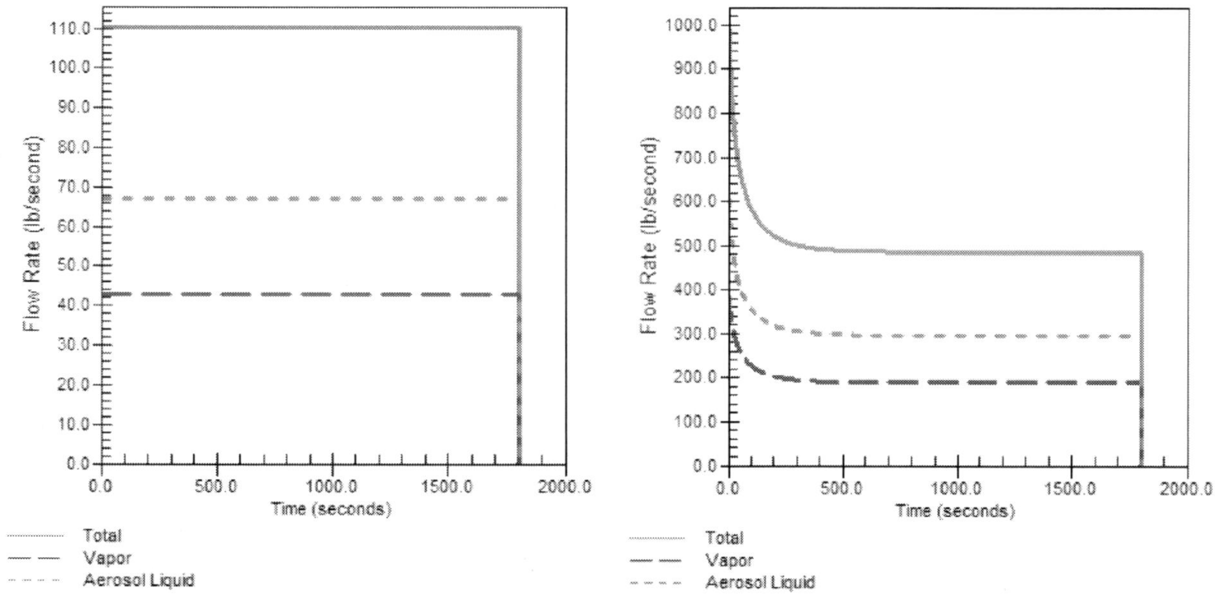

Figure 4: Ethylene Release – 2-inch (left figure) & 6-inch (right figure)

Consequence modeling for ethylene release was performed at specific operating process conditions for 2-inch and 6-inch hole size. As shown in the above figures, the release is flashing into aerosol liquid and vapor phase. No liquid pool is generated from the refrigerant release i.e. Ethylene release.

3.6 LNG Release at Low Operating Pressure condition

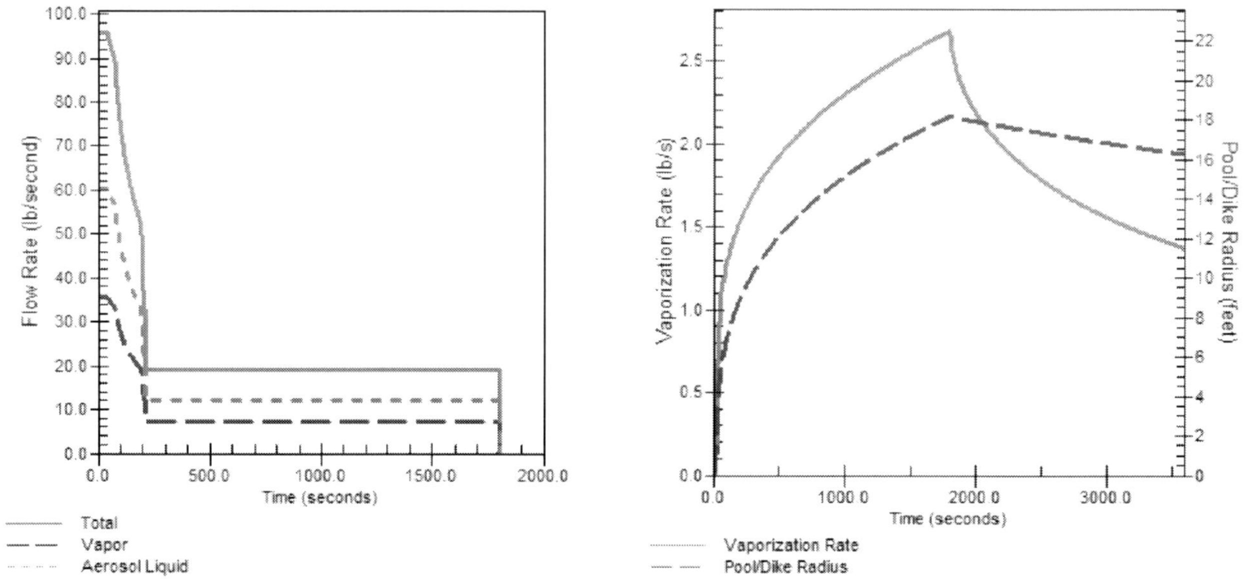

Figure 5: Liquefied Natural Gas Release – 2-inch (left figure) & 6-inch (right figure)

Consequence modeling for a liquefied methane was performed at the process conditions of an equipment for 2-inch. and 6-inch hole size. As shown in the above left side figure, the release for 2-inch hole size is flashing into aerosol liquid and vapor phase, without creating any pool. However, as shown in the right side figure, a release from 6in. hole size from the same equipment is creating a liquid pool of 16 ft. radius.

4 Conclusions

Consequence modeling results showed that out of the total twenty-eight (28) equipment of LNG liquefaction plant that were normally considered as Fire Potential Equipment (FPE) based on specific threshold volume, thirteen (13) equipment that handles methane, propane, ethylene at some specific operating conditions (e.g. low or medium or high pressure) did not create any liquid pool. Instead they flash along forming a large fraction of aerosol liquids, which never creates liquid pool.

The pool fire modeling was performed for a 2-inch release and a sensitivity analysis was carried out for 6-inch release to check the possibility of liquid pool formation for larger size releases for all the 28 equipment. The 6-inch release did not make a major difference in pool fire modeling results except for one equipment which contain LNG at low operating pressure condition. As stated earlier, thirteen (13) out of twenty-eight (28) equipment had not produced a liquid pool irrespective of the release size.

One (1) out of the remaining fifteen (15) equipment, that handles LNG at lower pressure condition did not generate liquid pool at 2-inch release, however it has generated liquid pool for a larger size (6-inch) release. Therefore, it is recommended to perform the consequence modeling for larger hole size to check the liquid pool formation.

The consequence modeling proved that 46% of the total equipment containing liquid such as methane or propane or ethylene or LNG had not produced a liquid pool. Therefore, considering the current economic trend, consequence modeling based approach for determining fireproofing requirements in conjunction with API 2218 requirement can be used in major LNG projects to justify considerable cost savings.

5 References

[1] J.D. Marx and A. Nicotra, "Is a Two-Inch Hole Adequate for Siting Study?", 12th Global Congress on Process Safety, American Institute of Chemical Engineers, Houston, Texas, 2016.

[2] API. "Fireproofing Practices in Petroleum and Petrochemical Processing Plants", American Petroleum Institute, 2013.

[3] CCPS (2003), "Guideline for Fire Protection in Chemical, Petrochemical, and Hydrocarbon Processing Facilities", Center for Chemical Process Safety of the American Institute of Chemical Engineers, 3 Park Avenue, New York, 10016-5991. (ISBN 0-8169-0898-2).

[4] CAN/CSA Z276, "Liquefied natural gas (LNG) - Production, Storage, and Handling", National Standard of Canada, 2015.

A Case Study in Human Reliability - Analysis to Support LOPA in Unloading Hazardous Liquids

Delmar "Trey" Morrison, Ph.D., P.E.
Principal Engineer, Exponent, Inc.
4580 Weaver Parkway, Suite 100, Warrenville, IL 60555, USA
tmorrison@exponent.com

Chris Aiken
Worldwide Process Safety Leader
Global Plant Operations, Cargill
Chris_aiken@Cargill.com

Sunil D. Lakhiani, Ph.D., P.E.
Exponent, Inc.
slakhiani@exponent.com

Pim van der Graaf
Global Manufacturing Technology Leader Starch & Maltodextrin
Cargill Starches, Sweeteners & Texturizers
Pim_van_der_graaf@cargill.com

Prepared for Presentation at
American Institute of Chemical Engineers
2017 Spring Meeting and 13th Global Congress on Process Safety
San Antonio, TX
March 26 – 29, 2017

A Case Study in Human Reliability - Analysis to Support LOPA in Unloading Hazardous Liquids

Delmar "Trey" Morrison, Ph.D., P.E.
Principal Engineer, Exponent, Inc.
4580 Weaver Parkway, Suite 100, Warrenville, IL 60555, USA
tmorrison@exponent.com

Chris Aiken
Worldwide Process Safety Leader
Global Plant Operations, Cargill
Chris_aiken@Cargill.com

Sunil D. Lakhiani, Ph.D., P.E.
Managing Engineer, Exponent, Inc.

Pim van der Graaf
Global Manufacturing Technology Leader Starch & Maltodextrin
Cargill Starches, Sweeteners & Texturizers

Keywords: Human Error Identification, Human Error Analysis, Human Reliability Analysis, Layer of Protection Analysis,

Abstract

Several of Cargill's operations involve transfer of liquids to or from transportation containers, such as portable tanks or truck tankers. Some liquids have benign characteristics while others have toxic or flammable properties. The risk of loss of containment (LOC) during transfer operations is typically evaluated through qualitative process hazard analysis (PHA) techniques. PHAs can be supplemented with Layer of Protection Analysis (LOPA) if the consequence or perceived risks are higher than an acceptable threshold. LOPA is a rule-based semi-quantitative risk analysis technique. A circumspective review of LOPA studies for unloading certain hazardous materials across Cargill revealed that although the final risks were a similar order of magnitude, the approaches to calculating these risks differed between LOPA studies. The initiating event frequency (IEF) was the primary parameter that varied among the studies. It was found that differing assumptions and different underlying failure rates were applied among the studies to estimate the IEF. The IEF varied by as much as three orders of magnitude for seemingly identical processes. This paper discusses the authors' approach to understanding and addressing this inconsistency between LOPA studies. The authors used Human Reliability Analysis (HRA) with the Human Error Assessment and Reduction Technique (HEART) method to develop uniform IEFs for LOPA. In addition, the HRA improved the organization's understanding of the potential initiating events, the critical procedural steps in the prevention of the scenario, and the most likely error producing conditions to cause operator error. The model for application of HRA to unloading procedures is described along with the assumptions and limitations of the approach.

1 Introduction

Cargill is a global organization providing food, agriculture, financial and industrial products and services. Cargill operates approximately 1,500 facilities in 70 countries with a multitude of processes. Many processes either produce or utilize hazardous liquids that must be loaded or unloaded via portable tank containers, tanker trucks, or rail tank cars. The risk of loss of containment (LOC) during these transfer operations is usually evaluated by qualitative process hazard analysis (PHA) techniques. In certain cases, the severity of the consequence or risk identified by a qualitative PHA may require further analysis. Layer of Protection Analysis (LOPA) may then be applied to develop an order of magnitude estimate for the risk posed by the scenario(s). LOPA is a rule-based semi-quantitative risk analysis technique that approximates the necessary elements of an incident scenario through a frequency or probability of failure for each element.

A circumspective review of LOPA studies for unloading certain hazardous materials across the global organization revealed that although the mitigated risks were a similar order of magnitude, the approaches to calculating these risks differed between LOPA studies. The initiating event frequency (IEF) was the primary parameter that varied among the studies. The authors found that differing assumptions and different underlying failure rates were applied among the studies to estimate the IEF. The IEF varied by as much as three orders of magnitude for seemingly identical processes. This paper discusses an approach to understanding and addressing this inconsistency between LOPA studies.

LOPA was originally developed as a simplified risk assessment tool, and it relies heavily upon standardized rule sets [1]. It is common for LOPA study teams to apply the representative failure rates, error probabilities, and probabilities of failure on demand from industry references in LOPA studies in lieu of undertaking independent development of the values. The authors found that the variability in the IEF values used in the example cases discussed below arose from a lack of solid rules upon which to assign the IEF for operator error in truck pull-away incidents, i.e., where the driver moves the truck with one or more hoses still connected to the container. The "operator" involved in specific scenarios may have been one of multiple plant operators involved in the transfer and/or the truck driver. To address this variability, the authors undertook a Human Reliability Analysis (HRA) to develop a consistent IEF basis to evaluate pull-away incidents for these specific processes. Two specific chemical loading/unloading operations were analyzed:

1. Unloading of pressurized liquefied flammable and toxic liquid into a stationary tank, and
2. Replacement of storage totes of a volatile reactive and toxic chemical connected to a process

A team consisting of process safety personnel and operations staff from Cargill and experts in PHA and human factors used a HRA framework to quantify the failure rates for critical steps in the procedures that required human actions or verifications using the 'Human Error Assessment and Reduction Technique' (HEART). The failure probabilities were then introduced into an event tree diagram to estimate the probability of a driver pull-away scenario in the two procedures described in the case studies.

The team reviewed the steps in the procedures and identified steps that were critical to the prevention of an inadvertent driver pull-away while the hoses are still connected to the tanker or portable container (i.e., a "tote"). Various assumptions and conditions were discussed to justify the failure rates and the effect of error-producing conditions on generic human unreliability conditions applicable to the critical steps.

2 Layer of Protection Analysis

LOPA is a semi-quantitative risk analysis technique that is used to augment qualitative PHAs. PHA study teams use their judgment, process safety information, consequence analysis results, and knowledge of past or similar incidents to qualitatively assign the risk to individual hazard scenarios. However, due to the nature of PHAs, the qualitatively-assigned risk level may not accurately represent the risk posed by the hazard scenario. Where the consequence for given scenario reaches a pre-determined severity, e.g., multiple fatalities, then companies often use LOPA to systematically evaluate the scenario in a more rigorous fashion.

The team conducting the PHA may not be the same team that is conducting the LOPA, although some personnel may participate in both analyses, and the LOPA may not be performed concurrently with the PHA. The LOPA team is a multi-disciplined group with skills required to analyze a given scenario more quantitatively. The LOPA team is expected to apply data from the manufacturing technology group, general industry and the specific site to identify the initiating events, enabling events, conditional modifiers, and independent protection layers (IPLs) for each scenario. These terms are defined in the Appendix A for the reader's reference.

LOPA is a simplification of fault tree/event tree analysis that allows the team to relatively quickly identify and evaluate the primary elements of an accident scenario and its preventative safeguards. The LOPA team applies order of magnitude estimates for the frequency of the initiating event and the probability of enabling events or conditional modifiers. The product of these values gives the intermediate value, unmitigated event frequency (UEF), which is the event frequency without applying the effects of IPLs. The probability of failure on demand (PFD) is assigned for IPLs. The product of the UEF and the PFDs for the IPLS is the mitigated event frequency (MEF) for the scenario. The MEF is then compared against the tolerable mitigated event frequency for the company (TMEF). If the MEF is greater than the TMEF, then the company requires further evaluation to reduce the MEF to an acceptable level by identification of additional IPLs, by modification of the process to reduce the IEF, or by modification of IPLs to reduce the PFD. In the examples below, the comparison stops at the UEF to illustrate the effects of assumptions applied to develop the IEF.

LOPA is well-suited to hazard scenarios caused by simple equipment failure, a single control loop failure, or field operator error. However, evaluating more complex scenarios, where multiple chances for recovery by the operator to prevent the accident exist, poses challenges for the technique. For example, if the procedure requires several steps and a subsequent step will naturally reveal the error in a prior step, then the operator must make at least two errors for the scenario to occur. One example that Cargill identified where LOPA proved to yield inconsistent results was the evaluation of potential accidents during hazardous liquid unloading operations. Loading and unloading activities are hands-on with a high level of human interaction to connect hoses, verify connections, open/close valves, actuate pumps or pressure transfer, and movement and monitoring of the truck or tank. The most common type of incident involving such transfer operations is a pull-away event where the vehicle or portable tank moves while still connected to the process. In such incidents, the transfer hose may be compromised leading to a catastrophic release of the hazardous chemical.

The LOPA team's definition of the "operator error" in the LOPA analysis may encompass just one step that leads to pull-away or it may be applied to a collection of steps. Regardless, the LOPA process assigns a frequency value (likelihood) to that initiating event (called IEF). Guidance values are provided for operator error frequencies in CCPS's Guidelines for Initiating Events and Independent Protection Layers in Layer of Protection Analysis [2]. The text defines a value for "human error during a routine task that is performed once per week or more often" as one error per year [pg. 68]. It defines a value for "human error for a task that is performed between once per month and once per week" as 0.1 error per year [pg. 70]. These general values are a reasonable starting point for the IEF in a loading operation, but Cargill found that this simplification did not adequately describe the frequencies of various human errors during the execution of complex procedure-driven unloading operations.

In a review of LOPAs for the same unloading process at different locations internationally, a manufacturing technology leader identified inconsistent application of IEF assumptions in the LOPAs. The LOPAs were facilitated by different third party professionals, and the net result of using different sets of assumptions were that the IEFs in two different studies on a similar unloading operation differed by three orders of magnitude.

The first team applied general values aligned with the table on page 71 in reference [1] and used an IEF of 10^{-2}/unloading operation. However, the team inappropriately applied a time-at-risk enabling condition modifier that reduced the overall event frequency. Table 1 is an example of the LOPA calculation illustrating the use of the time-at-risk enabling condition modifier. CCPS's Guidelines for Enabling Conditions and Conditional Modifiers addresses the use of time-at-risk conditional modifier in Chapter 2 and Appendix B [3]. The cause of the hose failure in the scenarios is a direct result of the operator making an error while executing the unloading procedure. Since the error is directly tied to the instance of unloading, the probability of time-at-risk is unity.

However, the removal of the modifier from the LOPA calculation resulted in a frequency of the initiating event to be significantly higher than the site or similar processes at sister facilities had experienced. The initiating event for the LOPA scenario required several errors in executing the procedure; thus, it became apparent that utilizing the general failure data for operator error was not sufficient to properly represent the IEF for this scenario.

Table 1. LOPA Scenario Calculation from Team 1.

Scenario: Driver pulls away during unloading with liquid hose still connected due to an error following unloading procedure. This results in a release of HazChem A from the stationary storage tank with a toxic consequence potential.			
	IEF (probability per opportunity)	0.01	
	Opportunities per year	120	
	IEF (per year)		1.2
Enabling Events or Conditional Modifiers	Probability of personnel present	1	
	Time-at-Risk: Scenario only valid during unloading, which occurs less than 10% of the time	0.1	
	Probability of fatality	1	
	UEF (per year)		0.12

The second team used data published in reference [3] (pg. 11.6-11.7) to estimate the IEF. The reference provides the failure rate for hoses as 1.3×10^{-5} per transfer based on UK HSD data from a 5 year period of hose and arm loading failures. The failure rate was then multiplied by the percentage of failures resulting from pull away events yielding the IEF in Table 2. There are several issues with using this approach to estimate the IEF but the most significant issue was that the frequency is mitigated (i.e. includes safeguards) and the team applied this frequency to the LOPA IEF which is unmitigated. The LOPA calculation is show in Table 2.

Table 2. LOPA Scenario Calculation from Team 2.

Scenario: Driver pulls away during unloading with liquid hose still connected due to an error following unloading procedure. This results in a release of HazChem A from the stationary storage tank with a toxic consequence potential.			
	IEF (probability per opportunity)	5.4×10^{-5}	
	Opportunities per year	100	
	IEF (per year)		5.4×10^{-3}
Enabling Events or Conditional Modifiers	Probability of personnel present	1	
	Probability of fatality	1	
	Unmitigated Event Frequency (per year)		1.6×10^{-3}

Upon further review of these two scenarios, it became evident that estimating IEF of the unloading procedures required a different analytical approach. The authors undertook an analysis of the potential operator errors during the unloading procedures to develop an IEF value for the LOPA scenario.

3 Background on Human Reliability Analysis

Human error has been deemed as a major contributor to process safety incidents resulting in fatalities, injuries and property damage [5]. Such errors are a result of actions or inactions by people handling the operations within a system such that it causes deviation in the normal functioning of the system, hence increasing the risk of failure within the system resulting in a negative consequence. Heinrich analyzed 75,000 industrial accidents and concluded that 88 percent of all industrial accidents were "caused primarily by the unsafe acts of persons" (p. 21) [6]. In assessing the risk imposed on a system or a subsystem due to the likelihood of people making errors, it is important to understand the roles and actions of people involved in running the system as well as the likelihood of human errors. Human reliability is defined as "the probability that a person (1) correctly performs some system-required activity in a required time period (if time is a limiting factor) and (2) performs no extraneous activity that can degrade the system" (p. 2-3) [7].

Human Reliability Analysis or Assessment (HRA) is a method to identify the human actions that can have an effect on the successful performance of the system and estimate their reliability. It comprises different quantitative and qualitative tools that "assess the human contribution to risk" (p. 1) [8]. A literature review conducted by the U.K. Health and Safety Executive (HSE) identified 72 potential human reliability tools and acronyms of which 35 HRA tools were reviewed and 17 of those (listed in Appendix B) were considered to be of use to the major hazards sector [8]. The HRA specialists summarized each of the 17 HRA tools based on the published literature and, in selected cases, communications with the authors of the tools and presented their opinions on the strengths and weaknesses of relating to different characteristics such as: potential application, availability, applicability, ease of use, validity, etc. A similar qualitative assessment was conducted by Kirwan [9] of seven main HRA techniques where they were qualitatively rated on six criteria: accuracy, validity, usefulness, effective use of resources, acceptability and maturity. Table 3 shows a summary of evaluation of the seven techniques identified in Kirwan.

While most of these shortlisted techniques have a reasonable degree of validity, the Human Error Assessment and Reduction Technique (HEART) was found to be the most acceptable and useful technique as it concentrates on the human factors that have a significant effect on human performance (see pg. 853 of Chapter 32 in [9]).

Table 3. Summary of evaluation techniques [9].

Technique	Accuracy	Validity	Usefulness	Effective use of resources*	Acceptability	Maturity
APJ	Moderate	Moderate/high	Moderate/high	Moderate	Moderate	High
PC	Moderate	Moderate	Low/Moderate	Low/Moderate	Moderate/high	Moderate
TESEO	(Low)**	Low	Moderate/high	High	Low	Low
THERP	Moderate	Moderate	Moderate	Low/Moderate	High	High
HEART	Moderate	Moderate	High	High	Moderate/high	Moderate
IDA	(Low)	Moderate	Moderate/high	Low/Moderate	(Moderate)	Low/Moderate
SLIM	Moderate	Moderate	High	Low/Moderate	Moderate/high	Moderate/high
HCR	(Low)	Low	Low/Moderate	Moderate	Low/Moderate	Low

*A rating of high on this criterion means the technique is favorable with respect to effective use of resources (i.e., resource requirements are low), and vice versa.
**Ratings in parentheses were based on the subjective opinions of the authors and contributions to the document, since insufficient empirical evidence was available to justify a rating from applications alone.
Note: This table was constructed prior to the experimental study assessing accuracy (Kirwan, 1988).

3.1 HEART

The HEART technique is one such method that was proposed in the 1980s (Williams, 1986, 1988, 1992) [10]-[12]. The approach is a valid [8] and simple approach that is aimed at assessing critical steps and determining the likelihood of failure of each step using the following methodology:

1. Identify the most appropriate task description (from a list of generic tasks) for the step to be analyzed. The generic tasks, listed in Table 4, have proposed nominal human error probability (HEP) values associated with them.

2. Identify all the Error Producing Conditions (EPCs) associated with each critical task. Each EPC has a modifier (or effect) associated with it. The 38 EPCs and corresponding modifiers/effect identified in the HEART methodology are listed in Appendix C.

3. Assess the effect proportion (or state) of the EPCs (through team discussion) and assign a value ranging between 0 and 1.

4. Calculate the final HEPs for the task using the equation shown below.

$$Final\ HEP = Nominal\ HEP \times \prod_{i=1}^{n}\left[\left(Effect_{EPC\_i} - 1\right) \times State_{EFC\_i} + 1\right]$$

Table 4. HEART's generic tasks and corresponding HEPs [11].

Generic tasks	Nominal HEP	$5^{th} - 95^{th}$ percentile HEPs
(A) Totally unfamiliar, performed at speed with no real idea of likely consequences	0.55	0.35 – 0.97
(B) Shift or restore system to a new or original state on a single attempt without supervision or procedures	0.26	0.14 – 0.42
(C) Complex task requiring high level of comprehension and skill	0.16	0.12 – 0.28
(D) Fairly simple task performed rapidly or given scant attention	0.09	0.06 – 0.13
(E) Routine, highly-practiced, rapid task involving relatively low level of skill	0.02	$7\times10^{-3} - 4.5\times10^{-2}$
(F) Restore or shift a system to original or new state following procedures, with some checking	0.003	$8\times10^{-4} - 7\times10^{-3}$
(G) Completely familiar, well-designed, highly practiced, routine task occurring several times per hour, performed to highest possible standards, by highly-motivated, highly-trained and experienced person, totally aware of implications of failure, with time to correct potential error, but without the benefit of significant job aids	4×10^{-4}	$8\times10^{-5} - 9\times10^{-3}$
(H) Respond correctly to system command even when there is an augmented or automated supervisory system providing accurate interpretation of system state	2×10^{-5}	$6\times10^{-6} - 9\times10^{-4}$
(M) Miscellaneous task for which no description can be found	3×10^{-2}	$8\times10^{-3} - 0.11$

4 Using HRA to Enable LOPA

The objective is to use HRA to define procedure-specific human error rate to replace the general error rate given in general LOPA guidelines. As we have discussed above, LOPA is a rule-based semi-quantitative analysis. LOPA error rates, often found in the CCPS tables or an individual organization's LOPA standard, are general examples. The order of magnitude error rates are not tied to the actual reliability of operators in correctly performing specific tasks or procedures at any individual facility. We have found that these general order of magnitude error rates may prove to be insufficient for describing the "operator error" associated with specific tasks or procedures. The single probability value for operator error is independent of the type of task or complexity of a series of errors that may represent the initiating event for a LOPA scenario.

The problem to be evaluated is not just to develop an individual operator error rate but to take a complex procedure and decompose it into a suitable form for analysis of human reliability. To do this, a procedure-driven activity can be thought of in terms of an event tree. Each step in the sequence of tasks composing the procedure may be a new decision rung in an event tree (see

Figure 1). The probability of a specific outcome is the sum of the probabilities of the multiple paths through the event tree that achieve that outcome. The cumulative probability of all potential outcomes (paths through the event tree) is one. In the case of LOPA, the event tree is used to identify the specific paths to failure for the specific scenario due to operator error.

Step 1	**Step 2**	**Step 3**	**Outcome**
Yes	Yes	Yes	OK
		No	**Failure**
	No	Yes	OK
		No	**Failure**

Figure 1. Example event tree.

For a complete HRA of a procedure, the full event tree can be analyzed to develop granularity in the types of operator error, but for LOPA, it is more useful to group all or similar error paths together for a given type of scenario. In the case of LOPA, the problem is simplified by evaluating the series of tasks in order to identify error-critical tasks. By error-critical tasks, we mean that these are the steps in the procedure that must be performed incorrectly, insufficiently, or omitted for the entire procedure to fail. Thus, the error state must not be recovered. For example, if an operator has to leave a manual valve in an open position for the undesired outcome to occur, but a later step in the procedure allows for identification of that error (recovery), then the product of the probabilities of failure of the two steps represents either part of or all of the procedure failure probability. Since the objective is to calculate an error rate, the probabilities of the successful outcomes in the event tree don't require quantification.

In the cases described here, the team evaluated each step in the unloading procedures to identify error-critical steps. The process will be discussed below for the two case studies.

5 Case Study 1 – HazChem A

A highly toxic flammable liquid is transferred from shipping tanker trucks into a stationary storage system. This chemical will be generally referred to here as HazChem A since its hazards and unloading procedure can be applied to numerous chemicals with similar properties. This chemical transfer utilized a detailed procedure including multiple, independent operator verifications. In this operation, two hoses were connected between the truck and the stationary equipment – a liquid transfer hose and a vapor return hose. Rupture of either hose could cause at the least a significant vapor release or at worst a large liquid spill. The problem to be analyzed was driver pull-away with one or more hoses remaining attached in order to determine appropriate IEF values for LOPA scenarios.

Three different IEF values were sought:

1. IEF for pull-away with liquid hose attached.
2. IEF for pull-away with vapor hose attached.
3. IEF for pull-away with both hoses attached.

Each of these scenarios will be inputs for two LOPA scenarios (LOC from tank and truck) which may call upon distinct IPLs; thus, the MEF values may not be the same. The revised LOPA table illustrating the UEF will be provided at the end of the section.

The team reviewed the detailed procedure and other less formal practices regarding the HazChem A unloading operation. In this review, the critical procedural tasks were identified. Each of these critical tasks was evaluated using the HEART approach to estimate the HEP(s). The set of tasks was then represented in an event tree approach with the final HEP for each step. Specific paths to LOC in the event tree were aggregated to develop the IEF for driver pull-away during the unloading activity. The event tree was used to quantify the IEF, but stopped short of including the IPLs.

The procedure is too detailed to provide here, but it can be summarized for the purpose of this paper. The unloading station is surrounded by a fence with a locked gate. The facility operator interacts with the truck driver to unlock and open the gate, position the truck inside the fence, close and lock the truck gate, and then secure the driver's keys in a lock box. During unloading, all non-essential personnel leave the area. A second operator, safety watch, independently places a lock on the lockbox containing the driver's keys and the safety watch observes the unloading activity from a safe distance. The operator performs a sequence of steps to unload the truck contents. After completing the unloading activity, the operator performs a sequence of steps to prepare and disconnect the hoses. The truck keys and truck gate are unlocked by the operator and safety watch allowing the driver to remove the truck.

5.1 Problem Definition & Task Analysis

The objective of the HRA was to evaluate only the scenario of driver pull-away with one or more hoses still connected to the truck. Many different scenarios caused by human error are possible, but only the scenarios with a pull-away as the initiating event were considered as part of this analysis. However, the techniques used here can also be applied to other error scenarios. The task analysis reviewed each step associated with line purge and disconnection and identification of critical tasks within the overall unloading process. For the driver pull-away error scenarios, the critical tasks identification was based on the following assumptions:

- Malicious acts and intentional violations are disregarded;
- All staff have appropriate training;
- Unloading process is complete;
- Unloading procedure has been followed up until the steps to purge and disconnect the line (except deviation in the lock out step);
- Driver is not present at the start of the line purge and disconnect sequence; and
- Only defined stressors are weather and timing.

The procedure steps/tasks were screened to identify the critical or necessary tasks to lead to a pull-away incident. Insignificant or non-related tasks to this scenario were not evaluated further.

The critical tasks that, if not successfully executed would lead to the driver pull-away error scenario, are listed in Table 5. Since access to the truck keys was related to the execution of the lock-out task that takes place prior to the unloading process, it was also considered as a critical task and is listed in the line purge and unloading steps. The final verification by the operator and the driver was split into two tasks since it is possible that these two individuals may not simultaneously undertake the task.

Table 5. HazChem A truck unloading procedure—Critical steps for prevention of driver pull-away error.

SOP section and step number	Step description	Explanation of step "FAILURE"
II.4.	Truck and Gate keys are locked in lock box by the Operator.	Operator fails to lock all sets of truck keys in the lock box.
V.10.	The Operator will disconnect the vapor unload fitting.	Operator fails to disconnect the vapor unload fitting.
V.18.	The Operator will disconnect the liquid unload fitting.	Operator fails to disconnect the liquid unload fitting.
V.30.	The Safety Watch will remove their lock from the lock box.	The Safety Watch removes the lock from the lock box without verifying hose disconnect.
V.31.	The Operator will remove their lock from the lock box.	The Operator removes the lock from the lock box prior to hose disconnections and allows drivers to have access to the keys.
V.35a.	The Operator will then complete a disconnect verification to ensure trailer is disconnected.	The Operator fails to verify complete disconnection of all hoses.
V.35b.	The Driver will then complete a disconnect verification to ensure trailer is disconnected.	The Driver fails to verify that hoses have been disconnected.

5.2 Applying HEART Approach to Critical Tasks

After task identification, the HEPs were evaluated for each critical task independently of associated tasks. Each critical task in Table 5 was assessed to identify the HEART generic tasks that closely associated with it, and the nominal HEP related to the generic task (as listed in Table 4) was assigned in Table 6. The team then assessed each task to identify the EPCs for each task and their effect rating (between 0-1). The task EPCs and their individual effect rating are summarized in Table 7.

Table 6. Generic task identification.

Step description	Generic task	Nominal HEP
Truck and Gate keys are locked in lock box by the Operator.	D/E	0.05 (Average of nominal HEPs for D and E)
The Operator will disconnect the vapor unload fitting.	F	0.003
The Operator will disconnect the liquid unload fitting.	F	0.003
The Safety Watch will remove their lock from the lock box.	F	0.003
The Operator will remove their lock from the lock box.		1 (Assumed to occur with the failure to lock keys in the lock box)
The Operator will then complete a disconnect verification to ensure trailer is disconnected.		1 (Assumed to occur with the failure to lock keys in the lock box)
The Driver will then complete a disconnect verification to ensure trailer is disconnected.	D	0.09

Table 7. EPC and affect modifier assignment for HazChem A truck pull-away.

Step Description	EPC Descriptions	Modifier	Assessed Effect	Comment
Truck and Gate keys are locked in lock box by the Operator.	6 – Mismatch	8	0.2	Mismatch - potential for Operator to not correlate "taking the keys" with the significance of allowing driver full control over the truck. Estimate 1 out of 5 operators doesn't get or understand the procedure.
The Operator will disconnect the vapor unload fitting.	17 – Little or no independent verification	-	0	Team felt that since the procedure requires an independent check that this doesn't apply.
The Operator will disconnect the liquid unload fitting.	17 – Little or no independent verification	-	0	Same as above.
The safety watch will remove their lock from the lock box.	15 – Operator Inexperience 1 – Unfamiliarity 16 – Impoverished 12 – Mismatch	3 17 3 4	0.1 0.25 1 0.1	Operator inexperience - Assume that few safety watch are inexperienced. Unfamiliarity - A plant employee can take on safety watch with no prior experience in this activity. But a simple task. Mismatch - High awareness of potential risk within operations staff, hence low affect rating.
The Operator will remove their lock from the lock box.	-	-	-	Assume that since the Operator has made an error in disconnecting hose, they believe they are disconnected. Dependent error.
The Operator will then complete a disconnect verification to ensure trailer is disconnected.	-	-	-	Assume that since the Operator has made an error in disconnecting hose, they believe they are disconnected. Dependent error.
The Driver will then complete a disconnect verification to ensure trailer is disconnected.	17 – Little or no independent verification	3	1	Assume driver will rely on Operator and would not do an independent verification.

The final HEPs, which were more conservative than the generic HEP values assigned, are listed in Table 8.

Table 8. Final HEPs for HazChem A driver pull-away error scenario.

Step description	Nominal HEP	Final HEPs
Truck and Gate keys are locked in lock box by the Operator.	0.05	0.05 × [{(8-1) ×0.2}+1] = **0.12**
The Operator will disconnect the vapor unload fitting.	0.003	**0.003**
The Operator will disconnect the liquid unload fitting.	0.003	**0.003**
The safety watch will remove their lock from the lock box.	0.003	0.003 × [{(3-1) ×0.1}+1] × [{(17-1) ×0.25}+1] × [{(3-1) ×1}+1] × [{(4-1) ×0.1}+1] = **0.702**
The Operator will remove their lock from the lock box.	1	**1**
The Operator will then complete a disconnect verification to ensure trailer is disconnected.	1	**1**
The Driver will then complete a disconnect verification to ensure trailer is disconnected.	0.09	0.09 x [{(3-1)x1}+1] = **0.27**

The orders of magnitude for these individual task HEPs are consistent with reference values provided in reference [9]. These values are also consistent with the general initiating event frequency recommended in LOPA [2]. Comparison values are provided in Table 9.

Table 9. Reference generic human error probabilities and initiating event frequency.

Category	Failure Probability	Source
Simple, frequently performed task, minimal stress	10^{-3}	[9]
More complex task, less time variable, some care necessary	10^{-2}	[9]
Complex, unfamiliar task, with little feedback and some distractions	10^{-1}	[9]
Highly complex task, considerable stress, little performance time	3×10^{-1}	[9]
Extreme stress, rarely performed task	10^{0}	[9]
Human error during a routine task that is performed once per week or more often. Initiating event frequency = 1/yr. $P \approx 1/52$	10^{-2}	[2]

5.3 Scenario Representation

An important step in the human reliability analysis is to establish a narrative description of the scenarios and to document relevant assumptions. Four paths leading to driver pull-away were identified from the event tree analysis. In one group of scenarios, the team assumed that all the steps prior to initiating the hose disconnection procedures were accomplished correctly. These three potential paths assumed that the succeeding operations could only occur if truck keys were successfully locked away by the operator and safety watch. The team also discussed a potential scenario where the truck driver retains a second set of keys and pulls the truck away at his/her own discretion. For this scenario, the team considered that the operator was not involved and both hoses would likely still be connected to the truck.

For the procedure, since the vapor and liquid hoses are disconnected and purged in non-sequential steps, the team assumed that each hose presented an independent error path. The first three paths are depicted in the event tree in Figure 2 as the red paths. From the event tree, it is apparent that the assumptions of independent hose errors lead to identical HEPs for each of the single hose errors and the product of the two hose error HEPs for the scenario with both hoses remaining connected (event probabilities of 5.69×10^{-5} and 1.71×10^{-7}, respectively). The complementary event tree where it was assumed that the driver's keys were not secured is provided in Figure 3. In that event tree, it was assumed that if the driver had the keys that he/she would pull away with both hoses connected leading to an event probability of 2.05×10^{-8}.

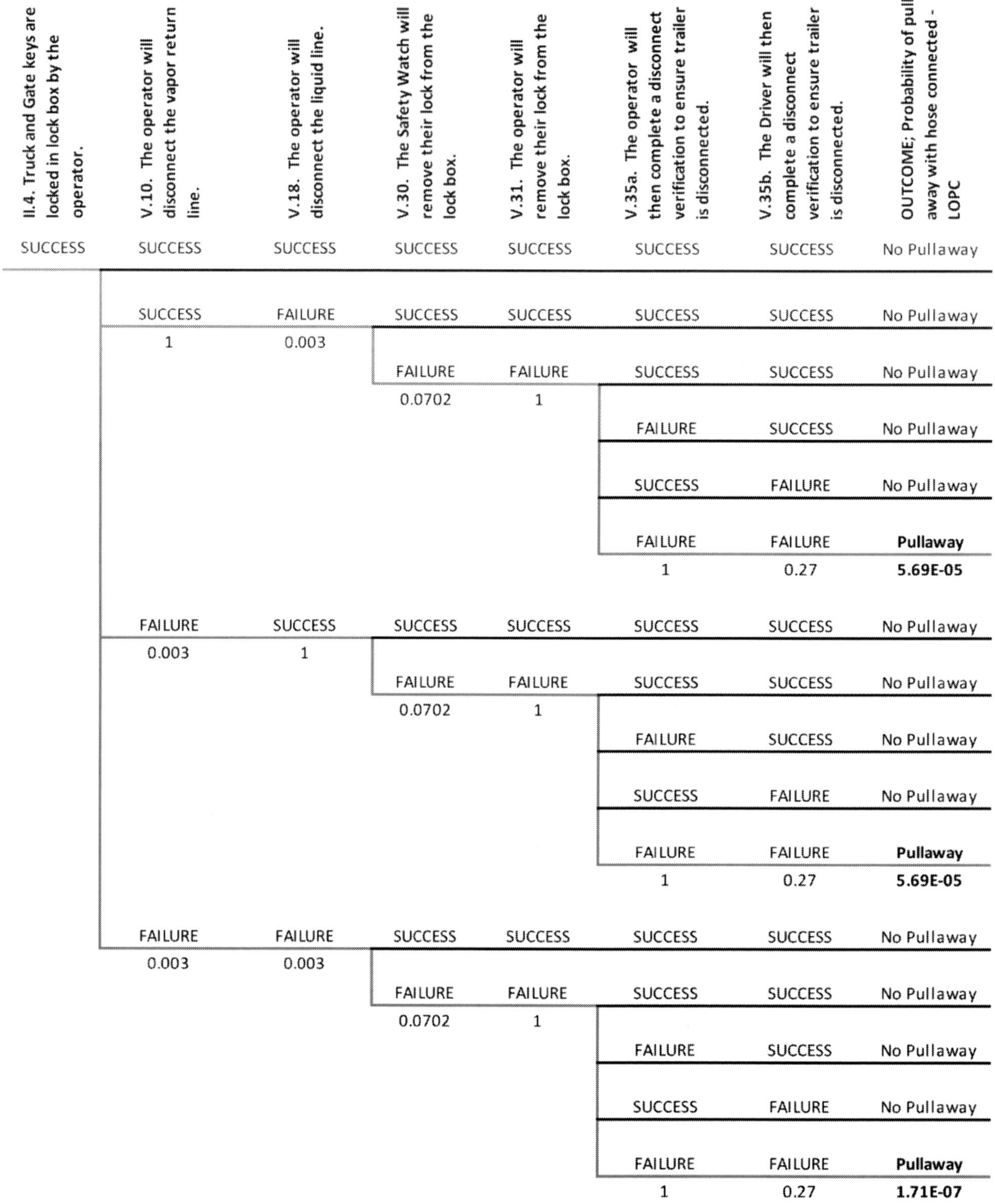

Figure 2. **Event tree displaying probability of three paths to driver pull-away. The procedural steps are listed and explanation of "FAILURE" is provided in Table 5.**

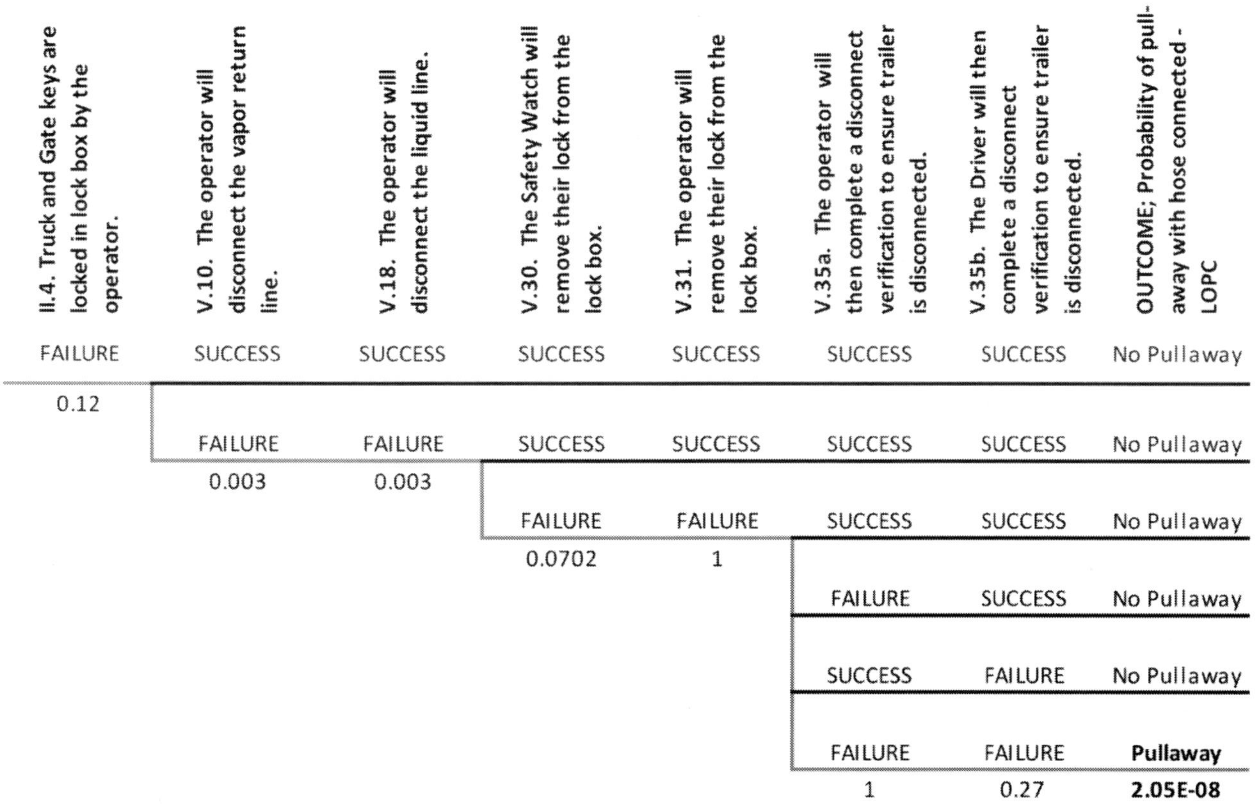

Figure 3. **Event tree displaying probability of driver pull-away if the keys are not secured. The procedural steps are listed and explanation of "FAILURE" is provided in Table 5.**

5.4 Driver Pull-Away Initiating Event Frequency

The probability of a driver pull-away from the event tree analysis is the sum of the probability of the independent error paths shown in Figure 2 and 3: 1.14×10^{-4}. Each path should be evaluated as to how it would apply to the distinct LOPA scenario because each pull-away LOPA scenario may contain different enabling events, conditional modifiers, or IPLs. For example, the IPLs may be different for toxic versus flammable considerations or for the vapor and liquid hoses. Thus, there needs to be careful consideration and evaluation when applying the new IEF to a scenario.

In addition, if the HRA result causes the LOPA team to separate the original scenario into multiple scenarios aligning with each path of the event tree, the TMEF should be re-evaluated. The separation will result in each scenario starting with an IEF for a single path which is lower than the previous cumulative frequency. Therefore, the TMEF needs to be adjusted appropriately or the MEF of similar scenarios may be aggregated. Given the frequency of receiving of one truck about every two to three days, the team used an estimate of 120 tanker unloading operations per year (as assumed in the original LOPAs). The calculated initiating event frequencies are listed in Table 10.

Table 10. Driver pull-away IEFs for LOPA.

Human Error Description	Opportunities per year	Event Probability	IEF (yr^{-1})
Driver pull-away with a single hose connected	120	5.7×10^{-5}	6.8×10^{-3}
Driver pull-away with both hoses connected	120	1.7×10^{-7}	2.1×10^{-5}
Driver's keys not locked away then driver pulls away with both hoses connected, may occur during unloading	120	2.1×10^{-8}	2.5×10^{-6}

A simplified example is provided in Table 11 to show how the HRA results can be applied in a LOPA calculation for the scenario involving pull-away with one or both hoses still attached. In the example, the conditional modifier for probability of fatality is taken to be unity. The LOPA team would then include the frequency reduction for existing IPLs to generate a MEF. The MEF would then be compared to a TMEL, and if necessary, additional IPLs would be added to reduce the risk further.

Table 11. LOPA Example Showing Calculation with HRA-derived Error Probability.

Scenario: Driver pulls away during unloading with **one or more** hoses still connected due to an error following unloading procedure. This results in a release of HazChem A from the stationary storage tank with a toxic consequence potential.			
	Event Probability - Liquid Hose Event Probability - Vapor Hose Event Probability - Vapor + Liquid Hose Event Probability - Driver second set of keys **Total Event Probability**	5.7×10^{-5} 5.7×10^{-5} 1.7×10^{-7} 2.1×10^{-8} $\mathbf{1.1\times10^{-4}}$	
	Opportunities per year	120	
	IEF (per year)		1.4×10^{-2}
Enabling Events or Conditional Modifiers	Probability of personnel present	1	
	Probability of fatality	1	
	Unmitigated Event Frequency (per year)		1.4×10^{-2}

6 Case Study 2 – HazChem B

The following case study addresses replacement of a portable container that is connected to the process, which involves a field operator, supervisor and a forklift driver. This second case study involves a smaller scale pull-away analysis involving a forklift operator who removes a portable container of hazardous chemical from a process unit while the container is still connected. A highly reactive and acutely toxic liquid chemical is supplied to a process directly from a stainless steel portable container (i.e., a tote) via pressurized transfer. This material will be referred to as HazChem B. The disconnection and removal procedure for the tote is much simpler than that for HazChem A. Essentially, a field operator isolates the tote from the process, uses nitrogen to purge the lines, isolates the tote valves, and disconnects the fittings. A supervisor then verifies the tote is disconnected and unlocks the room allowing access by the forklift operator. The forklift driver verifies the tote is disconnected and removes the tote and replaces with a full tote. Residual hazardous chemical vapor in the tote and from the process connections pose the hazard for the scenario if the tote is pulled away while still connected and the piping or hoses are ruptured. Three critical tasks were identified that would be associated with forklift driver pull-away error during the removal of an empty storage tote. Although the three tasks are performed independently, the HRA team determined it was appropriate to combined all three tasks to determine the IEF. The HEART generic task that closely represents each step and the nominal HEPs are listed in Table 12. The EPCs are listed in Table 13.

Table 12. HazChem B critical and generic tasks and associated HEPs.

Step description	Generic task	Nominal HEPs
Operator disconnects hose(s)	F	0.003
Supervisor verifies disconnection	F	0.003
Forklift driver removes tote	D	0.09

Table 13. EPC and affect modifier assignment for HazChem B storage tote pull-away.

Step Description	EPC Descriptions	Modifier	Assessed Effect
Operator disconnects hose(s)	HEPs are assumed to be nominal	-	0
Supervisor verifies disconnection	HEPs are assumed to be nominal	-	0
Forklift driver removes tote	12-Mismatched risk perception	4	0.5
	16-Impoverished quality	3	0.1

Table 14. Final HEPs for HazChem B forklift driver pull-away error scenario.

Step description	Nominal HEP	Final HEPs
Operator disconnects hose(s)	0.003	**0.003**
Supervisor verifies disconnection	0.003	**0.003**
Forklift driver removes tote	0.09	$0.09 \times [\{(4\text{-}1) \times 0.5\}+1] \times [\{(3\text{-}1) \times 0.1\}+1] =$ **0.27**

The calculated HEPs were used in the event tree analysis shown below and yielded a probability of 2.4×10^{-6} per opportunity.

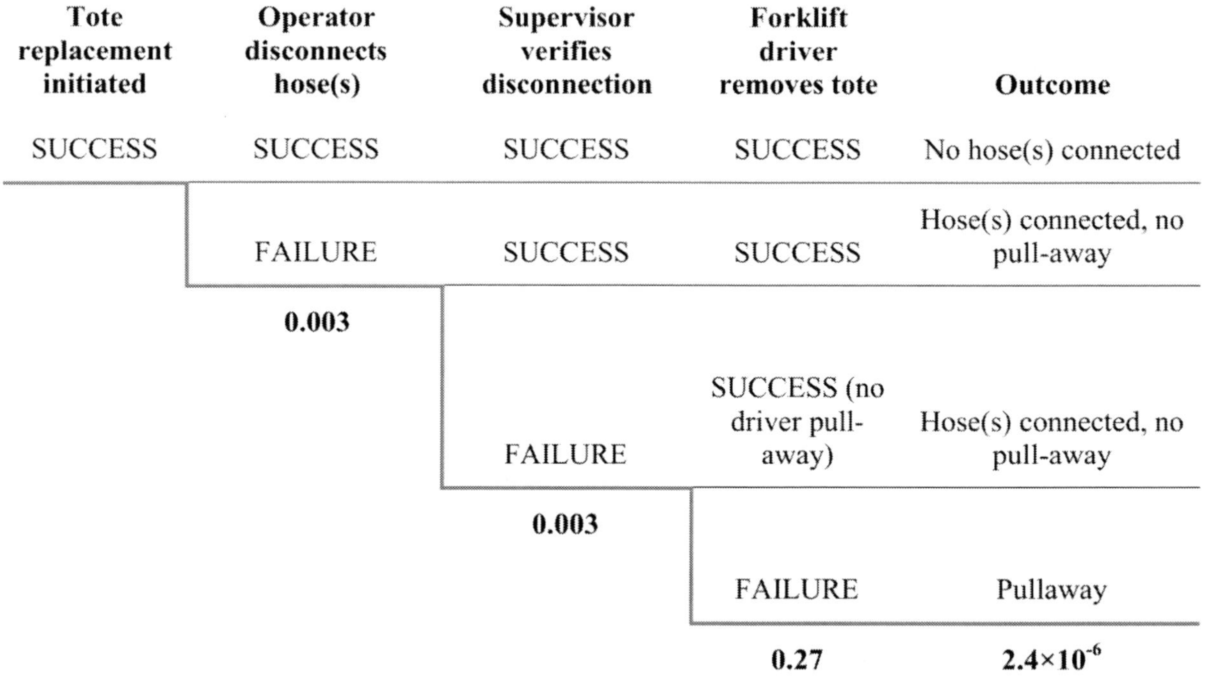

Figure 4. Event tree displaying probability of forklift driver pull-away.

6.1 Driver Pull-Away Initiating Event Frequency

A simplified example is provided in Table 15 to show how the HRA results can be applied in a LOPA calculation for the scenario of a forklift driver change a tote while still connected to the process. Assuming 12 tote replacements per year, the LOPA initiating event frequency of forklift driver pull-away for HazChem B storage tote replacement is 2.9×10^{-5} yr^{-1}. If the guidance value of 0.1 errors per opportunity was used instead of the HRA-developed error probability, then the IEF would be orders of magnitude higher and the supervisor and operator disconnect would possibly be taken as an administrative IPL. The HRA allowed the team to analyze the entire administrative task as an entire sequence of activities versus each procedure separately. This enabled the team to evaluate more clearly the level of independence and possible interaction of these seemingly separate activities.

Table 15. LOPA Example Showing Calculation with HRA-derived Error Probability..

Scenario: Driver pulls away during unloading with tote still connected to process due to an error in procedure. This results in a release of HazChem B from the tote and process vessel with a toxic consequence potential.			
	IEF (probability per opportunity)	2.4×10^{-6}	
	Opportunities per year	12	
	IEF (per year)		2.9×10^{-5}
Enabling Events or Conditional Modifiers	Probability of personnel present	1	
	Probability of fatality	1	
	Unmitigated Event Frequency (per year)		2.9×10^{-5}

7 Conclusions

This evaluation was initiated to address the inconsistency between different LOPA studies of a similar tank truck unloading process, which relied upon different assumptions. As a result, the initiating event frequencies differed significantly. The objective of this study was to employ a HRA technique to better understand the initiating event, the critical steps in the prevention of the scenario, the primary drivers that would cause operator error, and finally to develop a more representative estimate for driver pull-away IEFs to be used in a LOPA. The team utilized the HEART methodology to estimate the individual human error probabilities for critical tasks in order to develop estimates of the probability of pull-away per unloading event. The method was described in detail for a complex procedure to unload a hazardous chemical from tanker trucks and illustrated in a simplified fashion for change-out of a hazardous chemical tote. A summary table comparing the estimates of UEF derived from the different approaches is provided in Table 16.

Although the team was able to determine an IEF for these scenarios that was more representative of the actual process, the greatest value was not realized by the HRA team until the analysis was completed. The HRA allowed the team of operators, line management, engineers and human factors SMEs to dissect the procedure in a systematic way, identify the steps that are critical for the prevention of the scenario, evaluate the reliability of each critical step, and finally, understand the error producing causes so that the site can focus on the management of these critical steps and error producing causes. The outcome of the HRA was more valuable than generating a representative IEF because it provided an aligned understanding that can be translated to risk reduction activities at the shop floor.

Table 16. Comparison of Unmitigated Event Frequencies for HazChem A and HazChem B when using alternate methods to estimate the IEF.

Scenario	Study ID	UMEF (yr^{-1})	Comment
HazChem A	LOPA Team 1	1.2×10^{-1}	Handbook error rate; inappropriate conditional modifier
	LOPA Team 2	1.6×10^{-3}	UK HSD hose failure rate as IEF
	LOPA (HRA)	1.4×10^{-2}	HRA-derived IEF
HazChem B	LOPA (Handbook)	1.2	Handbook error rate
	LOPA (HRA)	2.9×10^{-5}	HRA-derived IEF

8 References

[1] Center for Chemical Process Safety, Layer of Protection Analysis - Simplified Process Risk Assessment, American Institute of Chemical Engineers, New York (2001).

[2] Center for Chemical Process Safety, Guidelines for Initiating Events and Independent Protection Layers in Layer of Protection Analysis, American Institute of Chemical Engineers, New York (2014)

[3] Center for Chemical Process Safety, Guidelines for Enabling Conditions and Conditional Modifiers, American Institute of Chemical Engineers, New York (2014)

[4] Taylor, J.R., Hazardous Materials Release and Accident Frequencies for Process Plant, Volume II, Process Unit Release Frequencies, Version 1 Issue 7, Taylor Associates AsP (September 2006)

[5] Center for Chemical Process Safety, Guidelines for Preventing Human Error in Process Safety, American Institute of Chemical Engineers, New York (1994).

[6] Heinrich, H.W., Industrial Accident Prevention: A Scientific Approach, McGraw-Hill, New York (1959).

[7] Swain, A.D. and H.E Guttmann, Handbook of Human Reliability Analysis with Emphasis on Nuclear Power Plant Applications, Final Report. NUREG/CR-1278 Prepared for U.S. Nuclear Regulatory Commission (1983).

[8] U.K. Health and Safety Executive (HSE), Review of Human Reliability Assessment Methods. Research Report RR679 (2009).

[9] Kirwan, B., "Human Reliability Assessment," Evaluation of Human Work, John R. Wilson and Nigel Corlett. 3rd ed. Boca Raton, FL: Taylor & Francis, pp. 833-875 (2005).

[10] Williams, J.C., "HEART - A Proposed Method For Assessing And Reducing Human Error". In Proceedings of the 9th Advances in Reliability Technology Symposium. University of Bradford, UK. 1986, pp. B3/R/1 – 14 (1986).

[11] Williams, J.C., "A data-based method for assessing and reducing human error to improve operational performance," Paper presented at the IEEE Conference on Human Factors in Nuclear Power, Monterey, CA, pp. 436-450 (1988).

[12] Williams, J.C., "Towards an improved evaluation analysis tool for users of HEART," In Proceedings of the International Conference on Hazard Identification and Risk Analysis, Human Factors and Human Reliability in Process Safety (Center for Chemical Process Safety, AIChE), Orlando, FL. pp. 261-271 (January 15-17, 1992).

Appendix A: LOPA Terminology [2]

Initiating Event – A device failure, system failure, external event, or wrong action (or inaction) that begins a sequence of events leading to a consequence of concern.

Enabling Condition – Operating condition necessary for an initiating cause to propagate into a hazardous event. Enabling conditions do not independently cause the incident, but must be present or active for it to proceed.

Conditional Modifier – One of several possible probabilities included in scenario risk calculations, generally when the risk criteria are expressed in impact terms (e.g., fatalities) instead of loss event terms (e.g., release, loss-of-containment, vessel rupture).

Independent Protection Layer (IPL) – A device, system, or action that is capable of preventing a scenario from proceeding to the undesired consequence without being adversely affected by the initiating event or by the action of any other protection layer associated with the scenario.

Initiating Event Frequency (IEF) – How often the initiating event is expected to occur; in LOPA, the IEF is typically expressed in terms of occurrences per year.

Mitigated Event Frequency (MEF) – How often the ultimate event is expected to occur when consideration of enabling conditions, conditional modifiers, and IPLs are taken into account.

Tolerable Mitigated Event Frequency (TMEF) – Maximum tolerable MEF for the given type of consequence based on corporate, regulatory, or other criteria.

Unmitigated Event Frequency (UEF) - The product of IEF and the probabilities of enabling events or conditional modifiers.to give an intermediate frequency value without applying the effects of IPLs.

Appendix B

List of the 17 tools reviewed by HSE and considered to be of potential use to the major hazards sector [8]:

1. THERP - Technique for Human Error Rate Prediction
2. ASEP - Accident Sequence Evaluation Programme
3. HEART - Human Error Assessment and Reduction Technique
4. SPAR-H - Simplified Plant Analysis Risk Human Reliability Assessment
5. ATHEANA - A Technique for Human Error Analysis
6. CREAM - Cognitive Reliability and Error Analysis Method
7. APJ - Absolute Probability Judgement
8. PC - Paired comparisons
9. SLIM-MAUD - Success likelihood index methodology, multi-attribute utility decomposition
10. HRMS - Human Reliability Management System
11. JHEDI - Justified Human Error Data Information
12. INTENT (Not an acronym)
13. CAHR Connectionism Assessment of Human Reliability
14. CESA Commission Errors Search and Assessment
15. CODA Conclusions from occurrences by descriptions of actions
16. MERMOS Method d'Evaluation de la Realisation des Missions Operateur pour la Surete(Assessment method for the performance of safety operation.)
17. NARA Nuclear Action Reliability Assessment

Appendix C: Error Producing Conditions (EPCs) in HEART [3]

Error-Producing Condition	Maximum predicted nominal amount by which unreliability might change going from "good" conditions to "bad"
Unfamiliarity with a situation which is potentially important but which only occurs infrequently or which is novel	17
A shortage of time available for error detection and correction	11
A low signal-noise ratio	10
A means of suppressing or over-riding information or features which is too easily accessible	9
No means of conveying spatial and functional information to operators in a form which they can readily assimilate	8
A mismatch between an operator's model of the world and that imagined by a designer	8
No obvious means of reversing an unintended action	8
A channel capacity overload, particularly one caused by simultaneous presentation of non-redundant information	6
A need to unlearn a technique and apply one which requires the application of an opposing philosophy	6
The need to transfer specific knowledge from task to task without loss	5.5
Ambiguity in the required performance standards	5
A mismatch between perceived and real risk	4
Poor, ambiguous or ill-matched system feedback	4
No clear direct and timely confirmation of an intended action from the portion of the system over which control is to be exerted	4
Operator inexperience (e.g. a newly- qualified tradesman, but not an "expert")	3
An impoverished quality of information conveyed by procedures and person/person interaction	3
Little or no independent checking or testing of output	3

A conflict between immediate and long-term objectives	2.5
No diversity of information input for veracity checks	2.5
A mismatch between the educational achievement level of an individual and the requirements of the task	2
An incentive to use other more dangerous procedures	2
Little opportunity to exercise mind and body outside the immediate confines of a job	1.8
Unreliable instrumentation (enough that it is noticed)	1.6
A need for absolute judgements which are beyond the capabilities or experience of an operator	1.6
Unclear allocation of function and responsibility	1.6
No obvious way to keep track of progress during an activity	1.4
A danger that finite physical capabilities will be exceeded	1.4
Little or no intrinsic meaning in a task	1.4
High-level emotional stress	1.3
Evidence of ill-health amongst operatives, especially fever	1.2
Low workforce morale	1.2
Inconsistency of meaning of displays and procedures	1.2
A poor or hostile environment (below 75% of health or life-threatening severity)	1.15
Prolonged inactivity or highly repetitious cycling of low mental workload tasks	1.1 (for 1st half hour) 1.05 (for each hour thereafter)
Disruption of normal work-sleep cycles	1.1
Task pacing caused by the intervention of others	1.06
Additional team members over and above those necessary to perform task normally and satisfactorily	1.03 (per additional person)
Age of personnel performing perceptual tasks	1.02

Current State of Bow Tie Risk Assessment Method

Presenter/Principal Author
Robin Pitblado, Oil & Gas, DNV-GL, Katy, TX,
robin.pitblado@dnvgl.com

and
Tim McGrath, Genentech, CA (mcgrath.timothy@gene.com)
Mark Manton, ABS Group, UK, (mmanton@absconsulting.com)
Charles Ian Cowley, CCPS, (charlescowley@btinternet.com)
Kiran Krishna, Shell, Houston, TX (Kiran.Krishna@shell.com)
Martin Johnson, BP, UK, (martin.johnson@uk.bp.com)
Ron McLeod, Ron McLeod Ltd., Glasgow, United Kingdom, (ron@ronmcleod.com)
Paul Haydock, CGE Risk Management Solutions, Australia
(p.haydock@cgerisk.com)
Tatiana Norman, DNV GL, Houston, TX (tatiana.norman@dnvgl.com)

Prepared for Presentation at
American Institute of Chemical Engineers
2017 Spring Meeting and 13th Global Congress on Process Safety
San Antonio, TX
March 26 – 29, 2017

Current State of Bow Tie Risk Assessment Method

Presenter/Principal Author
Robin Pitblado, Oil & Gas, DNV-GL, Katy, TX,
robin.pitblado@dnvgl.com

and
Tim McGrath, Genentech, CA (mcgrath.timothy@gene.com)
Mark Manton, ABS Group, UK, (mmanton@absconsulting.com)
Charles Ian Cowley, CCPS, (charlescowley@btinternet.com)
Kiran Krishna, Shell, Houston, TX (Kiran.Krishna@shell.com)
Martin Johnson, BP, UK, (martin.johnson@uk.bp.com)
Ron McLeod, Ron McLeod Ltd., Glasgow, United Kingdom, (ron@ronmcleod.com)
Paul Haydock, CGE Risk Management Solutions, Australia
(p.haydock@cgerisk.com)
Tatiana Norman, DNV GL, Houston, TX (tatiana.norman@dnvgl.com)

Keywords: risk assessment, methodology

Abstract

Bow tie analysis has been in use for many years, with applications in both the upstream and the downstream petrochemical industries, as well as other major hazard industries. The method is qualitative and uses a diagrammatic representation of major accident threat and consequence pathways showing the hazard, top event, threats, and potential consequences, with intervening barriers and degradation factor pathways linked to the main pathway barriers. Bow ties can be compiled from Process Hazards Analysis, Hazard and Operability Study, Layers of Protection Analysis and Safety Integrity Level safety studies to summarize and more effectively communicate the controls (barriers and safeguards) organizations rely on to protect against threats and to mitigate against the consequence should the top event occur.

One of the primary aims of bow ties is enhanced communication of how major hazards are controlled. A problem with bow ties has been that there is no widely accepted methodology or definitions and this has resulted in inconsistent diagrams often with structural errors. Additionally, diagrams frequently show many barriers that are actually not 'effective, independent and auditable', misleading the organization into believing they have a greater level of protection than is actually present.

CCPS has developed a concept book to better define the bow tie analysis method and to promote best practice ideas. The book aims to ensure consistent application of the bow tie technique by defining structural elements together with good and poor examples for clarification. For example, the book provides guidance on what constitutes a barrier and what differentiates barriers from measures that merely prevent a barrier degrading. The aim of this guidance is to avoid the overrepresentation of controls, which may otherwise lull the workforce into a false sense of security that an event can never occur.

The biggest area for inconsistency is the treatment of HOF - human and organizational factors. No current guidance addresses this topic well and the CCPS Committee along with the Energy Institute developed novel guidance for this area.

1 Bow Tie Introduction

The bow tie method has been in use for over thirty years for major hazard process industry, both upstream and downstream petro-chemical industries. It has also found application in aviation, railways, shipping and other high hazard industries as well as in business processes and strategic risks. The bow tie is a qualitative methodology that addresses major accident events and shows the full range of barriers and safeguards deployed to stop initiating events from becoming major accidents. It is under-pinned by the Swiss cheese model of accident causation developed by Reason, and displays the major accident events in a graphical format resembling a bow tie. A unique feature of the bow tie is its ability to communicate complex major hazard events in a simple format that is easily communicated to all members of staff, contractors, regulators and other stakeholders.

The first recorded application of bow ties was at a hazard assessment training course at the University of Queensland in the 1980's by ICI. It showed the development of a major accident as a combination of fault trees to the left and event trees to the right passing through a top event in the center. The methodology was soon developed and further refined by major industry users (e.g. Shell globally) and by bow tie software vendors. They added additional features and created a formal terminology.

As no recognized formal methodology statement has existed, many distinct formats of bow ties have developed, often reflecting corporate standards or software vendor guidance. This lack of standardization has led to bow ties that appear very different depending on the author, creating some confusion and weakening the benefits of the approach. Further, the treatment of Human and Organizational Factors (HOF) within bow ties has been inconsistent and not reflecting the best current thinking on this more difficult topic.

CCPS in conjunction with The Energy Institute has developed a concept book on bow tie risk assessment to provide guidance, a common vocabulary and a best practice approach to developing, assessing and communicating effective risk management through bow tie diagrams. This guidance will surely promote consistent application of the bow tie technique; however the ultimate goal of any such work is to enhance the quality and analysis of high hazard risk models, with the aim of reducing residual risk for a range of industries. The concept book aims to ensure consistent application of the bow tie technique by defining structural elements together with good and poor examples for clarification. For instance, the book differentiates measures which prevent or mitigate major accident events named 'barriers' from measures that prevent barriers degrading named 'safeguards'. This ensures controls are not overrepresented which could lull the workforce into a false belief that an event can never occur.

The concept book was written by DNV GL and CGE Risk Management Solutions, with a Committee of twenty independent authors including practitioners with experience in technical and human factors as well as those soundly conversant with the use of bow tie software tools.

2 Bow Tie Terminology

The Committee regarded current terminology as having some potential for improvement, either because the terms were confusing or because different types of controls with different characteristics were all called barriers. The basic diagram is shown in Figure 1. This shows the 8 main bow ties elements: 1) hazard, 2) top event, 3) consequence, 4) threat, 5) prevention barrier and 6) mitigation barrier which are connected by the main pathways and 7) degradation factor, and 8) degradation factor safeguard which are connected by defeating factor pathways. Real bow ties can be much larger than shown here and the concept book provides techniques to concisely filter the information for each target audience.

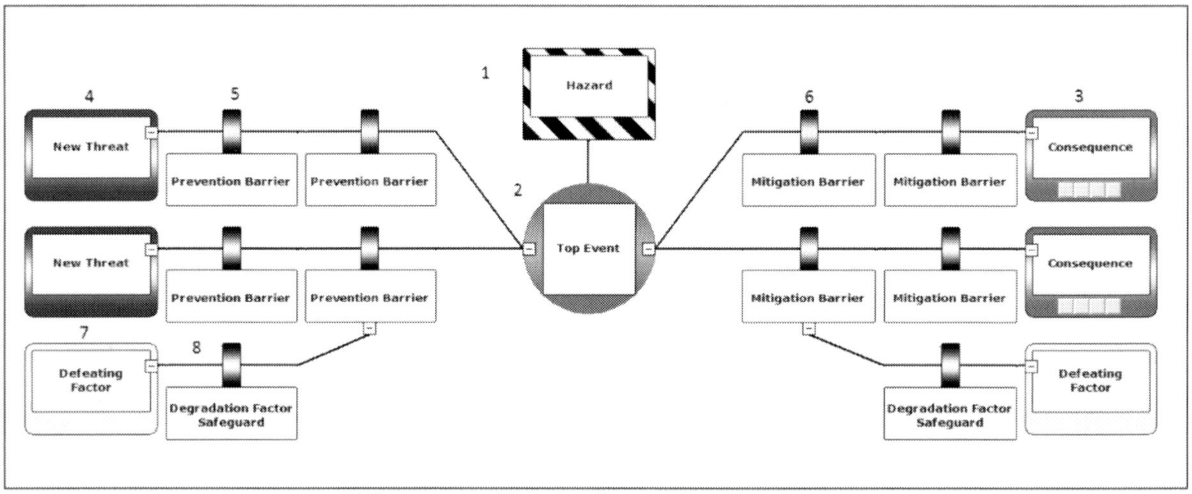

Figure 1. Basic Bow Tie Format

The concept book provides full definitions for these elements in the Glossary, along with good and poor examples for each. Over 30 pages are used to cover the basic elements of a bow tie, and while space in this paper prevents the inclusion of all terms, some important differences are provided here.

Barrier: A risk reduction measure that on its own can prevent a threat developing into a top event (prevention side) or can mitigate the consequences of a top event once it has occurred (mitigation side). The Committee believed that for a barrier to be valid, it must be effective, independent and auditable, similar to the requirements for an independent protection layer in LOPA although not all barriers will meet the LOPA criteria. This provides confidence that it will be able to act when required and as intended, without any action or intervention external to the barrier, and that its degradation will be prevented. It is represented by a single icon, such as 'emergency shut down' (ESD) – which is: a) effective because it prevents further loss of containment, b) independent due to back up power for use in emergency, and c) auditable – since it is on a risk based inspection, maintenance and repair regime.

The Committee decided that the barrier must be the complete system and so active barriers must contain all three elements of detect, decide and act. For example, an emergency shut down: a) detects – gas sensors for the loss of containment, b) decides – a logic solver determines which part of the process equipment requires isolation, and c) acts – actuators close valves.

There are different types of barriers and defining these can be helpful to teams in assessing the quality of barriers deployed against each threat or consequence. The Committee defined five types as shown in Table 1.

The Committee introduced the type 'active - hardware+human' to replace other terms in use, such as socio-technical, believing the term to be clearer to potential users of the combination involved. For example, most ESD systems are a combination of gas detection, operator decision, and actuation of the ESD valve. If the operator decision is not made explicit, many teams might categorize this system as purely hardware and omitting the important contribution of humans. Similarly, 'continuous hardware' (e.g. a ventilation system) was added to complement 'passive hardware' (e.g. firewall), as while both are always present the former involves motion or energy of some type (e.g. fans for a ventilation system or electrical current for a cathodic protection system).

Table 1. Types of Barriers

Barrier type	Description	Detect	Decide	Act	Examples
Active Hardware	All elements of the barrier are executed by technology	Technology (e.g. pressure sensor)	Technology (e.g. logic controller)	Technology (e.g. emergency shutdown valve)	Process control systems and Safety Instrumented Systems
Active Human	The barrier consists of human actions interacting with technology	Human (e.g. operator walk around detect leak)	Human (e.g. decides to shut down and isolate the equipment)	Human - but acting on technology (e.g. operator presses stop button or manually shuts a valve)	Operator detection and response (e.g. during structured walk arounds)
Active Hardware+Human	The barrier is a combination of human behavior and technological execution	Technology (e.g. high-high level indicator and alarm)	Human (e.g. operator hears and responds to alarm)	Technology (e.g. emergency shutdown valve) OR Human (e.g. operator manually shuts valve)	Operator-activated ESD valve Gas alarm and decision by human to evacuate
Passive Hardware	The barrier works by virtue of its presence	N/A	N/A	N/A	Bund, Blast wall, crash barrier, anti-corrosion paint
Continuous Hardware	The barrier is always operating	N/A	N/A	Technological	Ventilation system, active corrosion protection

<u>Degradation factor</u>: A situation, condition, defect or error that compromises the function of a main pathway barrier, either through defeating it or reducing its effectiveness.

The Committee also believed that 'degradation factor' was a clearer term than escalation factor as it clearly indicates the impact on the barrier to which it connects. Escalation factor is commonly used to describe a situation which may increase the severity of a consequence, for example the close proximity of a tank farm to process units. In principle, whilst all barriers can degrade and thus can have degradation factors, too many degradation factors can make diagrams overly complex and defeat the communication objective - a key use of bow ties. Generally, it is a good strategy to include degradation pathways for critical barriers (those which when defeated would dramatically increase the likelihood of the event or worsen the consequences) or where the degradation factor frequently reduces the effectiveness of the barrier.

<u>Safeguard</u>: safeguards lie along degradation pathways into that barrier where they help prevent the degradation factor from occurring or impacting the main pathway barrier.

The Committee believed that the term 'safeguard' was useful to differentiate only those controls that appear along degradation pathways extending from the main pathway barriers and to acknowledge that they frequently do not meet formal requirements of barrier validity. Examples of these are "training and competence", "inspection and maintenance" and many similar elements that are defined in management systems; as these do not contain all the essential components of a barrier, because they cannot terminate a threat pathway on their own. Safeguards can also be included to help the users place a control within the bow tie where there is a temptation to include the measure as a barrier when in fact it only qualifies as a degradation control.

In the interests of barrier management (Section 4), a group such as barrier owners does need to know all the safeguards supporting a barrier so they can be maintained, repaired and have operational criteria developed when barriers are degraded, missing or are of unknown status. However, degradation pathways and their complexity can be a distraction so these pathways can be displayed or hidden to allow specific groups of stakeholders to focus on the main pathway barriers.

3 Developing Bow Ties

Bow ties are best developed in a workshop format with a similar team structure to a HAZID or HAZOP (CCPS, 2008). The study is driven by specific terms of reference as this will help the team identify appropriate scenarios and the depth of treatment required (e.g. types of consequence to consider etc.). Typical factors to consider are shown in Table 2.

Table 2. Factors to Consider Prior to Bow Tie Workshop

Issue	Possible answers
Intended audience	Operators, maintenance technicians, top management, etc.
Study scope	Major hazards only, full hazards range, whole plant, specific unit, etc.
Study purpose	Design review, operations, activity review, incident investigation
Scenario prioritization	High consequence, high risk, manual controlled activities, etc.
Human factors issues	Strong dependence on reliable human performance
Rule set for study	CCPS concept book, company guidance, software vendor
Consequences	Harm to people, environment, asset, company reputation
Means to record	Paper notes, word processor, graphics tool, specialist bow tie software

Workshop Team

The Facilitator and Recorder would have similar qualifications to those listed in CCPS (2008) but with specialist knowledge of the bow tie method and terminology and the software being used for recording. Other team members are most affected by whether the study is for a design review or for operations. In the design phase, the team will be dominated by design staff and by experienced operations staff from a similar facility. For an operations phase review, team members will likely include process engineers, operators and maintenance technicians, safety and environmental specialists, instrument and control specialists, etc. The HAZID / HAZOP guidance (CCPS, 2008) about team size and composition applies to the bow tie team as well: a balance exists between having sufficient expertise present to address all likely issues, or having too many people for efficient facilitation. Bow tie workshops may last a few days and it is desirable to keep the team constant, rather than changing people each day. This helps maintaining a consistent approach through the study and being able to reuse bow tie fragments (e.g. threat pathways or degradation pathways) from a prior bow tie to ensure the management system is applied consistently.

Study Procedure

The suggested study procedure is shown in Figure 2. It is suggested that the hazard and top event be identified first: getting these correct to match the terms of reference is essential as they drive the team in a certain direction. Consequences should be identified next to confront them with the seriousness ("How bad could it be?") and motivate them to avoid the event rather than avoid threats. Once defined, the consequences and the related top event can be considered as the effect or outcome of certain causes, or threats. The team then turn their attention to discovering (via brainstorm and reference to other sources) these threats.

If steps 2 and 3 were reversed, as might seem more intuitive, then the team may spend wasted time generating a long list of threats that cannot actually generate the required magnitude of the consequences.

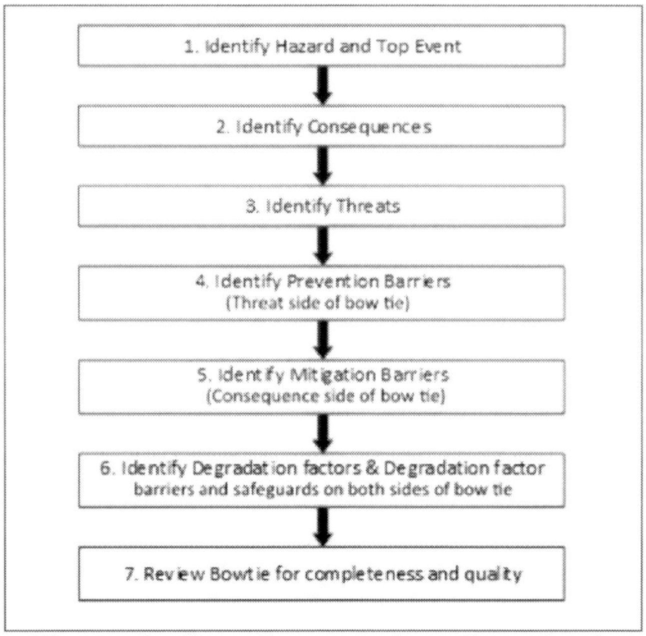

Figure 2. Bow tie workshop flowchart

It is unlikely that a team can generate a structurally correct bow tie in the workshop. Their time is better spent identifying the major accident events and associated hazards, top events, all relevant threats and consequences, barriers and safeguards, and any relevant metadata (e.g. barrier type, barrier owner, known historical reliability, criticality, etc.). It is the post-workshop task of the Facilitator and the Recorder to address bow tie structure and quality.

<u>Post-workshop checklist:</u>

The Committee identified many matters to check, post-workshop, in order that the final bow ties are useful and structurally correct. A list of 31 checks is provided in Table 3.

Table 3. Post- Bow Tie Workshop Quality Checks

Overall
1. Has an adequate number of bow ties been developed to meet terms of reference and prior PHA's
2. Did the bow tie workshop include the right mix of people and was adequate time available
3. Do all elements of the bow tie meet the required rule set (e.g. CCPS concept book)
4. Is the terminology correct and consistent across all bow ties
5. Is the bow tie structurally correct (e.g. no degradation factor safeguards located on main pathways)
6. Where multiple bow ties have similar threats / consequences, are the pathways similar
7. Does the complexity and content match the intended purpose and audience
Hazard and top events
8. Is the hazard clearly expressed and contain sufficient detail
9. Does the top event describe a loss of containment or loss of control
Threats and consequences
10. Are multiple very similar threats identified that might be combined to convey the same information
11. Do the threats lead directly to the top event
12. If a threat is a human error, can this be better defined as a degradation factor threat
13. Do consequences cover the range of types and magnitude as specified in the terms of reference

14. Where multiple bow ties exist, do the consequence categories match as appropriate
Barriers and safeguards
15. Do all barriers meet the validity requirements (e.g. effective, independent, auditable) and type
16. Have barriers been checked for less obvious dependencies (e.g. 2 barriers rely on the same operator)
17. Have safeguards been incorrectly included on a main pathway rather than a degradation pathway
18. Are barriers displayed in an appropriate order (e.g. time sequence of operation)
19. Do mitigation barriers appear on the prevention side (e.g. ignition prevention system)
20. Have too many barriers been included on one pathway (e.g. safeguards or trivial barriers)
21. Have barriers been cross checked against regulatory required barriers
22. Does the bow tie list only technical barriers and omit important human and organization safeguards
23. Are barriers included that cannot be measured or tested - if so they should be redefined as safeguards
24. Are all elements defined with sufficient words so that their meaning is clear on a stand-alone basis
25. Do barriers and safeguards include all required metadata as per company requirements
26. Are there too many barrier responsibilities relying on a single job role
27. Has the number and type of barriers been assessed for sufficient prevention or mitigation control
28. Has barrier past performance and incidents been considered in judging effectiveness
29. Have critical barriers and safeguards been documented and linked to the mechanical integrity system
30. Have trivial barriers been incorrectly included
31. Have advances in engineering practice, since initial operation, invalidated any barriers

4 Barrier Management

The Committee was careful to ensure the concept book emphasizes the importance of active management of barriers and safeguards to make sure they:
- are capable of their intended duty (adequate design and management of change);
- are available when demanded (adequate maintenance); and
- function as required (adequate operation).

To achieve these requirements a management system element is required to define and resource inspection, testing, calibration and maintenance of hardware barriers and safeguards, and the selection, training and work conditions of human barriers.

Bow ties can assist an organization with the critical task of barrier management. This is achieved by storing metadata for each barrier, and by enabling search and filter functions to produce customized views to aid in operational decision making. Bow tie diagrams are particularly useful for communicating important attributes such as the following:

- Barrier owner. This will identify the single individual on site, or rather their role, who is responsible to assure that the barrier functions as it should.
- Barrier effectiveness/strength. Whilst every barrier should fulfil the criteria of being effective, independent and auditable some barriers are better than others. This can be seen as analogous to the size and thickness of the cheese within the 'Swiss cheese model'.
- These are important discussion points during the development of the bow tie. Factors affecting the inherent, or as designed, strength of a barrier may include some or all of these sub-categories:
    - Barrier reliability: This can be assessed qualitatively or quantitatively.

    - Barrier effectiveness: whilst the goal is for all barriers to be fully effective, specific barriers might be included due to their importance even if not meeting a desired effectiveness target, particularly on the mitigation side of the bow tie (e.g. an ammonia water curtain might only address 50% of wind directions and reduce ammonia impacts by only 70%).
    - Barrier criticality: In principle, all barriers are important but some are more important than others. Factors determining criticality can include if the barrier is working to prevent a particularly prevalent threat or a critical consequence, or if a barrier is used across multiple threat or consequence pathways.
- Barrier condition. This is important during operation, to give an indication of the status of the barrier against the design intent and whether the barrier has degraded over time. A three level system lends to the use of 'traffic lighting':
    - Green: in place, available and effective,
    - Amber: in place and available but operating below its intended functionality
    - Red: not available or significantly degraded.

  Some companies have found it has been useful to also add white for a barrier that is yet to be assessed or no operational performance is available; and black to designate either a barrier is not installed or where a barrier has been removed or not available on a long term basis (deactivated).

Occasionally the bow tie review can lead to the recommendation for additional barriers. Care should be taken in adding barriers unless the original design intent and assessment is shown to be deficient (also considering any incidents). If the current barriers are significantly degraded, this may indicate the existence of a structural problem with deploying the organization's maintenance, inspection and testing; the additional burden of any new barrier is also likely to suffer degradation. Justification to add barriers may also include where new technologies are advised, performance standards for controls are to be met, or if there is a change in the environment or process and throughput.

Due to the complexities of different barrier strengths, criticalities and current condition, the Committee were unanimous in recommending not to create a barrier scoring system for a 'go, no go' process. A simple system of assigning the condition of the barrier allows focus on what to do next and avoids lengthy analysis and discussion that can result from using complex ways to grade the condition of barriers. Furthermore, assigning the condition to barriers gives an information source that management can use to decide these key questions: a) if operations should continue or be shut down; b) if immediate actions to strengthen (or add) barriers and safeguards are required to allow continued operation; and c) to help priorities and decide the longer-term actions to strengthen (or add) barriers and safeguards.

The Committee believes this guidance on utilizing bow ties for barrier management will assist organizations to better meet the requirements of Process Safety Management program elements of Leadership, Asset Integrity, and Management of Change.

5 Human Factors

The CCPS and the Energy Institute recognized the need to address the current inconsistencies in the treatment of HOF in bow ties as this could significantly improve process safety. A separate Human Factors Sub-Committee reviewed best practices to develop this novel guidance.

As described above, a threat is something that can lead to a top event if not prevented by a barrier. In bow ties HOF will generally be degradation factors which stop a barrier working as intended, rather than a threat on the main pathway because by themselves the HOFs generally do not have the ability to combine with the hazard to lead to a top event. HOF will also be present as safeguards to prevent other degradation factors affecting a barrier. They cannot be hazards or top events.

The Sub-Committee found many organizations developed bow ties where HOFs were represented as a threat for a main pathway. For example, Figure 3a shows adding catalyst to an exothermic reactor before proper mixing is established can lead to a runaway reaction and an explosion. Treating human error as a threat results in an incorrect analysis because:

1) The HOF fails to meet the definition of a threat as it lacks the ability to combine with the hazard to lead to a top event.
2) The resulting barriers "Training" and "Supervision" on the main pathway fail to meet the definition of barriers as they lack the ability on their own to prevent a threat developing into a top event and cannot detect-decide-act
3) The resulting four barriers on the main pathway lead to overconfidence in control over the threat

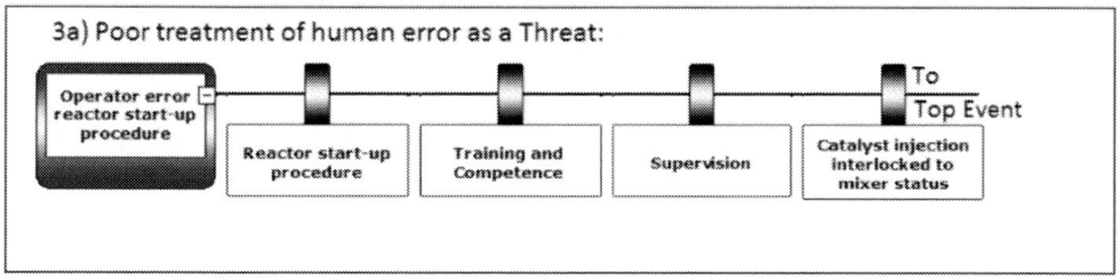

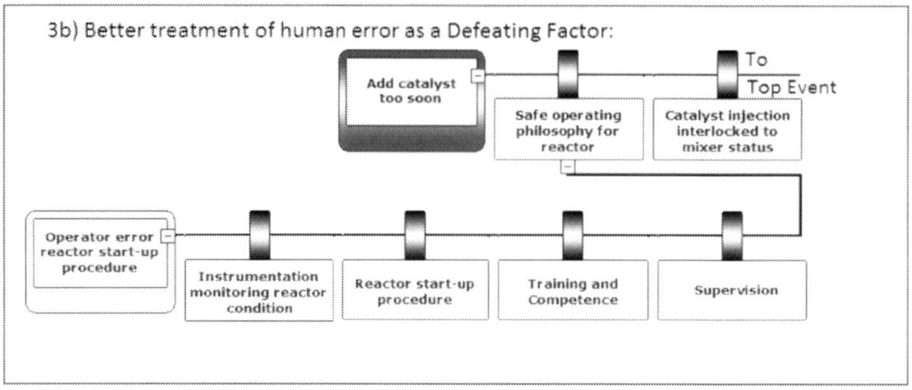

Figure 3. Poor and Better Treatment of Human Error

Treating human error as a degradation factor for a main pathway barrier is shown in Figure 3b which results in a correct analysis because:

1. The HOF "Operator error startup procedure" as a Defeating Factor for the barrier "Safe operating philosophy for reactor" meets the definition of being a situation, condition, defect, or error that compromises the function of a main pathway barrier, either through defeating it or reducing its effectiveness.
2. The resulting barriers "Training" and "Supervision" on the Defeating Factor pathway meet the definition of safeguards as they support the main pathway barrier and lie along degradation pathways into that barrier where they help defeat the degradation factor
3. The resulting two barriers "Safe operating philosophy for reactor" and "catalyst injection interlocked to mixer status" on the main pathway more correctly represent the control over the threat and the resulting four safeguards on the degradation pathway more correctly represent the level of attention paid to maintaining the barrier "Safe operating philosophy for reactor".

"Multi-Level Bow Tie" Approach

Up to this stage there has been nothing substantially new in the treatment of HOF, although there has been a refinement and clarification in the thoughts and processes associated with developing bow ties and the definitions around threats and barriers. The useful question may now be to consider in more detail as to how and why, for example, a safeguard may not work as desired, i.e. are there degradation factors for the safeguard?

The Human Factors Sub-Committee developed a novel preferred approach to address this additional analysis of HOF as a separate bow tie. The proposed approach is to call the main, top level bow tie including the degradation factors and safeguards 'Level 0' and for these next bow ties that are delving in to the details of what lies underneath the main bow tie as lower levels or Level -1. In theory then it would also be possible to produce bow ties showing the degradation factors on the degradation factors, or Level -2 bow ties, etc. but with diminishing returns of value to the organization.

The concept of multi-level bow ties is shown schematically in Figure 4. The upper part of this figure is the Level 0 comprising all the basic bow tie main threat and consequence pathways and degradation factors and safeguards. Level-1 duplicates the degradation factor pathway shown here for Preventative barrier 2 (only), but built out again as a basic bow tie. Thus, the level 0 degradation pathway threat (degradation factor A) becomes the level-1 main pathway threat, the top event becomes the hazard, and the new top event is the failure of Preventative barrier 2. However, safeguards remain safeguards as they do not meet the requirements for a barrier.

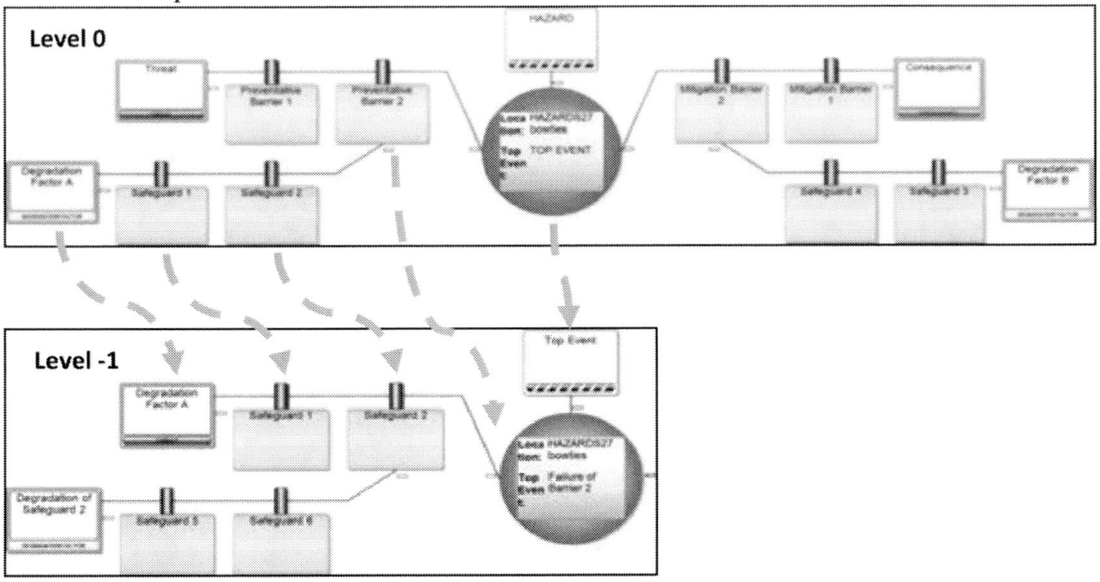

Figure 4: The concept of multi-level bow ties for Level 0 and Level -1

Taking the above into account, the recommended approach for addressing human factors in bow tie analysis is based on four important premises:

1. In bow ties human error should not be modeled as a threat that can lead to a top event. Instead, attention should be focused on identifying the barriers that human error can defeat or degrade and on ensuring those barriers are as robust as they reasonably can be, both to human error and to other degradation factors.
2. Human error will however frequently be identified as a potential degradation factor. When that is the case, it is important to understand both how the error can arise, and to ensure there are sufficient controls in place to mitigate the risk and to make the human error specific to the barrier being degraded (or safeguard being degraded in the lower level bow ties).
3. Organizations can decide for themselves whether to develop bow ties at a basic level (Level 0) where the first level of HOF can be included as degradation factors and safeguards or if they see value in investigating the further layers of possible degradation in more detail of Level-1, Level-2, etc.

4. Most of the Human and Organizational controls used to mitigate human error cannot meet the criteria necessary to be considered barriers in their own right (for Active barriers to detect, decide and act and to also be effective, independent and auditable). These factors are therefore considered as "safeguards", rather than barriers.

6 Case Study

To ensure consistent application of the bow tie technique good examples are provided for clarification. The concept book provides two case studies – one for a cross-country pipeline loss of containment with primarily technical barriers (Figure 5a Pipeline Loss of Containment Bow Tie – threat Pathways and 5b Pipeline Loss of Containment Bow Tie – consequence Pathways) and a more complex bow tie for a tank overfill event with both technical and human error issues.

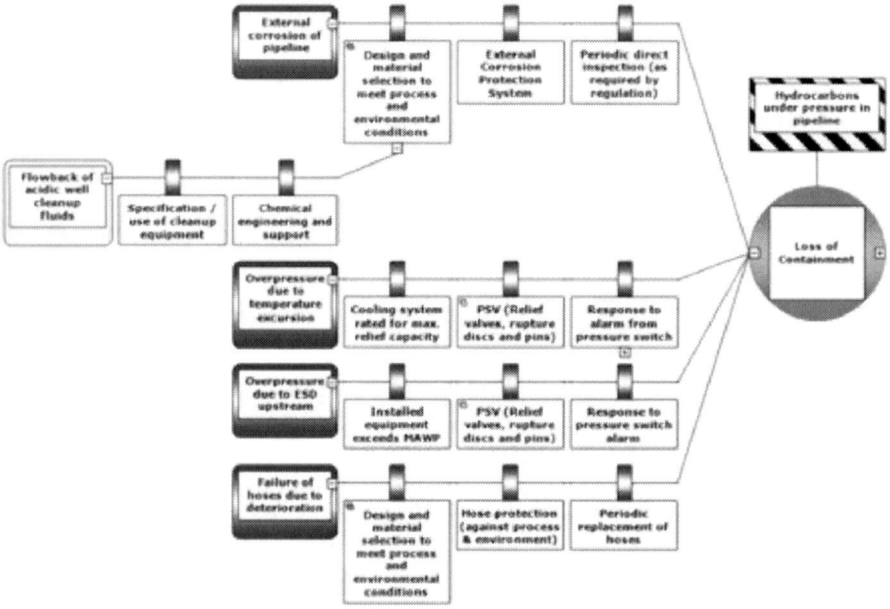

Figure 5a. Pipeline Loss of Containment Bow Tie – Threat Pathways

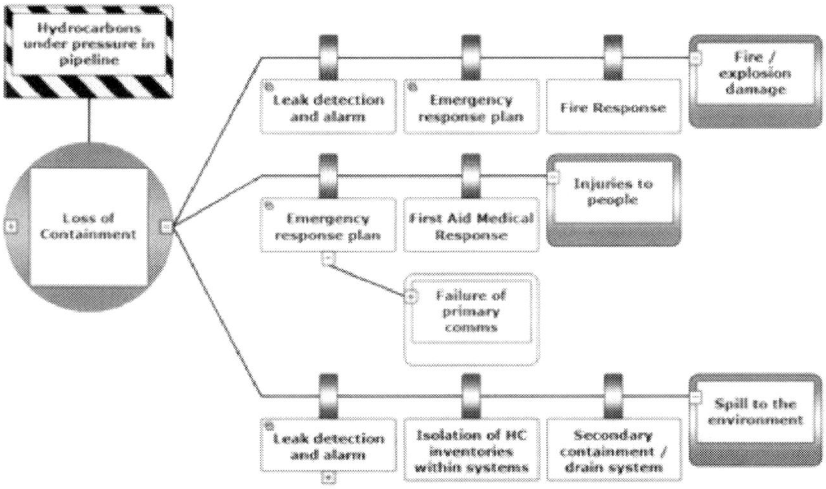

Figure 5b. Pipeline Loss of Containment Bow Tie – Consequence Pathways

7 Additional Uses

The bow tie has many uses beyond simply communicating major accident events and associated barriers and safeguards. The concept book lists these along with discussions how it achieves these other uses. This paper simply lists these other uses, including:

- Linking bow ties to Hazard and Operability Studies, Layers of Protection Analysis and Safety Integrity Level evaluation
- Integrating bow ties into As Low As Reasonably Pracicable demonstrations
- Operationalizing bow ties (Manual of Permitted Operations / Scope of Operational Boundaries) by listing required functional barrier before specific activities can be initiated
- Incident investigation
- Real-time process safety dashboards
- Barrier and safeguard verification
- Bow tie chaining
- Enterpriser wide analysis and windows onto systemic risks

Thus, bow tie analysis can be much more than a simple communication tool.

8 Conclusions

The bow tie method is a qualitative risk analysis method addressing major accident events and the key barriers and safeguards used to manage these. The method is growing in use in the process industries, for both upstream and downstream petrochemical industries, as well as other major hazard industries such as aviation, railways, and shipping.

An important feature is that the method, although useful in the design phase, is particularly valuable for the operations phase. It shows hazards, top events, threats, consequences, barriers, safeguards and degradation factors in a clear graphic format resembling a bow tie, with options of fully detailed or summarized formats. This diagram is easy to communicate to top management; to operators and maintenance technicians; contractors; and other stakeholders. A bow tie not only enhances the operators' understanding of all major accident events and their role in managing the barriers and safeguards that control the hazards, but also enhances management confidence in the operators' readiness.

Decision-making on any required actions is facilitated when degraded barriers or safeguards are clearly identified in the diagram. The bow tie also provides a very clear visual representation of cumulative risk where multiple barriers or safeguards are degraded, for example if all barriers on a single threat pathway are degraded.

An issue limiting the full value of the method has been a lack of an accepted clear methodology statement and terminology. This has led to bow ties with structural errors and inconsistent terminology. The CCPS concept book, authored jointly with the Energy Institute in the UK, has been created to address this, and has been created by a very experienced team with wide exposure to bow tie development. The Committee found that HOF were often poorly treated in current bow ties and the Committee developed a novel multi-level approach to deal with these issues. While this represents current best practice, it is recognized this is a developing field and that over time these approaches may evolve further.

9 References

In this Text:

CCPS (2008) Hazard Evaluation Procedures, 3rd Ed., Wiley-AIChE; New York.

General References:

Hollnagel, E., & Leonhardt, J. (2013). From Safety-I to Safety-II: A White Paper. European Organization for the Safety of Air Navigation (EUROCONTROL).

HSE. (2016, December). Human factors and ergonomics. Retrieved from HSE Guidance: http://www.hse.gov.uk/humanfactors/

Barrier Thinking Using Bow Ties for Offshore Production Platforms – A Case Study

Patricia Grant
Hess Corporation
Houston, Texas, United States of America
pgrant@hess.com

Krishna Bala
Hess Corporation
Houston, Texas, United States of America
kbala@hess.com

Stuart Chia, Fiona Wong, Michell Liu
Sherpa Consulting
Sydney, Australia
stuart.chia@sherpaconsulting.com.au

Prepared for Presentation at
American Institute of Chemical Engineers
2017 Spring Meeting and 13th Global Congress on Process Safety
San Antonio, TX
March 26 – 29, 2017

Barrier Thinking using Bow Ties for Offshore Production Platforms – A Case Study

Patricia Grant
Hess Corporation
Houston, Texas, United States of America
pgrant@hess.com

Krishna Bala
Hess Corporation
Houston, Texas, United States of America
kbala@hess.com

Stuart Chia, Fiona Wong, Michell Liu
Sherpa Consulting
Sydney, Australia
stuart.chia@sherpaconsulting.com.au

Keywords: barrier based thinking, bow tie, major accident event, risk management

Abstract

At Hess Corporation an extensive program involving 'barrier-based thinking' to demonstrate effective process hazard management is well underway. Hess is piloting the use of simplified bow ties to improve barrier thinking on its Offshore Production platforms. These bow ties are simple, visual tools used by Production Operations to manage process safety risk and demonstrate they have sufficient controls in place for daily operation and maintenance.

Traditional bow ties can be difficult for operators to use since they can be large and complicated. Hess has taken a measured approach in developing simplified bow ties that only include threats and barriers over which Operations has control.

The cornerstone of this approach was developing 'Major Accident Event (MAE) Bowtie' templates that outline the minimum barriers to be in place. At Hess, a Major Accident Event is an incident that could result in multiple fatalities. These templates are used to develop 'asset-specific' bowties that are tailored to include only the threats, barriers and consequences relevant to the asset. The templates simplify the process of creating the bow ties and also provide consistency across the assets that use them.

Another enhancement in the approach is the use of a barrier index. This index contains all the barriers on the bow ties and lists which bow ties the barrier appears on and how many times it appears on each bow tie. This index has been the key to operators finding the bow ties easy to use.

This case study reviews:

- How the MAE (Major Accident Event) templates were developed,
- How asset-specific bow ties are used when a barrier (e.g. safety critical equipment) is unavailable or taken out of service for maintenance
- How lessons learned from the successful pilot at one of the Hess Gulf of Mexico production platforms will be applied to other platforms.

1 Introduction

In 2015, Hess Senior Management posed the challenge to the EHS team 'How do we demonstrate assurance that our assets have and continue to maintain integrity?' In response, the EHS team worked with Offshore Production Management to develop a program improving overall risk management. by integrating barrier-based thinking into daily operations through the application of bow ties. This paper outlines the Hess Barrier Thinking journey and the lessons learned from a bow tie pilot at a Gulf of Mexico offshore facility.

2 The Hess Barrier Thinking Journey

2.1 Defining the Challenge

A barrier thinking task force team of integrity, engineering, production, EHS and process safety specialists was established. The team refined the challenge of assuring the continued integrity of assets into the following questions:

- Are the facilities continuing to be operated safely and with acceptable risks?
- Are the existing critical control measures (or barriers) effective in preventing and mitigating a potential major accident event (MAE)?
- Are the barriers sufficient? Do we need to strengthen or provide additional barriers?

Generally, some form of Safety Case or Safety Dossier was prepared for each production platform. Within these studies a barrier-based analysis such as Layer of Protection Analysis (LOPA) and/or bow tie analyses had been performed to demonstrate risk acceptability.

Interestingly, it was observed that there was a variation in the approach, depth and scale of assessing barriers that in turn influenced the nature of the bow ties. In some instances, for the same MAE type present at differing platforms, there were inconsistencies and differences in information and bow tie structure. The team concluded, after further investigation, that the barriers and key element structure of a bow tie needed to be better defined.

2.2 *Developing a Barrier Thinking Approach*

The definition of an MAE is well understood across Hess Offshore Production. However, the team realized that a sound and practical Barrier Thinking philosophy and approach was required to address the challenge of assuring continued asset integrity and provide consistent and effective risk management to reduce the likelihood of an MAE.

Hess characterized the Barrier Thinking as the understanding of two points:

- The role of barriers and their relationships to hazards, threats, events and consequences
- The management systems required to manage and maintain the barriers

Barrier thinking can be used to produce a range of visualizations for risk, depending on the needs of an organization. For example, barrier thinking can be used as a brainstorming tool (producing a basic Swiss cheese model) or a site work planning tool (producing a job safety analysis).

This case study focuses on using bow ties to control Loss of Primary Containment (LOPC) MAE on a production platform.

<u>Barrier Definition and Types</u>

After careful consideration, the definition of a barrier was adapted from Norway's Petroleum Safety Authority (PSA) principles for barrier management in the petroleum industry. At Hess, a barrier is a measure taken to prevent the release of the hazard and to limit the consequence of the event. This includes technical, operational and organizational elements which are intended individually or collectively to reduce/prevent the possibility for a hazard or accident, or, which limits its harm.

Hess further categorized barriers as either passive or active.

Passive barriers do not require any action or energy to manage a hazard. Examples of passive barriers include bunds/dikes, fire/ blast walls, and fire proofing.

Active barriers do require action or energy. To manage a hazard, they use the Detect-Decide-Act principle comprising the following aspects:

- Detector – can be a person or equipment
- Decide – a logic solver to make a decision based on the detected parameter (can be a person, hardware, or control system).
- Act – a final element to take the appropriate action (this may be a manual valve closed by an operator or an automated isolation valve).

Active barriers must include all three aspects of detect-decide-act. These barriers can be a combination of elements performed by people or technology. Examples of active barriers are a high pressure automatic trip, pressure safety valve (PSV), and an operator responding to a low-level alarm.

Barrier Thinking - use of bow ties to determine 'enough control'

Hess chose to operationalize barrier thinking through the use of bow tie diagrams. The bow tie is a diagram that visualizes risk. This tool should be intuitive and easy to understand, so it is accessible to personnel throughout the organization. See Appendix 1 for the bow tie structure adopted at Hess and a summary of key bow tie elements.

Hess requires that MAE risks are managed to an acceptable level using the Hess Risk Matrix. Using Asset-specific bow ties, risk tolerability can be demonstrated through the barrier counting technique. This relates to the number of valid barriers required to control a MAE.

Barrier counting is carried out through the Bow Tie from threat to consequence via the top event, for every combination of threat to consequence. The guidance on the number of recommended valid barriers for risk tolerability is equivalent to five (three prevention barriers and two mitigation barriers). Normally a bow tie barrier is equivalent in concept to a single independent protection layer (IPL) in a Layers of Protection Analysis (LOPA).

It must also be stressed that the barrier emphasis should be on the prevention side of the Bow Tie. This implies that one prevention barrier and four mitigation barriers is not the intent.

Barrier Validity

For the control of an MAE, a barrier is considered valid if it is effective, independent and auditable, as shown in Figure 1.

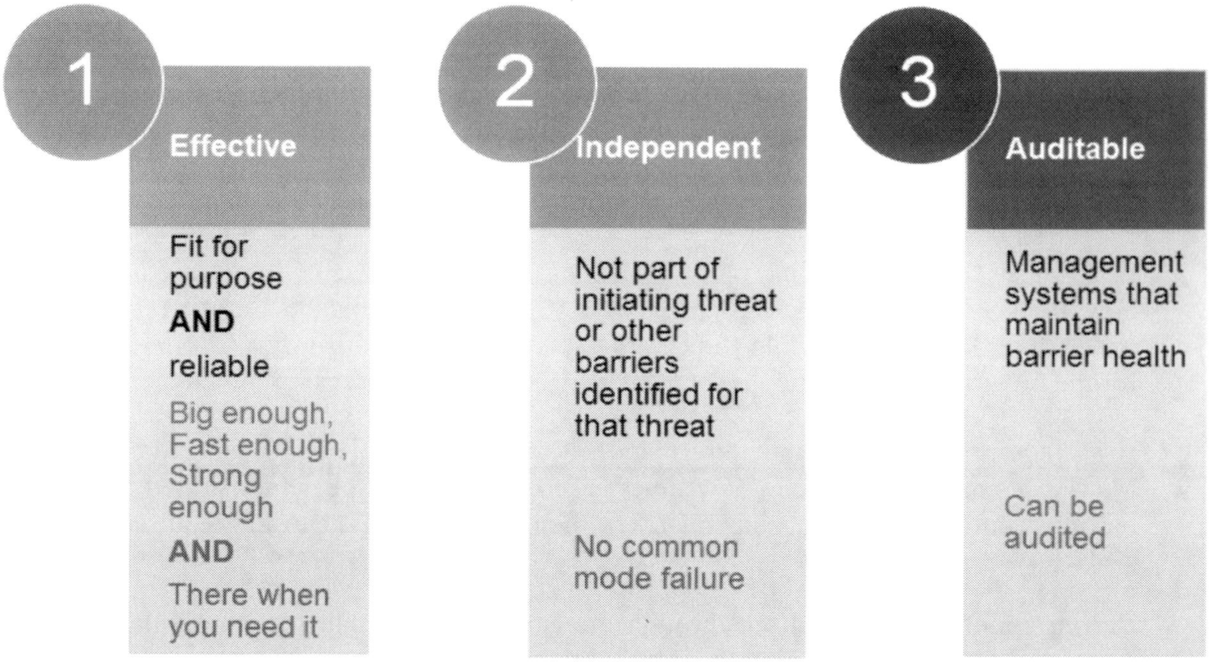

Figure 1. - Barrier Validity

Effective: A barrier must be fit for purpose and reliable in order to prevent the threat from leading to a top event or mitigate the consequence should the top event occur. For example, a PSV must be properly designed and maintained for it to be an effective barrier, Having a properly sized PSV is not enough, it must work when called upon. It must also be reliable,

Independent: The performance of the barrier cannot be compromised by the threat or the failure of another barrier to work.

Auditable: Barriers must be audited against procedures, records, previous validation assessments and other documented information to provide that design, testing, maintenance and operation of the barrier continues to conform to expectations. This includes auditing the performance of the barrier elements that support maintaining the "health" of the barrier. Hess requires that all MAE barriers are supported by reference to the maintenance system and/or operating procedures and/or design standard with a responsible party.

<u>Barrier Thinking - MAE Model Bow Tie Templates</u>

To provide that barrier thinking is consistently applied at Hess offshore facilities, the team developed a set of easy-to-use MAE model bow tie templates that are specific to operators and production operations. The Top Event in all these bow tie templates is LOPC (Loss of Primary Containment).

The benefits of adopting this approach are:

- Consistency across the global assets
- Simplification of the creation of asset-specific bow ties
- Validation by Hess Technical Authorities
- Minimum expected barriers defined
- Demonstrated risk tolerability through barrier counting
- Identification of barrier shortfalls

Working with the Facilities Engineering and Integrity subject matter experts, an MAE model bow tie template was developed for each major unit operation on a typical offshore Gulf of Mexico (GoM) platform, as shown in Figure 2. These templates reflected the most typical threats (inherent in the process and that an operator can control) and the minimum expected prevention and mitigation barriers.

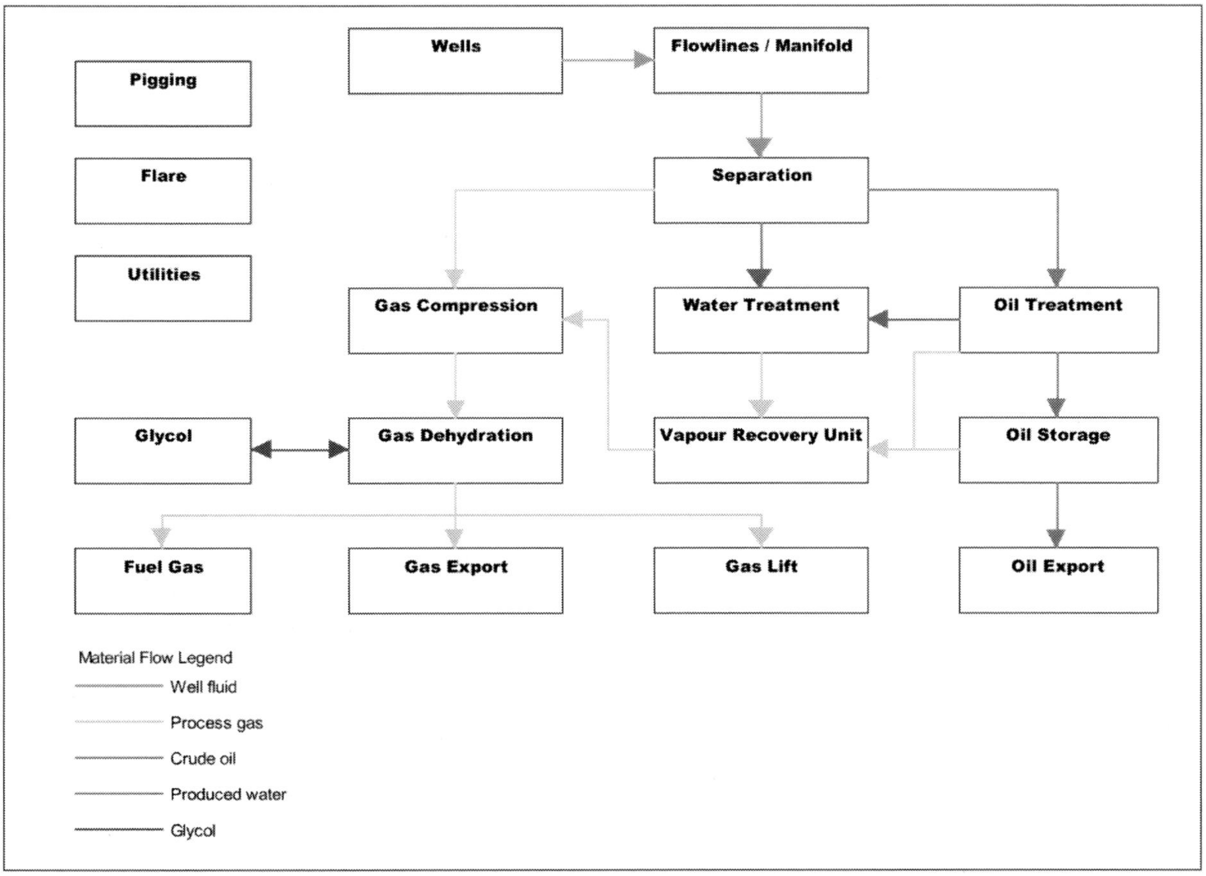

Figure 2. Typical Process Systems for a Hess Offshore GoM Facility

Threats: The bow tie templates show the most common threats leading to a LOPC within a process system that operations personnel can control. These are process-related threats and equipment integrity failures. However, threats such as external corrosion are not included because Operations is not directly responsible for controlling external corrosion.

By including only those threats that are the responsibility of Operations, the bow ties are streamlined and much easier to understand and use. The bow tie templates had up to a maximum of 6 common threats. The threats are different depending on the unit operations that the template covers. For example, the threats on a separation template will not be identical to the threats on the gas compression template.

In addition, each asset is also able to customize the bow tie to include additional threats specific to their facility or delete those that are not relevant. The simplified templates are easier to manage than the previous bow ties, which often listed more than 15 threats most of which the operator had no control over (e.g. external corrosion). Having so many threats on a bow tie makes them complicated and difficult to use.

Barriers: The bow tie templates show the minimum recommended barriers for a production asset or for the design of a new development. These were based upon relevant codes and standards (e.g.. API 14C), as well as industry best practices and Hess-specific design requirements

3 Case Study – Hess GoM Platform

3.1 Application of MAE Model Bow Tie Template

To pilot the program, Hess applied the barrier thinking approach to a GoM offshore facility, testing the efficacy of the LOPC suite of MAE model bow tie templates. Appendix 2 includes diagrams for developing asset-specific bow ties from the MAE model template.

During the case study, an engineer with less than five years' experience was able to successfully create the asset-specific bow ties for a platform by using:

- Site knowledge and operating experience from production personnel to confirm asset-specific hazards, threats and barriers
- The asset's Process Safety Information (PSI) and safety studies to further develop the hazards, threats, consequences and barriers
- Standard Operating Procedures (SOPs) and maintenance planning and execution activities (including performance standards) to underpin the barriers

The bow ties he created were also thoroughly reviewed and approved by the platform facilities engineers and the Bow Tie subject matter expert to validate they were complete and correct.

See Appendix 2 for an example of Oil Storage Tank MAE Template and Asset-Specific Bow Tie

3.2 The Bow Tie Barrier Index – the key to making the bow ties user friendly

The pilot platform generated 25 asset-specific bow ties from the templates. To simplify finding barriers across the bow ties, a barrier index was created. The index makes the bow ties more user friendly because it lists every barrier on the bow ties, and shows on which bow tie(s) each barrier can be found.

Some barriers appear on a single bow tie multiple times, against different threats, or they appear on multiple bow ties. The barrier index allows users to quickly and efficiently locate where a barrier is referenced on the bow ties. This is important because there is a natural tendency to assume that a barrier will only appear once on a single bow tie against a single threat, which is not always the case.

The barrier index assists operators in performing a thorough risk review by making it easier for them to review all the threats and consequences and showing them the whole risk picture.

3.3 Day to Day use of the bow ties to operate the platform

The bow tie pilot also assessed the usefulness of asset-specific bow ties for risk assessments when a barrier on the barrier index is bypassed, inhibited or unavailable (e.g. requires maintenance).

A simple process was put in place to structure how the bow ties would be used to perform the risk assessment

- When an instrument or other safety device, such as a safety valve, needs to be removed from service, the operator checks the barrier index to see if the equipment is on a bow tie.
    - If no – No additional bow tie review is required.
    - If yes – A bow tie review must be performed and documented.
- A bow tie review requires the operator to use the barrier index to identify all the bow ties and threats or consequences where the barrier provides protection.
- The operator reviews each identified bow tie. If there are at least two additional healthy barriers on each threat or consequence line, no additional action is required.
- If there are not at least two healthy barriers on the threat line, the Offshore Installation Manager (OIM) must review the bow tie and approve the removal of the barrier. The OIM may require an additional temporary barrier(s) be put in place to provide additional protection against that threat until the original barrier is returned to service.

3.4 Experiences and Lessons Learned

The following experiences were gained from the pilot:

- **Ease of creating asset-specific bow ties**: The availability of the bow ties templates allowed the production team to quickly construct bow ties for their asset. It also provided focus for the production team to verify their threats and barriers.
- **Barrier Shortfall Identification**: In developing the asset-specific bow ties, barriers from the MAE model template that were not present on the asset were initially treated as barrier shortfalls. This allowed the production team to identify alternative barriers, develop mitigation plans to provide new barriers, or strengthen existing barriers.
- **Leadership**: Training the production management on barrier thinking and use of bow ties allowed for a better appreciation of the risks of a potential MAE and a better understanding of whether these risks are being managed to a tolerable level.
- **Communication and Knowledge Sharing**: The visual nature of the bow ties engaged personnel and improved awareness of each hazard and the intent of associated control barriers. The bow ties allowed the distillation of hazard knowledge to production personnel.
- **Ease of use**: The asset-specific bow ties and barrier index display the inter-relationship of specific barriers to potential MAEs. Operations and maintenance personnel clearly understand the impact of an unavailable, degraded or modified barrier. They could also more easily determine whether the facility might be compromised (i.e. incur increased risk) or exceed its operating limits without interim barriers.
- **Improved Barrier Thinking and expanded use of bow ties** - Discussing barriers and adequately addressing risk has gained traction through the use of bow ties which was the original goal of the program. In addition, the bow ties are now also being used to assist in incident investigations to understand where barrier failures occurred. Now that the organization has bow ties their use is expanding beyond the original intent.

4 Conclusion

Operators such as Hess are working within a challenging energy environment and are still required to demonstrate assurance of 'continued integrity of assets'. The Hess Offshore Production and EHS team has developed a strategic approach to managing hazards and minimizing the risk of a potential major accident event (MAE) at their global assets utilizing barrier thinking through the application of bow ties.

The development of MAE bow tie templates for a typical Hess offshore facility outlining the minimum barrier requirements, allows a consistent approach for risk management.

By focusing the MAE bow tie templates on threats that operators could influence, the team created diagrams that are easier to read, understand and adapt to operations. Application of the MAE templates to a specific Hess offshore facility to create asset-specific bow ties affords the flexibility to:

- Demonstrate acceptable risk mitigation
- Identify additional improvements to strengthen potential barrier shortfalls
- Perform thorough risk assessments by the production team of mitigation measures in the event of degraded or unavailable barrier(s)

The pilot of the barrier thinking program at a Gulf of Mexico platform yielded positive results. Lessons learned from this pilot have been included in further program rollout scheduled for 2017-2018.

Appendix 1: Key Structure and Key Elements of a Bow Tie

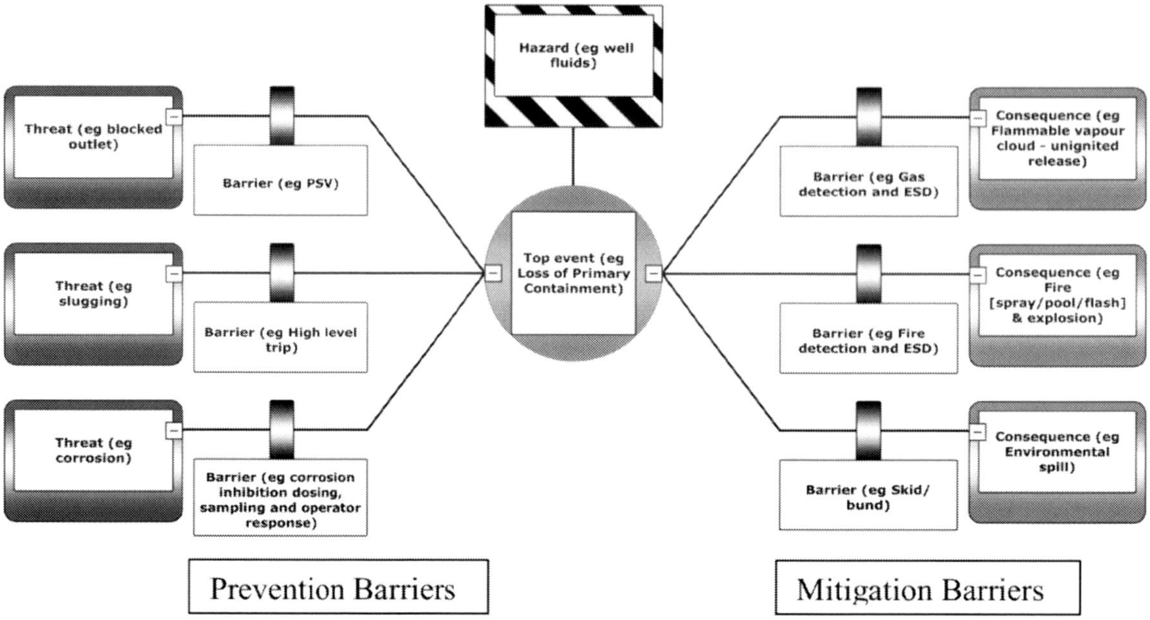

Typical Bow Tie Structure

Bow Tie Symbol	Bow Tie Item Description
	A substance (e.g. hydrocarbon under pressure) or a physical situation (e.g. overhead equipment) with the potential to lead to harm to people and/or environment. Examples include process gas, crude oil.
	The way a hazard can be released, (e.g. loss of containment), loss of control. Examples include loss of primary containment of gas during processing.
	Reason why, and how, the top event could occur (also referred to as 'cause'). The threats are the 'Causes' of the potential hazardous scenarios. Examples include blocked outlet, corrosion.
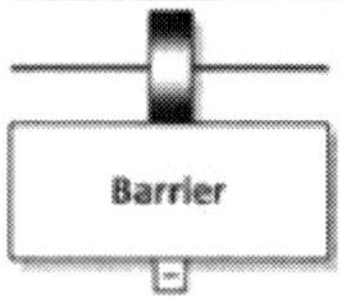	A barrier is a measure taken to prevent the release of the hazard and to limit the consequence of the event. This includes technical, operational and organizational elements which are intended individually or collectively to reduce/prevent the possibility for a hazard or accident, or, which limits its harm.
	The results that follow from the occurrence of the top event. This is normally harm to people, environment, property and/or business. Examples include fire, explosion.

Appendix 2 – MAE Bow Tie Template and Asset-Specific Bow Tie Example

Step 1

- Identify MAE Bow Tie Template for unit operation (eg oil storage tank)

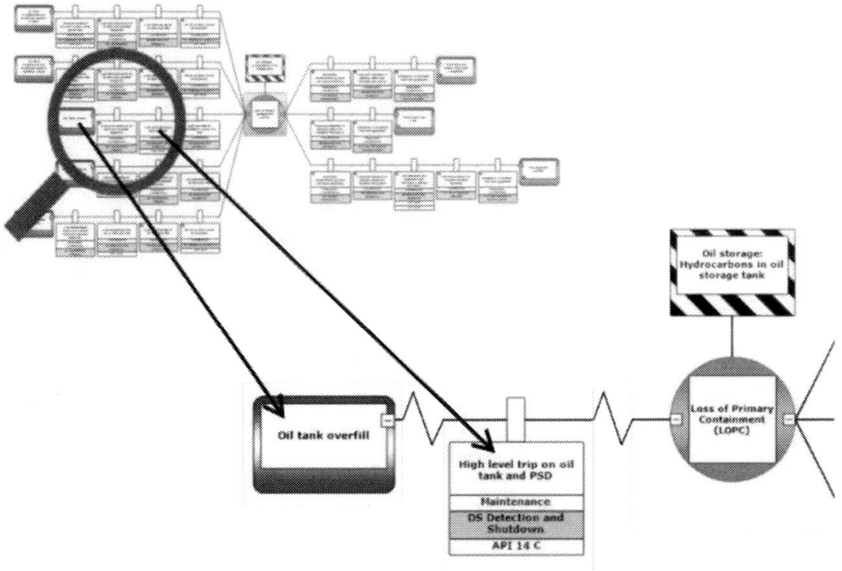

Step 2

- Provide asset-specific details for threats, barriers and consequences

MAE Bow Tie Template

- Generic threats & barriers
- Barriers linked to standards/codes

Asset-Specific Bow Tie

- Equipment tags
- Barriers linked to maintenance system & SOPs

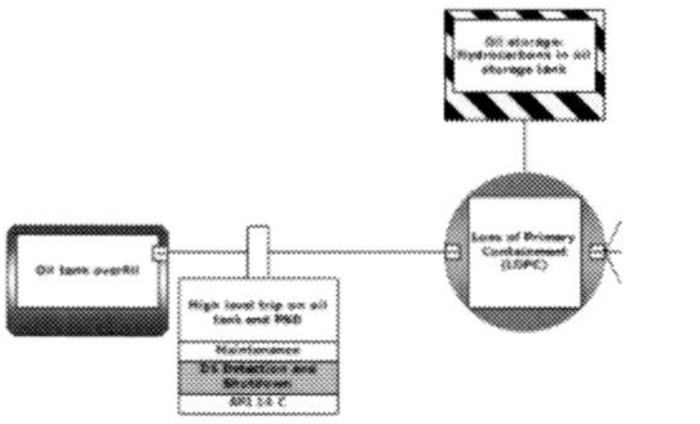

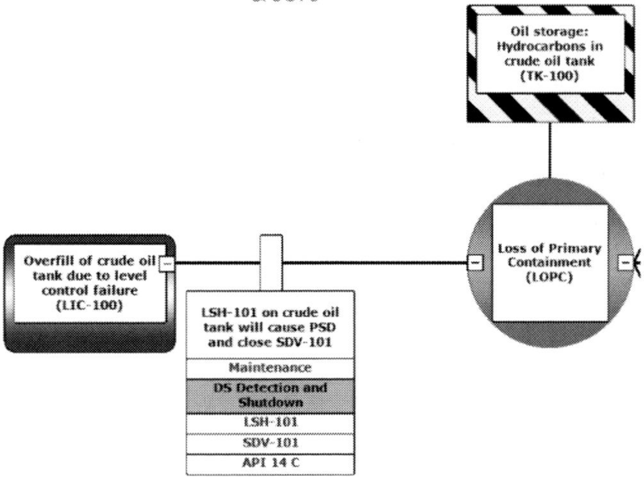

Addressing Process Safety Challenges in Batch Tolling

Brenda Seggerman
Hydrite Chemical Co.
114 N. Main Street, Cottage Grove, WI 53527
brenda.seggerman@hydrite.com

Prepared for Presentation at
American Institute of Chemical Engineers
2017 Spring Meeting and 13th Global Congress on Process Safety
San Antonio, TX
March 26 – 29, 2017

Addressing Process Safety Challenges in Batch Tolling

Brenda Seggerman
Hydrite Chemical Co.
114 N. Main Street, Cottage Grove, WI 53527
brenda.seggerman@hydrite.com

Keywords: Batch, Process Hazard Analysis, PHA, Process Safety Information, PSI, Operating Procedures, Management of Change, MOC, Human Factors

Abstract

Batch operations play a significant role in the Chemical Process Industries, but can face different process safety challenges than continuous processes, especially for batch tolling. A batch tolling facility will produce many different products, in multiple vessels, and with chemicals in varying amounts. The associated hazards of the raw materials and the products may range from highly hazardous to little to no hazard. In looking at the 14 elements of the Process Safety Management regulation, batch tolling faces increased challenges in successfully implementing and managing operating procedures, training, management of change, pre-start up safety reviews, process hazard analysis, emergency response, contractors, mechanical integrity, and process safety information due to the ever changing chemicals used and their hazards, types of reactions that occur, and the temperatures and pressures at which the processes are run. This paper will highlight some of the challenges and how one company addresses these challenges for process safety, while maintaining market competitiveness.

1 Introduction

Batch operations in the United States can be subject to the Federal regulation Process Safety Management (PSM) through the Occupational Safety and Health Administration (OSHA) and the Risk Management Plan (RMP) through the Environmental Protection Agency (EPA). While each regulation is based on the type and amount of chemicals onsite, a facility can be subject to one or both of these. Once a facility is subject to either of those rules, the facility must manage those chemicals and the processes those chemicals are run in per the regulation. While each rule has specific elements a facility must adhere to, this paper will look at adherence to OSHA's PSM regulation, the elements of which are listed in Table 1.

Table 1. Elements of OSHA's PSM Regulation

Program Elements	
Employee Participation	Safe Work Practices
Process Safety Information (PSI)	Emergency Planning & Response
Process Hazard Analysis (PHA)	Mechanical Integrity
Management of Change (MOC)	Training
Operating Procedures	Contractors
Compliance Audits	Incident Investigation
Pre-start Safety Review (PSSR)	Trade Secrets

The PSM regulation is a performance-based regulation, and as such OSHA does not specify exactly how a facility must comply with the regulation. This allows a facility flexibility to determine the best manner to comply. While interpretations exist regarding compliance, there still is room for interpretation. This ambiguity can lead to a lot of frustration on the part of a facility working to comply, especially when there exists a myriad of best management and industry practices in use throughout industry.

2 Batch Operations

Process safety can be applied to many different Chemical Process Industry (CPI) operations. While most CPI processes are continuous, batch operations play an integral role in the industry. Figure 1 shows the world-wide consumption of specialty chemicals in 2014 that were made using batch and/or semi-batch operations. This amounted to a market revenue of $345.5 billion in 2014 ((SOCMA), 2016) illustrating the importance of batch operations in the CPI.

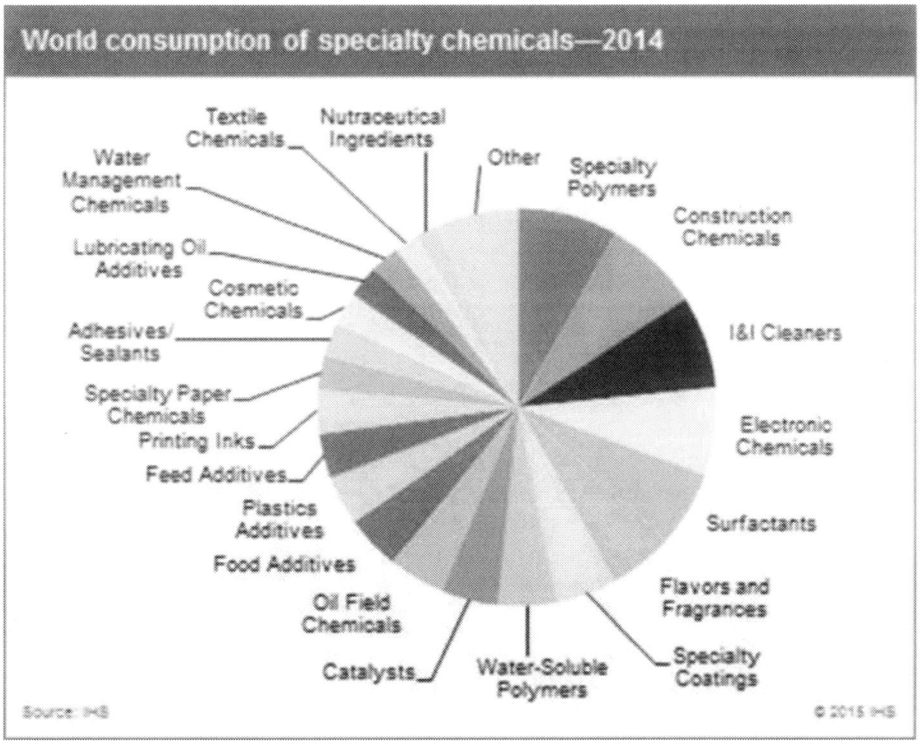

Figure 1. Specialty Chemical Consumption ((SOCMA), 2016)

Batch processing can be used where there is limited demand to make a product continuously or where running a product in a continuous process may not deliver the quality that is achieved in running the same process in a batch operation. Batch processing can also be more practical for multi-step processes while allowing for producing a maximum yield, yet at the same time potentially reducing waste generated. ((SOCMA), 2016)

While batch operations have its set of operating and PSM compliance challenges, batch tolling has a unique set of additional challenges. Toll operations consist of a facility producing products for a customer, based on contractual obligations and the customer's specifications. The customer can and will specify the raw materials to be used, the temperatures, pressures, flow rates, or other critical process parameters along with product and quality specifications for the product. In order for a batch toll facility to maintain profitability and competitiveness in the industry, a plant must be able to quickly produce a new product while maintaining the flexibility to produce current products and future products.

In tolling operations, products are continually changing along with the raw materials, processing conditions, chemical hazards, and vessels where the products are processed. One product may be run in one or multiple vessels or may require the use of different or even additional equipment. Due to this continual change, a facility first must determine if a product can be made within the equipment at the facility along with determining if the hazards of the raw materials, processing conditions, and products are within the hazard allowances of the facility.

3 Process Safety Information

Process Safety Information (PSI) is highly important in batch tolling operations. Per the OSHA PSM regulation, 29 CFR 1910.119 (d), PSI includes information pertaining to:

- The highly hazardous chemicals in the process
- The technology of the process
- The equipment in the process

With the constant changing nature of a batch toller, it is critical this information is properly reviewed to prevent a catastrophic incident.

3.1 Process Screening

Considering PSI in the screening process, a batch toller must review the raw materials that will be used and the equipment those materials will be used in, the hazards of the primary reaction and those of any side reactions, and the heating and/or cooling capabilities of the equipment. At one Hydrite Chemical Co. (Hydrite) facility this is performed by a process screening process.

3.1.1 Screening Process

When a toller is approached by a customer to produce a product, an initial request for information is sent to the customer to obtain information on:

- Chemical reactions and any subsequent side reactions involved in the process
- Safety Data Sheets (SDSs) for raw materials and the final product
- Reaction kinetics and stability data
- Analytical methods
- Raw material and final product specifications

This technology transfer from the customer is critical in helping the plant to determine what hazards may be associated with the products being run, especially if the product requires the use of monomers.

Once the initial screening request has been completed, the process safety evaluation continues with a Level 1 screen that is performed before a process lab study can begin. This screen is reviewed by site lab and safety personnel to determine if the process can safely be handled in the lab. If a process passes a Level 1 screen or if the process does not require a process lab study, then the process moves to a Level 2 screen. A process safety lab study, if performed, will look at the reaction(s) in the process and determine if the process can be scaled to a trial batch or to full production. The data from the lab study is then used to evaluate the process in the Level 2 screen.

In the Level 2 screen, site safety and process engineering personnel perform a full assessment on the process and determine if the product can be produced within the guidelines of the facility's permits and within the capabilities of the plant equipment. Within the Level 2 screen, the hazards of the chemicals are evaluated for their effect on:

- Employees that will work with or around the new chemicals
- Personal Protective Equipment (PPE) that should be used and whether existing PPE will adequately protect employees
- Fire protection equipment and whether existing equipment is adequate for a fire involving the new chemicals
- Foam systems for fire protection - to make sure the chemical is compatible with the foam chemicals, to prevent additional hazards
    - Storage capabilities and onsite movement and handling of raw materials, including whether employees can be exposed during transfer of materials in and out of the plant
- Emergency response equipment or procedures – determining if current equipment is adequate or new equipment needs to be purchased, and whether emergency response procedures and action plans must be revised to accommodate new chemical hazards.

During the screening process, chemical compatibility with the process equipment is evaluated. This is a very important step in preventing poor product quality and/or potential process safety incidents. One example of where the Level 2 screening process successfully prevented an accident involved a process that could only be run in a glass-lined vessel. In reviewing the chemical product information and associated Safety Data Sheets (SDSs), it was determined that the main raw material to be used in the process was incompatible with Hastelloy. Previous repairs made to the vessel included the use of patches made of Hastelloy and this incompatibility prevented the product from being manufactured at the plant. If this incompatibility had not been identified in the screening and the process was run at the plant, the vessel could have been seriously compromised potentially leading to a catastrophic release.

During the Level 2 screen, a preliminary process flow diagram and mass and energy balance, if not already received from the customer, is created. At the same time, raw materials, intermediates, by-products, wastes, equipment that will be used, rates, and process conditions are determined if not already supplied by the customer. Once it is determined the product can be safely made and is economically feasible, the product then goes through the Management of Change (MOC) process.

3.2 Materials of Construction

In order to be an economically productive facility, flexibility in switching from product to product is a key characteristic in process and equipment design. While this flexibility aids in quick response to new product trials and production, this characteristic leads to inherent process hazards.

Many times hoses and air operated transfer pumps are used to transfer materials from tanks to reactors and other equipment within a plant. While this reduces the cost of having to run piping to and from a process and can increase process flexibility, this also introduces chemical incompatibility hazards to the process. Hoses, pumps, valves and their internal components must be compatible with a wide range of temperature, chemicals, and material conductivity. One Hydrite facility standardized the type of hose used to handle a wide range of temperatures, chemicals, and material conductivity. This was done to ensure proper material of construction potential and reduce damage to hoses along with reducing operator injury and chemical releases. This same type of standardization was applied hose gaskets and gaskets in piping to cover most chemicals. A gasket that is specific to a raw material or chemical is specifically called out in the Standard Operating Procedure (SOP) for that product.

When reviewing the materials of construction of process equipment, one Hydrite facility chooses to use stainless steel (SS) in grades 304L and 316L in piping and equipment. The majority of the piping and equipment is 316 SS, which is fairly compatible with most materials. Designing processes from the onset to be as compatible with a wide range of chemicals can prevent dangers and having consistent material of construction throughout a facility will make reviewing processes simpler and reduce the risks.

3.3 Chemical Incompatibilities

Chemical incompatibilities can be very challenging to manage in batch toll operations due to the nature of batch processing and ever changing raw materials with very different hazards. One facility receives material into a main warehouse where operators then pull raw materials from specific designated locations in the warehouse and move them to the vessel or reactor where the materials will be used. An operator, who may be managing multiple reactors at one time, may stage multiple raw materials in the same area at the same time.

While there may be hundreds of chemicals in the warehouse, to properly evaluate the potential chemical incompatibilities of charging the wrong material, one Hydrite facility found it easiest to identify the chemicals that will be in the processing area during the time of staging and charging. The review then considers chemicals with similar naming such as glacial acetic acid and glacial acrylic acid and chemicals in like containers (e.g. same colored, plastic, closed-head drums).

One method of reviewing the incompatibilities of chemicals that may be present in the processing area is to use the Chemical Reactivity Worksheet (CRW) program developed by the National Oceanic and Atmospheric Administration (NOAA). With this program, one facility is able to input all of the chemicals that are currently used in the batch itself, the process area, and the new chemicals that will be introduced with a new or existing process in a new vessel. Given some processes use highly reactive or potentially reactive chemicals such as organic peroxides and monomers, the use of incompatible chemicals can result in an uncontrolled runaway reaction without any proper warning because the SOP would be written for correct material charges. In that case, alarms and indicators may not catch a high temperature or pressure until it was too late. This may also affect the relief capability of a rupture disc or other relief device on the vessel.

In order to reduce the likelihood of chemical incompatibilities, as the operations are so dependent on operators and human factors, a number of simple preventative actions can be taken. Chemicals in similar colored drums can be separated by a larger distance in a warehouse, additional signage added onto drums denoting chemical names, second operator and/or supervisor cross-checks used to confirm the proper material and amount to be charged, and inclusion of practices within the process SOP for robust calculations of charge quantities all aid to reduce the risks due to potential chemical incompatibilities.

Another option is to only stage materials when they are about to be charged instead of staging all the materials that will be used in the batch all at once. It may also be possible to use a bar coding system if the facility inventory system can accept the use and programming of a bar code scanner, and only after its use in the plant is permitted considering hazardous area classifications.

While raw materials present hazards, residues left in equipment and hoses after a batch or after a chemical transfer also present an incompatibility hazard. Not only does proper equipment and hose cleaning help to prevent poor product quality, it also prevents potential accidents from occurring. Proper vessel clean-out with water, appropriate solvents, or other clean out solutions and subsequent contaminant testing of post clean-out solutions help reduce the risk of potential incompatibilities with subsequent batches. In the case of cleaning hoses, tagging hoses with tags to notify an operator when a hose was last cleaned and what chemical was transferred last also helps to reduce incompatibility issues.

3.4 Equipment Compatibility

When reviewing a new process or producing an existing product in a new vessel, it is paramount to understand the reaction kinetics and chemical hazards to determine the heating and cooling capability of the particular vessel along with the relief vent sizing. In order to do that, correct data on the reaction kinetics and hazards from a customer is very important to have available when reviewing equipment compatibilities. Unfortunately, catastrophic accidents have occurred in the industry when relief and heating/cooling equipment capabilities were not considered in batch reactors. The Chemical Safety Board (CSB) has investigated numerous incidents ((U.S. Chemical Safety Hazard and Investigation Board, 2000), (U.S. Chemical Safety and Hazard Investigation Board, 2006), (U.S. Chemical Safety and Hazard Investigation Board, 2007)) that have occurred due to relief equipment or reactor cooling systems not being sized appropriately.

4 Management of Change

Management of Change (MOC) is critical in batch processing. Due to the very nature of batch processing, there are many changes on a regular basis, a situation often amplified in batch tolling. The changes may be in the process, the equipment used, raw materials (or quantities of them), operating parameters (e.g. temperatures, pressures, and their associated set points), operating procedures, and others. It is important that all changes are managed and communicated appropriately to prevent a serious incident. As there are so many changes in batch tolling facilities, and of varying degree of change, one important hurdle is to identify and address what constitutes a change in terms of "process safety" and how to evaluate MOCs using a formal, well defined process. When a facility logs an MOC, due to equipment changes, product trials, or new product production, every week or almost every week, it is essential the MOC process is well understood and efficient to use.

While the PSM regulation requires an MOC when there is a change associated with a PSM covered process, Hydrite has decided to manage all new products under the MOC system whether the process is covered under PSM. While this approach increases the number of MOCs performed, it helps the facility to review all changes and any potential impacts on PSM covered and non-covered processes, along with any impacts to the facility environmental permits. Even if the change is deemed minor, e.g. a ball valve changed to a gate valve, the impacts of that valve change must be evaluated for every product. When one vessel can run upwards of twenty different products at any given time that evaluation can be considerable. Operators must also be trained on not only the valve change itself, but also any changes on the use and/or position of the valve for every product. When changes involve monomers, it is critical that all aspects of the change are thoroughly reviewed. A root cause in a CSB incident investigation determined a lack of a formal review for changes made to a process that involved increasing amounts of monomer and solvents. These increases led to an increase in the heat of reaction that exceeded the cooling system capacity leading to a catastrophic accident. (U.S. Chemical Safety and Hazard Investigation Board, 2007)

5 Process Hazard Analyses

Process Hazard Analyses (PHAs) can also be a challenge with batch tolling operations. When the need for a "new" process arises, it could either be a new process altogether or a variant of a current process. While these new processes will be covered under an MOC, depending upon whether it is actually a completely new process or a variant of an existing one, a determination of the proper safety and health hazard review method will be made, including if a PHA is required.

In a batch facility, processes can be run in multiple or dedicated vessels. When the number of new trials and processes can reach 100 a year, this can put a strain on a facility's resources in the conduct of PHAs and managing the subsequent five-year revalidations. In the OSHA Process Safety Management Guidelines for Compliance document (Administration, 1994), OSHA notes that similar processes can be evaluated under one PHA. In some cases the only similarity between processes are the vessels they are run in. In order to streamline the PHA process, one approach is to create a baseline PHA of a vessel and perform a PHA to cover all the products that are run in that vessel along with the vessel itself and other associated equipment. (Center for Chemical Process Safety, 2008) Then, when a new process or existing process needs to be run in that vessel, the baseline PHA for that vessel can be reviewed and it can be determined if the hazards introduced by running a new or existing process in that particular vessel will be more hazardous or less hazardous than the products and hazards that were already covered in the PHA. Due to the nature of different products being run one after another, this also allows for the opportunity to review any potential hazards of residual material being left in the vessel or system and any potential issues of clean-out hazards or reactivity hazards with chemicals in the next batch.

5.1 Type of PHA

Another aspect to conducting PHAs for batch operations is the type of PHA to perform and the number of deviations to consider. Due to the discontinuous nature of batch operations, the hazard and operability study (HAZOP) method for PHAs is considered by many to be a better method as compared to a What-If type of analysis. (Bridges & Clark, 2011) However, due to the detailed, line-by-line nature of HAZOPs, they tend to take longer to complete and are more involved than a What-If type of PHA.

Due to the sequential nature of batch operations the process may have hundreds of steps with each step serving a specific and intended purpose. In one Hydrite facility, the typical steps for processes can include system preparations that can consist of: verifying the system is clean and ready for the next batch or campaign, verifying material charge lines are routed properly, and performing flushes and/or nitrogen purging if needed. The next step can include the charging of raw materials followed by re-inertions. Then the reactor may be heated and/or raw materials are fed and metered into the vessel and the heat of reaction closely monitored. Other materials may then be added or adjustments made to the batch. The process may then go through a decanting or cooling followed by the new product being transferred out of the vessel to another vessel, a storage tank, a tanker truck or rail car, totes, or drums.

In setting up a PHA for this type of process, one Hydrite facility found it easiest to define the nodes of the PHA per the process. For example, the first node included the system preparations, node two was for the charging of raw materials, node three for the heating and/or metered feeds, node four for any additional raw material charges and/or adjustments, node five to include decanting and/or cooling, and node eight the packing out of the vessel. Within each node there may be many deviations to cover. The deviations considered for a step where a monomer is charged, depending on the process conditions for a particular product, could consist of the following deviations as given in Table 2.

Table 2. Possible Deviations for a Batch Operation Involving Monomers

Deviation	Intent
High Flow	Flow is higher than desired
Low/No Flow	Flow is lower than desired, no flow at all
More Material	More material than desired is added
Less Material	Less materials than desired is added
Out of Sequence	Operator performs the step too soon or too late
Other Than	Operator charges in a material other than what is desired
Skip	Operator skips step and does not charge the desired material
High Agitation	The speed of the agitator is set too high or the agitator malfunctions and is running at too high of an amperage
Low/No Agitation	The speed of the agitator is set too low or the agitator malfunctions and is running at too low of an amperage or not at all

When charging monomers, or their subsequent catalysts or initiators, all of these deviations have the potential to cause a runaway reaction, depending on the nature of the reaction, so it is imperative to evaluate them fully in a PHA.

As can be seen from Table 2, just one charge can lead to an extensive, yet imperative, review, but also take a lot of time during a PHA. If the nature of the reaction is similar to that of a process that has already been reviewed in a reactor vessel baseline PHA, the time to review this charge is greatly reduced, especially if the causes, consequences, and safeguards are similar.

6 Operating Procedures and Supporting Information

Continuous operations in steady state require few operating changes to be performed on the system once running, unless a process upset occurs. The same is not true for batch operations as operators are continually making changes to processes based on raw materials being fed to the vessel, reaction rates of the processes, and the steps involved in the process. Operating procedures for batch operations must be written to reflect the constant operator involvement, supervision, and interaction with control systems. (Center for Chemical Process Safety, 1999)

The PSM regulation states what information must be contained in operating procedures, but do not offer any methods on how to comply with the regulations and rightfully so, since different operations require different types of procedures.

6.1 Batch Procedure Example

One Hydrite facility chose to tailor operating procedures for each individual product as shown in the in Figure 2 below. These operating procedures represent how a batch is produced. In other companies this procedure may be referred to as a batch record or batch tickets. These operating procedures act as a training document to train an operator on the process, the batch record, and procedure on how to manufacture the product. Depicted is the cover page for an SOP that can be tailored to the product produced. It contains six sections that all provide differing amounts and types of information.

Figure 2. Batch Procedure Cover Page

SOP NUMBER ________ **S.O.P. TITLE** Vessel Customer Product Name Production/Distillation

PSM REQUIRED Yes or No **REVISION** 65

ORIGINATOR: ________ (signature required) – Required for original SOP only

(Add title here) Date

MOC #: ________

APPROVALS FOR NEW SOP

Signature	Date	Title
		(Health & Safety Manager)
		(Plant Manager)
		(Plant Engineer)
		(Building Manager / Process/Project Engineer)
		(Quality and Operations Support Manager)
		(QC Lab Supervisor)
*		(R&D Manager Graphic Arts)

*Required for specific customer products

REVISED BY: ________ (signature required) – Required after major review

(add title here) Date

MOC #: ________

APPROVALS FOR REVISED SOP

Signature	Date	Title
		(Health & Safety Manager)
		(Building Manager / Process/Project Engineer)
		(Plant Engineer)
		(QC Lab Supervisor)
*		(R&D Manager Graphic Arts)

*Required for specific customer products

RE-CERTIFIED BY: ________ (signature required) – Required for annual Re-certification

Process/Project Engineer Date

APPROVALS FOR RE-CERTIFICATION

Signature	Date	Title
		(Health & Safety Manager)
		(Building Manager)

Air Operating Mode:	**OCPSF:**	**Hazardous Waste Treatment:**
0		

ANNUAL RE-CERTIFICATION DUE DATE: (Not later than 1 year after last signatures)

The first section lists the SOP number that is an internal facility reference, the SOP title that references the vessel the process is being run in and the name of the product, a designation of whether or not the process is covered by PSM, and the current SOP revision number.

The second section lists the person who was the original creator of the SOP, when it was created, and the MOC number the product was created under. A new product SOP must then be reviewed and signed off by the company officials listed, while certain products may require more review and sign-offs.

If a product has to be changed per an MOC, then the third section is completed. This details whom the SOP was revised by, the MOC number the revision was done under, and who reviewed and approved of the revised SOP.

SOPs are then recertified annually and this is documented in section four of the SOP cover page. Process/Project Engineers are assigned to particular process areas and are considered "owners" for any product that is run in a vessel in their corresponding process area. When the product SOP is due for recertification, the process engineer, Health & Safety Manager, and the business manager of the process area (i.e. Building Manager), all three review the SOP and certify it is current with how the process is being run at the time of recertification. Section six is filled out once the SOP has been recertified and the annual recertification due date is noted.

Because the facility referenced above is also regulated by a Title V air permit, a Part B hazardous waste permit, and a wastewater treatment permit, the SOP also contains information and documentation for running the process within the constraints of those permits as well. The Air Operating Mode distinguishes what type of air emissions control equipment must be used in conjunction with the process being run. This also specifies what must happen to the process if the plant's emission control equipment malfunctions. The OCPSF (Organic Chemicals, Plastics and Synthetic Fibers) Effluent Guidelines designation is due to whether or not any water discharge from the process must be sampled and tested before being discharged to the sewer. The Hazardous Waste Treatment designation is filled in if wastes from the process fall under hazardous waste treatment standards at the facility.

6.1.1 Process Chemistry, Summary, and Quality Specifications

Figure 3 displays the process chemistry for the reaction and can also include a narrative on the specifics of the reaction(s) and process. This narrative can also be tailored to provide the operator with any pertinent chemical reaction hazards, as well as any side reactions that could occur and the associated hazards.

While the product specification table may seem to be benign in its information other than from a product quality standpoint, if a product is out of specification, depending on the process, it may also be an indicator of a potential runaway reaction or some other problem (e.g. incorrect raw material, contamination, equipment malfunction). This indicator can help an operator also identify an issue before it becomes a more serious problem.

Figure 3. Process Chemistry, Process Summary, Quality Specification Information

Vessel Customer Product Production/Distillation

Process Chemistry:
Reaction:

Raw Material	+	Raw Material	→	Product	+	Product
	+				+	

Process Summary:

Product Specifications and Manufacturing Targets:

		PRODUCT SPECIFICATIONS		MANUFACTURING TARGETS	
Spec 1	=	XX	MIN	N/A	
Spec 2	=	XX	MAX	XX	MAX
Spec 3	=				
Spec 4 (Appearance)	=			N/A	

		PRODUCT SPECIFICATIONS		MANUFACTURING SPECIFICATIONS		MANUFACTURING TARGETS
BDO	=	xx	MIN	N/A		N/A
Water (PPM)	=	xx	MAX	N/A		xx
Color (APHA)	=	xx	MAX	xx	MAX	N/A

Materials Summary:

Chemical Name Here: Appearance:
n-Boiling Point: Mol. Formula: Water Solubility:
Melting Point: Mol. Weight: Vapor Density:
Flash Point: Density: Auto Ignition:
Other: Vapor Pressure: mmHg @ °F

Chemical Name Here: Appearance:
n-Boiling Point: Mol. Formula: Water Solubility:
Melting Point: Mol. Weight: Vapor Density:
Flash Point: Density: Auto Ignition:
Other: Vapor Pressure: mmHg @ °F

N/A=Not Applicable **ND**=No Data **NonD**=Non Detectable **SADT**=Self-Accelerating Decomposition Temperature

The materials summary contains information primarily taken from the SDS for the chemicals involved in the process and is a very handy reference for an operator without having to pull the actual SDS. If there are critical parameters, such as a self-accelerating decomposition temperature (SADT) or auto ignition temperature, that needs to be noted – typically noted here, in the critical process parameters, and within the relevant step of the procedure.

6.1.2 Alarms and Interlocks

Figure 4 displays a customizable table that illustrates the alarms for a particular vessel, the tag name corresponding to the computer screen in the control and the field, the type of alarm and associated units, and the action when the alarm is triggered. If an alarm also triggers an interlock, that information is noted below the alarm table.

Figure 4. Identification of Alarms and Interlocks

Vessel Customer Product Production/Distillation

Alarm Tables:

Reactor 1 Alarm Table				
Process Parameter	*Tag*	*Alarm*	*Value*	*Action*
R1 Reactor Temperature	TIC_R1_102	High-High	°F	Bannerlist, SPR Horn, See NOTE 1.
		High	°F	Bannerlist
		INC ROC	°F/min	Bannerlist, SPR Horn
		Low	°F	Bannerlist
			Note: Do Not enable Low alarm until Step ___.	
R1 Reactor Pressure	PIC_R1_103	High-High	"Hg	Bannerlist, SPR Horn
		High	"Hg	Bannerlist
		INC ROC	"Hg/min	Bannerlist
R1 Hot Oil Supply Temperature	TIC_R1_500	High-High	°F	Bannerlist, SPR Horn, See NOTE 2.
		High	°F	Bannerlist
		High Dev	°F	Bannerlist, SPR Horn, See NOTE 2.
R1 Coil Steam Pressure	PIC_R1_600	High-High	"Hg	Bannerlist, SPR Horn, See NOTE 3.
		High	"Hg	Bannerlist
		Low	"Hg	Bannerlist
R1 Receiver Level	LIC_R1_401	High-High	%	Bannerlist, SPR Horn
		High	%	Bannerlist
		Low	%	Bannerlist
R1 Agitator Motor Current	II_R1_104	High-High	AMPS	Bannerlist, SPR Horn
		High	AMPS	Bannerlist
		Low	5 AMPS	Bannerlist, SPR Horn, See Note 3.
			Note: Do Not enable Low alarm until Step ___.	
R1 Column Differential Pressure	DPI_R1_150	High-High	mm Hg	Bannerlist
		High	mm Hg	Bannerlist
N2:O2 % Oxygen	R1R2_N2O2_ALM	Low	%	Bannerlist, SPR Horn
N2:O2 Flow	R1R2_FLOW_ALM	Low	SCFM	Bannerlist, SPR Horn

NOTE 1: Sets TIC_R1_500 to Manual = 0.0 in standby sync (SS) mode. Sets TIC_R1_600 to Manual = 0.0 in standby sync (SS) mode.
NOTE 2: Sets TIC_R1_500 to Manual = 0.0.
NOTE 3: Sets TIC_R1_600 to Manual = 0.0 in standby sync (SS) mode.

6.1.3 Critical Process Parameters

Figure 5 is a page in the SOP that is used to inform the operator of specific critical process parameters or other important aspects of the process to adhere to. This page includes pertinent PSI, as well as information that pertains to how to operate the process per the facility's air, hazardous waste, and wastewater permits. This page is meant to be a good introduction to the critical parameters before the operator gets into the individual steps of the process.

It is also on this page that critical information is given as to when to perform a shutdown of the process, as compared to an emergency shutdown, and includes the triggers for each type of shutdown. Process engineers fill in specific temperatures, pressures, or temperature rate of change information indicating a need to perform an emergency shutdown, along with information regarding how to perform the emergency shutdown and how to start up after an emergency shutdown. Consequences of deviating from parameters such as pressure, temperature, flow, order of addition, or others can also be added to the particular parameter in this section or in individual steps in the process.

Figure 5. Critical Process Parameters

Critical Process Parameters:

- *For simple blends only: state:*
    - Minimum batch size ___________ lbs.
    - Maximum batch size ___________ lbs.
- ***Agitation*** –
- ***Alarms*** - Alarms must be set prior to start-up based on Alarm Table.
- *Capacity* –
- *Clean Outs* –
- *Cooling/Heating* –
- *Material Handling/ Transfer* –
- *Pressure/Vacuum* -
- *Rate of Addition/Order of Addition* --
- *Regulatory* –

  Choose one
    - This process must be vented to the thermal oxidizer. This process must be shut down if the thermal oxidizer malfunctions. The _____ Scrubber should be included in the vent line routing. Document actual time process was shut down and/or blocked from venting due to Control Device Malfunction in the Control Device Malfunction Log. If process cannot be shut down immediately due to safety reasons, document reason for delay lasting longer than 1minute in Control Device Malfunction Log.
    - For South Side processes (not running formaldehyde or a very odiferous compound): This process must be vented to the **South Scrubber**. This process may use the oxidizer. This process must be shut down if the thermal oxidizer malfunctions and the **South Scrubber** fluid exit temperature exceeds 60°F. Document actual time process was shut down and/or blocked from venting due to Control Device Malfunction in the Control Device Malfunction Log. If process cannot be shut down immediately due to safety reasons, document reason for any delay lasting longer than 1minute in Control Device Malfunction Log.
    - For R6 and centrifuge when running ammonia: The pH of the fluid in the scrubber that removes ammonia from the suction-vent-gas stream should be kept between 5 and 9. Samples must be taken every 6 hours to confirm that the pH is within this range. This process must be shut down if the scrubber pH exceeds 10.2. Document actual time process was shut down and/or blocked from venting due to Control Device Malfunction in the Control Device Malfunction Log. If process cannot be shut down immediately due to safety reasons, document reason for any delay lasting longer than 1minute in Control Device Malfunction Log.
    - For R6 when running an acid HAP (HCl, Sulfuric Acid, Nitric Acid): The pH of the fluid in the scrubber that removes an acid HAP from the suction-vent-gas stream should be kept at greater than 10.5. Samples must be taken every 6 hours to confirm that the pH is acceptable. This process must be shut down if the scrubber pH drops below 10.1. Document actual time process was shut down and/or blocked from venting due to Control Device Malfunction in the Control Device Malfunction Log. If process cannot be shut down immediately due to safety reasons, document reason for any delay lasting longer than 1minute in Control Device Malfunction Log.
    - For B4 only: Document whenever the process is shut down and/or emissions are blocked from venting on the "B4 Batch Log." The B4 process must/must not be routed through the C29 scrubber. If the C29 scrubber is being used, samples must be taken to the lab every 6 hours to monitor pH.
    - For OCPSF regulated processes: Water is OCPSF regulated and cannot be discharged without prior regulatory approval. The amount of recovered water must be tracked and recorded for offsite disposal or pre-approved discharge.
- ***Safety*** – Wear goggles or FF Respirator whenever blowing out lines or disconnecting pressurized lines of any kind.
- ***Shipping*** –
- ***Stoichiometry*** –
- ***Temperature*** –
- ***Vacuum Pump*** –
- ***Waste*** –
- ***Other*** - Operators must fill in all gray boxes before this SOP becomes valid to run per Building Manager's notes.
- ***Causes for Shutdown*** – Weather, fire, leaks, loss of controls, loss of Agitation
- ***Shutdown Procedure*** –
    1. Disengage heat.
    2. Deadhead vacuum pump.
- ***Emergency Shutdown Triggers*** - large exotherm, loss of agitation, high amps, temperature spike or pressure spike
- ***Emergency Shutdown Procedure*** -
    1. Disengage heat.
    2. Deadhead vacuum pump.

6.1.4 Batch Procedure Steps

Figure 6 shows an example of what the individual procedure steps in a batch might look like. Each process section and step is numbered and each step is to be initialed by the operator once completed.

Figure 6. Batch Procedure Steps

No.	Procedure Description	Initials
II.	**Initial Charges:**	N/A
15.	**Record start date and start time on Prism Log (Charge Time).**	
16.	Route lines to from location C to vessel D. **Have a 2nd Operator Check Line Routing: ________ (Initials).** **PPE: FF(MG), 48RJ, etc.**	
17.		
18.	Route lines to charge Material A (Resource Number, i.e. XX-####. Only include the last two digits if only one container type is available for charge) from location C to vessel D. **Have a 2nd Operator Check Line/Material Routing: ________ (Initials).** **PPE: FF(MG), 48RJ, etc.**	
19.	Charge X lbs. of Material A (Resource Number, i.e. XX-####. Only include the last two digits if only one container type is available for charge) into Vessel Z **Record appropriate information in the Prism Log. Lot Number Required.** **PPE: FF(MG), 48RJ, etc.**	
20.	**For monomer charging (or other material per building manager/engineer) via bulk:** **Initial Load Cells: ______________ lbs. Final Load Cells: ______________ lbs.** **Amount Charged: ____________ lbs. by Load Cells** **Amount Charged: ____________ lbs. by Scale Weight** **Amount Charged: ____________ lbs. by Mass Meter** **Initial Tanker Weight ______________ lbs. Final Tanker Weight ______________ lbs.** **Amount Charged ____________ lbs. by Tanker Weight** **Amount Charged: ____________ inches by Level Transmitter**	
21.	**For monomer charging via drums:** **Initial Load Cells: ______________ lbs. Final Load Cells: ______________ lbs.** **Amount Charged: ____________ lbs. by Load Cells** **Amount Charged: ____________ lbs. by Scale Weight**	
22.	**Other Charging:** **T-________ Initial Level: __________ lbs. T-________ Final Level: __________ lbs.** **Amount Charged: ____________ lbs. by Level Change**	
23.	Below are the standard formats that must be used for 2nd Operator/Supervisor Checks of various items. • The standard format for line routing when transfer of material is involved: **Have a 2nd Operator Check Line/Material Routing: ________ (Initials).** • The standard format for line routing of equipment, for exampling, route Column 1 vent to the Thermal Oxidizer, etc. when materials are not involved: **Have a 2nd Operator Check Line Routing: ________ (Initials).** • The standard format for checking the alarm settings: **Second Operator's Alarm Check: ________ (Initials).** • The standard format for a Supervisor check in an SOP: **Supervisor Check: ________ (Initials).** • The standard format for a verifying the location of a raw material or charge calculation: **Have Second Operator Confirm: ________ (Initials).**	
24.		
25.		
26.		
27.		
28.	**Record stop date and stop time on Prism Log (Charge Time).**	

The amount of information can vary in each step depending on the operation the operator is to perform. As can been seen in steps 18 and 19 above, the resource number for the raw material is referenced, just as it was in Figure 4, which is another means for the operator to verify the correct material is being charged.

The facility also employs second operator and supervisor cross-checks on raw material charging for the correct chemical, amount, and routing to help prevent mischarges. Any safety notes particular to a step are also added when warranted.

With the high number of changes that are made within batch tolling operations, operating procedures must continually be updated to account for changes and the specific steps of the process. The SOPs must also be able to accurately convey the correct operating conditions at each step along with the sequence of materials and any changes that are to be made that are inherent to the process such as adjusting a temperature alarm in a particular step in a process. Operator error in not properly adjusting a temperature alarm can lead to out of specification products or result in a catastrophic consequence such as a runaway reaction. A robust MOC program must be able to capture the nuances of batch systems and make sure all changes are covered along with their potential effects.

Figure 8 illustrates the log that is used to convey what changes have been made to the SOP for a revision number, when the change was made, and who made the change. The MOC number will be added to a change entry when applicable. The facility has strict requirements on when an MOC is required for a change to an SOP as denoted by the "Risk Level" column. An operator can use this log to see when the SOP was last revised and what changes were made. If an MOC was involved with the change to the SOP, the operator would also receive training or notification of that change depending on the complexity of the change.

Figure 7. Change History Log

CHANGE HISTORY LOG				
Rev #	**Risk Level**	**Date**	**Description or reference MOC#.**	**Initials**
64	N/A	xx/xx/xx	1. Removed DO NOT INCLUDE WEEKEND DOWN TIME from Vessel Usage table per.	ABC
65	N/A	xx/xx/xx	2. Updated Alarm Table per xx/xx/xx email from ABC.	ABC

7 Human Factors

Reliance on operators is also inherent to batch operations. There clearly are potential issues due to human factors as well as risks from human error in batch operations that are much more difficult to minimize as compared to continuous operations. Operators play a key part in safe operations of batch operations, especially in batch toll operations due to the many changes in the processes that are being run.

Operators can be expected to monitor processes for changes in temperatures, pressures, flow rates of raw materials, and other critical process parameters. Not only are they monitoring them, they are also diagnosing any problems that may arise and how to properly respond to those problems. Because of the ever changing chemical compositions and states reactions can be in, this adds difficulty in an operator's job to monitor, diagnose, and address any problems that arise during the batch. (Center for Chemical Process Safety, 1999)

Due to the manual nature of batch operations, operators are also located closer to the process hazards. Proper location and design of piping manifolds, headers, or other areas where there are multiple hose connections or valves is paramount to enhancing the operators' ability to choose the correct connection or open or close the correct valve and reduce human error. Error can be reduced with the aid of proper marking, unique connections, color-coding valve handles, and other methods all the way up to batch sequencing in a control system. (Center for Chemical Process Safety, 2012)

8 Conclusion

While batch operations play a major role in providing needed products to society, inherent process safety challenges exist. Careful and detailed processes can be put in place to help streamline process safety evaluations and management of risks from conception and screening to production - to ensure process safety hazards are being reviewed and evaluated and continually evaluated for the life of the process. Based upon industry experience, importance must be placed on reviewing chemical reactive hazards and the potential for runaway reactions in batch operations. Addressing and managing these hazards utilizing experienced personnel and formal process safety management processes can put batch manufactures ahead on the process safety curve.

9 References

(SOCMA), S. o. (2016, May 16). Batch Processing and Process Safety. *Presentation to the EPA.* Washington, D.C.: SOCMA.

Administration, O. S. (1994). *OSHA.* Retrieved from OSHA: https://www.osha.gov/Publications/osha3133.html

Bridges, W., & Clark, D. T. (2011). How to Efficiently Perform the Hazard Evaluation (PHA) Required for Non-Routine Modes of Operation (Startup, Shutdown, Online Maintenance). *American Institute of Chemical Engineers, 2011 Spring Meeting, 7th Global Congress on Process Safety.* Chicago.

Center for Chemical Process Safety. (1999). *Guidelines for Process Safety in Batch Reaction Systems.* New York: American Institute of Chemical Engineers.

Center for Chemical Process Safety. (2008). *Guidelines for Hazard Evaluation Procedures.* New York: John Wiley & Sons, Inc.

Center for Chemical Process Safety. (2012). *Guidelines for Engineering Design for Process Safety, Second Edition.* New York: John Wiley & Sons, Inc.

U.S. Chemical Safety and Hazard Investigation Board. (2006). *Investigation Report - Toxic Chemical Vapor Cloud Release - MFG Chemical, Inc. - Dalton, GA - April 12, 2004.*

U.S. Chemical Safety and Hazard Investigation Board. (2007). *Case Study - Runaway Chemical Reaction and Vapor Cloud Explosion - Synthron, LLC - Morganton, NC - January 31, 2006.*

U.S. Chemical Safety Hazard and Investigation Board. (2000). *Investigation Report - Chemical Manufacturing Incident - Morton International, Inc. - Paterson, NJ - April 8, 1998.*

13TH GLOBAL CONGRESS ON PROCESS SAFETY

A HAZOP Gap Query Approach for Hazard Assessment in Multi-Product Facilities

James Torres, FSP
Albemarle®
P.O. Box 341
Baton Rouge, LA 70821
james.torres@albemarle.com

Nick Ashley, Ph.D., P.E.
Albemarle®
451 Florida Street
Baton Rouge, LA 70801
nick.ashley@albemarle.com

Prepared for Presentation at
American Institute of Chemical Engineers
2017 Spring Meeting and 13th Global Congress on Process Safety
San Antonio, TX
March 26 – 29, 2017

A HAZOP Gap Query Approach for Hazard Assessment in Multi-Product Facilities

James Torres, FSP
Albemarle®
P.O. Box 341
Baton Rouge, LA 70821
james.torres@albemarle.com

Nick Ashley, Ph.D., P.E.
Albemarle®
451 Florida Street
Baton Rouge, LA 70801
nick.ashley@albemarle.com

Keywords: PHA, HAZOP, LOPA, Batch, Multi-Product, Hazard Assessment

Abstract

Batch, multi-product chemical processing facilities present special challenges for conducting thorough & timely hazard assessments. New or improved methods are needed to perform quality hazard analyses while reducing the time and resource demands on facility personnel. A technique has been developed to expand the traditional HAZOP methodology to include additional parameters for process operational intent, chemicals, and affected upstream/downstream equipment. Integrated Layer of Protection Analysis (LOPA) is applied as required to assess existing or proposed Independent Protection Layers (IPLs), and to compare resulting mitigated risk to established risk criteria limits. The resulting HAZOP/LOPA for each unit then becomes a baseline hazard assessment, to which subsequent processes/configurations can be compared, without having to repeat the full HAZOP for each new process. For new products, a process intent template is completed by selecting the chemicals and equipment to be used, as well as the pertinent unit operations (e.g., reaction, stripping, product isolation, etc.) and conditions for each step in a semi-batch process. The unit HAZOP file is electronically queried using programmed search logic to identify matching process intent/equipment/chemicals parameters that were considered during the base HAZOP exercise. Finally, a condensed Process Hazard Review (PHR) is conducted to review the query results, ensure the identified items are applicable to the proposed new process, and to evaluate new items that were not previously covered under the base HAZOP.

1 Introduction

The Hazard and Operability (HAZOP) methodology based on "critical examination" techniques was first introduced at ICI in the 1960s and 1970s with the first publicized reference work by Lawley appearing in 1974 [1]. Since then, HAZOP methodology has been improved and standardized by Kletz and many others [2-4] using a traditionally recognized process parameter-guideword-deviation format. The HAZOP technique has become the most commonly accepted hazard analysis methodology and HAZOP studies have subsequently been extensively conducted world-wide to identify potential hazards in chemical process units. This structured approach results in a thorough unit hazard analysis, but requires significant time and personnel resources. HAZOP methodology is often employed for hazard analysis of dedicated operating units producing singular products. Other hazard analysis techniques such as "What-If" or "What-If/Checklist" methodologies [5,6] are often used for simple processes but various methodologies can be to varying degrees somewhat tedious and/or redundant when chemicals, equipment, and human failure scenarios are common across typical process unit operations. Conduct of thorough Process Hazard Analyses (PHAs) can become an onerous exercise to meet regulatory and/or company-specific process safety requirements for multi-product facilities with frequent new product introduction or multiple related processes producing different, though very similar products with changes in raw materials or process conditions.

Batch and semi-batch chemical processes are frequently employed in the specialty chemicals, pharmaceutical and electronic materials industries [7]. Special challenges associated with batch processing include the flexibility and re-configurability of process equipment, manual operations & control, time-dependent component additions, large-volume static heating and cooling, and frequent sampling to determine process endpoints. These process conditions lead to other challenges associated with the Process Safety Information (PSI) and other fundamental tenets of risk-based process safety. The same flexibility of connection and reconfiguration of the process equipment which makes multi-product facilities versatile also taxes other resources to maintain up-to-date Piping and Instrumentation Diagrams (P&IDs), review numerous Management of Change (MOC) requests, and conduct frequent hazard analyses for new chemicals, unit operations, and process configurations. For these reasons, batch and semi-batch PHAs are often more complex and time consuming as compared to studies for continuous processes [8]. New or improved methods are needed to perform quality, thorough hazard analyses for batch chemical processes in multi-product facilities, while reducing the time and resource demands on facility personnel. Because the type of chemicals used at the facility may include pyrophorics, toxic inhalation hazard materials, compressed or highly flammable gases/liquids, and Active Pharmaceutical Intermediates (APIs), any new or improved methods must be shown to be rigorous and thorough, while not compromising assessment quality for the sake of efficiency.

1.1 Proposed Solution

The novel approach presented herein extends typical HAZOP methodology past a singular-focus view of specific item deviation-hazard-consequence consideration to a robust assessment tool for new common-theme process consideration. Individuals with previous HAZOP experience are cognizant that the exercise is thorough in identifying potential hazards and configuration deficiencies as evident by the typical summary output, often comprised of hundreds to thousands of line item entries. While the exercise is initially useful in identifying areas for pre-commission mitigation activities, usefulness for existing operating units may be overshadowed by the sheer volume of data. How can one use the generated data to effectively identify hazards of a slightly different process using the same equipment without having to repeat the entire exercise? A logic-searchable relationship between existing hazard analysis data and proposed intent can be established if additional parameters are assigned for defined unit operations, chemicals employed, and affected upstream/downstream equipment associated with the line item under consideration within a HAZOP. This approach parametrically associates multiple potential individual deviations with defined unit operations within the HAZOP.

Proposed defined unit operations, chemicals used, and affected upstream/downstream equipment are then specified for a new process and a subsequent search match between the proposed process and existing HAZOP can then quickly identify all deviations previously considered for the specified operations. The hazard analysis for the new process becomes more efficient by allowing the PHA team to spend more time on hazards unique to the proposed new process, with a concise review of line items that were previously considered in the base HAZOP. The method is able to simultaneously preserve a thorough analysis of the hazards, controls, and consequences of failure associated with the new process, while reducing the time required to review operational steps and configurations which have already been comprehensively assessed in the base HAZOP.

1.2 Comparison to Traditional HAZOP

The differences between traditional HAZOP and the novel approach presented here can be illustrated as follows. By way of analogy, standard output from a HAZOP exercise can be compared to a container with a large number of identically colored spheres, as depicted in Figure 1. Other than simple Node-item identification tags, of which there may be hundreds to thousands, it is not possible to easily query the data to identify specific items associated with particular unit operations, chemicals, and equipment configurations.

Figure 1. Typical HAZOP Output without Unique Data Identifiers

When a slightly different process or equipment configuration is proposed, no method exists for querying the existing data to determine which operating steps, chemicals, and equipment configurations have been previously addressed by the unit HAZOP. Currently, the only options are to exhaustively review the unit HAZOP manually, line by line, to look for matching operation steps, or completely repeat the HAZOP for the new proposed process. Either choice places a significant burden on facility resources when new processes may enter the multi-products facility several times per month. This dilemma is graphically illustrated in Figure 2.

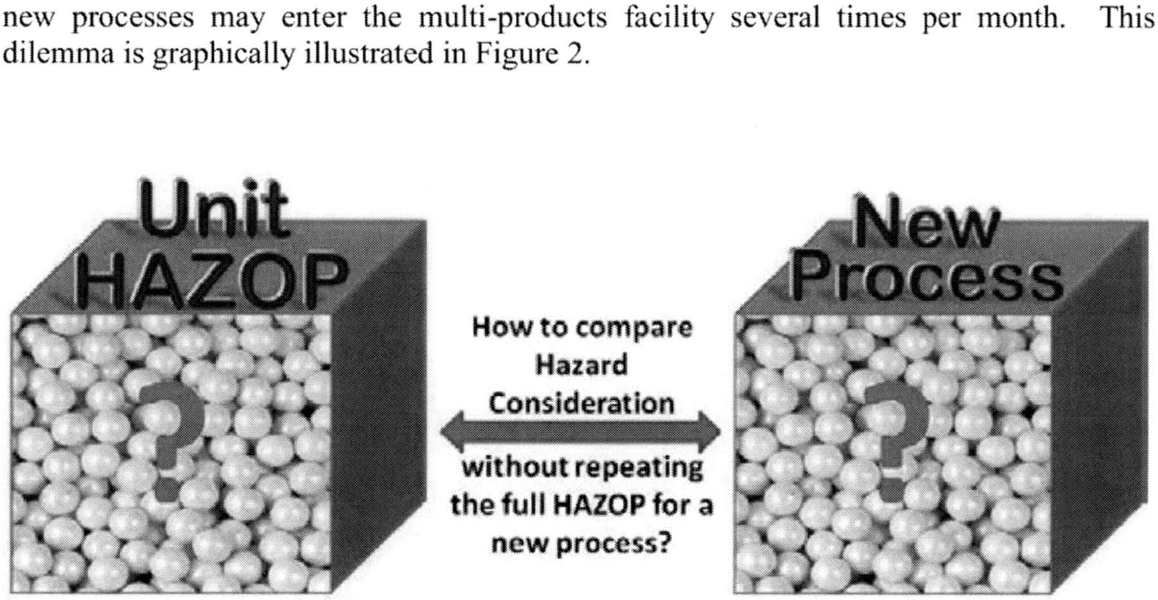

Figure 2. New Process Hazard Comparison Dilemma

By employing the method proposed in this paper, the base HAZOP template is modified to include additional parameters for defined unit operations, chemicals used, and affected upstream/downstream equipment. The individual HAZOP line-items are now marked with unique, searchable identifiers. This is analogous to tinting the spheres inside the

HAZOP output with distinguishable colors, which are easily identified and can be quickly sorted, as shown in Figure 3. Each unique color then represents a specific combination of unit operation, chemical component, and configuration of several pieces of process equipment. Individual line items within the HAZOP may be made up of single operations, and therefore single colors, or multiple operations, and therefore multiple colors as shown in Figure 3.

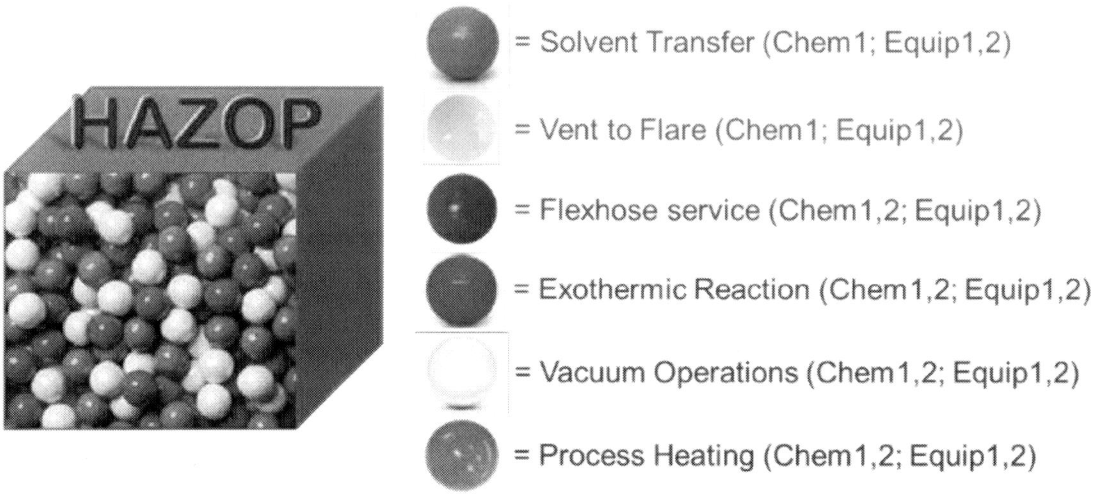

Figure 3. HAZOP Output with Unique Data Identifiers for Each Line-Item

Querying the data then allows identification of singular or multiple parameters which may be common across several different processes or configurations. For example, consider a process in which an exothermic reaction occurs between two chemicals with a specific overhead condenser configuration. In Figure 3, this would be analogous to selecting all of the red spheres from the base HAZOP. Consider a different operating step in the process, in which a solvent is charged to a vessel through a flexible hose. In Figure 3, this would be analogous to selecting all of the green spheres which are also associated with blue spheres. By specifying new parameters of defined operation, affected chemicals, and affected equipment in the unit HAZOP, a logic searchable parametric association becomes possible. By applying these parameters within the base HAZOP and then applying the same parameters to a process intent template for the proposed new process, the unit HAZOP data can be electronically queried to quickly identify operating conditions for the new process that have already been covered by the base HAZOP, as shown in Figure 4.

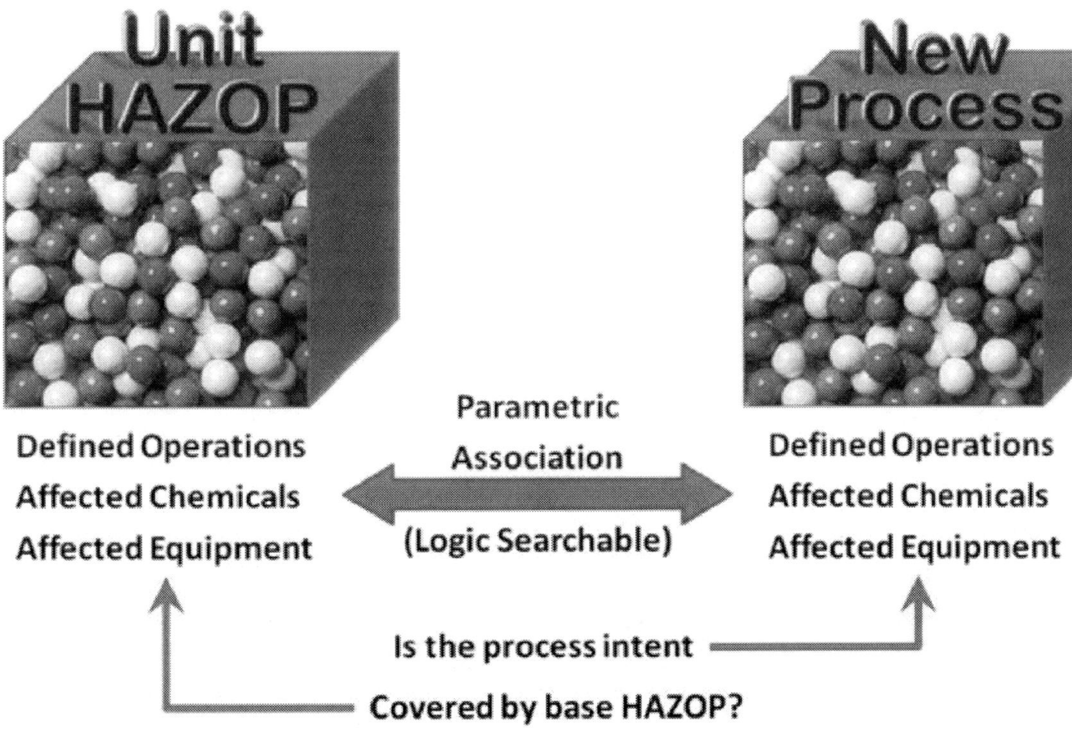

Figure 4. Establishing Parametric Association between HAZOP and New Process

1.3 Scope of the New Method

The gap query hazard assessment methodology consists of four major efforts: (1) one-time modification of the HAZOP platform to allow inclusion of additional searchable parameters during execution of the unit HAZOP, (2) creation and periodic population of a proposed process intent survey template for new processes, (3) importation of the proposed process intent data into the base HAZOP file and execution of the gap query, and (4) condensed Process Hazard Review (PHR) to compare proposed process parameters to items previously considered in the base HAZOP. To ensure thorough coverage of all hazards associated with the new process, the PHR additionally includes critical reviews of:

- specific chemical hazards
- process chemistry and reactive hazards
- inadvertent mixing hazard/potential
- synthesis reaction energy balances
- existing or proposed alarms/interlocks
- previous incident history

Each of the four steps in the above methodology will be described in detail below.

2 HAZOP Modification by Addition of New Parameters for Batch Process Hazard Analysis

The crux of the improved HAZOP method for batch process hazard analysis is based upon the addition of several key parameters to the individual line-items within each node of the HAZOP. These additional parameters are: (1) process operational intent, (2) chemical utilization, and (3) affected upstream/downstream equipment. In typical batch chemical processes, the major fixed process equipment (i.e. reactors, mixing tanks, distillation columns, large filters, dryers, etc.) are permanently located within the process structure of each unit. Smaller equipment, such as cartridge or Nutsche filters, may be brought in and located within designated areas within the unit footprint. Specific reactants or intermediates are often received in small, portable containers such as portable tanks, sample bombs, or pre-weighed sterile bags, and are charged to the process from the original containers.

While the physical location of the fixed process equipment may not change, the order and interconnectivity among equipment may differ from process to process. Overhead condensers or pump-around heater/coolers may be re-piped to be utilized by one of several different vessels depending upon the specific process needs. It is this flexibility of equipment configuration which makes the facility versatile and readily adaptable for the manufacture of a wide variety of products that directly contributes to the company's success. At the same time, the flexibility and near constant changing of chemical processes and equipment configuration creates an environment where new batch process hazard analyses become increasingly complex, frequent, and resource-intensive. Thus, the primary objective of the modified HAZOP methodology described in this paper is to maintain the ability to perform a thorough, high-quality risk assessment for each process that comes through the facility, while reducing the time and resource demands on the facility personnel due to frequent process configuration changes or new product introduction.

2.1 Process Operational Intent

Process Operational Intent describes the primary unit operation(s) being performed in an individual step or line-item within the HAZOP. These may include: reactant charge from fixed tank or portable container, exothermic/endothermic chemical reaction, stripping/distillation of byproducts and impurities, crystallization or filtration, solids drying, pneumatic conveying, process heating or cooling, and many others. By capturing the major unit operations occurring within a process as a dedicated parameter, future processes which utilize identical or nearly identical operations can be quickly screened and identified. Common or recurring steps such as solvent charges, and similar reactions or separations, may be imported into the HAZOP for the newly proposed process and re-used without the need to repeat the exercise for each common step of each new process.

2.2 Chemical Utilization

Chemical utilization involves definition of both the specific chemical employed in each step of the process, as well as the primary hazard(s) associated with that material. For example, if hydrogen was being charged from compressed gas cylinders to a reactor, the chemical utilization would list hydrogen, and flammable compressed gas. Charging toluene to a vessel would list toluene and flammable liquid solvent. By capturing as many details as possible about the specific chemicals and associated major hazards, the query for a new process can identify steps where similar hazards may be present, even if the chemicals are not exactly identical. For example, charging ethylene from a compressed gas cylinder instead of hydrogen would generate a match on the flammable compressed gas parameter, and charging hexane instead of toluene would match on the flammable liquid solvent parameter. The PHA team is then able to evaluate the matched operational steps and compare the properties of the new flammable compressed gas or flammable solvent with those of the original material, along with the equipment operating limits, to determine if the initial hazard assessment remains adequate for the new chemicals.

2.3 Affected Upstream/Downstream Equipment

As previously discussed, due to the flexibility of equipment configuration for a process even amongst the fixed process equipment, the specific configuration employed is a critical parameter to capture when evaluating HAZOP scenarios across multiple processes. Affected equipment may include heating/cooling, material collection and storage, venting to a flare or other destruction device, and many others. For example, an exothermic reaction in which a knockback condenser is used to help remove heat from the reactor and an exothermic reaction in which all heat is removed by the reactor jacket or a pump-around exchanger are not similar configurations. Therefore, when identifying scenarios in which process unit operations and equipment configurations are similar from one process to the next, the connectivity of both upstream and downstream equipment is a key parameter in the expanded HAZOP methodology proposed herein.

With the three new HAZOP parameters described, the discussion now turns to the details of using the modified HAZOP method within a batch process PHA.

3 PHA Platform Modification

Microsoft® Excel® was chosen as the Process Hazard Analysis (PHA) software platform for the diverse customization capabilities afforded by the integrated Visual Basic for Applications (VBA) programming functionality. The author has developed and refined an Excel template that provides both depth-of-analysis and flexibility of use in PHAs, by incorporating the ability to perform What-If, HAZOP, and/or LOPA within the same worksheet. The existing integrated platform incorporated standard HAZOP data fields for process parameter, guideword, and resulting process deviation. Additional data fields were added for defined process operational intent, chemical utilization, and affected upstream/downstream equipment, as shown in Figure 5.

Node	Item	**Traditional HAZOP**			**Additional Parameters for Modified HAZOP**			
		Process Parameter	Guide Word	Deviation	Process Operation	Process Chemical	Source Equipment	Downstream Equipment
#	#							

Figure 5. Addition of New Parameters for the Modified HAZOP Method

Data entry Userforms containing data selection Listboxes populated from master data lists were created for selection consistency and data input time reduction. Userforms are custom-built dialog areas containing enhanced data entry and user interaction control options that can be displayed on demand by code execution. Userforms can be used to collect, display, manipulate, and verify information from multiple source options before being transferred to a worksheet. Listboxes are configurable controls placed in a Userform to let users make multiple choices of items from a specified list or to enter their own value in a list [9]. Master data lists were created for defined standard process operations, typical chemical usage, and equipment by unit operating area. Prior to performing a HAZOP study, the facilitator pre-configures the platform by utilizing Userforms to select the specific operating area and utilized equipment, applicable chemicals, and defined operations from the master data lists. The limited selection lists then become the specified selection lists displayed when populating the extended HAZOP parameters by Userform guidance. A unit HAZOP study is subsequently performed per standard methodology practice with the additional tasks of completing the data fields for defined operations, chemical utilization, and affected upstream/downstream equipment associated with the line item of interest. An example of HAZOP execution with the newly added parameters is shown in Figure 6.

4 Process Intent Survey Template

A standardized process intent template was created for site deployment. Unit process engineers input process intent data for any new process introduced to the respective operating unit. The master data lists utilized in the HAZOP platform were duplicated in the process intent survey file for search consistency. Engineers then simply manipulate a series of form control buttons (aka "push buttons") executing macro code to display Userforms with Listbox data selections to choose existing or specify new proposed chemicals and equipment usage by operating area. Targeted process steps with operating conditions are entered and a form control button displays a Userform/Listbox to select pre-defined specific operational steps that match targeted operations for the proposed process. Other form control buttons are used to select chemicals and upstream/downstream equipment globally specified for the process to associate those items with the current process step.

HAZOP Process Parameter	Deviation Guide Word	Process Deviation & Generic Hazard	Defined Unit Operations	Affected Process Chemicals	Affected Upstream Equipment	Affected Downstream Equipment
Flow	More	High Flow	Solvent feed from header (Iso source), Vent to flare header	Toluene, Flammable Liquid	T-100, P-100, 6,000 gal Iso	R-001, D-500, FS-500
		Erroneous flow totalizer indication, Excessive level condition, Overflow to dry flare header, Overflow secondary containment, Overflow elevated flare stack, Release of flammable liquid from elevated position, Snuff out flare pilot flame, Potential Vapor Cloud Explosion (VCE) consequence				

Figure 6. Example of Modified HAZOP with New Parameters Applied to a Flow Deviation

5 Data Import/HAZOP Gap Query

The HAZOP software platform was modified to include a user selectable data import option where proposed process intent template data can be automatically imported into the base HAZOP file. The lead process engineer fills in the template for each step of the semi-batch process. The template includes all pertinent steps of the process, including chemical charges or downloads, process heating/cooling, product purification, filtering/drying, and many others. In addition, expected operating conditions for each step are also included, such as pressure, temperature, mass and volumetric charge/flow, and expected duration. These parameters enable comparison with specific equipment safe operating limits during the gap query and condensed Process Hazard Review. An example illustrating a condensed portion of the process intent data after it has been imported into the HAZOP file is shown in Figure 7.

Specific Process Step Input Data:

Step-#	Sub-#	Process Step Intent	Defined Process Operations Associated with Intent	Process Chemical	Upstream Equipment	Downstream Equipment
1	1	Charge toluene to R-001	Solvent feed from header (Iso source), Glass in hazardous service (flammable vapor)	Toluene, Flammable Liquid	T-100, 6,000 gal Iso	R-001, D-500, FS-500
	2	Charge Reactant-1 to R-001	Material feed from portable tank, Material feed with exothermic reaction in jacketed vessel, Process Utility cooling, Glass in hazardous service (flammable vapor)	Reactant-1, Flammable Liquid	430 gal PT	R-001, D-500, FS-500
	3	Hold R-001 at 30°C for 0.5 hours	Isothermal process hold, Glass in hazardous service (flammable vapor)	Flammable Liquid	R-001	R-001

Figure 7. Process Intent Data after Import

Once the data has been imported, a gap query macro command can be executed to query the base HAZOP file. The heart of the gap methodology is based on programmed hierarchal search logic to automatically perform the gap query. The query search proceeds through three phases: First, proposed process chemicals are queried against global chemical usage specified during the base HAZOP. Exact match instances are noted with affirmation versus a "no match" indication. Second, proposed equipment usage is queried against global equipment usage specified during the base HAZOP. Exact equipment match instances are noted with affirmation versus a "no match" indication. These initial search match steps are used to globally identify instances of proposed chemicals and equipment that were not included during the base HAZOP. Noted exclusions will fail to match during the subsequent specific operation step/HAZOP line item comparison search, but indicating the exclusions in this area results in a time saving benefit during the condensed Process Hazard Review by immediately identifying chemicals and equipment that may require review. An example of global parameter search match results for affected chemicals and equipment is shown in Figure 8.

Finally, proposed defined operations are queried against specified defined operations assigned in each line item during the base HAZOP. Progressively specific sub-searches are performed during this step. First, the HAZOP is queried line-by-line for a general match to the specified proposed defined operation. Existing match instances are further queried for additional matches to the specified chemical(s) utilized during the operational step and then for a match to both affected upstream and downstream equipment. Resulting positive matches are respectively noted with the Node-Item number match reference identifier.

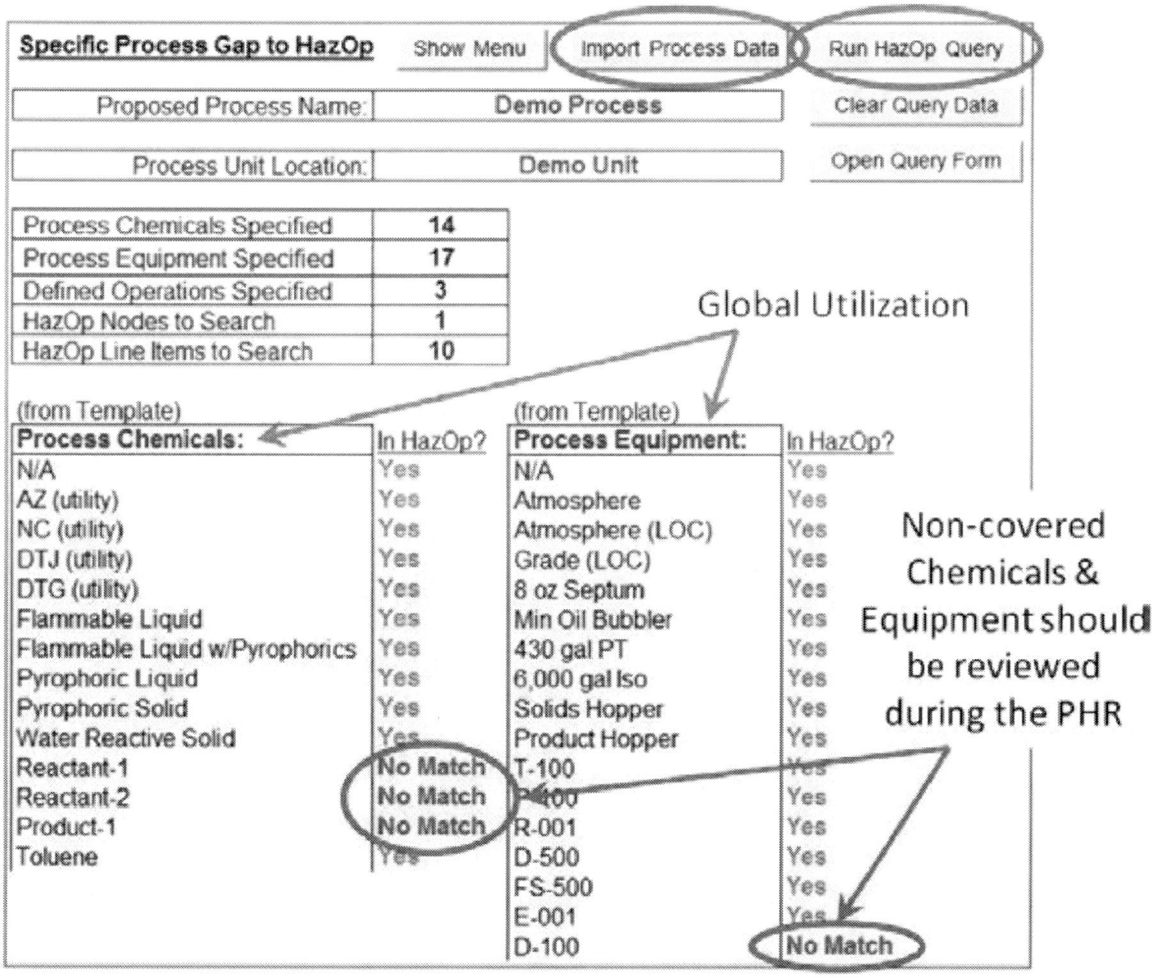

Figure 8. Global Search Match Results for Affected Chemicals and Equipment

Double-clicking on a valid Node-Item reference in the generated Gap Summary table executes code to automatically redirect spreadsheet focus to the HAZOP summary page and places the cursor on the referenced Node-Item for line review of HAZOP assessment specifics. This option significantly reduces review time requirements by eliminating the need to manually search through multiple nodes to locate item match instances. A condensed example of the resulting Gap Summary indicating search match results is shown in Figure 9.

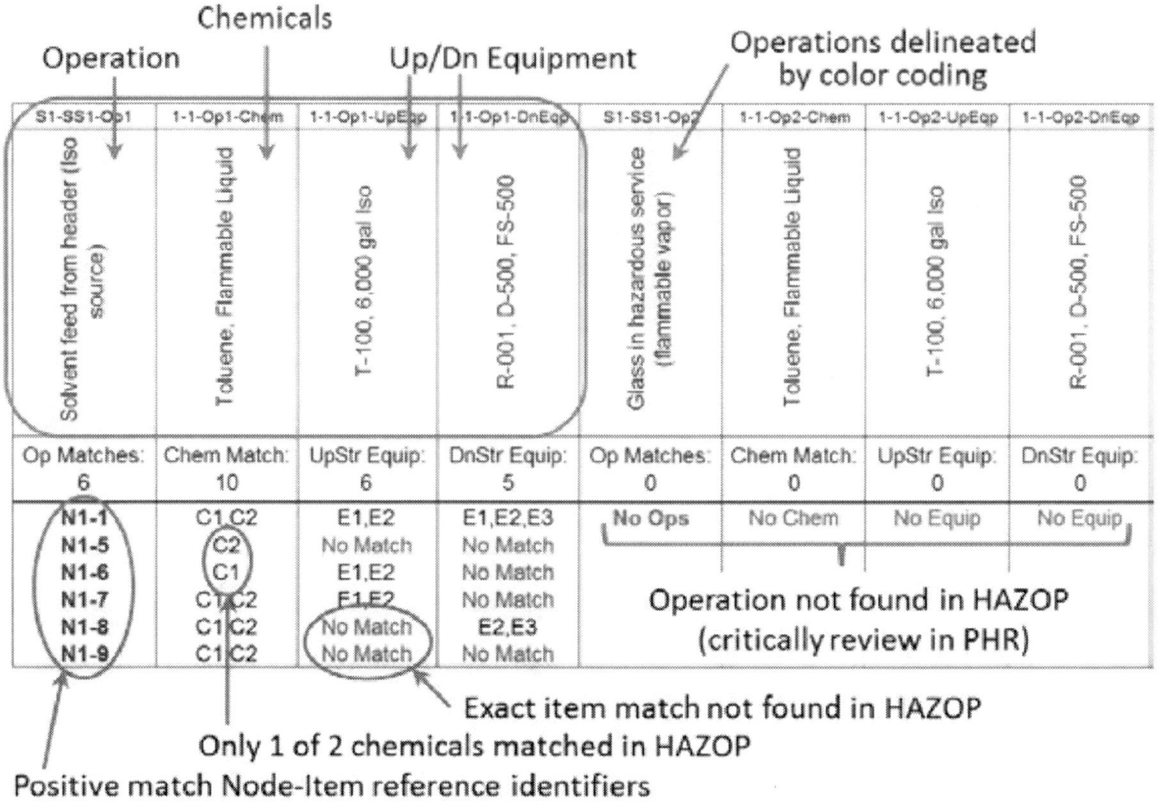

Figure 9. Gap Summary Table with Search Match Results

6 Condensed Process Hazard Review

The intent of the modified HAZOP gap methodology is to perform a thorough hazard analysis of newly introduced processes without having to repeat redundant formal assessments of typical generic failures previously considered for common unit operations, chemicals, and equipment. The underlying assumption is that generic equipment/human failures and hazards should be identified during a base HAZOP. Once the base exercise is complete, all that remains is consideration for any incrementally specific hazards introduced by a new process.

A condensed Process Hazard Review (PHR) is conducted to critically review specific hazards introduced by a newly proposed process to a level that can be proven to meet governing agency process safety or company-specific hazard identification consideration requirements. To consistently accomplish this goal, the PHR is guided by a checklist specifying standard review requirements with sections for review acknowledgement and observation notations. Much like a full PHA, the PHR begins with an overview of methodologies used to conduct the HAZOP and gap query. The newly proposed process is then subsequently reviewed starting with a process scope description, process flow diagram, markup Piping & Instrument Drawings (P&IDs) associated with the process, and proposed operating steps. The HAZOP gap query results are then reviewed for gap deficiencies to determine if the proposed process operational steps, chemicals, equipment, and conditions were adequately considered during the base HAZOP. Any gap result items noted as "not found" are critically reviewed and the standard HAZOP methodology may be employed to assess non-covered items as required.

Once the gap query assessment and any new items have been appropriately addressed, normal PHR activities are continued with critical reviews of specific chemical hazards and the process chemistry/Reactive Hazard Evaluation (RHE) report. The RHE process summarizes measured or calculated adiabatic temperature rise of synthesis reaction steps assuming instantaneous stoichiometric addition and temperature rise adjusted for total mass dilution effects. Differential Scanning Calorimetry (DSC) screening results of all raw materials, intermediates, side streams, and final products with summarization of total integrated exothermic/endothermic energy content and established Maximum Safe Temperature Limits (MTLs) for each stream are also provided. Additionally, any advanced calorimetry evaluation efforts and upset stream consideration for identified inadvertent stream mixing potential are included. Critical reviews of the process synthesis reaction energy balances are conducted and compared to the RHE report for rate and magnitude of temperature/pressure increase and heat generation by reactive material feed including hydraulic assessment of maximum feed capability by failure scenarios and dynamic simulation of semi-batch reactant feed versus heat removal capability. Existing and proposed Basic Process Control System (BPCS) alarms/interlocks are reviewed for set point determination compared to process requirements. Critical hardwire Safety Instrumented System (SIS) interlocks are reviewed for type/trigger/response action and current set point versus proposed process limitations and requirements. Opportunities for inadvertent mixing potential and proposed mitigation are assessed based on proposed process intent and equipment configuration. Previous incident history is reviewed for process applicability, corrective action response, and repeat potential. Once the PHR is completed, standard wrap-up activities are conducted such as: management reviews, corrective action item assignment, and completion of a final documentation report summarizing the review content and any identified action items.

7 Demonstrated Benefits

Typical semi-batch process PHA efforts at the author's facility historically require 2-8 days depending on the number of steps, chemical complexity, and equipment configuration. Initial limited usage of the modified HAZOP gap approach has demonstrated that a thorough gap assessment/review can be completed in 2-6 hours for standard common-theme processes. Aside from the initial time investment for the base HAZOP study and assuming comparison of a 40 hour full PHA versus a 4 hour gap assessment, it is evident that usage of the gap assessment approach can result in a 90% reduction in time resource demand. For a facility that has a historic demand of roughly 25-30 PHAs per year, this represents a significant resource benefit by allowing more effective time utilization for process turnaround preparation and more time availability to focus on other process safety demands. Additionally, the frequent review process effectively converts the base HAZOP into a living document, which ensures unit process hazard assessment is thorough to current operations instead of a buried data-vault that gets reviewed at regulatory or company defined intervals.

8 Conclusion

This paper introduces a modified HAZOP technique for performing thorough hazard analyses of multi-product facilities without the need to repeat redundant studies for unit operations, chemicals, and equipment used in typical processes with only minor theme variations. The technique described in this paper is being actively employed with encouraging results. Current usage of this method has produced significant reductions in time resource demands without sacrificing hazard analysis thoroughness for newly introduced batch or semi-batch processes.

The challenges associated with batch or semi-batch process hazard analyses are many; and the novel approach presented herein presents one method for reducing time requirements, while preserving high-quality and comprehensive analysis of the hazards associated with new processes. It is hoped that others will find this approach useful, and that the method will provide an additional tool for process safety practitioners to improve hazard analyses and process safety performance.

9 References

[1] H. Lawley. "Operability Studies and Hazard Analysis," *Chemical Engineering Progress* 70 (1974), p. 45.

[2] T. Kletz. *A Guide to Hazard and Operability Studies*. ICI and Chemical Industries Association, 1977.

[3] T. Kletz. *HAZOP and HAZAN, 4th ed.* IChemE, UK, 2006.

[4] F. Crawley, M. Preston and B. Tyler. *HAZOP: Guide to Best Practice, 2nd ed.* IChemE, UK, 2008.

[5] J. Gould. *Review of Hazard Identification Techniques*. HSE, 2000.

[6] F. Lees. *Loss Prevention in the Process Industries, 2nd ed.* Butterworth-Heinemann, Oxford, U.K., 1996.

[7] W. D. Seider, J. Seader, and D. R. Lewin. *Product and Process Design Principles, 2nd ed.* Wiley, New York, NY, 2004.

[8] CCPS. *Guidelines for Hazard Evaluation Procedures*. Center for Chemical Process Safety, American Institute of Chemical Engineers, New York, NY, 1985.

[9] C. Frye, W. S. Freeze and F. K. Buckingham. *Microsoft Office Excel 2003 Programming Inside Out*. Microsoft Press, Redmond, WA, 2004.

13TH GLOBAL CONGRESS ON PROCESS SAFETY

Inherently Safer Reactor Design: Challenges in the Changeover from Semi-batch to Continuous

Jingyao Wang

Mary Kay O'Connor Process Safety Center, Artie McFerrin Department of Chemical Engineering, Texas A&M University, College Station, Texas 77843-3122, USA, +1(732)910-0929, jingyao@tamu.edu

Maria Papadaki[1], Noor Quddus[2], M. Sam Mannan[2]

1 **Department of Environmental & Natural Resources Management, School of Engineering, University of Patras, Seferi 2, Agrinio GR30100, Greece, marpapadaki@upatras.gr**

2 **Mary Kay O'Connor Process Safety Center, Artie McFerrin Department of Chemical Engineering, Texas A&M University, College Station, Texas 77843-3122, USA, +1(979) 862-3985, mannan@tamu.edu**

Prepared for Presentation at
American Institute of Chemical Engineers
2017 Spring Meeting and 13th Global Congress on Process Safety
San Antonio, TX
March 26 – 29, 2017

Inherently Safer Reactor Design: Challenges in the Changeover from Semi-batch to Continuous

Jingyao Wang[1a], Maria Papadaki[2], Noor Quddus[1], M. Sam Mannan[1b]

[1]Mary Kay O'Connor Process Safety Center, Artie McFerrin Department of Chemical Engineering, Texas A&M University, College Station, Texas 77843-3122, USA,
[a]jingyao@tamu.edu, [b]mannan@tamu.edu
[2]Department of Environmental & Natural Resources Management, School of Engineering, University of Patras, Seferi 2, Agrinio GR30100, Greece

Keywords: Inherently Safer Design, Process Intensification, Runaway Reaction, Computational Fluid Dynamics

Abstract

Multipurpose batch and semi-batch reactors are widely used in the chemical process industries. Highly exothermic reactions are performed with large amounts of thermally unstable and/or toxic reactants, products and intermediates. While better heat control can be achieved through gradual reactant dosing in semi-batch mode, the reactors still contain large amounts of hazardous chemicals under critical reaction conditions, which may lead to severe consequences if an incident occurs. Semi-batch processes are subject to low efficiencies due to longer residence time and larger reactor volumes. In this paper, the process intensification inherent safety design principle was applied, and a novel continuous tube-in-tube jet mixing reactor was proposed in place of traditional batch and semi-batch reactors to perform the picoline *N*-oxidation reaction by hydrogen peroxide. The five most widely used turbulence models were used to predict jet mixing phenomena in the reactor. Among them, realizable k- ε with enhancement wall treatment shows good agreement with the experiments in literature [1]. The reactor was simulated using an eddy-dissipation model and the mixing controlled kinetics mechanism was successfully predicted. Different hydrogen peroxide feed conditions were modeled; as expected, the results show that the risk of runaway reaction is strongly affected by hydrogen peroxide dosing rate.

1 Introduction

The chemical incident that occurred at T2 Laboratories, Inc. has drawn significant concerns from public and scientific communities for the risk of catastrophic thermal runaway reaction in batch process operation[2]. The Chemical Safety Board (CSB) issued a study identifying 167 serious reactive chemical incidents between 1980 and 2001 that involved more than 100 fatalities. Even after 2001, another 24 incidents were identified by CSB [3]. Multipurpose batch and semi-batch reactors like those in the incidents are extensively utilized in the fine chemicals, pharmaceutical, and agrochemical industries. Typically, thermally unstable and toxic chemicals are involved and the reaction types include oxidation, epoxidation, and halogenation reactions of organics. Most reactions are performed in semi-batch mode, which is inherently safer than batch operation, by adding one of the reactants gradually. Although most of the primary root causes from the CSB investigation reports have been attributed to human operating errors,[4] semi-batch operations still may pose a high risk of thermal runaway reaction due to the large amount of hazardous reactant materials contained under unstable conditions, and potentially with insufficient cooling systems and limited mixing capabilities [5].

To lower risks associated with traditional batch processes, the best and most effective approach is to tackle the problem in the early design phase by applying inherently safer design principles and utilizing safer and more efficient manufacturing technologies. Considering the major hazards introduced by the large amount of reactive chemicals at critical state in a batch reactor, Trevor Kletz proposed the concept of process intensification, aiming to reduce the size of the operation unit, inventories of hazardous materials, and the consequences of process failures [6]. Therefore, an intensified, inherently safer reactor design topology was investigated, in substitution of the current batch reactors, towards more reliable and efficient chemical manufacturing.

Among potential alternatives, continuous reactors of significantly smaller size compared to batch reactors of the same production rate, have raised interest in academia, industry, and government agencies as a promising solution. Bristol-Myers Squibb developed the continuous plug flow process for preparing a key intermediate for Brivanib Alaninate, partially achieving the continuous production of a key Active Pharmaceutical Ingredient (API) intermediate [7]. Boukouvala & Ierapetritou proposed the modeling and optimization framework for continuous pharmaceutical tablet manufacturing that proves to be profitable with high product quality guarantee [8]. In 2012, the Food and Drug Administration (FDA) also advocated the perspective of continuous manufacturing with the aim of cleaner, more flexible, and efficient manufacturing practices [9]. In addition to enhancing profitability and product quality of continuous processes, many researchers have focused on the safety perspective of continuous operations by studying liquid phase highly exothermic catalytic oxidation reactions in plug-flow reactors.

Hydrogen peroxide is a widely used oxidant as it is both a homogeneous liquid reactant and is a green chemistry oxidant. It has broad applications in many industry sectors [10]. However, the potential for thermal and catalytic decomposition is always a process safety concern and can compromise the final product selectivity as well [11]. Vernières-Hassimi *et al.* studied reacting sodium thiosulfate solution with hydrogen peroxide in a tubular reactor. The location with maximum temperature in the reactor was identified both experimentally and theoretically [12]. Following those publications, Vernieres-Hassimi et al. conducted different incident scenarios by introducing operation errors on reactant concentrations and/or feed temperature. The thermal runaway induced by the operational error was mitigated by direct injection of quench solvents along the reactor [13]. Benaissa *et al.* designed a "heat-exchanger/reactor" open plate continuous reactor to perform the propionic anhydride esterification by 2-butanol. HAZOP studies were applied in identifying the most critical runaway scenarios. Both the experimental and theoretical studies revealed an intrinsically safer behavior of the continuous reactor [14]. However, one of the common issues among the previous plug-flow design is excess dosing rate of hydrogen peroxide (H_2O_2) at the inlet of the reactor which compromised both the safety and the product quality. The heat generation is proportional to the concentration of the oxidants and the reactants. Direct mixing of the two generates highly concentrated reactant solutions and may trigger temperature excursions at the beginning of the reactor and lead to thermal runaway.

In this research, we focus on consequence analysis of different H_2O_2 dosing rates in a tubular continuous reactor. An intensified tube-in-tube jet mixing tubular reactor was designed to achieve fast mixing of the reactants with H_2O_2. Suitable turbulent and reaction models are used with convincing agreement with experimental results. Different oxidant injection scenarios and potential impacts on thermal runaway are also considered. The reaction of interest is the synthesis of pharmaceutical intermediate 3-picoline *N*-oxide from 3-picoline and H_2O_2. It is rapid and highly exothermic with -160 kJ/mol reaction heat and strong runaway reaction potential [15]. At the Corden Pharmachem facility in Cork, Ireland, its product picoline *N*-oxide decomposition was involved in a runaway reaction incident resulting in one operator death and severe injuries of another. Dramatic building damage and ejection of debris also occurred [13]. Apart from picoline *N*-oxide decomposition, hydrogen peroxide may also decompose in a high temperature regime, generating incondensable oxygen which can cause reactor explosion. The consequences could be more catastrophic with flammable organic reactants and products contacted with the generated oxygen.

2 Chemistry Mechanisms

2.1 Calorimetry Test

Thermal calorimetry of the *N*-oxide reaction has been intensively studied by many researchers [15-18]. Usually, the reaction is carried out isothermally below 100 °C in an open system in semi-batch mode with a complex tungsten catalyst dissolved in the liquid homogeneous reaction [15]. The appropriate amount of H_2O_2 is dosed into the reactor until the reaction is complete. The general reaction can be simplified to two reactions as shown:

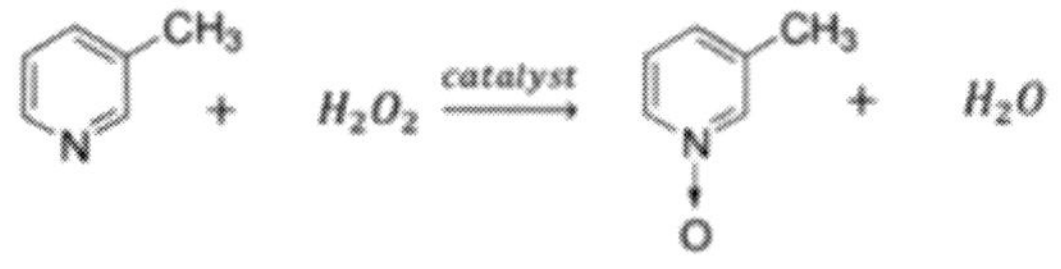

(1)

$$2H_2O_2 \xrightarrow{catalyst} 2H_2O + O_2 \quad (2)$$

Reaction (1) is the main reaction with picoline oxidized by H_2O_2 to generate picoline *N*-oxide. Reaction (2) is the side competitive reaction with H_2O_2 is decomposed into water and oxygen. Several hazards are related to the reactions: reaction (1) is highly exothermic with significant heat generation. Without effective heat removal, the product picoline *N*-oxide may reach decomposition onset temperature and generate the explosive and flammable gas; the H_2O_2 may also decompose and form incondensable oxygen; meanwhile, the generated O_2, once contacted with the flammable organic reactants or products, may lead to fires and/or explosions. Therefore, the reaction can potentially trigger a significant pressure surge, fires and explosions in the reactor. However, the reaction is more efficient and safer under intensified conditions in a closed system at high temperature (110 °C to 125 °C) and using a more concentrated catalyst [16, 18]. Higher temperature can result in a significant increase in the main reaction rate and suppress the H_2O_2 decomposition reactions. The *N*-oxidation reaction is a first order reaction and proportional to the concentration of the picoline. However, the H_2O_2 decomposition reaction rate is second order and proportional to the square of H_2O_2 concentration. Therefore, the concentration of the H_2O_2 should be minimized and H_2O_2 decomposition can be nearly eliminated under fast dosing rate conditions, allowing rapid *N*-oxidation reaction with little H_2O_2 decomposition potential. The incondensable O_2 generation also can be minimized, making the process inherently safer [16]. This is a good example of process intensification as an inherently safer design principle. In this work, the intensified condition will be applied in the reactor design - with fast H_2O_2 dosing rates generating turbulent mixing, higher temperatures at 110 °C to 125 °C, and higher catalyst concentration conditions to achieve inherently safer and more efficient production.

In terms of rate expression of the *N*-oxidation reaction, the complete power-law kinetics model was proposed by previous researchers describing the kinetic control regime of the reactions with good agreement [15, 18]. However, as shown in the calorimetry results of the picoline *N*-oxidation reaction in Figure 1, the kinetics regime for this reaction is very limited and model simulation failed to capture the dosing controlled region. The majority of the reaction is controlled by H_2O_2 dosing rate, suggesting the reaction is dosing controlled instead of kinetics controlled until the picoline reaches 95% conversion under intensified conditions [18].

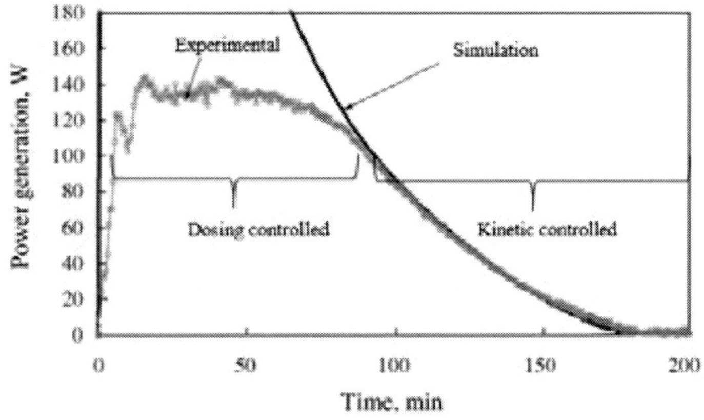

Figure 1. Picoline *N*-oxidation Calorimetry Experimental Results [15]

1.2 Eddy-dissipation Model

The eddy-dissipation model proposed by Magnussen [19] is used to describe the mixing controlled *N*-oxidation reaction. It is particularly suitable for the rapid turbulent-chemistry reaction, where the overall rate of reaction is controlled by turbulent mixing. The unknown and complex kinetics can be approximately neglected without significantly compromising the accuracy for rapid mixing controlled reaction. It is assumed that the rate of reaction is much faster than the turbulent mixing rate of reactants, which is related to the timescale of turbulent eddies in the flow. Therefore, the rate of reaction can be expressed by the dissipation eddies of the species *i* with the smallest time-mean concentration and the most marked intermittency [19]. The general formula of eddy-dissipation is shown in the following [20]:

$$R_{i,r} = v'_{i,r} M_{w,i} A \rho \frac{\varepsilon}{k} \min_{R} \left(\frac{Y_R}{v'_{R,r} M_{w,j}} \right) \quad (3)$$

$$R_{i,r} = v'_{i,r} M_{w,i} A B \rho \frac{\varepsilon}{k} \frac{\sum_P Y_P}{\sum_j^N v''_{j,r} M_{w,j}} \quad (4)$$

Where k is the turbulence kinetic energy, ε is the turbulence dissipation rate, Y_p and Y_R are the mass fractions of the product and reactants, respectively. A and B are the Magnussen constants for reactants and products, respectively. $M_{w,i}$ is the molecular weight. The net rate of production of species i is given by the limiting value of the two equations. $v'_{i,r}$ and $v''_{j,r}$ represents stoichiometric coefficient for reactant species i and product species j. The density of the mixture (ρ), and the molecular weights for reactant i and products j ($M_{w,i}$ and $M_{w,j}$)are multipliers in both equations.

3 Tube-in-Tube Jet Mixing Reactor design

For this study, we adopted and revised the preliminary reactor design proposed by Pineda-Solano *et al.* [17]. The original continuous tubular reactor design is simplified and shown in Figure 2 (a). The picoline is in plug flow in the reactor at 110 °C to 125 °C. The H_2O_2 solution is added to the picoline feed flow through an injection pipe that allows the H_2O_2 to inject and spread axially and mix with the pure picoline flow through rows of jet holes. The amount of H_2O_2 dosed is controlled by the number and the size of the jet flows. The whole reaction regime is expected to be turbulent with optimum mixing of the H_2O_2. The coils of hot product are circulating around the tubular reactor and supply heating utility. Additional cooling coils are also supplied around the reactor to satisfy the cooling requirements. The reactor is designed to operate at inherently safer intensified conditions to minimize the potential for H_2O_2 decomposition, and enhance the selectivity towards the *N*-oxidation reaction. The reactor size is significantly reduced compared with the traditional semi-batch reactor and the potential consequences are also less severe. However, there is one major drawback of the previous design. The H_2O_2 solution is highly unstable above 70 °C [11]. And the H_2O_2 injection pipe is in direct contact with the high temperature reactants. Therefore, there is a risk that the H_2O_2 may decompose in the injection pipe and trigger a sharp pressure rise and explosion. An alternative design is proposed in this work as shown in Figure 2 (b). The H_2O_2 is fed through the outer tube and isolated with a cooling agent layer from the hot inner reactor tube. The temperature of the H_2O_2 is maintained below the decomposition onset temperature. A cooling tube is also injected into the reactor inner tube to maintain the isothermal reaction conditions. In this work, only one row of jets is modelled without cooling agents to study the worst-case scenarios of the novel reactor design in trying to determine the maximum amount of H_2O_2 jet flows allowed for safety concerns.

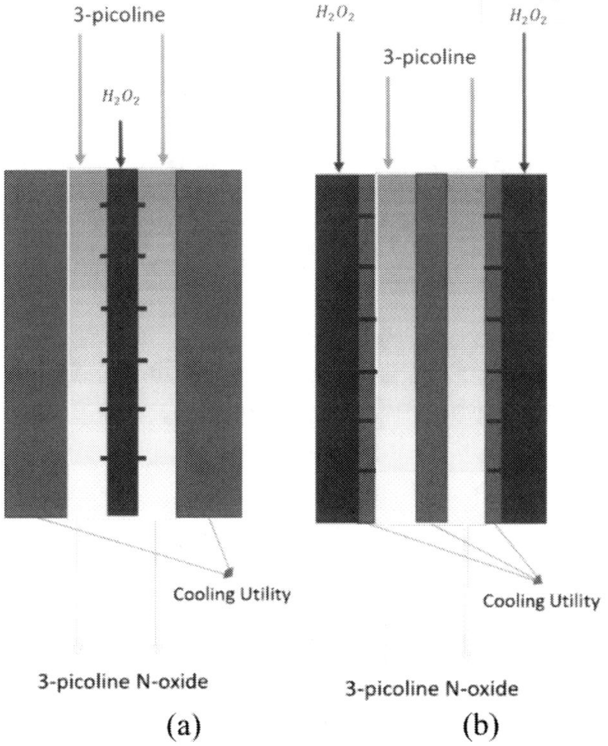

Figure 2. (a) Simplified Sketch of a Preliminary Reactor Design from Pineda [17], (b) Improved Tube-in-Tube Jet Mixing Reactor.

4 Model Development

4.1 Model Formulations

The major challenge is to model the liquid turbulent jet mixing process in the tube-in-tube reactor. Among the three major types of turbulence models available, Direct Numerical Simulation (DNS) model numerically solves the N-S equation directly at an extremely high computation cost and is not applicable for practical usage [21]; the Large Eddy Simulation (LES) model solves the large eddies directly but applies a subgrid scale turbulent model to solve the smaller eddies. However, the computational efforts are still quite significant [22]. The most widely used current approach is the Reynolds Averaged

Navier-Stokes Simulation (RANS) model, where all turbulent length scales are modeled using a Reynolds-Average approach with economically acceptable computational times [23]. Since in this case we only consider the steady state design, the RANS model is selected in modeling time-averaged transport phenomena. The general formulation of the RANS models are listed in the following equations [24]:

$$u = \bar{u} + u' \ (5)$$

$$\frac{\partial \bar{u}_i}{\partial t} + \bar{u}_j \frac{\partial \bar{u}_i}{\partial x_j} + \frac{\partial \overline{u'_j u'_i}}{\partial x_j} = -\frac{1}{\rho}\frac{\partial \bar{P}}{\partial x_i} + v\nabla^2 \bar{u}_i \ (6)$$

In the RANS model, the fluid velocity u is modeled using the mean velocity $\bar{u}$ and fluctuating velocity u'. Substituting the expression (5) into the instantaneous momentum equations yields the RANS equation (6). v is the kinetic viscosity, ρ is the fluid viscosity. Reynolds stresses $\overline{u'_j u'_i}$ must be solved with additional equations. Turbulent-viscosity-based models are considered here with the following hypothesis to model the Reynold Stress:

$$\overline{u'_j u'_i} = \frac{2}{3} k \delta_{ij} - v_T \left(\frac{\partial \overline{u_i}}{x_j} + \frac{\partial \overline{u_j}}{x_i}\right) \ (7)$$

v_T is called turbulent viscosity, an unknown yet to be modeled for solving the $\overline{u'_j u'_i}$. k is turbulent kinetics energy. In this paper, we consider introducing turbulence frequency ω as $k - \omega$ in the model and turbulence dissipation rate ε as $k - \varepsilon$ to solve the turbulent kinetic viscosity v_T using the following two equations:

$$v_T = C_\mu \frac{k^2}{\varepsilon} \ (8)$$

$$v_T = \frac{\alpha^* \rho k}{\omega} \ (9)$$

α^* and C_μ are empirically determined constants. Equations for solving the turbulent kinetic energy k, turbulence frequency ω and turbulence dissipation rate ε are the following:

$$\frac{\partial k}{\partial t} + \bar{u} \cdot \nabla k = \nabla \cdot \left(\frac{v_T}{\sigma_k} \nabla k\right) + \mathrm{P} - \varepsilon \ (10)$$

$$\frac{\partial \varepsilon}{\partial t} + \bar{u} \cdot \nabla \varepsilon = \nabla \cdot \left(\frac{v_T}{\sigma_k} \nabla \varepsilon\right) + \frac{\varepsilon}{k}(C_{\varepsilon 1}\mathrm{P} - C_{\varepsilon 2}\varepsilon) \ (11)$$

$$\frac{\partial \omega}{\partial t} + \bar{u} \cdot \nabla \omega = \nabla \cdot \left(\frac{v_T}{\sigma_\omega} \nabla \omega\right) + C_{\omega 1}\frac{\omega}{k}\mathrm{P} - C_{\omega 2}\omega^2 \ (12)$$

P is the turbulent kinetic energy production term, σ_k, σ_ω, $C_{\varepsilon 1}$, $C_{\varepsilon 2}$, $C_{\omega 1}$ and $C_{\omega 2}$ are empirical model constants. Combining equations (6), (7), (8), (10), and (11) is the standard $k - \varepsilon$ equation and the $k - \omega$ model includes equations (6), (7), (8), (10), and

(12). The standard $k-\varepsilon$ is widely used and very robust, but performs poorly in complex boundary layer flow. The standard $k-\omega$ model performs much better in low Reynolds number and boundary layer flow. In addition to standard $k-\varepsilon$ and $k-\omega$ models, we also applied the realizable $k-\varepsilon$ model and SST $k-\omega$ model for simulations. The realizable $k-\varepsilon$ model introduces an alternative equation in modeling the turbulent viscosity and derives the dissipation rate ε using mean square vorticity fluctuation. It provides superior benefits in simulating more complex flows. The SST $k-\omega$ model is a hybrid two-equation model which combines the advantages of both $k-\varepsilon$ and $k-\omega$ models [23, 24].

The convection-diffusion equation (13) and the energy equation (14) are used for the mass and energy balances, where R_i is the net rate of production for species i by chemical reaction and expressed as an eddy-dissipation reaction model using equations (13) and (14). S_i is the source term defined by the inlet streams. $\vec{J_i}$ is the diffusion flux of species i. Y_i is the mass fraction of the species i. $\vec{v}$ is the fluid velocity vector. k_{eff} is the effective conductivity. S_h includes the heat of chemical reaction. E is the total energy and h_j is the enthalpy of species j.

$$\frac{\partial}{\partial t}(\rho Y_i) + \nabla \cdot (\rho \vec{v} Y_i) = -\nabla \cdot \vec{J_i} + R_i + S_i \quad (13)$$

$$\frac{\partial}{\partial t}(\rho E) + \nabla \cdot \left(\vec{v}(\rho E + p)\right) = \nabla \cdot \left(k_{eff} \nabla T - \sum_j h_j \vec{J_j} + \left(\bar{\bar{\tau}}_{eff} \cdot \vec{v}\right)\right) + S_h \quad (14)$$

4.2 *ANSYS-Fluent Set-Up*

The simulations from four RANS models are compared with the experimental results [1] to find the best fitted RANS model describing the liquid mixing. The geometry of the mixing tube in the experiments is shown in Figure 3 (a) and (b). Stream A consists of dye solution and stream B contains only water. Water stream B is injected into the mixing tube through four small holes along the tube wall. A high-speed camera continuously takes pictures of the concentration profile of the dyes in the mixing process. The mixing tube in the experiments is reproduced and meshed in ANSYS Fluent 17.0 as shown in Figure 3 (c). Only half of the mixing tube as symmetric geometry is simulated to reduce the computational cost. The fine mesh is applied at the jet inlet region to more accurately capture transport phenomena of the complex fluid flows. Coarse mesh is used in the downstream, where the mixed fluids are steady. Inflation layers are used in modeling the surface of the mixing tube to generate a very fine near-wall mesh. The enhanced wall treatment (EWT) option in ANSYS Fluent was enabled which provides better prediction of complex near-wall phenomena in the mixing tube. The hydraulic diameter of the inlet condition is assumed to be the same size as the tube diameter and turbulent intensity of the jet and pipe flow is assumed to be 5%.

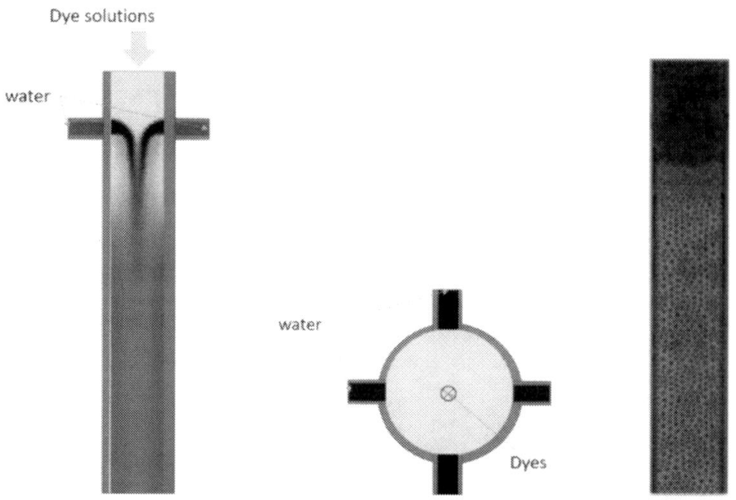

(a) (b) (c)

Figure 3. (a) Front View of Simplified Mixing Tube Geometry from the Experiments[17]; (b) Top View of the Mixing Tube; (c) Meshed Mixing Tube in Fluent.

5 Results and Discussion

5.1 Model Validation

The experimental results are simulated and reproduced with five RANS models, which are standard $k-\varepsilon$, realizable $k-\varepsilon$ with EWT, realizable $k-\varepsilon$ without EWT, standard $k-\omega$, and the SST $k-\omega$ model. The concentration profile is shown in Figure 4 for the dyes. In case 1 (a), the experiment shows that backwards mixing happened above the jet mixing plate. However, the standard $k-\varepsilon$ model, (b), showed poor agreement with the experimental concentration, and showed bad predictions on large streamline curvature, where jets are mixing. The realizable $\mathrm{k}-\varepsilon$ with enhanced wall treatment model, (c), manages to predict the backwards mixing. The other models failed to mimic the experimental results. The realizable $\mathrm{k}-\varepsilon$ without enhanced wall treatment model, (d), managed to capture the concentration profile below the mixing region, but it still failed to predict the mixing behavior above the jet injection plane. The standard $k-\omega$ and SST $k-\omega$ models (e) and (f) take longer computation time to converge, and diverge significantly from the experimental profile. This is because the formulation of the $k-\omega$ model is highly dependent on the wall distance, which causes inaccurate prediction in the far wall jet mixing zone. Among the five models, only the realizable $k-\varepsilon$ model with EWT successfully predicts the backwards mixing inside the tube. It shows advantages of accurate description of both complex jet mixing and near-wall phenomena. Therefore, it has been selected as the turbulent model in the reactor modeling parts.

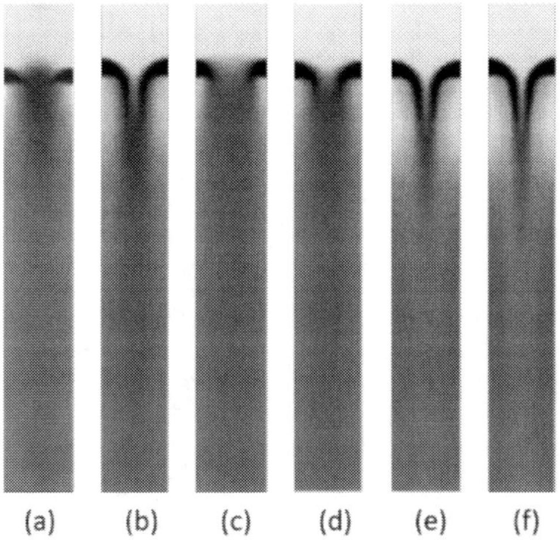

Figure 4. (a) Concentration profile from experiments [17]; (b) Simulation results from standard $\mathbf{k-\varepsilon}$; (c) Simulation results from realizable $\mathbf{k-\varepsilon}$ with EWT ; (d) Simulation results from realizable $\mathbf{k-\varepsilon}$ without EWT ; (e) Simulation results from standard $\mathbf{k-\omega}$; (f) Results from SST $\mathbf{k-\omega}$.

5.2 Rector Modeling

The tube-in-tube reactor with a single row of jets is simulated under different H_2O_2 dosing rates. The reactor is operated adiabatically to test the worst-case scenarios and determine the maximum allowable H_2O_2 dosing amount. The realizable $k-\varepsilon$ model with EWT is selected as the turbulence model, showing convincing agreement with liquid jet mixing experiments. An eddy-dissipation reaction model is applied to simulate the H_2O_2 dosing controlled reaction. The different simulated feeding cases are summarized in Table 1.

Table 1. Feed Conditions for Different Study Cases.

Case	H_2O_2 Jet Velocity (m/s)	Picoline Pipe Flow Velocity (m/s)	Dosing Rate $\dot{n}_{H2O2}$ (mol/s)
1	3	1	2.66
2	4.5	1	3.98
3	5.5	1	4.868
4	6	1	5.311
5	7	1	6.196

Figure 5 shows the picoline concentration inside the reactor in case 1, 2, 3, 4, and 5, and Figure 6 describes the H_2O_2 concentration profiles. From Figure 5 (a) and Figure 6 (a), excess picoline is dosed and the H_2O_2 dosed is completely consumed. According to the H_2O_2 and picoline profile of cases 4 and 5, H_2O_2 is over-fed and picoline reaches 100%

conversion. Case 3 has almost equal stochiometric molar feeds of H_2O_2 and picoline, and all the reactants are completely converted to the products. It is observed that the H_2O_2 profile in Figure 5 matches almost exactly the shape of the picoline profile in Figure 6, which means H_2O_2 and picoline can hardly coexist, and the reaction rate is so fast that the reactants are already consumed before mixing. Therefore, the reaction simulated by an eddy-dissipation model is completely dosing controlled.

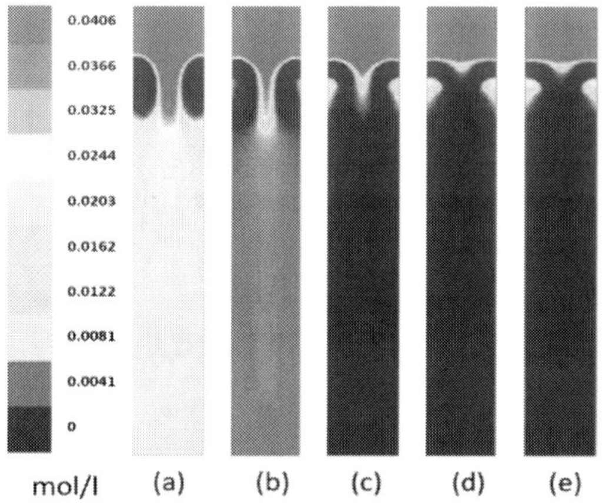

Figure 5. Picoline Concentration Contour: (a) Case 1; (b) Case 2; (c) Case 3; (d) Case 4 ; (c) Case 5

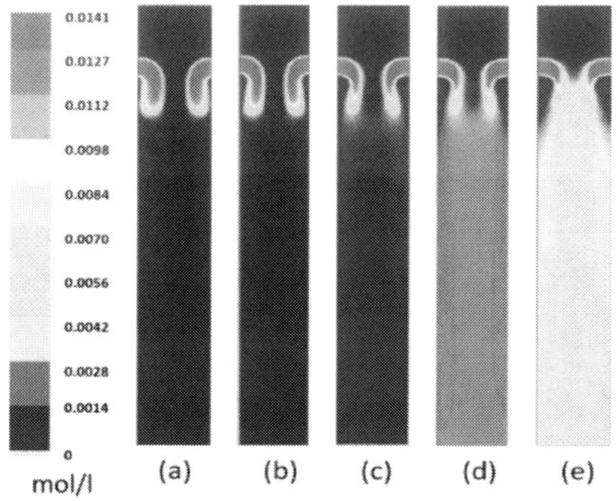

Figure 6. H_2O_2 Concentration Contour: (a) Case 1; (b) Case 2; (c) Case 3; (d) Case 4 ; (e) Case 5

The temperature and 3-picoline *N*-oxide concentration profiles of each case are shown in Figure 7 and Figure 8, respectively. It is interesting to see the shapes of the two figures being quite similar. This result is because the reaction is highly exothermic and a significant amount of heat will be produced along with the generated products. Therefore, the concentrated products area indicates excess reaction heat accumulated. As discussed in the chemistry mechanism section, the temperature of the reactor should be maintained between 110 to 125 °C (383.15 K to 398.15 K). A temperature too low or too high may lead to the need for an increased reactor size, lower selectivity for the desired products, and H_2O_2 decomposition. The temperature in case 1 is below 383.15 K, where the picoline oxidation reaction is not fast enough and long residence time is required for reaction completion. While in case 3 where equal molar concentrations of H_2O_2 and 3-picoline exist, the maximum temperature exceeds 410 K. It represents a particularly dangerous scenario, where the picoline *N*-oxide may decompose into incondensable side products and trigger an explosion. Cases 4 and 5 seem to be operating in a safer and more efficient temperature zone. However, as shown in the H_2O_2 concentration profile in Figure 6 (e), there is a considerable amount of unreacted H_2O_2 left in the reactor. That unreacted H_2O_2 may be subject to decomposition at high reaction temperature. Based on the above analysis, it is concluded that dosing equal molar or excess amounts of H_2O_2 at once may cause severe thermal runaway consequences and trigger fires and explosions. On the contrary, insufficient dosing is adverse to the process efficiency. An optimum dosing rate between the two may exist, enabling the reactor to operate within design temperatures. Multistage tube-in-tube reactors can be designed with several rows of H_2O_2 jets and maintain the reactor under design intensified conditions throughout the process. It should also be noted that hot spots exist in Figure 7 (a), (b), and (e) and are usually located in the dead zone below the jet trajectory. In future designs, a proper cooling system should be applied to avoid local temperature excursions.

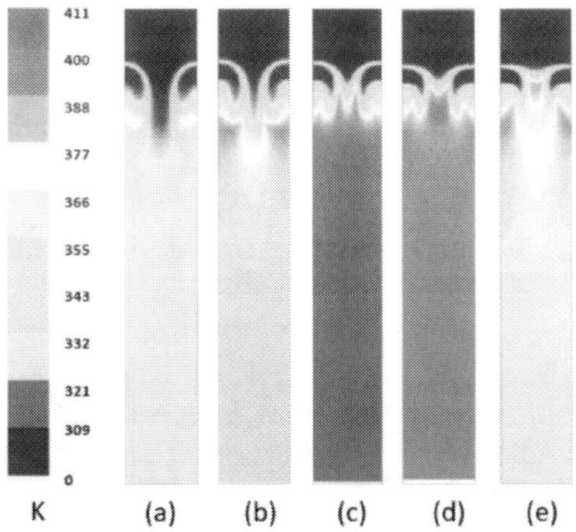

Figure 7. Temperature Profile: (a) Case 1; (b) Case 2; (c) Case 3; (d) Case 4 ; (e) Case 5

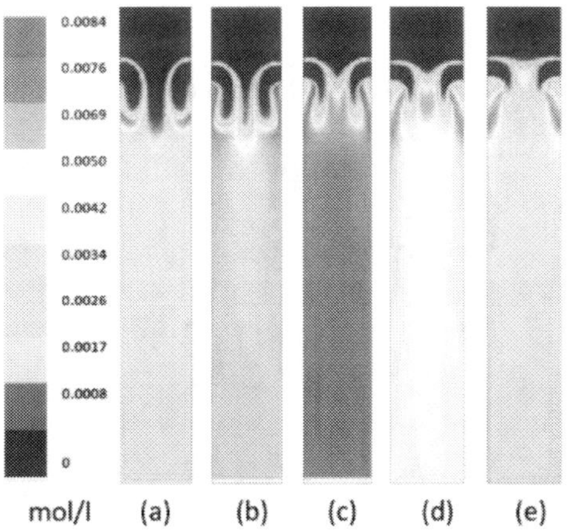

Figure 8. 3-picoline N-oxide Concentration Contour: (a) Case 1; (b) Case 2 ; (c) Case 3; (d) Case 4 ; (e) Case 5.

6 Conclusions

In this paper, we applied the process intensification inherently safer design principle and proposed a novel tube-in-tube jet mixing reactor to perform the picoline *N*-oxidation reaction. To simulate the jet mixing transport phenomena within the reactor, five RANS turbulence models were tested and it was found that only the realizable k- ε model with Enhanced Wall Treatment (EWT) successfully predicts the backwards mixing at the jet mixing plane. After model validation, the realizable k- ε model was coupled with eddy-dissipation to simulate the reactor operation under different feed conditions. The results show that complete injection of equal or excess molar amount of H_2O_2 will generate significant high temperatures and may lead to catastrophic picoline *N*-oxide and H_2O_2 decompositions. There should be an optimum feeding condition that allows the reactor to operate within the design temperature. The results of this work facilitate multistage reactor design and hotspot prevention.

7 References

1. Luo, P., et al., *An experimental study of liquid mixing in a multi-orifice-impinging transverse jet mixer using PLIF.* Chemical engineering journal, 2013. **228**: p. 554-564.
2. Theis, A.E., *Case study: T2 Laboratories explosion.* Journal of Loss Prevention in the Process Industries, 2014. **30**: p. 296-300.
3. Joseph, G., M. Kaszniak, and L. Long, *Lessons after Bhopal: CSB a catalyst for change.* Journal of Loss Prevention in the Process Industries, 2005. **18**(4): p. 537-548.
4. Barton, J. and P. Nolan, *Incidents in the chemical industry due to thermal runaway chemical reactions.* Hazards X: Process Safety in Fine and Speciality Chemical Plants, 1989(115): p. 3-18.
5. Leveneur, S.b., et al., *From kinetic study to thermal safety assessment: application to peroxyformic acid synthesis.* Industrial & Engineering Chemistry Research, 2012. **51**(43): p. 13999-14007.
6. Kletz, T.A. and P. Amyotte, *Process plants: A handbook for inherently safer design.* 2010: CRC Press.
7. LaPorte, T.L., et al., *Development of a Continuous Plug Flow Process for Preparation of a Key Intermediate for Brivanib Alaninate.* Organic Process Research & Development, 2014. **18**(11): p. 1492-1502.
8. Boukouvala, F. and M.G. Ierapetritou, *Surrogate-based optimization of expensive flowsheet modeling for continuous pharmaceutical manufacturing.* Journal of Pharmaceutical Innovation, 2013. **8**(2): p. 131-145.
9. Chatterjee, S. *FDA perspective on continuous manufacturing.* in *IFPAC Annual Meeting, Baltimore, MD.* 2012.
10. Ventura, M. and S. Yuan. *Commercial production and use of hydrogen peroxide.* in *36th AIAA/ASME/SAE/ASEE Joint Propulsion Conference and Exhibit.* 2000.
11. Mackenzie, J., *Considerations for the safe design of processes using hydrogen peroxide and organics.* Plant/Operations Progress, 1991. **10**(3): p. 164-170.
12. Vernières-Hassimi, L., M. Abdelghani-Idrissi, and D. Seguin, *Experimental and theoretical steady state maximum temperature localization along an exothermic tubular chemical reactor.* Open Chemical Engineering Journal, 2008. **2**: p. 57-65.
13. Vernières-Hassimi, L. and S. Leveneur, *Alternative method to prevent thermal runaway in case of error on operating conditions continuous reactor.* Process Safety and Environmental Protection, 2015. **98**: p. 365-373.
14. Benaïssa, W., et al., *Evaluation of an intensified continuous heat-exchanger reactor for inherently safer characteristics.* Journal of Loss Prevention in the Process Industries, 2008. **21**(5): p. 528-536.
15. Sempere, J., et al., *Modelling of the reaction of N-oxidation of 2-methylpyridine using hydrogen peroxide and a complex metal catalyst.* Chemical Engineering and Processing: Process Intensification, 1998. **37**(1): p. 33-46.
16. Pineda-Solano, A., et al., *Inherently safer reactors: Improved efficiency of 3-picoline N-oxidation in the temperature range 110–125° C.* Process Safety and Environmental Protection, 2012. **90**(5): p. 404-410.
17. Pineda-Solano, A., et al., *Toward an inherently safer design and operation of batch and semi-batch processes: The N-oxidation of alkylpyridines.* Journal of loss prevention in the process industries, 2012. **25**(5): p. 797-802.
18. Papadaki, M. and J. Gao, *Kinetic models of complex reaction systems.* Computers & chemical engineering, 2005. **29**(11): p. 2449-2460.

19. Magnussen, B.F. and B.H. Hjertager. *On mathematical modeling of turbulent combustion with special emphasis on soot formation and combustion*. in *Symposium (international) on Combustion*. 1977. Elsevier.
20. Fluent, A., *14.0: Theory guide, ansys.* Inc., Canonsburg, PA, 2011.
21. Moin, P. and K. Mahesh, *Direct numerical simulation: a tool in turbulence research.* Annual review of fluid mechanics, 1998. **30**(1): p. 539-578.
22. Pope, S.B., *Ten questions concerning the large-eddy simulation of turbulent flows.* New journal of Physics, 2004. **6**(1): p. 35.
23. Fluent, A., *14.5 Theory Guide.* Canonsburg, PA, USA: ANSYS Inc, 2012.
24. Fox, R.O. and H.L. Stiles, *Computational models for turbulent reacting flows*. Vol. 419. 2003: Cambridge university press Cambridge.

PSM Challenges in an Aging Cyanide Plant

John L. Gohres*, William R. Zecha, Hans E. Thullesen, Joel F. Singleton

Evonik Corporation, 4201 Evonik Rd. Theodore, AL 36582

*E-mail: john.gohres@evonik.com

Abstract

A 40-year old hydrogen cyanide (HCN) plant is examined as a case study. The plant is approximately 90% PSM-covered, predates the OSHA Process Safety Management (PSM) standard by over a decade, and has a large distribution in operator-level experience. Despite the challenges of being designed before the OSHA PSM standard was written, the plant has an outstanding safety record. Heightened company safety standards have required the unit to implement additional engineering safeguards and new technologies, particularly Safety Instrumented Systems (SIS). The handover of new technology to a workforce with changing age and experience demographics can be especially challenging. This study examines the plant's practices and execution through periodic turnarounds, scheduling of Mechanical Integrity inspections, and other topics.

1.0 Introduction

Evonik is a major producer of hydrogen cyanide (HCN) worldwide. Although HCN is highly toxic and requires hazardous raw materials (ammonia, sulfuric acid, sulfur dioxide, natural gas) to produce, it is a key raw material for a variety of useful end products (Pesce, 2010). Evonik promotes knowledge exchange for handling HCN and sponsors intra-company exchanges between the European, Asian, and domestic sites. The plant also participates and contributes to an annual HCN Industrial Safety Conference. Such programs promote best safety practice and lessons learned exchanges to help ensure safe HCN production. Process Safety Management (PSM) information exchange specific to HCN can be challenging since the plant is the company's lone HCN producer in the United States. The foreign sites have similar governmental requirements and must meet company process safety standards; however, they may not be directly related to all PSM elements. Many PSM requirements are clearly defined; however, others are performance-based and provide flexibility for the company to manage. The challenges these topics present and some of the strategies which have been implemented are shared in this case study. Most of the 14 PSM elements (OSHA 29 CFR 1910.119, 1992) are discussed directly or indirectly; however, this article focuses on personnel-related elements (e.g. training, procedures, and employee involvement), mechanical integrity and process safety information since these tend to present challenges for an aging facility.

1.1 Overview of the Unit

The HCN unit was built between 1976 and 1977 with start-up in early 1978. The capacity has increased to roughly double the initial capacity through several expansion projects. The technology is the Blausaure von Methan und Ammoniak (BMA, or Degussa) process and is one of three of its kind in the world. A simplified block-flow diagram is provided in Figure 1. A BMA plant presents unique challenges in that it depends on a highly endothermic reaction requiring a furnace to achieve very high temperatures. The materials of construction of the reactor are sensitive to thermal shock, and consequently the reactor raw material feeds and furnaces are rarely shut down to prevent reactor damage. The downstream process can be bypassed for maintenance while the reactor is in operation; however, this is undesirable due to high operating costs. Consequently, mechanical integrity planning is critical and execution must be performed during short turnaround windows.

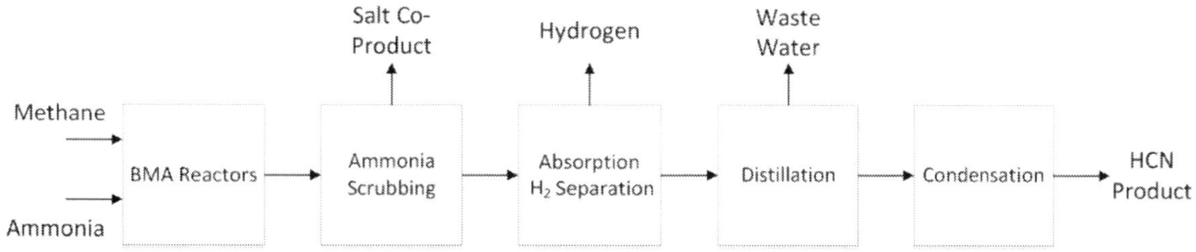

Figure 1: Simplified block-flow diagram of the BMA process.

2.0 Personnel Development and Training

2.1 Change in Experience Level

The HCN unit started operations in 1978, and until 2013 had operations personnel working in the unit who were original to the plant start up. Additionally, several more first generation employees (those hired within five years of start-up) were working in the plant at that time. Between 2012 and 2015, there were eleven retirements and five attrition losses, representing a turnover of more than one third of the workforce. Additionally, two experienced operators spent a year-long plant start-up assignment at a new facility in Asia. The changes are best reflected in Figure 2, where nearly 700 years of total HCN experience dropped to just over 500 years of total experience within a year. A loss of that much experience has the potential to impact safety and production performance if the change in personnel is not managed. Consequently, the company chose to add two full-time equivalents (FTE's) in middle of 2011 when the first retirees indicated their retirements were within 12 months. An age and experience distribution graph, like the one shown in Figure 3 is used to demonstrate impending loss of knowledge to justify additional new hires.

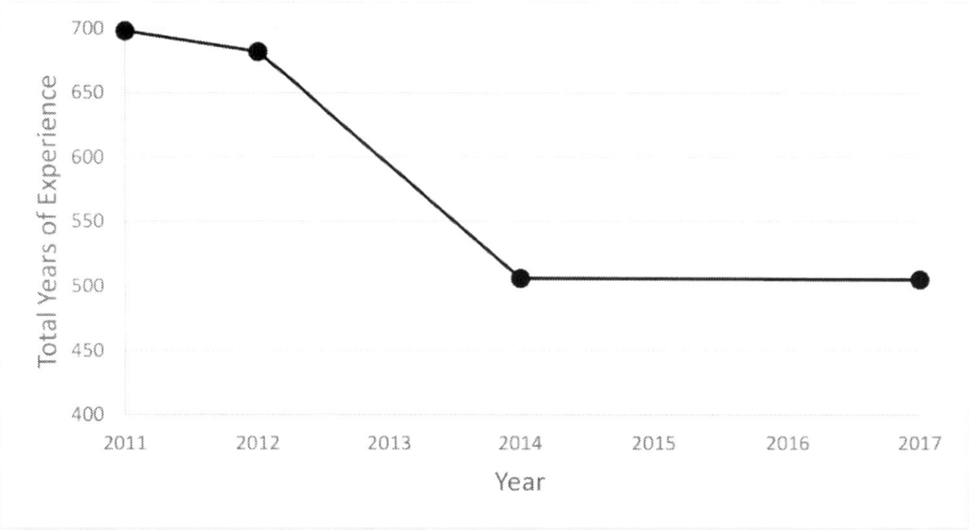

Figure 2: Total years of experience progression for the unit.

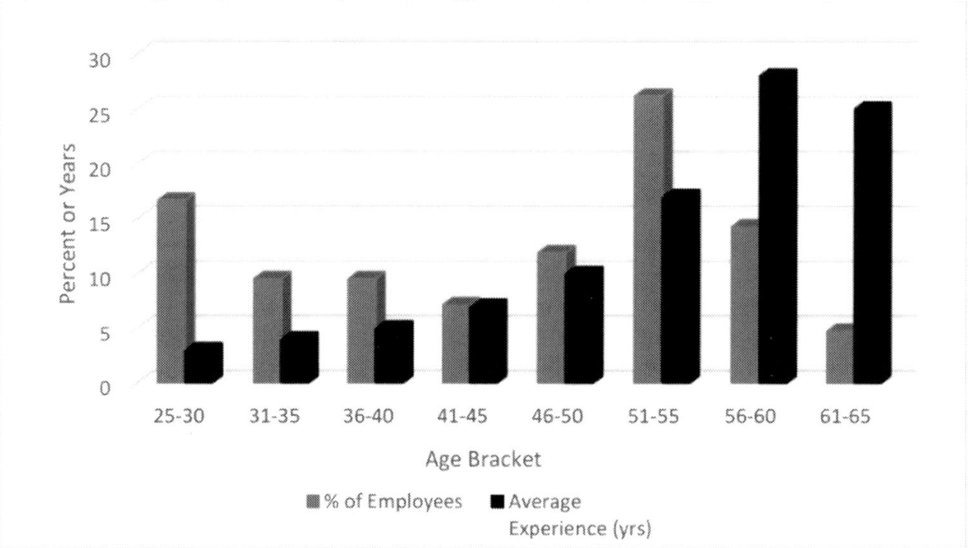

Figure 3: Age and experience distribution (as of 2017).

The practice of hiring additional employees (beyond staffing requirements) is a common tool within Evonik, especially when used to manage periods of high attrition. Evonik in North America has even launched an incentive program to promote early announcement of a retirement. A potential retiree can be paid a one-time payment if the announcement is made at six in advance. Additional payout is given if twelve months of notice are given. The program has guidelines for situations if the employee changes his/her mind. The program helps plant managers better manage attrition and ensure personnel are hired early enough so they can be trained before the attrition occurs.

2.2 Training and Shift Structure

The operator training program has become important as a result of increasing personnel turnover. Only sixty percent of personnel remain from early 2011. This has caused the high percent of inexperienced personnel (less than six years of experience) that is depicted in Figure 4. Turnover has also affected specialty skills like distributed control system (DCS) certified operators and maintenance coordinator positions.

Shifts are self-directed (no on-shift supervision) and are organized to include at least two DCS operators; one normally has over 15 years of experience and at least one field operator has 10-20 years of experience. All shifts have at least one member of the emergency response team (ERT). The objective is to evenly distribute experience level and provide mentorship to newer operators. A sample shift structure, including roles and years of experience is provided in Figure 5.

All operations staff undergo frequent training and must recertify on all critical job tasks every three years or when an edit is made or a management of change (MOC) request is approved. Recertification consists of a written test and a field review with a member of the production Leadership Team (generally a 10+ year employee). The operator must explain the key equipment, basic operations, critical interlocks and safety precautions to the senior operator. Training for minor job tasks, simple MOC's, and procedural edits are conducted via computer based training and are documented in a database. Operating procedures are kept current and accurate using the MOC program and are continuously scrutinized for improvement. Operations personnel are encouraged to point out inaccuracies through the course of training, and the entire production team periodically reviews operating procedures for accuracy. For example, in 2015, the production team performed and reviewed one "high hazard, high frequency" procedure each month to further optimize and assure the procedures reflect current operating practice.

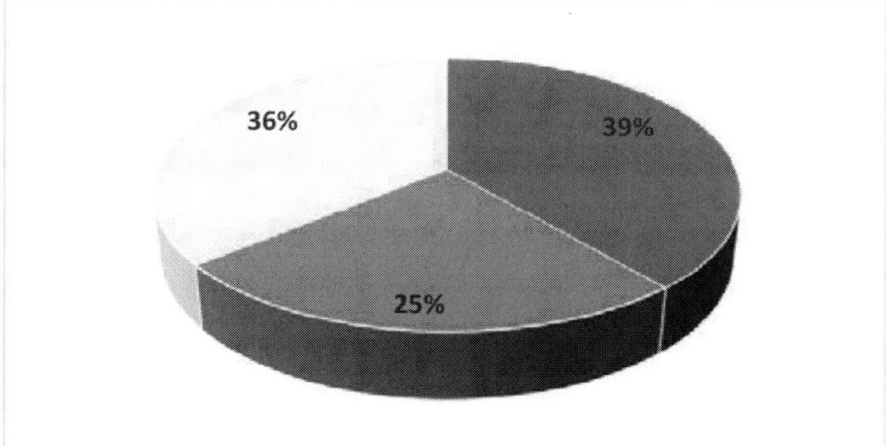

Figure 4: Current breakdown of operations workforce experience

Figure 5: Example of a shift structure. Black indicator represents a member of the plant Leadership Team, orange indicator designates a DCS certified operator, red indicator is a reactor maintenance specialist, and red circle indicates a member of the ERT. Hash marks represent an employee in training.

2.3 Employee Involvement

A challenge of a work force with multiple experience levels is engaging both veterans and new employees in PSM. Communication is perhaps the most important way to involve the workforce, and several platforms are used to provide information to everyone in the plant. One successful concept has been the introduction of a variance database and notification tool where employees can enter any deviation from normal operation. Entries include near miss reporting, safety suggestions, mechanical or quality issues, and the database automatically generates a notification e-mail to all personnel in the unit. Incident investigations can be initiated directly from the database interface and the reports are accessible to anyone in the unit. HCN personnel are required to enter one safety-related variance per quarter. Variances have been a great way to involve everyone in occupational and process safety.

The number of variances in the HCN unit averages 600-700 per year and approximately 40% of all variances are EHS related. Variances are classified as Corrective (Example: "Pump seal is leaking"), Continuous Improvement ("Installed a new decontamination station"), or Preventative ("Stopped a contractor working without proper PPE"). The latter two classifications, particularly Preventative, are considered leading indicators since they demonstrate a proactive approach to safety while Corrective variances are considered lagging indicators since it requires a reactive measure. A breakdown of variance type by year is shown in Figure 6. In recent years, there has been a shift from Corrective based to Preventative, due to increased emphasis on quality variances.

Another forum is weekly communication meetings in the control room to discuss topics directly with shift personnel. Topics can include site news, production issues and safety items, it is also a setting to share lessons learned and outcomes from incident investigations, MOC's, and upcoming process changes.

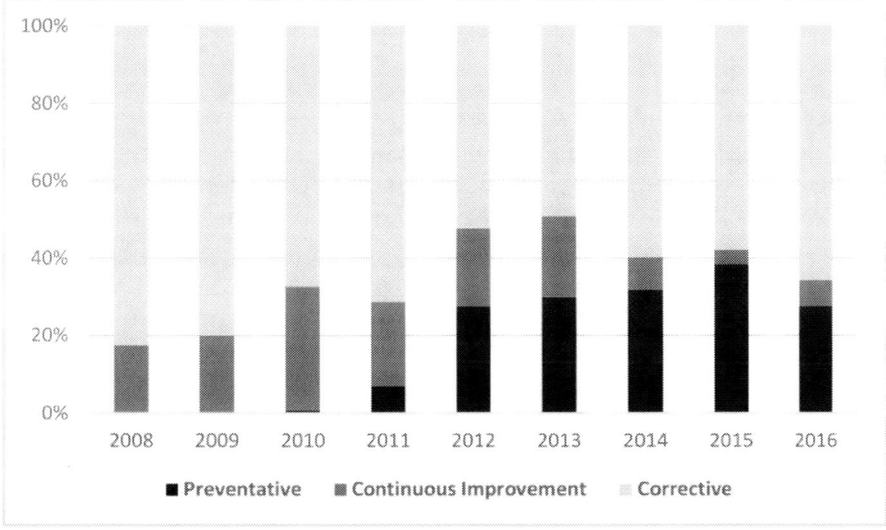

Figure 6: Variance entries into the database by type, since 2009.

3.0 Process Safety to the Forefront: 2009-2014

Recently, Evonik has made a push to improve process safety and has started an initiative to harmonize process safety requirements worldwide. Many of the PSM program enhancements in the HCN unit pre-date this initiative and often require additional resources to meet all of the aspects of PSM associated with these changes.

3.1 Implementation of LOPA and Procedure Changes

Layer of Protection Analysis (LOPA) was added to the existing Hazard and Operability studies (HAZOP) format as our Theodore site PHA methodology starting in 2002. The addition of LOPA added a semi-quantitative analysis to the PHA. As a result, the number of recommendations within one revalidation grew from twelve in 1997 to ninety-eight in 2002, most of which required new instrumentation and Safety Instrumented System (SIS)/Basic Process Control System (BPCS) interlocks. The LOPA methodology created more recommendations due to its semi-quantitative nature and conservative risk tolerance approach. Further improvements to the LOPA procedure were made in 2010 when a work team revised the procedure. The most significant change was the elimination of redundant DCS credit. Before this change BPCS alarms and interlocks were treated as independent protection layers (IPL); however, the DCS provides a common cause failure. When the next revalidation PHA was conducted in 2012, most of the BPCS interlocks used as IPL's were converted to SIS interlocks. Currently there are 40 SIL1 and 12 SIL2 SIF's. In addition to installation and process challenges, there were PSM hurdles to overcome. The most significant items were ensuring adequate training of personnel and adjusting the mechanical integrity program to include additional proof tests to meet SIL ratings.

In December of 2014, Evonik's Process Safety Competency Center issued a binding technical document to provide corporate guidelines for performing hazard identification and evaluation. The purpose of the standard is to provide global harmonization of safety and risk evaluation, so that all sites worldwide perform risk assessments that meet a base level of requirements. A new risk matrix will be rolled out in June 2017 to compliment the technical standard and provide a consistent methodology to evaluate risks.

3.2 Renewed Emphasis on MOCs, PSSRs

Rollout of improvement projects starts with a Management of Change (MOC) request to ensure that all components of Process Safety Information (PSI) are modified and personnel are appropriately trained. MOCs have been written since PSM was first introduced. The current MOC database allows for responsibility to be assigned, providing accountability for PSI changes and ensuring effective documentation and training. MOC's always require a safety review, the type of review is based on the level and type of change. Significant changes (new equipment, major piping changes) require the site PHA procedures, while minor changes require a reference to the most recent revalidation PHA to determine if a formal PHA is needed. The MOC database provides the suggested type of review based on the user input.

In 2012 the site hired a FTE to assist the production units with MOC's, PSSR's, and incident investigations. This person's main responsibility is to provide process safety support on these items and ensure timely completion of responsibilities and MOC closure. Additionally, emphasis was increased on pre start-up safety reviews (PSSRs) following a 2012 incident at a European facility. PSSRs have always been conducted before starting up a change related to an MOC or shut down; however, a new procedure was developed to specify the minimum attendees and the agenda. PSSRs include a review of all edited PSI, ensuring PHA action items are complete, and a field review of the change. A PSSR template is included in the MOC database to simplify the process and ensure proper completion.

4.0 Enhanced Mechanical Integrity: 2014-Present

4.1 Mechanical Integrity Challenges with Infrequent Shutdowns and Aging Equipment

The BMA reactors are sensitive to thermal shock, so an unexpected loss of fuel or inadvertent shutdown to the furnace can damage the reactor and lead to production losses. As a result, high plant and reactor reliability is needed and total shutdowns are periodic. Some vessels, like storage tanks, can be taken out of service while the plant is in operation, for example where redundant pieces of equipment are present. Most process equipment like columns or heat exchangers can only be inspected in a shutdown. In HCN service, it takes additional time and resources to ensure equipment is decontaminated before a work permit can be issued. Reactors must also be shut down in a controlled manner to minimize damage, which adds to the total downtime. As the equipment ages, the inspection frequency may need to increase to ensure fitness for service. The unit requires a thorough inspection to understand deficiencies and develop repair or replacement plans to prevent equipment failures. In some instances, MI issues have forced rescheduling of a shutdown. A heat exchanger inspection during a 2014 shutdown revealed cracks to the tubesheet of an exchanger in HCN service. This required a repair plan to weld a crack and plug deficient tubes in order to place the equipment back into temporary service, but also required an intermediate turnaround several months later to replace the equipment.

4.2 Mechanical Integrity, Fixed Equipment

One of the more rewarding PSM developments has been an update to the MI program. Inspections have always been performed; however, there have been site-level initiatives to improve the quality of inspection reports to better manage the program. Before 2011, the report format was at the discretion of the testing company. The reports generally contained comments, an occasional picture, and data such as eddy current measurement numbers and standard information, thickness reading and locations. The current reports follow a template, developed in-house by the Reliability Department (a sample template is included in Appendix A) and includes much more detailed information and recommendations. They are also reviewed by in-house personnel with certifications (API). The new reporting style includes an executive summary, pictures of deficiencies, and recommendations for repair and changes to inspection frequency. Some reports include a welding repair plan.

The reporting upgrades have allowed for better management of the MI program instead of simply performing inspections and correcting deficiencies. This has allowed us to better justify an inspection frequency basis and has improved our asset integrity management since we can prevent mechanical integrity issues before they occur. Our improved program allows us to better manage our older assets by providing a solid baseline and subsequent inspections.

4.3 Mechanical Integrity, Instrumentation

Technology improvements have pushed instrumentation and automation to the forefront. Just like fixed equipment, critical instruments and interlock functions must be tested at a frequency required to meet the safety integrity rating of the function. The PHA procedure changes that were adopted in 2010 have dramatically increased the number of instruments to deliver the requisite number of protection layers. Generally, the instruments are used in an SIS function; however, many additional BPCS alarms and interlocks were added as layers of protection. The reactor design makes it difficult to test all critical alarms and interlocks without risking an unplanned shutdown. This has required us to proactively manage our instrumentation MI program with better diagnostics, planning, and evaluation of failure data to extend testing intervals.

5.0 Emerging Technology: The Future

Our process safety concept relies heavily on safety instrumentation. With the newly adopted PHA procedures, emphasis has shifted from mechanical devices (e.g. relief valves and operator intervention) to SIS and BPCS interlocks. Consequently, safety instrumentation must be well maintained to function as designed when required. While automation is not a new concept in our unit, implementation of new devices and interlocks requires training, documentation, and a quality MI program for safe operation. Safe and reliable operation is life cycle management: smart instruments and valves, online testing with partial stroke technology, and asset management platforms. As older instrumentation phases out, it is being replaced with newer models that offer better diagnostic coverage. Advanced technologies have allowed us to reevaluate interlock and alarm testing frequencies which are important for ensuring safety integrity level (SIL) are met. They also facilitate management of failure data and failure modes. A recent example of emerging technology is installation of smart valves with partial stroke capability in liquid HCN service.

The partial stroke test (PST) is considered a diagnostic test and not a proof test, but adds value in the reliability management of safety functions and can lead to reduced proof test intervals. A proof test must always be performed at a calculated interval since a PST cannot detect all failure modes. For example, a proof test can detect packing seizure, over-tightened packing, damaged instrument gas lines (crimped or blocked), and stem stick, seat scarring, debris blockage, or plugged seat. Unfortunately the partial stroke diagnostic test cannot determine these failure modes.

A smart PST valve increases the diagnostic coverage of a SIF. A diagnostic credit is applicable for failure modes performing a PST test, which can improve SIL average probability of failure on demand (PFD_{avg}) and reduce proof testing frequency (IEC 61508, Lundteigen and Rausand, 2008; Gruhn, et al, 2008). Additionally, the PST can be performed automatically and while the plant is running. Of course the whole SIF has to be designed and evaluated to determine the benefits of the partial stroke valve. The process and the results can be transferred into an asset management database for storage and evaluation. In a highly hazardous, yet highly reliable plant, emerging technologies are needed to minimize risk of personnel exposure while maintaining maximum uptime and PSM compliance.

An example of a partial-stroke application is a SIL2 storage vessel low-level shutoff SIF. The SIF consists of 1-out-of-2 (1oo2) voting on level transmitters that closes two shutoff valves on the feed line and a SIL3 rated logic solver. With failure data and proof test intervals that include a daily PST, a PFD_{avg} of 3.1E-3 and SIL2 rating are achievable. Removing the daily PST for only one of the shutoff valves reduces the PFD_{avg} to 1.2E-2, and the SIL2 rating is no long achieved. Similar examples have been employed in the HCN unit to ensure SIL ratings are achieved at reasonable proof test intervals.

6.0 Conclusions

An effective PSM program is critical in a hydrogen cyanide plant from a compliance and safety standpoint. There are too many hazards with the end product and raw materials to take shortcuts. Perhaps the most challenging PSM components in our unit are employee participation, training and mechanical integrity. While there are difficulties, the challenges are not impossible to overcome if sufficient resources are utilized.

The most important resource in our unit is personnel, and the training component of PSM is critical to ensure safe operation and effective emergency response. As the unit approaches 40 years of operation, the first generation production staff has phased out. There are several effective ways to transfer knowledge to newer employees before the information is lost to retirement or attrition:

- Encourage open discussion about potential retirement dates and incentivizing early notice.
- Consider investing in “Supernumeraries” if a potential retirement may occur. This can provide hands-on training with a veteran and allows a new employee to phase into the job.
- Ensure new employees are paired with a mentor on shift and develop training programs that require a veteran operator to approve successful completion of a training module.
- Periodic review and evaluate training materials and development programs to ensure they are accurate and effective.

Changing standards and company requirements provide the opportunity to implement the latest safety technology. The biggest effect has been modifications to the site PHA procedure. The end result has been an upgrade of critical instrumentation. We have implemented several large projects to improve SIS capabilities and install new safety instrument functions and BPCS capacity to meet the new PHA recommendations. Consequently the MI program has grown to ensure compliance and safe operation. While the capital projects and maintenance work load has increased, there was the opportunity to exchange dated instrumentation with state-of-the-art technology, namely smart devices and final control elements and asset management programs. With new technology, effective management of the MI program is achievable and reliability of BPCS and critical safety systems is ensured, while maintenance costs are reduced.

Acknowledgements

The authors would like to thank Steven Emerson of Emerson Technical Analysis and Mike Broadribb of Baker Engineering and Risk Consulting, John Traynor of Evonik Corporation, and Ivan Pelgrims of Evonik Nutrition & Care, GmBH for their critical review and helpful comments. Additional thanks to Kimberly Holt of Evonik Corporation for compiling data for the figures, and Mary Kilgore of Evonik Corporation for providing the inspection report template.

References

29 CFR Part 1910.119 (2014) Process Safety Management of Highly Hazardous Chemicals.

Pesce, L. D. (2010) Cyanides. Kirk-Othmer Encyclopedia of Chemical Technology. 1–29.

IEC 61508 (2010) Functional Safety of Electrical/Electronic/programmable electronic safety-related systems. Geneva: International Electrotechnical Commission.

Gruhn, P., Pittman, J., Wiley, S., LeBlanc, T. (1998) Quantifying the impact of partial stroke valve testing of safety instrumented systems. *ISA Transactions* 37, 87-94.

Lundteigen, M., Rausand, M. (2008) Partial stroke testing of process shutdown valves: How to determine the test coverage. *Journal of Loss Prevention Process in the Process Industries* 21, 579-588.

Appendix A: Sample fixed equipment inspection report template

EVONIK
POWER TO CREATE

Fixed Equipment Inspection Report

Current Inspection Information		
Work Order(s)		
Equipment Number & Plant		
Process/Content		
Inspection Date		
Previous Inspection Date		
Inspection(s) Method(s)		
NDE Performed by Examiner		
Directed by Contract API Inspector	John Doe	
Review by Evonik API Inspector	John Doe API Cert# XXXXX	
Equipment Information		
Construction Code/Standard		
National Board Number		
Manufacturer		
Serial Number / Year Built		
Design Pressure		
Max Allow Work Pres. (MAWP)		
SRV Tag No. & Set Pressure		
Design Temperature		
Temperature at Time of Inspection		
Shell Height /Length/Diameter		
Shell and Head Materials		
Corrosion Allowance		
Subject to or Evidence of Corrosion Under Insulation (CUI)		
Joint Efficiency		
Flange & Coupling Class		

Overall View

Executive Summary

Important point for the equipment owner to consider (repair, replace, change inspection interval)

Recommendations

Change inspection interval, repair damage, address a corrosion concern, repair grounding/insulation

Next Inspection – Per API XXX Inspection Code (RAGAGEP)

RAGAGEP= Recommended and Generally Accepted good Engineering Practices

Inspection Type	Current Frequency	Recommend Frequency	Max. Frequency Per Code
External Inspection	years	years	5 years
Internal/On-Stream Inspection	years	years	10 years

EVONIK
POWER TO CREATE

Fixed Equipment Inspection Report

Summary of Thickness Measurements

Component	Average $t_{previous}$ (in inches)	$Date_{previous}$	Average t_{actual} (in inches)	$Date_{actual}$	Corrosion Rate (inches Per Year)	Established TFlag (in inches)	Remaining Life (To TFlag Thickness) (in Years)
Shell 1"	0.766	10/10/11	0.768	08/01/16	0.00042	0.705	Greater Than 40 Years
Shell 6"	0.773	10/10/11	0.773	08/01/16	0.00187	0.705	36.37
Shell 12"	0.773	10/10/11	0.771	08/01/16	0.00083	0.705	Greater Than 40 Years
Shell 18"	0.774	10/10/11	0.768	08/01/16	0.00021	0.705	Greater Than 40 Years
Shell 24"	0.772	10/10/11	0.769	08/01/16	0.00000	0.705	No Wear Indicated
Shell 30"	0.770	10/10/11	0.765	08/01/16	No Wear	0.705	No Wear Indicated
Shell 35"	0.772	10/10/11	0.773	08/01/16	No Wear	0.705	No Wear Indicated
Top Head 1"	0.779	10/10/11	0.760	08/01/16	0.00208	0.705	26.48
Top Head 6"	0.743	10/10/11	0.743	08/01/16	0.00145	0.705	26.13
Top Head 11"	0.747	10/10/11	0.747	08/01/16	No Wear	0.705	No Wear Indicated
Bottom Head 1"	0.770	10/10/11	0.772	08/01/16	No Wear	0.705	No Wear Indicated
Bottom Head 2"	0.772	10/10/11	0.772	08/01/16	0.00166	0.705	Greater Than 40 Years

Tprevious = Actual thickness measurements from the previous inspection.

Date previous = Date of previous thickness readings

Tactual = The actual thickness of a CML, in inches, measured during the most recent inspection.

Dateactual = Date of Current Ultrasonic Thickness Reading of Material.

Tflag – Nominal thickness minus corrosion allowance (.0625) substituted for tmin.

External Inspection

General condition of equipment (foundation, shell, heads, insulation, grounding, etc.)

Internal Inspection

General condition of the heads/shell/nozzles, presence of cracks, pits, visible leaks, condition of liner (if applicable). Results of specialty testing (magnetic particle, dye penetrant, eddy current/IRIS).

We Built a (Process Safety) Ship – Who is Going to Sail it in 5 Years?

Monica Stiglich
3M
Maplewood, MN USA
mrstiglich1@mmm.com

Prepared for Presentation at
American Institute of Chemical Engineers
2017 Spring Meeting and 13th Global Congress on Process Safety
San Antonio, TX
March 26 – 29, 2017

We Built a (Process Safety) Ship – Who is Going to Sail it in 5 Years?

Monica Stiglich
3M
Maplewood, MN USA
mrstiglich1@mmm.com

Keywords: Process Safety Competency, Workforce Involvement, Process Knowledge Management

Abstract

3M has recognized the criticality of ongoing personnel development in a public sustainability goal of "100% participation in employee development programs to advance individual and organizational capabilities". One input to an individual's development plan is a company-developed structure of competency models for each major job function at 3M. Competencies are defined as a set of behaviors that can be identified, measured and developed. Defining and communicating global competencies enables shared understanding of the capabilities valued by 3M. An overview of the 3M competency structure will be compared to the CCPS "Guidelines to Process Safety Competency Requirements".

1 Introduction

The topic of "Aging Assets" which typically deals with the integrity of legacy equipment and systems is re-focused in this paper on the current state and subsequent loss of legacy human knowledge and practices which have been developed and implemented over the last several decades. Asset Integrity may use preventive maintenance and inspections to identify deficiencies which can alert an organization to malfunctions and deterioration in operating systems. Another type of analysis is needed to identify deficiencies in organization resources for both current and future operations.

3M has developed a business case for career development based on both individual satisfaction and organizational needs. The key points of this business case are:

- 3M believes an individual's career development is important
- 3M has a vested interest in developing their employees

- People are 3M's best asset and to continue 3M's success into the future, we need to work with our employees and to maximize their strengths, improve their weaknesses, and develop them into the leaders they want to be and the company needs them to be.

There is also an understanding that, in order to stay competitive, 3M needs employees who:

- Are dedicated to the idea of continuous learning
- Regularly assess their interests, values and skills
- Understand the skills and behaviors the company will need in the future
- Can easily move across functional boundaries and are able to switch between regular duties and special projects.

In contrast, the concept of "process safety competency" is an element of the RBPS pillar "committing to process safety" as described in the CCPS "Guidelines for Risk Based Process Safety" [1]. A supplementary CCPS publication "Guidelines for Defining Process Safety Competency Requirements" [2] was published to "help organizations design and implement the Process Safety Competency element of Risk Based Process Safety". A method was provided in the later publication to define job roles with necessary proficiency levels within an organization and to assign knowledge and skills to each job role proficiency level through a Process Safety Competency Matrix.

Both the 3M Competencies Models and the CCPS Process Safety Competency Matrix system are more fully described in the following sections of this paper.

2 Competency Frameworks

2.1 3M Competencies Model

Developed in 2015, the 3M Competencies Model has the goal of a shared understanding of capabilities valued by 3M. Competencies are identified for each major job function based on behaviors which can be identified, measured, and developed. "Behaviors" are the observable actions which shape the organization culture and demonstrate the implementation of 3M strategies and corporate vision. Individuals review the job function competencies and create development plans for capability and career growth. Personnel development is included in the "Goals and Progress" reported on the 3M.com/Sustainability webpage [3]:

Education & Development

Support empowerment and enrichment to provide people everywhere the opportunity to live life to its fullest.

- **Goal: Invest cash and products for education, community and environmental programs.** We will continue to invest in increasing access for people worldwide to science, technology, engineering, math and business skills, and to improving lives through investments in environment, health and safety programs. Cash, in-kind donations, sponsorships, and volunteerism will remain core to our 3Mgives strategy. Since 1953, we have invested $1.4 billion in cash and product donations in 3M communities.

- **Goal: 100% participation in employee development programs to advance individual and organizational capabilities.** Our employees are encouraged and supported to pursue areas of interest and develop career paths and opportunities aligned with those interests.

- **Goal: Double the pipeline of diverse talent in management to build a diverse workforce.** Increasing women and diversity in every region in which we operate is a core part of our Sustainability strategy. We recognize that doubling our diverse talent in management will help build a diverse pipeline for executive leadership as well.

Figure 1. Education and Development Goal from 3M.com Sustainability webpage

Two terms related to job roles and functions are used in the 3M system:

- Job Descriptions: Describe duties and responsibilities of common positions

- Functional Competencies: Describe key capabilities valued by leaders to be developed over a career and linked to overarching goals of the function within 3M and define expectations globally, across every business segment.

The Functional Competencies describe knowledge, skills, and abilities necessary to establish the capabilities to accomplish the duties and responsibilities of a given Job Description.

Major 3M job functions which have been identified with specific competency models include:

3M Job Functions with Competency Models	
Design	Finance
Human Resources	Information Technology
Legal Affairs	Marketing
Sales	Research and Development
Logistics	Technical / Engineering
Sourcing	Environmental, Health and Safety

Table 1. 3M Job Functions with Competency Models

Each competency model is linked to job function goals with five to seven competencies which describe unique function capabilities. Each of the competencies list key components which describe five levels of proficiency listed as Foundational, Basic, Proficient, Advanced, and Expert. Note that there is NOT an expectation for individuals in any job function to achieve "Expert" proficiency in all competencies and components.

As an example, Technical / Engineering competencies for people working in a technical role – typically, engineers in a Division Engineering, Manufacturing, Packaging, Design, Quality or other function are:

- Analyzing Manufacturing Operations
- Creating Packaging Solutions
- Developing and Applying Manufacturing Technology
- Driving Technical Improvements
- Leveraging the Supplier Network
- Managing Technical Projects

- Operating and Maintaining Facilities
- Sustaining Product / Operations Compliance.

The Technical / Engineering job functions "Sustaining Product / Operations Compliance" competency has the following components:

- External Compliance
- Internal Standards
- Management of Internal and External Audits.

For example, the graphic below contains the component details and descriptions of proficiency for Management of Internal and External Audits.

Foundational	Basic	Proficient	Advanced	Expert
• Recognizes the scope of auditing that is likely in assigned work areas	• Prepares for and participates in audits • Communicates the pertinent work processes and information flows to the auditor	• Develops and implements strategies to prepare for upcoming audits in the functional area • Identifies root cause and implements Corrective Action and Preventative Action (CAPA) to address audit findings across the functional area • Trains others on audit protocol	• Develops and implements strategies to prepare for upcoming audits across the local value stream • Identifies root cause and implements Corrective Action and Preventative Action (CAPA) to address audit findings across the Local Value Stream • Coaches and mentors others to proactively identify and resolve potential audit concerns	• Develops and implements strategies to prepare for audits across the value stream • Identifies root cause and implements Corrective Action and Preventative Action (CAPA) to address audit findings across the value stream

Figure 2. Management of Internal and External Audits Details

Another example is the Competency Model identified for the Environmental, Health, and Safety job functions:

- Identifying and Reporting Hazards and Releases
- Assessing Risks and Compliance Obligations
- Developing Control and Compliance Strategies and Solutions
- Implementing Controls and Compliance Requirements
- Developing, Managing and Improving Programs
- Engaging and Influencing Externally

Notice that, for both the Technical / Engineering and the Environmental, Health, and Safety grouped job functions, "process safety" is not an explicitly stated area in the details and descriptions in their respective Competency Models. Several facts and other systematic approaches address the visibility and presence of "process safety competency" within 3M:

- Of more than 230 facilities worldwide, about half are identified for process safety implementation based on regulatory requirements or internal risk assessment
- The 3M Global Safety and Health Plan implementation requires specific and measurable safety and health (including process safety) expectations and performance evaluation for all jobs. Specific expectations for demonstration of environmental, health, and safety leadership are also stated for facility management.
- Corporate, business and regional process safety organizations have the responsibility to continually monitor, audit and develop process safety resources at all levels of the organization.

A self-evaluation tool has been developed for each job function to identify gaps and create an individual development plan based on desired or target competencies. Development can be through formal methods such as classes, or informal methods such as on the job experiences or learning through others. This tool can also be used for future career growth by identifying and developing proficiencies for another job function.

Competency	Component	Actual Proficiency Level	Target Proficiency Level	Development Ideas
Assessing Risks and Compliance Obligations	Assess Compliance Obligations	Basic	Proficient	Form work group to study and develop plan for implementation of EPA RMP changes

Table 2. 3M Self-Evaluation Tool Example

As a further example, my personal development plan for 2016 included the following items:

- Coaching and mentoring others
  - Specifics: provide guidance on PHM implementation, PHA facilitation, combustible dust and flammable liquids attributes / hazards, engineering control methods
  - Related Competency: Developing control and compliance strategies and solutions
- Schedule time to complete assigned projects
  - Specifics: projects to have scheduled times include fire analysis and reports, updates to PHM Guidance Manual, individual facility PHM applicability analysis completed
  - Related Competency: Developing control and compliance strategies and solutions
- Conference attendance

  - Specifics: Attend 2016 Global Conference on Process Safety and 3M Environmental Conference
  - Related Competency: Developing, Managing and Improving Programs
- Training courses
  - Specifics: Arrange and attend PHAST course (external trainer brought to 3M), attend PSM audit course (external trainer brought to 3M), complete 3M required legal courses
  - Related Competency: Assessing Risks and Compliance Obligations

Completion of individual annual development plans and documentation of completed tasks and activities is recorded and monitored on a corporate basis to demonstrate conformance with the 3M Sustainability Goal for Education and Development. Feedback may be provided to both individuals and their direct supervision for absence of personal development information in a corporate database.

2.2 CCPS Process Safety Competency Matrix

The purpose of the CCPS publication "Guidelines for Defining Process Safety Competency Requirements" is to "help organizations design and implement the Process Safety Competency element of Risk Based Process Safety". A method was provided to define job roles with necessary proficiency levels within an organization and to assign knowledge and skills to each job role proficiency level through a Process Safety Competency Matrix as documented in an Excel ® workbook available from the CCPS website. Generic job roles and corresponding process safety competencies for that role are given for:

- Facility Manager
- HSE Manager
- Operations Manager
- Inspection, Testing and Maintenance Manager
- Supervisor
- Operator
- Technician
- Engineer
- Project Manager
- Project Engineer

The Process Safety Competency Matrix workbook is arranged with expectations for each Pillar of Risk Based Process Safety with its corresponding elements:

Commit to Process Safety

- Process Safety Culture
- Compliance with Standards
- Process Safety Competency
- Workforce Involvement

- Stakeholder Outreach

Understand Hazards and Risks

- Process Knowledge Management
- Hazard Identification and Risk Analysis

Manage Risk

- Operating Procedures
- Safe Work Practices
- Asset Integrity and Reliability
- Contractor Management
- Training and Performance Assurance
- Management of Change
- Operational Readiness
- Conduct of Operations
- Emergency Management

Learn From Experience

- Incident Investigation
- Measurement and Metrics
- Auditing
- Management Review and Continuous Improvement

And an additional element for **Engineering Design.**

Proficiency levels for the Competency Matrix are described as:

Level Name	Level Description
Awareness	Be generally aware of this topic and associated terms. May not know the answer, but knows where to get more information.
Basic Knowledge	Has general working knowledge of the topic. Has basic training required to carry out general tasks related to their role
Practitioner	Can execute specific tasks within the topic with minimal direction. Has the experience levels to complete assigned tasks. Typically an engineering or manager role.
Expert	Is a recognized expert with extensive knowledge and skills

	Has specialized training or certification which may be required for certain tasks.
Leader	Is responsible for implementation of a PSM element or activity. Knowledgeable of the subject matter, may be a subject matter expert, but that isn't required. Coordinates all resources and assigns tasks to ensure effective implementation of the topic.

Table 3. CCPS Proficiency Level Descriptions

The recommended use of this Competency Matrix is for customization to corporate, divisional, or facility organizations with adjustment of job titles and assigned responsibilities.
Note that there is NOT an expectation for individuals in an individual job function to achieve "Expert" or "Leader" proficiency in all competencies. The recommended proficiency levels for each Pillar, Element, and participating job role have been tabulated in the Matrix. The graphic below from the RBPS worksheet displays the Typical Proficiency Levels by Competency for the RBPS Pillar "Commit to Process Safety" and Element "Process Safety Culture".

		Typical Proficiency Levels 2 = Basic Knowledge; 3 = Practitioner; 4 = Expert													
		Operations				Inspection, Testing & Maintenance				Projects		Health Safety & Environmental		Others (HR, Training, Security, Purchasing, R&D, Marketing, Sales,	Corporate or Contractor
Competencies	**Responsibility Assigned To (Leader -**	Operator	Supervisor	Engineer	Manager	Technician	Supervisor	Engineer	Manager	Engineer	Manager	Engineer	Manager		
	Facility Manager		2		2		2		2			2	3		4
Establish a vision statement for the facility with regard to process safety															
Develop a training program for process safety which highlights the desired culture, vision, expectations, standards, roles, and responsibilities															
Establish process safety goals/measurable objectives for the facility and monitor performance															
Develop and maintain process safety policies and mission statement for the															
Monitor process safety incidents within the industry and share key learnings with the															
Ensure that adequate resources are assigned to comply with process safety															

Table 4. Competencies for the Pillar "Commit to Process Safety" / "Process Safety Culture" Element

The Process Safety Competency Matrix workbook also has a Skills and Knowledge worksheet with recommended proficiency content for each Pillar and Element combination:

Basic	**Practitioner (Basic Level Plus the Following)**	**Expert (Practioner Level Plus the Following)**	**Leader (Basic level Plus the Following)**
Knowledge of the implications of actions and how they reinforce or violate the vision statement -Knowledge of how the progress towards the process safety goals and objectives is measured	Knowledge of resources required to implement process safety elements. -Experience with sharing learnings related to process safety, including how incidents from other industries may be used to improve the processes at this facility. - Experience with monitoring typical leading and lagging indicators associated with process safety - Familiarity with RAGAGEPs which may impact process safety policies.		Knowledge of how process safety goals and objectives and progress towards them are communicated throughout the facility. - Authority to assign responsibilities for process safety to other personnel on-facility. - Familiarity with procedures necessary to obtain outside resources to implement process safety elements.

Table 5. Skills and Knowledge for the Pillar "Commit to Process Safety" / "Process Safety Culture" Element

An assessment of current proficiency can be compiled and opportunities for improvement identified using the RBPS worksheet for either an individual in a job role or for an organization. The assessment can also show gaps where elements do not have assigned personnel.

3 Comparison of 3M Competency Framework and CCPS Process Safety Competencies

Both the 3M and the CCPS frameworks for competencies have the goal of describing key attributes and knowledge for a variety of jobs in an organization and the ultimate goal of sustained, uninterrupted operations. Other similarities and differences include:

- Job Roles vs Job Functions and the related Competency Needs vs Function Capabilities
  - CCPS focuses on the essential job role process safety content and competency needs by job role
  - 3M focuses on the job function knowledge base necessary for essential business function capabilities with limited safety and health aspects; other corporate structures address process safety specific job content and continually monitor, audit and develop process safety resources at all levels of the organization.
- Proficiency Levels: Both CCPS and 3M provide descriptions for similar levels of increasing job competency with the provision for targeting proficiency level based on job demands.
- Assessor vs Self-Evaluation
  - Both CCPS and 3M competency systems can be used as an individual's assessment for their job function or role in an organization.
  - The CCPS system can also be used to assess an organization, by an evaluator either internal or external to an organization, to determine if process safety roles and functionality are met.
- Gap Closure Plan vs Development Plan
  - The CCPS system specifically identifies gaps either in personnel assignments or knowledge base for process safety roles, with the expectation that a plan be developed to close those gaps.
  - The 3M system focuses on individuals evaluating the competencies related to their job function and building a personal development plan for career growth.

4 Conclusions

Who is going to sail the process safety ship in 5 years?

3M has recognized the criticality of ongoing personnel development in a public sustainability goal of "100% participation in employee development programs to advance individual and organizational capabilities" and in the communicated business case "to continue 3M's success into the future". Competency Models were developed for use across the enterprise for the major business and operational job functions. Specific Process Safety competency is addressed by corporate, business and regional process safety organizations which have the responsibility to continually monitor, audit and develop process safety resources at all levels of the organization.

The CCPS Process Safety Competency Requirements also focus on critical content of various job functions aligning with the concepts of Risk Based Process Safety. Outcomes from use of this information are improved organizational learning and a knowledgeable process safety culture.

Both the CCPS and 3M Competency frameworks were built based on knowledge and experience with job functions within organizations. The frameworks describe the current state of individuals and organizations while identifying development and resource gaps for the purpose ensuring business continuity – keeping "the ship" afloat and moving forward.

5 References

[1] *Guidelines for Risk Based Process Safety*. Center for Chemical Process Safety, American Institute of Chemical Engineers, New York, NY, 2007.

[2] *Guidelines for Defining Process Safety Competency Requirements*. Center for Chemical Process Safety, American Institute of Chemical Engineers, New York, NY, 2015.

[3] 3M.com/Sustainability

Changes to industry guidance for relief and blowdown system design and the impact on existing infrastructure

Paul Frey, MIChemE, C.Eng
Process Systems Enterprise Limited
26/28 Hammersmith Grove, London,W6 7HA
p.frey@psenterprise.com

Sathish Natarajan,
Process Systems Enterprise Limited
Two Allen Center, 1200 Smith Str, Suite 1600, Houston, TX
s.natarajan@psenterprise.com

Prepared for Presentation at
American Institute of Chemical Engineers
2017 Spring Meeting and 13th Global Congress on Process Safety
San Antonio, TX
March 26 – 29, 2017

Changes to industry guidance for relief and blowdown system design and the impact on existing infrastructure

Paul Frey, MIChemE, C.Eng
Process Systems Enterprise Limited
p.frey@psenterprise.com

Sathish Natarajan,
Process Systems Enterprise Limited
s.natarajan@psenterprise.com

Keywords: Incidents, verification, brittle fracture, acoustically induced vibration, flow induced vibration, Low Temperature, fire survivability, blowdown, relief, flare

Abstract

As the last line of defense for process and equipment integrity, relief, blowdown and flare systems need to be adequately designed and maintained. Systems need to be periodically re-assessed so that they operate safely throughout the life of the facility that they help protect. This talk highlights the importance of ensuring flare and relief systems remain fit for purpose throughout their lifetime maintaining accountability for plant modifications, changes to operations, and changes to industry guidance and regulation. Recognizing high profile incidents involving flare systems from Westlake to Texas City and to a number of recent incidents that have led to loss of containment due to excessive vibrations and brittle fracture of the flare piping, we show that the root cause of these incidents can be linked to failure to recognize hazards and perform adequate analysis, which is fundamental to Process Hazards Analysis programs under OSHA 3133

In recognition of these & other incidents, we describe how relief and blowdown adequacy assessment has tightened in recent years in regards to significant changes to API 521, and in particular:

A. to ensuring the suitability of materials of construction so as to avoid brittle fracture risks during depressurization
B. the adoption of the more rigorous analytical methodology for assessing vessel survivability under fire attack for blowdown system design
C. to assessing the likelihood of failure due to acoustic and flow induced vibration in flare piping
D. the importance of verification of relief and flare systems to ensure they comply with latest Industry Standards

Through a number of case studies we describe a methodology for assessing and analyzing existing infrastructure for these risks and explain how a detailed model-based analysis can ensure an inherently safe relief and blowdown system design that is compliant with the latest industry guidance.

1 Introduction

Processing facilities in offshore assets and chemicals refining, share common technologies for the safe venting and disposal of hydrocarbons. Depressuring valves and Pressure relief valves prevent overpressure in various systems such as slug catchers, gas liquid separators, vessels, distillation columns, compressor systems, packed beds and distribution piping. The undesirable effects of rapid depressuring of gas systems in upstream facilities, may be the generation of very low metal embrittlement temperatures. Conversely, in fire events, the depressuring may not be rapid enough to prevent vessel wall stresses causing rupture and loss of containment of the vessel.

Relief valves on the other hand, have to operate under a wide range of overpressure scenarios and may not have been incorrectly sized or not adequate for a case which later turns out to be governing. Plant modifications are often made which can invalidate existing overpressure protection systems.

It is important for Contractors as part of design, and Operators maintaining compliance of existing assets, to take full responsibility for applying the most appropriate analysis techniques. Whilst little has changed in hardware terms with blowdown valves, relief valves, rupture discs and flare systems, our understanding has changed of the parameters which cause relieving events, the calculation of the required capacity and device sizing methods. Recent changes in the API Standards [3,6] have also meant that new techniques are now recommended to reduce un-certainty and over- conservatism.

Driving the improvement of Standards, has been a number of well publicized Industry Incidents. The fire and explosion at Texas City, resulted in a wide ranging review of plant design safety, operational factors and industry codes of practice. Consequently, API Standard 521 was reviewed and updated following the 1987 Hydrocracker explosion at Grangemouth, Texas City and more recently, following the Williams Geismar Olefins Plant fire in 2013.

The impact changes to Standards on existing and ageing facilities is critical. Older facilities however well maintained from a mechanical perspective, often do not maintain well documented process design basis, relief and blowdown philosophy documentation and up to date calculations, especially where assets have changed ownership over their lifetime.

Experience in performing relief system verifications on operational facilities, have shown large numbers of significant faults due to impact of plant modifications after installation, or changes to the device itself, or operational changes to the process. Therefore periodically, Process Hazards Analysis (PHA) and independent third party verifications, are necessary to assess blowdown, flare and relief systems to ensure the records and sizing is compliant with latest standards.

In the USA, OSHA 3133 provides the framework for process safety management and the impetus to organizations to comply with the standards. In other regions of the world the National Offshore Petroleum Safety and Environmental Management Authority (NOPSEMA) in Australia, and the Health and Safety Executive (HSE) in the UK adopt similar schemes.

Formal and mandatory programs of assessment (OSHA) are an important part of maintenance of safety integrity. The PHA should be conducted on a new facility or after a significant process modification prior to start up, and then re-validated on a 5 year cycle. The re-validation cycle is also supported by the general requirement to update and re-issue API Standards every 6 years.

The Operator is also responsible for the analysis and documentation for existing equipment designed to standards and codes which are no longer in use, to determine that the equipment design is still appropriate, and operating in a safe manner.

Whilst a PHA must be periodically re-validated, there is no reason why a more proactive approach to hazard evaluation should not be the basis of process safety management. When major incidents occur, the root causes are analysed to ensure that an Operators own facility does not share similar risks.

2 Major Incidents involving overpressure protection systems and case studies of advanced analysis

2.1 Texas City Fire and Explosion – De-isohexaniser overfill, March 2005- Overpressure Protection

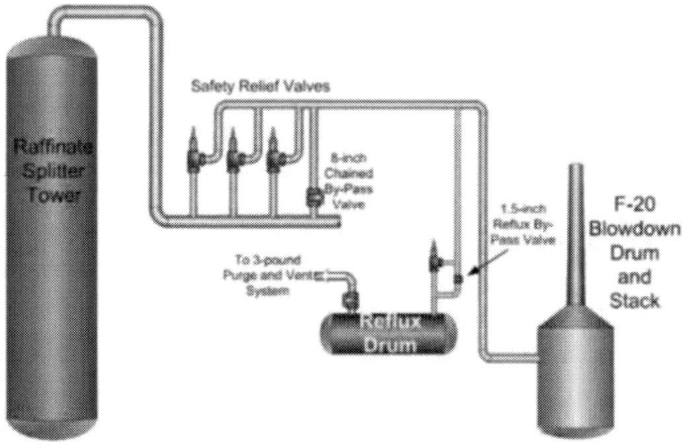

Figure 1. Schematic of relief and blowdown system

The incident investigation conducted by the Chemical Safety and Hazards Board (CSB) of the Refinery fire and explosion at Texas City [1], resulting in 15 Killed and 180 Injured identified the following root causes and actions related to overpressure protection systems:

- Overfilling of the Raffinate Splitter tower resulted in overpressurisation of the relief valves allowing liquid to overflow to the blowdown drum.

- The blowdown drum overfilled, discharging a large quantity of flammable liquid and vapour which engulfed the ISOM unit area and was ignited resulting in the explosion and fire.
- The potential for liquid overfilling had been determined as not credible and the relief system was not designed for liquid duty.
- "BP was unable to provide the CSB with any documents from the historical files concerning the original design basis of the blowdown drum and stack."

As part of the incident investigation, from plant records, it was apparent that the level controller on the column base was faulty, as was the high level switch which failed to trigger the event. However, due to conflicting operator / witness statements, it was difficult to identify the sequence of actions taken leading up to the overfilling event, and there was insufficient instrumentation which could measure and record the physical flows, pressures and temperatures in key locations of the system.

2.2 Changes to the API Standards

- The API Standard 521 did not have specific guidance on the potential overpressure hazard of vessel liquid overfill
- There was no guidance on the hazard of a large liquid release to a disposal drum that vents directly to the atmosphere.
- In light of the ambiguities and gaps discovered in reviewing API 521, CSB issued a recommendation to API on October 31, 2005 to revise its recommended practice.

In the latest revision, API have re-emphasised that care should be taken using the general ideal gas equations in all circumstances, and there are many examples of existing facilities which have undersized or grossly oversized valves due to mis-interpretation of the guidelines. In particular verification studies often identify:

- Relief valves undersized due to use of compressibility factor below 0.8, and use of simulated real gas Cp/Cv as approximation of isentropic coefficient.
- Inlet/ Outlet piping losses exceeded reducing the valve capacity
- Undersized valve due to un-accounted for relief scenario – e.g gas breakthrough, or liquid overfill.

2.3 Application of Dynamic Simulation to incident investigation

PSE performed the model based investigations into the incident, analyzing the hydraulics of the column system. Dynamic simulation was used to determine the overpressure in the column and quantify the flows to the blowdown drum, and mass of liquid and vapor discharged as part of the consequence modelling.

The dynamic simulation showed accurate representation against the column pressure transmitter, and could corroborate when the manual interventions were made by operators, including at 12:41 hrs – opening of the 8" chain bypass valve, and the timings when the relief valves opened and at 13:20 hrs when the explosion occurred.

The benefit of using a detailed modelling approach was:

- Simulations confirmed the timings of the events
- Operator interventions and changes to process setpoints were confirmed
- Tower overpressure beyond the maximum allowable accumulation confirmed previous inspection warnings that the relief valves were undersized.

3 Brittle Fracture Risks

3.1 Westlake Ethylene Cracker flare Fire, 2002 – brittle fracture of flare header

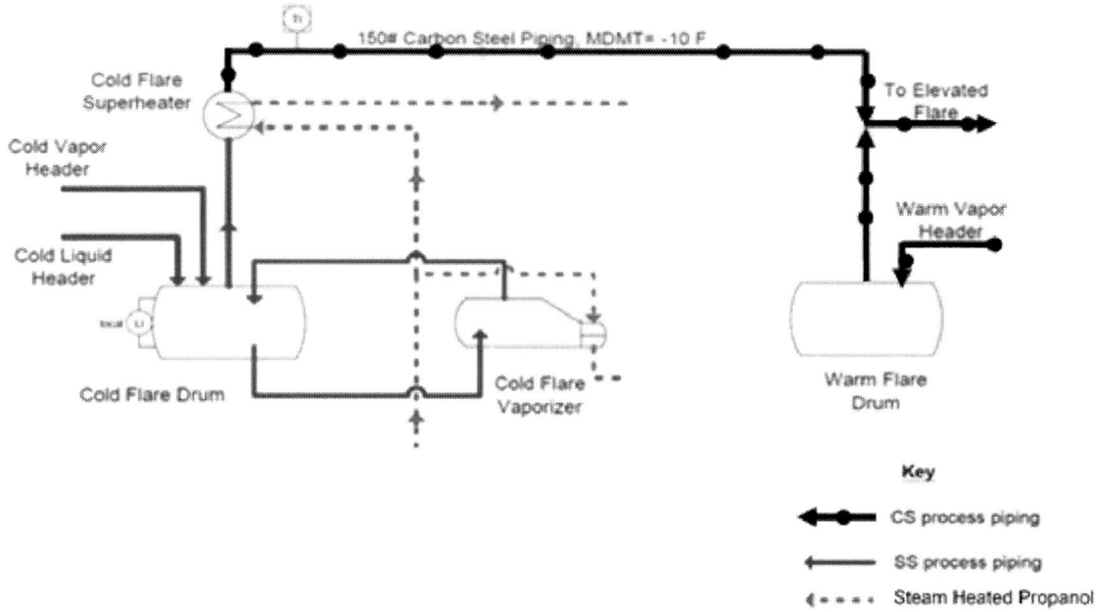

Figure 2. Simplified schematic of Cold flare vaporizer and Flare Header

This is an example of a loss of hydrocarbon containment, explosion and fire resulting from brittle fracture of a flare header. Fortunately, no one was injured, however the fire was not extinguished for a week and the plant was shut down for a month. Corrective actions to replace/ upgrade systems had to be scheduled in for the 2003 turnaround.

Sequence of events:

- Westlake Petro 1 ethylene plant started to produce off spec product, resulting in the flaring of liquid ethylene into the cold flare drum.
- The minimum design metal temperature of the flare piping was -10F.
- The cold flare drum vaporizer and outlet superheater failed to provide sufficient duty and the flare temperature dropped below -13F which was the minimum temperature recorded before the temperature instrument failed.

- The low temperatures propagating into the carbon steel headers caused brittle fracture of the header and loss of containment.

A number of root causes were determined by Westlake Group:

- Fouling of the vaporizer and superheater had reduced the available duty. It was determined that heavy oil could remain in the cold flare drum for long periods leading to increased risk of fouling.
- The risks to the flare line safety of mal-operation of the heat exchangers were not recognized
- The brittle fracture required a coincident stress which although not discovered may have been pipe contraction, or external stress from heavy rain at the time.
- The material was impact tested to -10F, subsequently parts of the flare header failed the charpy v-notch testing for -20F.

Corrective Design Actions:

Inherent safety was improved to reduce the principal risk to the carbon steel piping through low performance of the heat exchangers, by replacing the header downstream of the vaporizer with stainless steel. This was continued to the warm flare drum, to prevent effects of reverse flow into the warm flare piping, and into the main combined header 100 ft downstream of the cold/warm flare drum tie in point. A new thermocouple was installed at this point and the thermocouple at the super heater re-ranged for -200F. Stress analysis was conducted for the new flare line and flare drum. Westlake also replaced the super heater with a 50% larger duty and uprated the propanol piping. A heavy oil removal system and new level indicators and alarms were installed on the cold flare drum to manage any liquid accumulation.

3.2 Changes to API Standard for low temperature analysis

The Westlake incident occurred prior to the revision of the API 521 Standard, which now highlights the importance of accurate assessment of low temperature/ brittle fracture risks.

There are numerous incidents and alerts involving brittle fractures. As Carbon Steel is cooled, the impact strength reduces rapidly as the material changes from a ductile to brittle zone making it less resistant to stresses which are present through implied loads, internal pressure or fluid flow reaction forces.

API 521[3] highlights a number of risks of brittle fracture:

"Materials exposed to temperatures below the specified minimum design temperature may suffer permanent damage or brittle failure, depending on the mechanical stresses present in areas subjected to low temperatures."

"...particularly if there is a possibility of low-boiling point liquids entering the disposal system. Autorefrigeration will occur as liquid boils at the reduced pressure. If the equilibrium temperature is sufficiently low, piping and drums fabricated of materials designed for low temperature may be required to eliminate the risk of brittle fracture" [3] Section 5.2.2.4.

Whilst this is invariably the root cause of the Westlake incident, it is also important to consider non-equilibrium processes occurring during gas depressuring

Gas depressuring systems are essential for protecting vessels from failure during fire events, and to minimize the consequences of a loss of containment by removing inventory safely to a flare system. However the phenomenon of liquids condensing as pressure is reduced and then re-evaporating in combination with the Joule-Thomson temperature reduction, is often overlooked by conventional design approaches. This is now recognized in the new edition of the API 521 standard [3].

3.3 Application of Dynamic Simulation

Research into depressuring of vessels by M.A Haque et Al [4] indicated the phenomenon that the gas cooled the vessel surfaces far less than the surfaces where liquids accumulated and evaporated. As can be seen from figure 3, where there is liquid, the wall temperature can be as much as 40°C lower than the gas temperature.

Furthermore, conventional approaches to simulation which do not model the non-equilibrium processes were shown to be highly inaccurate. In the case of gas depressuring, they were generally conservative however where liquids were present, highly in-accurate.

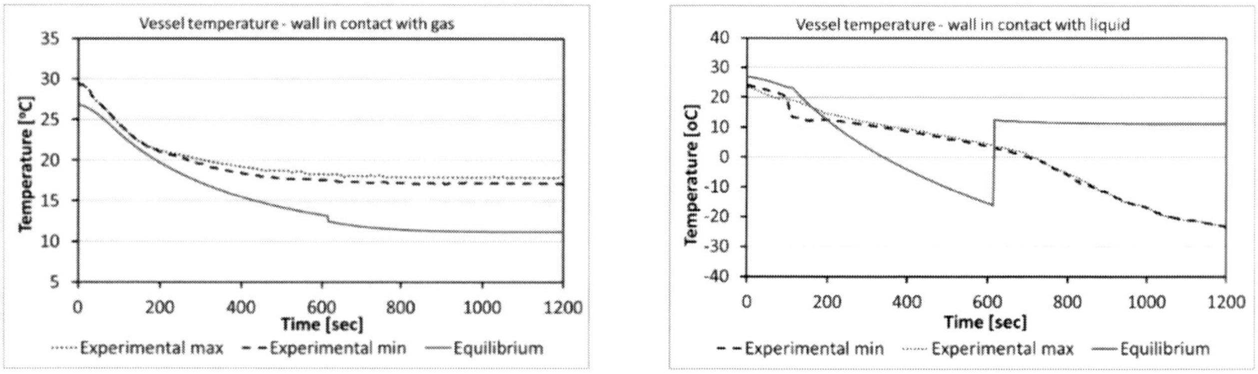

Figure 3 Experimental data showing wall temperatures for the depressuring of process vessels compared with conventional equilibrium simulation approaches.

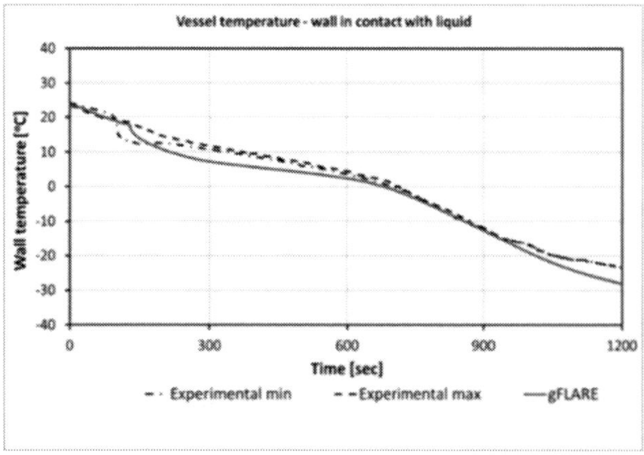

Figure 4. Comparison between g-FLARE® dynamic simulation and experimental data

For more complex systems of interconnected vessels and piping, typically encountered with compression systems, gas sweetening and absorption systems, it is important to be able to represent the actual topography of the plant. This means the equipment orientations, elevations, and the detail of the piping falls and pockets where liquid may accumulate need to be included within a simulation model, as well as addressing the complex thermodynamics of the process. This allows decisions on equipment metallurgy to be based on specific locations within the plant.

In one recent example a multi-national Oil & Gas operator performed a preliminary blowdown analysis on a high pressure gas plant that indicated a requirement for 80% of the plant to be constructed from stainless steel which would have made the economics un-feasible. To minimize the requirement for stainless steel construction, the operator commissioned a more detailed analysis study compliant with the latest API guidance to determine the minimum metal temperatures in the process equipment and piping during system depressurization. The resulting optimized design decreased the stainless steel construction requirement from 80% of the facility to just 30%, with total project savings of up to $1.8bn.

Because pressure drops throughout the process control the gas flowrates via different flow paths, including reverse flows and mixing of streams, certain locations within the same depressurizing section exhibited different pipe wall temperatures during the depressuring process. For example, due to the location of the blowdown orifice, some segments could not depressurize until after the inlet line and separator pressure had reduced. The low gas flow allowed heat gain from ambient to keep the pipe wall from falling below the limit of -29°C for non-impact tested carbon steel.

However within the separator the starting gas temperature was lower due to evaporating condensate and non-equilibrium cooling of the gas stream. Due to the significant gas volume and much higher flowrate the separator outlet pipe wall exceeded the MDMT of -46°C for low temperature carbon steel (LTCS). This was also a long process line of large diameter which made changes to this system un-desirable. To minimize changes to materials and cost impact, the separator was fabricated in LTCS however the Orifice was relocated to another part of the system so that the separator did not depressurize through the process line.

At the orifice, the extremely high velocity in the downstream piping caused increased heat loss from the gas due expansion and conversion to kinetic energy, and a very low temperature of -120°C. In this case it was necessary to change the tailpipes material to stainless steel, and review how far this cold front extended to the flare header.

The benefit of using a detailed modelling approach was:

- Piping low points and areas with liquid accumulation such as the separator, and areas with high kinetic energy such as the tailpipe were confirmed to be much colder than expected.
- Re-design of the blowdown locations and segment configurations (removal of low points) assisted with low temperature design

- The precise location of the temperatures were discovered to aid material selections.

4 Fire Survivability

4.1 Piper Alpha Offshore Platform fire and explosion, 1988 - Fire Depressurisation

Piper Alpha ranks as the worst offshore human catastrophe with the loss of 167 lives and insurance claims of approximately US$ 1.4 billion. Whilst the root causes of the initial fire were due to operator error in starting up a standby pump, fire spread un-checked throughout the platform. The majority of the deaths occurred within the accommodation module in which personnel had sought refuge and was overwhelmed by fire and smoke. The Cullen report made 106 separate recommendations to improve offshore safety.

API 14C mandates ESD and depressurization systems as one of the Emergency Support Systems required. Depressurization systems increase the survivability time of vessels and equipment to reduce the risk of premature rupture and uncontrolled release of hydrocarbons.

4.2 Changes to API Standards for Fire Depressurization analysis

In 2014, API 521 was significantly changed to include a more advanced analytical methodology which had been in use by Statoil / Norsk Hydro to provide protection to offshore vessels exposed to fire. Recently, offshore safety guideline API 14C was re-issued to align with API 521.

API recognized that the previous guidelines were based on empirical fire heat input equations, suitable for on-shore installations. The basis of these equations were from a series of experiments on storage tanks exposed to pool fires in a well ventilated and open / un-congested plant layout.

Depressurization guidelines for typical refinery vessels were found not to be applicable for thin wall vessels and congested equipment layout as found offshore. In offshore facilities there is also the real risk of impingement of high pressure gas jet fires, which can produce a localized peak heat flux higher than the surface average heat flux applied in the empirical model. The fire heat fluxes typically encountered in an open pool fire are between 50 to 150 kW/m2, and this increases up to 400 kW/m2 [3] for the largest Jet fires.

Subsequently the API 521 guidance was revised to include the following:

"For pool fire exposure, this generally involves reducing the equipment pressure from initial conditions to a level equivalent to 50% of the vessel's design pressure within approximately 15 min. This criterion is based on the vessel wall temperature versus stress to rupture and applies generally to carbon steel vessels with a wall thickness of approximately 25.4mm or more" and

"The vessel material and thickness influences the depressurization rate. For vessels other than 25.4mm carbon steel the user may choose to apply the 50%/15 min criterion or some other criterion, or may choose to perform more specific calculations" [3] Section 4.6.6

Prior to the revised guidelines, depressuring design was generally achieved by a dynamic simulation of the vessel pressure with time, by varying the discharge line orifice sizes to achieve the 15minute depressuring criterion. There was no associated check of the internal stress and the ultimate tensile stress of the vessel which varies with the wall temperature. Fundamentally, if a pressurized vessel wall is heated the material strength reduces to the point where internal stress (principally pressure) exceeds the tensile strength of the vessel wall and rupture occurs, as indicated by Figure 5. The API 521 Standard (in Annex A) as well as other guidelines provides a revised methodology for calculating the fire heat flux based on the stefan-bolzmann equation and details of the calculation of internal stresses due to pressure within piping and vessels.

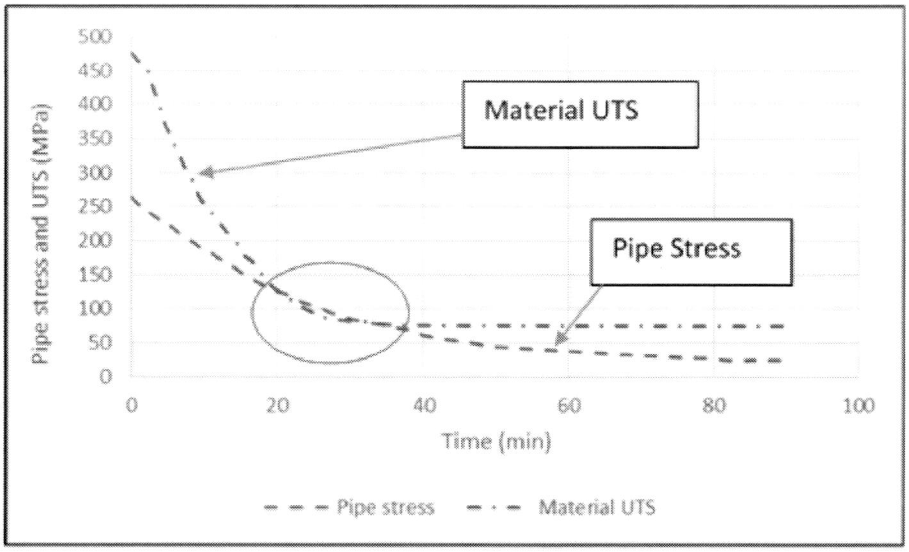

Figure 5. Graph illustrating rupture condition when depressurization is not sufficiently rapid.

4.3 Application of Dynamic Simulation

In the following example, a significant on-shore facility consisted of four trains, each train had four separators (ID=4m, L=20m)

- In event of fire on one separator, adjacent separators were blown down simultaneously to prevent escalation

Analysis of above scenario indicated that:

- Steady state simulations showed flare capacity was exceeded
- In order not to exceed the flare design capacity, the survivability criterion was violated for the vessel exposed to fire
- In order not to exceed the flare capacity and not violate the fire survivability criterion Passive Fire Protection (PFP) would have to be used.
- Operator decided to investigate option of more accurately determining the extent of flare capacity utilization to answer the following:

- Is the flare capacity limit actually exceeded?
- How much thickness of PFP, if any, is required?

Results:

A fully representative dynamic model was configured, to include the process vessel, blowdown lines, orifice plates, and flare system headers including the knock out drum and flare riser. The simulation solved both the pressure-stress calculation and the wall temperature due to fire heat input simultaneously using the analytical equation for heat flux [3].

The following results were obtained :

- Flare capacity limit was suitable for the maximum load when considered dynamically (figure 7)
- Orifice size optimised to avoid risk of rupture as shown in Figure 6
- No requirement for PFP

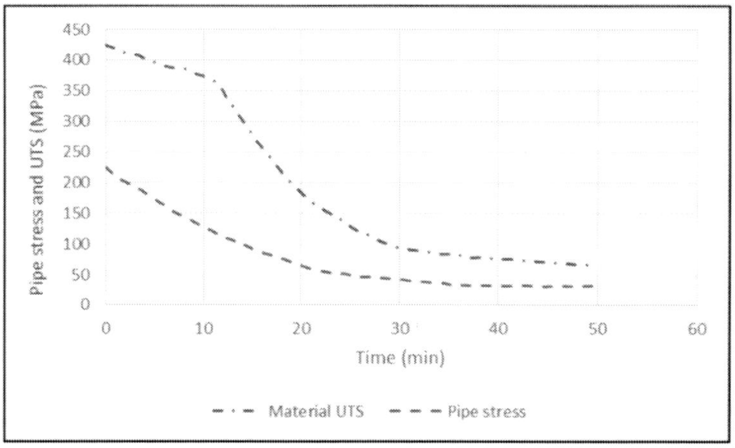

Figure 6. Case Study: graph indicating separator depressurization meeting stress requirements

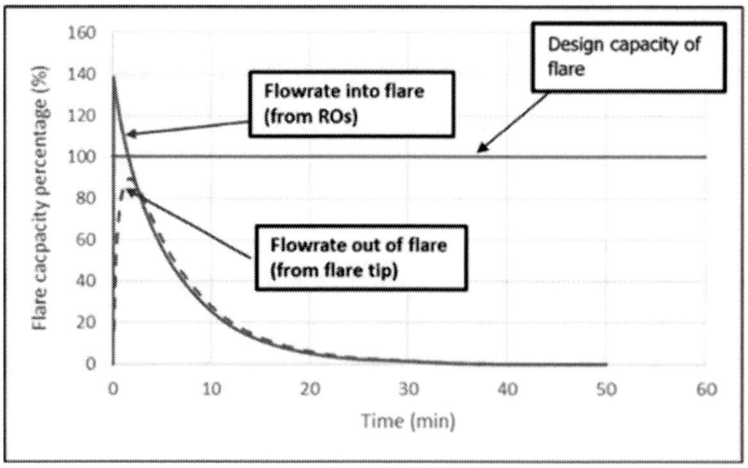

Figure 7. Case Study – graph comparing depressuring flow to flare tip flow

4.4 Insala Gas Plant, Algeria 2005, Flare Piping Failure from vibration

This incident [7] involved the failure of a reactor vent line at the flare header. Gas had been flowing into the header from the first train, and when the second train was started up, noise and excessive vibration was seen and the tailpipe failed at the junction with the main flare header.

80,000kg of gas released, fortunately no ignition source was encountered, and there was no consequent explosion or loss of life.

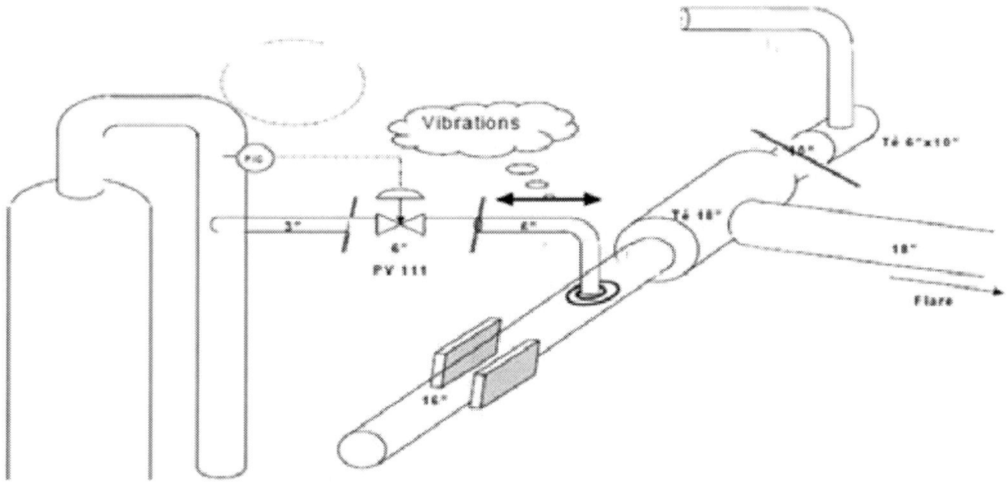

Figure 8. Schematic showing location of 6" tie downstream of PV-111

Recommendations:

- The complete flare header was reviewed and >20 pipe modifications were made to limit excessive velocities.
- 'Set on' welded branch connections were found to be in-adequately re-enforced. These were addressed by providing a full circumferential reinforcement of the pipe branch and wall thickness increases for tailpipes entering the flare header.
- The main flare header was found not to have been sized for the governing case and was replaced.

4.8 Changes to API Standard 521 – Vibration failures

Industry has been discussing vibration failure modes based on an increasing awareness of specific failures and forced shutdowns due to discovery of weld failures and cracks during routine plant inspections. Data published by the UK Health and Safety Executive for the offshore industry have shown that in the UK sector of the North Sea piping vibration and fatigue accounts for over 20% of hydrocarbon releases. Similar figures between 10% and 15% are reported for onshore plants in Western Europe. [5]

We are also seeing examples of recent designs where no assessment of vibrational risks has been made.

The two principal vibrational risks for piping systems are high frequency Acoustically Induced Vibration (AIV), and a lower frequency Flow Induced Vibration (FIV). The former is caused by high flow and pressure drops across control and relief valves and is now recognized in API521 (2014). FIV is principally caused by flow turbulence or vortex shedding effects downstream of branches which causes flow pulsation.

Both of these are excitation modes which when coupled with the piping natural frequency create amplified oscillatory response and potential failure of insufficiently supported piping due to the concentration of stresses at branches or fabricated sections, as shown on the example figure 8 above.

The recommended approach [3,5] is a screening of the Sound Power Level, and application of the Carucci/ Mueller Guidelines. This research established a correlation between an SPL >155dB(A), pipe diameter and wall thickness for a number of failures associated with AIV.

The API guidelines do not include any reference to the other principal mode of pipe vibration, however, the Energy Institute Guidelines [5] detail the approach to be taken for both AIV and FIV analysis.

The Likelihood of Failure (LOF) is a screening approach using Qualitative or Quantitative Analysis. The latter method for calculating LOF for Flow Induced Vibration requires a knowledge of the stiffness of the piping section based the support span, the kinetic energy and composition of the fluid. An LOF below 0.29 is acceptable, with no further action required. LOF above 0.5 for a main header requires remedial investigation of the pipe supporting and flexibility, and special focus on branch connections where there is greatest failure risk.

Often in complex systems, dynamic simulation can be used to model the response to transient flow conditions. For example a choke valve and production piping manifold design, was modeled to check the vibrational risk during start up. The flowrate from the well affected the 2-phase flow regime in the piping, and ultimately the LOF.

The mitigation approach followed was to control the start up flowrate and ramp rate until the pipe manifold downstream of the choke was under stable flow conditions and piping design was improved to provide stiffer pipe supporting.

In other cases, the piping backpressure develops as the flow establishes, and the worst case for vibration may exist at the beginning of a ramp flow change. These scenarios are readily tested using a dynamic model, which allows the exact fluid properties to be used in the calculation of AIV or FIV screening during transient flow conditions.

5 Conclusion

Prior to the issue of the revised API Standard 521, and Standard 520 Parts 1&2, these Standards were written around the downstream sector, but were applied also for certain areas of upstream design. Changes to industry practices in upstream sectors affecting depressuring design and better understanding of how dynamic simulation analysis tools can be applied has brought pressure on the need to revise the guidelines.

The key changes to API Standard 521 which affect the maintenance of both ageing as well as the design of new plants are:

- Relief valve sizing and importance of maintaining adequate records through periodic verification
- Low temperature depressurization risks and avoidance of brittle fracture through detailed analysis
- Vessel Depressurisation to achieve adequate Fire Survivability may not be rapid enough for thin walled vessels - requiring a faster blowdown rate, and a detailed dynamic analysis.
- Screening of vibration risks in flare headers due to high flow or pressure drops is now recommended using the Energy Institute guidelines with particular focus on small bore connections.

6 References

[1] U.S. Chemical Safety and Hazards Board report no 2005-04-TX available at web link *www.csb.gov/assets/1/19/csbfinal**report**bp.pdf*, Accessed 03/2017

[2] "Flare line failure case, what have we learned" Presentation prepared by Westlake Group for 16th Annual Ethylene Producers' Conference, New Orleans, LA, April 2004

[3] Pressure-relieving and Depressuring Systems, API Standard 521, Sixth Edition, July 2014

[4] M.A Haque et al., Blowdown of pressure vessels, Trans IChemE Part B Process Safety Environmental Protection, 70 (BI) 1 and 10, 1992

[5] Guidelines for the avoidance of vibration induced fatigue failure in process pipework, 2nd Edition, Energy Institute, London, January 2008

[6] Sizing Selection and Installation of Pressure-Relieving Devices, API Standard 520 Part 1, Ninth Edition July 2014, Part 2 Sixth Edition March 2015

[7] "Acoustic Fatigue –Turbulence Induced Fatigue Failure of Relief System Piping", E. Zamejc given at spring 2006 API Refining Meeting, May 3, 2005

Limiting OSHA's Authority to Issue Citations: Court Rejects OSHA's Efforts to Evade OSH Act Statute of Limitations by Alleging Continuing Violation

Mark S. Dreux
Arent Fox LLP
1717 K Street, NW
Washington, DC 20006
Mark.Dreux@arentfox.com

Alexandra M. Romero
Arent Fox LLP
1717 K Street, NW
Washington, DC 20006
Alexandra.Romero@arentfox.com

Prepared for Presentation at
American Institute of Chemical Engineers
2017 Spring Meeting and 13th Global Congress on Process Safety
San Antonio, TX
March 26 – 29, 2017

Limiting OSHA's Authority to Issue Citations: Court Rejects OSHA's Efforts to Evade OSH Act Statute of Limitations by Alleging Continuing Violation

Mark S. Dreux
Arent Fox LLP
1717 K Street, NW
Washington, DC 20006
Mark.Dreux@arentfox.com

Alexandra M. Romero
Arent Fox LLP
1717 K Street, NW
Washington, DC 20006
Alexandra.Romero@arentfox.com

Keywords: OSH Act, OSHA, PSM, process safety management, statute of limitations

Abstract

In recent years, the Occupational Safety and Health Administration ("OSHA") has adopted an expansive interpretation of the six-month statute of limitations contained in the Occupational Safety and Health Act (the "OSH Act"). According to OSHA, violations of the agency's health and safety standards continue until the employer has abated them, and only at that point does the six-month limitations period begin to run. OSHA has sought to use this "continuing violation" theory to justify a number of recent enforcement actions that the agency did not bring until years, and in some cases more than a decade, after the alleged violation first occurred.

However, OSHA has not yet been successful in defending its position in court. In 2012, a federal appeals court in the District of Columbia rejected the application of the continuing violation doctrine to citations issued for alleged violations of OSHA's recordkeeping regulations. In response to the decision, OSHA has sought to amend its recordkeeping regulations to "clarify" that the duty to maintain correct injury and illness records is ongoing and to authorize the issuance of citations up until six months after the mandatory retention period has expired. OSHA issued a new final rule to that effect in December 2016.

However, a federal appeals court in Texas issued an opinion a few weeks after the final rule came out that again rejected OSHA's continuing violation theory and vacated two citations for alleged violations of OSHA's Process Safety Management ("PSM") standard as untimely. Since then, the House of Representatives has passed a resolution invalidating OSHA's final rule clarifying its recordkeeping regulations, and a similar resolution has been introduced in the Senate. This presentation will explain OSHA's interpretation of the OSH Act's statute of limitations, the reasoning of the two appellate decisions rejecting that interpretation, and how the PSM standard will subsequently be enforced if OSHA's clarifying rule is not invalidated by Congress.

On December 29, 2016, a three-judge panel from the U.S. Court of Appeals for the Fifth Circuit unanimously vacated two Occupational Safety and Health Administration ("OSHA") process safety management ("PSM") citation items issued to a Delek Refining Ltd. ("Delek") facility for alleged safety violations that occurred years prior to its ownership of the refinery.[1] The decision was a significant victory for employers because it marked the second time that a federal appeals court has rejected OSHA's attempt to allege the existence of a "continuing violation" that suspends the six-month statute of limitations contained in the Occupational Safety and Health ("OSH") Act.[2]

In 2008, OSHA issued citations to Delek under its process safety management ("PSM") standard for the alleged failure to resolve recommendations from process hazard analyses ("PHAs") and compliance audits that were conducted in 1994, 1998, 1999, 2004, and 2005 — all prior to Delek's acquisition of the refinery. The PSM standard requires employers to conduct a PHA to "identify, evaluate, and control the hazards involved" in covered processes, and then establish procedures to "promptly" address the PHA's findings, resolve them in a "timely" manner, and document the manner in which they were resolved. [3] Of particular significance to the Fifth Circuit's ultimate decision, the PSM standard does not require employers to *correct* each PHA finding. Instead, all that the standard requires is that each finding be "resolved," which in some circumstances could mean that the employer determines that no corrective action is necessary. Each PHA must be updated and revalidated every five years.

Similarly, employers with processes covered under the PSM standard must conduct compliance audits every three years to "verify that the procedures and practices developed under the standard are adequate and are being followed."[4] After the audit has been completed, employers must create a report and "promptly determine and document an appropriate response to each of the findings of the compliance audit, and document that deficiencies have been corrected."[5] Accordingly, as with PHAs, the PSM standard does not contain any requirement that every audit finding must be corrected. In some circumstances, the "appropriate response" that is required by the standard may be to do nothing. Regardless, covered employers must retain records from their two most recent compliance audits confirming that each finding has been adequately addressed.[6]

The OSH Act states that no citations may be issued more than six months following "the occurrence of any violation."[7] In the Fifth Circuit case, OSHA argued that Delek's alleged failures to address the PHA and audit findings conducted by its corporate predecessor were "continuing violations" that would not trigger the OSH Act's statute of limitations until they were abated. Accordingly, taken to its logical conclusion, OSHA's theory, if adopted, would have permitted the agency to issue citations based on the allegedly unresolved PHA and audit findings indefinitely. In response, Delek argued that conducting the PHAs and the audit were the "occurrences" that triggered the six-month statute of limitations, meaning that the limitation period had expired years ago and thus OSHA's citations should be vacated as untimely.

The Fifth Circuit ultimately agreed with Delek and expressly rejected OSHA's efforts to apply the continuing violation doctrine to the PSM standard's PHA and audit requirements. In doing so, the Fifth Circuit held that OSHA's position was contrary to both the plain language of the OSH Act and the PSM standard, as well as "the basic purposes of a statutory limitations period."[8] First, the Court noted that the PSM standard used the words "promptly" and "timely" to describe a covered employer's obligation to resolve and respond to PHA and audit findings. As such, the Court held that a single violation occurs when an employer "does not 'promptly' or 'timely' do as [the PSM standard] directs."[9] To permit OSHA to bring enforcement actions at any time would be entirely inconsistent with the commonly understood meaning of those words.

Second, the Fifth Circuit held that OSHA's interpretation was "at odds" with the language of the six-month limitations period that Congress "deliberately wrote into [the OSH Act]."[10] If OSHA were allowed to bring enforcement actions in perpetuity after the occurrence of a safety violation, which the agency conceded would be the case if their interpretation were adopted, the Court held that such an outcome would be in direct conflict with clear congressional intent.

And finally, the Fifth Circuit panel stated that permitting OSHA to issue citations in perpetuity would be "utterly repugnant to the genius of our laws." Accordingly, the Court held that the failure to address a PHA or audit recommendation is a single violation for the purposes of the statute of limitations, reasoning that a strict interpretation of the OSH Act's six-month statute of limitations in this context would "provide security and stability." It would also prevent OSHA from instituting enforcement actions long after the relevant witnesses' memories had faded and the available evidence had gone stale or disappeared entirely.

The Fifth Circuit's decision in the Delek case relied heavily on and extended the U.S. Court of Appeals for the District of Columbia Circuit's reasoning in a case presenting similar issues that involved violations of OSHA's injury and illness recordkeeping regulations. In Volks Constructors v. Secretary of Labor (the "Volks decision"), the D.C. Circuit rejected OSHA's claim that violations of its recordkeeping regulations were continuing violations that prohibited the statute of limitations from expiring until they were corrected.[11] As did the Fifth Circuit, the D.C. Circuit based its decision on the Court's interpretation of the plain meaning of the OSH Act as well as the principles of fairness and finality that are the primary reasons why statutes of limitation are so important.

Nonetheless, both courts left open the possibility that a continuing violation that tolls the statute of limitations could exist where an employer fails to correct a persistent hazard to employee health or safety. As such, although the Delek decision is hugely important because it is a clear repudiation of OSHA's long-standing claim that the Volks decision applied only to recordkeeping violations, it does have its limits. The Fifth Circuit's decision relied heavily on the fact that employers were not obligated to "*actually remedy* the issues addressed in a PHA or audit recommendation."[12] Accordingly, the possibility remains that OSHA may ultimately be successful in persuading a court that a continuing violation exists where the agency can show that the violation at issue permits an unlawful safety risk to continue unabated. However, the Fifth Circuit's opinion in Delek signals that courts will nonetheless continue to place the burden on OSHA to justify citations issued outside of the six-month limitations period, and heralds a growing consensus that exclusions from the OSH Act's statute of limitations should be strictly limited.

The Delek decision is also significant given OSHA's effort to achieve by regulation what it could not in court. Following the Volks decision, the agency issued a proposed rule "clarifying" that the obligation to maintain accurate and complete injury and illness records is a continuing one, that extends for the length of the mandated retention period. The rule was perceived as an effort by the agency to "overturn" the Volks decision and provide regulatory authorization for the agency to treat recordkeeping violations as continuing ones. OSHA issued its final rule on December 19, 2016. However, a number of commentators have observed that the issuance of the regulation does not address the courts' reasoning in Volks and Delek. In both cases, the Fifth Circuit and the D.C. Circuit held that OSHA's adoption of a continuing violation theory was in direct conflict with the language of the OSH Act — a statute passed by Congress. Many have observed that congressional intent cannot be circumvented simply by administrative action.

As a result, a number of employers and industry trade groups have been successful in persuading members of Congress to pursue a challenge to OSHA's final rule under the Congressional Review Act. That statute authorizes Congress to review, on an expedited basis, new federal regulations issued by executive branch agencies and overrule them upon passage of a joint resolution of both houses of Congress. On March 1, 2017, the House of Representatives passed H.J. Res. 83, invalidating OSHA's final rule by a vote of 231-191. The Senate introduced its version of the resolution, S.J. Res. 27, on March 2, 2017. Nonetheless, even if Congress is unable to invalidate the rule, the Delek and Volks cases suggest that OSHA's continuing violation theory will continue to face skepticism in the courts.

Mark Dreux, head of the OSHA practice at Arent Fox LLP, led the team of partner, James H. Hulme, and associate, Alexandra M. Romero, that represented Delek Refining Ltd. in the Delek case. Mr. Dreux argued the appeal in the Fifth Circuit.

PSM/RMP Modernization Programs in California

(2016 Developments and Correlation to Evolution at the Federal Level)

Steven T. Maher, PE CSP
Risk Management Professionals
Steve.Maher@RMPCorp.com

Aleksandar Metulev
Risk Management Professionals
Aleksandar.Metulev@RMPCorp.com

Cho Nai Cheung
Contra Costa County – Health Services Division
ChoNai.Cheung@hsd.cccounty.us

Michael A. Vigliarolo
Alon USA
MVigliarolo@ppcla.com

Prepared for Presentation at
American Institute of Chemical Engineers
2017 Spring Meeting and 13th Global Congress on Process Safety
San Antonio, Texas
March 26-29, 2017

PSM/RMP Modernization Programs in California

(2016 Developments and Correlation to Evolution at the Federal Level)

Steven T. Maher, PE CSP, Risk Management Professionals

Aleksandar Metulev, Risk Management Professionals

Cho Nai Cheung, Contra Costa County, Health Services Division

Michael A. Vigliarolo, Alon USA

Keywords: PSM, RMP, CalARP, SMS, Cal/OSHA, 5189.1, OSHA, EPA, Process Safety

Abstract

Paper 441463, presented at the 2016 Global Congress on Process Safety, provided a snapshot of the significant changes being made to the PSM/RMP Regulatory Framework in California and at the federal level. Post-publication, additional changes were proposed by Cal/OSHA (July 2016), CalEPA/CalOES (July 2016), and U.S. EPA (March 2016). Cal/OSHA's requirements will include a separate PSM regulation focused on petroleum refineries, and CalEPA/CalOES requirements will introduce a new program level (Program 4), initially applying only to petroleum refineries. These two California requirements are very similar, and differences will be identified in this paper. The changes to the U.S. EPA's Risk Management Program (RMP) Rule are distinct from the California requirements, but apply to a wider range of facilities.

The final Cal/OSHA language for the new California petroleum refinery requirements was issued on February 10, 2017, with a comment deadline of March 3, 2017, and promulgation for both the CalEPA/CalOES requirements and Cal/OSHA requirements is anticipated for Summer/Fall 2017. Federal OSHA and U.S. EPA had been considering updates to Process Safety Management (PSM) and Risk Management Program (RMP) requirements as part of the PSM/RMP Modernization/Expansion Programs designed to address significant accidents that occurred in the recent past. Updates to the U.S. EPA RMP Rule were published in the Federal Register on January 13, 2017[44] and are currently being reviewed/validated as part of the Congressional Review Act[45].

The changes being proposed for these Safety Management System (enveloping term for PSM and RMP) programs are extensive, still being formulated, and not synchronized between the various agencies. The focus of this paper will be to provide relevant background information, a summary of the proposed changes, and tips for addressing key technical and regulatory challenges, as well as providing the reader with a "roadmap" for effective implementation. This paper addresses the regulatory environment in March 2017 and will be updated periodically to reflect this "moving target". Key topics include:

- Background – Pre-2012 PSM/RMP Universe
- Brief Overview of Key Safety Management System (SMS) Program Modernization Initiatives, Focused on 2016 Changes

- Key Elements of the Various SMS Regulatory Initiatives
- SMS Modernization Status and Potential SMS Program Impact
- What Should I Be Doing Now?

1. Background – Pre-2012 PSM/RMP Universe

The December 2, 1984 Methyl Isocyanate (MIC) release from the Union Carbide Bhopal Facility is considered a pivotal event in catalyzing the application of SMS approaches to enhance process safety. Referencing the analogies in Figure 1.1, the MIC release resulted from the concurrent alignment of several "holes," and the magnitude of the tragedy (3928 fatalities and over 100,000 permanent injuries are estimated)[1], drew the attention of industry, the public, and the regulatory community to the potential consequences associated with process safety events (Figure 1.2). Industry was quick to realize the significance of the event, with respect to the need to create and implement SMS mechanisms at highly-hazardous facilities, and the importance of developing mechanisms to control process safety.

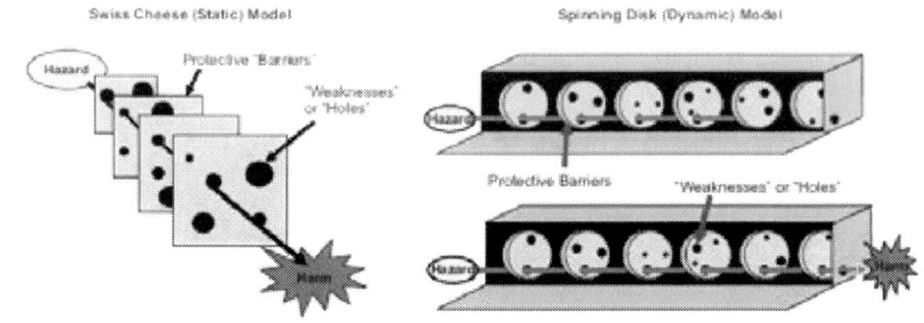

FIGURE 1.1 – Swiss Cheese and Spinning Disk Models for Accident Causes

API RP 754

FIGURE 1.2 – Tragedies to Avoid

Industry's response was swift and definitive. The American Institute of Chemical Engineers (AIChE) founded the Center for Chemical Process Safety (CCPS) in 1985, recognizing that the most effective mechanism for addressing process safety was not the application of additional prescriptive mechanisms, or by addressing any specific action, but by effecting changes in the way business is done (i.e., safety culture and management systems). CCPS Guidebooks are currently considered key references in conveying the technologies needed for process safety, and the very first guidebook ("Guidelines for Technical Management of Chemical Process Safety"[2]) published in 1987 was designed to address this pressing need. Shortly thereafter, the American Petroleum Institute (API) distilled its version of SMS and issued Recommended Practice 750, "Management of Process Hazards"[3] in 1990. The following characteristics have remained consistent from the onset:

- All segments of the process industries correctly identified SMS as the primary and most effective mechanism for addressing core issues associated with process safety incidents.
- Although some details differ between the above two documents, and also between the current requirements of OSHA's Process Safety Management (PSM) Program[4], U.S. EPA's Risk Management Program (RMP)[5], and the Bureau of Safety and Environmental Enforcement's (BSEE's) Safety and Environmental Management Systems (SEMS) Program[6], **the same key SMS elements are at the core of PSM, RMP, and SEMS, spanning a spectrum of facility types and geographic applications**.

Figure 1.3 illustrates the parallel evolution of PSM, RMP, and SEMS programs. It's interesting to note that although PSM and RMP were catalyzed by an onshore tragedy and focus on onshore facilities in the United States, and SEMS was catalyzed by an offshore-US tragedy and focuses on offshore facilities in waters off the coast of the United States, the key prevention program elements are nearly identical (see the key elements of PSM in Figure 1.4). This parallel evolution and resultant overlap are important to note for several reasons:

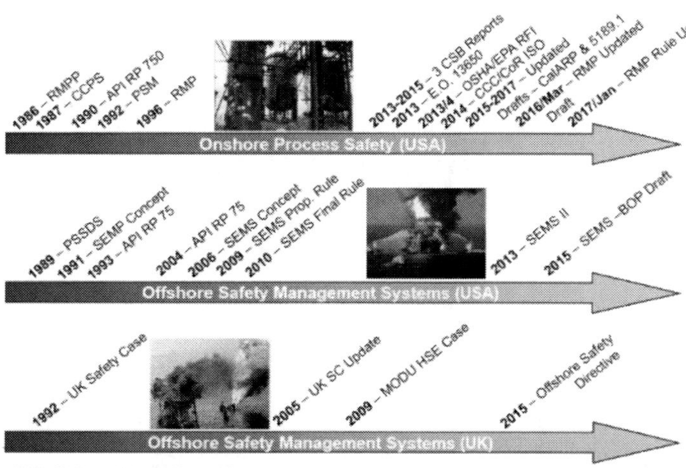

FIGURE 1.3 – Evolution of Select SMS Guidelines

- Although these regulatory programs were developed independently, at different times, and in different locations, industry and the regulatory community noted the importance of SMS application across all facility types.
- As **performance-based** regulatory requirements continue to evolve, the same phenomenon associated with the effectiveness of key SMS elements will tend to gravitate the programs towards synchronization.

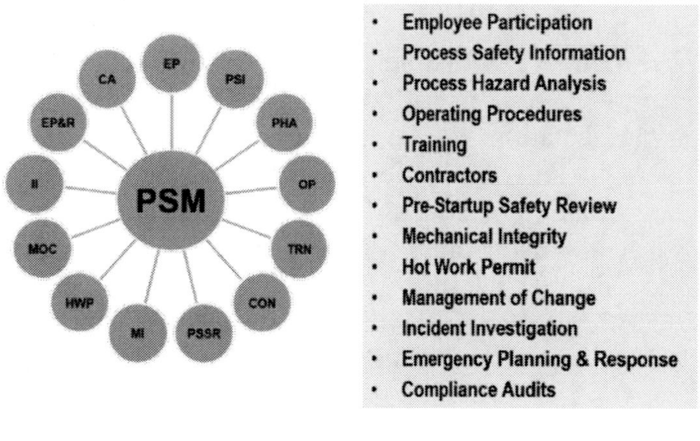

FIGURE 1.4 – Key PSM Elements (2016)

The Richmond (California) Refinery Fire on August 6, 2012 and West (Texas) Ammonium Nitrate Explosion (April 17, 2013) triggered a fresh look at the different SMS programs and resulted in several proposals for the modernization of PSM and RMP. The remainder of this paper will address the regulatory changes that resulted from this fresh look, which were promulgated in early-2017 or are scheduled for promulgation in Summer/Fall 2017. Figure 1.5 identifies the geographic span of the facilities affected by these regulatory changes. Changes to the RMP Rule apply throughout the United States and its territories, whereas the more significant changes associated with the California-specific SMS regulatory requirements primarily affect approximately a dozen facilities in California. It is worth noting that California agencies are chartered to administer the PSM and CalARP Programs independent of federal requirements. This is an important distinction at the time that this paper is being written, because the current review/validation of the RMP Rule Updates as part of the Congressional Review Act do not apply to California-specific regulatory requirements.

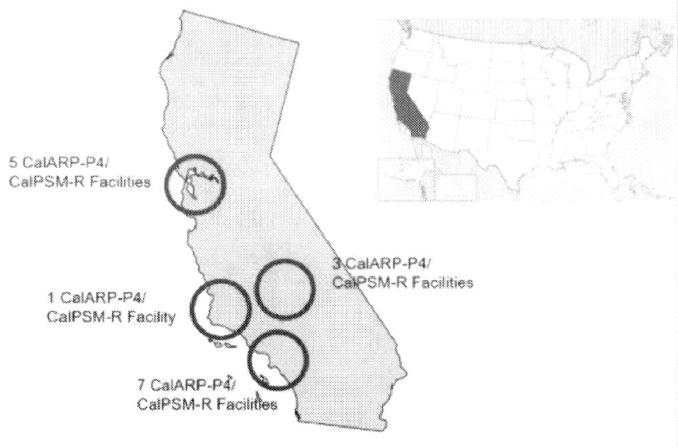

FIGURE 1.5
CalARP-P4/CalPSM-R Affected Facilities

2. Brief Overview of Key SMS Program Modernization Initiatives, Focused on 2016 Changes

Since August 2012, significant changes (Figure 2.1) have been proposed by a number of agencies to SMS Programs. After various high-profile incidents occurred, Executive Order 13650 was signed by President Obama on August 1, 2013 to emphasize the importance of various agencies working together towards the common goal of improving process safety, as well as modernizing the current standards[7]. In addition, the Chemical Safety Board (CSB) investigated the Richmond Refinery Fire, and identified improvement areas[8,9,10] for both PSM/RMP implementation at the facility, as well as agency oversight and cooperation (which essentially resulted in the formation of the California Governor's Interagency Refinery Task Force). As a result, there were changes proposed for the following programs: Federal OSHA PSM[11], U.S. EPA RMP[12], Cal/OSHA PSM[13], and CalARP[37, 43]. Table 2.1 identifies the key elements of the California SMS regulatory elements that were in-place (January 1, 2017), prior to the January 13, 2017 RMP Rule Update.

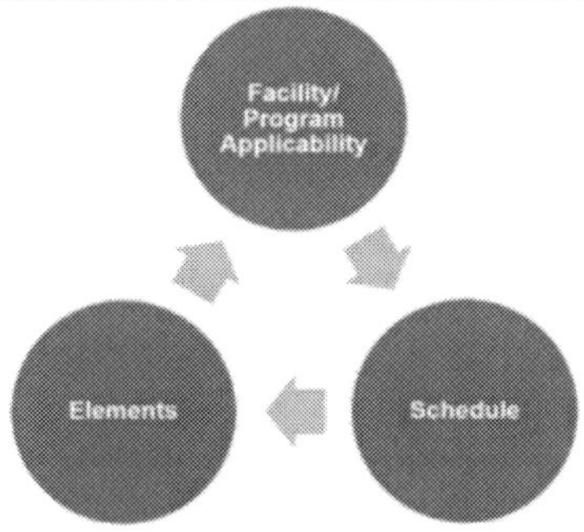

FIGURE 2.1
Types of SMS Program Changes

TABLE 2.1 – SMS Program Overlap Matrix (January 1, 2017)				
Element	**Fed-OSHA (PSM)**	**Cal/OSHA (PSM)**	**U.S. EPA (RMP)**	**CalOES/CalEPA (CalARP)**
(Process) Safety Information	1910.119 (d)	5189 (d)	68.65	2760.1
Process Hazard Analysis (PHA)	1910.119 (e)	5189 (e)	68.67	2760.2
Operating Procedures	1910.119 (f)	5189 (f)	68.69	2760.3
Training	1910.119 (g)	5189(g)	68.71	2760.4
Mechanical Integrity	1910.119 (j)	5189 (j)	68.73	2760.5
Management of Change	1910.119 (l)	5189 (l)	68.75	2760.6
Pre-Startup Safety Review	1910.119 (i)	5189 (i)	68.77	2760.7
Compliance Audits	1910.119 (o)	5189 (o)[a]	68.79	2760.8
Incident Investigation	1910.119 (m)	5189 (m)	68.81	2760.9
Employee Participation	1910.119 (p)	5189 (p)	68.83	2760.10
Hot Work Permit	1910.119 (k)	5189 (k)	68.85	2760.11
Contractors	1910.119 (h)	5189 (h)	68.87	2760.12
Emergency Planning and Response	1910.119 (n)	5189 (n)	68.90-95	2765
Off-site Consequence Analysis	---	---	68.20-42	2750

[a] – Captured as part of the "Injury and Illness Prevention Program" Subsection.

Table 2.2A summarizes and compares the various changes included in the January 13, 2017 RMP Rule Update and those being proposed (March 2017) for the California SMS Regulatory Program Modernization Initiatives, CalARP-P4 and CalPSM-R. Table 2.2A also delineates the various topics by the type and magnitude of the impact on RMP and PSM programs.

TABLE 2.2A – Current CalPSM-R, CalARP, and U.S. EPA RMP Initiatives					
Subset	**Topic**	**Cal/OSHA CalPSM-R** (Draft)[13]	**CalOES/CalEPA CalARP-P4** (Draft)[37,43]	**U.S. EPA RMP-P2/3** (Update)[44]	
1	Process Safety Information (PSI) (including confirmation of RAGAGEP adherence)	5189.1(d)	2762.1	68.65	
2	Process Hazard Analysis (PHA) Safeguard Protection Analysis (SPA) (CA-Only)	5189.1(e)	2762.2 2762.2.1	68.50/67	
2	Operating Procedures (OP)	5189.1(f)	2762.3	68.69	
1	Training (TRN)	5189.1(g)	2762.4	68.54/71	
1	Contractors (CON) (including safety research & documentation)	5189.1(h)	2762.12	68.87	
1	Pre-Startup Safety Review (PSSR)	5189.1(i)	2762.7	68.77	
3	Mechanical Integrity (MI) Damage Mechanism Review (DMR) (CA-Only)	5189.1(j) 5189.1(k)	2762.5	68.73	
3	Inherently Safer Technology (IST) / Hierarchy of Hazard Control Analysis (HCA) / Safer Technology and Alternatives Analysis (STAA)	5189.1(l)	2762.13	68.67	
1	Hot Work (HW)	5189.1(m)	2762.11	68.85	
2	Management of Change (MOC)	5189.1(n)	2762.6	68.75	
2	Incident Inv. – Root Cause Analysis (II-RCA)	5189.1(o)	2762.9	68.81	
3	Emergency Planning and Response (EP&R)	5189.1(p)	2765.2/3	68.95/96	
1	Employee Participation (EP)	5189.1(q)	2762.10	68.83	
3	Process Safety Culture Assessment (PSCA)	5189.1(r)	2762.14	---	
3	Human Factors (HF)	5189.1(s)	2762.15	---	
3	Management of Organizational Change (MOOC)	5189.1(t)	2762.6	---	
2	Compliance Audits (CA) (Fed Rqmt. for 3rd Party Auditor that is familiar with RAGAGEP)	5189.1(u)	2762.8	68.58/59, 68.79/80	

TABLE 2.2A – Current CalPSM-R, CalARP, and U.S. EPA RMP Initiatives				
Subset	Topic	Cal/OSHA CalPSM-R (Draft)[13]	CalOES/CalEPA CalARP-P4 (Draft)[37,43]	U.S. EPA RMP-P2/3 (Update)[44]
2	Process Safety Management Program (PSMP)	5189.1(v,w)	2762.16	---

Description of Subset Categories	
Subset	Categories
1	Minimal changes to regulation or minimal effort needed for compliance
2	Moderate changes to regulation or moderate effort needed for compliance
3	New element or significant effort needed for compliance

Table 2.2B lists Fed-OSHA PSM initiatives that are being or may be handled by Letters of Interpretation. Fed-OSHA has not identified a current intention to update 29 CFR 1910.119. References 11, 12, 13, 37, and 43 may be reviewed for more complete details. Section 3 provides additional background on the proposed changes listed in Table 2.2A. Many of these additional requirements overlap, and with some planning, existing PSM, RMP, and CalARP efforts can be slightly modified to address some of the more significant new requirements with minimal effort. The focus of this paper will be to summarize the more significant changes proposed by the various agencies, and provide readers with a "roadmap" of how to effectively implement these changes in as efficient a manner as possible.

TABLE 2.2B – Fed-OSHA PSM Initiatives Potentially Handled by Letters of Interpretation (LoI)		
Status	Topic	Fed OSHA[11]
LoI 02Feb03	Atmospheric Tank PSM Exemption Clarification	Topic 1
Pending	Inclusion of Oil/Gas-Well Drilling & Servicing	Topic 2
Pending	Inclusion of Oil/Gas-Production Facilities	Topic 3
Pending	Expand Coverage & Requirements for Reactivity Hazards	Topic 4
Pending	Updating the List of Applicable Chemicals / Threshold Quantities	Topic 5
LoI 05Jun13 & 05May16	Adding a Definition for Recognized and Generally Accepted Good Engineering Practices (RAGAGEP) – This has already been incorporated into the CalPSM-R and CalARP draft updates.	Topic 8
Pending	Expanding the Requirements of §1910.109 to Cover Dismantling and Disposal of Explosives, Blasting Agents, and Pyrotechnics	Topic 13
Pending	Updating §1910.106 and 1910.107 Based on the Latest Applicable Consensus Standards	Topic 14
Pending	Regulations Addressing the Storage, Handling, and Management of Ammonium Nitrate	Topic 15
LoI 22Jul15 LoI 23Dec15	Enforcement of the Exemption for Retail Facilities	Topic 16
LoI 18Jul16	Enforcement Policy for Highly Hazardous Chemicals Without Specific Concentration Thresholds (The 1% rule)	Topic 17

3. Key Elements of the Various SMS Regulatory Initiatives

The key elements of the various SMS regulatory initiatives identified in Table 2.2A have been categorized into three subsets to facilitate understanding and the development of implementation strategies. Subset 1 topics in Table 2.2A were judged to have only minimal changes to regulation or minimal effort needed for compliance. The only measureable difference is for Contractors, where some additional requirements are placed on both the facility and the contract owner regarding training verification and hazard identification.

3.1 SMS Program Modernization Initiatives – Subset 2

Subset 2 topics in Table 2.2A were judged to have a potential for moderate changes to regulation or moderate effort needed for compliance.

1. **Process Hazard Analysis (PHA) / Safeguard Protection Analysis (SPA)** – Key changes proposed for CalPSM-R and CalARP-P4 include recommendation closure timelines and the performance of Safeguard Protection Analysis (SPA). The SPA is a new element with the objectives of assessing the effectiveness of existing and combined safeguards for each failure scenario identified in the PHA, and assuring that safeguards are independent of the initiating event and each other. Passive, Active, or Procedural Safeguards are evaluated by the SPA using LOPA or an "equally effective method". The SPA report can be standalone or part of the PHA. Specific implementation tips are offered in Section 5.2.2 of this paper. Timing:
   - Conduct SPA and complete report within 6 months of finalizing a PHA.
   - SPA reports must be retained for the life of the process.

2. **Operating Procedures (OP)** – The primary changes to the Operating Procedures requirements for CalPSM-R and CalARP-P4 address additional details associated with Emergency Operations. No implementation timing changes are identified.

3. **Management of Change (MOC)** –The primary changes to the MOC element for CalPSM-R and CalARP-P4 address the depth of assessment performed during the MOC process:
   - Performance of a Hierarchy of Hazard Control Analysis (HCA) prior to implementing a major change, with the findings and recommendations of the HCA included in the MOC documentation
   - Use of qualified personnel and appropriate methods for all MOCs, based upon hazard, complexity, and type of change
   - Provisions for employee participation throughout the MOC process

4. **Incident Investigation (II) – Root Cause Analysis (RCA)** – The primary changes to the II element for CalPSM-R, CalARP-P4, and U.S. EPA RMP is the requirement for the application of a RCA technique for more serious events (i.e., resulted in or could reasonably have resulted in a catastrophic release). Other changes:
   - **CalPSM-R/CalARP-P4** – Complete a HCA in a timely manner for all recommendations that result from the investigation of a major incident.
   - **CalPSM-R/CalARP-P4** – Additional employee participation requirements
   - **CalARP-P4** – "The local Unified Program Agency [UPA] may perform an independent PSCA, Incident Investigation, and Human Factors Analysis after a major incident. The owner or operator shall assist the UPA in conducting the independent analysis. The costs of the independent analysis shall be paid by the owner or operator." Note that the acronym PSCA stands for a Process Safety Culture Assessment.
   - **CalARP-P4** – UPA to post reports from investigations of major incidents on the agency's website within 30 days of receipt

New timelines:

- **RMP** – Incident Investigation report completion within 12 months of the incident
- **RMP** – Implementation of RCA requirements within 4 years of the effective date of the regulation
- **CalPSM-R/CalARP-P4** – Investigation report completion within 90 calendar days of the incident. If the team demonstrates in writing that additional time is needed due to the complexity of the investigation, the team shall prepare a status report within ninety (90) calendar days of the incident and every thirty (30) calendar days thereafter until the investigation is complete. The team shall prepare a final investigation report within five (5) months of the incident.
- **CalPSM-R/CalARP-P4** - Incident investigation reports shall be retained for the life of the process unit.
- **CalPSM**-R – Each corrective action from an incident investigation shall be completed within (18) months after completion of the audit, unless it is demonstrated in writing that it is infeasible to do so.

5. **Compliance Audits (CA)** – The changes proposed by CalPSM-R and CalARP-P4 are relatively minor (e.g., retaining the 3 most recent compliance audit reports, instead of 2); however, the RMP Update requires the use of a third-party auditor, for both Program 2 and Program 3 facilities, if an applicable accidental release has occurred, or if justified by the implementing agency. The RMP Update includes specific auditor qualifications, independence and impartiality requirements, specifications for the audit report, and specifications for recordkeeping. Timing for audit, report, and submittal to the implementing agency:
   - Within 12 months of request/requirement
   - Within 3 years of completion of the previous compliance audit

6. **Process Safety Management Program (PSMP)** – A PSMP is a new element required of CalPSM-R and CalARP-P4. Both have specifications for recommendation implementation timing (typically 2.5 years) and detailed criteria for rejecting recommendations. CalARP-P4 also includes provisions for a written Stop Work Procedure, a written Hazard Reporting Plan, a system for documenting work refusals for process safety concerns, and for the development and tracking of Process Safety Performance Indicators. Timing:
   - Immediate – Completion of corrective actions addressing process safety hazards
   - 18 Months – Completion of corrective actions from a CA or PSCA
   - 18 Months – Completion of corrective actions from a II (or two years after the incident, whichever is earlier)
   - 30 Months – Completion of PHA corrective actions not requiring a process shutdown
   - Triennial – PSMP review and update
   - First Regularly-Scheduled Turnaround – Completion of corrective actions requiring a process shutdown
   - **CalARP-P4** – 14 Days After Completion – Findings/Recommendations from PHA, DMR, HCA, II, PSCA, CA, and SPA must be provided to the facility owner

- **CalARP-P4** – 30 days - time to respond in writing to written hazard reports submitted by employees or employees of contractors.
- **CalARP-P4** – 90 days from effective regulation date – Develop a system to effectively document work refusals for process safety concerns, written employee reports of process safety hazards, recommendations to shut down an operation or process, and the actual shut down of an operation or process that occurs pursuant to a Stop Work procedure.

3.2 SMS Program Modernization Initiatives – Subset 3

Subset 3 topics in Table 2.2A identified a new element or changes that were judged as having a potential significant effort needed for compliance.

1. **Mechanical Integrity (MI) / Damage Mechanism Review (DMR)** – DMR is a new element for CalPSM-R and CalARP-P4, with the objective to identify any vulnerabilities associated with "damage mechanisms," defined as the "mechanical, chemical, physical, or other process that results in equipment or material degradation." This effort should involve a broad team, including inspection and damage/failure mechanism experts, using process flow diagrams, materials of construction, process conditions, and chemical substances to determine potential damage/failures. Specific implementation tips are offered in Section 5.2.1 of this paper. At a minimum, the following should be evaluated:
    - Mechanical Loading Failure – Ductile fracture, brittle fracture, buckling, mechanical fatigue
    - Erosion – Abrasive wear, adhesive wear, fretting
    - Corrosion – Uniform corrosion, localized corrosion, pitting
    - Thermal-Related Failure – Creep, thermal fatigue, transformation
    - Cracking – Stress-corrosion cracking
    - Embrittlement – High Temperature Hydrogen Attack (HTHA)

    Timing – In conjunction with any PHA and prior to changes affecting chemistry, metallurgy, or operating limits, and the initial DMR to take place at the earliest of the following:
    - Within 5 years (50% within 3 years and all within 5 years of CalARP-P4/CalPSM-R promulgation)
    - Revalidated every 5 years or prior to a major change
    - Reviewed as part of an incident investigation
    - DMR reports must be retained for the life of the process.

2. **Inherently Safer Technology (IST), Hierarchy of Hazard Control Analysis (HCA), and Safer Technology and Alternatives Analysis (STAA)** – For contemporary hazard reviews, IST evaluations are performed as a best-practice[18]. CalPSM-R and CalARP-P4 require the performance of HCA, and the RMP Update requires the performance of STAA, to evaluate hazard prevention and control measures, in priority order, to eliminate or minimize a hazard, ranked from most preferred to least preferred:

- First Order Inherent Safety – Prevents a major incident by eliminating or reducing the hazard (e.g., changing to a safer chemical)
- Second Order Inherent Safety – Reduces the severity or likelihood of a release without the use of add-on safety devices (e.g., changing operating parameters to safer levels, such as lower pressures or temperatures)
- Independent Protection Layers – Reduce the likelihood or consequence of a major incident through the application of add-on safety devices, from most preferred to least:
    - Passive safeguards (e.g., double-walled vessels)
    - Active safeguards (e.g., PSVs or automated shutdowns)
    - Procedural safeguards (e.g., emergency response plans)

Specific implementation tips are offered in Section 5.2.3 of this paper. Applicability of the CalPSM-R and CalARP-P4 HCA is currently focused on California refineries, whereas the RMP Update broadens application to any Program 3 NAICS 322/324/325 Facility (i.e., Paper Manufacturing, Petroleum and Coal Products Manufacturing, Chemical Manufacturing). The following timing applies for conducting the HCA/STAA:

- **CalPSM-R/CalARP-P4 (HCA)** – All existing processes (50% within 3 years and all within 5 years of CalARP/CalPSM-R promulgation), with a 5-year revalidation
- **CalPSM-R/CalARP-P4 (HCA)** – Applicable for all PHA recommendations addressing a "major incident potential"
- **CalPSM-R/CalARP-P4 (HCA)** – Applicable for all major changes (e.g., as part of MOC)
- **CalPSM-R/CalARP-P4 (HCA)** – Applicable for all recommendations from the incident investigation report for any "major incidents"
- **CalARP-P4 (HCA)** – Applicable for new processes (including making the HCA available to the public)
- **RMP (STAA)** – 4 years from effective regulation date

3. **Emergency Planning and Response** – CalPSM-R has not yet identified any significant changes to these requirements; however, the proposed CalARP and RMP do contain significant additional requirements for:
    - **CalARP-P4** – Annual update and submittal of the Emergency Response Program to the UPA
    - **CalARP-P4** – For a designated "geographic zone," development of a community outreach/notification program
    - **CalARP-P4** – Provisions for the establishment of a Unified Command
    - **CalARP-P4** – Provisions for ensuring sufficient emergency response resources
    - **CalARP-P4** – Emergency response personnel training program
    - **CalARP-P4** – Annual drills/exercises, with a triennial full-scale exercise
    - **CalARP-P4** – Fixed and mobile, routine and emergency, air monitoring and detection systems
    - **RMP** – Emergency response coordination with local emergency planning and response organizations

- **RMP** – For responding Program 2/3 facilities, annual review and update of Emergency Response program, including notification procedures and coordination with community emergency response plans
- **RMP** – For all Program 2/3 facilities, annual notification exercises
- **RMP** – For responding Program 2/3 facilities, annual tabletop exercises and field exercises once every five years

Timing:

- **CalARP-P4** – Within 12 months of the effective regulation date
- **RMP** – Compliance dates:
  - ➢ Emergency Response Coordination Activities (1 year)
  - ➢ Local Emergency Planning Committee (LEPC) Requires Compliance with §68.95 (3 years)
  - ➢ Emergency Response Exercises (4 years)

4. **Process Safety Culture Assessment (PSCA)** – The PSCA is a new element of CalPSM-R and CalARP-P4 and requires an assessment of cultural safety practices at a facility. The team should include one person knowledgeable in refinery operations and at least one employee representative. CalARP-P4 requirements also allow the UPA to perform its own PSCA after a major incident or any incident that could reasonably have led to a major accident, with the costs paid by the refinery owner/operator. Timing:
   - Initial PSCA – Within 18 months after the effective regulation date
   - PSCA Updates – Every five years
   - Evaluation of Recommendation Effectiveness – Within three years of PSCA report completion

5. **Human Factors (HF)** – HF is a new element of CalPSM-R and CalARP-P4 that requires, at a minimum, that the facility accounts for the complexity of tasks, levels of training and expertise required for employees, the human-machine and human-human interfaces are effective, and that the physical challenges of the work environment and effects of shift work and overtime are accounted for.

6. **Management of Organizational Change (MOOC)** – MOOC is a new element of CalPSM-R and CalARP-P4 that requires the employer to designate a team to conduct a MOOC assessment prior to reducing staffing levels, reducing experience levels of employees, changing shift duration, or increasing employee responsibilities at or above 15%. The employer shall provide for employee participation. The MOOC assessment is required for changes affecting operations, engineering, maintenance, health and safety, and emergency response.

4. SMS Modernization Status and Potential SMS Program Impact

4.1 Status of Current Initiatives

Figure 4.1 identifies Recent SMS Regulatory Activities, and Figure 4.2 depicts the different agencies involved in the current SMS modernization initiatives. Even though the ultimate performance objectives of each agency include accident prevention and consequence minimization, each of these agencies have a slightly different assigned mission and perspectives on the role they should play in SMS regulation implementation. Thus, at the time that this paper is written (March 2017), different agencies use different approaches, and apply different timelines. The following summarizes the current status and anticipated outcome.

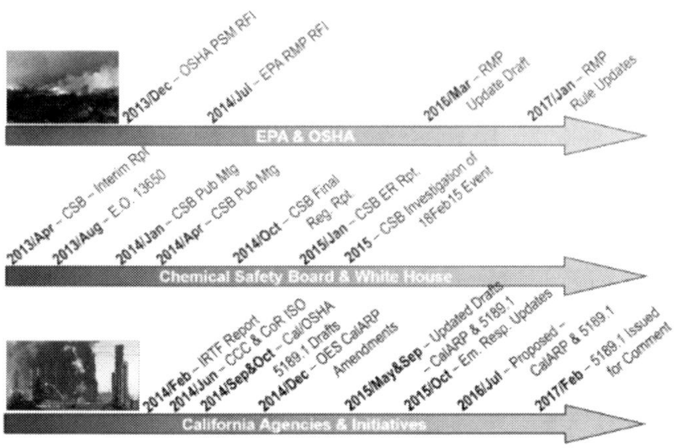

FIGURE 4.1 – Recent SMS Regulatory Activities

Chemical Safety Board (CSB)[19] – Three reports[8,9,10] associated with the August 6, 2012 Richmond Refinery Fire were created by the CSB. Each report has specific recommended actions for industry and the regulatory community. No further CSB investigation efforts regarding the Richmond Refinery Fire are currently planned. CSB will continue tracking the implementation of their recommendations by industry and other agencies.

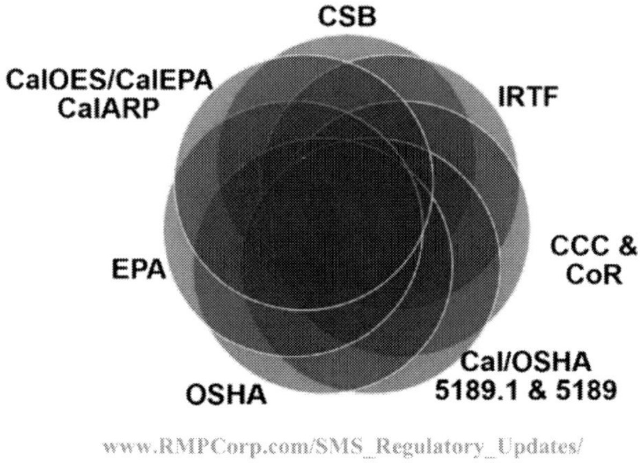

FIGURE 4.2 – SMS Modernization Initiatives

California Interagency Refinery Task Force (IRTF)[20] – The task force is composed of various federal, state, and local agencies with jurisdiction over refineries in California.

- **Publication** – "Improving Public and Worker Safety at Oil Refineries," February 2014.[21]
- **Status** – Task Force meetings continue. Additional reports are not currently planned.

Contra Costa County and City of Richmond

- **Rule-Making**
  - June 2014 – Contra Costa County Issued ISO 450-8 ("Risk Management")[38]
  - June 2014 – The City of Richmond Issued ISO 6.43 ("Industrial Safety")[39]
- **Status** – Current initiative to increase involvement of Community Oversight Group in Process Safety Culture Assessments.

U.S. Environmental Protection Agency (EPA)[22]Risk Management Program (RMP)

- **Rule-Making** – Final Amendments to the Risk Management Program (RMP) Rule, January 13, 2017.[44]
- **Status** – The RMP Rule Amendments are currently subject to review/validation by the Congressional Review Act Implementation and, if not disapproved, are scheduled for implementation on March 21, 2017.
- **Anticipated Outcome** – If implemented on March 21, 2017 the following compliance dates (years after effective regulation date) apply[44]:
  - Third-Party Audit (4 years)
  - Root Cause Analysis (4 years)
  - STAA (4 years)
  - Emergency Response Coordination Activities (1 year)
  - LEPC Requires Compliance with §68.95 (3 years)
  - Emergency Response Exercises (4 years)
  - Information Sharing (4 years)
  - Update RMP (5 years)

California Division of Occupational Safety & Health (Cal/OSHA)[24]Process Safety Management (PSM) Program

- **Rule-Making** – Proposed General Industry Safety Order (GISO) §5189.1, Process Safety Management for Petroleum Refineries, February 10, 2017.[13]
- **Status** – Updates addressing petroleum refinery safety are in the final stages of rulemaking. The February 10, 2017 GISO provided a "Notice of Proposed Modifications to California Code of Regulations" with comments due on March 3, 2017.
- **Anticipated Outcome**
  - Final Regulation Promulgation – Estimated Summer/Fall 2017

  - Initially, 2 separate PSM regulations in California – §5189.1 for Petroleum Refineries (CalPSM-R), §5189 for Others
  - Initially, harmonization with CalARP, but no planned synchronization with Federal PSM/RMP Requirements

Federal Occupational Safety and Health Administration (OSHA)[23] Process Safety Management (PSM) Program

- **Publication** – Request for Information on Proposed Revisions to Federal OSHA's Process Safety Management Program Regulations, December 9, 2013.[11]
- **Status** – OSHA stated that it will process comments received by the March 31, 2014 deadline; however, specific Letters of Interpretation have been issued to address some of the issues identified in its Request for Information.
- **Anticipated Outcome** – Promulgation of updated PSM requirements and timing is unclear.

California Office of Emergency Services (CalOES)[25] and California Environmental Protection Agency (CalEPA)[40] California Accidental Release Prevention (CalARP) Program[26]

- **Rule-Making**
  - CalARP Regulation Amendments 2015, January 1, 2015.[27] These requirements are relatively minor updates to the existing program, but include specified timeframes and deadlines for addressing corrective actions.
  - California Accidental Release Prevention (CalARP) Program Proposed Updates, February 14, 2017 and October 26, 2015.[37,43]
- **Status** – Although CalARP regulatory requirements were updated for January 1, 2015 implementation, updates addressing petroleum refinery safety are in the final stages of rulemaking.
- **Anticipated Outcome**
  - Final Regulation Promulgation – Estimated Summer/Fall 2017
  - Single CalARP Regulation with additional requirements (Program 4) for "High Hazard Facilities" (initially petroleum refineries)
  - Initially, harmonization with CalPSM-R, but no planned synchronization with Federal PSM/RMP requirements

4.2 Anticipated Near-Term Impact

At this point, the most significant near-term impact on SMS programs for California facilities will be the promulgation of §5189.1 by Cal/OSHA and the new Program 4 elements of the CalARP Program. Both of these focus on petroleum refineries in California. The current updates to the RMP requirements will have a smaller impact, but some elements may apply to facilities other than refineries. California will have two PSM regulations (one for petroleum refineries and one for non-refineries). For California facilities, Federal OSHA (along with Federal PSM requirements) does not have direct jurisdiction; however, for refineries that have "sister facilities" that are part of the same company, but in other parts of the United States, PSM Program synchronization within the company may no longer be practical. In addition, although there is harmonization, the current organization of the PSM elements in §5189.1 is not quite the same as the Prevention Program 4 elements of CalARP, thus desynchronizing the programs and making implementation more challenging. In addition, §5189.1 and the CalARP Program 4 elements are not synchronized with the January 1, 2017 U.S. EPA RMP requirements nor the January 13, 2017 RMP Rule Updates. Significant differences exist for both the element-specific requirements, implementation timeframes, and facility applicability. Figure 4.3 depicts the anticipated near-term regulatory framework, and highlights the updates to the U.S. EPA requirements in red lettering.

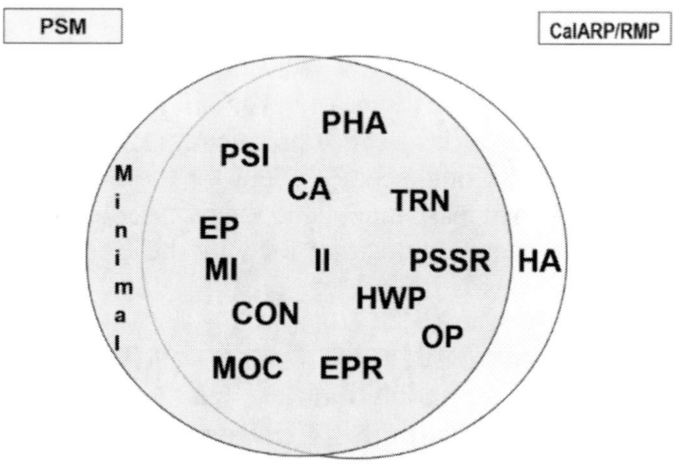

2017 CA SMS Regulatory Environment

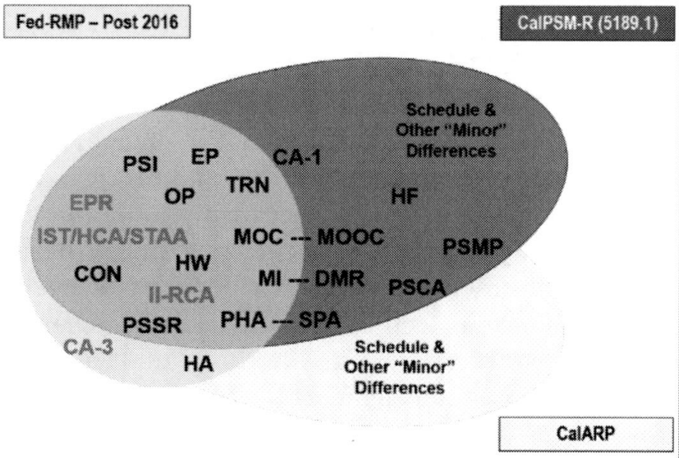

Post-2017 Near-Term CA SMS Environment

FIGURE 4.3 – Current Modernization Plan

In addition to SMS program implementation challenges at a plant site, the existing decentralization of regulatory oversight in California for CalARP can present additional challenges. In California, CalARP oversight is provided by the UPA with jurisdiction over the facility. Commonly, the UPA is either a municipal-level or county-level fire agency, and there are 97 UPAs in the State of California (7 of these have petroleum refineries in their jurisdiction). Each UPA has one or more individuals involved in the CalARP Program, along with their other program responsibilities, so if the UPA is to retain their jurisdictional authority, significant training will be required to bring all of these responsible individuals up-to-speed with updated CalARP regulatory requirements.

Another potential concern with multiple agencies having different programs with nearly identical objectives is the general nature of regulations in the United States, and especially in California, to gain momentum and become more complex. Although there are examples of regulations becoming less complex over time (e.g., the transition of the Risk Management and Prevention Program (RMPP) requirements to CalARP in 1999), the general tendency is for the regulatory environment to evolve towards increased complexity. As the application of the additional program elements being proposed for §5189.1 becomes more common-place for refineries, there is going to a natural inclination for regulators to encourage application to other hazardous materials facilities or require justification from the facility for not applying them.

Another potential complexity for industry is SB-612[42] for California facilities and the U.S. EPA General Duty Clause[28]. If a non-petroleum-refinery highly-hazardous facility does not apply elements that are required (and implicitly considered a best practice) for other types of highly-hazardous facilities (e.g., a petroleum refinery), could there be increased liability issues? At a minimum, it would seem prudent for the SMS Specialists at all facilities encompassed by PSM and RMP to at least understand the requirements for petroleum refineries in California and at least briefly document a position for why these requirements do not need to be implemented at their facility.

It should be noted that a complex SMS regulatory environment can create implementation challenges for not only industry, but also for regulators providing oversight. It is an important objective for regulators to always minimize unnecessary complexity, because unnecessary complexity can erode some of the safety benefits desired from the promulgation of modernized PSM/RMP regulations.

4.2 Anticipated Long-Term Impact

Challenges with overlapping regulations from multiple agencies that are designed to achieve similar objectives is not a new phenomenon. For SMS regulatory requirements, this environment existed in California in the 1990s. In California, RMPP requirements were promulgated in 1986[29] that were designed to provide a platform for the formulation of a "prevention program" to minimize the potential for accidents at highly-hazardous facilities. When PSM was promulgated in 1992, it created a similar challenge with overlapping, but desynchronized, SMS regulatory requirements that were essentially trying to accomplish the same thing. This environment existed until 1999, when the CalARP Program was created to replace RMPP and synchronize key prevention program elements with U.S. EPA RMP requirements, which went into effect in 1997. The net result were programs that were easier to manage, both for industry and for the regulatory community, thus becoming more effective in achieving their objectives of protecting workers, the community, and the environment. Although there will likely be several iterations in the current SMS modernization program initiatives, it would be appropriate to anticipate eventual program synchronization. In the long-term, this will provide a SMS regulatory environment that is easier for facilities to manage, more effective in its implementation, and provide a platform for facilitated regulation and enforcement.

5. What Should I Be Doing Now?

5.1 General Regulatory Program Compliance, Monitoring, and Awareness

The January 1, 2015 CalARP Program Updates are promulgated and currently applicable for all CalARP-covered facilities, refinery and non-refinery. Thus, facilities must adhere to these relatively minor changes in the program at this time.

Although the focus of the current SMS modernization program is on petroleum refineries in California, it is important for all U.S. "highly-hazardous facilities" to be carefully monitoring the proposed changes to CalARP, Federal PSM, California PSM, and RMP, and as changes are made, understanding the requirements and any potential for interpreting them as applying to your facility. Even if these new requirements do not directly affect your facility at this time, it is possible that:

- Applicability could be later expanded to other highly-hazardous facilities in California.
- Applicability could be later expanded to non-California petroleum refineries and other facilities.
- Inference could be made to the new requirements in California as "best practice."
- General Duty Clause covenants could point to the new requirements in California.

5.2 Specific Program Implementation Tips

At this time, the SMS regulatory program modernization initiatives are still a "moving target." Many of the specific requirements are being formulated/finalized, and there are likely to be changes; however, there are several common elements between the different programs, which are also directly linked to CSB-identified SMS program gaps and which are likely to be included in the final SMS regulatory requirements. This section will focus on these higher-priority elements and suggest "high-value" implementation strategies, such that substantive progress can be made on addressing important aspects of recommendations made by the CSB (i.e., those which could be considered best practice) in a focused manner. The specific tips offered will focus on streamlined approaches for adapting efforts currently expended on PHA to address and make progress on some of these new requirements. Although the initial focus of the SMS regulatory modernization programs will be on California petroleum refineries, applying these high-value elements to refineries in other U.S. locations and to other types of facilities may be prudent.

In addition to the need for the SMS Program Coordinator to structure facility programs to efficiently accommodate desynchonized program requirements, a critical challenge that has been identified by the regulatory community is a shortage of capable resources for some of these specialized areas to deal with the anticipated tight schedules, once the proposed regulations are promulgated. One key regulatory agency recommended applying "High-Value/Priority" elements now. Even if schedules are tight, if substantial progress has been made addressing the spirit of the new requirements, "selective-'grandfathering' may be allowed."

5.2.1 Damage Mechanism Review (DMR)

The complete implementation of DMR can require extensive resources, and Figure 5.1 depicts the range of approaches that can be used to address DMR requirements. The following resources clarify the challenge and provide some focused/practical approaches for implementation:

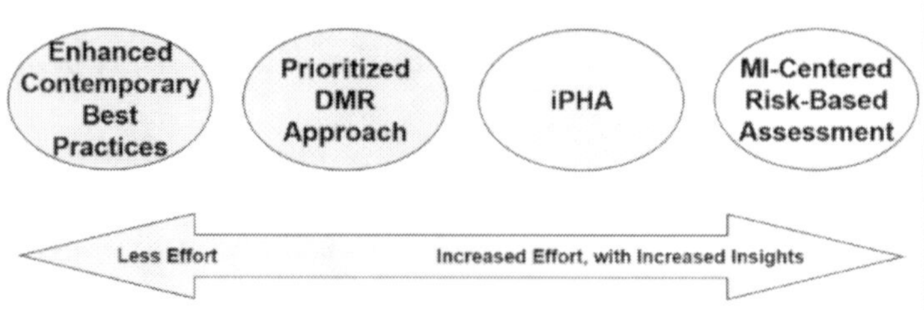

FIGURE 5.1 – DMR Implementation Spectrum

- Maher, Nour, Schultz, "Using PHA as a Framework for Effectively Addressing Evolving PSM/RMP Guidelines, Such As Damage Mechanism Hazard Reviews," Global Congress on Process Safety 2015[30].
- RMP/PSM Series Educational Webinars (March 26, 2015 and August 27, 2015)[41]

5.2.2 Safeguard Protection Analysis (SPA)

Safeguards effectiveness was identified as an important issue in CSB Recommendations 2012-03-I-CA-6 and 2012-03-I-CA-12, and SPA is specifically required by the proposed CalPSM-R §5189.1(e) and CalARP Sections 2762.2/2762.2.1 to validate the effectiveness of safeguards for petroleum refineries. The CSB recommendations focus on documenting "recognized methodologies, rationale, and conclusions used to claim that safeguards intended to control hazards will be effective," and a key objective of §5189.1(e) is that "All safeguards for each failure scenario shall be independent of each other and independent of initiating causes."

Best practices for PHA implementation already include the Facilitator challenging the Team and only crediting effective, and typically independent, safeguards during the PHA. This objective can be achieved by careful application of PHA best practices by a qualified Facilitator during a PHA, and this can be accomplished especially well during an integrated HAZOP/LOPA exercise. The SPA is designed to validate that these best practices are adhered to. Thus, the balance of this subsection will provide tips on the performance of a quality PHA (which, of course, addresses the requirements of a SPA). Tips on maximizing the quality of your PHA effort can be found in a free on-line webinar series that addresses HAZOP/LOPA Facilitation Best Practices.[31] Module 11 is especially well-tuned to best practices that support SPA objectives.

Practical PHA Facilitation Tips to Apply Best Practices and Address SPA Objectives:

- Clearly identify and document transmitters/controllers affecting causes & safeguards
- Frequently refresh the team on risk ranking
- Frequently challenge the team on safeguard effectiveness, availability, & reliability
- List/partition the safeguards as Independent Protection Layers (IPLs)
- Group alarms as a single safeguard, where appropriate

- Apply Risk-Graph and LOPA to provide additional insights
- Apply Risk-Graph and LOPA for high consequence/risk scenarios to identify if a Safety Instrumented Function (SIF) is needed or to validate that a SIF is not needed
- Considerations for reliability & timing of Operator response to alarms
  - Present to hear the alarm
  - Alarm prioritization & diagnosis
  - Permission for corrective action
  - Initiating the corrective action
  - Time for the corrective action to mitigate the event

5.2.3 Inherently Safer Technology (IST) / Hierarchy of Hazard Control Analysis (HCA) / Safer Technology and Alternatives Analysis (STAA)

The evaluation of IST was identified as an important issue in CSB Recommendations 2012-03-I-CA-7 and 2012-03-I-CA-13, HCA is specifically required by the proposed §5189.1(l) and 2762.13, and STAA is specifically required by the proposed RMP Section 68.67 to identify the inherent safety features associated with the process and evaluate the feasibility of applying IST. CSB's recommendations apply to all CalARP facilities, the proposed CalPSM-Rand CalARP-P4 requirements are initially focused on California refineries, and the proposed RMP Section 68.67 applies to all U.S. refineries and other identified highly hazardous facilities. The specific CSB guidance is to minimize risk to the public, personnel, and the environment by performing an Inherently Safer Systems (ISS) Analysis triggered by all MOC and PHA reviews. The stated objective of §5189.1(l) is to:

- "Eliminate hazards to the greatest extent feasible using first order inherent safety measures;
- Reduce any remaining hazards to the greatest extent feasible using second order inherent safety measures;
- Effectively reduce remaining risks using passive safeguards;
- Effectively reduce remaining risks using active safeguards; and
- Effectively reduce remaining risks using procedural safeguards."

Best practices for PHA implementation already include the application of a module to perform an ISS Analysis. This is done during the course of the PHA, and there are many guidance documents and resources available. Reference 33 provides a technical paper from GCPS-2011 that summarizes the use of a HAZID study, HAZOP study, and LOPA to generate ideas for Inherently Safer Designs, and identifies (Figure 5.2) optimal timing for introducing the concepts into the design process. It should be noted that the opportunity to cost-effectively make significant changes to the process to introduce inherently safer design concepts decreases over time. References 14, 19, 34, 35, and 36 provide guidelines and agency checklists for evaluating the feasibility of implementing ISS concepts into a process.

Practical PHA Facilitation Tips to Apply Best Practices and Address IST/HCA/STAA Issues:

- Provide a briefing during PHA Synchronization Training.
- Provide a list of IST Considerations in the PHA Quick Reference provided to the team.
    - ➢ Minimize
    - ➢ Substitute
    - ➢ Moderate
    - ➢ Simplify – More Robust
    - ➢ Simplify – Human Factors
    - ➢ Simplify – Facility Siting
- Apply a "Deviation" to discuss IST/HCA/STAA issues in each node.
- At the end of the PHA, apply an IST Checklist and document as separate node.

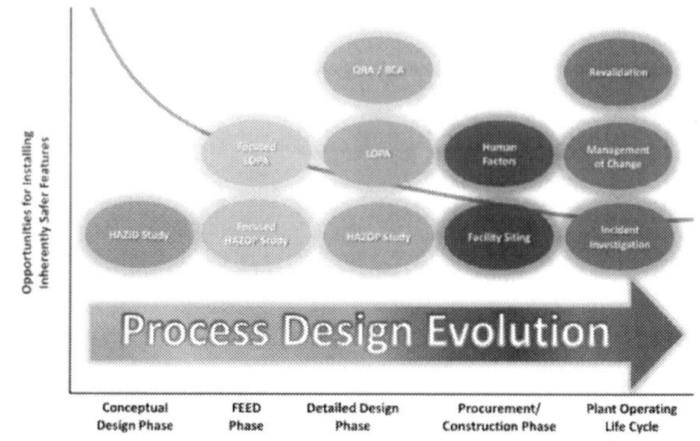

FIGURE 5.2 – ISS Life Cycle

6. Summary and Conclusions

The aftermath of the August 2012 Richmond Refinery Fire resulted in the CSB and other agencies reflecting on industry's application of PHA and MI, as well as the agencies' oversight of PSM and RMP. This has resulted in a number of agencies proposing ideas for improvements in both the specific areas associated with the root cause(s) of the event and the application of other SMS elements. These SMS Program modernization initiatives have focused on PSM, RMP, CalARP, and Cal/OSHA's PSM, with a special emphasis on petroleum refineries in California. Although California refineries are the current focus of CalPSM-R and CalARP-P4, and some elements are already included in the RMP Update, it would be prudent to anticipate that many of these initiatives will eventually affect PSM/RMP facilities throughout the United States.

The months to come are likely to see significant efforts by industry to implement changes, and even more significant will be the challenges to both industry and the regulatory community due to differences in implementation, timing, and the SMS elements included in the different regulatory requirements. The only certainty is the uncertainty in what will be the final outcome and when implementation will be required. Although there will likely be significant near-term variances in the different SMS regulatory requirements, it would be appropriate to anticipate a long-term movement towards synchronization.

This paper has provided relevant background information, summaries of the various SMS regulatory program modernization initiatives, and tips for practical actions to take as part of current PSM/RMP/CalARP efforts. These tips are also conducive towards maximizing the value associated with the implementation of these best-practices and minimizing the later impact of these updated SMS regulatory requirements.

It is critical that the process safety and risk management industrial and regulatory community carefully monitor the modernization concepts being proposed and potential implementation timing. Process safety and risk management professionals at the plant sites should focus on charting the course for long-term success for their programs. Being aware of these upcoming changes will facilitate planning a strategy for effective implementation that will best serve the interests of those being regulated (and regulators), as well as minimizing risks to plant personnel, the public, and the environment.

7. References

[1] http://en.wikipedia.org/wiki/Bhopal_disaster – Accessed on March 20, 2015.
[2] CCPS – "Guidelines for Technical Management of Chemical Process Safety," 1987.
[3] API RP 750, First Edition – "Management of Process Hazards," 1990.
[4] PSM – 29 CFR 1910.119, "Process Safety Management (PSM) of Highly Hazardous Chemicals, Explosives and Blasting Agents," 1992.
[5] RMP – 40 CFR Part 68, "Risk Management Programs (RMP) for Chemical Accidental Release Prevention," 1996.
[6] SEMS Final Rule – Federal Register – Title 30, Code of Federal Regulations (CFR) Part 250 – "Oil and Gas and Sulphur Operations in the Outer Continental Shelf – Safety and Environmental Management Systems," Federal Register, Vol. 78, No. 66, April 5, 2013.
[7] https://www.osha.gov/chemicalexecutiveorder/, "Executive Order 13650: Actions to Improve Chemical Facility Safety and Security – A Shared Commitment".
[8] http://www.csb.gov/assets/1/19/Chevron_Interim_Report_Final_2013-04-17.pdf, "Interim Investigation Report, Chevron Richmond Refinery Fire," April 2013.
[9] http://www.csb.gov/assets/1/19/Chevron_Regulatory_Report_11102014_FINAL_-_post.pdf, "Regulatory Report, Chevron Richmond Refinery Pipe Rupture and Fire," 2012-03-I-CA, October 2014.
[10] http://www.csb.gov/assets/1/19/Chevron_Final_Investigation_Report_2015-01-28.pdf, "Final Investigation Report, Chevron Richmond Refinery Pipe Rupture and Fire," 2012-03-I-CA, January 2015.
[11] https://www.OSHA.gov/chemicalexecutiveorder/section6.html, Request for Information on Proposed Revisions to Federal OSHA's Process Safety Management Program Regulations, December 9, 2013.
[12] https://www.gpo.gov/fdsys/pkg/FR-2016-03-14/pdf/2016-05191.pdf, 40 CFR Part 68, Accidental Release Prevention Requirements: Risk Management Programs Under the Clean Air Act; Proposed Rule, March 14, 2016.
[13] http://www.rmpcorp.com/wp-content/uploads/2014/08/15-day-Notice-Process-Safety-Management-for-Petroleum-Refin.pdf, Proposed General Industry Safety Order (GISO) §5189.1, Process Safety Management for Petroleum Refineries, February 10, 2017.
[14] http://www.nj.gov/dep/rpp/brp/tcpa/downloads/Reactive_HA_Guide_2014.pdf, "Guidance on Hazard Assessment for Reactive Hazard Substances (RHS) and RHS Mixtures," New Jersey Department of Environmental Protection: Bureau of Release Prevention. Toxic Catastrophe Prevention Act (TCPA) Program.

[15] http://www.aiche.org/ccps/topics/elements-process-safety/learn-experience/measurement-and-metrics/overview, "Overview of Measurements and Metrics," Center for Chemical Process Safety.
[16] http://www.aiche.org/ccps/topics/elements-process-safety/learn-experience/management-review-and-continuous-improvement/introduction, "Introduction to Management Review and Continuous Improvement," Center for Chemical Process Safety.
[17] http://www.aiche.org/ccps/topics/elements-process-safety/commitment-process-safety/process-safety-competency/introduction, "Introduction to Process Safety Competency," Center for Chemical Process Safety.
[18] http://www.nj.gov/dep/enforcement/tcpa/downloads/IST_workshop_concepts_methodologies.pdf, "IST Concepts and Methodologies for Implementation," Inherently Safer Technology Workshop for NJDEP, June 2008.
[19] http://www.CSB.gov – Source website for the Chemical Safety Board.
[20] http://www.CalEPA.CA.gov/Refinery/ – Source website for the Interagency Refinery Task Force.
[21] http://www.calepa.ca.gov/publications/Reports/2014/RefineryRpt.pdf, "Improving Public and Worker Safety at Oil Refineries," February 2014.
[22] http://www.EPA.gov – Source website for the United States Environmental Protection Agency.
[23] http://www.OSHA.gov – Source website for the Federal Occupational Safety & Health Administration.
[24] https://www.dir.ca.gov/dosh/ - Source website for the California Division of Occupational Safety & Health (Cal/OSHA).
[25] http://www.CalOES.ca.gov – Source website for the California Office of Emergency Services.
[26] http://www.caloes.ca.gov/cal-oes-divisions/fire-rescue/hazardous-materials/california-accidental-release-prevention – Source website for the California Accidental Release Prevention Program.
[27] http://www.caloes.ca.gov/FireRescueSite/Documents/CalARP%20Regulations%2001012015.pdf, CalARP Regulation Amendments 2015, January 1, 2015.
[28] Clean Air Act (CAA) Section 112(r)(1) – General Duty Clause.
[29] Section 25534 of Chapter 6.95, Risk Management and Prevention Programs, 1986.
[30] Maher, Nour, Schultz, "Using PHA as a Framework for Effectively Addressing Evolving PSM/RMP Guidelines, Such As Damage Mechanism Hazard Reviews," Global Congress on Process Safety 2015.
[31] http://www.RMPCorp.com/HAZOP-Study-series-module, HAZOP/LOPA Facilitation Best Practices Webinar Series.
[32] http://www.RMPCorp.com/SMS_Regulatory_Updates/ - Website tracking Safety Management Systems U.S. Regulatory Updates.
[33] Surmeli, Norton, Maher, "Using the HAZID Study, HAZOP Study and LOPA to Generate Ideas for Inherently Safer Designs," Global Congress on Process Safety 2011.
[34] http://cchealth.org/hazmat/pdf/inherently_safer_systems.pdf, "Inherently Safer Systems Checklist," Contra Costa County Health Services.
[35] http://www.nj.gov/dep/enforcement/tcpa.html - Source website for the New Jersey Toxic Catastrophe Prevention Act (TCPA) Program.

[36] "Inherently Safer Chemical Processes: A Life Cycle Approach," 2nd Edition, Center for Chemical Process Safety (CCPS), ISBN: 978-0-471-77892-9, December 2008.
[37] http://www.caloes.ca.gov/cal-oes-divisions/fire-rescue/hazardous-materials/california-accidental-release-prevention, California Accidental Release Prevention (CalARP) Program Proposed Updates, February 14, 2017.
[38] http://cchealth.org/hazmat/calarp/ – Source website for the Contra Costa County Health Services CalARP Program.
[39] http://cchealth.org/hazmat/pdf/iso/RISO-Chapter-6-43-INDUSTRIAL-SAFETY.pdf – City of Richmond Industrial Safety Ordinance, Chapter 6.43.
[40] http://www.CalEPA.ca.gov/ – Source website for the California Environmental Protection Agency.
[41] http://www.RMPCorp.com/rmppsm-series/ - RMP/PSM Series Educational Webinars.
[42] http://leginfo.legislature.ca.gov/faces/billNavClient.xhtml?bill_id=201520160SB612 – Senate Bill 612 (SB-612), "Hazardous Materials."
[43] http://www.caloes.ca.gov/FireRescueSite/Documents/CalARP%20Refinery%20Emergency%20Response%20Program%20Draft%20Regulations%2010-26-15.pdf, "CalARP Regulation Draft Proposed Amendments- applicable to Petroleum Refineries," October 26, 2015.
[44] https://www.epa.gov/rmp/final-amendments-risk-management-program-rmp-rule, "Final Amendments to the Risk Management Program (RMP) Rule," January 13, 2017.
[45] https://www.law.cornell.edu/uscode/text/5/part-I/chapter-8, "5 U.S. Code Chapter 8 - CONGRESSIONAL REVIEW OF AGENCY RULEMAKING."

Existing Challenges in Incorporating Process Safety Management in Developing Countries and Ways to Overcome Them

Yogesh Koirala
Mary Kay O'Connor Process Safety Center, Texas A&M University
College Station, 77843-3122, USA
ykoirala@tamu.edu

Andres Hernandez[1] and M. Sam Mannan[1,2]
[1] Mary Kay O'Connor Process Safety Center, Texas A&M University
[2] Artie McFerrin Department of Chemical Engineering, Texas A&M University
College Station, 77843-3122, USA
andreshdez8@hotmail.com
mannan@tamu.edu

Prepared for Presentation at
American Institute of Chemical Engineers
2017 Spring Meeting and 13th Global Congress on Process Safety
San Antonio, TX
March 26 – 29, 2017

Existing Challenges in Incorporating Process Safety Management in Developing Countries and Ways to Overcome Them

Yogesh Koirala
Mary Kay O'Connor Process Safety Center, Texas A&M University
College Station, 77843-3122, USA
ykoirala@tamu.edu

Andres Hernandez[1] & M. Sam Mannan[1,2]
[1] Mary Kay O'Connor Process Safety Center, Texas A&M University
[2] Artie McFerrin Department of Chemical Engineering, Texas A&M University
College Station, 77843-3122, USA
ykoirala@tamu.edu
andreshdez8@hotmail.com
mannan@tamu.edu

Keywords: Process Safety Management; PSM; Process Safety Challenges; Developing Countries; Strategies to Improve Process Safety; Process Safety Regulations; CCPS Vision 20/20

Abstract

Many of the products that we use or consume in our daily life are produced in developing countries because of socio-economic factors. Most of the manufacturing industries in these countries are an important part of the international markets, providing products and raw materials of desired specifications for other industries around the world. Demands from developed countries and their own local needs should be met with safer production. Various incidents that have occurred in these manufacturing industries indicate the lack of comprehensive process safety programs and regulations. Because of this weakness, the rigorous standards and regulations that the multinational companies promise to follow become less effective when implemented in developing nations. Most employees working in these industries are aware of the occupational hazards, but identifying and managing the process hazards requires proper process safety management systems. Hence, developing countries need to create strategies to improve the process safety management systems.

This paper presents the resulting root cause analyses for incidents that could be attributed to improper, inadequate and/or ineffective implementation of process safety programs in developing countries, and compares their existing practice/regulations with the Occupational Safety and Health Administration's Process Safety Management Program. The key issues for incorporating good management systems are identified. Based on these identified challenges, recommendations for best practices [16] are provided to improve the current situation with regard to process safety and risk management programs in developing countries.

1. Introduction

The American Institute of Chemical Engineers (AIChE) defines process safety as "a disciplined framework for managing the integrity of operating systems and processes handling hazardous substances by applying good design principles, engineering, and operating practices. It deals with the prevention and control of incidents that have the potential to release hazardous material or energy"[1]. In the United States, the Occupational Safety and Health Administration (OSHA) defines process safety management (PSM) as "the proactive identification, evaluation and mitigation or prevention of chemical releases that occur as a result of failures of process, procedures, or equipment" [2]. In the last 50 years, socio-economic development, along with scientific and technical progress, has brought an unprecedented amount of research and knowledge regarding risk management and process safety. This knowledge is practiced by many industries through their own internal standards or through standards and regulations imposed by law, but there are aspects that must be considered. Existing literature on process safety and the challenges of implementing it correctly shows that most of the research, conferences, and symposia are primarily about North American and European industry, and appear to overlook the developing countries. Despite availability of information and resources regarding process safety management, adapting to these PSM standards and good engineering practices is slow for companies in developing regions where many process industries are growing at much higher rates due to the globalization of world economies. This has resulted in higher risk for the employees and for the environment in these developing nations [3].

Reports from the International Labour Office (ILO), International Association of Oil and Gas Producers (IOGP), and the Ibero-American Social Security Organization in the last decade conclude that process safety programs are not fully implemented or given a high priority in companies in developing countries due to socioeconomic factors [3-4]. For example, IOGP published a report about fatal incidents around the world, from which information concerning regions with developing countries was extracted [4b]. It is important to note that this information comes from the reported incidents within the IOGP members, but according to the Economic Commission for Latin America and the Caribbean (ECLAC), South and Central American countries report only 25% of incidents at work and 5% of occupational diseases [5], which means that the statistics do not necessarily reflect the reality of the situation. The study indicated by Renton *et al.* in 2012 estimated that in developing countries people are exposed to more than 80% of global hazards [6]. The World Bank also conducted a study in 1995, which indicated that a manufacturing worker in a developing country (*e.g.*, Pakistan) is almost eight times more likely to die at work than a manufacturing worker in a developed country (*e.g.*, France). Similar differences have remained for many developing countries even today [7]. This paper highlights ways to overcome the challenges in improving process safety in developing countries.

2. Current Process Safety Status

The primary objective for petrochemical companies in developing countries should be to maximize profit while making the process(es) safer. This requires additional safety barriers, which can often be costly. Without strong regulations that require companies to demonstrate a safer process, companies in developing countries could easily reduce the number of safety barriers and in turn reduce the cost associated with the processes. Costs normally considered within the industry are those inherent in the production process and its management, while the social value of human life, health and environmental quality are considered external costs. Human life is not an effective expression in terms of monetary value; however, the relative comparison of human life between developing and developed countries is a topic that deserves more attention and often leads to higher risk operating conditions in developing countries. Furthermore, legal systems utilizing fines or sanctions are considered more stringent in developed countries as compared to developing countries, affecting the inherent issue of valuing life and the environment [8].

It has been observed that the quantitative approach to risk analysis is not too evident in the process safety management programs of many industries in developing countries. Some adopt a management system based on static objectives, (*e.g.,* adherence to minimum legal requirements), where continuous improvement is not sought, demonstrating the continued need for work in the areas of education, collaboration, and commitment. The needed action to make a process safer is to reduce the process hazards in the plant, but this action will differ depending of the country and the company. According to the reports of the ILO, it is common in developing countries for the cost of compliance to fall heavily on the employees, who are often not aware of the risk associated with their work [3].

Normally, the authorities impose one of two types of requirements on the industry. The first is prescriptive; these requirements are standards, norms, and criteria that the industry must meet to guarantee a minimum level of safety, but that do not contribute to the continuous improvement of the process. The second type is performance-based requirements, which include the use of tools for continuous improvement, best available techniques, as low as reasonably achievable/practicable (ALARA, ALARP), and can be more costly. Thus, for many developing countries, only prescriptive requirements are considered, while performance-based are viewed as optional and their use will depend on the company. The company may decide to implement different program to meet the different markets needs [8].

3. Standards and Regulations in Developing Countries

Literature shows that developing countries are looking for a broad and/or general treatment for safety issues. They often handle process safety within the framework of occupational safety and health (OSH), referring to ILO policies; leaving the selection of PSM programs to the companies who want to embrace international practices and guidelines. The OSH-related organizations that provide guidelines in some developing regions include the Andean Community (CAN) and Southern Cone Common Market (MERCOSUR) in Latin America, and the Southern African Development Community (SADC) in Africa. But taking a look at the statistics of incidents in the industry, we can conclude that the measures taken have not been sufficient, and because of that, a more process safety focused framework is required[4c, 9].

Many developing countries are in the process of creating a framework or updating their national policies on process safety and occupational health. There are a number of such cases in Latin America, Africa and Asia. Countries such as Dominican Republic, Costa Rica, Guatemala, and Nigeria are adopting new policies, while others are improving the existing ones (Colombia) or creating new OSH institutes (Algeria and Serbia). Mali has recently requested assistance to initiate a national OSH policy and Sri Lanka has integrated it into their policies [3-4, 10].

4. Factors Driving the Need of Change

In 2008, Joseph Louvar reviewed the U.S. Chemical Safety and Hazard Investigation Board (CSB) incident investigations, and reported that 50% of the incidents occurred in facilities covered by the PSM standard, and the other 50% occurred in small or mid-sized plants that were not covered by this standard. One clear conclusion from this review is that the OSHA PSM regulation and recommendations by Center of Chemical Process Safety (CCPS), CSB are not being used as detailed [11]. For those countries where the regulations are weakly enforced, we can extrapolate that the number of incidents would be much higher. Large multinational companies are recognized for going beyond the minimum regulatory requirements of the countries where they are active. These companies comply with the requirements that the larger markets demand. For mid-sized or small local companies in developing countries, this is not always the case, and often they take advantage of the socio-economic conditions and lack of law enforcement to continue operating in a sub-standard fashion without fear of being sanctioned.

To understand some of the shortcomings that can lead to PSM implementation problems in developing countries, the results of specific studies were used. In one study conducted in China in 2013 by Zhao *et al.* regarding the safety performance of small- and medium-sized enterprises (SMEs), it was found that management knowledge was one of the most common problems when implementing PSM [12]. Similarly, a survey was sent to owner/managers of SMEs in Ghana, with a response rate of 32%. The findings revealed that only a few of the SMEs adopted proactive health and safety practices such as incident reporting procedures, incident investigation procedures, documentation of method statements, health and safety inductions, and use of health and safety posters.

In order to improve PSM implementation in developing countries, it is important to learn about the incidents that have occurred and oftentimes it is difficult to obtain this information due to the lack of published data and records [9]. In-depth reviews of the existing situation of the different developing regions found the following problems affect the development of good process safety management programs;

- **Safety culture**: Perception and perspectives about the advantage of good process safety culture in developing countries show they are still struggling [11, 13]. Stakeholders and top management are not engaged unless there is a public outcry.
- **Laws and regulations***:* Most countries have laws and regulations to protect the safety of workers and citizens and developing countries are no exception. However, in developing countries these laws and regulations are not strongly enforced. As a result, owners do not feel the need to put many resources toward training and workshops; therefore, workers are not always aware of the hazards associated with their work, and even if they are, they might accept the risks because they are limited in employment opportunities [7, 14].
- **Training:** Most developing countries are not capable of providing state-of-the-art training for their employees. These training facilities are often expensive to run or are not available locally. Employees could be sent to developed countries for training which

would increase their competency; however, this is considered expensive and is done minimally.

- **Contractors:** Contractors are an integral part of process industries and play a significant role in managing risk throughout the process lifecycle. One great challenge in developing countries is to find experienced local contractors to perform tasks. Inexperienced workers performing specialized tasks could lead to unsafe practices which could lead to incidents. Furthermore, if contractors are being brought in from other countries, it is often an expensive and lengthy process; long periods of time to hire contractors could lead to increased risk.
- **Mechanical Integrity:** PSM rules mandate training for process maintenance employees, written procedures for operation, and inspection and testing for process equipment. Not every company has a competent workforce to perform these activities. These activities include inspection and testing of process equipment such as piping systems; pumps; relief and vent systems and devices; pressure vessels and storage tanks; emergency shutdown systems; and controls (including monitoring sensors, alarms and interlocks). In addition, equipment used in the process industries must meet certain design codes and standards. It should also be mentioned that for economic reasons, some companies overuse their equipment above its capabilities or service life.
- **Incident Investigation:** If a developing country does not have a well-established regulatory compliance structure, it could significantly affect the investigation of incidents. Furthermore, lax or non-stringent regulations set by the regulators could impact the number of investigations performed. In some countries, incident investigations are often avoided by management where weak management systems could be to blame. The incidents which were not investigated often had similar root causes to incidents that occurred later. Hence, not doing an incident investigation or having a report would make it impossible for others to learn [3].
- **Compliance Audits:** Compliance audits depend upon the regulatory requirement. The biggest challenge in meeting the regulatory requirement is that auditors are more susceptible to be bribed to keep facilities running. The recent incident in Tianjin is an example where it was found that, despite having some regulations, the company had illegally built and operated a freight-yard of hazardous materials with inadequate process safety management. Local officials were also deemed to have failed to enforce regulations [15].
- **Process Hazards Analysis (PHA):** The PHA must address previous hazardous incidents, the hazards of the process, administrative and engineering controls that are in place and the consequences of the failure of the control measures. However, regulations do not always define the set of criteria for conducting PHAs. Also, PHAs are often prioritized lower or ignored as they require a significant number of personnel and expertise. When PHAs are conducted, recommendations that need to be addressed are made. If there are no strict regulations, these recommendations could easily be delayed or not implemented, as they could be costly.
- **Data management:** Poor data management of incident records and the lack of adequate programs to use them for lessons learned activities makes improvement of the PSM programs difficult [11, 13a]. Robust documentation systems cost money and industries are often reluctant to spend money without a business case.

- **Corruption:** Even with the well-written laws and regulations, corruption can impinge on those responsible to perform the job honestly. Some organizations with a selfish agenda like to exploit this to save money [13a].
- **Poverty:** There are some regions with low standards of living in developing countries, where just having a decent paying job means a lot for a considerable amount of people. This can lead to exploitation of the work force by paying low salaries, working long hours and bad workplace conditions [13b, 14].

5. Path Forward in Implementing PSM in Developing Countries

To overcome the problems and challenges previously discussed, CCPS' Vision 20/20[16] was used as a reference to develop the following recommendations for improving the implementation of PSM in developing countries.

- **Management committed to safety:** Leaders should be connected to the employees within the company, have the capacity to involve them in the idea of "it can happen here too", and promote the commitment of senior executives showing the importance of safety. This commitment must be documented and shared with local industries to promote safer actions and routines in accordance with existing policies and procedures.

- **Basic but continuous education for all employees:** The objective is to continuously educate all employees with safety knowledge in line with their safety critical roles. This helps employees to clearly understand the goals and expectations of the company and its commitment for process safety. It enhances safety knowledge for all employees and creates a strong safety culture.

- **Development and pursuit of competency:** It does not matter how strong the company's commitment is or how good the management systems are if the company does not have competent personnel to safely carry out critical tasks. It is necessary that all employees who have an impact on process safety have the skills and knowledge to execute the technical and cultural requirements that their position requires.

- **Willingness to adapt:** There should be constant evolution of the standards and best practices of safety through the use of excellent management of information. This allows for continuous improvement in updating the equipment, the personnel and the procedures driven by the idea of learning from many sources.

- **Cooperative work:** Government regulatory bodies, labor organizations, industry associations/consortiums, research institutions, universities and industries should work together with the intent of sharing information on best practices and lessons learned to improve standards and regulations. These groups should produce guidelines for the safe design, maintenance and operation of equipment, and harmonize them for the sake of a better understanding and implementation on a regional level. Governments should encourage industries to provide basic training and seminars, inviting employees to discuss regulatory issues and open a dialogue on how to work together to create safer processes.

- **Systematic implementation of PSM programs:** Many developing countries do not have PSM requirements that cover all companies. Large multinational companies comply with international standards while small ones do not have the resources or the technical knowledge to do the same. The use of a Plan-Do-Check-Act cycle implemented in a gradual and systematic way could start to change this situation. Quantitative and qualitative indicators must be implemented as defined by the companies. It is important to be able to evaluate the structure, process and outcome of the PSM program as they are aligned with and are part of the strategic platform. If companies do not have the resources to implement this new PSM program, a consortium can be formed to share the cost and resources to create a platform that has the capabilities to provide service to these industries.

  Companies should be granted a transition period to strengthen the PSM programs gradually and systematically. To make the transition to employers less stressful, labor organizations should be empowered to provide advice and technical assistance to affiliated companies and to submit quarterly reports of the progress made.

  Developing countries such as Colombia, which is in the process of implementing a new occupational safety and health management system, have a transition period as follows:

  1. 18 months for companies with less than 20 workers
  2. 24 months for companies with 21 to 100 workers
  3. 30 months for companies with 101 to 200 workers
  4. 36 months for companies with 201 or more workers

- **Academic support:** Another aspect that must be addressed is the importance of proper education and research in the process safety area. In developing countries, especially in Africa, most OSH and PSM education is limited to inspectors and managers. However, efforts to make this education more inclusive are being made in countries such as India and Ecuador, where the subject has been included in their new drafts of OSH policies, or Tobago and Trinidad, where the OSH agency is currently working with the Ministry of Education to include OSH in the curriculum. Other countries like Bangladesh and Malaysia have started safety programs in some of their leading national institutions. Brazil is providing OSH courses within the framework of specialized studies for safety engineers and technicians, whereas OSH issues are included at all levels of the education system in Cuba, and Mexico. To what extent PSM education is being addressed has yet to be seen [3].

  The implementation of these provisions would create an environment to educate new and old generations of employees by providing education and training within their reach. The inclusion of PSM at all levels of education and training, including higher technical and professional education, is a way to produce an early familiarization with PSM knowledge and principles, and is the most effective way to ensure compliance, while increasing awareness among future employers and workers in hazard and risk reduction strategies. In this way, the changing industrial world and the new generation working population can both be dealt with.

This also opens the door for industries to reach out to academic institutions for any technical assistance they may need in solving major problems. Further, a country can allocate money to support such institutions, generate research agendas, and promote cooperation between the industry and academia.

6. Conclusions

This paper presented the resulting root cause analyses for incidents that could be attributed to improper, inadequate and/or ineffective implementation of process safety programs in developing countries, and compared their existing practice/regulations with the Occupational Safety and Health Administration's Process Safety Management Program. The key issues for incorporating good management systems were identified. Based on these identified challenges, recommendations for best practices were provided to improve the current situation with regard to process safety and risk management programs.

References

1. Center of Chemical Process Safety (CCPS) Process Safety Glossary. http://www.aiche.org/ccps/resources/glossary/process-safety-glossary/process-safety (accessed October 10th 2016).
2. Occupational Safety and Health Administration (OSHA), Occupational Safety and Health Standards. In *1910*, Occupational Safety and Health Administration (OSHA), Ed. 1992; Vol. Compliance Guidelines and Recommendations for Process Safety Management (Nonmandatory) 1910.119 Appendix C.
3. International Labour Office Report III (Part 1B) ILO standards on occupational safety and health Promoting a safe and healthy working environment. In *International Labour Conference*, Geneva, 2009.
4. (a) International Labour Organization Salud y seguridad en trabajo en América Latina y el Caribe. http://www.ilo.org/americas/temas/salud-y-seguridad-en-trabajo/lang--es/index.htm#banner (accessed September 14th); (b) International Association of Oil & Gas Producers *Safety performance indicators – 2014 data*; 2015; (c) Ibero-American Social Security Organization, Recopilacion de los Principales Indicadores de Siniestraliedad Laboral y Enfermedad Ocupacional Utilizados en Iberoamerica. **2012**.
5. CEPAL Social Panorama of Latin America. http://www.cepal.org/en/publications/social-panorama-latin-america-2013 (accessed September 14th 2016).
6. Renton, N. C.; Baker, J. M.; Tan, H. Tackling occupational hazards in developing countries, Topic 29 lectured at Safety Engineering and Risk Management Debate.
7. World Bank, World Development Report 1995 : Workers in an Integrating World. New York: Oxford University Press. © World Bank1995. https://openknowledge.worldbank.org/handle/10986/5978.

8. Duijm, N. J.; Fiévez, C.; Gerbec, M.; Hauptmanns, U.; Konstandinidou, M., Management of health, safety and environment in process industry. *Safety Science* **2008,** *46* (6), 908-920.

9. Ghettas, S. The Evaluation of Process Safety Management Effectiveness in the Oil and Gas Sector: Case of Sonatrach, Algeria. Rochester Institute of Technology, 2015.

10. Ministry of Labor, Sistema de Gestion de Seguridad y Salud en el Trabajo. Mintrabajo: 2015.

11. Louvar, J., Improving the Effectiveness of Process Safety Management in Small Companies *Process Safety Progress* **2008,** *27* (4).

12. Zhao, J.; Joas, R.; Abel, J.; Marques, T.; Suikkanen, J., Process safety challenges for SMEs in China. *Journal of Loss Prevention in the Process Industries* **2013,** *26* (5), 880-886.

13. (a) Khan, M. I., Developing a Safety Culture in Developing Countries. In *International Conference on Safety, Construction Engineering and Project Management*, 2013; (b) Abu-Khader, M. M., Impact of Human Behaviour on Process Safety Management in Developing Countries. *Process Safety and Environmental Protection* **2004,** *82* (6), 431-437.

14. Iunes, R. F. Occupational Safety and Health in Latin America and the Caribbean: Overview, Issues and Policy Recommendations 2002. http://www.iadb.org/wmsfiles/products/publications/documents/354332.pdf (accessed September 14th 2016).

15. Lee, S. (2016, November). Businessman gets death sentence for Tianjin chemical disaster. Retrieved from https://chemicalwatch.com/asiahub/51260/businessman-gets-death-sentence-for-tianjin-chemical-disasterchina/

16. Center for Chemical Process Safety, Vision 20/20 Process Safety: The Journey Continues. **2013.**

Proven Approaches to Ensuring Operators Can Respond to Critical Process Deviations in Time (Human Response IPL)

William G. Bridges, President
PROCESS IMPROVEMENT INSTITUTE, INC. (PII)
1321 Waterside Lane, Knoxville, TN 37922
Phone: (865) 675-3458
Fax: (865) 622-6800
e-mail: wbridges@piii.com

Prepared for Presentation at
13th Global Congress on Process Safety
San Antonio, TX
March 27-29, 2017

UNPUBLISHED

13TH GLOBAL CONGRESS ON PROCESS SAFETY

Proven Approaches to Ensuring Operators Can Respond to Critical Process Deviations in Time (Human Response IPL)

William G. Bridges, President
PROCESS IMPROVEMENT INSTITUTE, INC. (PII)
e-mail: wbridges@piii.com

Keywords: Risk control, process safety management, PSM, LOPA, Layer of Protection Analysis, Independent Protection Layers, IPLs, human error prevention, Human Factors, human response

Abstract

Humans can be the cause on an accident scenario (the Initiating Event [IE]) or humans can serve or participate as an independent protection layer (IPL). Validating Human IPLs has been a show stopper for many companies considering the use of human response as an IPL. Human IPLs include preventative steps that may stop a scenario from progressing once it is initiated, but more typically the human IPLs are responses to alerts or alarms or troubling readings and sample results.

This paper first describes the fundamentals of clear alarms, practical actions, and having enough time to perform the action, all without being in harm's way at the end of the action. This paper builds upon earlier studies (based on similar papers from 2010 and 2011)[13, 14] of how to collect the data needed for directly measuring the probability of failure on demand (PFD) of the human response. The preferred method for data collection covers the training requirements that should be met, proof drills for response to alarms, simulations and tests, and frequency of proofs, and of course the effect of human factors on human error rates. An example is provided of how a simple data collection and validation method can be set up within a company. This paper also provides an overview of alternative methods for estimating the PFD of a Human IPL, based on plant and scenario specific factors (such as stress factors, complexity, and communication factors); and the paper evaluates and compares alternative approaches to validating human IPLs, including expert judgment based on frequent practice of trouble-shooting by operators. All of these methods were available in Appendix B and C of the initial full draft of the CCPS book, *Guidelines for Initiating Events and Independent Protection Layers, April 2012 (unpublished), but unfortunately o*nly a small fraction of one of the Appendices was retain in the version that was published in January 2015. This paper provides that missing information.

Response to Critical Process Deviations – Fundamentals

Humans response to an alarm or troublesome reading or sample result can be a great safeguard layer against major accidents because the human has the capability to diagnose false alarms and in other ways prevent a spurious shutdown of a process. Many organizations are reluctant to use human response as a protection layer because "humans are human" and make mistakes. Of course, instrumented systems fail as well. Nothing is perfect. The key to using human response as a layer of protection is to follow the general guidelines for qualifying a response as a Human Independent Protection Layer (IPL).

There are 5 steps to using a human to respond to a critical process deviation as a Human Response IPL:

1. Determine which parameters limits (announced by alarms or other triggers) should have human response, and why human response is best
2. Ensure human response action meets the definition of an IPL
3. Develop a trouble-shooting guide (general steps for the operators to take) for each response
4. Perform initial training on each human response IPL
5. Validate that human response success rates are high enough

This paper explains what we have found to be best practices for each of these steps

1. Determine which parameters limits (announced by alarms or other triggers) should have human response, and why human response is best

This important first step is difficult for many organizations since they either have a prejudice against using a response by a human as an IPL or because they do not trust their hazard evaluation teams to make such judgments. The CCPS book on LOPA (2001)[1] and the follow-on book, *Guidelines for Initiating Events and Independent Protection Layers,* CCPS (2015)[2] both allow the use of Human IPLs and these books given criteria on their use. Neither book tells how to determine if and when a Human IPL should be used, but instead state that the PHA (HAZOP, What-If, etc.) is the setting for determining what layers of protection are appropriate and if that fails, then LOPA itself allows the use of a Human IPL as one IPL in a scenario, but it is up to the LOPA analyst to determine that the Human IPL is the best selection from alternatives for risk reduction.

If qualitative methods such as HAZOP or What-If brainstorming teams are used to determine when a Human IPL is best, then it is recommended that the team follows the protocol in Bridges & Dowell, 2016.[3] This approach provides for qualitatively determining when safeguards (including human response safeguards) meet the definition of an IPL. Once determined, the Human IPL should be documented as such in the PHA analysis as shown on in Table 1.

Table 1. Documentation of IPLs in a PHA (in this case HAZOP method) Analysis Table[3]

No.: 2 XXXX storage spheres xxx-T-XX A/B/C/D/E/F/G/H/I/J/K/L (1 of 12)					
#	**Dev.**	**Causes**	**Consequences**	**Safeguards**	**Recommendations**
2.1	High level	Too much flow to one sphere from XX Plant (through their pump; about 40 bar MDH)	High pressure (see 2.5)	High level SIF with level sensors voted 2oo2, to close inlet valve - **SIL 1** Overflow thru pressure equalization line to other spheres (through normally open [NO] valve) - **IPL**	
		Misdirected flow - Liquid from xxx Plant(s) to spheres (see 1.4)	Overpressure of sphere not credible from high level, for normal operating pressure of the column (which is 1.75 MPa), unless all spheres are liquid filled and then thermal expansion of the liquid could overpressure the spheres	High level SIF with level sensors voted 2oo2, to close inlet valve - **SIL 1** Overflow thru pressure equalization line to other spheres (through normally open [NO] valve) - **IPL** Spheres rated for 1.95MPa (19.5 Bar, approx) and the highest pressure possible from the column feeding the spheres is 1.75 MPa Level indication and high level alarm in DCS, used by operators to manually select which tank to fill – **Human IPL**	
			Overflow into the equalization line will interfere with withdrawal from the column, but this is an operational upset only		
			Excessive pressure on inlet of high pressure liquid pumps, leading to excess load on pumps and trip of pumps on high pumps, causing trips of xxx, xxx, etc. - significant operability issue		
2.2	Low level	Failing to switch from the sphere with low level in time (based on level indication)	Low/no flow - Liquid from spheres through high pressure product pumps to the vaporizer (see 4.2)	Level indication and low level alarm, inspected each year, per government regulation **(not IPL; part of the cause)** 9 other spheres with possibly enough level to switch to Feeding from two spheres at all times, so unlikely for BOTH spheres to have low level at the same time - **IPL** Two level indication from SIS level transmitter, with low level alarm, with more than 60 min available to switch tanks (SIF driven alarm and response) - **possible IPL, if action of the operator is quick enough**	Rec 4. Make sure the Human IPL of response to low level in all spheres and tanks is described in a trouble-shooting guide (like an SOP) and practiced once per year per unit operator. This will make this response a valid IPL.
			Low/no flow - Unqualified liquid from spheres back to Plant (see 6.2)		
2.3	High temp.	Large area of damaged insulation Loss of cooling. when the tank is isolated from column	High pressure - vapor from spheres through condenser and return to liquid pump out line (only used when plant is shutdown) (see 3.7)		

2. Ensure Human Response Action Meets the Definition of an IPL

The prior step should have already performed this check when determining if a Human Response IPL is the best choice of protection layer for the scenario. Regardless, before going to the further effort on ensuring proper human response, the organization should ensure all of the criteria for a human response IPL has been met. Per the *Guidelines for Initiating Events and Independent Protection Layers, 2015*[2] this includes:

A. The Human IPL (and any associated alarm) must be truly independent of the other protection layers. That is, there must be no failure that can deactivate two or more protection layers.

The IPL (also applies to IE) includes the ENTIRE sub-system, including any root valves, impulse lines and bypasses. The other IPLs cannot share any of these or other components (except for the mother board of the BPCS loops).

A device, system, or action is **not** independent of the initiating event and cannot be credited as an IPL for either approach if either of the following are true:

- ➢ Operator error is the initiating event and the candidate IPL assumes that the same operator must act to mitigate the situation. Human error is equivalent to the failure of a system and once a human has committed an error it is not reasonable to expect the same operator to act correctly later in the sequence of events. This approach is justified because the error may be due to fatigue, illness, incapacity (drugs or alcohol), distraction, work overload, inexperience, faulty operating instructions, lack of knowledge, etc., that are still present later when the action is required.

Examples where the Human IPL is not independent include

- ➢ Assuming that the same operator acts correctly after operator error initiated the event.
- ➢ Alarms that are annunciated on the BPCS are not independent of the BPCS; if the BPCS is counted as an IPL, then such alarms cannot be counted as an IPL (again, see the exception discussed later).

B. The Human IPL is specifically designed (capable) to prevent or mitigate the consequences of a potentially hazardous event.

- Is the Human IPL valid for the mode of operation for the scenario (startup, shutdown, normal, batch, etc.)?
- Is the Human strong enough to perform the required action, such as closing a manual isolation valve
- Is the Human fast enough (discussed in a little more detail later in this section)?
- What is the maintenance/reliability practices and plant/company history for any related equipment that the Human must use the complete the desired action? How much likelihood reduction credit will you take for the alarm working?
- How good are the procedures and related training (and drills)? Were the operators trained in specifics of how to respond to this alarm/indication? Are they test often enough?

C. **The Human IPL must be *Maintained, Tested,* and *Validated* periodically; it must be proven that the Human IPL can be relied upon to do what it was intended to do.**

The IPL must be periodically maintained and it must be proven or validated. The site must have data that supports the reliability factor. The frequency and test method must comply with best industry practices for such IPLs. Also, the site must maintain a database for each IPL that statistically supports the PFD stated. For a component or instrumentation IPL, this requires maintaining a statistical failure rate database that justifies the PFD listed for each IPL. For a human IPL, the site must maintain data from "drills" of the action of the worker that statistically demonstrates that the worker(s) can indeed implement the required action (of the IPL) with the time specified in the IPL, or else they must use another means of validation as discussed in Section 5 of this paper.

D. **The Human IPL maintenance and validation must be *Audited.*** Auditing is required to ensure the validation, procedures, training, and resulting data are adequate. This is an administrative check. This auditing cycle is set frequent enough (typically 1 year for the first audit and then 5 year frequency after that) to ensure that validation is being carried out as planned and is sufficient to justify the IPL and its PFD.

Specific Criteria on Speed of Response versus Process safety Time.

The criteria for setting the alarm level (that sets the time available for response) should be true before going to the effort of developing a procedure for response (before developing a trouble-shooting guide explained in Section 3):

- The response is typically still possible, but it is time dependent. The time available is called the process safety time (PST). The operator must complete the diagnosis, make the necessary change(s), and make sure they are out of harm's way by the end of the Maximum Allowable Response Time (MART). [5]
- We usually set an alarm or a pre-alarm to trigger this action. This is usually before the shutdown triggers (ESD occur automatically) or release points (PSV set points) are reached
- The Min and Max shown in a Trouble-shooting Guide are not the absolute safety limits for a system, but are instead some values that leave us some time to take action to prevent from reaching the absolute limits.
- There is still time to prevent or avoid the final consequence that could occur if we reach the ultimate limits of the process. Usually, we want the MART to be ½ or less of the PST, and we want MART > 10 minutes for trouble shooting in the field/plant and MART > 5 min for trouble-shooting only from the control room.[5]

See the PII database of IPL Datasheets for detailed criteria on qualifying a human IPL[4]. Similar datasheets are also available in *Guidelines for Initiating Events and Independent Protection Layers,* CCPS (2015)[2]

3. Develop a Trouble-Shooting Guide (TSG) for Each Response

TSGs are a special form of operating procedure. They are written for the actions we want the operators to take to recover from a process deviation, *before an emergency situation occurs.* They are called guides since rarely can we predict the process conditions at the time the action is required. Trouble-shooting guides (and necessary training and drills) are required for any action that is considered a Human IPL. The Action Limit is what we show as the Min or Max in a Trouble-Shooting Guide. The action limit triggers the demand to use the trouble-shooting guide.

If the unit has a good PHA/HAZOP, then it is best to extract information from the HAZOP (or What-If) analysis tables to start the development of each guide. (Table 2 shows examples of the conversion of HAZOP entries into TSG entries.) The guide is then finished with input from the process experts.[6]

Table 2 Examples of Creating a TSG from a HAZOP Table

HAZOP Table Entry	Trouble-Shooting Guide Entry
Cause: *Bypass valve is open or passing*	Make sure the bypass is tightly closed
Safeguard: *Isolation valves for the vessel*	Isolate the vessel, if necessary
Safeguard: *Relief valve*	Make sure the relief valve block valves/relief path are open

The key categories of information needed in a trouble-shooting guide are:[6]

- IMMEDIATE ACTION (by system or by operator)
- DECIDE IF ALARM is REAL
- FINDING and FIXING the CAUSE
- FIX or BYPASS PROBLEM

Figure 1 provides an example of a trouble shooting guide for one critical alarm/action. This guide is in portrait format and follows best human factors practices for formatting.

Optimal Presentation of Trouble-Shooting Information

Ideally trouble-shooting information should be imbedded in the basic process control system (DCS) so that the operators can access the information on demand, with the click of mouse or key. Using the DCS for display of the steps for response to alarms (trouble-shooting), to be displayed "on demand" is becoming more of the norm each year. So, for more than 15 years, many companies have been taking the TSG information, such as that shown in Figure 1, and imbedding it in the DCS; the operators can then access this reminder of the proper response with a click of the human system interface (such as a click of the mouse).

Figure 1: Example of Trouble-Shooting Guide that Follows Best Practices

Trouble-Shooting Guide

Alarm or Indicator:	**PAL 4446 – Low Pressure Alarm for Suction of Organic Feed Pump 40-PM-18.445**

Action Limit:	5 kPa		
Consequence:	Possible pump seal failure, releasing or spraying organic waste into the berm.		
Process Area:	FB&D Incinerator; Liquid Organic Liquid Feed	**Oper. Mode:**	Normal
Drawing #s:	D-400-PI-013		

IMMEDIATE ACTION (by system or by operator)

- DCS should shut down the organic feed pump (40-PM-18.445).
- From the DCS display, MAKE SURE the organic feed pump is shutdown.
- HAVE the field operator check for leaks near the organic feed pump.
- IF there is a large leak/release, THEN use the ESD switch to shutdown the unit and then follow/complete the shutdown and isolation procedure, OPS-ESD-117.
- IF there is a minor leak or no leak, THEN:
  - COMPLETE the rest of the trouble-shooting,
  - and DECIDE how to contain the leak for now,

DECIDE IF ALARM is REAL

- From the DCS, CHECK the pressure and feed tank level trends. IF the trends indicate the alarm if valid, THEN continue with finding the cause or fixing or bypassing the problem.

FINDING and FIXING the CAUSE

- CHECK valves upstream of the organic feed pump to see if any are closed too far, including checking ESD valves.
- CHECK, by feel with hand, if the heat tracing is on; IF Not, then TURN ON or open heat trace valves
- MAKE SURE nitrogen to the pump seal is at the normal operating pressure.
- CHECK if the line if plugged or frozen (skill)

ABC Chemical Company **Prepared by:** ***Printed copy of this procedure is good for one job task duration.***	**OPS-76-TSG-233**	**Incinerator Unit** **Revised 8/24/2015** **Printed 2/21/16** **Page 1 of 2**

Figure 1: Example of Trouble-Shooting Guide that Follows Best Practices (continued)

Trouble-Shooting Guide

Alarm or Indicator:	**PAL 4446 – Low Pressure Alarm for Suction of Organic Feed Pump 40-PM-18.445**

- CHECK if the line if plugged or frozen (skill)
- CHECK if the level is actually low, use the Organic Feed Tank.
- IF the cause is low level in the feed tank, THEN resolve the problem if necessary based on cause that is found (skill)

FIX or BYPASS PROBLEM

- IF necessary, SHUT DOWN the Unit to allow fixing of the problem.
- FOLLOW the proper procedure to resolve problems (repair procedure, line clearing procedure, etc.)
- IF the decision is made to continuing operation without all equipment in normal condition, THEN:
    - FOLLOW Temporary Operating mode, if there is a temporary procedure already written for this possible problem/condition
    - FOLLOW MOC procedures to obtain approval for any non-standard temporary operating procedure or mode

END

4. Perform Initial Training on Each Human Response IPL

Once the TSG is developed then initial training of operators can start. Initial training is straight forward and includes a progression of activities such as:

- Reading the TSG
- Classroom or simulator training on the response action
- Discussions to increase understanding on the importance of the action
- Discussions to ensure operators know the options and judgments needed with the TSG approach
- Tools needed (if any, such as a valve wrench to get more leverage on a valve)
- Optional Written Exam on the task

As with all other activities in life, the real "training" occurs in practice in the control room or unit. With a Human Response IPL, the hands-on practice is doubly important, since if the triggering alarm sounds, the operator will many times Not have time to refer to a procedure; the response must be practiced enough to make it second nature to the operator. The amount of practice should be enough to ensure the probability of failure to accomplish the task in time is less than 0.1.

Repeat practice necessary to maintain adequate response capabilities (see Section 5 for details on Human IPL Validation)

As discussed in Section 5.3, this may require one drill per operator per alarm per year. But after the first year or two, the organization may decide to use a different method of Human IPL validation, such as Expert Judgment.

5. Validate that Human Response Success Rate is High Enough

The values for failure rates and probabilities of failure on demand (PFDs) used in LOPA should be conducted to ensure that the selected values are appropriate. Validation of the values used in risk analysis can follow any of the four methods used for initially establishing failure rates. The validation method used for any particular value can differ from the original method used to determine a failure rate or probability of failure on demand.

As mentioned earlier, much of Section 5 was originally included in the *Guidelines for Initiating Events and Independent Protection Layers and Initiating Events, 2015*[2], but was subsequently cut before publication of the final book. This paper provides that useful information once again.

Validation approaches

The four methods of *validation* are presented in increasing order of robustness. As values to be used in a risk analysis increase in the reliability claimed, consideration should be given to using more robust validation methods. In particular, site-specific data can be used to justify extending the reliability claimed for initiating events or independent protection layers beyond the values derived from generic data.

- *Expert Judgment* is often used as a validation method for all of the other methods of generation of values for failure rates and probabilities of failure on demand. The critical evaluation of values derived from other methods by one or more experts is a means of validating the values used in a risk analysis.
- *Generic Data.* Validation by the use of generic data is through monitoring of any updates to the data used in an analysis along with other sources of generic data to ensure that improved values developed by industry sources do not invalidate the values originally selected. Validation by the use of generic data can be used for original values derived from expert judgment, generic sources, and predicted values. Normally, values for failure rates derived from analysis of site data are more accurate than any generic source, and generic source data would not be used to change the value derived from analysis of site data.
- *Predicted Reliability Values* can be used for validation in situations where the original values selected were based on expert judgment or generic data and predicted values become available. As an example, a level transmitter might be assumed to fail once in 10 years based on expert judgment or generic data. If a predicted value of failure is later reported by the manufacturer for the particular level transmitter of one failure in six years based on calculation from component failure rates, the lower reliability may invalidate the LOPA based on the less conservative value.
- *Site-specific data.* The most robust means of validation of the values used in LOPA is through the collection and analysis of failure rate data in the area being analyzed. Site-specific data can be used to validate failure rates developed from expert judgment, generic data, and predicted values. The process of validation may support the values developed from the original method, show that the actual reliability of the systems are not as good as that predicted from the other methods, or reveal that the systems are actually more reliable than anticipated. When the site-specific data reveals that systems are not as reliable as originally developed, steps can be taken to improve the reliability or correct any potential deficiency in risk management by taking additional mitigating actions. In situations where the site-specific data reveals that the reliability of the systems in a particular application are better than originally derived, the better values can be used in the design of future applications.

Choosing the Approach for Validation of Human IPLs

Figure 2 describes the decision making necessary to choose Expert Judgment, Prediction/Estimation, or Site-Specific Data approaches for validation. (Note that CPQRA in Figure 2 refers to *Guidelines for Chemical Process Quantitative Risk Analysis* [CCPS 2000].)[7]

Figure 2 Example Decision Path for Validating Human Response IPL; ***copyright PII, 2017***

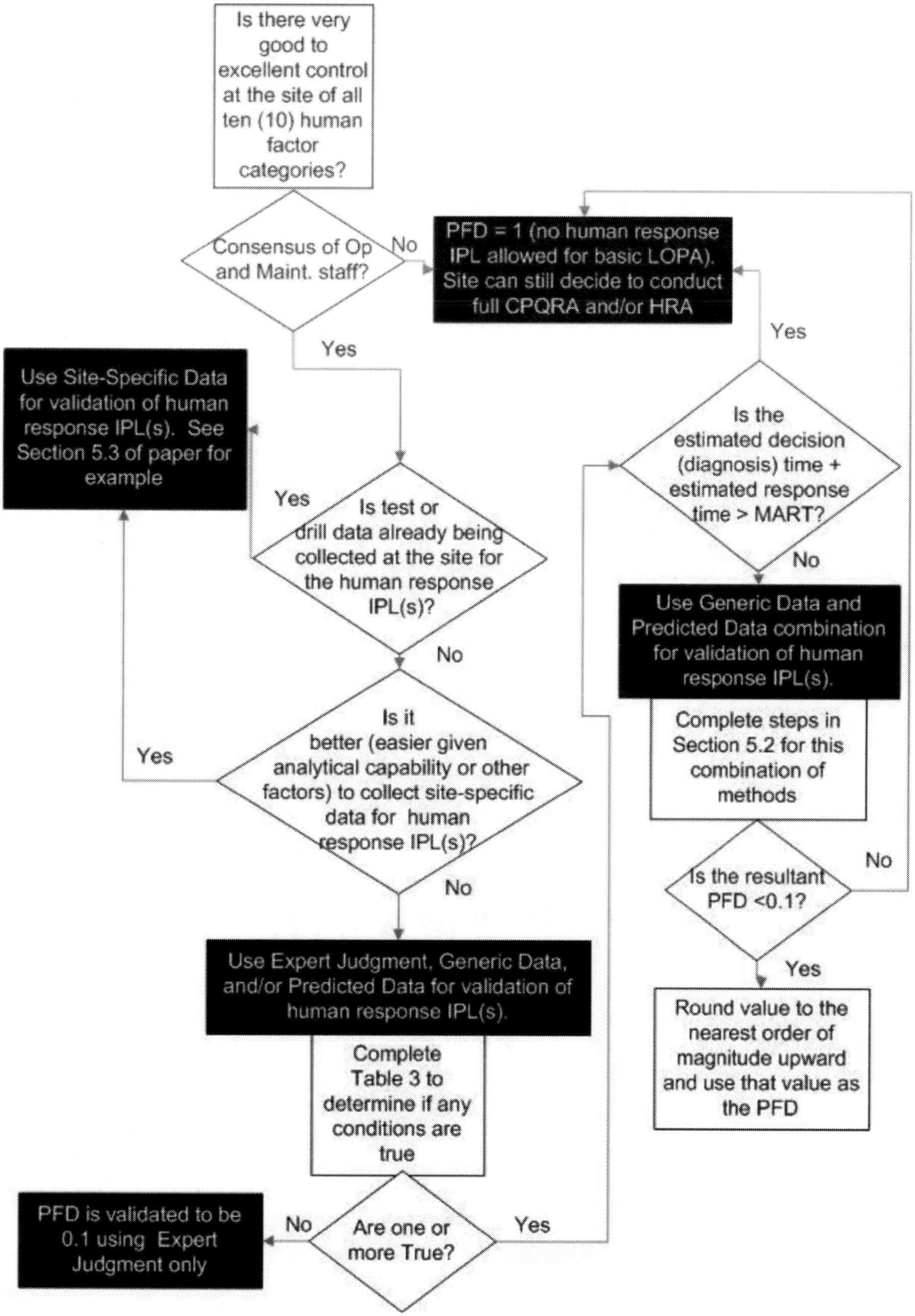

5.1 Determine if Validation by "Expert Judgment Only" Is Allowed

The questions in Table 3 are used in the initial screening process. If ALL are False, *then a PFD of 0.1 is valid without further calculation*. If ANY of the statements in Table 3 are true, then the validation calculation approach (section 5.2) or the Site-Specific Data (Section 5.3) must be applied to verify that the required reliability is achieved.

TABLE 3 Determine *(Using Expert Judgment)* if Validation by Predicted Data Is Required

#	Criteria	True	False
1	Based on consensus of expert opinion, the operator has less than 15 minutes to successfully detect, diagnose, and perform the required action		
2	Operator response is required without explicit criteria and response instructions		
3	Critical or emergency responses involve multiple people		
4	Response actions provide no feedback that they are effective		
		If any are true, go through Predicted Data method for validation of the PFD	If all are false, use PFD = 0.1 (stop: no further calculation needed; Expert Judgment is sufficient for validation)

5.1.1 Determine if the Expert Judgment Estimate of the Baseline Human IPL Time Is Less than MART

The time available is critical to the reliability of any response activity. A shortage of time leads to hurrying and increased stress. In addition, under these conditions, any errors that occur may not be correctable. Ultimately, the action must be accomplished within the MART.

For a human response IPL, MART is the time from when the sensed parameter reaches the set point (and then perhaps a few moments later the alarm sounds, if alarmed) to the point of no return where the corrective action can no longer prevent the incident. This value is determined from process dynamics independent of any hoped-for human response. It includes any time delay in alarm activation and any time for automated actions (initiated by the operator) to occur.

This example of validation by predicted data requires three different time elements versus MART (maximum time available to stop the event):

A. Detection time. Time from when the parameter of interest exceeds the "safety" limit until the deviation is noticed by the human.

- Detection and annunciation could be via a sensor and alarm, followed by sensory perception of the annunciation by the operator.
- Detection could be by the operator taking a sample to the lab and then subsequent analysis and reporting of results by the lab technician. The detection time in this case includes time between sampling (at least one cycle) plus the time to take the sample plus the time to wait for analysis and perform analysis, plus the time to report the results to the appropriate operating staff.
- Detection could be the operator noticing a problem during routine operator rounds, in which case the time since the previous rounds is the major portion of the time consideration. So, for rounds every four hours, the detection time is greater than or equal to four hours; but note that it is best to rotate operators every round to enhance vigilance.

Use Expert Judgment for this estimate.

B. <u>*Decision time*</u> (time to decide what action to take; also called *diagnosis* time in HRA). The decision time was identified as a source of variability when people assessed the reliability of these activities. Consequently, the decision time is fixed within this validation method based upon the activity type. For purposes of alarms that a site would allow for LOPA, the decision time is normally less than one minute. But some HRA data developed for diagnosis time in control rooms (Swain, 1983) suggests that there is 90% chance the diagnosis will be correct if the worker in a nuclear power plant control room has at least 10 minutes to diagnose, and a 99% chance of correct diagnosis if they have 40 minutes. Because of these traditional values, the decision time is typically set at 10 minutes. However, for actions that require no or very little diagnosis or in simple process units, this value can reasonably be set to five minutes. Use Table 4.

C. <u>*Response time*</u> (time to complete all the alarm response activities). This is the time required to complete the tasks that will prevent the undesired event, as detailed in the alarm response procedure (e.g., after the procedure has been chosen as the correct course of action). Use Expert Judgment for this estimate. (For comparison, this is the time that is measured directly by testing/drills in validation using Site-specific data; see Section 5.3 for details.)

Estimate the task response time: Using solicitation of expert opinion (including at least two senior operators from the unit and one senior process engineer or equivalent), develop an Expert Judgment estimate of the time to complete the response activities, given that the diagnosis is performed correctly.

IF: Detection Time (including any delays in a related instrument system) + *Decision time* + *Task response time* > *MART*

THEN: The human response IPL is not valid.

TABLE 4 The Decision-Time Factor Assigned to the Different Activity Types

Activity Type	Decision Time (minutes)
Unambiguous cue in a continuously staffed control room or similar staffing near an alarm annunciation location, with simple process and little or no diagnosis (with a decision tool, such as a troubleshooting guide).	5
Unambiguous cue in a continuously staffed control room or similar staffing near an alarm annunciation location, with complex process unit that requires diagnosis to deduce the failure cause and the proper action to take (with a decision tool, such as a troubleshooting guide).	10
Requires diagnosis of a novel fault situation (cannot be used for IPL in LOPA).	Beyond LOPA

If the total human IPL time is too great, then the site may:

- Decide to use other methods to validate the human response IPL
- Decide to redesign the human response IPL so that it can be done in less than the MART.
- Decide to redesign the system to eliminate or reduce the risk.
- Decide to install or upgrade other types of IPLs (such as IPS, which are faster to respond) in lieu of the human response IPL not being available (because it is currently invalid).

IF: Detection Time + Decision time + Task response time < MART

THEN: Proceed to the steps in the next section

5.2 Approach for Validation of PFD of HUMAN IPL Using Combination of GENERIC Data and PREDICTED Data

The approach shown below is based largely on the methods described in SPAR-H (NUREG /CR 6883)[8], which is mainly a simplification of the much more complicated and detailed HRA methodology developed for the analysis of critical tasks (Swain, 1983)[9]. The approach starts with a baseline human error rate (0.0008 per year) is the lowest human error rate and then corrects that rank based on the multipliers related to good or bad human factors. See Table 5 for PII's method of how to estimate the PFD of a human response based on adjustment for human factors at the site and adjustment for the number of practices per operator per year (total practice is the combination of drills and actual responses to that same alarm by the same operator).

TABLE 5 Estimation of the PFD for Human Response IPL based on Basic Human Factors

Human Factor Category	Human Factor Issue/Level	Multiplier for Cognitive & Diagnosis Errors	USED
		Absolute Baseline at practice multiple times per day	0.0008
Available Time (includes staffing issues) – ***for responses only***	Inadequate time	P(failure)=100%	1
	Barely adequate time (≈2/3 x nominal)	10	
	Nominal time (1x what is expected)	1	
	Extra time (at least 2x nominal and >20 min)	0.5	
	Expansive time (> 4 x nominal and > 20 min)	0.1	
Stress/Stressors (includes staffing issues)	Extreme stress (threat stress; unloading ship with crane non-stop for more than 2 hours, imminent hazard nearby)	5	1
	High stress (time pressures such as during a maintenance outage; issues at home, etc.)	2	
	Nominal	1	
Complexity & Task Design	Highly complex task. Or very low complexity/boring task that requires 100% attention for more than 45 min.	5	1
	Moderately complex (requires more than one staff)	2	
	Nominal	1	
	Obvious diagnosis	0.2	
Experience/Training* (see the practice rate adjustment in at end of table)	Low experience relative to complexity of task; or poor/no training	10	1
	Nominal	1	
	High	0.5	
Procedures	Not available in the field as a reference, but should be	20	1
	Incomplete; missing this task or these steps; or untrustworthy (< 85% accuracy)	8	
	Available and >90% accurate, but does not follow format rules ***(normal value for process industry)***	3	
	Good, 95% accurate, follows >90% of format rules	1	
	Diagnostic/symptom oriented	1	
Human-Machine Interface (includes tools)	Missing/Misleading (violates populational stereotype; including round valve handle is facing away from worker)	20	1
	Poor or hard to find the right device; in the head calc	10	
	Some unclear labels or displays	2	
	Good	1	
Fitness for Duty	Unfit (extreme fatigue level at >80 hrs/wk or >17 hr/day, no day off in 6-day period; or illness, legally intoxicated, etc.)	20	
	Highly degraded fitness (high fatigue such as >15 hr/day or >72 hr/wk, no day off after 4 shifts of 12 hours or more, illness, injury, legally barely intoxicated, etc.)	10	

	Moderately Degraded Fitness (≥12 hr day or ≥ 60 hours/wk; 1 day off [break] or more per week)	5	2
	Slight fatigue (more than 8 hr per day, but not more than 12 hr day; up to 48 hrs per work week, 1 day off [break] or more, after 48 hours of work; *normal value for process industry)*	2	
	Nominal	1	
Work Processes & Supervision	Poor	2	1
	Nominal	1	
	Good	0.8	
Work Environment	Extreme (in temp, humidity, noise, lighting, vibration, etc.)	5	
	Good	1	1
Communication	Communication system/interference damaged; poor communication environment	10	3
	No standard for verbal communication rules *(normal value for process industry); use this value if coordinated task for response*	3	
	Well implemented and practiced standard	1	
		Product	6.0
*** adjustment for practice frequency**	Number of times task performed and/or practiced per year	2	16.5
		Revised Product	98.7
		Product	0.0790
		PFD to use for this Human IPL (must be equal to or less than 0.1 to allow this Human IPL)	**0.10**

Simplifying Assumptions

Not all inputs for a full HRA evaluation are likely to be readily available to the LOPA team or analyst. Some HRA inputs are often team or analyst judgments that would lead to continued variability in the results. For HRA inputs that could be reasonably held constant, human factors are set at the expected norm or standard. If the team or analyst feels that the human factor are not "up to standard" for the IPL (or related process unit) being validated, a recommendation for improvement to the particular constant human factor would be made with the validation being contingent upon the plant or site completing the recommendation.

It is ideal if all of the human factors (except for practice) can be set to 1 (no negative effect), but the analyst should use their judgment for the task, mode of operation, etc., for the value of each human factor. ***It is very rare for a site to have all human factors at a value of 1.***

As mentioned earlier, the approach shown below is based largely on the methods described in SPAR-H (NUREG/CR 6883)[8]. A somewhat similar calculation approach to human IPL validation has been used by Dow (Stack, 2010),[10] but it does not allow correction of the estimate for poor human factors.

For a given human response IPL (triggered by an alarm or some other call for action), this validation approach consists of the following steps:

1. Estimate the time required for successful response (which must be less than MART, as defined earlier). This is estimated from the following aspects of the response:
    a. Estimate the time for instrumentation or/or operator detection.
    b. Estimate the time for operator decision making (this is normally small compared to operator action).
    c. Estimate the time for operator action and verification
    d. Fill in the appropriate factor in the *USED* column in Table 5 for the "Available Time" row at the table
2. Rate or scale the other Human Factors based on the descriptions in the columns. Fill in the appropriate factor in the *USED* column for each Human Factors
3. Fill in the number of practices per year, which includes number of actual alarms responded to per year and number of times this alarm or a very similar alarm is drilled/practiced. The formula used to get to the right Practice Factor, starting with a value of 0.0008 as the lowest baseline error rate possible, is:

    = (250/(number of practices per year))^0.6+0.4*((number of practices per year)/1000)^2
4. Calculate the overall human unreliability which gives the PFD (normally 0.1) for the human IPL. If the number is > 0.1 then no PFD is allowed.

This PII method takes about 15 minutes for validation of a PFD for one human IPL. This assumes the analysis is performed by a Subject Matter Expert (SME) who is trained in this method, is a human factors expert, and has access to the site data needed, which may in turn require solicitation of expert judgment (such as for the estimate of "time to respond"). Note that the time invested for validation by this method is comparable to validating one human IPL using Site-specific data, presented in Section 5.3, indicated that simply validating the actual responses to alarms (in drills) is likely the better approach.

NOTE: Approach and calculation method above is copyrighted to PII, 2017.

5.3 Validation of Human IPLs by Site-Specific Data

A 0.1 PFD value for a human response indicates that the correct response occurs at least 9 out of 10 times (or no more than 1 wrong response for every 10 attempts). Most organizations will have identified many human responses involving a number of personnel, as part of their LOPA studies. Some organizations believe that if they have a procedure and a training program in place, that they can claim the PFD value of 0.1 for a Human IPL. This is no truer for a human IPL than it is for an active component-based IPL.

As required for all IPLs, a human IPL must be validated. The Preferred approach to validation is direct measurement or testing of the human response (under controlled conditions or drills); but other methods of validation can include Expert Judgment, using data from other comparable settings (Generic Data method), and estimation of the PFD of human IPLs by mathematical modeling (Predicted Data method, see an example of this method later in this paper).

On the next few pages are the options for validating human IPLs with **site-specific data**. These include:

- 100% testing for each human responder and for each action to be taken
- Sample plan testing of random combinations of human responders and actions.

One key focus of this paper is discussion of practical means for collecting raw data in a plant setting for substantiating the error rates for the site, and especially for crediting a human IPL. The method for data collection covers the training requirements that should be met, proof drills for response to alarms, simulations and tests, and frequency of proofs, and of course the effect of human factors on human error rates. *Actual plant data and tests are included in this paper to provide the reader with some examples of how a simple data collection and validation method can be set up within their companies.*

This appendix provides an example of the data needed for adequately counting the human in a LOPA (and other risk assessments) using Site-specific data for validation. One key focus of the appendix is discussion of practical means for collecting raw data in a plant setting for substantiating the error rates for the site, and especially for crediting a human IPL. The method for data collection covers the training requirements that should be met, proof drills for response to alarms, simulations and tests, and frequency of proofs, and, of course, the effect of human factors on human error rates. *Actual plant data and tests are included in this appendix to provide the reader with some examples of how a simple data collection and validation method can be set up within their companies.*

If a site has a very good system for reporting and investigating near-misses, then this system can be used to find site-specific data for human errors (including both IEs and failure of human IPLs). Getting high near-miss reporting rates is covered in other research and papers (Bridges 2008, 2012, etc.)[11]; note the ratio of near-misses to loss events likely needs to be higher than 15 to provide sufficient data for validation using near-miss data alone.

Another way to collect site-specific data on error rates is to measure error rates with tests or drills of the action; and for human IPLs discussed later, the results may need to be adjusted to account for actual stress levels of a response. This practice is not commonplace in the chemical industry, but the US Nuclear Regulatory Agency (NRC) in 10 CFR 55.45 and 55.59[12] requires that all power plant operators be tested once per year on abnormal procedures. This testing is mostly related to humans involved as IPLs. 10 CFR 55.45 and 55.59 also allude to the possibility of using a test of a representative sample of human responses, but we address this option later in this appendix.

EXAMPLE OF SITE-SPECIFIC DATA FOR HUMAN RESPONSE IPLs

As required for all IPLs, a human IPL must be validated. The preferred approach to validation is direct measurement or testing of the human response (under controlled conditions or drills); but other methods of validation can include expert judgment, using data from other comparable settings (Generic Data method), and estimation of the PFD of human IPLs by mathematical modeling (Predicted Data method), see Sections 5.1 and 5.2 of this paper for a discussion of these alternate approaches.

On the following pages are the options for validating human IPLs by direct measurement including (1) 100% testing for each human responder and for each action to be taken and (2) a sample-plan testing of random combinations of human responders and actions.

Approach to Using a 100% Individual Test Plan for Validation of Human IPLs

One method to prove the operators will reliably respond for each human IPL trigger is to have each operator demonstrate they can individually respond to each alarm (or other trigger). This response can be demonstrated by walk-throughs in the field or using simulators of the process unit. The nuclear power industry uses this approach for validating response by control room operators, but many in the chemical industry perceive this will take "too much time." As an example of the effort required, one nuclear power plant allocates about 200 hours per year per operator for refresher training activities, including 60 hours per year per operator for demonstration of skill in responding to critical alarms (from Tennessee Valley Authority [TVA] internal requirements to meet the performance-based requirements of NRC regulations 10 CFR 55.45 and 55.59). This equals about 3% of the work-year for an operator. (The example below and also the example of the sample-plan approach discussed later shows how this investment of time to measure response effectiveness can be reduced by a factor of 10 or more.)

Consider the following as an example of the application of this method of testing responses for 100% of the triggers by 100% of the operators at a chemical plant:

Background: The operating area being evaluated has 20 operators (spread across 4 rotating shifts of work) and 130 identified (from a hazards analysis and LOPA) human response IPLs.

Validation Test: The "tests" are documented on a set of index cards that call out various alarm conditions; these are the events (triggers) for a scenario identified in the hazards analysis (and LOPA) to be protected by the given human response IPL.

Demonstrated Result: A correct answer (success of the IPL) is the desired response to the alarm scenario. A wrong answer (failure of the IPL) is any response other than the one desired or a response that takes too long (longer than the MART, defined earlier).

Estimation of Resources Requirements for 100% Testing Scheme: Below is an estimate of how much test effort is needed to ensure that training and retraining programs are sufficient to validate a 10^{-1} value for the PFD of all of these identified human response IPLs:

1. *Determine the number of tests to be performed.* This would be the number of human response IPLs multiplied by the number of people who are expected to respond (perform as the human IPL) at some point in the future during their own shift. This example would yield 2600 discrete tests (20 operators X 130 human response IPLs = 2600 tests) for one test of each combination of trigger and human responder (which makes up the 2600 IPLs).
2. *Determine the test frequency.* It is difficult to get consensus on this value. One documented example is from the nuclear industry. The U.S. NRC regulation for control room operators (10 CFR 55.45 and 55.59)[12] requires annual demonstration of proficiency in response to critical alarms and signals. This in fact would likely not be enough testing to give a 90% chance of proper response to every alarm, except for the fact that response to one alarm is normally similar to response to other alarms, so the operators are in essence getting more **practice on similar alarms** as they perform each demonstration. Operators under this regimen of recertification have shown a 90% chance or better of proper response (detection, diagnosis, and action) within 10 minutes of an annunciation of an event of interest (from internal power plant records and also inferred from Swain (1983), since the basic control room regimen of testing each operator with each action each year was essentially the same in the 1970s as it is now). For this example, a frequency of 1 validation per year is chosen.
3. *Determine the time required to perform each test.* Assuming 10 minutes of allowed response time for success (per alarm/trigger), then the test time for the organization would be 26,000 minutes or about 430 staff-hours (22 hours per operator per period; most likely the period would be once per year). With a frequency (test period) of once per year per alarm, this equates to about 1% of the normal staff-hours for a worker in the USA. *(NOTE: An equal amount of time would also be required to record/document the training record for the test [once the tracking system for such demonstrations is set up], but this is likely not a load the operator will have to bear.) (Note that testing only a sample of these responses, discussed later, would reduce the load considerably.)*

ACTUAL EXAMPLE of Using Tests/Drills (Site-Specific Data Collection) to Validate Human Response IPLs

Until now, the actual effort to collect such data has not been well documented in the literature, though many chemical companies, refineries, and nuclear power plants do in fact validate human response using this method. Recent research (Bridges, 2010 and 2011)[13, 14] by three chemical companies documented the effort required to validate human response IPLs using Site-specific data. The following is an excerpt of the research results.

Validation Setup: A simple test was used in measuring the response to an alarm condition. The test was not meant to measure the probability of detection of the alarm, but rather was meant to measure the time and success in determining and accomplishing the proper response to critical alarms as part of human IPLs. Two chemical plants belonging to large organizations (one in Malaysia and one in the USA) performed the test.

The test involved having multiple operators in one unit of one plant/site (one for each company) perform responses to critical process alarms. These alarms were related to human IPLs. The actual response and time of response was measured, but essentially the tests were setup as a "pass or fail" – in other words, the tests were to determine if the operators were able to respond as desired/expected, within the allotted MART.

To run each test, the plants printed a data card (the size of an index card) and handed it to an operator chosen at random. Below is an example of such an index card.

Figure 3 Example of card used to administer validation of a single human IPL

Human IPL Validation Test/Drill			
Response Task:	Max. Allowable Resp. Time (MART)		Response Time:
LAH for Tank 105	*15 minutes*		5:20 minutes
Date of Test:	Time/Shift:		Employee Number:
1/23/10	07:35/A		23122
	Pass/Fail:	Pass	

Note that the card contains an estimate of the MART – as defined earlier in this appendix and elsewhere in this guide; this is the time an operator has to perform the task once the alarm is received until it is too late to take any further action. The time it took to print and hand out the index card was minimal.

Validating/Testing: A human response IPL "failed" if the operator could not perform the required action to prevent the hypothetical outcome within the MART (defined earlier). The person administering the test timed the operator response and recorded the results. (Again note that these tests did not validate the probability of an operator failing to detect an alarm.) Each test took 10-15 minutes to administer and less than one minute to record the data. The validation was performed by multiple operators on multiple shifts. The tests were administered by a shift supervisor, a shift engineer, or in some cases, a process safety coordinator. It is likely another operator could administer most proof tests (validations) and then the site management could audit some percentage of the tests to help ensure against bias. If the operators test each other, then the time to administer a test is likely not significant enough to measure, since they have to be there, regardless of their other duties. The total time for the test varied, but the two sites that performed the test considered the time to administer the test to be minimal; the largest effort was simply for someone other than the operator to be there to "independently" measure the operator's response time (i.e., time to administer the test).

For the most part, the tests came with little warning and occurred on all shifts. Several critical alarms were tested using various operators, all randomly selected. (It is anticipated that, unless a sample plan is used, each operator will perform roughly one response related to each human IPL each year.) The time to respond was recorded on the index card.

Based on such raw data, the site was able to (1) evaluate the degree to which they were controlling human factors for the units, (2) identify which human responses qualify as IPLs, and (3) validate that the response is accurate enough and quick enough to qualify for the PFD used as the credit for the IPL (which for LOPA, the PFD is limited to a value of 0.1).

Table 6 provides a sample of the site-specific data for several similar human IPLs from the three sites (one in Malaysia, one in Canada, and one in the USA). For the Malaysia and Canada sites, the data was from the same operating area consisting of multiple operators across four rotating shifts of eight hours per shift. For the USA site, the shift was 12 hours.

All of the IPLs passed (all operators performed each action correctly within the allotted time) during these tests/drills. The labor to perform the test took less than 15 minutes per test (including documentation time). After the initial resistance at each site to performing the test, the subsequent tests were not resisted and in fact the operations staff embraced the testing/drills since they saw many side benefits from the test/drills, including the re-enforcement of "what to do" with all parties involved (the person doing the test, the person recording the test, and the other operators who noticed the test in progress). Lead operators and supervisors administered the test; very little training or coaching was necessary to have the drills done properly.

All three sites believe it is possible to use a sampling of human error for similarly qualified humans doing similar response or proactive tasks. (This is because the responses for all IPLs were the same when the same operator acts on different alarms or when different operators act on the same alarms.) A sample plan of perhaps only 5% to 10% of the number of human-task pairs may be necessary to have a valid statistic for human error for a "type of action." Sampling is valid for human actions because of how the mind processes information and how humans take the necessary actions for similar situations. Obviously, sampling can greatly reduce the measurement and documentation load for validation of human error rates. If sampling is used, the sites suggested that:

- The site should first screen which responses can be grouped together into general types of response IPLs. Then a lesser percentage will need individual validation drills.

Table 6: Site-Specific Validation of Human Response IPLs

IPL No.	Response Task	Number of Test Performed	Average Response Time (minutes)	Maximum Average Response Time (minutes)	Number Failures	PFD (average)	LOPA PFD
Company A (USA), Site 1							
IPL 1	ABP: High Temp in Generator	6	2.3	10	0	0	0.1
IPL 2	ABP: Loss of Acid Flow to Generator	12	2.2	10	0	0	0.1
Company B (Canada), Site 1							
IPL 1	Low Seal Gas Pressure to Turbo-Exchanger Bearings	5	5.7	15	0	0	0.1
IPL 2	High Lube Oil Temperature – XX Compressor	5	6.3	30	0	0	0.1
IPL 3	Low Level Emergency Alarm -- Steam Drum	5	4.9	15	0	0	0.1
IPL 4	High-High Lube Oil Temperature – XX Compressor	5	6.1	30	0	0	0.1
Company C (Malaysia), Site 1							
IPL 1	High Level on Ammonia Absorber Column	10	5.1	15	0	0	0.1
IPL 2	Low Temperature Alarm on Ammonia Compressor Discharge	10	4.9	15	0	0	0.1
IPL 3	Low Level in CO_2 Compressor Interstage KO drum	10	7.0	15	0	0	0.1
IPL 4	High Level on High Pressure Carbamate Heat Exchanger	10	6.1	15	0	0	0.1
IPL 5	High Pressure High Pressure Carbamate Condenser	10	5.0	15	0	0	0.1
IPL 6	High Pressure on Steam Controller to Rectifying Column Recirculation Heater	10	5.4	15	0	0	0.1

[a]Data provided by 3 companies (in the USA, Canada, and Malaysia)

- Perhaps do just one or two drills per shift per group per year for the simpler ones on some periodic basis; that gives a chance to test feasibility (make sure valves are not locked, make sure valve wrench or other tools are available, etc.).

Human performance sampling is discussed in more detail later in this section.

ADJUSTMENT for STRESS: As mentioned, these data (as with all drills) are collected during a simulation of a call for action. In a real event, *the stress to perform the task correctly would increase the average error rate.* NRC estimates (Gertman, 2005)[8] the stress for this type of pre-emergency response action (versus emergency response and evacuation) will likely not be "extreme" but it will be "high," in which case a conservative estimate is that error rates would double from the test/drill case. It is likely not possible to get a drill that accurately mimics the stress of a real alarm event, so there will likely always be a need to adjust data for increases in errors due to stress. It is likely more appropriate to **double the observed error rates** (observed PFDs) rather than doubling the observed response time. But in either case, the IPL data collected above must still "pass" when adjusted for stress for an IPL to be validated using Site-specific data.

Approach to Using a Statistical Sample Plan for Validation of Human IPLs

It is important to ensure that a 10^{-1} value is indeed valid before using a human IPL. Rather than testing 100% of the responses by 100% of the operators, it is normally valid to use a sampling plan. This is especially true for groups of responses that are similar in action and response time. U.S. NRC alluded to a "sampling" of human response in 10 CFR 55.59, indicating that this may be acceptable for validating human response to triggers (i.e., for validating human IPLs).

There is resistance by some to the idea of only testing a sample of the human response pairs, with the argument that not all humans are the "same." Of course all humans are not "exactly the same," but every site eventually agrees that an individual is competent enough to be allowed to perform a task on their own (without direct supervision). This is the level of sameness that is sufficient for sampling; in other words, all operators in a unit who are competent enough to be allowed to operate independently can likely perform troubleshooting and others aspects of a human IPL within 90% of the skill of other operators.

Another argument is that valuable drills could be missed by many individuals if the sample plan is too small. This argument is focused on the training benefit derived while doing the validation drills for a human IPL. But the learning achieved by doing many drills is much greater than the training of the individuals/alarms included in the sample size alone. Unlike machines, humans learn by (1) doing similar activities, (2) giving the tests to others and watching/scoring them, and (3) by watching a test if they are neither the official observer nor the one being tested.

Statistical techniques developed decades ago are used to establish a basis of acceptance of all kinds of products, raw materials, components, etc. (See Walpole, 2006[15], for typical approach on statistics and sampling.) These methods can also be used to validate human IPLs.

The sample plan approach must group similar type of actions and similar response time requirements. For a sample plan approach, choice of responder and trigger must be chosen at random. The lot size is the product of the number of responders multiplied by the number of similar response actions in the group.

As a rough rule of thumb, the sample should be about 10% to 5% of your total population of data, but not smaller than 30 and not greater than 350 to 500. The sample size and number of failures before the PFD is invalid is related to the confidence level and margin of error that is acceptable to the organization. A confidence level of 95% and an error margin of 5% indicate that the result of your testing (validation of a human IPL) will be within +/-5% of the true PFD 95% of the time the validation testing is performed.

The correct sample size is a function of those three elements – your universe (how many people multiplied by the number of actions; each pairing makes up a human IPL), the desired error margin, and the preferred confidence level. For IPL validation purposes, it is likely reasonable to use a 5% error margin at 95% confidence. Below are typical sample sizes (the first at a 10% error margin, the second at 5%):

50 in the population, sample 33 or 44
100 in the population, sample 49 or 80
200 in the population, sample 65 or 132
500 in the population, sample 81 or 217
1000 in the population, sample 88 or 278

The trend above approaches a limit that hardly moves above 350 in the sample size no matter how large the population, for a 10% error margin (and approaches a limit of 500 for a 5% error margin). The sample size can also be approximated using the equation below:

Sample Size (SS):

Equation 1: SS (for infinite population) =

$$\frac{Z^2 * (p) * (1-p)}{c^2}$$

Where:

- Z = Z value (e.g. 1.96 for 97.5% confidence level for single-sided, normal distribution; note that though human action is pass/fail and so is typically described using binomial distributions, for large populations/groups, a normal distribution approximates a binomial distribution)
- p = percentage picking a choice, expressed as decimal (0.9 used for human IPL sample size determination)
- c = confidence interval, expressed as decimal (e.g., ±5%)

This calculation then must be corrected to account for a finite population (validation SS):

Equation 2: Validation SS (finite population) =

$$\frac{SS}{1 + \frac{SS - 1}{\text{population}}}$$

Meeting acceptance criteria for a human response means that the procedures, training, retraining, communication control, and all other human factors are achieving the desired result of no more than 1 wrong response in 10 demands. Rejection means that the PFD of 10^{-1} is not valid (or that the control of human error needs improvement for the PFD to remain valid). When applying this sampling plan to a validation of administrative IPLs, the "acceptance" criteria (the desired response) is the response that prevents the consequence that is being considered in the LOPA, e.g., *the operator response to the alarm prevents the overflow*.

Sampling plan schemes rely on developing a valid definition of a "group." For human IPLs, the group is a combination of similar operators (likely all "qualified, independent" operators can be treated equally for this purpose, since the company has likewise made the judgment that the operators are "qualified") and similar responses to similar triggers. Creating this grouping takes careful consideration by a multi-disciplinary team (with heavy emphasis on the operators in the team composition), to ensure the grouping of IPLs makes sense (i.e., use expert judgment). Although it is likely that all operators can be lumped into the same statistical group (if they are selected at random for validation drills of IPLs), the triggers and responses will need to be grouped to ensure that validation of one trigger/response is essentially the same as validating other triggers/responses within the same group.

Next, the sample size and pass/fail targets must be estimated. This is based on the size of the groupings, the confidence level desired in the result, and the expected error margin (distribution) of the results. One method for determining sample size is to use Equation 1 and 2 discussed earlier. After the sample size is determined, then the pass/fail target (to prove if the hypothesis of a PFD of 0.1 is valid or invalid) can be estimated (refer to statistical text, such as Walpole, 2006, for this derivation). Another recognized standard for determining the sample size and the pass/fail targets is to use ANSI Z1.4[16]; this standard is now incorporated by reference with U.S. MIL standards for lot sampling and pass/fail determination. Examples calculations and the resulting savings achievable by use of sample plans are also provided in ANSI Z1.4.

> *Example Calculations:* Using the same theoretical example as before with 20 operators and 130 alarms that require response, the total population of combinations of responses and operators is 2600.
>
> *Case A:* For one case, assume that all actions and all operators are equivalent, so the population makes up one group. In this case, at a confidence level of 97.5% and a confidence interval of 5%, the sample size adjusted for the finite population would be 131. So, only 131 combinations of operators and alarms/actions would have to be validated (tested) each period (and a typical period is each year). This is about 5% of the total population of human IPLs and so the validation would likely only take .05% of a staff year per operator, or 1 hour per year per operator (a very small investment in time for validation of IPLs). For this sample of 131 validation test, if 8 or more fail to accomplish the action in the required time, then ALL human IPLs have failed their validations. In addition, this is without accounting for the stress of an actual response. If the error rate is doubled to account for stress, then the number of acceptable failures is cut in half, so for the sample size of 131, if 4 or more fail, then ALL human IPLs have failed their validations. If the validation failed, then the site would likely enlarge the sample size and re-test and/or find out where the problems are occurring and work to improve the response success. Regardless, the workload would be less with sampling and the company would obtain valuable insights into where to focus improvement efforts for human IPLs.

Case B: For this case, assume that not all alarms/actions are equivalent, but still assume as in Case A that operators are about 95% equivalent with respect to alertness and to experience with such response actions. Further assume that the alarms and actions can be separated into 10 groups of 13 similar alarms/actions each. In this case, the population of any one group is 20 operators multiplied by 13 alarms or actions for a total of 260. In this case, at a confidence level of 97.5% and a confidence interval of 5%, the sample size adjusted for the finite population would be 91. So, 91 combinations of operators and alarms/actions from each of the 10 groups would have to be validated (tested) each period (and a typical period is each year). This is about 35% of the total population of human IPLs, so the validation would likely only take .35% of a staff year per operator, or six hour per year per operator (a small investment in time for validation of IPLs). For this sample of 91 validation tests, if six or more fail to accomplish the action in the required time, then all human IPLs in this grouping of 13 alarms or actions have failed their validations. In addition, this is without accounting for the stress of an actual response. If the error rate is doubled to account for stress, then the number of acceptable failures is cut in half, so for the sample size of 91, if 3 or more fail, then ALL human IPLs in this grouping have failed their validations. If the validation of a group of alarms failed, then the site would likely enlarge the sample size for that group and re-test and/or find out where the problems are occurring and find ways to improve the response success. Regardless, the workload would be less with sampling and the company would obtain valuable insights into where to focus improvement efforts for human IPLs.

Table 7 provides various sample sizes versus confidence values and relates these to the maximum number of failures (pass/fail target) using standard statistical methods (such as Equation 1, 2, and those from standard statistical handbooks). This table includes the data for Case A and B above.

Similar values can be obtained using the look-up tables and graphs in ANSI/ASQC Z1.4[16]. This approach is based on a binary distribution, and the operating curves and acceptable quality levels have slightly different meanings than confidence levels and confidence intervals for normal distributions, such as used for Case A and B above.

Also, using MIL-STD-105E (1989) adaptation of ANSI Z1.4, a repeat of Case A requires similar steps and produces identical results (not shown here).

Table 7: Sample Size and Acceptance Criteria Estimation for Human IPL Validation (single-sided Z test; for normal distribution)

Employee	Alarms	pop = actual size of population	SS = sample size based on infinite population	Z value	Confidence Level	p = anticipated PFD	c = confidence interval (related to expected range of results; +/-)	Validation SS = sample size adjusted for actual population size	% of actual population sampled	Expected average number of failures per test period using the sample size	Acceptable failures to have Z confidence (95% if Z = 1.645) that the entire population has less than 1-p error rate (accounting for one side of the confidence interval [+%] of the test plan)	**Therefore, the PFD of the human response IPL is not 0.1 or better if the number of failures are equal to or greater than:**
20	130	2600	59	1.28	90%	0.9	0.05	58	2	5.8	3	4
20	13	260	59	1.28	90%	0.9	0.05	48	19	4.8	2	3
20	130	2600	97	1.645	95%	0.9	0.05	94	4	9.4	5	6
20	13	260	97	1.645	95%	0.9	0.05	71	27	7.1	4	5
20	130	2600	138	1.96	97.5%	0.9	0.05	131	5	13.1	7	8
20	13	260	138	1.96	97.5%	0.9	0.05	91	35	9.1	5	6

6. Conclusion

Operator response to an alarm can be a valid safeguard and even an IPL, of the approach for identifying the necessary action is valid, and if there is a procedure (trouble-shooting guide) and training. But in addition, the proper response action and speed must be proven in the field as well; otherwise the alarm response will not happen as hoped. The steps outlined in this paper will ensure the probability of proper operator response is high.

7. Acronyms Used

AIChE– American Institute of Chemical Engineers
ASM – Abnormal Situation Management consortium
CCPS – Center for Chemical Process Safety (a division of AIChE)
CMA – Chemical Manufacturer 's Association, now American Chemical Council (ACC)
HAZOP – Hazard and Operability Analysis
IPL - Independent Protection Layer
JSA – Job Safety Analysis
LOPA – Layer of Protection Analysis
MART – Maximum Allowable Response Time
MIL – Military
MOC – Management of Change
NRC – Nuclear Regulatory Commission (USA)
NUREG – Nuclear Regulation, US Nuclear Regulatory Commission
OSHA – Occupational Safety and Health Administration, US Department of Labor
PHA – Process Hazard Analysis
PFD – Probability of Failure on Demand
PII – Process Improvement Institute, Inc.
P&ID – Piping & Instrumentation Diagram
PSI – Process Safety Information
PSM – Process Safety Management
PST – Process Safety Time
SOP – Standard Operating Procedure
STD – Standard

8. References

1. "Layer of Protection Analysis (LOPA) Guideline," CCPS/AIChE, 2001.
2. "Guidelines for Initiating Events and Independent Protection Layers and Initiating Events," AIChE/CCPS, 2015.
3. Bridges, W., & Dowell, A., "Identify SIF and Specify Necessary SIL, and other IPLs, as part of PHA/HAZOP," *12th Global Congress on Process Safety*, AIChE, 2015.

4. *Online Database of IPLs and IEs*, Process Improvement Institute, Inc., 2017 [pending].

5. Bridges, W., "LOPA and Human Reliability," *6th Global Congress on Process Safety*, AIChE, 2010.

6. Bridges, W and Tew, R., "Best Practices for Writing Operating Procedures and Trouble-Shooting Guides;" *13th Global Congress on Process Safety*, AIChE, 20106

7. "Guidelines for Chemical Process Quantitative Risk Analysis," AIChE/CCPS, 2000.

8. Gertman, D.; Blackman, H.; Marble, J.; Byers, J. and Smith, C., "The SPAR-H Human Reliability Analysis Method," NUREG/CR-6883, U.S. Nuclear Regulatory Commission, Office of Nuclear Regulatory Research, Washington, DC, August 2005.

9. Swain, A. D., Guttmann, H. E., "Handbook of Human Reliability Analysis with Emphasis on Nuclear Power Plant Applications, Final Report," NUREG/CR-1278, 1983, US Nuclear Regulatory Commission.

10. Stack, R. and Delanoy, P., "Evaluating Human Response to An Alarm for LOPA or Safety Studies," *6th Global Congress on Process Safety*, AIChE, 2010.

11. Bridges, W., "Gains in Getting Near Misses Reported," *8th Conference, ASSE-MEC*, Bahrain, 2008; and *8th Global Congress on Process Safety*, AIChE, 2012.

12. "Training Requirements for Nuclear Power Plant Operators," 10 CFR 55.45 and 55.59, US Nuclear Regulatory Commission.

13. Bridges, W., "LOPA and Human Reliability – Human Errors and Human IPLs*;" 6th Global Congress on Process Safety*, 2010.

14. Bridges, W., Clark, T, "LOPA and Human Reliability – Human Errors and Human IPLs (Updated)*;" 7th Global Congress on Process Safety*, 2011

15. Walpole, R, et. al., *Probability & Statistics for Engineers & Scientists (8th Edition),* Prentice Hall; March 5, 2006.

16. MIL-STD-105E. US Department of Defense; Military Standard: *Sampling Procedures and Tables for Inspection by Attributes*, 10 May, 1989. MIL-STD-1916 supersedes MIL-STD-105E, 1996.

"Closing the Holes in the Swiss Cheese Model" – Maximizing the Reliability of Operator Response to Alarms

Todd Stauffer, PE
Director Alarm Management, exida consulting
80 N. Main Street, Sellersville, PA 18960
tstauffer@exida.com

Nicholas P. Sands, CAP, PE
Manufacturing Technology Fellow, DuPont Protection Solutions
nicholas.p.sands@dupont.com

David Strobhar, PE
Chief Human Factors Engineer at Beville Engineering, Inc
dstrobhar@beville.com

Prepared for Presentation at
American Institute of Chemical Engineers
2017 Spring Meeting and 13th Global Congress on Process Safety
San Antonio, TX
March 26 – 29, 2017

"Closing the Holes in the Swiss Cheese Model" – Maximizing the Reliability of Operator Response to Alarms

Todd Stauffer, PE
Director Alarm Management, exida consulting
80 N. Main Street, Sellersville, PA 18960
tstauffer@exida.com

Nicholas P. Sands, CAP, PE
Manufacturing Technology Fellow, DuPont Protection Solutions

David Strobhar, PE
Chief Human Factors Engineer at Beville Engineering, Inc

Keywords: Conduct of Operations, Human Factors & Culture, Compliance with Standards, alarm management. Safety alarms, alarm rationalization, ISA-18.2, situation awareness, IPL Alarms, operator response to alarms, classification

Abstract

Layers of protection for abnormal event management can be modeled as slices of swiss cheese according to James Reason [1]. An operator's response to an alarm is one of the first layers of protection to prevent a hazard from escalating to an incident. This paper will present best practices for maximizing the operator's reliability for understanding and responding to abnormal situations as adapted from the alarm management standards ANSI/ISA-18.2-2016 and IEC 62682. Examples include alarm rationalization to ensure all alarms are meaningful and to capture "tribal knowledge", prioritization to help operators determine which alarms are most critical, and creation of alarm response procedures. The treatment of safety alarms, which are those that are deemed critical to process safety or to the protection of human life or the environment, will be specifically highlighted.

The paper will also discuss key human factors considerations for maximizing operator situation awareness (SA) by preventing SA "demons"; such as developing an errant mental model of the process, attention tunneling, data overload, and misplaced salience. As such the resolution of issues which inhibit operator performance, such as nuisance alarms and alarm floods, will also be discussed.

1 Introduction

The purpose of an alarm is to notify the operator of an equipment malfunction, process deviation or abnormal condition that requires a timely response [2]. Alarms help the operator keep the process within normal operating conditions and play a significant role in maintaining plant safety. Alarms are one of the first layers of protection for preventing a hazard from escalating to an incident or accident. They work in conjunction with other IPLs such as relief valves, dikes, and safety instrumented systems (SIS), as shown in Figure 1 [3].

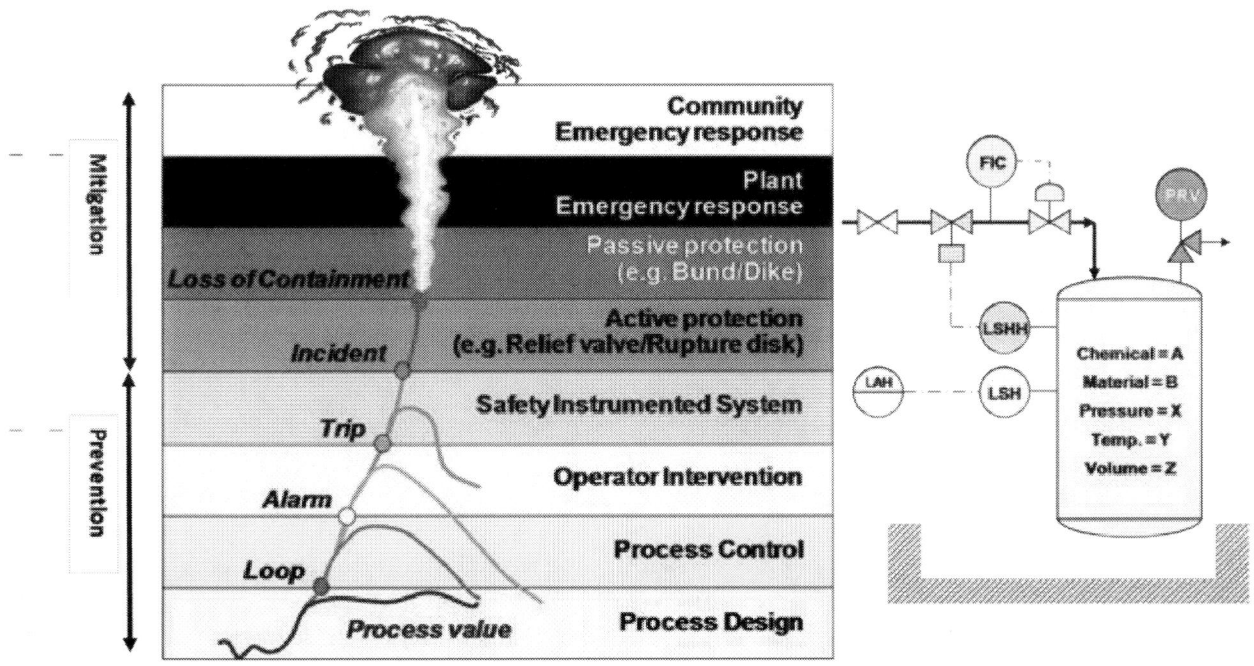

Figure 1. Layers of Protection and Their Impact on the Process [3]

Unlike other safeguards or layers of protection, such as a relief valve or safety instrumented system (SIS), the operator's response to an alarm relies on human intervention. There are numerous potential failure modes for operator response to an alarm including hardware, software, and human behavior. Failures in human behavior become more likely with poor alarm system design and performance (nuisance alarms, stale alarms, redundant alarms, and alarm floods). These failures are often improperly labeled as "operator error"; but are often more appropriately characterized as alarm management failures.

All valid alarms provide a measure of risk reduction to prevent an unwanted consequence. Risk reduction can be qualitative, such as for a safeguard in a Hazard and Operability Study (HAZOP), or semi-quantitative as for an independent protection layer (IPL) in a Layer of Protection Analysis (LOPA). This paper discusses techniques that can be applied to maximize the risk reduction from operator response to alarms and to minimize the chance of "operator error".

2 Notation

BPCS	Basic Process Control System
HAZOP	Hazard and Operability Study
HMI	Human Machine Interface
IEC	International Electrotechnic Committee
IPL	Independent Protection Layer
ISA	International Society of Automation
KPI	Key Performance Indicator

LOPA	Layer of Protection Analysis
MOC	Management of Change
OSHA	Occupational Safety & Health Administration
PFD	Probability of Failure on Demand
PHA	Process Hazard Analysis
PSM	Process Safety Management
RAGAGEP	Recommended and Generally Accepted Good Engineering Practice
SIS	Safety Instrumented System

3 Purpose / Problem Description

3.1 What is the Swiss Cheese Model?

Investigation of industrial incidents has shown that most include multiple independent failures. Thus it can be useful to invoke the metaphor of slices of swiss cheese to represent the layers of protection as proposed by James Reason [1]. As shown in Figure 2, each slice of swiss cheese represents an opportunity to prevent the hazard from escalating to an incident. No layer of protection is 100% reliable, so the holes in the swiss cheese represent failures. For an incident to occur, the holes in the swiss cheese must be aligned. The area of the holes in the swiss cheese would represent the un-reliability of the layer of protection (the more holes that exist and the larger the holes, the higher the probability of failure on demand).

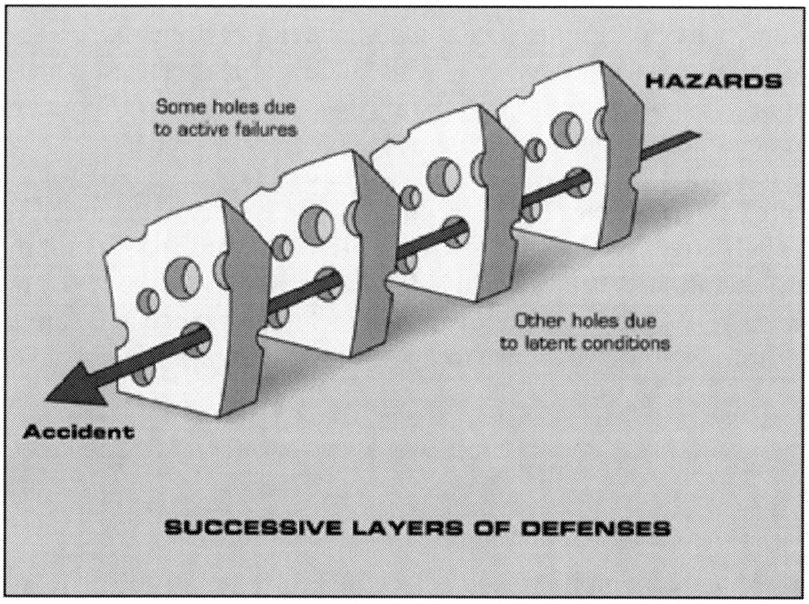

Figure 2. Layers of Protection – Swiss Cheese Model [4]

Typical reliability used in a Process Hazard Analysis (PHA) for an operator response to alarm is 0.9 (PFD = 0.1) assuming the action is simple and well-documented, with clear and reliable indications that the action is required [5]. Applying the swiss cheese model, would indicate that the area of the holes in the slice of swiss cheese is 10%. Keeping with the analogy, to improve performance of the operator response to alarm layer, techniques should be applied to reduce the area of the holes and to make sure the holes don't line up from one layer to the next. For an alarm system with poor performance, the size of the holes would be greater than 10%.

3.2 Operator Response Model

To analyze the failure modes for operator response to an alarm it is helpful to define a model for evaluation. For this study the operator response model is defined to consist of the following three components, as shown in Figure 3.

- **Detect** – the operator becomes aware of the deviation from the desired condition
- **Diagnose** – the operator uses knowledge and skills to interpret the information, diagnose the situation and determine the corrective action to take in response
- **Respond** –the operator starts and completes corrective action in response to the deviation [2]

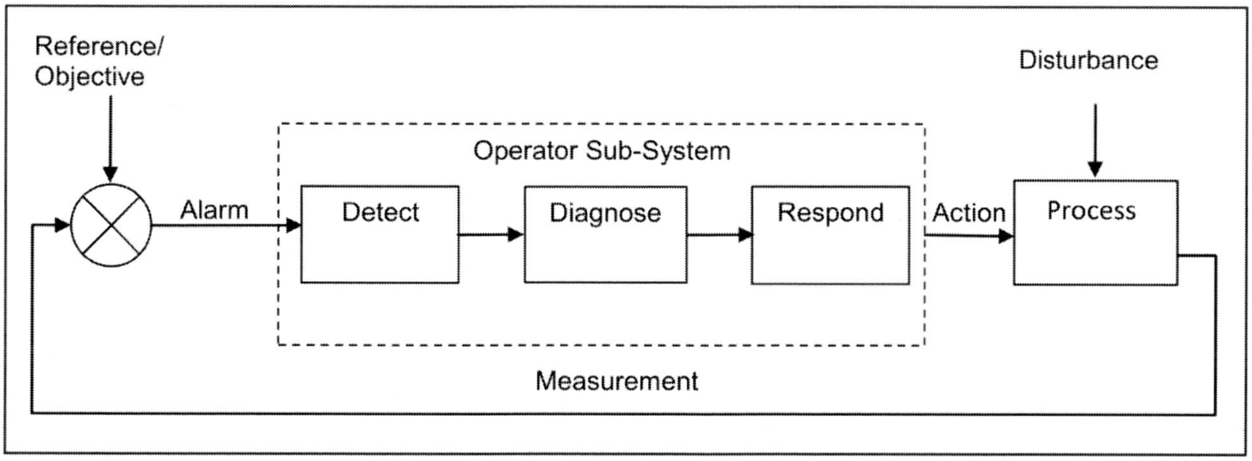

Figure 3. Feedback Model of Operator Process Interaction [2]

D.G. Dunn, et al. performed a study of eleven vessel overflow incidents to analyze the mechanisms where operational discipline broke down and the alarm protection layer failed. The mechanisms studied in the review are summarized in Figure 4 [6].

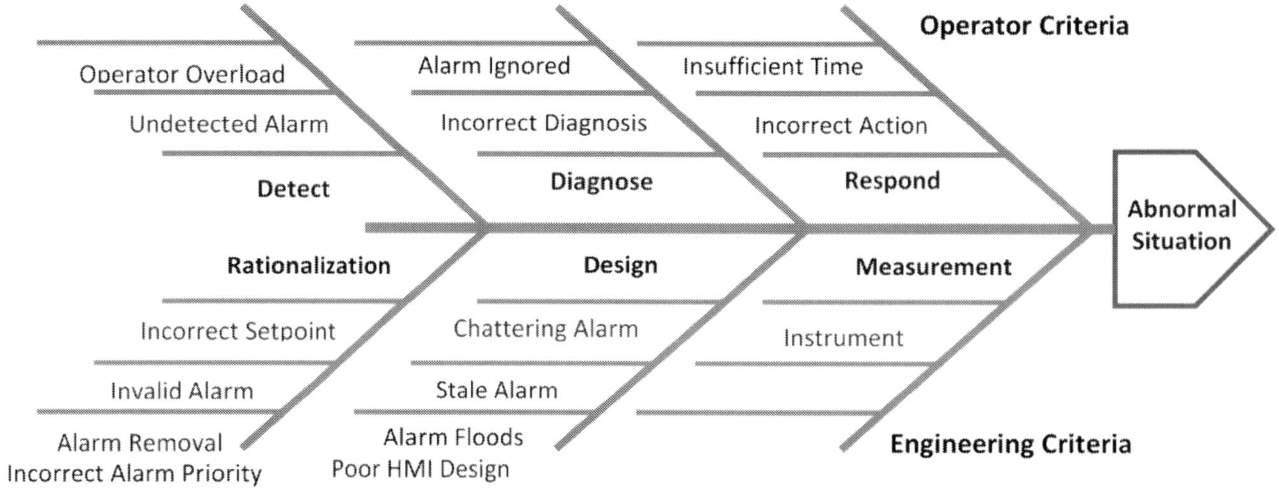

Figure 4. Failure Mechanisms Fishbone [6]

The failure mechanisms in the **"Detect"** component of operator response, include:

- **Undetected Alarm** - the alarm is annunciated but the operator does not notice the alarm
  - Stale alarms can indicate an alarm condition existed previously
  - Alarm floods can hide an alarm in a large number of alarms
  - Poor alarm human machine interface (HMI) design may make alarm annunciation difficult to detect
- **Operator Overload** - Rate of alarm actuation exceeds human capacity for signal detection

The failure mechanisms in the **"Diagnose"** component of operator response include:

- **Alarm ignored** – the operator receives the alarm indication and fails to take any action.
  - Chattering alarms can cause an alarm to be ignored because of frequent annunciation
  - Stale alarms can cause an alarm to be ignored because it is accepted as normal
- **Insufficient training** – the operator does not have enough knowledge to take the corrective action.
- **Incorrect Diagnosis** – the operator does have sufficient knowledge but fails to identify the corrective action.
- **Poor Descriptor** – Alarm information does not convey the real problem.

The failure mechanisms in the **"Respond"** component of operator response include:

- **Incorrect Action** – the operator determines the correct response but fails to take the correct action
- **No Action** – the operator determines the correct response but fails to take the corrective action.
- **Untimely Action** – The operator takes action but not quick enough to prevent the consequence from occurring.

An improper output from the "Respond" component of the operator subsystem, either the wrong or no output, is often referred to as "operator error". Operator error can be thought of as:

- not doing something that should be done (errors of omission),
- doing something in the wrong sequence,
- not doing the action in time, or
- doing something that shouldn't be done (errors of commission).

While failures can occur in all three components (Detect – Diagnose – Respond), most operator response failures in process plants are caused by failures in Detection or Diagnosis. Either the operator failed to notice the problem (Detection) or they incorrectly identified the cause and applied the associated (incorrect) action (Diagnosis). Instances where the operator knew what to do, but performed the incorrect action (such as turning the wrong valve), are much less frequent [7].

3.3 Situation Awareness

The concept of Situation Awareness will be applied to help evaluate and categorize failures in the operator subsystem (Detect-Diagnose-Respond). Situation Awareness (SA), which comes from the study of human factors and is more widely known in the airline industry, can be defined as "being aware of what is happening around you and understanding what that information means to you now and in the future [8]." As such, SA drives effective decision making and performance.

As a framework for categorizing SA failures, eight (8) factors, called SA Demons, have been identified which undermine effective situation awareness [8]. Several of these factors, as described below, will be referenced in this paper [6]:

- **Attentional Tunneling** – Focusing on one area or issue to the extent that alarms from another area or issue are excluded.
- **Misplaced Salience** – Incorrect alarm priority or HMI representation of alarm importance and other status information.
- **Errant Mental Models** – Thought process that incorrectly interprets alarms or mistakenly discounts relevant alarms.

3.4 The Impact of Human Factors

To improve upon operator performance requires an understanding of how humans process information. The discipline of human factors has become increasingly important with the evolution of technology. Numerous authors have commented on how the span of control (responsibility) of the operator has grown with the adoption of distributed control systems – as indicated by the number of control loops and alarms / operator [7, 9].

One of the challenges to applying human factors is that how well a person performs a task cannot be attributed to a single factor. Instead human performance is the product of several variables and their interaction. As shown by Strobhar this multidimensional aspect of human performance presents a couple of significant challenges when trying to understand it. First, deficiencies are often not addressed in the most direct and appropriate manor. In some cases the resolution contributes further to the problem. The response to an operator ignoring a nuisance alarm might be disciplinary action, additional training, more alarms, or longer procedures, instead of determination and resolution of the true root cause. Second, human performance issues rarely have simple and absolute answers [7].

To help understand the causes and resolutions, the above problems will be looked at from the point of view of signal detection theory. Chattering alarms, standing alarms, and alarm floods can all be thought of as visual "noise" to the operator obscuring their ability to correctly detect the alarm signal. According to signal detection theory, as the level of "noise" increases, the ability to discriminate a true alarm from a false alarm is reduced.

3.5 Typical Alarm Failure Example

The continuing review of vessel overflow incidents by Dunn et al and Chidambaram et al, attempts to identify the factors contributing to the failure of alarms using both the failure mechanisms shown in Figure 4 and the SA demons described above. In a typical incident where the alarm response layer of protection failed to prevent the overflow incident, the detect step of the measurement and control system worked as designed and presented the alarm to the operator. The failure occurs in the diagnose step.

In some cases, the operator was overloaded with alarms (alarm flood) or with actions related to a process upset. The most common case was that the operator dismissed the alarm as not requiring action either because it has not required action in the past, or because they do not believe the condition indicated by the alarm.

In one of the incidents reviewed, a product was to be transferred to a dedicated storage tank. A high-level alarm annunciated for a second tank because a valving error sent the product to the wrong tank. No action was taken by the operator in response to the alarm because he mistakenly believed no material was going to the second tank [6, 10].

3.6 Failure Example – Three Mile Island

One of the most famous and costly cases of operator error was the meltdown of the reactor at the Three Mile Island Nuclear Generation Station. The details below come from documented information and first-hand experience by one of the authors. Numerous factors contributed to the operators misdiagnosing the event and taking the incorrect action (it was correct for what they thought was occurring, but totally inappropriate for the actual event). The alarm flood quickly resulted in the operators abandoning the alarm system. This was one of two nuclear reactors on site. The previously constructed reactor was identical in design, but had about one-half as many alarms as its sister plant that melted down. In addition, an alarm critical to diagnosing the event was located on a remote panel, not given enough salience, and was not viewed until after a rupture disk had blown and returned the alarm to normal.

3.7 Failure Example – Petrochemical Plant in Sarnia

As witnessed first-hand by one of the authors, the impact of an alarm cannot be viewed in isolation, it is part of a system that the operator is using to control the plant. A process plant in Canada had a level controller freeze on its treated water tank. While the frozen controller alarmed high, the redundant level indicator alarmed low a few minutes later. The operator took no action for eight hours. When the treated water pumps began to cavitate, the operator realized there was no more water available to feed the plant's boilers. Steam was lost to the plant in January in Canada. The event had been alarmed, but the operator failed to react due to poor practices (did not compare the original high level alarm to the redundant indicator to identify the frozen transmitter), poor alarm management (the low level alarm was one of 15 on a page of standing alarms), and poor salience (the low level alarm differed from the high in only two letters, PVHI vs PVLO).

4 State of Alarm Management

The discipline of alarm management has evolved significantly over the last ten years. In 2009, the standard ANSI/ISA-18.2, "Management of Alarm Systems for the Process Industries" (ISA-18.2) was released. It provides guidance that can help users design, implement and maintain an alarm system in order to optimize performance for an operator response to alarm [2]. ISA-18.2 was used as the starting point for the creation of an international standard, IEC 62682, which was released in 2014 [11]. The ISA-18.2 standard was updated in 2016 based on lessons learned during the adoption phase and based on input from the IEC committee. The ISA-18.2 standard is considered a recommended and generally accepted good engineering practice (RAGAGEP) by insurance and regulatory agencies.

These standards provide a framework for the successful design, implementation, operation and management of alarm systems. They contain guidance to help prevent and eliminate the most common alarm management problems, as well as a methodology for measuring and analyzing performance of the alarm system. As shown in Figure 5, alarm management activities are structured to follow a lifecycle approach wherein the key activities are executed in the different stages of the lifecycle [2, 11, 12]. The products of each stage are the inputs for the activities of the next stage.

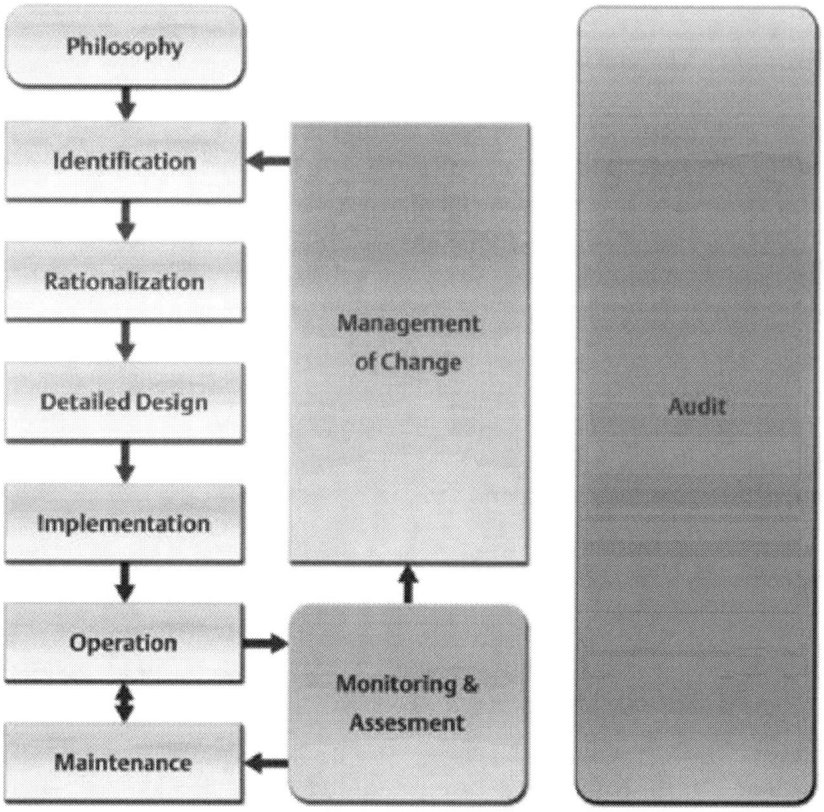

Figure 5. Alarm Management Lifecycle

For the purposes of this paper, the following activities will be highlighted:

- Philosophy
- Rationalization
- Detailed Design
- Operation
- Monitoring & Assessment.

5 Addressing the Problems by following Alarm Management Principles from ISA-18.2

Implementation of an alarm management program is a journey that requires an ongoing commitment. A detailed discussion on how to implement a program is available in the July 2012 issue of Chemical Engineering Progress (CEP) [13]. Key elements of an alarm management program that have a significant impact on operator performance are summarized here below.

5.1 Alarm Philosophy

A necessary and often first step is creating an alarm philosophy document, which is the cornerstone of an effective alarm management program. It establishes the guidelines for how to address all aspects of alarm management, including the criteria for determining what should be alarmed, roles and responsibilities, human machine interface (HMI) design, management of change (MOC), and key performance indicators (KPIs). This document is critical for helping plant staff maintain an alarm system over time and for driving consistency.

Establishing the methodology for alarm prioritization and classification is particularly important before beginning rationalization. Priority is used to indicate criticality and to help the operator understand the relative importance of each alarm. To ensure consistency, alarms should be prioritized based on the severity of the potential consequences and the time available for the operator to respond. Alarm classification organizes alarms based on common requirements (e.g., testing, training, MOC, reporting). Certainly an alarm that is identified as safeguard in a hazard and operability study (HAZOP) or as an independent protection layer (IPL) will have more stringent requirements for testing and operator training than the "average" process alarm. A good philosophy will provide a listing of relevant alarm classes (e.g., personnel safety, quality critical, environmental critical; safety critical, OSHA PSM critical), and their requirements.

5.2 Alarm Rationalization

To maximize dependability the operator must believe that every alarm is valid and requires their response. Alarm rationalization is the process for ensuring that every alarm configured in the system is valid and justified. Rationalizing the alarms in the system helps to improve the operator's trust in the alarm system, and serves to document the cause, consequence, corrective action, and time to respond. It also defines the priority of the alarm, which is a measure of the alarm's criticality. Priority, which is typically assigned based on the severity of the potential consequences and the time available for the operator to respond, tells the operator which alarm they should respond to first. While rationalization only requires that a single operator action and consequence for each alarm be identified, the process can become a rich source of training material and decision aids if a thorough documentation of all causes, responses, and consequences is conducted.

5.3 Classification of Alarms

Along with prioritization, the activity of classification is important to the effectiveness of alarms. Classification is process of separating alarms into alarm classes based on common requirements (e.g., testing, training, monitoring, and auditing requirements). Not all alarms are created equal and for alarms that need higher reliability, a higher level of management is needed. In fact, ISA-18.2 uses the term Highly Managed Alarms (HMA) for the classes of alarms that need higher reliability. Most of the requirements for highly managed alarms are derived from OSHA PSM. These requirements target:

- Training and documentation of training, both initial training and refresher training for both operations personnel and maintenance personnel
- Testing and documentation of testing
- Procedural control of shelving, including authorization and documentation
- Procedural control of out-of-service suppression, including authorization, documentation, and alternate risk reduction.

The HMA requirements apply to alarms classified as safety alarms; alarms critical to process safety for the protection of human life or the environment.

Alarm classification is usually based on the consequence the alarm is designed to prevent and the method used to identify the alarm (e.g., PHA, LOPA). It is worth noting that no risk reduction needs to be explicitly assigned to a safety alarm. If the alarm is specifically designed to detect a life-threatening condition, it is a safety alarm. All alarms on chlorine detectors, for example, might be assigned to a safety alarm class. The testing, calibration, training and suppression would all be documented.

5.4 Operation – Alarm Response Procedures

For an operator response to alarm to be dependable, it is critical that the operator know what to do in the event of the alarm. This is best achieved through training and by making alarm response procedures available. The alarm response procedure, which contains key information documented during rationalization, can be provided in context to the operator from within the HMI. Use of alarm response procedures can reduce the time it takes the operator to diagnose the problem and determine the appropriate corrective action, as well as promote consistency between operators.

According to ISA-18.2 [2], alarm response procedures "shall_be readily accessible to the operator as specified in the alarm philosophy." ISA-18.2 recommends that the following content be included in an alarm response procedure:

- Tag name for alarm
- Tag / alarm description
- Alarm Type
- Alarm Setpoint
- Potential Causes
- Consequence of Inaction
- Operator Action
- Allowable Response Time
- Alarm Class.

Training on how to respond to safety alarms is especially important as these alarms will not occur frequently and because they are most likely to occur during stressful situations such as a major plant upset. Providing operators with alarm response procedures is a recommended practice that should be considered mandatory for safety alarms according to some practitioners [3].

Well trained operators should possess knowledge necessary to diagnose an alarm; however, as the Deepwater Horizon incident illustrates, in times of stress, operator decision processes can be slow and/or incorrect. Making this type procedure available to the operator will support operational discipline by providing information that can improve the speed and accuracy of diagnosis, especially in times of stress [14].

5.5 Monitoring & Assessment

Monitoring the performance of an alarm system provides an indication as to whether the holes in the swiss cheese are getting smaller or larger. In a perfect plant, the alarm summary is a blank screen in the control room. The only time a message appears on the screen is when the operator needs to take an action to prevent an undesired consequence. The message disappears after the corrective action is taken and the screen is blank again. In such a plant, the alarm system does not negatively impact the ability of the operator to respond to an alarm, though other factors may. There are some plants that have achieved this level of alarm system performance.

In some plants the alarm system may overwhelm the operator, significantly increasing the probability the operator will fail to take the corrective action in time to prevent the consequence. Metrics allow categorization and trending of the alarm system performance.

The most commonly used performance metrics are average alarm rate and time-in-flood.

- Average alarm rate is the average number of alarms to an operator position in a unit of time (e.g., alarms per operator per hour).
- Time-in-flood is the percentage of 10-minute time intervals with greater than 10 alarms (per operator).

These metrics only roughly measure system performance. Generally, if the alarm rate is less than 6 per hour and the time in flood is less than 1 percent, the alarm system is performing well. Achieving this performance may also indicate that nuisance alarms are not a significant problem.

Another useful performance metric is the alarm priority distribution. When alarms have been prioritized based on the urgency for the operator to respond, the priority distribution can be an indication of the risk the operator is managing. The general guidance is ~80% of alarm annunciations should be in the lowest priority. Better guidance might be the higher the better. In a 3-priority alarm system, the general guidance is ~5% of alarm annunciations should be in the highest priority. Better guidance might be the lower the better. These numbers vary on the method of prioritization and type of process.

More focused metrics measure the performance of a subset of alarms, like the alarms in a safety alarm class. These are not system performance metrics, but can still be safety KPIs. Some class performance metrics could be alarm rate, time in alarm, and the number of stale alarm occurrences.

- Alarm rate could be used to infer how often the operator response to alarm layer of protection (the slice of swiss cheese) is challenged by an escalating hazard.
- Time-in-alarm is the measure of the average time from alarm annunciation to alarm return to normal.
- Number of stale alarm occurrences is the count of the number of times the stale alarm threshold, typically 24 hours, is exceeded.

These metrics indicate the amount of time alarms in the measured class, for example a safety alarm class, are annunciated to the operator. If safety alarms remain active for an extended period of time, the alarm is not effective and the alarm state will be normalized, making future alarms less effective. Again, the holes in the swiss cheese grow larger. With poor alarm system performance or poor alarm class performance, a safety alarm cannot be considered an effective layer of protection at all.

5.6 Design

The detailed design of alarms also has a significant impact on reliability, or minimizing the area of the holes in the cheese. Alarm design includes several aspects, as reflected in Figure 3 including:

- Selection and design of the process measurement
- Selection of alarm attributes that impact alarm behavior
- Designed suppression of the alarm when the alarm is not relevant
- Annunciation of the alarm to the operator (HMI)

It is a given that the process measurement must be reliable for the alarm to be reliable. Yet in some plants, measurements become faulty over time and the alarms become ineffective. With the monitoring described above, this can be detected and corrected.

Certain alarm attributes greatly impact alarm behavior, especially alarm deadband, on-delays, and off-delays. These attributes can be adjusted to reduce nuisance alarm behaviors like chattering or fleeting.

Designed suppression is used to hide alarms from the operator when the alarms are not relevant to the current mode of operation. An example is suppressing a low pump discharge pressure alarm if the pump is not running.

The alarm annunciation is also critical to reliability. If the alarm is hidden on a graphic in the HMI it will be less effective. Use of alarm priorities increase the salience of the most urgent alarms. For true safety alarms, an indication independent of the basic process control system (BPCS) may be needed to achieve the desired reliability. Often field horns and lights are used to indicate the need to evacuate an area due to a severe hazard. There is a significant difference in reliability between an alarm passing through the BPCS HMI for the operator to take action, and an alarm in the field that indicates the action is immediate evacuation. This was demonstrated in the Deepwater Horizon accident where the general evacuation alarm was NOT automated, but instead had to be initiated by the control room operator. The delay in was a contributing factor to the fatality and injury count [14]. Independent field alarms may have a reliability of 0.99 (PFD = 0.01) as long as the system performance is good as described above.

6 Addressing the Problems by Applying Human Factors Principles

6.1 Decision Theory applied to Operator Response to Alarms

Operator detection of an abnormal event, like the detection of any event or signal, has been studied by the human factors community since WWII. They have found that performance is based upon two key variables: (1) difference in the signal from the surrounding noise and (2) the willingness of the operator to accept false alarms versus missing a true event. The latter can be highlighted by torpedo lookouts on Navy vessels; they were willing to have more false alarms in order to ensure that they did not miss a true event.

All judgements are made in an environment of uncertainty. The signal indicating the event has some degree of noise associated with it. That signal then is conveyed to the operator in an environment with its own degree of "signal" noise (Figure 6). The greater the difference in the nature of the noise in the environment and the characteristics of the signal (D'), the more likely is that the signal will be detected. As the environmental noise (audible or visual) more closely matches the signal (e.g., red is used for both warning and pump status), the more likely it is that the signal will be missed. As D' approaches zero (e.g,, large number of standing alarms or false alarms), the ability of the operator to detect a valid alarm is little better than 50%, a coin toss.

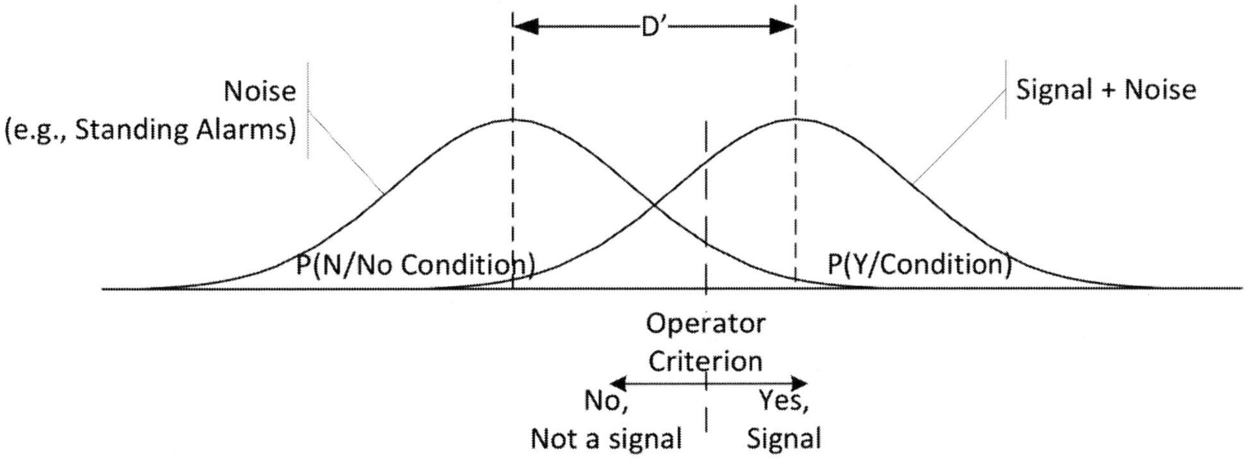

Figure 6. Signal and Noise Conditions

While the difference between the noise and signal affects the detection of the signal, it is not the only factor. The person, the operator, creates their own criterion for what signal to accept as true or valid. As the signal and noise become more alike, the probability of mistaking a false condition for a true event increases (e.g., a poorly designed alarm system). If experience has placed little importance on missing the actual event, then the user will generally opt to reduce false alarms at the expense of missing a true event – the "cry-wolf" phenomena. In reducing the number of false alarms, the operator increases the probability of missing a valid alarm. Ensuring a large difference in alarm signal/noise is important not only for the ability to detect the signal, but the operator's criterion for doing so.

6.2 Human Factors Consideration for Nuisance Alarms

It does not take a high false alarm rate to result in the operator essentially abandoning their alarm system. A 25% false alarm rate, one alarm out of every four, is enough to have the operator no longer rely on the alarm system to detect an abnormal event. A false alarm is does not necessarily indicate that the condition is not true (e.g., level is high), but simply that no action is needed on the part of the operator ("It's not high enough to warrant my doing something about it").

Ignoring nuisance alarms is not a behavior that can be changed by training or disciplining the operator. According to Endsley "A person's reluctance to respond immediately to a system that is known to have many false alarms is actually quite rational. Responding takes time and attention away from other ongoing tasks perceived as important [8]".

Shifting the operator's reliance on the alarm system is often a byproduct of the alarm rationalization process. As redundant alarms are eliminated, the operators begin to realize that they must pay attention when an alarm actuates. They will not get a second, third, and fourth alarm prior to their need to take action. Several operators have commented during rationalization, "No more ignoring alarms, we're going to have to act when they come in".

Additionally design of HMI displays should support situation awareness. As stated by Endsley "Visual displays need to do more than simply provide a visual indication of the alarm; they should help people rapidly confirm that the alarm is a real indication [8]".

6.3 Human Factors Consideration for Training

Helping operators identify valid alarm conditions from invalid ones should be part of the basic operator training program. Expert operators know what to look for in order to determine if the alarm condition is real. Rarely does a process variable move in isolation. Something else will have changed as well, either another variable or the control system. As noted in the earlier example, had the operator in the Canadian refinery compared the redundant level transmitters, one would have been identified as failed and the facility would not have lost steam in January. A key part of operator training needs to be how to cross-check alarms with other process variables to determine their validity. For example, if a low flow alarm occurs, is there a change in level in the up or down stream vessels.

6.4 Human Factors Consideration for Alarm Response Procedures

A general guideline in alarm rationalization is DO NOT alarm the normal or expected. That provides no real information to the operator. So it is with alarm response procedures. Providing the operator with information from the rationalization that is obvious for a trained operator just creates noise, not value. While the entire output of the alarm rationalization is useful for training, only a subset of that information should be provided to the operators as a real-time decision aid. Alarm response procedures should contain or highlight that which is unique about this particular alarm, either in cause, response, or consequence of inaction. Alarm response procedures are an aid for the operator, not a replacement for good operating procedures.

6.5 Human Factors Consideration to best deal with alarm floods

Even well designed alarm systems can generate a large number of alarms as a result of a major process upset, such as a loss of power. Humans have a relatively limited capacity for processing information, so the potential to exceed that capacity is high. However, there are several techniques that can be employed to maximize the potential that alarms will be processed.

Two of the techniques reduce the number of alarms that must be processed. The easiest is to automatically suppress low priority alarms for some period of time. If the alarm prioritization has been done correctly, this will reduce the number of alarms by 80% with limited risk in the short term, given that low priority alarms generally do not need to be responded to with the same urgency as high priority. A more complicated technique is to utilize state based alarming such that those alarms that are expected to occur during such upsets do not actuate. For example, once a unit has shutdown, alarms due to low energy (e.g., low temperature, pressure, flow) will likely be "normal". Alarm for these conditions should therefore be suppressed when entering the low-energy state, prior to annunciating.

The third technique is to aggregate the alarm information into higher order alarms for entire systems, thereby reducing what needs to be processed by the operator. For example, all alarms for a tower or set of towers operating in series can be combined so the operator need not process them individually, but qualitatively as the tower is either "okay" or "upset".

6.6 Human Factors - How to prevent attention tunneling

Due to the limit of human processing capabilities, it is possible for the operator to become overly focused on the task at hand and miss events in other parts of the system. This is a loss of the "big picture" or situation awareness. A single display for the operator's entire span of control is needed to ensure that they can, at all times, assess the status or health of all the equipment for which they are responsible. Included in this display should be the status of the alarm system – what units / areas have alarms and what is their priority/importance.

6.7 Human Factors - How to prevent misplaced salience

People are very poor at detecting changes, resulting in a phenomenon known as change blindness. Overcoming change blindness requires that changes be highlighted when they happen. This is the purpose of having new alarms flash until acknowledged directing the operator's attention to the variables that have changed. It would be very difficult for any individual to notice that some values on a screen had changed color if that were the only indication that an alarm had occurred, particularly if the display was very colorful.

6.8 Human Factors – How to Improve Use of Mental Models

Mental models are an important mechanism for interpreting new information. They are thinking tools that provide a construct for a person to combine disparate pieces of information, interpret the significance of that information, and to develop reasonable projections of what will happen in the future [8]. Operators work with mental models of how the process works (**if** the reactor feed flow rate is increased, **then** the reactor temperature will increase without additional coolant).

Application of mental models run into problems when the operator uses an incomplete or incorrect mental model. One of the keys to using mental models is to be able to realize when you are using the wrong one. Operators may misinterpret alarms or events as fitting into their current mental model without realizing these cues indicate that they should be thinking differently (using a different mental model). Unfortunately people tend to explain away conflicting cues to their current mental models (confirmation bias) and can be slow to realize this mistake.

One way to improve operator response is to have them develop multiple mental models so that they can pick the right one for the situation. To help in the development of more and better mental models one recommendation is to apply a pre-mortem strategy [15]. This involves analyzing / brainstorming how a process or operation could fail through the creation of if – then scenarios and discussion of how to rectify the situation. Experienced operators have more and more varied models of plant operation, developed after years of experience plant upsets. These models need to be transferred to new operators without them having to experience the upsets for themselves. This could be especially useful before starting up or shutting down equipment or changing its mode of operation. Rationalization can be effective in identifying where multiple models exist, and training is how we share with the entire staff.

6.9 Human Factors – Conditioning / Normalization

Standing alarms are a negative in alarm metrics as they create noise in the visual environment, in this case the alarm display. In the case cited earlier of the Canadian Refinery, the alarms for the faulty and valid level indications were present over eight hours before water was lost to the boilers. How does an operator go eight hours with alarms in their visual field and not notice them? It happens because some 15 or more alarms already existed on the alarm summary display. The two alarms of the anomalous condition simply blended into the other alarms. More standing alarms results in greater visual noise in the environment and increasing difficulty to detect a signal when it occurs.

7 Conclusion / Summary

This paper has discussed alarm management and human factors techniques for improving an operator's response to an alarm; thus reducing the size and area of the holes in its "swiss cheese" layer and making it more reliable. These techniques can be applied to all alarms, but are particularly important for safety alarms. Some of the key takeaways are summarized below.

- Performing a thorough rationalization is important to ensure that each alarm is actionable and has a purpose.
- Prioritizing alarms based on consequences and time to respond helps the operator know which alarm to respond to first.
- Classification in (conjunction with prioritization) identifies safety alarms so that they can be managed and maintained appropriately to their risk reduction and can be displayed on dedicated HMIs or annunciated uniquely.
- Alarm system performance is monitored and assessed to ensure that operators are not being overloaded or flooded with too many alarms. Performance of individual safety alarms should be examined in more detail to ensure that they are functioning acceptably.

- Alarm response procedures, created from the results of rationalization, can be presented to the operator as a real-time decision aid.
- Alarms should be designed for reliability from the sensor to the HMI with an appropriate level of salience for quick and easy detection by the operator.
- A nuisance alarm rate of more than 25% can cause operators to ignore alarms or delay their response. Thus significant effort should be put into eliminating nuisance alarms and creating an environment where the operator can quickly and easily confirm the validity of an alarm.
- Training – first believe what the indication is, look for confirmation to prove your mental model…Be careful discounting an alarm without corroborating evidence. Instincts can lead you awry…Challenge when closing of possibilities…When I see an alarm I look for a confirmatory set of actions to confirm that it is real….Factors that correlate with making an error
- Operators should work on developing more and better mental models to provide alternate scenarios for dealing with plant upsets and should be wary of slipping into an attention tunneling episode.

While it is important to follow the alarm management best practices in this document and other references (e.g., ISA-18.2 / IEC 62682), the importance of changing the operator's mindset (behavior) cannot be overemphasized. As discussed, the majority of failures occur in the diagnosis step of the operator response model; thus addressing this failure becomes one of the key actions to improving reliability. No amount of alarm management can make up for operators who mistakenly think that information is not "real", eliminate potential causes too quickly, or exhibit confirmation bias when responding to prevent the escalation of a hazard.

8 References

[1] J. Reason, *Managing the Risks of Organizational Accidents.* Burlington, VT: Ashgate Publishing, 1997.

[2] ANSI/ISA–18.2–2016. Management of Alarm Systems for the Process Industries. June 2016.

[3] Stauffer, T., Clarke, P., Using Alarms as a Layer of Protection – AICHE 8th Global Congress on Process Safety, April 2012.

[4] J. Reason, (1990) *Human Error.* Cambridge, Cambridge: University Press, 1990.

[5] CCPS. *Guidelines for Safe and Reliable Instrumented Protective Systems.* Center for Chemical Process Safety, American Institute of Chemical Engineers, Hoboken, NJ, 2007.

[6] D.G. Dunn, N.P. Sands, and T. Stauffer, When Good Alarms go Bad: Learning from Incidents, 70th Annual Instrumentation and Automation Symposium – Texas A&M University, January 2015.

[7] D. Strobhar, P.E., *Human Factors in Process Plant Operation*, Momentum Press, New York, NY, 2013.

[8] M. Endsley, *Designing for Situation Awareness: An Approach to User-Centered Design*, CRC Press, Boca Raton, Fl, 2012.

[9] B. Hollifield, E. Habibi, *Alarm Management: A Comprehensive Guide, Second Edition*, International Society of Automation, Research Triangle Park, NC, 2011.

[10] Chidambaram, P., Sands, N., Analysis of Overflow Incidents: Searching for Situational Awareness Demons – AICHE 12th Global Congress on Process Safety, March 2016.

[11] IEC62682-2014 "Management of Alarm Systems for the Process Industries". October 2014.

[12] Stauffer, T., Sands, N., and Dunn, D., "Alarm Management and ISA-18 – A Journey, Not a Destination", Texas A&M Instrumentation Symposium (2010).

[13] T. Stauffer, P.E., Implement an Effective Alarm Management Program, Chemical Engineering Progress, pp. 19-27, July 2012.

[14] Chastain-Knight, D., Stauffer, T., Managing Alarms to Support Operational Discipline – AICHE 12th Global Congress on Process Safety, March 2016

[15] G. Klein, *Sources of Power: How People Make Decisions*, The MIT Press, Cambridge, MA, 1998.

13TH GLOBAL CONGRESS ON PROCESS SAFETY

Adapting cause and effects methodology to your Safety instrumented system (SIS) to reduce human errors from engineering, operations and beyond

Charles M. Fialkowski, CFSE
Siemens Industry Inc.
Spring House, PA
Charles.Fialkowski@siemens.com

Prepared for Presentation at
American Institute of Chemical Engineers
2017 Spring Meeting and 13th Global Congress on Process Safety
San Antonio, TX
March 26 – 29, 2017

Adapting cause and effects methodology to your Safety instrumented system (SIS) to reduce human errors from engineering, operations and beyond

Charles M. Fialkowski, CFSE
Siemens Industry Inc.
Spring House, PA
Charles.Fialkowski@siemens.com

Keywords: Process Safety, safety instrumented systems, IEC 61508, ANSI/ISA 84, IEC 61511, Safety PLC, safety lifecycle, cause and effects, Safety Integrity Level, SIL, Probability of Failure on Demand, PFD, Risk Reduction Factor

Abstract

A safety instrumented system (SIS) is designed to reduce the likelihood of hazardous risk by decreasing the frequency of unwanted accidents. The amount of risk reduction that an SIS can provide is represented by its safety integrity level (SIL). The SIS is designed to detect when the process reaches a hazardous condition and respond accordingly to move the process to a safe state, thus preventing the unwanted accident from occurring. Studies indicate however, that over 50% of all SIS failures are related to human error. While many SIS systems boast having SIL 3 certification, it's often the human interactions that render many of these well intended systems to be essentially non-existent. A cause and effects methodology is an approach many in the industry are exploring to help reduce human errors throughout the entire safety lifecycle of the SIS.

1 Introduction to Cause and Effects Methodology

Logic is defined as the science of correct reasoning, or necessary connection or outcome, as through working of cause and effect.

Anyone who has ever watched Star Trek knows that any problem, if approached logically, will either yield a solution, or verify that a solution is not possible. By examining all eventualities with respect to cause and effect a solution can usually be found.

The cause and effect diagram is used for defining how and when actions are executed in a safety system. This methodology involves organizing process events into categories of causes and effects and then linking them together logically via the intersection which will indicate what effect(s) will result from which active cause(s).

There are three key fields of information contained in the cause and effect diagram. A cause occupies a row, an effect occupies a column and the intersection is the cell that is common between a cause row and an effect column. These fields may contain and represent the following functional requirements for a safety instrumented system as shown below in Figure 1:

- Cause - This field reflects a process deviation. When the cause tags meet certain user-configured conditions, it becomes active.
- Effect - This field reflects a process action. When the effect is active, the effect tags will be set to their failsafe values.
- Intersection - This field determines how the effect responds to the cause. If the intersection is empty, the cause does not influence the effect. If the intersection is configured, an active cause will trigger the associated effect and the effect will become active.

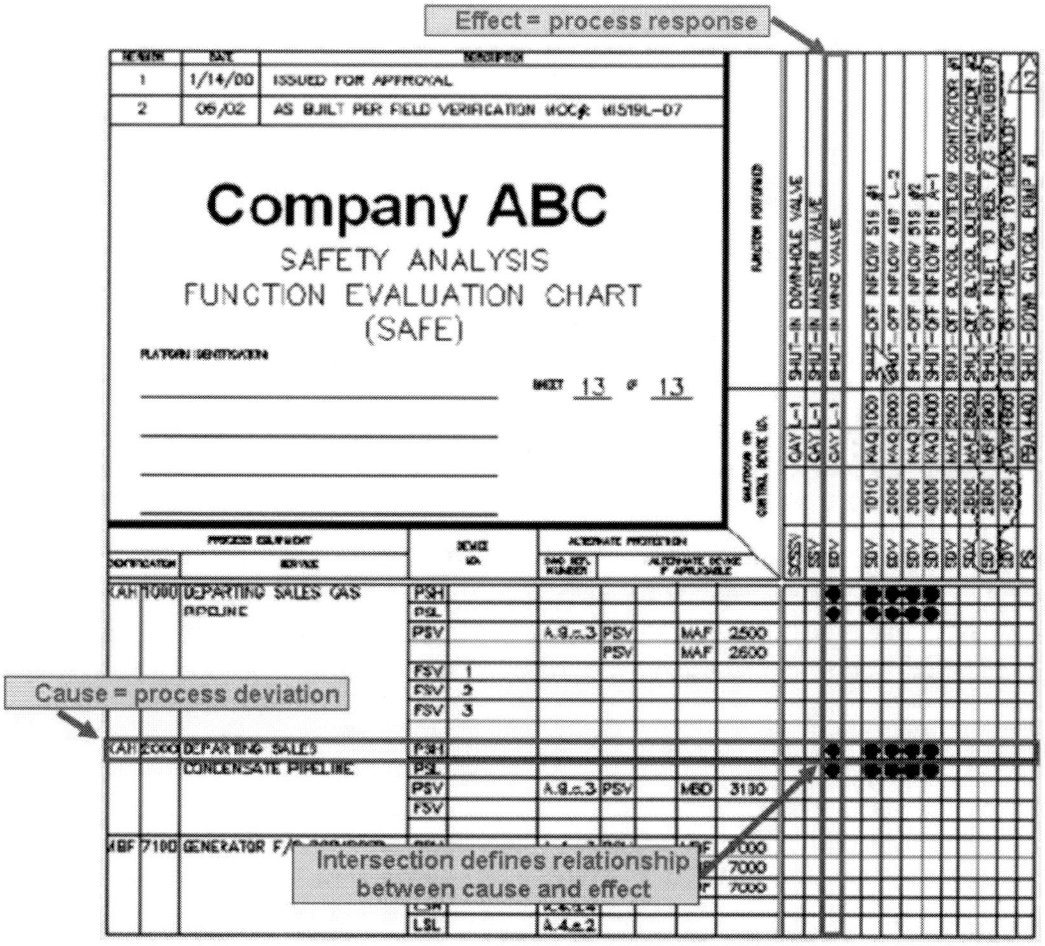

Figure 1. Cause and Effect Diagram

2 The Safety Lifecycle

In 1995, the U.K. Health and Safety Executive (HSE) published a report titled "Out of Control: Why Control Systems Go Wrong and How to Prevent Failure" [1] where they analyzed the root cause of several industrial accidents that were initiated by failure of the control system and published the chart shown in Figure 2, showing where failure occurred. They updated their publication in 2003 where they noted the analysis of incidents remained unchanged from the first edition.

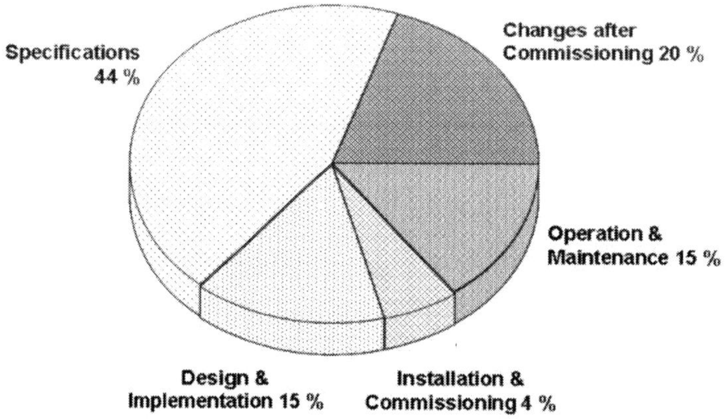

Source: *"Out of Control: Why Control Systems Go Wrong and How to Prevent Failure,"* U.K.: Sheffield, Health and Safety Executive, 1995

Figure 2. Root Cause Of Control System Failures

In many ways, it was the findings of this study that led to the development of the safety design lifecycle which serves as the basis for international standards on functional safety such as the IEC 61511:2016 standard [2]. The safety design lifecycle is a practical methodology that defines the process necessary to ensure overall plant safety by defining a sequence of steps and deliverables from each phase to help prevent the failures that were identified in the HSE report.

Figure 3, shows the safety life-cycle as described in the IEC 61511 standard. The lifecycle can be divided into three main phases; the analysis phase, the realization phase and the operation phase. The analysis phase is focused on determining and documenting how much safety is needed. The realization phase is focused on the design and implementation of the system and safety achieved. The operation phase is focused on the activities and documentation required in operating and maintaining the system to ensure the performance is maintained.

The safety life-cycle phases are meant to address the root causes of failure as identified by the HSE. The analysis phase is focused on addressing the percent of failures caused by improper specification by requiring quantitative risk analysis and risk reduction techniques resulting in a safety requirements specification (SRS). The realization phase is focused on minimizing the failures caused by design & implementation and installation & commissioning. It does so by requiring documented verification that the design meets the targets defined in the analysis phase. The operation phase is focused on addressing the percentage of failures caused by improper operation & maintenance and changes after commissioning. Again, this is accomplished through mandatory procedures and documentation.

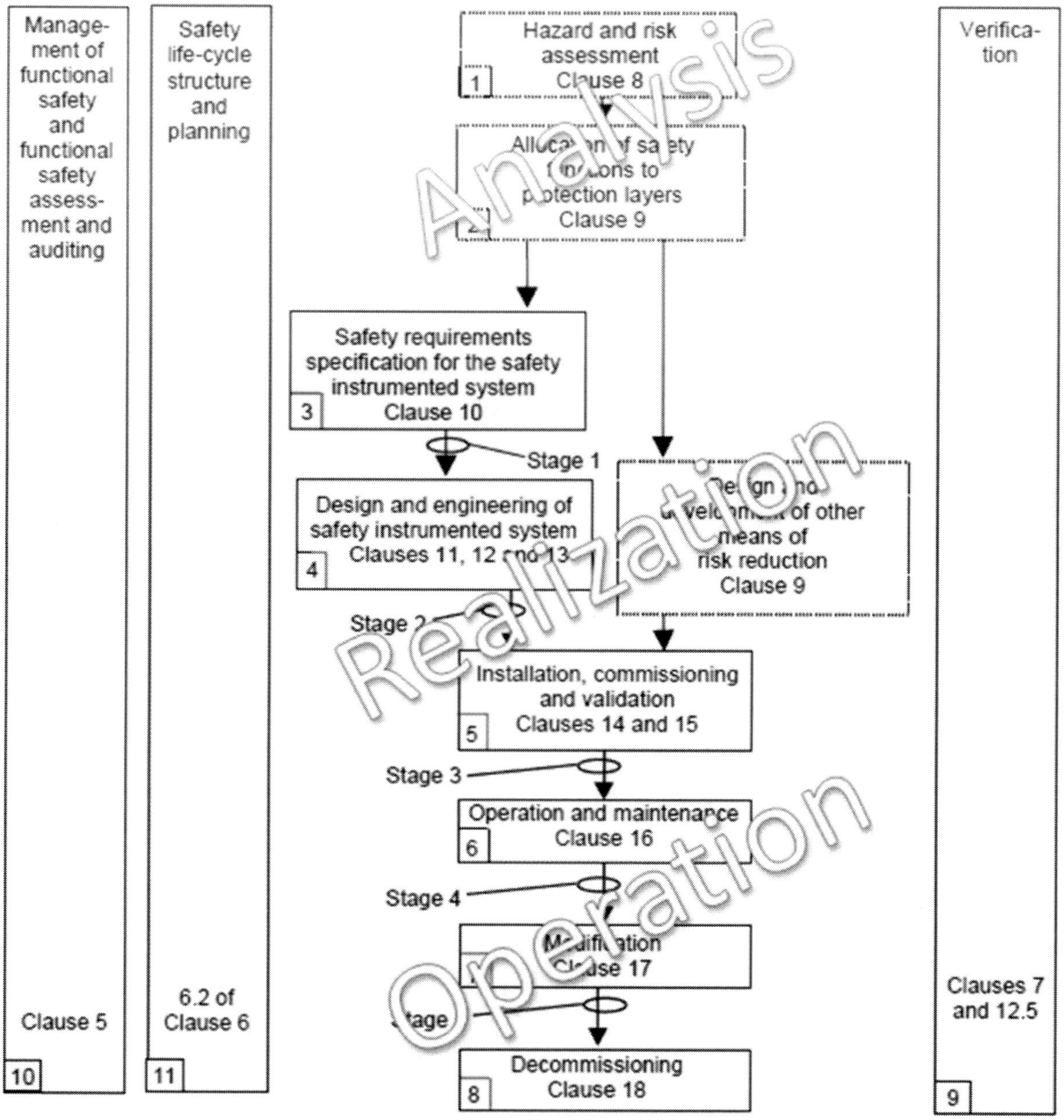

Figure 3. SIS Safety Life-Cycle Phases

3 The IEC 61511 standard requirements

In June 2016, the IEC technical committee released the 2nd edition of their functional safety standard titled "Safety instrumented systems for the process sector". This edition cancels and replaces the first edition which was published in 2003, and includes a number of editorial improvements (definitions, etc.), and technical changes specifically around the requirements to improve the management of functional safety.

Safety instrumented systems (SISs) have been around for many years to perform safety instrumented functions (SFIs) in the process industries. If instrumentation is to be effectively used for SIFs, it is essential that they can achieve certain minimum performance levels. The IEC 61511 standard drives towards two basic concepts which are fundamental to its application, the SIS safety life-cycle and safety integrity levels (SILs).

Fundamentally, the goals and purpose of the 2nd edition of the IEC 61511 standard is identical to the first edition which maintains the same general title "Functional safety: Safety Instrumented Systems for the process industry sector" as well consisting of the same 3 parts:

1. Part 1: Framework, definitions, system, hardware and software requirements - normative
2. Part 2: Guidelines in the application of IEC 61511-1- informative
3. Part 3: Guidance for the determination of the required safety integrity levels - informative

The principle behind this standard is simple. Follow the activities of the safety lifecycle, make sure you achieve the defined minimum requirements, and document everything. This approach is intended to direct a process that is rational and consistent. To facilitate this approach, this standard requires the following:

- a hazard and risk assessment is carried out to identify the overall safety requirements
- The allocation of the safety requirements to the safety instrumented system(s)
- works within a framework applicable to all instrumented methods of achieving functional safety
- Details the use of certain activities, such as safety management, which may be applicable to all methods of achieving functional safety

Specifically this standard addresses all safety life-cycle phases from initial concept, design, implementation, operation and maintenance through to decommissioning. In order to claim conformance to this standard the owner/operator would have to demonstrate that the requirements outlined in clause 5 through 19 have been satisfied. This paper will discuss how adopting a cause and effects methodology, may be used to satisfy most of these requirements.

Clause 5: Management of functional safety

The objective of clause 5 is to identify the management activities (organizational, planning, assessment, etc.) that are necessary to ensure the functional safety for the SIS is met. In general it contains the policy and strategy for the organization on how to achieve functional safety. Sub-clause 5.2.7 requires SIS configuration management techniques have to be developed specifically for the application program of the SIS where proper revision control shall be maintained. An example of revision control is shown in Figure 4. This process requires that if further changes are made (after properly reviewed and approved) by the SIS designers, the revision of the cause and effects diagram may be incrementally advanced forward allowing for appropriate descriptions of the changes to be captured as well.

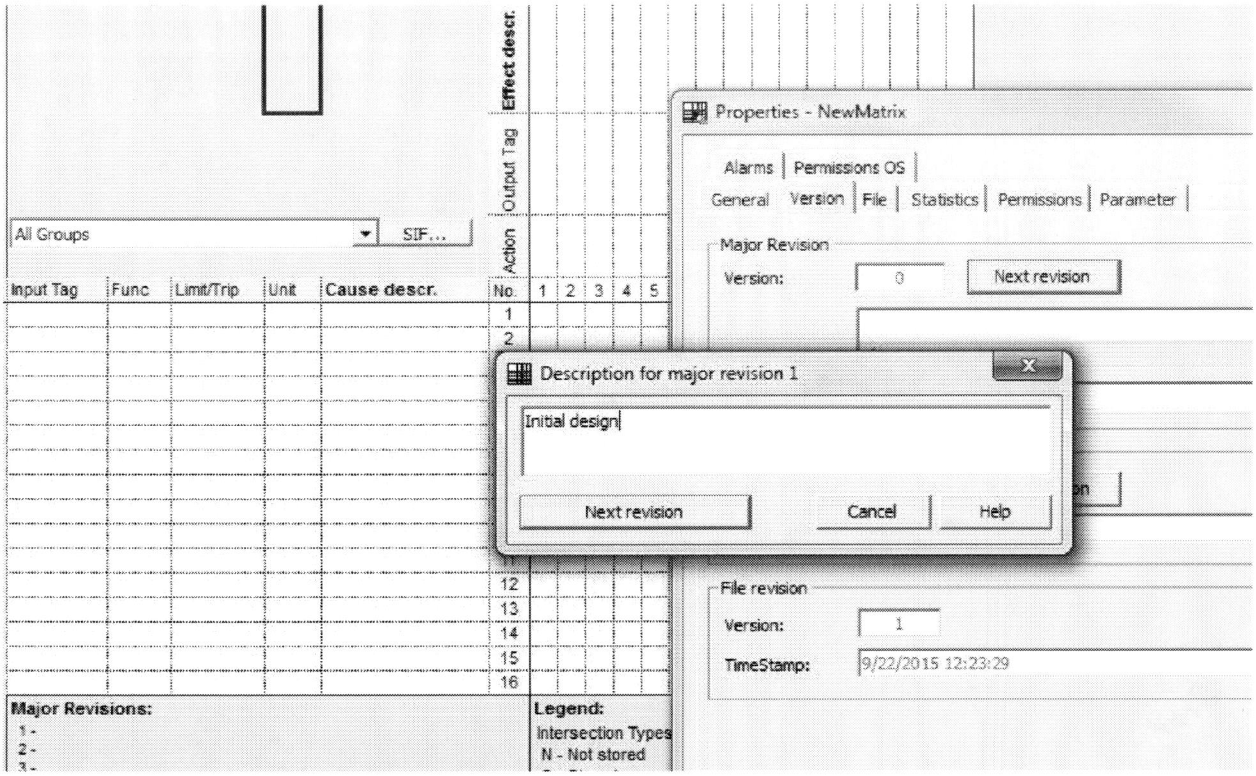

Figure 4. Example Of Revision Control

Clause 6: Safety life-cycle requirements

The main objective of clause 6 is to define the phases and establish the requirements of the SIS safety life-cycle activities and to ensure that adequate planning exists to make certain the SIS meets the safety requirements.

A cause and effects diagram will not only help management organize and define their safety strategy, but will also help to present their safety philosophy in a clear concise manner that various levels can read and understand. Prior to this, a control narrative would have to be derived from reading complex logic charts. This process often took many man hours to compete as it might require the attention of multi-disciplined personnel.

The cause and effects diagram will significantly reduce the time it takes to organize the technical functions of the safety system. It is noted in the standard that the SIS requirements should be expressed and structured in such a way that they are clear, precise, verifiable, maintainable, and written to aid comprehension by those who are likely to utilize the information at any phase of the lifecycle.

Traditionally, safety logic would be in the form of high level programming language such as Boolean algebra or ladder logic. It would consist of several defined "logical connections" or relationships between variables that depend on, or follow each other logically. Either of these procedures are fine for Safety PLC programmers, but additional work would still be required to meet the criteria for properly documenting the safety functions.

Take for example the logic shown below in Figure 5. This indicates how a fired heater shutdown system needs to be operated. This form is again fine for programming but would not be acceptable for documentation. From this logic diagram, a control narrative would still need to be developed to clarify the operation of this "de-energized to trip" logic circuit as many would find this counter-intuitive to the way it is diagramed below where a logic 0 at the final control element would equal an active state (i.e. closed).

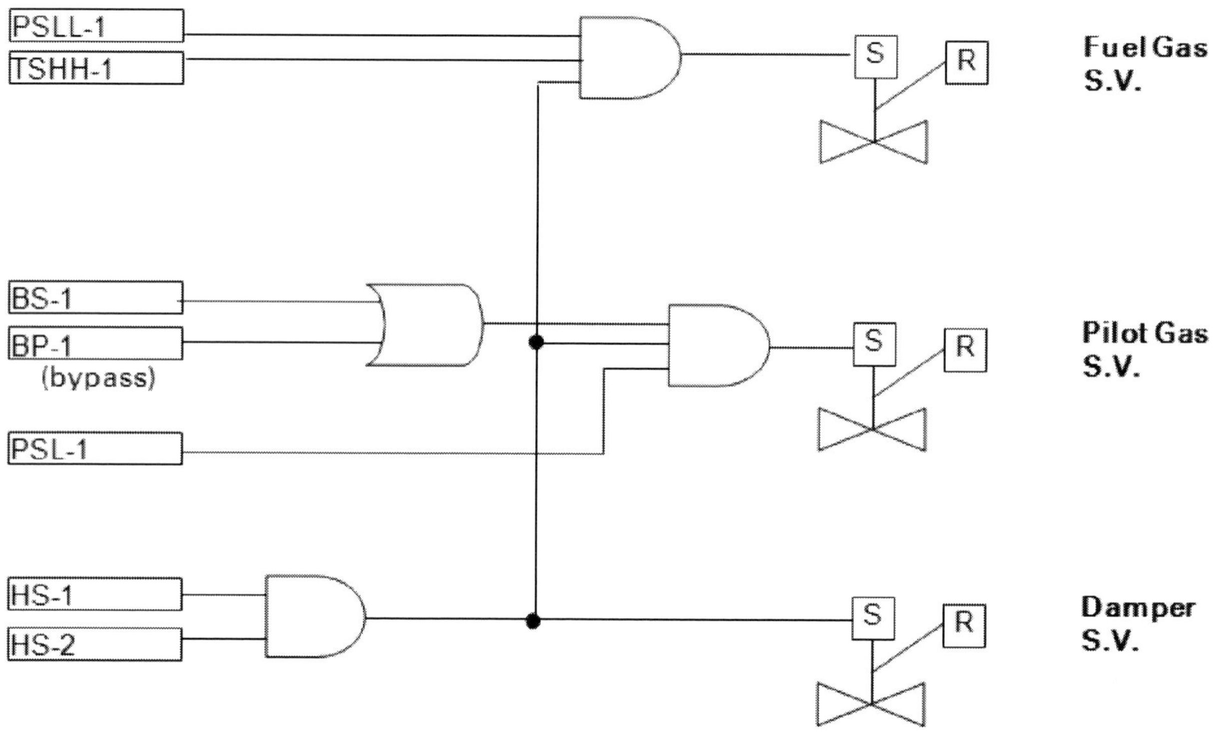

Figure 5. Sample Shutdown Logic

Figure 6, shows the same fired heater shutdown logic, but illustrated in a cause and effects diagram. This view shows a more concise and descriptive way that when either emergency trip pushbuttons HS-1 or HS-2 are activated, as defined by the relationship of the intersection, they would logically demand that all three control elements (valves and damper) close. In addition, either PSL-2 or BS-1 would demand "close" the pilot gas shutoff valve, and either PSLL-1 or TSHH-1 would demand "close" the main fuel gas shutoff valve.

Controller Name:

Matrix Name:

Effects

Num	Action	Output Tag	Description	Type	SIL
1	Close	SV-3	Main Air Damper		1
2	Close	SV-2	Pilot Gas shutoff valve		2
3	Close	SV-1	Main Fuel Gas shutoff valve		2
4					
5					

Causes

Input Tag	Func	Limit/Trip	EngUnit	Description	SIL	Num	1	2	3	4	5
HS-1 HS-2	OR	FALSE		Burner Emergency Trip PB		1	N	N	N		
PSL-2 BP-1	AND	FALSE		Pilot Gas Pressure Low		2		N			
BS-1		FALSE		Burner switch (No Pilot Flame)		3		N			
PSLL-1		FALSE		Fuel Gas Pressure Low		4			N		
TSHH-1		FALSE		Furnace Stack Temperature High		5			N		
						6					
						7					

Figure 6. Cause and Effects Diagram

Clause 7: Verification

The main objective of clause 7 is to demonstrate by review, analysis, and/or testing that the required output for each phase of the safety life-cycle meets the requirements.

Verification planning requires a number of items to be properly addressed throughout the safety life-cycle. One area of concern is the management of change (MOC) of the SIS logic. Most modern day software tools provide an option that allows your cause and effect diagram tool to perform a compare function. This utility would allow the engineer to compare the current program to one that is stored offline, thus being able to identify and record any functional logic changes that might have taken place during the verification phase. Figure 7 below illustrates an example of compare utility; where the output is in a comprehensive report that describes in great detail the differences between the two documents.

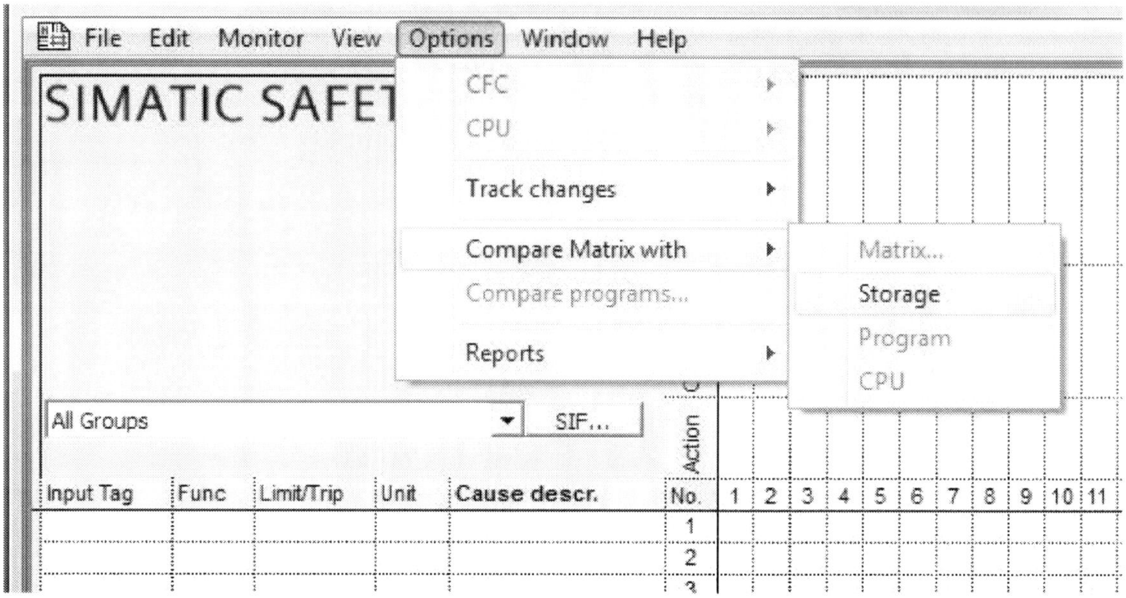

Figure 7. Management of Change (MOC) Requirements

Clause 8: Process Hazard And Risk Assessment

Identifying the hazards and/or hazardous events of the process along with determining the risk associated with these hazards is the main objective of clause 8. This would also include determining if any safety functions are necessary to reduce the risk and if so, are any of these safety functions SIFs?

While there maybe a number of engineering documents needed to conduct a proper process hazard analysis (PHA) to gain a thorough understanding of the process such as sequential function charts, process flow diagrams and piping and instrumentation diagrams, the output of the PHA could be in the form of a cause and effect diagram to list the identified SIFs as shown in Figure 8.

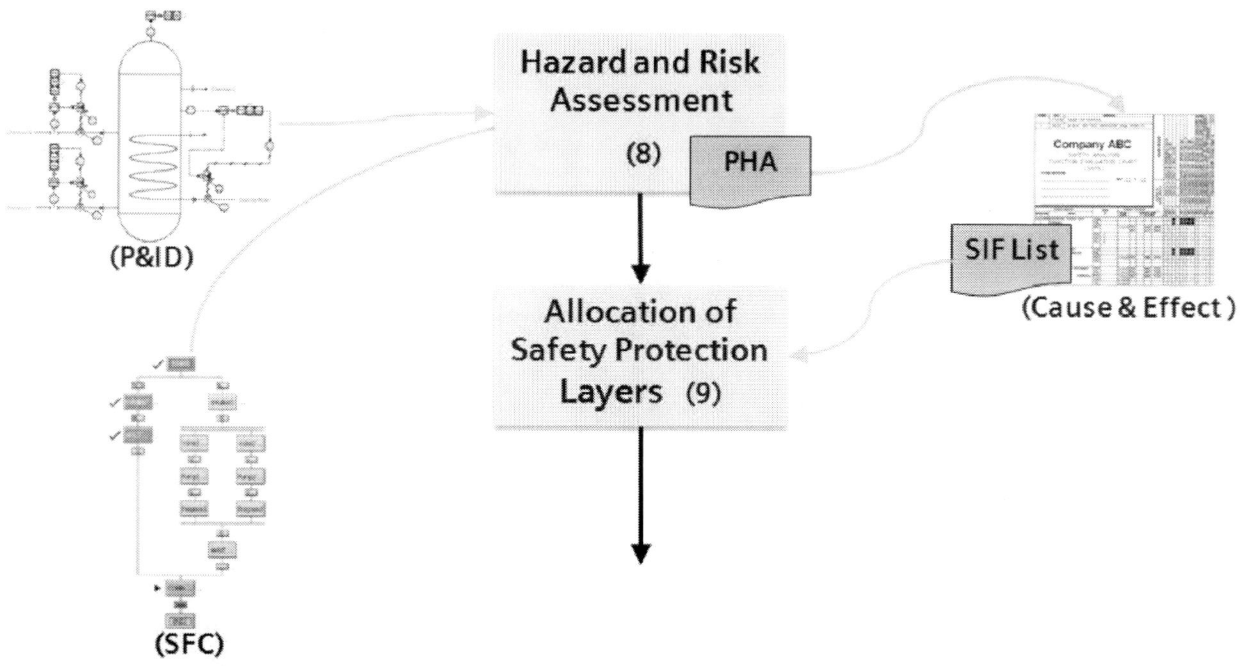

Figure 8. Hazard And Risk Assessment To Protection Layers

Clause 9: Allocation of safety functions to protection layers

In this clause, safety functions are allocated to specific protection layers. It should be noted that not all safety functions will be part of the SIS (e.g. mechanical relief device), however those identified as SIFs could be managed and documented within a cause and effects diagram. In addition one of the requirements of this phase is to determine the necessary risk reduction or SIL for each SIF. This may also be recorded and documented within the same cause and effect diagram during the configuration of each specific effect as shown below in Figure 9.

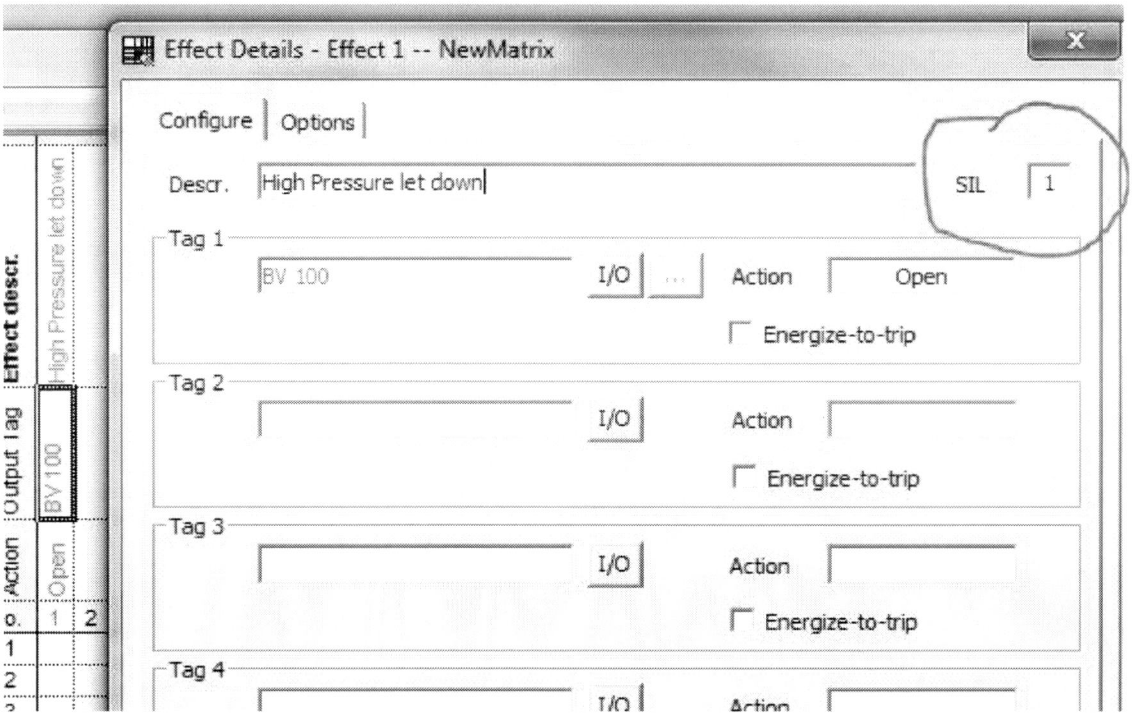

Figure 9. Recording The SIL Requirement For Each SIF

Clause 10: SIS safety requirements specification (SRS)

The objective of this clause is to specify the requirements for the SIS including a clear and concise description of the application program as well as the architecture of the SIS.

Cause and effects diagrams play a crucial role in providing an intuitive documentation format for the application program. Conventional programming tools such as ladder logic, or function block diagrams are not considered adequate since they do not convey the ease of comprehension, therefore it would typically require a separate control narrative to verbally describe the safety functions that are being executed in the application program. Cause and effects diagrams will generally contain the configuration details (architecture voting, input and output tags, etc.) along with the options (bypass tags, time delays, etc.) to fully satisfy the requirements of the SRS.

Clause 11: SIS design and engineering

The objective of this clause is to design and engineer the SIS to meet the specified integrity requirements (e.g. SIL, risk reduction, PFD, etc.). In many cases SIS designers have found success leveraging cause and effect diagrams to help meet some of clause 11 requirements for operability, maintainability, diagnostics, inspections and testability to reduce the likelihood of dangerous failures. The level of implementation will vary from using the cause and effects as the design document to a full replication of an active cause and effects diagram running in the SIS logic solver, where the latter would be preferred as it would remove potential human errors during the translation from to the application program. In some cases end-users have reported that this feature alone can reduce their detailed design effort by 50%, not to mention the reduction in translation errors.

Clause 12: SIS application program development

The objective of this clause is to define the requirements for the development of the application program. In the first edition of the standard this clause was 17 pages, it is now 3 and a half pages. This clause has been significantly rewritten to remove life-cycle steps covered in other sections, and now only provides useful information around the requirements for proper application design, development and implementation.

Using a cause and effects tool that is capable of developing the application program will vary from SIS vendors, thus to avoid commercial conflicts this clause will not be covered.

Clause 13: Factory acceptance test (FAT)

The objective of this clause is to test the devices of the SIS to ensure that the requirements defined in the SRS are met. Historically this process involved comparing the functional operation of the application logic (what it is supposed to do) to the traditional programmable logic diagrams (what it's going to do) that are designed and programmed in the system such as ladder logic diagrams, function blocks, etc. A cause and effect diagram could further support the program validation depending on the system and its level of integration using cause and effect diagrams. In some cases it may be available to test the application logic, via an intuitive interface that illustrates the logic in a simple and concise manner such as the cause and effect diagram rather than having to trace the logic thru one of the other conventional methods. This would require that the system being used is capable of supporting both the ability to monitor online visualization of the logic, as well as a means to 'record' the interaction to generate a validation report to prove that the logic provided the correct outputs as it was designed.

Clause 14 & 15: SIS installation and commissioning and safety validation

The objective of these two clauses are to ensure that the SIS is properly installed, commissioned and validated according to the requirements as stated in the SRS. Much of the requirements listed in clause 14 are geared toward the field devices (properly installed, grounded, calibrated and configured, etc.) however a detailed cause and effect diagram could be vital in recording and identifying these devices as well as documenting the interactions between the sensors (inputs) all the way thru to the final control elements (outputs).

Clause 15 defines all the required validation planning activities for the entire SIS which is sometimes referred to as the site acceptance test (SAT).

Clause 16: SIS operation and maintenance

The objective of this clause is to ensure that the required SIL for each SIF is maintained during operation and maintenance of the SIS. Depending on how integrated the cause and effect diagram is to the safety PLC operating the SIFs, will vary to what functions are available to meet the requirements of this objective. Owner/operators would need to verify what their current system is capable of covering, but below are a few examples of how a cause and effect diagram maybe used to meet the operation and maintenance requirements:

- System diagnostics that impact the performance of specific SIFs might be displayed on a cause and effect diagram to quickly draw the attention to the operator to reduce the time to repair.
- Inspection and proof testing maybe conducted, recorded and documented using the online capability of the cause and effect diagram to meet SIL verification requirements.

Clause 17 & 18: SIS modification and decommissioning

The objectives of clauses 17 and 18 are to ensure that any changes (modification or decommissioning) to the SIS are properly planned, reviewed, approved and documented. Today it's well understood that depending on the plants circumstances, modifications may be necessary either online or offline and in either case rigorous revision tracking would be required, with the ultimate goal of maintaining the performance of the SIS. In the case where a cause and effect diagram is well integrated with the SIS logic solver, much of the required documentation can be automatically updated as the system goes thru modifications as well as automatically updating the SRS to keep it current.

Clause 19: Information and documentation requirements

The objective of this clause is to ensure that all of the necessary information is available and documented to support all phases of the SIS safety life-cycle. Figure 10 below summarizes the typical workflow between the safety life-cycle and a cause and effect program to minimize human errors through all phases.

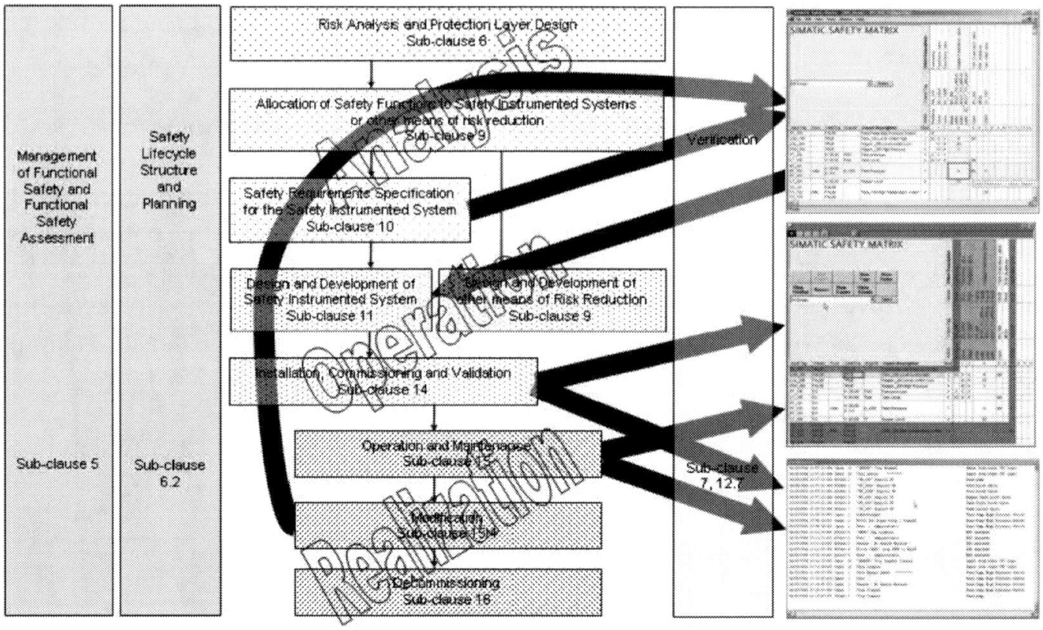

Figure 10. Safety Lifecycle Workflow

4 Summary

The safety life-cycle is regarded as the foundation of the international functional safety standard for the process industries. It recognized as an important engineering methodology, but documentation intensive process that is prone to errors. A number of tools have been developed throughout the industry to try and address some of the leading contributors to failures (specifications, installation and commissioning, operations and maintenance, etc.) which are largely tied to human errors. One approach the industry is looking towards is how a comprehensive cause and effect methodology that covers the SIS design, programming, operations, and maintenance tool tying together all phases to address the impact of human error through the life-cycle.

5 References

1. U.K. Health and Safety Executive, Out of control: Why control systems go wrong and how to prevent failure, Available at http://www.hse.gov.uk/pubns/books/hsg238.htm Accessed on December 8, 2016
2. IEC 61511-1: 2016, Functional Safety – safety instrumented systems for the process industry sector
3. Permissive Sequencing and ISA 84 -- The Shape of Things to Come By Gene Cammack, PE; Francisco Sanchez, PDVSA and Luis M. Garcia G. CFSE Siemens Energy & Automation, Houston, Texas 2008
4. SIMATIC Safety Matrix V 6.2 Configuration Manual, Copy right 1995-2015, Siemens AG
5. Innovative new tools for automating SAFE Charts as required for API RP 14C Charles M. Fialkowski: Siemens Energy & Automation; Spring House, PA USA Presented at IEEE – PCIC technical conference 2006

A simplified process hazards analysis (PHA) approach for non-covered processes

Jonas Duarte
Chemtura
Ontario, Canada
jonas.duarte@chemtura.com

Prepared for Presentation at
American Institute of Chemical Engineers
2017 Spring Meeting and 13th Global Congress on Process Safety
San Antonio, TX
March 26 – 29, 2017

A simplified process hazards analysis (PHA) approach for non-covered processes

Jonas Duarte
Chemtura
Ontario, Canada
jonas.duarte@chemtura.com

Keywords: hazard identification, risk analysis, manage risks and process safety

Abstract

Often times plant managers and engineers discuss how they should address hazard identification of non-covered processes units with the same quality of covered processes at chemical plants. This paper will review a simplified PHA approach applied for non-covered processes in need of hazard identification. The proposed approach is to use a PHA screening checklist with several level questions ranging from a regulatory check to basic risk assessment. The screening checklist uses questions applied to categories such as process, health, safety, environment, and security. The intent is to provide guidance to manufacturing facilities to determine if there are any processes that may not require a detailed PHA (i.e. Hazop, what-if, etc.) based on the inherent hazards of the raw materials, intermediates, and finished goods. The general PHA screening might contain specific sections.

Local regulatory requirements other than Process safety management (PSM): -environmental compliance, building code, electrical code, nuclear, or other specific code;

Hazard screening; -reactive hazard (i.e. water, air, acid, bases, etc.); -flammable materials (i.e. any quantity of liquid, gas, solid below limit of local regulatory threshold;

-dust material (i.e. dealing with fugitive emissions or transportation of raw material, waste or by product; -hazardous dangerous goods including waste; -hazardous operating condition (i.e. pressure, temperature used for storage or transfer); -proximity to adjacent covered process (i.e. potential risk effects from other process); -toxic or corrosive materials (i.e. health and safety issues as result of potential release or exposure).

1 Introduction

Often times plant managers and engineers discuss how they should address hazard identification of non-covered processes units with the same quality of covered processes at chemical plants. This paper will review a simplified PHA approach applied for non-covered processes in needs for hazard identification. The proposal approach is to use a PHA screening checklist with several level questions from regulatory check to basic risk assessment. The screening checklist uses questions applied to categories such as process, health, safety, environment, and security. The intent is to provide guidance to manufacturing facilities to determine if there are any processes that may not require a PHA (i.e. Hazop, what-if, etc.) based on the inherent hazards of the raw materials, intermediates, finished goods. The general PHA screening might contain specific sections.

2 Definitions

For purpose of limiting scope of this paper the below definitions apply:

"Non-Covered Process" is a process which **does not** fit in to a definition of "Covered Process" which means a process subject to regulatory requirements established under the OSHA PSM Standard[1] or the EPA RMP[2] Rule.

"Hazardous Chemical" means a material that is toxic, reactive, or flammable and is capable of causing a process safety incident if released. [6]

"Hazardous Material", in a broad sense, any substance or mixture of substances having properties capable of producing adverse effects to the health or safety of human beings or the environment. Material presenting dangers beyond the fire problems relating to flash point and boiling point. These dangers may arise from, but are not limited to, toxicity, reactivity, instability, or corrosivity. [6]

"Process Hazard Analysis", means an organized effort to identify and evaluate hazards associated with processes and operations to enable their control. This review normally involves the use of qualitative techniques to identify and assess the significance of hazards. Conclusions and appropriate recommendations are developed. Occasionally, quantitative methods are used to help prioritized risk reduction. [6]

"Process safety information", means physical, chemical, and toxicological information related to the chemicals, process, and equipment. It is used to document the configuration of a process, its characteristics, its limitations, and as data for process hazard analyses. [6]

3 Suggested process safety information

The process safety review of any process should start with collecting data about the process including trade secret if needed. Then the screening process starts such as:

- List of process chemicals, chemistry, and maximum intended inventory of feedstock and in-process.

- Compile safety data sheet information about chemicals present at the site such as flammable, combustible, reactive, toxic, and corrosive.

- List process units, connective, and supply chain activities throughout of the facility.

- Overview of facility siting layout of what chemical process operations are performed indoors, or in close proximity to occupied buildings.

- Process flow diagram, piping and instrument diagrams, plot plants, electrical classification, etc.

- Operating procedures, safe operating limits, and safety systems.

This PHA practical approach intend to use series of checklist to provide guidance to manufacturing facilities in order to determine if there are any processes that may require a PHA (process hazards analysis) based on the inherent hazards of the raw materials, intermediates, waste, and finished goods.

4 Risk assessment of major hazards

The risk assessment might be required, depending of complexity and detail of review. It is suggested to document potential impacts to health, safety, environmental, security, and public matters. For the purposes of this paper it is suggested to determine minimum evaluation if risk is acceptable or not acceptable. If necessary a more complex risk matrix can be used for specific hazards, but for this approach use following decision making:

- Risk is acceptable with existing safeguards. Document and validate administrative controls in place; or

- Risk is not acceptable and mitigation is required. Document actions, responsible person, and timeline.

Identify potential for failures events that could lead to incident classified as API Tier 1 or 2 considering UN dangerous good chemical list. See detail on sections 7 and 8.4 assessment checklist.

5 – Validate exemptions from PSM / RMP regulatory requirements

Most important step and sometimes difficult is the determination of each process unit is a non-covered process as per OSHA PSM Standard[1] or EPA RMP Rule.[2]. It all starts with detailing analysis of chemical safety data sheet and its inventory in storages and inventory in process. Further, a critical analysis on definitions of connectivity of a process unit to other process units or process and utilities is also important.

It is highly recommended to document the rational for regulatory exceptions before a determination of non-covered process of a facility or process unit. Moreover, extensive validation for inventory of chemicals including waste and by products that doesn't exceed the threshold of chemicals listed on OSHA PSM Standard[1] or EPA RMP Rule.[2].

5.1 - PSM / RMP regulatory checklist

This simplified PHA screening approach assumes that exemption from PSM \ RMP regulatory have been determined. Possible conclusions are:

No - PHA Non-Covered Process;

Yes - PHA Covered Process Required.

Specific Regulatory Applicability	Conclusion
a) Are storages of hazardous chemical **at or above** the threshold quantities specified in 29CFR 1910.119 Appendix A? [1]	<Select>
Provide rationale & supporting information to justify overall conclusion: (List Chemicals and Quantity):	
b) Are storages of flammable liquid or gas (see specific definition for flammable liquid or gas in the Introduction) onsite in one location in a quantity **at or above** 10,000 lb. (4,535.9 kg)? [1] OSHA PSM 1910119 /[2] EPA RMP :	<Select>
Provide rationale & supporting information to justify overall conclusion: (List Chemicals and Quantity):	
c) Are storages of regulated substances **at or above** threshold quantity? The Risk Management Plan (RMP) rule provides a List of Regulated Substances under section 112(r) of the Clean Air Act. This includes their synonyms and threshold quantities (in pounds) to help assess if a process is subject to the RMP rule. [2]	<Select>
Provide rationale & supporting information to justify overall conclusion: (List Chemicals and Quantity):	

6 – Local regulatory or other requirements

What if scenario / question	**Impact**	**Risk**	**Mitigation**
a) Have there been any local Environmental regulatory update (air emissions, hazardous waste, soil, noise) that		<Select>	

impact the facility?			
b) Have there been any local regulatory updates such as fire code, building code, electrical code that impacts the facility?		<Select>	
Have there been any changes to City zoning by law within proximity to the facility that requires update of emergency response plan?		<Select>	
		<Select>	

No: Risk Acceptable with existing safeguards; Yes: Risk not acceptable; needs mitigation

7 - Incident and lessons learned

Hazard recognition and identification is continuous improved by reviewing lessons learned from industry incidents throughout public domain such as media, newspaper, standards organizations, professional associations, and networks.

What if scenario / question	Impact	Risk	Mitigation
b) Has there been new industry incident information (e.g. internal Process Safety Alerts, Insurance claims, NFPA, and US Chemical Safety Board) such that the process is considered hazardous that could result in multiple injuries, and or off-site impact?		<Select>	
c) Has there been any incident resulting in multiple injuries, serious near miss, or recordable		<Select>	

that have occurred in the process?			
d) Has there been any significant loss primary to environment and potential for soil or ground water contamination, or offsite pubic impact?		<Select>	
e) Are there potential for failures leading to incident classified as API Tier 1 or 2?		<Select>	
f) ?		<Select>	

No: Risk Acceptable with existing safeguards; Yes: Risk not acceptable; needs mitigation

8 – Steps for screening process safety risk

8.1– Reactive hazards

Assess the potential impacts of reactivity hazards, and evaluate safeguards and administrative controls. Reactive hazard (i.e. inadvertent mixing, water, air, strong acid, strong bases, oxidizing, etc.).

What if scenario / question	**Impact**	**Risk**	**Mitigation**
a) Are there any materials that if improperly mixed will result in an exothermic reaction adequate to heat the mass 30°C or more?		<Select>	
b) Are there any materials that if heated, could result in a decomposition of the material / run away reaction resulting in high pressure or vapor emission?		<Select>	

c) Are there any chemical reactive with water or air?		<Select>	
d) Are there any unstable (reactive) liquid shall mean a liquid which in the pure state or as commercially produced or transported will vigorously polymerize, decompose, condense, or will become self-reactive under conditions of shocks, pressure, or temperature? [3] 1910.106(a)(20)		<Select>	
e) ?		<Select>	

No: Risk Acceptable with existing safeguards; Yes: Risk not acceptable; needs mitigation

8.2 – Flammable and combustible liquids

Assess the potential impacts of fire or explosions with handling flammable materials, and evaluate safeguards and administrative controls. Flammable materials (i.e. quality of liquid, gas, solid below limit of local regulatory threshold).

What if scenario / question	Impact	Risk	Mitigation
a) Are there any materials that can / could be processed within 10C of it flashpoint? [3]		<Select>	
b) Any material with a flash point less than 140°F (60°C)? [4]		<Select>	
c) Are there any pyrophoric as raw materials or catalyst?		<Select>	
d) Any flammable material listed as a Class 1 flammable per		<Select>	

NFPA in a quantity greater than 2500 lbs.? [4]			
e) Have there been any significant management changes that potential created hazards leading to a fire or explosion? (1) materials in process (2) in process equipment (3) in process control (4) in operating procedures or assignments. NFPA30 [4]		<Select>	
f) ?		<Select>	

No: Risk Acceptable with existing safeguards; Yes: Risk not acceptable; needs mitigation

8.3 – Dust hazards [5]

Assess the potential impacts of handling powders with potential dust explosion, and evaluate safeguards and administrative controls. Dust material (i.e. dealing with fugitive emissions or transportation of raw material, waste or by product).

What if scenario / question	Impact	Risk	Mitigation
a) Is there a combustible dust with Minimum Ignition Energy (MIE) < 200 mile-joules that is susceptible to air borne dispersion?		<Select>	
b) Is there any dust that can be dispersed in concentrations above its Minimum Explosive Concentration (MEC)?		<Select>	
c) Is there a powder material that requires anti-static control handling recommended by manufacturer?		<Select>	

e) Are there any dust materials with St-2 or St-3 explosion classification?		<Select>	
d) Are there anti-static control required powders to be added into vessel with flammable atmosphere?		<Select>	
f) ?		<Select>	

No: Risk Acceptable with existing safeguards; Yes: Risk not acceptable; needs mitigation

8.4 – Hazardous material

Assess the potential impacts of handling hazardous material including waste, and evaluate safeguards and administrative controls. Hazardous dangerous goods including waste; -hazardous operating condition (i.e. pressure, temperature used for storage or transfer).

What if scenario / question	**Impact**	**Risk**	**Mitigation**
a) Any UNDG (United Nations Dangerous Goods) listed material that is stored or handled in quantities greater than the threshold quantity?		<Select>	
b) Any material listed as a marine pollutant in quantity that could adversely impact a local or adjacent waterway?		<Select>	
c) ?		<Select>	

No: Risk Acceptable with existing safeguards; Yes: Risk not acceptable; needs mitigation

8.5– Operating condition

Assess the potential impacts of uncontrolled process conditions, and evaluate safeguards and administrative controls.

What if scenario /	**Impact**	**Risk**	**Mitigation**

question			
a) Can the process exceed pressure of greater than 15 PSI? (a utility such as water / air / nitrogen may not apply)		<Select>	
b) Can the process exceed temperatures in excess of 60°C that could result in serious injury or hazardous vaporization or impact to environment / community?		<Select>	
c) Does the process require onsite pressurized loading or unloading of DOT\IMG Bulk or Portable Container?		<Select>	
d) Does the process handle equipment with radiation source?		<Select>	
e) Does the process handle compressed gas as raw material or refrigerant media?		<Select>	
f) Are relief devices sized, and installed properly on pressure retaining equipment?		<Select>	
g) ?		<Select>	

No: Risk Acceptable with existing safeguards; Yes: Risk not acceptable; needs mitigation

8.6– Siting & business continuity

Assess the process unit siting layout for potential impacts to business continuity or impact on other units in proximity, and evaluate safeguards and administrative controls. Proximity to adjacent covered process (i.e., potential risk effects from other process).

What if scenario / question	Impact	Risk	Mitigation
a) Does the process have any indirect feedstock / supply chain to\from a process that currently is required to have a PHA for covered process?		<Select>	
b) Could the process impact an adjacent process and result in the release of a hazardous material causing injuries or significant damage?		<Select>	
c) Are there safe exit routes from process unit?		<Select>	
d) Are there adequate fire protection and emergency equipment?		<Select>	
e) Are the inventory of flammable and toxic held to a minimum?		<Select>	
f) ?		<Select>	

No: Risk Acceptable with existing safeguards; Yes: Risk not acceptable; needs mitigation

8.7– Toxic material (Reference: Safety Data Sheet)

Assess the potential impacts of handling toxic substances, and evaluate safeguards and administrative controls for toxic or corrosive materials (i.e., health and safety issues as result of potential release or exposure).

What if scenario / question	Impact	Risk	Mitigation
a) Are there potential for acute toxicity LD50 or LC50 exposure above ACGIH limits? (e.g. Lethal Dose (LD) 50,Lethal Concentration (LC) 50)		<Select>	

b) Are any materials handles classified as a Carcinogenic, Mutagenic or Reproduction-toxic substances?		<Select>	
c) Are there potential for offsite impact such ERPG? AIHA[8] at exposure limits? (e.g. Emergency Response Planning Guideline-ERPG)		<Select>	
d) ?		<Select>	

No: Risk Acceptable with existing safeguards; Yes: Risk not acceptable; needs mitigation

8.8 – Human factors

Assess the human factors and employee interaction that could impact employee ability to make an informed decision and potential create workplace hazards, and evaluate safeguards and administrative controls.

What if scenario / question	**Impact**	**Risk**	**Mitigation**
a) Is the ambient temperature normally within comfortable bounds?		<Select>	
b) Is noise maintained at a tolerable level (<80dBA)?		<Select>	
c) Is the lighting sufficient for all process unit operations?		<Select>	
d) Could heat stress and fatigue on its impact on alertness?		<Select>	
e) Is ventilation adequate to prevent exposure to fumes \ vapors?		<Select>	
e) Is there safety shower available?		<Select>	

f) ?		<Select>	

No: Risk Acceptable with existing safeguards; Yes: Risk not acceptable; needs mitigation

8.9 – Other "what if" scenarios

Assess the overall process design, procedures, and technology for potential failures. Identify impact, document risk assessment, and evaluate safeguards and administrative controls.

What if scenario / question	Impact	Risk	Mitigation
a) What if high level on a tank\vessel overflows?		<Select>	
b) What if high pressure causes rupture of a tank\vessel?		<Select>	
c) What if high temperature causes process upset or unexpected hazards?		<Select>	
d) ?		<Select>	

No: Risk Acceptable with existing safeguards; Yes: Risk not acceptable; needs mitigation

9 Suggested resources

Resources depend on head count of the Facility, and complexity of process.

- Whenever possible use a cross-functional, and interdisciplinary team having knowledge of the activities, processes or products.
- Possibly a team consisting of engineer, operating personnel, and health & safety staff.
- Others resources where appropriate such as maintenance personnel, chemist, industry expert, etc.
- Management, and supervisors may fill in some of these required positions at small Facilities.

10 Conclusion

As per below example a decision making process can be documented to make sure the process unit is evaluated and assessed for regulatory and associated risk identified.

☐	**PHA Simplified Approach for Non-Covered Process** *This PHA screening satisfies minimum hazard identification. Choose this option, if no local regulatory applies to the process.*
☐	**PHA as per Site Policy** *Local jurisdiction supersedes these requirements. Local jurisdiction supersedes these requirements. If any question in the Section 1 is answered "YES", then "Regulatory PHA" might be required. Therefore, choose this option to comply with the Site PHA Policy.*

Certify the information has been accurately stated to the best of knowledge available at the time, and make sure document must be kept for the life of the process.

This paper assumed that similar potential hazards of a covered process might occur on non-covered process but with much less degree of risks and potential severity. Therefore, the screening checklist helps to identify hazards, evaluate impact severity, and document mitigation if required.

This simplified PHA approach might be useful for hazard identification on pilot plant, laboratories, warehouses of small packaging and portable containers, small chemical and manufacturing facilities

11 References

[1] Occupational Safety and health administration 29 Code of Federal Regulations (CFR) 1910.119, Process Safety Management of Highly Hazardous Chemicals (PSM). Appendix A

[2] Risk Management Program for Chemical Accidental Release Prevention, U.S. Environmental Protection Agency, 40 CFR Part 68.

[3] Occupational Safety and health administration 29 Code of Federal Regulations (CFR) 1910.106 Flammable and combustible liquids

[4] National Fire Protection Association, 30 flammable and combustible liquids code, 2015 edition. Chapter 6 Fire and Explosion Prevention and Risk Control 6.4.2 Management of Change.

[5] National Fire Protection Association, 652 Standard on fundamentals of combustible dust, 2016 edition

[6] Center for Chemical Process Safety. CCPS Process Safety Glossary

[7] U.S. Chemical Safety and Hazard Investigation Board

[8] American Conference of Industrial Hygienist-(ACGIH)

Beyond Compliance – Using PSM to Enhance the Grain Handling Facility Standard

Michael S. Schmidt
Bluefield Process Safety
3101 Cherokee St.
St. Louis, MO 63118
bluefieldsafety@gmail.com

Luca Clemons
Bluefield Process Safety
3101 Cherokee St.
St. Louis, MO 63118
luca.clemons.bluefieldsafety@gmail.com

Prepared for Presentation at
American Institute of Chemical Engineers
2017 Spring Meeting and 13th Global Congress on Process Safety
San Antonio, TX
March 26 – 29, 2017

Beyond Compliance – Using PSM to Enhance the Grain Handling Facility Standard

Michael S. Schmidt
Bluefield Process Safety
3101 Cherokee St.
St. Louis, MO 63118
bluefieldsafety@gmail.com

Luca Clemons
Bluefield Process Safety
3101 Cherokee St.
St. Louis, MO 63118
luca.clemons.bluefieldsafety@gmail.com

Keywords: OSHA, Process Safety Management, PSM, Grain Handling, Combustible Dust

Abstract

The purpose of OSHA's Process Safety Management standard, 29 CFR 1910.119, is to prevent fires, explosions, and toxic releases. The PSM standard covers processes based on chemical properties, but it is well recognized that not all facilities that are vulnerable to fires, explosions, and toxic releases are covered under the PSM standard. Moreover, fire, explosion, and toxic hazards based on physical properties are not addressed at all by the PSM standard. One industry segment that is not covered is that handling combustible dusts. Instead, combustible dusts are addressed, in part, by the Grain Handling Facility standard, 29 CFR 1910.272.

Recognizing the contribution that process safety management (PSM) has to make to improving safety in the grain handling industry, there is a movement by some in the industry to apply the principals of PSM to the design, operation, and maintenance of grain handling facilities. Nonetheless, even as they incorporate the elements of PSM into grain handling, these facilities are still compelled to comply with the requirements of the Grain Handling Facility (GHF) standard. To the extent that grain handling facilities can use PSM to comply with 29 CFR 1910.272, the effort will be more efficient and easier to sell to other parts of their organization.

This paper compares and contrasts the two standards and shows how the elements of the PSM standard can be used to comply with the GHF standard. It also shows the gaps in the grain handling standard addressed by the PSM standard and how PSM strengthens the overall process safety effort. Finally, it discusses the implementation of this approach in a recently constructed grain handling facility.

1 Introduction

Following a series of catastrophic incidents at facilities that handle grain and seeds, the Occupational Safety and Health Administration (OSHA) promulgated the Grain Handling Facilities (GHF) standard—29 CFR 1910.272—in 1988 to address combustible dust explosions. [1] As is common to OSHA regulations, the GHF standard begins with introductory paragraphs that establish and limit the scope of the standard as well as define the facilities to which it applies. These introductory paragraphs are followed by 14 sections that address the elements of the GHF standard. The GHF standard concludes with three non-mandatory appendices: guidance on how to comply with the standard, a list of NFPA consensus standards that can be followed in lieu of following certain sections of the GHF standard, and a list of references that may be helpful to covered facilities.

Similarly, following a series of catastrophic incidents in the chemical process industries, OSHA promulgated the Process Safety Management (PSM) standard—29 CFR 1910.119—in 1992 to address fires, explosions, and toxic releases in facilities that handle or use highly hazardous chemicals. [2] Again, the PSM standard begins with introductory paragraphs that establish and limit the scope of the standard as well as define the facilities to which it applies. These introductory paragraphs are also followed by 14 sections that address the elements of the PSM standard. The PSM standard concludes with four appendices: a mandatory list of toxics and highly reactive hazardous chemicals and their threshold quantities and three non-mandatory guidance documents. The three non-mandatory guidance documents give suggestions on what flow diagrams are, guidance on how to comply with the standard, and a list of references that may be helpful to covered facilities.

Both standards are similar in their intent: preventing catastrophic events associated with hazardous materials. However, the GHF standard was included in Subpart R, Special Industries, and so has a very specific industry focus, while the PSM standard was included in Subpart H, Hazardous Materials, and is intended for any facility that has defined process safety hazards.

It is simply a coincidence that both standards have fourteen elements. More interesting is that there are five elements that are common to both:

- Training
- Contractors
- Mechanical Integrity
- Hot Work Permits
- Emergency Planning and Response

There are nine elements in the GHF standard that are unique to that standard. They are specific to the grain handling industry, focusing on grain handling and grain handling equipment, and so have little to offer to other industries in the way of insights toward operating more safely. On the other hand, there are also nine elements in the PSM standard that are unique to the PSM standard. Because the PSM standard directs its attention to material hazards rather than to a specific industry, the nine elements that are unique to the PSM standard have the potential to contribute to the safety of any industry dealing with material hazards.

2 Common Elements

While the two regulations have five elements in common—training, contractor safety, mechanical integrity, hot work permits, and emergency planning and response—their approaches to these elements are not necessarily the same. In general, the GHF standard is less sophisticated in its approach. An element-by-element comparison between the two standards will be useful in understanding the differences and similarities.

2.1 Training

Both standards have training requirements. Both duplicate training requirements called for in other regulations, such as emergency action plans [3], confined space entry [4], lockout/tagout [5], and hot work [6], thus giving OSHA inspectors an opportunity to cite a facility twice for failure to comply with a training requirement stipulated elsewhere.

Each standard requires that employees be trained on the material hazards in their workplace; workers in grain handling facilities are required to be trained on dust explosion hazards and engulfment hazards, while workers in process facilities are required to be trained on the hazards of the process. Additionally, both standards require that workers be trained on the maintenance or mechanical integrity tasks they are expected to perform.

Both standards require refresher training. The GHF standard requires annual refresher training while the PSM standard requires refresher training at least once every three years for operating personnel.

Where the PSM standard goes beyond the GHF standard is that it also requires that operating personnel must be trained on operating procedures, and requires that mandated training be documented with specific information.

2.2 Contractors

Both standards require that employers inform contract employers of potential fire and explosion hazards related to the contractors' work. Both standards also require that employers explain the applicable provisions of the emergency action plan to contractors. The PSM standard exempts certain categories of contractors from its requirements, for example janitors, food services, laundry services, and delivery services.

In regard to contractor safety, the PSM standard goes beyond the GHF standard. It requires that employers evaluate the safety record of contract employers. It requires that employers assure that contract employees are trained, and it lists the responsibilities of contract employers as well as the responsibility of employers. Also, the PSM standard requires that access by contractors to covered processes be restricted and controlled.

2.3 Mechanical Integrity/Preventative Maintenance

Both standards require programs to assure the mechanical integrity of specific types of equipment in the covered process. Under the GHF standard, the specific types of equipment are

- ❖ Dryers
- ❖ Grain stream processing equipment
- ❖ Dust collection equipment including filter collectors
- ❖ Bucket elevators

Under PSM, the specific equipment types are

- ❖ Pressure vessels and storage tanks
- ❖ Piping systems (including piping components such as valves)
- ❖ Relief and vent systems and devices
- ❖ Emergency shutdown systems
- ❖ Controls (including monitoring devices and sensors, alarms, and interlocks)
- ❖ Pumps

Clearly, the differences in equipment types reflect OSHA expectations about which types of equipment will be critical to the safe operation of each type of facility.

Nonetheless, both standards require that equipment be inspected, tested, and maintained in accordance with manufacturer's recommendations or other recognized and generally accepted good engineering practices (although it only the PSM standard that invokes that awkward phrase that has come to be known as "RAGAGEP") and that records of the inspection, testing, and maintenance be kept. Likewise, both standards require that equipment deficiencies be corrected promptly.

The principal difference between the two regulations in regard to mechanical integrity and preventative maintenance is that the PSM standard requires a quality assurance program to assure that equipment is installed properly and more importantly, that maintenance materials and spare parts be suitable. The GHF standard does not. This is understandable. Maintenance materials and spare parts in grain handling facilities are far less vulnerable to hazardous incompatibilities than similar parts in chemical processes.

2.4 Hot Work

Both standards require hot work procedures and compliance with the Hot Work standard, 29 CFR 1910.252(a), which are required of facilities regardless of whether or not the Hot Work standard was invoked.

The two regulations differ in that the GHF standard is less restrictive in its requirements of how to comply with the Hot Work standard. Specifically, the Hot Work standard requires an authorization procedure, and suggests a written permit as a means of implementing this authorization procedure. The GHF standard requires a written permit in covered areas except when the employer's representative will be present. Under the PSM standard, written hot work permits are required for all hot work in covered areas.

2.5 Emergency Planning and Response

Both standards require employers to have emergency action plans that are also compliant with the Emergency Action Plan standard, 29 CFR 1910.38, which again, facilities would be required to have regardless of whether or not these required it.

The PSM standard goes beyond the Emergency Action Plan standard, however, in that it requires that a facility's emergency action plan include procedures for handling small releases. It also reminds employers that if employees will be responding to releases, they may also be subject to the provisions of the Hazardous Waste Operations and Emergency Response standard, 29 CFR 1910.120 (HazWOpER). [7]

It makes sense that grain handling facilities would not be subject to the additional requirements for small releases and spill response imposed by the PSM standard. There is no hazard imposed by small release of grain and a response to a grain spill will not be life threatening.

3 Elements Unique to PSM

In addition to the five elements of the PSM standard that are already addressed by the GHF standard, there are nine elements of the PSM standard that are not:

- Employee Participation
- Process Safety Information
- Process Hazard Analysis
- Operating Procedures
- Pre-Startup Safety Review
- Management of Change
- Incident Investigation
- Compliance Audits
- Trade Secrets

The following sections review the intent of these elements in terms of improved safety, summarize the regulatory requirements of compliance, and discuss the applicability of each element to grain handling facilities.

3.1 Employee participation

The intent of the employee participation element is to create a system that promotes a cooperative flow of process safety knowledge from employer representative to employee and vice versa. Consulting with employees and representatives on process safety issues aids an employer in addressing their concerns and suggestions. Employees' knowledge of how the process runs and of their own experiences provide relevant information to the facility.

The PSM standard requires a written plan, participation by employees in hazard review, consultation with employees on the development of all the elements of process safety, and access to safety information.

These requirements, when done voluntarily and in a spirit of cooperation, can enhance the safety culture of an organization. When "employee participation" is compulsory, either on the part of the employer or the employee, the effect is to breed resentment and undermine the safety culture. Implementing employee participation in the grain industry on a voluntary, cooperative basis is a worthwhile part of developing a positive safety culture and a valuable means sharing safety information, both top down and bottom up.

3.2 Process Safety Information

The intent of the process safety information element is to assure that data about the key aspects of the process is available, so that adequate hazard reviews can be performed.

The PSM standard requires that specific information be compiled in three areas: hazards of the substances in the process, the technology of the process, and the equipment in the process.

Some of the process safety information required by the PSM sandard is not relevant to grain handling facilities. For instance, grain is not toxic, reactive, corrosive, unstable, or hazardous when mixed with other grain. On the other hand, grain dust can pose both a respiratory hazard and an explosive hazard. So, it is still reasonable to compile all of the information that is relevant. This would include data about dust combustibility, process flows, operating limits, and equipment specifications. Having the relevant information available is essential to performing meaningful hazard analyses.

Where many grain handling facilities are likely to balk is at the requirement to document that equipment complies with RAGAGEP. It is not that the industry is opposed to using codes and standards, where they are available, but that to retroactively document compliance with non-existent standards is burdensome. Moreover, there is a sense that OSHA is more interested in invoking obscure RAGAGEP as a bludgeon than in using RAGAGEP to promote safety.

3.3 Process Hazard Analysis

The intent of the process hazard analysis (PHA) element is to assure that a facility has identified all of the specific hazards in its process, understands the consequences associated with those hazards, and has addressed the risk of those hazards with the appropriate safeguards.

The PSM standard has requirements about the make-up of the hazard review team, about the PHA methods that are considered acceptable, about the frequency at which reviews must be conducted and validated, about the topics that must be addressed, about resolving recommendations, and about recordkeeping.

The need for PHAs in the grain handling industry has been recognized for some time. The National Fire Protection Association (NFPA) recently developed and published a national consensus standard, *Standard on the Fundamentals of Combustible Dust*, NFPA 652. Among other things, NFPA 652 requires a dust hazard analysis (DHA). The goal of a DHA is to identify and document hazards related to fires, deflagrations, and explosion hazards of combustible dusts. The requirements for a DHA spelled out in NFPA 652 are not as thorough as the requirements for a PHA listed in the PSM standard. While some of the requirements in the PSM standard may be questioned, such as retaining every PHA document ever created for the entire life of the process, there is much to recommend from the PSM standard. Safety at grain handling facilities will benefit from going beyond the focus on explosion hazards that DHAs have.

3.4 Operating procedures

The intent of the operating procedures element is to prevent operators from "winging it," to assure that there are clearly considered written procedures for normal operations and abnormal operations, that they reflect what is actually being done in the facility, and that they remind operators of the hazards they face so they can perform the procedures safely.

The PSM standard requires operating procedures that include all of this information for a specific set of normal and abnormal operating phases, that written procedures be readily accessible to operators, and that procedures be reviewed and certified for accuracy at least annually. In addition, the PSM standard requires procedures for safe work practices, most of which are procedures required by other regulations. A safe work practice that is unique to the PSM standard is the requirement for a procedure to control entrance into covered processes by contractors and non-operating personnel.

Grain handling facilities should have accurate operating procedures. Whether procedures to control entrance to the facility and whether annual certification in and of themselves provide any benefit are questions to be considered, but there should be no question that written procedures are required. It is well accepted that the probability of an error while performing a task without a written procedure is at least an order of magnitude higher than the probability of an error while performing an accurate, written procedure on which an operator has been trained, even if the written procedure is not in front of the operator when the procedure is performed.

3.5 Pre-startup safety review

The intent of the pre-startup safety review (PSSR) element is to ensure that a new or modified facility is ready before hazardous substances are introduced. It is much simpler and far less dangerous to address a deficiency before hazardous substances are present.

The PSM standard requires a PSSR for any new covered process or before starting an existing covered process that has been modified sufficiently to require changes to the process safety information. The review must assure that the process is built and installed, or modified, according to specifications, that all required procedures have been written, that a hazard review has been completed and recommendations resolved, and that everyone involved in the process has been trained. For a new process, the hazard review will be a PHA; for an existing process the hazard review will be through the MOC procedure.

Grain handling facilities, like any kind of facility, should be ready before being started.

- ❖ The facility should be built, installed, or modified according to specifications.

- All required procedures should be written before they are needed, and anyone using those procedures should be trained on them.
- The facility should be reviewed for hazards and any recommendations to address those hazards should be resolved.

The question for the grain handling industry, however, is about the urgency of a PSSR. Unlike processes covered by the PSM standard, the mere introduction of grain to a grain handling facility is very unlikely to create a hazard. The hazards in a grain handling facility are typically the result of accumulation, of running for an extended period of time. It could be argued that while all of the items of a PSSR should be done, some may be concurrent with startup, rather than pre-startup. The counter argument is that only by holding startup hostage to the PSSR is completion of all the PSSR items assured. This is a question that each organization needs to resolve for itself, as a matter of policy.

3.6 Management of change

The intent of the management of change (MOC) element is to assure that the hazards of any change are understood and taken into account before introducing the change, to assure than new hazards are not inadvertently introduced, and to assure that existing safeguards are not inadvertently removed.

The PSM standard requires a written MOC procedure and that a number of items be addressed as part of considering a potential change before the change is made. It requires authorization, that any process safety information or operating procedures impacted by the change be updated, and that anyone affected by the change be trained on the change before being exposed to it. The standard exempts "replacements-in-kind" from this requirement.

It is hard to argue that changes to a potentially hazardous process should not be managed. All of the specific requirements of the MOC element are applicable to grain handling facilities.

3.7 Incident investigation

The intent of the incident investigations element is to ensure that catastrophic incidents and potentially catastrophic incidents are understood, in order to put in place measures that will make similar incidents less likely in the future.

Except for facilities that had 10 or fewer employees throughout the prior year and facilities in certain industry classifications [9] considered particularly safe (which do not include industries covered by either the PSM standard or the GHF standard), all facilities are required to complete incident reports using OSHA Form 301 for any recordable injury within seven days of the incident.[10] The PSM standard requires an incident investigation for any catastrophic release of a highly hazardous chemical or any incident that reasonably could have resulted in such a catastrophic release (a near miss). A PSM incident investigation is required whether or not a recordable injury occurred.

Completing OSHA Form 301 is not a substitute for a PSM incident investigation. A PSM incident investigation is required to begin within 48 hours of the incident. It must result in a report containing specific types of information. Recommendations from the investigation must be resolved, the report must be reviewed with all affected personnel, and the report must be retained for five years.

Incident investigations that lead to changes that make catastrophic events less likely in the future are to be commended under any circumstances. While the timing requirements of the PSM standard should probably be viewed as guidelines rather than as essential to the purpose of incident investigations, the other PSM requirements make just as much sense for grain handling facilities as they do for PSM-covered processes.

It is important for grain handling facilities to define in advance the kinds of consequences that warrant incident investigations. Fires and explosions above a certain size certainly qualify. More challenging will be the defining in advance the kinds of events that will qualify as a near miss requiring investigation.

3.8 Compliance audits

The intent of the compliance audit element is to ensure that the rest of the PSM standard is being followed. It is the "check" step in the plan-do-check-act cycle common to most quality management methods.

The PSM standard requires that compliance audits be conducted at least every three years, that reports on each audit be prepared and certified, that the last two PSM audits be retained, and that recommendations from audits be tracked and resolved.

Philip Stanhope, the 4th Earl of Chesterfield said, "Whatever is worth doing at all is worth doing well." Process safety management is worth doing. Quality management is a way of assuring that it is done well and compliance audits are essential part of the "check" step of the quality cycle. The hesitation that any organization may have in conducting compliance audits is not in the audit, but in a possible OSHA inspection. There is belief that OSHA inspectors request copies of compliance audits when they are required or are believed to exist, not simply to confirm that audits exist, but to use the findings of the self-audit as a road map to violations and citations. That would suggest that voluntary compliance audits only be conducted by grain handling facilities that are prepared to respond immediately to any findings.

3.9 Trade secrets

The intent of the trade secret element of the PSM standard is to assure that all information relevant to the safety of the facility is available to employees that may need it, regardless of its trade secret status. Additionally, however, the PSM standard allows employers to require confidentiality agreements in exchange for sharing trade secret information.

The primary hazards associated with grain handling—combustible dust explosions and engulfment—are not subject to any information that grain handling facilities might consider trade secret. The effect, then, is that this element is simply not relevant.

4 A Case Study

The authors recently had an opportunity to work with a diverse group of safety, operations, and engineering professionals at a grain handling company on the occasion of the construction of a new grain handling facility. The review team came from the new facility, other facilities, and from the corporate offices. There were three purposes for the review:

- To comply with a corporate mandate to conduct a PHA consistent with the requirements of the PSM standard.
- To complete a general review of the facility while it was still under construction, in order to identify hazards and assure compliance with relevant regulations and standards so that any gaps discovered could be addressed before the facility was commissioned.
- To establish how personnel would conduct similar reviews at other facilities, whether new or existing, and to determine how the elements of PSM would be applied.

4.1 PHA

The grain handling facility had no processes covered under the PSM standard, so a PSM-compliant PHA was not required by regulation. Nonetheless, the hazard review was conducted in a manner consistent with the requirements of 29 CFR 1910.119(e), including

- The use of a PSM-allowed methodology (checklists),
- Addressing the hazards of the process, previous incidents, engineering and administrative controls, consequences, facility siting, human factors, and the possible safety and health effects of failure, and
- Assembling a team with expertise in engineering and process operations, including at least one employee who has experience and knowledge specific to the process being evaluated, and including a facilitator knowledgeable in the specific process hazard analysis methodology being used.

The review addressed the entire facility, including the grain dryer and grain elevators, the belt conveyor from the docks, the shop, and the roadway for truck traffic within the facility. It did not address the aspects of the maritime loading operation covered under OSHA's maritime terminal regulations found in 29 CFR 1917, or the administrative building on the site.

4.2 Compliance Review

In addition to the specific concerns of complying with the GHF standard, which was mandatory for this facility, this review addressed the broader issues of complying with OSHA general industry regulations and NFPA standards relevant to the design of the facility. The review did not extend to administrative and procedural compliance issues, except incidentally.

This review considered the following regulations and standards:

- 29 CFR 1910 Subpart D Walking-Working Surfaces, especially 1910.21-.30
- 29 CFR 1910 Subpart E Means of Egress, especially 1910.33-.39

- 29 CFR 1910 Subpart F Powered Platforms, Manlifts, and Vehicle-Mounted Work Platforms, especially 1910.67 Vehicle-mounted elevating and rotating work platforms
- 29 CFR 1910 Subpart G Occupational Heath and Environmental Control, especially 1910.94 Ventilation and 1910.95 Occupational noise exposure
- 29 CFR 1910 Subpart H Hazardous Materials, especially 1910.119 Process safety management of highly hazardous chemicals
- 29 CFR 1910 Subpart J General Environmental Controls, especially 1910.146 Permit-required confined spaces and 1910.147 Control of hazardous energy (lockout-tagout)
- 29 CFR 1910 Subpart L Fire Protection, especially 1910.155-.165
- 29 CFR 1910 Subpart M Compressed Gas and Compressed Air Equipment, especially 1910.169 Air receivers
- 29 CFR 1910 Subpart N Materials Handling and Storage, especially 1910.178 Powered industrial trucks
- 29 CFR 1910 Subpart O Machinery and Machine Guarding, especially 1910.211-.219
- 29 CFR 1910 Subpart P Hand and Portable Powered Tools and Other Hand-Held Equipment, especially 1910.241-.244
- 29 CFR 1910 Subpart Q Welding, Cutting, and Brazing, especially 1910.252(a) General requirements [for welding, cutting, and brazing]
- 29 CFR 1910 Subpart R Special Industries, especially 1910.272 Grain handling facilities
- 29 CFR 1910 Subpart S Electrical, especially 1910.301-.399
- 29 CFR 1910 Subpart Z Toxic and Hazardous Substance, especially 1910.1200 Hazard communication
- NFPA 1 Fire Code
- NFPA 10 Standard for Portable Fire Extinguishers
- NFPA 13 Standard for the Installation of Sprinkler Systems
- NFPA 61 Standard for the Prevention of Fires and Dust Explosions in Agricultural and Food Processing Facilities
- NFPA 69 Standard on Explosion Prevention Systems
- NFPA 70 National Electric Code
- NFPA 499 Recommended Practice for the Classification for Combustible Dusts and of Hazardous (Classified) Locations for Electrical Installations in Chemical Process Areas
- NFPA 654 Standard for the Prevention of Fire and Dust Explosions from the Manufacturing, Processing, and Handling of Combustible Particulate Solids

4.3 Applying the Elements of PSM

The review team considered each of the fourteen elements of the PSM standard and determined how they wanted to apply them at the facility under construction, and in the future, to other facilities, new and existing.

4.3.1 Employee Participation

The review team struggled to understand the distinction that OSHA makes between employees and employer representatives. Accept for the authors, each member of the review team considered himself or herself an employee. They were dismayed to learn that as far as the PSM standard was concerned, OSHA would not consider them employees, but employer representatives. Nonetheless, the review team agreed that employee participation is an important indication of a healthy safety culture and wanted to encourage it as much as possible.

They were reluctant, however, to mandate it. For example, the facility being reviewed, since it was still under construction, had not hired any "employees." So, if employee participation was mandated, none of the work by the review team would count. The review team did not want circumstances like this to prevent progress.

The review team concluded that employee participation is to be encouraged, but not mandated.

4.3.2 Process Safety Information

The review team agreed that process safety information was valuable and determined that facilities should keep certain types available and accessible at each facility.

Regarding process safety information about the hazards of the substances in the process, the team determined that the information in the safety data sheets collected in compliance with the Hazard Communication standard, 29 CFR 1910.1200 [11], was sufficient.

Regarding the technology of the process, the review team saw the value in block flow diagrams, the safe upper and lower limits for temperatures, and an evaluation of the consequences of deviations. Otherwise, the team concluded that the information required by the PSM standard was not relevant to a grain handling facility.

Regarding equipment in the process, the team was satisfied that it was important to keep the specifications for equipment, including specification for materials of construction, and to the extent possible, design codes and standards employed. Likewise, they agreed that electrical classification of equipment and electrical classification drawings were essential, as were ventilation system design. To the extent they are used, the team agreed that it was important to keep documentation on relief system design and design basis and on safety systems.

The PSM standard also mentions P&IDs and material and energy balances, neither of which are commonly developed for grain handling facilities and for which no need was seen.

As long as they had equipment specifications, the review team saw no value in documenting that equipment complied with RAGAGEP, nor in reverse engineering equipment to complete equipment files with RAGAGEP.

4.3.3 Process Hazard Analysis

The decision to conduct PHAs consistent with the requirements of the PHA standard was one that had already been made at a corporate level. The review team did not question the wisdom of the decision.

4.3.4 Operating Procedures

The review team agreed that written operating procedures were required and identified operation, safe startup, shutdown, and response to upset conditions as necessary. Additionally, the team recognized the need for safe work practices, especially for housekeeping, hot work, lockout/tagout, and confined space entry, since regulations specifically require these.

A procedure to control entrance into a facility by maintenance, contractor, laboratory, or other support personnel was not one the review team deemed practical or necessary.

4.3.5 Training

The review team developed an extensive list of training requirements:

- General safety precautions
- Hazards of the workplace
- Hazards of accumulated dust
- Common ignition sources
- Safety practices/procedures
- Choked leg clearing
- Housekeeping procedures
- Hot work procedures
- Preventative maintenance
- Lockout/tagout
- Confined space entry
- Handling flammables/toxics
- Emergency action plan
- Process description
- Equipment operation, safe startup, shutdown, and response to upset conditions
- Fire and explosion protection systems
- Maintenance requirements and practices
- Explosion prevention system arming and disarming procedures
- Management of change procedures
- System impairment reporting procedures
- Fire extinguisher use

Initial training would include all of the topics listed above. The topics that require refresher training was still to be determined, as was the frequency of the refresher training. As for the training required by various regulations, the refresher training would be as required by the regulations, at a minimum.

4.3.6 Contractors

The review team determined that the contractor safety program required by the GHF standard are sufficient to assure the safety of contract employees in their facilities.

4.3.7 Pre-Startup Safety Review

The review team concluded that the commissioning procedures that were in place for capital projects were sufficient for new projects. For existing facilities modified under MOC, the MOC procedure they have decided to develop should be sufficient.

4.3.8 Mechanical Integrity/Preventative Maintenance

The review team determined that the additional requirements of the mechanical integrity element of the PSM standard offer no safety benefits for a grain handling facility, and so decided that compliance with the preventative maintenance element of the GHF standard was more than adequate.

4.3.9 Hot Work Permit

The review team confirmed that it would continue to use hot work permits, accepted as allowed for a designated welding areas. As required by the standards, the permits would be kept on file "until the completion of the hot work operations."

4.3.10 Management of Change

The review team determined that it would develop and implement a management of change procedure consistent with the requirements of the PSM standard. It would apply to changes to technology, equipment, and procedures. Process chemicals are not relevant. Other changes to facilities, such as changes to organizational structure, would not be addressed, except on a case-by-case basis.

4.3.11 Incident Investigation

The review team determined that incident investigations of any catastrophic event would be completed, whether or not the incident involved the release of material or a recordable injury. Any fire, including incipient fires, would be subject to investigation, as would any explosion (deflagration or detonation). The review team was unable to establish a working definition of a near miss that would require incident investigation, leaving it for a facility to determine on a case-by-case basis.

The review team was unwilling to accept the timing dictates of the PSM standard. They believed it was more important to assemble the right investigation team than to be driven by a 48 hour deadline. As for record retention, the team concluded that it should comply with the record retention policies developed by corporate counsel.

4.3.12 Emergency Planning and Response

The review team determined that it was sufficient for a facility to comply with 29 CFR 1910.38, as stipulated by the GHF standard. They determined that the additional measures called out by the PSM standard were unnecessary.

4.3.13 Compliance Audits

The review team recognized the value in audits and concluded that they wanted to conduct them periodically, at a frequency to be determined later. Concerned about the liability associated with audits, however, they determined that audits should not be retained at the facility at all. Instead, they are to be addressed to a corporate official, who, after evaluating the findings and recommendations, would determine an appropriate course of action. Once an appropriate course of action was determined, action plans would be communicated but the audit itself would not be retained.

4.3.14 Trade Secrets

The review team determined that the trade secret element of PSM was not applicable.

5 Summary and Conclusions

A diverse team from the grain handling industry, most with no previous exposure to formal process safety management, considered the fourteen elements of the PSM standard as a way to supplement compliance with the GHF standard. The GHF standard shares five elements with the PSM standard, although the implementation is not exactly the same in the two regulations. The PSM standard has an additional nine elements that are not shared with the GHF standard.

After considering the five elements that the GHF standard and the PSM standard have in common—training, contractors, mechanical integrity/preventative maintenance, hot work, and emergency response—the team concluded that the programs that they had put in place to comply with the GHF standard were appropriate and sufficient. In the case of training, they had already concluded that training on operating procedures was necessary, and that documentation of training was simply good management, neither of which is a requirement of the GHF standard. As for the other four, the team concluded that the additional requirements of the PSM standard were either not applicable to a grain handling facility, or not helpful.

Of the nine elements unique to the PSM standard, the element of PHA was accepted in its entirety. In part, that is because of a corporate mandate. In part, it was because of their conclusion that the DHA called for in NFPA 652, with which they were already familiar, was too narrowly focused. Mostly, though, it was because they recognized the value of the element.

On the other hand, there were two elements unique to the PSM standard that the team rejected. One was the element of trade secrets, which they concluded was not now nor ever likely to be relevant. The other was the element of PSSR, which they believed was adequately addressed by their existing commissioning procedures.

Of the remaining six elements unique to the PSM standard—employee participation, process safety information, operating procedures, management of change, incident investigation, and compliance audits—the team accepted that all had value, was worth the effort, and incorporated each, with limitations.

Other facilities in the grain handling industry are likely to come to similar conclusions and find similar value in enhancing their compliance with the GHF standard with the elements of process safety management.

6 References

[1] OSHA. *29 CFR 1910.272, Grain handling facilities.* Occupational Safety and Health Administration, U.S. Department of Labor, Washington, DC. Last revised 27-Dec-2011.

[2] OSHA. *29 CFR 1910.119, Process safety management of highly hazardous chemicals.* Occupational Safety and Health Administration, U.S. Department of Labor, Washington, DC. Last revised 08-Feb-2013.

[3] OSHA. *29 CFR 1910.38, Emergency action plans.* Occupational Safety and Health Administration, U.S. Department of Labor, Washington, DC. Last revised 07-Nov-2002.

[4] OSHA. *29 CFR 1910.146, Permit-required confined spaces.* Occupational Safety and Health Administration, U.S. Department of Labor, Washington, DC. Last revised 27-Dec-2011.

[5] OSHA. *29 CFR 1910.147, Control of hazardous energy (lockout/tagout).* Occupational Safety and Health Administration, U.S. Department of Labor, Washington, DC. Last revised 02-May-2011.

[6] OSHA. *29 CFR 1910.252(a), General requirements-Fire prevention and protection.* Occupational Safety and Health Administration, U.S. Department of Labor, Washington, DC. Last revised 26-Mar-2012.

[7] OSHA. *29 CFR 1910.120, Hazardous waste operations and emergency response.* Occupational Safety and Health Administration, U.S. Department of Labor, Washington, DC. Last revised 08-Feb-2013.

[8] NFPA. *Standard on the Fundamentals of Combustible Dust-NFPA 652, 2016 Edition.* National Fire Protection Association, Quincy, MA. 2015.

[9] OSHA. *Non-Mandatory Appendix A to Subpart B – Partially Exempt Industries*, Occupational Safety and Health Administration, U.S. Department of Labor, Washington, DC. Last revised 18-Sep-2014.

[10] OSHA. *OSHA's Form 301, Injury and Illness Incident Report*, Occupational Safety and Health Administration, U.S. Department of Labor, Washington, DC. Last revised Jan-2004.

[11] OSHA. *29 CFR 1910.1200, Hazard Communication,* Occupational Safety and Health Administration, U.S. Department of Labor, Washington, DC. Last revised 26-Mar-2012.

Common Hurdles, Benefits, and Costs for Fully Implementing Process Safety Worldwide – Especially in Countries without PSM Regulations

William G. Bridges, Art Dowell III, Jeff Thomas, & Paul Casarez
PROCESS IMPROVEMENT INSTITUTE, INC. (PII)
1321 Waterside Lane, Knoxville, TN 37922
Phone: (865) 675-3458
Fax: (865) 622-6800
e-mail: wbridges@piii.com

Prepared for Presentation at
13th Global Congress on Process Safety
San Antonio, TX
March 27-29, 2017

UNPUBLISHED

Common Hurdles, Benefits, and Costs for Fully Implementing Process Safety Worldwide – Especially in Countries without PSM Regulations

William G. Bridges, Art Dowell III, Jeff Thomas, & Paul Casarez
PROCESS IMPROVEMENT INSTITUTE, INC. (PII)
e-mail: wbridges@piii.com

Keywords: PSM, process safety, Risk control, process safety management

Abstract

Process safety is implemented around the world and most of those sites do not have government regulations for compliance to push them along. The hurdles for effective (full) implementation appear to be roughly common from country to country, and site to site. This paper summarizes the lessons learned from multiple companies/sites around the world. Specifically, the paper compares hurdles to effective implementation and how company crossed these hurdles. We also update earlier papers on the costs and benefits of effective implementation of process safety. Each of these implementations is an example of process safety implementation at a non-covered process and in many of the cases mentioned, the facilities implementing process safety outside of countries with process safety regulations do so better than those in regulated countries and extend process safety to all processes (including to processes such as steel making).

Background

"Process safety management (PSM) is the application of management principles to the identification, understanding, and control of process hazards to prevent process-related incidents"[1]. PSM entails development and implementation of programs or systems to ensure that the practices and equipment used in hazardous processes are adequate and are maintained appropriately. The primary categories of programs or systems have come to be called elements of PSM. However, the basic elements of PSM have been defined by many groups in a number of ways. Figure 1 lists the elements of PSM systems from various industry and government groups. Many of the elements with different names have essentially the same meaning. For instance, "maintenance and inspection of facilities," together with some aspects of "personnel" practices, both under ACC's (formerly CMA's) Process Safety Code of Responsible Care™ are essentially the same as the single element, "mechanical integrity," under 29 CFR 1910.1190).

Figure 1. Comparison of PSM Systems

OSHA 29 CFR 1910.119 and EPA 40 CFR 68

- Management System
- Employee Participation†
- Process Safety Information
- Process Hazard Analysis
- Operating Procedures
- Training
- Contractors†
- Pre-startup Safety Review
- Mechanical Integrity
- Hot Work Permit†
- Management of Change
- Incident Investigation
- Emergency Planning and Response
- Compliance Audits
- Trade Secrets

AIChE/CCPS Risk-Based Process Safety (RBPS) standard

Commitment to Process Safety
- Process Safety Culture *
- Compliance with Standards
- Process Safety Competency
- Workforce Involvement *
- Stakeholder Outreach

Understand Hazards and Evaluate Risk
- Process Knowledge Management
- Hazard identification and Risk Analysis

Manage Risk
- Operating Procedures
- Training and Performance *
- Safe Work Practices
- Asset Integrity and Reliability
- Contractor Management
- Management of Change
- Operational Readiness *
- Conduct of Operations *
- Emergency Management

Learn from Experience
- Incident Investigation
- Measurement and Metrics
- Auditing
- Management Review and Continuous Improvement

* Contains some of the 9 key sub-elements of a comprehensive Human Factors element

Responsible Care® Process Safety Code

Management Leadership
- Commitment *
- Accountability *
- Performance Measurement
- Incident Investigation
- Information Sharing
- CAER Integration

Technology
- Design Documentation
- Process Hazards Information
- Process Hazards Analysis
- Management of Change

Facilities
- Siting
- Codes and Standards
- Safety Reviews
- Maintenance and Inspection
- Multiple Safeguards *
- Emergency Management

Personnel
- Job Skills *
- Safe Work Practices
- Initial Training *
- Employee Proficiency
- Fitness for Duty *
- Contractors

Note that the newest definition of process safety is CCPS's *Risk Based Process Safety (RBPS)*; this replaces their earlier process safety definition. In the older definition from CCPS, there was an element on Human Factors, which brought strong focus to this necessary element; in RBPS, the human factors sub-elements are now spread across 6 different elements.

Although nearly the entire industry agrees that implementing PSM is the right thing to do, interpreting and converting the PSM requirements into practices is unique to each company, and even unique to each plant site. Not only can the requirements be interpreted differently for each site based on local needs, but each site also starts from a different point when they begin to implement a system that is consistent with this regulation. These starting points, and the hurdles and advantages inherent to each, are important factors when estimating the cost of implementing each element. Many companies, especially those that are in countries with process safety regulations such as in the USA, UK, EU, Japan, and a few other countries, are choosing to interpret the process safety implementation in a minimalistic manner (e.g. just meeting the regulatory requirements). Though they may survive a regulatory audit, they are also minimizing the benefits their company will receive from implementing a more thorough process safety system. Other companies in and outside of the USA are going well beyond the minimal interpretation of the OSHA PSM requirements; therefore their cost to implement and sustain process safety will be higher than average, but these companies should (and do) have better PSM performance (fewer process safety incidents).

Very few countries have a process safety regulation. These include:

- USA
- UK (COMAH)
- EU (Seveso II)
- Japan
- Korea
- Peru (just issued)
- Australia
- Canada (sort of)
- Norway
- Perhaps others

But, in addition to the countries listed above, PII has helped oil, gas, petrochemical, fertilizer, chemical, and even steel companies implement process safety in the following countries:

- Saudi Arabia
- Bahrain
- Qatar
- Kuwait
- UAE
- Yemen
- Mexico
- Costa Rico
- Peru (pre-regulations)
- Argentina
- China
- Malaysia
- Thailand
- South Africa
- Nigeria
- Egypt
- Pakistan
- India
- Kazakhstan
- Azerbaijan
- Trinidad/Tobago
- Curacao

These countries do NOT have specific process safety regulations and there is essentially no government impetus to implement process safety. Yet, in many cases, the implementation in these countries is superior to implementation of process safety in the USA or other countries with regulations. As an illustration of this point, in the USA, clients often ask:

> *What is the minimum required to comply with the US OSHA and US EPA regulations on PSM and RMP?*
>
> *Or, (especially in the case of the oil refining industry)*
>
> *What are my peers doing to implement process safety?*

In non-regulated countries, we of course do not ever hear the first question above, but in most cases we are instead asked:

> *What are best practices for process safety and how do we go about implementing those best practices?*

Certainly, many clients in the USA, Canada, and UK also ask this same question about best practices, but a higher percentage of clients from outside of the regulated countries listed above want to only know and implement best practices.

The next sections recap some fundamental data on the cost and benefits of implementing process safety.

COST of Implementing Process Safety (general to all countries)

To understand the cost of PSM, we must realize that there are two main phases of implementing a PSM program: 1) installing a program, and then 2) maintaining the quality of the program. What industry is finding is that to get buy-in throughout a facility so that the PSM maintenance phase becomes feasible, you first have to make PSM a part of the facility's "culture". This does not happen overnight – it is a gradual process requiring sincere management commitment and constant nurturing. A good PSM program should provide the sound foundation for maintaining the ongoing program.

The costs of implementing process safety include:

- Costs for PHAs (process hazard analyses), LOPA (layer of protection analysis), and other risk assessment studies (if needed).
- Physical improvements to the facility to address PHA and LOPA recommendations (including appropriate relief, SIFs (safety instrumented functions) compliance with international standards, facility siting, dikes, etc.).
- Maintenance and testing for the additional physical facilities to manage process safety (relief, SIFs, dikes). Remember if you don't test your independent protection layers, you'll find out they are broken after an incident.
- Management of change studies to manage risks that may be introduced as changes are made to the facility.
- Training of operations, maintenance, technical personnel, supervision, and management in the principles of process safety and in the specific tasks for ongoing process safety management.

- Writing and maintaining procedures for operations, maintenance, and other process safety activities.

Costs of implementing a process safety system is not trivial, but the costs are more than balanced by the benefits of strong process safety systems.

Costs of implementing process safety can be significant – generally in the form of resources (people), and equipment (safety systems, gas/fire detection, automation, etc.). Also, maintenance is a critical cost component of process safety (inspections, overhauls of machinery, calibrations, etc.) that is often neglected. Again it takes management leadership and commitment to ensure there are sufficient resources to properly implement process safety and sustain good process safety performance.

Implementation Planning and Cost Estimation:
The installation phase is typically viewed as a one-time project, and includes activities such as developing concepts for each element, planning the work, training the PSM resources, writing draft programs and practices, pilot testing major programs, implementing the finished product (usually with more training), and responding to findings and recommendations of certain programs (such as recommendations from PHAs, MOC hazard reviews, and incident investigations). Most of the cost related to the development and initial implementation of a PSM program is labor (people resources), whereas the major cost related to responding to recommendations is typically capital improvements. The labor cost and out-of-pocket costs for developing and implementing a PSM element can be accounted for in one or more of the following categories:

- Meetings
- Writing documents
- Reviewing
- Revising
- Training/ orientation
- Pilot testing
- More revising
- Initial implementation

Where PSM is being implemented for the first time, it is best to begin with a gap analysis, followed by an implementation plan. Table 1 shows an excerpt from a data analysis table from a PSM Implementation Plan, where each Process Safety Gap is dissected and then a small project estimated for the closure of each gap. From such analysis, the cost to develop and initially implement PSM can be estimated and managed, like any other project. The costs to implement PSM can wary widely depending on the size of the organization. For example, the costs may be relatively small for a company with one or two small sites as opposed to a large global company with many sites in many countries.

Cost from Prior Paper:

But, where does a company get the cost factors for such implementation? The answer is from the experience of others. One source of overall cost data is from the paper on PSM Cost & Benefits from 1994.[2] This paper provides the overall cost of implementing process safety are a site, the data is scalable, and in some cases, the data is provided for the cost of individual process safety activities.

Table 1. Excerpt from a Detail Implementation Plan Analysis Table

Element 8: Mechanical Integrity (j) Responsibility: Joe Doe Support: Maintenance Manager, Consultant/Contractors

Main Activity: Develop and implement a program, procedures, and training for ensuring mechanical integrity Revision Date: March 31, 2004

Task	Staff-hours (E-Engineer/management, H-Hourly/nonexempt, C-Clerical)									Expenses & Capital
	Code	Meet	Write	Review	Revise	Pilot	Train	Implement	Total	
1. MI Program: Develop overall MI program	E	32	160	40	20				252	$10,000
	H	32	40	40	20				132	
	C		40		10				50	
2. MI Program: Improve and implement the proposed quality assurance program	E	16	80	40	20				156	
	H	16	80	40	20				156	
	C		24		10				34	
3a. MI Procedures: Write standard maintenance procedures (SMPs) for repair and preventive maintenance of mechanical equipment (if not provided by vendor information)	E	24	40	40			64		168	See Op. Pro. element for course costs
	H	24	400	40	24		64		552	
	C		40		8				48	
3b. MI Procedures: Collect and input vendor supplied procedures for repair and preventive maintenance of mechanical equipment into the MI procedures	E		80						80	$2,000
	H								0	
	C		300						300	
3c. MI Procedures: Improve and implement the proposed inspection, testing, and calibration procedures	E	80	400	72	40				592	
	H	240	200	48	40				528	
	C	40	80						120	
4. MI Training: Determine how to ascertain and document that each employee has received and understood the training. (Some labor included in Task 3 above, and some in the Operating Procedures element)	E								0	
	H								0	
	C								0	
5. MI Training: Train personnel on MI issues and standard maintenance practices	E		80				480	480	1040	$200,000
	H						7200		7200	
	C		20				80		100	
Total Resources for This Element	E	152	840	192	80	0	544	480	2288	$212,000
	H	312	720	168	104	0	7264	0	8568	
	C	40	504	0	28	0	80	0	652	

In the 1994 paper, 84 facilities, representing a total of 25 companies, completed surveys. Dow Chemicals and Olin Chemicals responded with helpful comments and data, but were unable to complete a survey, since their programs are so mature and they did not gather data on major elements of PSM during each program's evolution. Both companies also shared some interesting insights and data on PSM benefits. The facilities that completed surveys together employ about 31,000 workers, which represent about 1 % of the workers which OSHA claims are protected by the PSM regulation.[2] The size of the survey sample and breakdown of respondents by industry segment are shown in Tables 2 and 3, respectively.

Table 2 Sample Size for 1994 PSM Cost/Benefit Survey[2]

- 25 Companies
- 84 Locations
- 25,000 Direct Hire Employees
- 6,000 Contract Employees
- 20,000 P&IDs
- 850 PHAs

Table 3 Industry Represented in 1994 PSM Cost/Benefit Survey[2]

Industry	Number of Sites
Gas Plant/Oil field	46
Oil Refinery	5
Petrochemical/Plastic	3
Pulp/Paper	13
Chemical	16
Iron & Steel	1
ALL	84

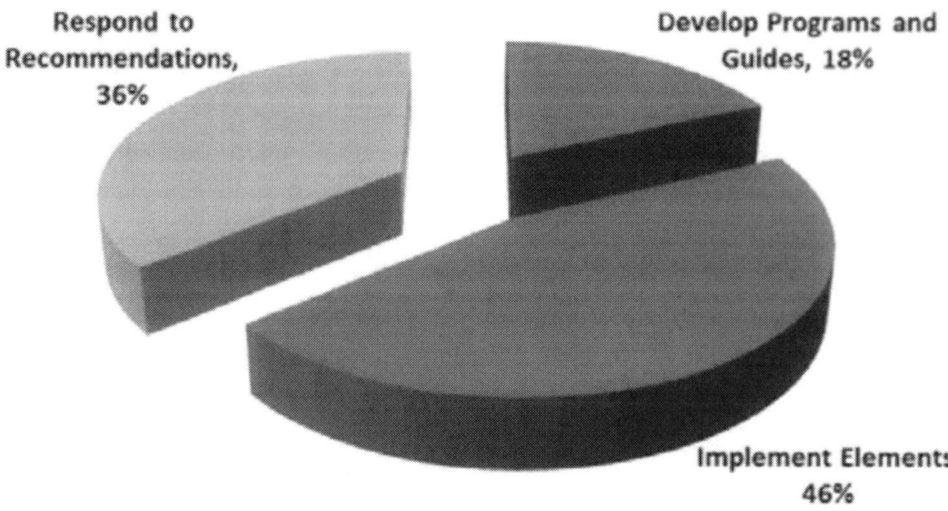

Figure 2. Breakdown of Cost for Initial Implementation of Process Safety from 1994 Cost/Benefit Survey[2]

Figure 2 illustrates the distribution of the cost of compliance based on an average of all survey data, versus the three primary cost categories:

- ***Developing PSM Programs***. The cost, primarily in equivalent labor costs, to bring the PSM program (and individual element programs) from the concept stage through the final design (such as developing an MOC or MI written program that your facility personnel are confident will work). This category also includes the cost of training personnel to be proficient in various PSM activities, such as leading PHAs, leading incident investigations, leading compliance audits, writing procedures, and leading employee training.

- ***Implementing PSM Programs.*** The cost (again primarily in equivalent labor costs) to do implementation tasks, such as writing operating and maintenance procedures, updating PSI, doing initial training of operators and maintenance personnel, and performing/documenting PHAs.

- ***Responding to Recommendations.*** The cost, primarily capital costs and expenses, to implement improvements to address recommendations from PHAs, MOC hazard reviews, incident investigations, and MI deficiency reports.

The total cost for years 1-10 for the 84 facilities surveyed was $484 million beginning at an average compliance of 40 %, or $592 million if extrapolated proportionally back to 0% compliance. As the prior paper warns, the averaged factors presented in that paper are not appropriate for estimating the cost of PSM at an individual facility. However, large, multi-facility companies will be able to use these averaged factors with more confidence and the factors can be used to estimate the cost for the industry as a whole. ***The average cost per facility to reach full compliance was estimated at $5.8 million (over a period of 10 years, beginning at 40% compliance) which is based on the average facility responding to this***

survey. For a continuous process plant, it was recommended that the scaling factor of $22,000 per P&ID be used (rather than the factor per employee) - more reliable results are produced)) to estimate the cost of reaching PSM compliance, starting a 40% compliance. So, if there are 100 P&IDs worth of equipment at the site, PSM implementation might take 4-5 years and might cost 2.2 million (in 1994 dollars), or about 3 to 4 million today, since the inflation adjustment between 1994 and 2016 is a multiplier of 1.6.[2]

The 1994 survey[2] also collected cost data for completing individual PSM activities. For example, the average cost to update a D-sized P&ID was $1,800. For comparison, an informal survey of over 500 attendees of a 3-day course on PSM indicated that the average cost is (1) $1,500 to field validate and then update an existing D-sized P&ID and (2) between $2,500 to $3,500 to create (or substantially revise) a P&ID, in 1994 dollars (multiply by 1.6 to get to 2016 dollars).[2]

A detailed analysis of the survey data (which is only summarized in this paper) indicates that the combined cost of (1) training personnel to lead PHAs (or contracting leaders), (2) performing and documenting PHAs, and (3) responding to PHA recommendations accounts for about 50% of the cost of PSM. Therefore, it is in every facility's best interest to maximize the benefit (value) of each PHA. Using the right PHA techniques, providing skilled (efficient) leaders and scribes, and allocating the highest qualified personnel to participate as subject-matter experts during the PHA meetings are fundamental ways to optimize the PHA efforts. Also, expanding the scope of the meetings to uncover more operability improvements can result in significant benefits that have proven to far outweigh (often by a factor of 10 or more) the incremental cost of longer PHA meetings to uncover and discuss operability problems. (There are many fine papers on how to perform high quality PHAs [3, 4, 5, 6, 7].)

Most respondents indicated verbally that MI and MOC were the most difficult elements for facilities' employees to develop and implement, since so much indoctrination and "culture" change were necessary.

BENEFITS of Implementing Process Safety (general to all countries)

For a moral and ethical company, it is not appropriate to tolerate higher risk in countries that do not have process safety regulations. It is not appropriate (and it is not good public relations) to "export" risk from a developed country to an emerging-market country. The value statement for many corporations emphasizes the importance of protecting the community, the employees, and the environment from adverse impacts from the company's operations. Implementing process safety worldwide, even in countries where it is not required by regulation, ensures compliance with the stated values of the corporation.

- Implementing process safety actually (documented; proven) improves the economic performance of the facilities. Improving the understanding of the chemistry, engineering, and reliability of the process can pay dividends in improved performance, improved quality, and reduced downtime. Lost production and repairs due to a process safety incident can be very expensive. Modifications made after an incident are typically 2 to 4 times as expensive as the same modifications made in normal times (the additional cost is attributed to overtime, urgent procurement, and overkill in corrective actions in the aftermath of a serious incident).

- Improving the knowledge and skills of employees to implement process safety can also improve the knowledge and skills to produce quality product with fewer upsets, thereby improving the economic performance of the facility
- The company can use the improved reliability of the facility that has implemented process safety as a selling point to their customers.
- We have observed that organizations that implement strong management of change programs to support process safety make fewer changes, but the changes are better engineered, better executed, better operated, and better maintained. The changes are less likely to be removed after startup and operation because potential issues were detected and addressed before the change was made.
- The same system that produces product also produces fires, explosions, injuries, and off grade product. Improving the system by implementing process safety also improves the performance of the system for on-time delivery and quality product.
- Process Safety and Process Reliability and Operational Excellence are all related and intertwined. And to a lesser extent, so is Occupational Safety. Having good process safety processes (proactive MI, good procedures, good control of human factors, excellent near miss reporting and incident investigation, thorough PHAs, etc.), will improve process reliability, profitability, and occupational safety, by having fewer incidents and less production downtime.
- A company's reputation and license to operate can be adversely affected by process safety incidents. So by implementing a good process safety system, these type of incidents can be reduced/eliminated thus improving: 1) company's bottom line (profit), and 2) improving how a company is perceived by various host governments which will help to get new agreements and licenses to operate. In most places safety performance (including process safety) plays a role in which company will be selected to build/operate a new facility.

Benefits from Prior Paper:

Very few respondents provided benefit data, since the cost avoided due to implementing PSM is so difficult to isolate from other ongoing safety and quality improvement efforts. Obviously, better data on the benefits of individual PSM activities are needed. Many companies have stated that an effective PHA program, especially when expanded to include hazard evaluations during the early phases of new projects, will produce benefits far beyond the cost of performing and responding to the PHAs. In the preamble to 29 CFR 1910.119, OSHA quoted one plant manager concerning PSM benefits:

> "Our small organization was quietly infused with a rebirth of innovative thinking. Process technology that was more than 35 years old was routinely being questioned .... This quickly led to the same questioning being applied to process improvement .... Ultimately, I believe that this thoroughness and training approach will result in cost savings to a small plant site on the order of 4 to 7 percent of an operating budget". [2]

Overall, the majority of companies responding said they feel PSM should be implemented, though most objected to some of the paperwork requirements (especially process safety information). Companies with more than 5 years of PSM experience unanimously echoed the comment in one survey response:

"It is indefensible not to implement PSM, and it's worth it."

But, the prior paper also concludes that due to the "fear of citations resulting from insufficient documentation on what we've done," the paperwork and associated labor to over-document PSM activities could outweigh many of the benefits that would otherwise be realized in the absence of regulatory pressure.

Table 4. PSM Compliance Benefits: Avoided Incidents (2016 dollars)

Type of Incident Avoided	*EPA USD Value per Incident*[8]
• Impact to Facility/Operations	
o Business/Production Interruption	13,000
o Equipment Damage	Not available (but more than 20,000,000 per significant PSE)
• Environmental Harm	
o Soil Contamination	$310,000
o Vegetation Damage	$80,000
o Groundwater Contamination	2,200,000
• Personal Injury (per person)	
o Injury/First Aid/Outpatient	28,000
o Hospitalization (excess over injury alone)	19,000
o Death (and Litigation)	18,000,000 (versus 10,000,000 for fatality on construction jobs)
• Other Expenses	
o Litigation per Environmental Action	90,000
o Evacuation and Sheltering in Place (per incident; effect on public only)	120,000

Note: Source is EPA's regulatory impact analysis related to RMP regulation (1993); corrected for inflation to 2016[8]

Table 5. PSM Compliance Benefits: Non-Incident related

- Quality and Productivity Improvements due to less process upsets and increased efficiency
- Lower Insurance Rates (or less increases in the rates in the future)
- Lower Workers' Compensation for Lost Time

Note: Some companies, such as Olin Chemicals, have reported that improvements in productivity and quality that were identified during their PHAs more than offset the total cost of the PSM implementation. Others, such as Eli Lilly and ALCOA have reported that just improvements in Near Miss reporting have paid for all of PSM implementation.

The expect return on investment is generally accepted as 100 to 1, if the site implements best practices.

Anecdotal Benefits Observed by Various Companies (outside and within the USA)

- Olin Chemicals found in the 1980s that the "reliability and operability improvements" that resulted from PHAs that were balanced between safety and operability paid for all of process safety. So the rest of the benefits were free!
- PII has found that doing excellent PHAs of all modes of operation has nearly a 1000 to 1 payback due to avoided cost of major accidents that occur during startup, shutdown, and online maintenance.
- AMOCO Oil found in the early 1990s that doing multiple PHAs (3 to 4) during a project, versus just one PHA/HAZOP at design freeze, saved enough in startup cost to pay for all project PHAs several times over.
- An undisclosed multi-national company (head-quartered in the USA) found that when the board of directors cut the budget of process safety in 2008 and 2009 due to the "economic" crisis, that the number of Significant Process Safety Events spiked, resulting in an increase of more than $30 million in actual losses and leading to the occurrence of two events that nearly cost more than $1 billion each, but luck (not good practices) saved the day in both cases. And yet the cost savings of the cuts was only about $2 million.

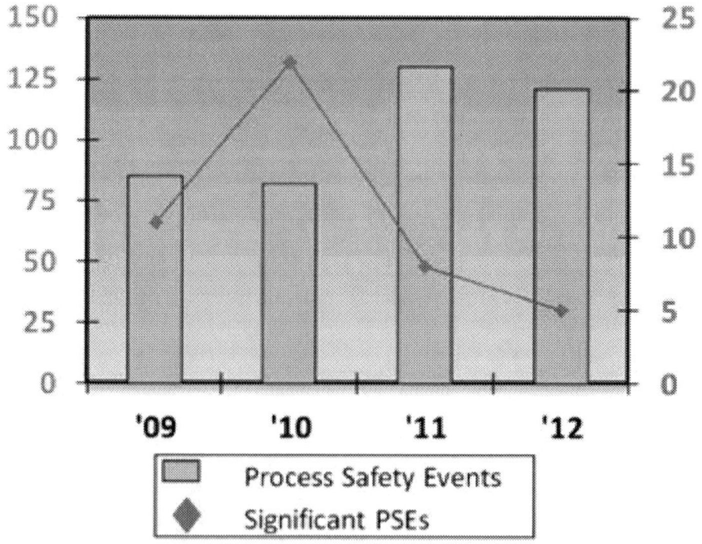

Figure 3. Significant PSE Trend

- The total implementation cost estimate for Excellent PSM in all 25,000 regulated facilities in the USA is about $100 billion initial implementation and $10 billion annual cost thereafter. This is about $4 to 5 million per site initially and $0.5 million per site thereafter. Just ONE significant process safety event averages more than $20 million but could be higher than the replacement cost of the facility (about $100 million replacement cost, on average), especially when legal liability is added.

The rest of this paper discusses the Hurdles and Advantages of implementing process safety, especially in sites that are not in countries with PSM regulations. This data is summarized from PSM implementation in more than 20 countries without PSM regulations, working with more than 50 companies.

HURDLES to Implementing Process Safety in Countries that Do NOT have a PSM Regulation

There are many hurdles to implementing process safety in countries that do not have regulations. Chief among these is the lack of in-country resources to help with the implementation since fewer companies may be implementing PSM in that country. One limitation is the lack of motivation because there is not regulation for a government to enforce; but on the other hand, we have seen that in many cases the quality of the implementation of process safety is better outside of regulated countries because once a company decides to implement process safety, the company is more inclined to follow best practices (best industry standards) rather than settling for minimal compliance. This may be in part because in the USA, some may confuse compliance with OSHA PSM as sufficient to prevent accidents. This is not necessarily true. So, some disadvantages become advantages.

Below are some of the hurdles we have found, in order of importance.

General resistance to Process safety:

- It is much more difficult to get interest started in process safety in countries that
  - do not have regulations that require implementation of a process safety system (like OSHA PSM or UK HSE process safety regulations), and
  - where human life may not seen as being as important/valuable as in developed countries (e.g., some third world countries, but not limited to just third world countries).
- Since it is not required by the government, our competitors are not doing it, and we cannot implement process safety and remain competitive. Further, our customers will not pay for process safety.
- Due to lack of regulations, most companies do not seem to participate in industry organizations (such as CCPS or API). Nor do they monitor the development of standards and recommended practices. So unless the technical firm responsible for the design or some good partner is still actively involved, this key PSM knowledge source is untapped.
- PSM is generally still a "Safety" responsibility, not part of the operational discipline of the facility.

- Process Safety is harder to understand and explain than personnel safety, especially in countries without process safety specific regulations. Most people understand personnel safety (slips, trips, falls, personnel injuries), but much fewer people understand process safety (preventing loss of containment, fires, explosions, etc.). The benefits of a good process safety system/program may not be easy to calculate on a cost benefit basis, and benefits are generally longer term (preventing large process safety incidents).
- Isolated, but still strong pockets of Senior Management insisting that "it hasn't happened yet". Or, "the company XYZ responsible for technology has not told us this is could be a problem."
- The knowledge, skills, and training of our workforce are not at a sufficient level to implement process safety. There is limited supply of appropriately-skilled personnel in the country to conduct studies, write procedures, and train personnel. There is a lower percentage of folks in the population of technical staff in such countries with process safety expertise, so the companies wanting to implement PSM according to best practices must start by importing labor (at first these being consultants) with this expertise.
- Many of the personnel who built, started-up, operated and maintained the facility since inception are at retirement age, and persons with sufficient experience are not available in-house to backfill these positions.
- Related to the two points above, there is no pool of qualified staff to facilitate a succession plan. So even if new hire persons are technically proficient, knowledge of the specific facility has just retired and walked out the door. Life experiences while working at the facility have not been intentionally captured through incident investigations, near misses, knowledge transfer, etc. Frequently technical and operating manuals are out of date.
- The point immediately above is exacerbated by a struggle to convince the technical personnel that their skills/experience is lacking in process safety issues.
- For companies that have started implementing PSM, they are implementing largely in silos within each element. Linking elements together to fulfill interlocking horizontal implementation is quite weak. One recurring weakness is these companies lack of near miss reporting and investigation with subsequent RCAs, making it difficult for management to see the importance of horizontal PSM integration/function.
- Establishing nomenclature, vocabulary, and definitions to insure complete communication (with outsiders and within organization) is even a problem, since process safety is not as widespread.

Hurdles related to Specific Process Safety Elements

- PHAs are out of date and have been conducted without using "best practices," particularly for the non-continuous modes of operations. This results in the remaining PSM program implementation being severely flawed/lacking. This is especially true for the PHAs (HAZOPs) conducted during the initial projects. In nearly all cases outside of the USA and UK, there was only one PHA/HAZOP conducted at the project design freeze, and changes since the design freeze were never incorporated into the Project PHA (rendering the original PHA/HAZOP essentially useless).

- All of the project PHAs are missing analysis of non-routine modes of operation, and many sites do not realize how important it is to perform a detailed analysis of these modes of operation.[3, 7]

ADVANTAGES to Implementing Process Safety in Countries that Do NOT have a PSM Regulation

In many ways, there are significant advantages in driving process safety forward in countries without regulations. The key is company leadership and site leadership. Below is list of the some of these advantages.

- Facilities tend to be newer; 5 to 30 years old or younger. Therefore MI is not quite as challenging (yet) and the design standards used are typically better than at older facilities and the materials of construction are usually better as well (if good vendors and EPCs were chosen). This is significant advantage compared to the USA in particular, where most of the facilities tend to be older.
- Companies that decide to implement process safety in such countries are far more receptive to "best practices" rather than a goal of "compliance" since there are no regulations to comply with. This is a very strong positive for implementation efforts since such companies are more focused on the business case for each facet of process safety, rather than considering a minimalistic approach for the sake of compliance. On the other-hand, this relatively stronger drive to be the best in process safety may reflect (1) the influence of joint venture partners who have high standards, (2) the desire to be best due to strong leadership in the company, or (3) the legal environment (less regulation → less lawsuits → less fear of doing what is best). Regardless of the reasons, it is certainly refreshing to work with clients who ask: "What are best practices for this element?"
- Global Companies operating outside of USA many times have a "core value" of process safety throughout their worldwide operations, and work to ensure that everyone in their organization (worldwide) understands the importance of process safety, and resources are provided to properly implement and sustain process safety systems and processes. In some instances, the affiliates outside of the regulated countries become the Best in process safety within such companies.
    - KEMYA, a petrochemical company in Saudi Arabia, is a JV of SABIC and EXXONMOBIL. They implemented process safety in a strong way before startup about 30 years ago and due to local leadership, they have never taken their eye off the ball. They have the best process safety performance of any facility in Saudi Arabia and we rank them in the top 5% of the world in quality of their process safety implementation (and this is echoed in their very low rate of process safety accidents).
    - HADEED, a large steel plant in Saudi Arabia and another affiliate of SABIC, implements process safety nearly as well as (perhaps even just as well) as KEMYA, just mentioned. HADEED implements process safety in every portion of the steel plant, not just the units with highly hazardous chemicals. This is not even done in many facilities in the USA.
- Operational skills (Operations and Maintenance) seem to be strong and enthusiastic in such countries, likely because the companies implementing process safety tend to be larger, more

profitable companies, or because they bring in labor from other countries. But, as mentioned earlier, major PSM competency gaps exist.

OTHER DRIVERS for Implementing Process Safety in Countries that Do NOT have a PSM Regulation

Besides the drivers mentioned above under "Advantages," there are several drivers for implementation of PSM outside of countries with PSM regulations:

- Companies with strong ties to or are affiliates of a company with strong process safety standards and culture, exhibit stronger desire to implement process safety correctly.
- Responsible Care™: Many international companies have made commitments at the executive level to fully and voluntarily comply with *Responsible Care*™ and this has driven compliance with the associated *Process Safety Code* shown in Figure 1.
- AIChE: The American Institute of Chemical Engineers has more than 50,000 members world-wide and there are many local chapters or sections of AIChE around the world. These sections work with the CCPS (of AIChE) to push for global implementation of best practices in process safety, with particular focus on the process safety model of RBPS.
- Economics: Implementing process safety correctly improves the competitiveness of a company primarily by improvements to operating efficiencies and reductions in downtime (or reduction of other losses)
- Companies with strong ties to a government or to a central agency (such as with strong enforcement by insurance companies) exhibit a stronger desire to implement process safety correctly. This is very evident in China.

Conclusions

Implementing process safety in facilities that are not regulated is no different than in regulated facilities, but there are some unique hurdles and advantages. Some of the best examples of process safety implementation can in fact be found in sites outside of countries with PSM regulations or in sites that are not covered by regulations. The keys to effective implementation are the same in all instances: effective and tangible leadership, active employee participation, solid technical competencies in process safety activities, keen attention to optimizing all controls of human factors, and thorough risk assessments.

The cost of implementing process safety is likely in the range of $9 million USD per average facility and will normally take 4-5 years to reach minimal (but full) compliance and 5-10 years to reach excellence. The benefits of PSM implementation likely outweigh the cost by a factor of about 100.

Acronyms Used

AIChE– American Institute of Chemical Engineers
ACC – American Chemical Council (formerly CMA)

API – American Petroleum Institute
CCPS – Center for Chemical Process Safety (a division of AIChE)
CFR – Code of Federal Registry (USA)
CMA – Chemical Manufacturer 's Association, now American Chemical Council (ACC)
COMAH – Control of Major Accidents Hazards
EPA – US Environmental Protection Agency
EU – European Union
HAZOP – Hazard and Operability Analysis
HSE – UK Health Safety Executive
IPL - Independent Protection Layer
JSA – Job Safety Analysis
JV – Joint Venture
LOPA – Layer of Protection Analysis
MI – Mechanical Integrity
MOC – Management of Change
NUREG – Nuclear Regulation, US Nuclear Regulatory Commission
OSHA – Occupational Safety and Health Administration, US Department of Labor
PHA – Process Hazard Analysis
PII – Process Improvement Institute, Inc.
P&ID – Piping & Instrumentation Diagram
PSE – Process Safety Event
PSI – Process Safety Information
PSM – Process Safety Management
RBPS – Risk Based Process Safety
RCA – Root Cause Analysis
SOP – Standard Operating Procedure
UAE – United Arab Emirates
UK – United Kingdom of Great Britain
USA – United States of America

References

1. *Guidelines for Technical Management of Chemical Process Safety,* CCPS/AIChE, 1985.
2. "The Cost and Benefits of Process Safety Management: Industry Survey Results," W, Bridges, *Process Safety Progress*, AIChE, January 1994.
3. "Necessity of Performing Hazard Evaluations (PHAs) of Non-normal Modes of Operation (Startup, Shutdown, & Online Maintenance)", W. Bridges and Mike Marshall (US OSHA), *18th Annual International Symposium, Mary Kay-O-Connor Process Safety Center, College Station, TX, October 2015.*
4. "How to Efficiently Perform the Hazard Evaluation (PHA) Required for Non-Routine Modes of Operation (Startup, Shutdown, Online Maintenance)," W. Bridges and T. Clark, *7th Global Congress on Process Safety*, Chicago, AIChE, March 2011.

5. "Optimizing Qualitative Hazard Evaluations for Maximized Brainstorming," W. Bridges and R. Tew, *5th Global Congress on Process Safety*, AIChE, March 2009.
6. Bridges, W.G., et. al., "Addressing Human Error During Process Hazard Analyses," *Chemical Engineering Progress*, May 1994
7. *Guidelines for Hazard Evaluation Procedures*, 3rd Edition, 2008, CCPS/AIChE.
8. *Regulatory Impact Analysis in Support of Listing Regulated Substances and Thresholds and Mandating Risk Management Programs for Chemical Accident Release Prevention, as Required by Section 112 (r) of the Clean Air Act*, Draft Report, Chemical Emergency Preparedness and Prevention Office, U.S. EPA, May 1992.

Dust-Explosion Hazards: Housekeeping Quantified

Richard W. Prugh, PE, CSP
Principal Process Safety Engineer
Chilworth Technology, Inc., a Dekra Insight Company
609-799-4449; richard.prugh@dekra.com

Prepared for Presentation at
American Institute of Chemical Engineers
2017 Spring Meeting and 13th Global Congress on Process Safety
San Antonio, Texas
March 26 to 29, 2017

Dust-Explosion Hazards: Housekeeping Quantified

Richard W. Prugh, PE, CSP
Principal Process Safety Engineer
Chilworth Technology, Inc., a Dekra Insight Company
609-799-4449; richard.prugh@dekra.com

Keywords: housekeeping; combustible dust; secondary explosion; dust accumulation; electrical classification; personal protection

Abstract

Limiting the accumulation of combustible dusts in the workplace is critically important in the prevention of primary and secondary dust explosions. Guidance on accumulation limits began in 1997 by the NFPA in the form of a layer-thickness criterion. As stated in NFPA 654, "the hazard area shall include areas where dust accumulations exceed 1/32 inch (0.8 mm)." Since that date, there have been several modifications of dust-layer guidance, and the relationship of dust accumulations to electrical classification was introduced in 2008. The 2013 issue of NFPA 654 introduced four additional criteria, for maximum weights of dust accumulations. These criteria can be expressed in terms of grams per square foot, and the results of calculations lead to very low "allowable" values. Two of these "mass" methods require an assumption of the extent of entrainment of dust from accumulations, as caused by a primary explosion; the available data indicate that the suggested entrainment fraction is not "conservative".

This paper compiles the current methods for determining the amounts of dust that would be hazardous as fuel for flash fire or secondary explosion. Also presented is guidance for housekeeping to minimize these hazards while providing an environment where general-purpose electrical equipment could be used, in an "unclassified" area. Further, guidance is given for the personal protective equipment that should be worn by persons who perform housekeeping, particularly where large spills are involved.

1 Introduction

The purpose of this presentation is to provide a review of previous and current standards and the guidance that these standards provide for "housekeeping". For combustible dusts, the term "housekeeping" is applied to the removal of accumulations in the workplace before they become hazardous [Reference 1]. Thus, the hazards of fires, flash-fires, and "secondary" explosions would be minimized [References 2 and 3]. An additional objective of this document is to compile the references to similar previous presentations, at Loss Prevention Symposia, and at other conferences. Thus, the overall objective is to assist persons who are tasked with establishing housekeeping policies and procedures, and for auditing the conformance with in-house and regulatory housekeeping standards.

Determining the magnitude of hazardous accumulations of dust has been mentioned previously, with descriptions of the available methods for [References 4, 5, and 6]. Also, the entrainment of such deposits – as the result of a "primary" event – to provide the fuel for a "secondary" explosion or flash-fire has been extensively investigated [References 7, 8, and 9].

2 Combustible-Dust Accumulation Limits, for "Organic" Dusts

2.1 The "Layer-Depth" Criterion

The National Fire Protection Association [NFPA] has included a dust-accumulation limit of 1/32 inch since 1994, in NFPA 654, with the following statement: "the hazard area shall include areas where dust accumulations exceed 1/32 inch [0.8 mm]." [Reference 10]. This requirement for prevention of secondary dust explosions was repeated in the 1997 issue [Reference 11], with supporting calculations.

The intention of these calculations was to show how a 1/32-inch layer of a dust having a density of 75 pounds per cubic foot could produce a dust cloud of "optimum" concentration that was 10 feet high over the area that was covered by the dust. A conclusion was that a 1/32-inch layer of combustible dust over 5% of the floor area "is capable of creating a hazardous condition". A companion observation was that the available surface area of overhead beams and ledges would be about 5% of the floor area. Thus, overhead accumulations would be hazardous even if the floor was kept clean. This issue of the standard [Reference 12] cautioned against the use of "compressed-air cleaning" since the smaller particles of powders and dusts can be lofted onto such overhead surfaces, where cleaning is difficult and hazardous accumulations can then occur.

In the 2000 edition of NFPA 654, the dust-layer criterion was modified to allow consideration of the bulk density of dusts, since the previous density would not apply to most organic dusts [Reference 13]. Thus, the allowable thickness would be given by:

$$t_{\text{ALLOWABLE}} = [1/32]\,[75/BD] \quad \text{inch, or} \quad (1)$$

$$t_{\text{ALLOWABLE}} = [\,0.8\,]\,[75/BD] \quad \text{millimeters} \quad (2)$$

where BD is the bulk density, in pounds per cubic foot. Thus, the following Table 1 can be constructed:

Table 1. Allowable Dust-Layer Accumulations

BULK DENSITY [pounds per cubic foot]									
BD	5	10	15	20	25	30	45	60	>75
Example		Flock	Most Organic Dusts						Metals
$t_{\text{ALLOWABLE}}$	0.5"	0.25"	1/6"	1/8"	0.1"	1/12"	1/20"	1/25"	1/32"
$t_{\text{ALLOWABLE}}$	12 mm	6.0 mm	4.0 mm	3.0 mm	2.5 mm	2.0 mm	1.3 mm	1.0 mm	0.8 mm

Bulk densities can cover a very wide range [Reference 14] and can also be a function of particle-size distribution.

The Occupational Safety and Health Administration [OSHA] has incorporated the dust-layer criterion in its guidance to compliance officers as: "Annex D of NFPA 654 contains guidance on dust layer characterization and precautions". It indicates that immediate cleaning is warranted whenever a dust layer of 1/32- inch thickness accumulates over a surface area of at least 5% of the floor area of the facility or any given room [Reference 15]. This criterion was based on a bulk density of 75 pounds per cubic foot, but the bulk-density modification was introduced in a later memorandum [Reference 16].

FM Global provides similar guidance, with a dust-accumulation limit of 1/16" inch for a dust having a bulk density of 36 pounds per cubic foot, and with this accumulation being limited to 5% of the floor area [Reference 17].

A measure of the combustible-dust "density" – in terms of grams per square foot – can be obtained by multiplying the layer depth by the bulk density:

$$D_{\text{ACCUMULATION}} = [453.5/(25.4*12)]\, t_{\text{ALLOWABLE}}\, BD = 1.49\, t\, BD \quad (3)$$

$$D_{\text{ACCUMULATION}} = 1.49\,[\,0.8\,(\,75/BD\,)\,]\, BD = 90 \text{ grams per square foot of floor area.} \quad (4)$$

Thus, based on the layer-depth criterion alone, the maximum allowable "density" – or the "threshold dust mass/accumulation" – of accumulated dust would be about 90 grams per square foot, as shown in Table 2.

Table 2. Allowable Dust-Layer Densities

BULK DENSITY [pounds per cubic foot]									
BD	5	10	15	20	25	30	45	60	>75
Example		Flock	Most Organic Dusts						Metals
$t_{ALLOWABLE}$	0.5"	0.25"	1/6"	1/8"	0.1"	1/12"	1/20"	1/25"	1/32"
$t_{ALLOWABLE}$	12 mm	6.0 mm	4.0 mm	3.0 mm	2.5 mm	2.0 mm	1.3 mm	1.0 mm	0.8 mm
g./sq.ft.	90	90	90	90	90	90	90	90	90

However, this high density of combustible dust would only be "allowable" in an area that is 5% of the total floor area. Thus, the averaged dust density over the whole floor area would be 5% of 90 grams per square foot or about 4.5 grams per square foot of floor area [about 10 pounds over 1,000 square feet], for all bulk densities. This corresponds to the guidance provided in an earlier presentation [Reference 18].

As will be shown later, this threshold dust mass/accumulation of 4.5 g./sq.ft. may not be sufficiently "conservative", particularly for prevention of flash-fire hazards.

2.2 Mass Methods

The following statements are from the "Origin and Development of NFPA 654", in the 2013 issue of NFPA 654 [Reference 19]:

The most important change is the inclusion of four methods to determine whether a dust fire or explosion hazardous condition exists in a facility. The methods include those based on mass accumulation and risk evaluation as well as the layer depth criterion, which was described in the 2006 edition. Each of the methods is viewed as equivalent in establishing that a hazardous condition exists for either fires or explosions, so the change offers options for users of the standard when determining this fundamental condition. [This] standard includes changes to the housekeeping requirements, [involving] the determination of a dust hazardous condition and prompts a cleaning frequency based on the nature of the dust layer or dust mass.

The four methods are [Reference 20]:

(1) Layer depth criterion method
(2) Mass method A
(3) Mass method B

(4) Risk evaluation method

The equivalency of the above four methods is stated as: "Each of the methods shall be deemed to provide equivalent levels of safety."

The following paragraphs present Mass Methods A and B.

2.2.1 Mass Method A, for Protection Against Flash-Fire Hazards

According to NFPA 654: "The threshold dust mass[/accumulation] establishing a building or room as a **dust flash-fire hazard** area, $M_{BASIC\text{-}FIRE}$, shall be determined by the following equation:

$$M_{BASIC\text{-}FLASHFIRE} = 0.02 \cdot A_{FLOOR} \qquad \text{kilograms"} \qquad (5)$$

where:
$M_{BASIC\text{-}FLASHFIRE}$ = threshold dust mass (kg.) based on personnel fire-exposure criterion; and
A_{FLOOR} = lesser of enclosure floor area (sq.m.) or 2000 sq.m.

The above relationship can be rearranged and converted to useful units:

$$M_{BASIC\text{-}FLASHFIRE} / A_{FLOOR} = 1.9 \text{ g./sq.ft., or about 4 lbs. per 1,000 sq. ft. of floor area} \qquad (6)$$

Note that this criterion is independent of bulk density – or other properties – of the dust, and is independent of the size or pressure-resistance of the room or building. However, the total allowable weight of dust in a room or building is directly proportional to the floor area.

2.2.2 Mass Method A, for Protection Against Explosion Hazards

According to NFPA 654: "The threshold dust mass[/accumulation] establishing a building or room as a **dust explosion hazard** area, $M_{BASIC\text{-}EXPLOSION}$, shall be determined by the following equation:

$$M_{BASIC\text{-}EXPLOSION} = 0.004 \cdot A_{FLOOR} \cdot H \qquad \text{kilograms"} \qquad (7)$$

where:
$M_{BASIC\text{-}EXPLOSION}$ = threshold dust mass (kg) based on building damage criterion;
A_{FLOOR} = lesser of enclosure floor area (sq.m.) or 2000 sq.m.; and
H = lesser of enclosure ceiling height (m.) or 12 m.

The above relationship can be rearranged and converted to useful units:

$$M_{BASIC\text{-}EXPLOSION} / A_{FLOOR} = 0.00025\,H = 0.11\,H \text{ g./sq.ft.,} \tag{8}$$

or 0.24 *H* lbs./1000 sq.ft. of floor area

where:
H = lesser of enclosure ceiling height (feet) or 39 ft.

Note that this criterion also is independent of bulk density – or other properties – of the dust, and is independent of the floor area or the pressure resistance of the room or building. For a ceiling height of 12 feet, the above relationship would limit the dust accumulation to 1.3 g./sq.ft., or about 3 pounds over a 1,000-square-foot area.

The limiting dust "densities" for flash-fire and explosion hazards would be equal – at about 1.9 g./sq.ft., or about 4 pounds per 1,000 square feet – for a room height of about 17 feet, or about 5.3 meters.

2.2.3 Mass Method B, for Protection Against Flash-Fire Hazards

To provide an estimate of the effects of flash fire on an exposed person, the NFPA standard apparently visualizes that dust accumulations having a mass $W_{SURFACE}$ on elevated surfaces are dislodged by a primary event, such as an explosion. The dust particles then descend essentially to floor level as a dense "shower" that has a "worst-case" concentration, C_{WORST}, at which the Deflagration Index, K_{ST}, and/or P_{MAX} values are determined. Thus, the initial volume of the dust cloud, before combustion occurs, would be given by:

$$V_{INITIAL} = W_{DISPERSED} / C_{WORST} \qquad \text{cubic meters} \tag{9}$$

where $W_{SURFACE}$ is the weight of dust that is dislodged from elevated surfaces, in grams;
C_{WORST} is the worst-case dust concentration, in grams per cubic meter.

The weight of the accumulated dust that is dispersed is related to the weight of accumulated dust by:

$$W_{DISPERSED} = W_{DUST}\,\eta_{DISPERSION} \qquad \text{grams} \tag{10}$$

where $\eta_{DISPERSION}$ is the fraction of the accumulated dust that is dispersed by the initial event.

Thus, the "initial" volume of the layer of dust/air mixture above the floor would be given by:

$$V_{INITIAL} = [\, W_{DUST}\, \eta_{DISPERSION} \,] / C_{WORST} \quad \text{cubic meters} \qquad (11)$$

To evaluate the flash-fire hazard to personnel in the enclosure, it would be appropriate to determine how a burning dust cloud could envelop those personnel. If ignition of the initial layer of dense dust/air mixture were to occur, this cloud would expand upward as a result of combustion. The volume of a cloud that would have a height, D, equal to that of an exposed person can be calculated. The cloud expansion is directly related to the increase in temperature, and the increase in temperature is closely related to the increase in the absolute pressure that results from confined combustion or explosion; that is:

$$V_{FINAL} / V_{INITIAL} = [\, P_{MAX} + P_{INITIAL} \,] / P_{INITIAL} = [\, A_{FLOOR}\, D \,] / V_{INITIAL} \qquad (12)$$

or:

$$V_{INITIAL} = [\, A_{FLOOR}\, D \,]\, [\, P_{INITIAL} / [\, P_{MAX} + P_{INITIAL} \,]$$
$$= [\, W_{DUST}\, \eta_{DISPERSION} \,] / C_{WORST} \quad \text{cubic meters} \qquad (13)$$

The above relationships can be combined to give:

$$W_{DUST} = C_{WORST}\, [A_{FLOOR}\, D]\, [\, P_{INITIAL} / [P_{MAX} + P_{INITIAL}] / \eta_{DISPERSION} \quad \text{grams} \qquad (14)$$

There is an assumption that the probability of one or more persons avoiding exposure to the flash fire is very low, and the standard suggested a probability of 5% or less [a]. Thus, the "allowable" weight of dust accumulations should be reduced by this factor. Thus, for a 5% probability of avoiding flash-fire exposure of 5%, the above relationship would become:

$$W_{DUST} = 0.05\, C_{WORST}[A_{FLOOR}\, D][P_{INITIAL} / [P_{MAX} + P_{INITIAL}] / \eta_{DISPERSION} \quad \text{grams} \qquad (15)$$

The above derivation yields the statement in NFPA 654: "The threshold dust mass[/accumulation] establishing a building or room as a **dust flash-fire hazard** area, $M_{FLASHFIRE}$ shall be determined by the following equation:

$$M_{FLASHFIRE} = P\, C_{WORST}\, [P_{INITIAL} / (P_{INITIAL} + P_{MAX})][A_{FLOOR}\, D / \eta_{DISPERSION}] \quad \text{kg."} \qquad (16)$$

where:

M_{FIRE} = threshold dust mass (kg) based on personnel fire exposure criterion;

P = probability of [a person avoiding] flame impingement ~~on a person~~, not to exceed 0.05 (5 percent probability) [a];

[a] This definition departs significantly from the definition in NFPA 654, paragraph 6.1.5.2; however, this definition meets the intention of the evaluation of flash-fire exposure.

C_{WORST} = worst-case dust concentration (kg./cu.m.) for the maximum rate-of-pressure-rise;
$P_{INITIAL}$ = 1 bar absolute;
P_{MAX} = maximum pressure (bar g) developed in tests on the accumulated dust;
A_{FLOOR} = enclosure floor area (sq.m.);
D = nominal height of a person (2 meters); and
η_{DUST} = entrainment fraction = 0.25 [non-conservative[b]]

The above relationship can be rearranged, with the above "default" values accepted (with a more-realistic nominal height of 1.75 meters or 5'-10"), and converted to useful units:

$$M_{FLASHFIRE} / A_{FLOOR} = [0.05][C_{WORST}][1/(1+P_{MAX})][(5.75)/35.3][1/0.25][2.205][453.5] \text{ g./sq.ft.} \quad (17)$$

$$M_{FLASHFIRE} / A_{FLOOR} = 32.5 \, [C_{WORST}] \, [1/(1+P_{MAX})] \text{ g./square foot of floor area.} \quad (18)$$

For typical organic dusts, the concentration C_{WORST} at which the maximum burning rate occurs – and the concentration at which the value of the Deflagration Index K_{ST} is obtained – is about 0.65 kilogram per cubic meter [References 21 and 22]. The maximum explosion pressure attained by organic dusts is about 9.0 barg [Reference 23]. Thus, for many organic dusts, the preceding equation becomes:

$$M_{FLASHFIRE} / A_{FLOOR} = 32.5 \, [0.65] \, [1/(1+9)] = 2.1 \text{ g./sq.ft., or about 4.5 lbs./1000 sq.ft.} \quad (19)$$

2.2.4 Mass Method B, for Protection Against Explosion Hazards

The NFPA standard for explosion venting [Reference 24] contains a relationship for "partial-volume" combustion in an enclosing structure, where the combustion would not rupture the confining walls and roof:

$$V_{CLOUD} / V_{TOTAL} = P_{REDUCED} / P_{MAX} \quad (20)$$

where V_{CLOUD} is the volume of the dust/air cloud at an explosible mixture, in cu.ft. (or cu.m.);
V_{TOTAL} is the total volume of the enclosure, in cubic feet (or cubic meters);
$P_{REDUCED}$ is the pressure that would result from the combustion, in psig (or barg); and
P_{MAX} is the explosion pressure when the mixture fills the volume, in psig (or barg).

[b] "Non-conservative", as explained later in this paper.

This relationship also can be shown as:

$$[\, W_{DISPERSED} / C_{WORST} \,] / V_{TOTAL} = P_{REDUCED} / P_{MAX} \qquad (21)$$

where $W_{DISPERSED}$ is the weight of dust in the suspended mixture of dust and air, in grams;
C_{WORST} is the concentration at which the Deflagration Index, K_{ST}, and/or P_{MAX} values are determined, in grams per cubic foot (or grams per cubic meter);
V_{TOTAL} is the volume of the enclosure, in cubic feet (or cubic meters).

The weight of dust in the suspended mixture of dust and air is related to the fraction of the accumulated dust that is raised and dispersed from horizontal surfaces (or from the walls of the enclosure): the "entrainment factor", $\eta_{DISPERSION}$; thus, $W_{DISPERSED} = W_{SURFACE}\, \eta_{DISPERSION}$. The total volume of the enclosure is the product of the floor area and the height of the enclosure: thus, $V_{TOTAL} = A_{FLOOR}\, H$. The value of $P_{REDUCED}$ is to be increased by a "Dynamic Load Factor", DLF, to obtain the strength of the enclosure walls and ceiling/roof, P_{ES}, in response to a brief-overpressure event: thus, $P_{ES} = P_{RED}\, DLF$, or $P_{RED} = P_{ES} / DLF$.

The preceding relationship would then become:

$$W_{SURFACE}/A_{FLOOR} = [\, P_{ES} / DLF \,]\, [\, C_{WORST} / P_{MAX} \,]\, [\, H / \eta_{DISPERSION} \,] \;\; g./m^2 \qquad (22)$$

where P_{ES} and P_{MAX} are in barg (or psig);
C_{WORST} is in grams per cubic meter; and
H is in meters.

or, for the enclosure height in feet, and the floor area in square feet:

$$W_{SURFACE} / A_{FLOOR} = 0.028\, [\, P_{ES} / DLF \,]\, [\, C_{WORST} / P_{MAX} \,]\, [\, H / \eta_{DUST} \,] \;\; g./ft^2 \qquad (23)$$

The above derivation yields the statement in NFPA 654: "The threshold dust mass[/accumulation] establishing a building or room as a **dust explosion hazard** area, $M_{EXPLOSION}$, shall be determined by the following equation [to contain a combustible dust explosion, and minimize damage to the walls and ceiling]:

$$M_{EXPLOSION} = [\, P_{ES} / DLF \,]\, [\, C_{WORST} / P_{MAX} \,]\, [\, A_{FLOOR}\, H / \eta_{DUST} \,] \qquad kg. \qquad (24)$$

where:
$M_{EXPLOSION}$ = threshold dust mass (kg) based on building damage criterion;
P_{ES} = enclosure strength evaluated based on static pressure calculations for the weakest building structural element not intended to vent or fail (bar g);

DLF = dynamic load factor; a [worst-case, conservative] value of 1.5;
C_{WORST} = worst-case dust concentration (kg./cu.m.) for the maximum rate-of-pressure-rise;
P_{MAX} = maximum explosion pressure (bar g) developed in tests on the accumulated dust;
A_{FLOOR} = enclosure floor area (sq.m.);
H = enclosure ceiling height (m); and
η_{DUST} = entrainment fraction; a [best-case, non-conservative[b]] value of 0.25.

The product of the above values of DLF and η_{DUST} is about 0.4 and represents a combination of "conservative" and "non-conservative" values. The value of DLF is very unlikely to be 1.0 or less, and the value of η_{DUST} is unlikely to be much above 0.6, for a product of 0.6. Thus, the "permitted" values of 1.5 and 0.25 appear to be reasonable.

The above relationship can be rearranged, with the above "default" values accepted, and converted to useful units:

$$M_{EXPLOSION} / A_{FLOOR} = [1000/35.3]\,[P_{ES}/P_{MAX}]\,[C_{WORST}/1.5]\,[H/0.25] \quad \text{g./ft}^2 \qquad (25)$$

$$M_{EXPLOSION} / A_{FLOOR} = 76\,[P_{ES}/P_{MAX}][C_{WORST}]\,[H] \quad \text{g./square foot of floor area} \qquad (26)$$

For typical organic dusts, the concentration C_{WORST} at which the maximum burning rate occurs – and the concentration at which the value of the Deflagration Index K_{ST} is obtained – is about 0.65 kilogram per cubic meter [References 21 and 22]. The maximum explosion pressure attained by organic dusts is about 9.0 barg [Reference 23]. Thus, for many organic dusts, the preceding equation becomes:

$$M_{EXPLOSION} / A_{FLOOR} = 76\,[\,P_{ES} / 9\,]\,[\,0.65\,]\,[H]\,[1/14.5] \quad \text{g./ft}^2 \qquad (27)$$

$$M_{EXPLOSION} / A_{FLOOR} = 0.38\,P_{ES}\,H \quad \text{grams per square foot of floor area} \qquad (28)$$

where:
P_{ES} = enclosure strength (psig); and
H = enclosure ceiling height (ft.).

[b] "Non-conservative", as explained later in this paper.

Guidance concerning the pressure resistance of wall and ceilings is limited, and there do not appear to be any "rules of thumb" for wall strengths. However, the following "data" might be useful:

1. The maximum wind speeds were probably considered in the design of metal or composite-material siding (on steel frames), and the required wind-pressure resistance would be given by: $P = 0.0025\ V^2$ pounds per square foot, where V is the wind speed in miles per hour [Reference 25]. For example, a wind speed of 120 MPH could produce a pressure of about 40 psf, or about 0.28 psig.

2. A steel-sided room within a building would suffer 50% damage if exposed to a pressure of about 0.55 psig [Reference 26].

3. Generally, buildings cannot withstand more than about 1 to 2 psig [Reference 27].

4. A brick wall would have very little ductility, and pressure resistance would be determined by the mass of the wall. An 18-inch wall could withstand about 10 psig [Reference 28], so a 4-inch wall probably could withstand about 2 psig], and 8 to 12-inch walls could withstand 7 to 8 psig [Reference 29].

5. Unreinforced concrete-masonry-unit [block] walls could withstand 2 to 3 psig [Reference 29]

6. A corrugated sheet-steel wall, 1/4-inch thick, could withstand 8 psig [Reference 30].

7. A concrete-masonry-unit [block] wall could withstand 10 psig if fully grouted and with double reinforcement [Reference 28].

8. Reinforced-concrete walls can withstand pressures of 10 psig or more [Reference 31].

As an example of the use of the preceding equation, the dust-accumulation limit for a room having a pressure resistance of 0.55 psig [about 80 pounds per square foot] and a 16-foot ceiling would be given by:

$$M_{EXPLOSION} / A_{FLOOR} = 0.38\ [0.55]\ [16] = 3.3\ \text{g./ft}^2\text{; about 7 lbs./1,000 ft}^2 \qquad (29)$$

For very large rooms of buildings, and for very strong rooms or buildings, the flash-fire limits on dust-accumulation weights will be smaller than the explosion limits. Thus, the flash-fire limits usually should be used as the housekeeping criteria.

2.3 Summary of Combustible-Dust Accumulation Limitations

The results of the calculations in the preceding sections can be tabulated as follows:

Table 3. Summary of Calculations

Properties	Method for Determining the Threshold Dust Mass/Accumulation [g./sq.ft.; lbs./1000 sq.ft.]				
	Flash-Fire Hazard		"Secondary" Explosion Hazard		
	Mass A	Mass B	Layer	Mass A	Mass B
Limit	1.9 g/ft^2	2.1 g/ft^2	4.5 g/ft^2	1.3 g/ft^2	3.3 g/ft^2
	4 lbs/1,000 ft^2	4 lbs/1,000 ft^2	10 lbs/1,000 ft^2	3 lbs/1,000 ft^2	7 lbs/1,000 ft^2
Variables	None	C_{WORST}; P_{MAX}	None	H	C_{WORST}; P_{MAX};P_{ES}; H

It may be of interest to note that equal dust-accumulation limits of 1.9 g./sq.ft. [or about 4 pounds per square foot of floor area] – are obtained only if:

a. The allowable area for the Layer method is only about 2% of the floor area, rather than 5%.

b. The room height is 17 feet, for the "Explosion-A" method.

c. The ratio of $C_{WORST}/(1+P_{MAX})$ is about 0.058, for the "Flash Fire-B" method; that is, for a value of C_{WORST} of 0.65 g./cu.m., the value of P_{MAX} would be about 10 barg (typical of a metal dust).

d. The ratio of $C_{WORST}/(1+P_{MAX})$ is about 0.058, and the room height is 17 feet, and the value of P_{ES} – the enclosure strength – is about 0.023 barg, or about 50 pounds per square foot, for the "Explosion-B" method.

2.4 Example of Dust Collection

In the recently-published "NFPA Guide to Combustible Dusts" [Reference 32], an example of hypothetical dust accumulation is provided. The "data" can be shown – with extended calculations that are based on a bulk density of 30 pounds per cubic foot – in the following Table:

Table 4. Summary of Housekeeping Results for a Segregated Room [Example]

Location	Area	Interval	Dust	Rate	Thickness	Density
Floor	500 sq.ft.	8 hours	25 lbs.	75 lbs./day	0.02 in.; 0.5 mm.	23 g./sq.ft.
Equipment	100 sq.ft.	7 days	10 lbs.	1.4 lbs./day	0.04 in.; 1 mm.	45 g./sq.ft.
Overhead	500 sq.ft.	60 days	30 lbs.	0.5 lbs./day	0.024 in.; 0.6 mm.	27 g./sq.ft.
Totals	1,100 sq.ft.	[22% of area]	65 lbs.	-	-	27 g./sq.ft.
Summary	5,000 sq.ft. [4,600 sq.ft. floor]		65 lbs.	-	-	5.9 g./sq.ft.

As stated by the authors, the layer-depth criterion would be 0.078 inch [2 mm.], but only over 5% of the total horizontal-surface area of 5,000 square feet. Thus, the total "allowable" dust accumulation would be limited to 1.6 cubic feet, or about 50 pounds of dust. Therefore, it is very likely that the layer-depth criterion would be exceeded, particularly prior to the 60-day cleanup.

Further, the authors state: "The total dust accumulation in this example exceeds the allowable for the room footprint area; thus, the current cleaning frequencies result in a dust deflagration hazard in the room."

The situation that is described by the preceding Table 4 is particularly hazardous because "Any dust that is elevated above floor level in joists, beams, equipment tops, etc., should be considered as being capable of being thrown into suspension." [Reference 33]. A 40-pound accumulation of combustible dust over a floor area of 4,600 square feet would represent a "density" of 4 grams per square foot. This exceeds the "Mass A" criteria for a flash-fire hazard and for an explosion hazard unless the ceiling was more than 35 feet above floor level. A density of 4 g./sq.ft. also would exceed the "Mass B" criteria for a flash-fire hazard and for an explosion hazard for most combustible dusts in typical enclosed process buildings.

As also stated by the authors, there are three alternatives:

1. Change the cleaning frequencies/intervals to limit the accumulation to 50 pounds;

2. Provide personnel protection from the potential flash-fire hazard, and protect the structure from overpressure/explosion damage; or

3. Provide engineering controls to reduce and/or capture fugitive dust releases from the equipment.

Facilities that have been experiencing "housekeeping problems" with accumulations of combustible dust should begin weighing the dust that is swept or vacuumed from floors; from elevated surfaces such as roof supports and ductwork; and from the surfaces of equipment, including motors, electrical panels, and process equipment. The purposes are:

1. The accumulated-dust "densities" – in terms of grams per square foot, or pounds per 1,000 square feet of floor area – could be used to establish and/or adjust the frequency of cleaning floors and elevated surfaces.

2. The trend in recovered-dust weight divided by the interval between such housekeeping activities – such as pounds per week – can serve as a "leading indicator" concerning the adequacy [or inadequacy] of the control of flash-fire and secondary-explosion hazards.

3. Guidance for minimizing the flash-fire and secondary-explosion hazards resulting from accumulations of combustible dust are:

   a. "Most-conservative" limitation for flash fire and explosion: 1.9 grams per square foot, or 4 pounds per 1,000 square feet of floor area.

   b. "Least-conservative" limitation for flash fire: 1.9 g./sq.ft., or 4 lbs./1,000 sq.ft., as based on the properties of "typical" organic dusts.

   c. "Least-conservative" limitation for explosion: 3.3 g./sq.ft., or 7 lbs./1,000 sq.ft., as based on (1) the properties of "typical" organic dusts, (2) 180-MPH wind-resistance walls, and (3) a 16-foot ceiling.

   d. The dust accumulations that would be "allowed" by the Layer method [about 4.5 g./sq.ft., or about 10 lbs./1,000 square feet of floor area] appear to be "excessive" when compared to the limitations that would be imposed by the above "Mass Methods".

There is another important aspect in the control of dust accumulations, and that is the relationship between the amount [particularly, the layer depth] of combustible-dust accumulations and the electrical classification that would apply to areas in which combustible-dust accumulations occur. Of concern regarding dust deposits on electrical devices is the hazard associated with the thermal insulating properties of the layer, which can limit the heat dissipation and allow powder temperature rise. If a dust layer gets too hot, it can self-ignite. Such smoldering dust layers present both an immediate fire hazard as well as a high energy ignition source for the expansion of the fire or ignition of dust clouds that could come into contact with the embers.

2.5 Housekeeping Guidance for "Chronic" or "Continual" Releases of Combustible Dusts

Where "episodic" releases of combustible dusts occur – frequently, or infrequently – guidance is provided in NFPA 654, NFPA 652, and NFPA 484 for "unscheduled" housekeeping [References 34, 35, and 36]; that is, removal of dust spillages or other types of "transient" dust releases. In the following Table 5, "TDMA" is the Threshold Dust Mass/Accumulation of about 90 grams per square foot, as determined from the "Layer Method" calculation that is described above. Since this dust "density" would be hazardous if covering a large area, other "more-conservative" values of "TDMA" can be selected, as justified by use of the "Mass A" or "Mass B" methods.

Table 5. Unscheduled Housekeeping Guidance [NFPA 654]

Accumulation Level [NFPA 652]	Accumulation on the Worst Single Square Meter [10 square feet] of Surface	Average Layer Depth at 30 Pounds per Cubic Foot [480 kg./cu.m.]	Max. Spill-Area-to-Room-Area Ratio [%}	Longest Time to Complete Local Cleaning of Accessible Surfaces	Longest Time to Complete Local Cleaning of Remote Surfaces
1	1X < TDMA < 2X	5/64" to 5/32" 2 to 4 mm.	2% to 1%, for 1.9 g./cu.m. average	8 hours	24 hours
2	2X < TDMA < 4X	5/32" to 5/16" 4.2 to 8.3 mm.	1.0% to 0.5%, for 1.9 g./cu.m. average	4 hours	12 hours
3	>4X TDMA	>5/16" >8 mm.	<0.5%, for 1.9 g./cu.m. average	1 hour	3 hours

The preceding Table 5 could be utilized to establish the maximum area – as a percentage of the total room area – that is "allowed" to be covered with dust at a given depth. Thus, if frequent spillages or deposits occur rather frequently – as at bagging stations – it might be advisable to provide floor markings that would show the maximum area, to prompt cleanup of that area.

3 Relationship Between Dust Accumulations and Electrical Classification of Dusty Areas

In the U. S., there are two NFPA publications that are concerned with the electrical classification of areas in which combustible dust might be present in the atmosphere or as accumulations on surfaces in the work area. The following publications provide dust-accumulation limits in the appendices that are "*for informational purposes only*":

1. NFPA 654, Table A.6.5.2 [Reference 37].

2. NFPA 499, Tables A.6.3.2(a) and A.6.7 [Reference 38].

The following Table 6 copies information from the NFPA 654 standard [in red], with the addition of calculated "Dust Densities" and "Maximum Ratio of Spill Area to Room Floor Area" (in percent).

Table 6. Guidance for Area Electrical Classification [NFPA 654; Table A.6.5.2]

Depth of Dust Accumulation	Dust Density [g./sq.ft.]	Max. Spill/Room Area %	Frequency	Housekeeping Requirement	Electrical Classification
Surface color discernible	Near zero; Negligible	100%	No spills or releases	Maintain negligible	Unclassified
0 to 1/32" [0 to 0.8 mm.]	0 to 35 g./sq.ft.	100% to 5%, for 1.9 g./cu.m. average	2 to 3 times per year	Clean up during same shift	Unclassified
			>3 times/year; 24-hr. accumulation	Maintain <1/64"/0.4 mm.	Unclassified, but dust-tight; NEMA 12
1/32" to 1/8" [0.8 to 3 mm.]	35 to 140 g./sq.ft.	5% to 1.5%, for 1.9 g./cu.m. average	2 to 3 times per year	Clean up during same shift	Unclassified, but dust-tight; NEMA 12
			>3 times/year; 24-hr. accumulation	Maintain <1/16"/1.6 mm.	Class II, Div. 2
> 1/8" [>3 mm.]	>140 g./sq.ft.	<1.5%, for 1.9 g./cu.m. average	2 to 3 times per year	Immediate shutdown; clean	Class II, Div. 2
			>3 times/year; 24-hr. accumulation	Minimize accumulation	Class II, Div. 1

The following Table 7 copies information from the NFPA 499 standard [in red], with the addition of calculated "Dust Densities" and "Maximum Ratio of Spill Area to Room Floor Area" (in percent). When Tables 6 and 7 are compared, it is clear that there are some significant differences between these two standards that have yet to be resolved by the NFPA.

Table 7. Guidance for Area Electrical Classification [NFPA 499; Tables A.6.3.2(a); A.6.7]

Depth of Dust Accumulation	Dust Density [g./sq.ft.]	Max. Spill/Room Area %	Frequency	Housekeeping Requirement	Electrical Classification
Surface color discernible	Near zero; Negligible	100%	No spills or releases	Not Specified	Unclassified
0 to 1/32" [0 to 1 mm.]	0 to 35 g./sq.ft.	100% to 5%, for 1.9 g./cu.m. average	Infrequent Release	Clean as appropriate	Unclassified
1/32" to 1/8" [0.8 to 3 mm.]	35 to 140 g./sq.ft.	5% to 1.5%, for 1.9 g./cu.m. average	Infrequent Release	Maintain less than 3 mm. [1/8"]	Class II, Div. 2
			Occasional Cloud Formation	Minimize additional accumulation or cloud formation	Class II, Div. 1 or Div. 2
>1/8" to 1/2" [>3 mm. to 12.5 mm.]	140 to 570 g./sq.ft.	1.5% to 0.3%, for 1.9 g./cu.m. average	Continuous; Frequently	Minimize additional accumulation	Class II, Div. 1
>1/2" [>12.5 mm.]	>570 g./sq.ft.	<0.3%, for 1.9 g./cu.m. average	Infrequent thick dust layer or extensive dust cloud	Immediate shutdown; clean	Class II, Div. 1

Despite the differences in the two standards – NFPA 654 and NFPA 499 – the above Tables 6 and 7 provide good guidance concerning minimum-hazard dust-layer thicknesses and maximum thick-dust areas. Thus, the preceding two Tables can be combined to give "minimum essential" guidance for protection against flash-fire and explosion hazards resulting from combustible-dust accumulations, as shown in Table 8.

Table 8. Suggested Guidance for Area Electrical Classification [Combined Standards]

Depth of Dust Accumulation	Dust Density [g./sq.ft.]	Max. Spill/Room Area %	Frequency	Housekeeping Requirement	Electrical Classification
Surface color discernible	Near zero; Negligible	100%	No spills or releases	Not Specified	Unclassified
0 to 1/32" [0 to 1 mm.]	0 to 35 g./sq.ft.	100% to 5%, for 1.9 g./cu.m. average	2 to 3 times per year	Remove during same shift	Unclassified
			>3 times per year; 24-hr. accumulation	Maintain <1/64"/0.4 mm.	Unclassified, but dust-tight; NEMA 12
1/32" to 1/8" [0.8 to 3 mm.]	35 to 140 g./sq.ft.	5% to 1.5%, for 1.9 g./cu.m. average	2 to 3 times per year	Remove during same shift	Unclassified, but dust-tight; NEMA 12
			>3 times per year; 24-hr. accumulation Occasional cloud	Maintain <1/16"/1.6 mm.	Class II, Div. 2
>1/8" to 1/2" [>3 mm. to 12.5 mm.]	140 to 570 g./sq.ft.	1.5% to 0.3%, for 1.9 g./cu.m. average	2 to 3 times per year	Immediate shutdown; clean	Class II, Div. 2
			>3 times per year; 24-hr. accumulation; Occasional cloud	Minimize accumulation	Class II, Div. 1
>1/2" [>12.5 mm.]	>570 g./sq.ft.	<0.3%, for 1.9 g./cu.m. average	Infrequent thick dust layer or extensive dust cloud	Immediate shutdown; clean	Class II, Div. 1

The above guidance for Class and Division electrical area classifications can also be applied to Zones, where Zone 22 corresponds to Class II, Division 2, and Zone 21 corresponds to Class II, Division 1. Zone 20 is essentially limited to equipment, where a hazardous concentration of dust occurs very frequently or continuously, and the nearest equivalent is Class II, Division 1.

4 Consideration of Non-Electrical Ignition Sources

There are several types of "non-electrical" ignition sources that are prevalent in chemical plants, and they also need isolation from combustible dusts, just as combustible dusts are isolated from electrical arcs and sparks. Typical "non-electrical" ignition sources – for "primary" dust fires, flash-fires, and explosions – include the following, in approximate descending order of importance [References 39 and 40]:

Table 9. Non-Electrical Ignition Sources

Open flames	[boilers, incinerators, air heaters]
Welding and cutting	[torches or arcs]
Explosions in equipment	[ignition of dust/air, gas/air, vapor/air mixtures]
Hot surfaces of equipment	[ovens, driers, piping for heating fluids]
Accidental Fire	[combustibles at hot spots]
Frictional hot spots	[bearings, metal-to-metal rubbing]
Frictional sparks	[grinding, polishing]
Vehicles	[exhaust systems]
Impact sparks	[thermite, mechanical systems]
Unstable materials	[peroxides, pyrophorics, spontaneous combustion]
Adiabatic compression	[diesel effects]
Electromagnetic radiation	[stray fields]
Ionizing radiation	[instrumentation]
Lightning	[storms]

A method for studying and correcting the undesired effects of these "non-electrical" ignition sources on combustible dusts [Reference 41] is the "Mechanical Equipment Ignition Risk Assessment" [MEIRA]. Assessment of mechanical ignition hazards is outside the scope of this presentation. However, the control of dust accumulations near such ignition sources is important in preventing fires, flash-fires, and explosions. Persons performing Dust Hazards Analysis should be aware of these other types of ignition sources and determine whether or not conformance to the area electrical classification adequately isolates these types of ignition sources from combustible dusts.

5 "Experimental" Data for "Organic" Dusts

During a recent Dust Hazards Analysis, an operator was observed sweeping-up dust in a powder-storage area, and he was asked to weigh the dust that he had gathered. The weight was about 5 pounds, and the floor area was about 2,500 square feet. Thus, the average dust "density" was about 2 pounds per 1,000 square feet, or about 0.9 gram per square foot. Thus, this amount of dust accumulation was unlikely to be sufficient to "fuel" a secondary dust explosion.

At another location, the site management was asked to collect the dust that was (1) on the ground floor at a bagging station and (2) on equipment and the building structure in an "upstairs" operating area. The weight that was collected at the first location was about 4 pounds, and the floor area that was swept-up was about 240 square feet. Thus, the average dust density was about 17 pounds per 1,000 square feet, or about 7.5 grams per square foot. This is a high dust density. However, the floor and surfaces in the rest of this 1,150-square-foot room did not contain significant dust; that is, the surface colors were not obscured. Thus, the average dust density was about 3.5 pounds per 1,000 square feet, or about 1.6 grams per square foot. Therefore, this amount of dust accumulation could be considered to be marginally acceptable, at the time of this "fortuitous" dust cleanup.

At the second location – which was isolated from the ground floor by solid flooring – about 10.2 pounds of dust were collected, and the floor area was 1,150 square feet. Thus, the average dust density was about 9 pounds per 1,000 square feet, or about 4.0 grams per square foot. This dust density exceeds the NFPA 654 "Mass Method" guidelines and indicated that more-frequent housekeeping – or better dust containment – was needed.

Figure 1, below, shows a 30-gram deposit of "dust" [sodium bicarbonate; baking powder] uniformly sifted onto a one-square-foot area; thus, 30 grams per square foot [g./sq.ft.]. The depth of the deposit is about one millimeter, as indicated by the "standard" paper clip. A wide-spread deposit of combustible dust having a "density" of 30 g./sq.ft. would significantly exceed the NFPA criteria for the prevention of flash fire and explosion for typical chemical-plant constructions.

Figure 2, below, shows a 3-gram deposit of baking powder sifted onto a one square foot area. As indicated by this photograph, a deposit of 3 g./sq.ft. would marginally obscure the color of the surface under the deposit. As indicated by the preceding discussion, a combustible-dust density of 3 g./sq.ft. would exceed several of the NFPA critera for flash-fire and explosion prevention.

6 Combustible-Dust Accumulation Limits, for "Metal" Dusts

6.1 The "Layer-Depth" Criterion

In the NFPA 484 standard for metal dusts, the Threshold Limit is defined as: "*An established value of dust accumulation, determined as part of the hazard analysis for a particular enclosure of facility and specific metal, for taking an appropriate action.*"

There have been several proposed methods for determining an appropriate threshold value for fugitive dust accumulations. It should be noted that some methods might not be appropriate in all situations. Fugitive dust shall not be allowed to accumulate to a level that obscures the color of the surface beneath it. It shall be permissible to establish, in a building or room, an alternate housekeeping dust accumulation threshold based on a documented hazard assessment acceptable to the Authority Having Jurisdiction [AHJ].

Unscheduled housekeeping should be performed to limit the time that a local spill or short-term accumulation of dust is allowed to remain before cleaning the local area to less than the threshold dust mass/accumulation. Table A.7.4(b) [of the NFPA standard] shows approximate equivalent depths for the accumulation values in Table A.7.4(a) [of the NFPA standard] when the threshold dust mass/accumulation is 1 kg/m².

It is likely that a double layer of dust particles would provide significant obscuration of the color of the surface under the layer of dust. For example, a double layer of dust particles having a uniform diameter of 100 microns and a density of 1 g./cu.cm. would have a "dust density" of about 10 grams per square foot, as based on:

$$W / A = 100\, d\, \rho \text{ grams per square foot} \tag{30}$$

where d is the diameter of the particles, in millimeters, and
ρ is the particle density, in grams per cubic centimeter.

The thickness of the dust layer would be about twice the particle diameter, or about 0.2 mm.

Since the above relationship indicates that the "dust density" is directly proportional to particle density, an obscuring layer of 100-micron steel particles – for a particle density of 7.8 g./cu.cm. – would have a "dust density" of about 78 grams per square foot, with the same dust thickness, near 0.2 mm. If the layer-depth criterion of NFPA 654 were to be used to determine a hazardous accumulation for iron dust, the result would be:

$$t_{ALLOWABLE} = [\,0.8\,]\,[75/BD] = [\,0.8\,]\,[75/180] = 0.33 \quad \text{millimeter} \tag{31}$$

and this is close to the "obscuring" depth. The following Table provides the results of similar calculations for other metal dusts:

Table 10. Dust-Accumulation Thresholds Based on Layer Depth

Dust		Aluminum	Copper	Iron	Magnesium	Titanium	Zirconium
Bulk Density	(lbs./cu.ft.)	44	101	180	46*	60	100*
Layer Thickness	(mm.)	1.4	0.6	0.3	1.3	1.0	0.6

* approximate bulk density [based on particle density]

6.2 *"Mass Method" Criteria for Metal Dusts*

The "Mass Method A" and "Mass Method B" calculations that are described above for organic dusts are derived from considerations of heat of combustion and the results of combustion-rate tests on dispersed dusts [and the effects of thermal radiation and blast pressures on humans]. It might be expected that similar "Mass Method A" and "Mass Method B" calculations will be available for metal dusts in the near future, also.

7 Estimation of Entrainment Fraction

In the NFPA 654 standard, a procedure for determining the maximum "allowable" dust accumulations is presented as "Mass Method B". In the equations for flash-fire and explosion hazards, an "Entrainment Fraction" defines the fraction of dust that is removed from the [top of] the dust deposits. An "entrainment fraction" of 0.25 is suggested, with the provision that *"an alternative value of η_D, based on a risk assessment that is acceptable to the AHJ is permitted"*. However, two sets of tests that were conducted to determine the extent of entrainment of dust deposits – when exposed to blasts of combustion gases [Reference 42] – indicate that a value of 0.25 probably is not "conservative" for thin layers of combustible [organic] dusts.

7.1 NGFA Tests

In the referenced study, the results of two sets of tests were reported. The results of five tests that were supported by the National Grain and Feed Association [NGFA] are presented in Table 11. Cornstarch was spread on the floor of an 8'x8'x80' gallery, and the layers of dust were subjected to an explosion of a cornstarch/air mixture at one end of the gallery. The bulk density of the cornstarch was about 42 pounds per cubic foot [0.67 g./cu.cm.], and the particle density of cornstarch is about 1.5 g./cu.cm. [94 pounds per cubic foot].

Table 11. Interpreted Results of NGFA Tests on Deposits of Cornstarch Powder

Test No.	Initial Dust Density	Initial Layer	End Dust Density	End Layer	Entrainment Fraction
50	17 g./sq.ft.	0.27 mm.	7 g./sq.ft.	0.10 mm.	0.62
56	21 g./sq.ft.	0.33 mm.	10 g./sq.ft.	0.16 mm.	0.53
53	23 g./sq.ft.	0.37 mm.	15 g./sq.ft.	0.25 mm.	0.34
51	35 g./sq.ft.	0.55 mm.	21 g./sq.ft.	0.33 mm.	0.40
52	64 g./sq.ft.	1.02 mm.	44 g./sq.ft.	0.71 mm.	0.30

The results of these tests show that the Entrainment Fraction increases as the thickness of the dust layer decreases. An equation that describes this behavior [approximately] is:

$$E = 1 - e^{-12.5 / D^{0.925}} \quad (32)$$

where D is the "dust density" in grams per square foot.

7.2 NIOSH Tests

The results of tests that were supported by the National Institute of Occupational Safety and Health [NIOSH] involved mixtures of coal dust and limestone dust. The mixtures were spread between two 1-inch rails in a 20-foot-wide, 8-foot-high, and 1600-foot long tunnel. The particle densities for coal dust and limestone dust are about 1.33 and 2.75 g./cu.cm., respectively [83 and 172 pounds per cubic foot], and the bulk densities of the mixtures were about 0.85 g./cu.cm. [about 53 pounds per cubic foot].

Ten tests are reported in the cited reference, with the coal-dust concentrations ranging from 20 to 35 weight-percent. The dusts were screened though a 74-micron screen, and the tests used dusts in which 20% to 80% passed through the screen. The dust-layer thicknesses were 1.0 inch [25.4 mm.] in all of the tests. Thus, the "dust density" was about 2,000 grams per square foot. The layers of dust were subjected to explosion of a methane/air mixture at one end of the tunnel.

The amounts of dust that were removed from the deposits ranged from 1.0 to 2.6 millimeters, with an average of 1.70 millimeters. Thus, the entrainment fractions ranged from 0.039 to 0.102. The results of the tests indicated that there were no relationships between the percentage of coal dust and the particle-size distribution – within the above ranges – and the amount of dust that was entrained by the explosion-generated flow of combustion gases.

7.3 Conclusions Resulting from the Tests

The results of these tests indicate that the Entrainment Fraction is a strong function of the amount of dust in the layer: high values of the Entrainment Fraction for thin layers of dust, and low values of the Entrainment Fraction for thick layers of dust. Thus, a conservative approach to the use of the "Mass Method B" procedures for determining the maximum "allowable" dust accumulation – in terms of grams per square foot – would be to calculate with an Entrainment Fraction of 0.25 and then recalculate with an entrainment fraction that would be obtained from the above equation (32). Repetitive calculations might be required to obtain the final result.

8 Other Quantifiable Aspects of Housekeeping

It is important that personnel who are expected to clean-up spills or releases of combustible dusts (1) be trained to avoid creating large dense [explosible] clouds of combustible dust and (2) be well-protected against the hazard of flash-fire that might result from ignition of smaller clouds of dust. This would include the members of the spill-response team, contractors, operators, and/or mechanics that would be involved in collection and removal of combustible-dust spills or released. Careful sweeping and cautious handling of the collected dust can minimize the likelihood of large dust clouds and the hazard of explosion. For dusts that could be ignited by electrostatic charges on humans – 25 millijoules or less [Reference 43] – the cleaning implements should be static-dissipative or semi-conductive, with a resistance to ground that is less than 100 megohms [Reference 44]. Such "grounding" could be through the person doing the cleaning, if gloves, footwear, and working surface have similar conductivities.

For protection against flash-fire, persons who clean-up dust deposits should be wearing static-dissipative footwear, to minimize the ignition hazard. Also, they should be wearing flame-resistant outer clothing, with face and head covering, for protection against possible flash fire, since the mortality likelihood is about 25% for ignition of "everyday" clothing [Reference 45]. Further, if the clean-up involves "pyrophoric" metal dusts – such as titanium, zirconium, aluminum, lithium, magnesium, cesium, or sodium and potassium "sand" [Reference 46] – a metallized "flash oversuit" should be worn. The flash-fire temperatures of metal dusts are much higher than those of most organic dusts. The adiabatic flame temperatures of metal dusts is limited only by the boiling points of the oxides: about 3,500°C for aluminum oxide and about 3,600°C for magnesium oxide [Reference 47]. This is in contrast with the adiabatic flame temperatures for organic dusts – about 2,000°C [Reference 48] – as limited by the decomposition of carbon dioxide and water vapor [References 49 and 50].

The preceding discussion concerning personal protective equipment would also apply to persons who are – or could be – exposed to dense clouds of combustible dust, either routinely or during "transient" activities. The former exposure would apply to slitting and dumping bags of small-particle powder into mixers, or for recovery or re-work of product. The latter exposure could involve cleaning of ductwork or emptying dust collectors. Work within dust collectors will also present additional hazards associated with confined space entry.

Cleaning of equipment and surfaces with compressed air should be prohibited, since such dispersion of dusts in air tends to "classify" the dust [References 51 and 52]. The larger particles fall rather rapidly to the floor, but the smaller particles can be lofted by air currents onto elevated surfaces, where they could become fuel for secondary explosions [Reference 17]. The settling velocity for 75-micron [dust] particles is about 0.3 meter per second [Reference 53], and this corresponds to an air velocity of about 0.7 MPH. Thus, a very small upward air current – such as a thermal updraft – could keep such particles suspended indefinitely.

9 Path Forward

There are some aspects of housekeeping that should be resolved in the NFPA standards:

1. Provide better guidance concerning the "Entrainment Fraction" that is used in the "Mass Method B" equations for maximum dust accumulations, and

2. Provide guidance for metal dusts concerning the flash-fire and explosion hazards in NFPA 484, similar to that which is given in NFPA 654 for "organic" dusts, and

3. Tests should be conducted to determine the "Entrainment Fraction" for metal dusts, which typically have bulk and particle densities that are significantly greater than those of "organic" dusts.

10 Conclusions

The publications of the National Fire Protection Association provide good guidance concerning limitations of combustible-dust accumulations, to avoid flash-fire and explosion hazards. A chemical plant – or other site where combustible dusts are handled – could use the results of the calculations based on the NFPA guidance to establish housekeeping policies and procedures.

Additional guidance is needed from the NFPA concerning the maximum "allowable" accumulations of metal dusts, to minimize the flash-fire and explosion hazards, like the "Mass Method B" procedures for "organic" dusts.

The NFPA publications also show the close relationship between combustible-dust hazard and electrical classification, to minimize the likelihood of electrical arcs, sparks, and high temperatures in electrically-powered equipment from igniting combustible dusts. With sufficiently-frequent cleaning, the powder-handling and dust-handling areas could be considered as "unclassified", and "general-purpose" electrical equipment could be used in these clean areas. Otherwise, the use of electrical equipment should be restricted to those types that meet the electrical classification of the area. Other types of ignition sources also should be isolated from dense clouds and layers of combustible dusts.

It is important that persons who respond to combustible-dust emergencies and persons who could be exposed to dense clouds of combustible dust be protected against flash-fire hazards. Good containment of dusts within equipment minimizes these hazards, and activities that could lead to formation of dust clouds or to the accumulation of dusts on elevated surfaces should be strictly limited.

Figure 1 30 grams of Sodium Bicarbonate Dust on One Square Foot

Figure 2 3 grams of Sodium Bicarbonate Dust on One Square Foot

11 References

1. W. L. Frank, S. A. Rodgers, and G. R. Colonna, *NFPA Guide to Combustible Dusts*, NFPA, Quincy MA, Pages 4, 52, 130, 132, 136, 150, 179, and 234 to 246 (2012).

2. W. L. Frank, S. A. Rodgers, and G. R. Colonna, *NFPA Guide to Combustible Dusts*, NFPA, Quincy MA, Pages 5, 15, 20, 144, 234, 242, and 311 (2012).

3. V. Babrauskas, *Ignition Handbook*, Fire Science Publishers, Issaquah WA, Page 144 (1992).

4. H. L. Febo, Combustible Dust Hazard Recognition – An Insurer's View, Process Safety Progress, 30 (1), Page 84 (March 2011).

5. T. J Myers and A. F. Ibarreta, Tutorial on Combustible Dust, Process Safety Progress, 32 (3), Page 305 (September 2013).

6. W. L. Frank G. R. Colonna, On-Going Developments in Addressing Combustible Dust Hazards, Process Safety Progress, 34 (1) Page 29 (March 2015).

7. E. A. Ural, Dust Entrainability and Its Effect on Explosion Propagation in Elongated Structures, Process Operations Progress, 11 (3), Page 176 (July 1992).

8. E. A. Ural, Towards Estimating Entrainment Fraction for Dust Layers, Fire Protection Research Foundation, NFPA, Quincy MA (June 2011).

9. R. K. Eckhoff, *Dust Explosions in the Process Industries*, 2nd Ed., Butterworth Heinemann, Woburn MA, Pages 226 to 234, 246, 490, 493 (Figure 7.10), and 572 (1999).

10. National Fire Protection Association, *Standard for the Prevention of Fire and Dust Explosions from the Manufacturing, Processing, and Handling of Combustible Particulate Solids*, NFPA 654, NFPA, Quincy MA (1994).

11. National Fire Protection Association, *Standard for the Prevention of Fire and Dust Explosions from the Manufacturing, Processing, and Handling of Combustible Particulate Solids*, NFPA 654, NFPA, Quincy MA, Paragraphs 2-2.3.1, 4-2.1, and A-4-2-1 (1997).

12. National Fire Protection Association, *Standard for the Prevention of Fire and Dust Explosions from the Manufacturing, Processing, and Handling of Combustible Particulate Solids*, NFPA 654, NFPA, Quincy MA, Paragraphs 4-2.2 and A-4-2.1 (1997).

13. National Fire Protection Association, *Standard for the Prevention of Fire and Dust Explosions from the Manufacturing, Processing, and Handling of Combustible Particulate Solids*, NFPA 654, NFPA, Quincy MA, Paragraphs 2.2.3.1 and A.2.2.3.1 (2000).

14. Tapco, Inc., Bulk Material Density Table, Available at http://www.tapcoinc.com/content/product_data/tapco_catalog_09_p88-94.pdf. Accessed on January 12, 2017.

15. U. S. Dept. of Labor, Occupational Safety and Health Administration, OSHA Instruction – Combustible Dust National Emphasis Program (Reissued), PL 03-00-008, GPO, Washington DC, Paragraph IX.E.3.c (Oct. 1, 2015).

16. U. S. Dept. of Labor, Occupational Safety and Health Administration, OSHA Memo – Evaluating Hazardous Levels of Accumulation Depth for Combustible Dusts, GPO, Washington DC, (June 15, 2015).

17. FM Global, Property Loss Prevention Data Sheet 7-76 – Prevention and Mitigation of Combustible Dust Explosion and Fire, FM Global, Johnston RI, Paragraph 3.1.1 (2014).

18. S. A. Rodgers, Application of the NFPA 654 Dust Layer Thickness Criteria – Recognizing the Hazard, Process Safety Progress, 31 (1), Page 24 (March 2012).

19. National Fire Protection Association, *Standard for the Prevention of Fire and Dust Explosions from the Manufacturing, Processing, and Handling of Combustible Particulate Solids*, NFPA 654, NFPA, Quincy MA, Page 654-1 (2013).

20. National Fire Protection Association, *Standard for the Prevention of Fire and Dust Explosions from the Manufacturing, Processing, and Handling of Combustible Particulate Solids*, NFPA 654, NFPA, Quincy MA, Paragraphs 6.1.1.3, 6.1.1.4, 6.1.4, 6.1.5, and 6.1.6 (2017).

21. W. L. Frank, S. A. Rodgers, and G. R. Colonna, *NFPA Guide to Combustible Dusts*, NFPA, Quincy MA, Page 66 (2012).

22. R. K. Eckhoff, *Dust Explosions in the Process Industries*, 2nd Ed., Butterworth Heinemann, Woburn MA, Page 40 (1999).

23. W. L. Frank, S. A. Rodgers, and G. R. Colonna, *NFPA Guide to Combustible Dusts*, NFPA, Quincy MA, Pages 66 to 70 (2012).

24. National Fire Protection Association, "Standard on Explosion Protection by Deflagration Venting", NFPA 68, Quincy MA, Paragraphs 7.2.7 and 8.3.2 (2013).

25. E. A. Avallone, T. Baumeister, and A. M. Sadegh, *Marks' Standard Handbook for Mechanical Engineers*, 11th Ed., McGraw-Hill, New York NY, Page 12-20 (2007).

26. Center for Chemical Process Safety, *Guidelines for Evaluating Process Plant Buildings for External Explosions, Fires, and Toxic Releases*, 2nd Ed., AIChE, New York NY, Figure 4.1, Page 48 (2012).

27. S. Mannan [editor], *Lees' Loss Prevention in the Process Industries*, 3rd Ed., Elsevier, Boston MA, Page 17/251 (2005).

28. D. O. Dusenberry [editor], *Handbook for Blast-Resistant Design of Buildings*, John Wiley & Sons, Edison NJ, Pages 285 and 430 (2010).

29. D. A. Crowl and J. F. Louvar, *Chemical Process Safety*, 3rd Ed., Prentice Hall, Upper Saddle River NJ, Page 289 (2011).

30. Safe Havens Enterprises, Inc., Blast-Resistant Modular Buildings. Available at http://SafeHavenEnterprises.com. Accessed on January 12, 2017.

31. C. Naito, Blast-Resistant Design Considerations, Ascent Publishers, Allahabad India, Page 31 (Winter 2011).

32. W. L. Frank, S. A. Rodgers, and G. R. Colonna, *NFPA Guide to Combustible Dusts*, NFPA, Quincy MA, Table 5.13 on Page 241 (2012).

33. E. A. Ural, Towards Estimating Entrainment Fraction For Dust Layers – Final Report, Fire Protection Research Foundation, NFPA, Quincy MA, Pages 34 to 42 (June 2011).

34. National Fire Protection Association, *Standard for the Prevention of Fire and Dust Explosions from the Manufacturing, Processing, and Handling of Combustible Particulate Solids*, NFPA 654, NFPA, Quincy MA, Paragraphs Tables A.8.2.1.3 (a) and (b) (2017).

35. National Fire Protection Association, *Standard on the Fundamentals of Combustible Dust*, NFPA 652, NFPA, Quincy MA, Paragraph 8.4 and Table A.8.4.6.3 (2016).

36. National Fire Protection Association, *Standard for Combustible Metals*, NFPA 484, NFPA, Quincy MA, Paragraph 5.4, Chapter 7, and Paragraph 8.8 (2015).

37. National Fire Protection Association, *Standard for the Prevention of Fire and Dust Explosions from the Manufacturing, Processing, and Handling of Combustible Particulate Solids*, NFPA 654, NFPA, Quincy MA, Table A.6.5.2 (2017).

38. National Fire Protection Association, *Recommended Practice for the Classification of Combustible Dusts and of Hazardous (Classified) Locations for Electrical Installations in Chemical Process Areas*, NFPA 499, NFPA, Quincy MA, Paragraph 6.3.2 and Table A.6.7 (2013).

39. S. Mannan [editor], *Lees' Loss Prevention in the Process Industries*, 3rd Ed., Elsevier, Boston MA, Pages 17/260 and 17/264 (2005).

40. V. Babrauskas, *Ignition Handbook*, Fire Science Publishers, Issaquah WA, Page 736 (1992).

41. British Standards Institute, *Methodology for the Risk Assessment of Non-Electrical Equipment and Components for Intended Use in Potentially Explosive Atmospheres,* BS EN 15198 (2007).

42. E. A. Ural, Towards Estimating Entrainment Fraction For Dust Layers – Final Report, Fire Protection Research Foundation, NFPA, Quincy MA, Pages 34 to 42 (June 2011).

43. V. Babrauskas, *Ignition Handbook*, Fire Science Publishers, Issaquah WA, Page 560 (1992).

44. L.G. Britton, *Avoiding Static Ignition Hazards in Chemical Operations*, AIChE/CCPS, New York NY, Pages 72 and 78 to 81 (1999).

45. S. Mannan [editor], *Lees' Loss Prevention in the Process Industries*, 3rd Ed., Elsevier, Boston MA, Page 16/251 (2005).

46. P. G. Urben [editor], *Bretherick's Handbook of Reactive Chemical Hazards*, 6th Ed., Butterworth Heinemann, Woburn MA, Pages 20, 1492, 1723, 1749, 1758, 1815, 1913, and 1924 (1999).

47. V. Babrauskas, *Ignition Handbook*, Fire Science Publishers, Issaquah WA, Page 356 (1992).\

48. V. Babrauskas, *Ignition Handbook*, Fire Science Publishers, Issaquah WA, Pages 1024 to 1053 (1992).

49. J. H. Perry, *Chemical Engineers' Handbook*, 4th Ed., McGraw-Hill, New York NY, Page 9-40 (1963).

50. E. A. Avallone, T. Baumeister, and A. M. Sadegh, *Marks' Standard Handbook for Mechanical Engineers*, 11th Ed., McGraw-Hill, New York NY, Page 4-30 (2007).

51. National Fire Protection Association, *Standard for Combustible Metals*, NFPA 484, NFPA, Quincy MA, Paragraph 7.6.1 (2015).

52. National Fire Protection Association, *Standard on the Fundamentals of Combustible Dust*, NFPA 652, NFPA, Quincy MA, Paragraph A.8.4.2.6.1 (2016).

53. R. K. Eckhoff, *Dust Explosions in the Process Industries*, 2nd Ed., Butterworth Heinemann, Woburn MA, Page 218 (1999).

Note: The opinions that are expressed in this presentation are those of the author and may not represent the opinions, or risk-avoidance criteria or guidance, of Chilworth Technology, Inc. or Dekra Insight.

13TH GLOBAL CONGRESS ON PROCESS SAFETY

Implementing an Effective Management of Change (MOC) System for E&P Business and Non-PSM Facilities

Sandipan Laskar
Risk Engineering Leader
Occidental Oil and Gas Corporation
5 Greenway Plaza
Houston, TX 77046
Sandipan_Laskar@Oxy.com

Prepared for Presentation at
American Institute of Chemical Engineers
2017 Spring Meeting and 13th Global Congress on Process Safety
San Antonio, TX
March 26 – 29, 2017

Implementing an Effective Management of Change (MOC) System for E&P Business and Non-PSM Facilities

Sandipan Laskar
Risk Engineering Leader
Occidental Oil and Gas Corporation
5 Greenway Plaza
Houston, TX 77046
Sandipan_Laskar@Oxy.com

Keywords: Process Safety Management (PSM), Process Safety Information (PSI), Management of Change (MOC), Safety Management Systems.

Abstract

The upstream oil and gas industry involves processing high volume of live data during different operations (such as drilling, production, well servicing etc.) and thereby generating high volume of process or procedure changes. It is very important to manage these changes as based on the past incidents in the industry it can be noted that accidents often occurs after changes. Any change that has safety implications should be carefully evaluated, including performing a hazard analysis, before it is allowed. One of the practicable approach in managing these changes is by implementing a Management of Change (MOC) system.

Most companies have policies for management of change, but the implementation and enforcement of these policies can vary greatly. An effective MOC work process includes a review and authorization process for evaluating proposed adjustments to facility design, operations, organization or activities prior to implementation to make certain that no unforeseen new hazards are introduced and that the risk of the existing hazards to employee, the public, or the environment is not unknowingly increased. It also includes steps to help ensure that potentially affected personnel are notified of the change and that pertinent document, such as procedures, process safety knowledge, and so forth, are kept up-to-date.

The main aim of this paper is to address the three basic questions (what? why? and how?) in implementing an effective MOC system. There are more robust MOC program available for the facilities covered under OSHA Process Safety Management program. However the management of change (MOC) work process mentioned in this paper is fit for purpose for upstream oil and gas operation. This paper discusses challenges in implementing MOC system for facilities not covered under OSHA Process Safety Management program and addresses common misconceptions of Management of Change system.

Introduction

Most oil and gas production, refining and petrochemical organizations have some kind safety management program that stems from the requirements outlined under OSHA's CFR 1910.119. Some have a very extensive program that covers most of the 14 elements of Process Safety Management (PSM) and is managed via an integrated platform while others follow an ad hoc approach that does not span all of the elements or is disjointed.

The OSHA PSM guidelines are enforceable throughout the United States and specific US holdings for facilities having a threshold volume of hazardous chemicals, but are not applicable where another federal agency has occupational safety and health jurisdiction, including in federal waters controlled by Bureau of Ocean Management Regulation and Enforcement (BOEMRE, formerly Minerals Management Service (MMS)). Oil and gas drilling and servicing operations and normally unmanned remote facilities are specifically excluded from the standard, but typical facilities in the upstream industry such as production facilities, gas oil separation plants, and gas plants are covered if the threshold chemical volumes, typically hydrocarbons, are present.

For offshore facilities, a more holistic approach applying to both production and drilling and without a threshold volume was published originally in 1998 by the API, with the most recent update in 2004 (API 2004). This recommended practice 75 (API RP 75) establishes methods to develop a Safety and Environmental Management Program (SEMP) covering 12 target elements, which addresses all the major aspects of an OSHA PSM program. Although the MMS and industry groups ran a number of audit surveys during the initial implementation period of API RP 75 (BOEMRE 2010), recent compliance status data are not publicly available.

The MMS published an Advance Notice of Proposed Rulemaking (ANPR) in the May 2006 US Federal Register related to the creation of an enforceable PSM/SEMP, and provided a report in 2009 on the comments and data gathered during the voluntary PSM/SEMP period (MMS 2009). Data analysis was performed on a number of MMS incident reports, noncompliance notices, and spill reports, showing a preponderance of incidents resulting from failures in one or more of four main program areas.

Historically the MMS has not been regarded as a strong enforcement agency, though with the reorganization announced by the Interior Secretary in May 2010 (Straub 2010) splitting the field development, enforcement, and revenue functions of the former MMS body, in combination with the promulgation of an enforceable PSM/SEMP and the recent major catastrophe in the Gulf of Mexico, one might have expected a more stringent regulatory environment with greater emphasis on design review and prevention as well as increased penalties and fines should incidents occur. Following a flurry of activity in the second half of 2010, the SEMP as described in API RP 75 became mandatory 15 November 2010 (BOEMRE 2010).

Management of Change – What?

Management of Change is defined as a formal process that systematically manages Changes to the organization, operations, process equipment, technology, procedures, chemicals, or for any temporary modifications that could impact the original design, operating & maintenance criteria, and/or environmental performance. The MOC process is intended to control the potential impact of any proposed Changes (other than for replacement-in-kind) that might impact personal safety, legal & regulatory compliance, safeguard the environment and/or the safety or emergency response of the facility or equipment. This shall also include personnel or organizational Changes which might impact the safety or emergency response provided to the Process. So in an essence management of changes, or MOC is a best practice used to ensure that safety, Health and environmental risks are controlled when a company makes changes in their facilities, documentation, personnel or operations.

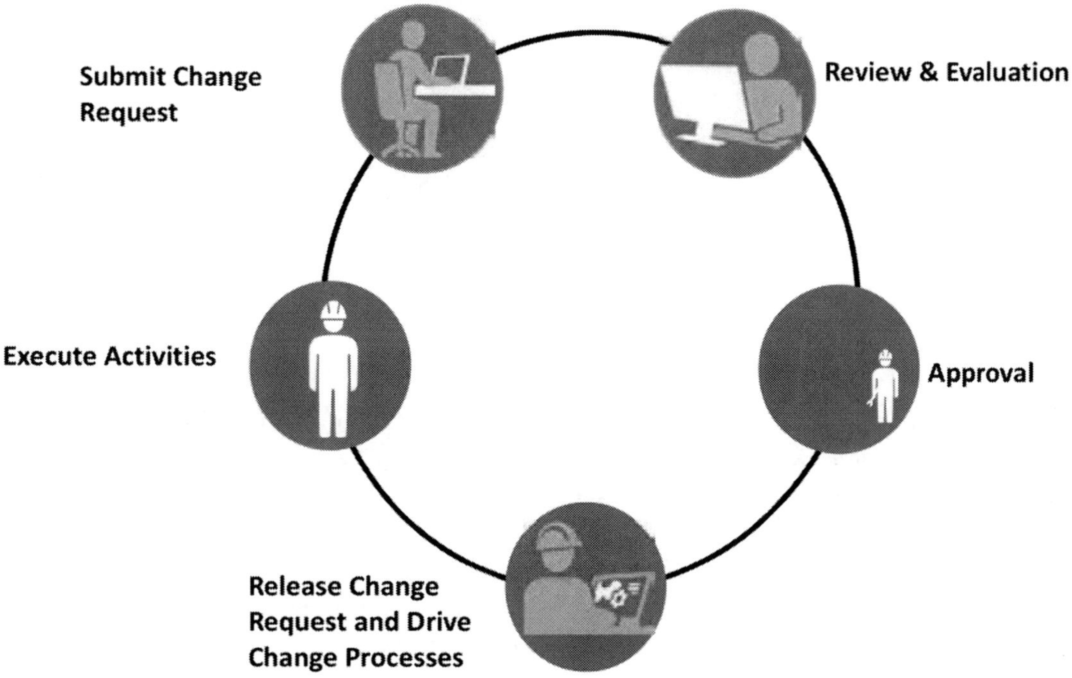

Figure 1- An Overview of Management of Change

The above figure illustrates a simple overview of the management of change work process and each of those activities could be further broken down into more detailed workprocess having multiple layers of activities attached to them.

The MOC process must address changes that range from minor changes to major capital projects as well as handling temporary and permanent changes. This can be accomplished in many ways. Some companies have developed "one-size-fits-all" MOC processes that are used for every change. Other companies find it easier to have multiple MOC systems. The multiple systems can be split based on the type of change (equipment vs. procedural), the duration (temporary vs. permanent) or the size of the change (minor vs. major). The decision to use an electronic MOC process or a paper process should also be considered. The important thing to remember is that the MOC process or processes should be flexible, easy to use and fit your facilities culture.

A typical management of change process should address following changes:

- Facility Changes: The scope of the facility changes include any modifications made to the equipment within a facility. For a piping system, this may include replacing a carbon steel pipe replacing 2-in. pipe with 3-in. pipe, replacing 150-lb flanges with 300-lb flanges, adding tracing or insulation, substantially changing the pipe routing, tying the piping into another line, or anything that changes the pipe class on a P&ID. Other facility changes include replacing a centrifugal pump with a diaphragm pump, re-tubing a heat exchanger with a different gauge tubing or material of construction etc.

  Facility changes also includes changes related to the control system. Some example could be changes to the programming or control logic, changes to who has access to the logic, resetting interlock timers, bypassing any circuit element or protective device, or changes to an alarm point that takes it outside of the safety or quality limits specified in the operating procedure.

- Procedural Changes: This category includes any changes to previously established safety, quality, or operating limits in the operating procedure.

- Organizational Changes: The Organizational MOC is intended to ensure that the site operating organization can provide safe operation during both normal and emergency situations. Examples of an Organization MOC include: changes in the number of operating personnel at the site, changes in hours worked (e.g. 5-day maintenance coverage to 4 day maintenance coverage), changes in delegation of authority where safety is impacted, changes in organizational structure (e.g. Elimination of a lead operator or HES position), changes from company to contract operators and/or maintenance personnel. It should be noted that the above examples are not exhaustive list and other Organizational MOC situations may exist.

Changes may be permanent or temporary. If a temporary change is made, it must be tracked and properly reversed. Companies should also have a procedure to handle emergency changes, which are those that must be made quickly during off-hours *(e.g.,* nights. weekends, and holidays) when the regular MOC approvers may not be available. For example, approval could be obtained verbally rather than in writing (as long as it is properly documented) or approval could be obtained from authorized on-call personnel rather than the regular approvers.

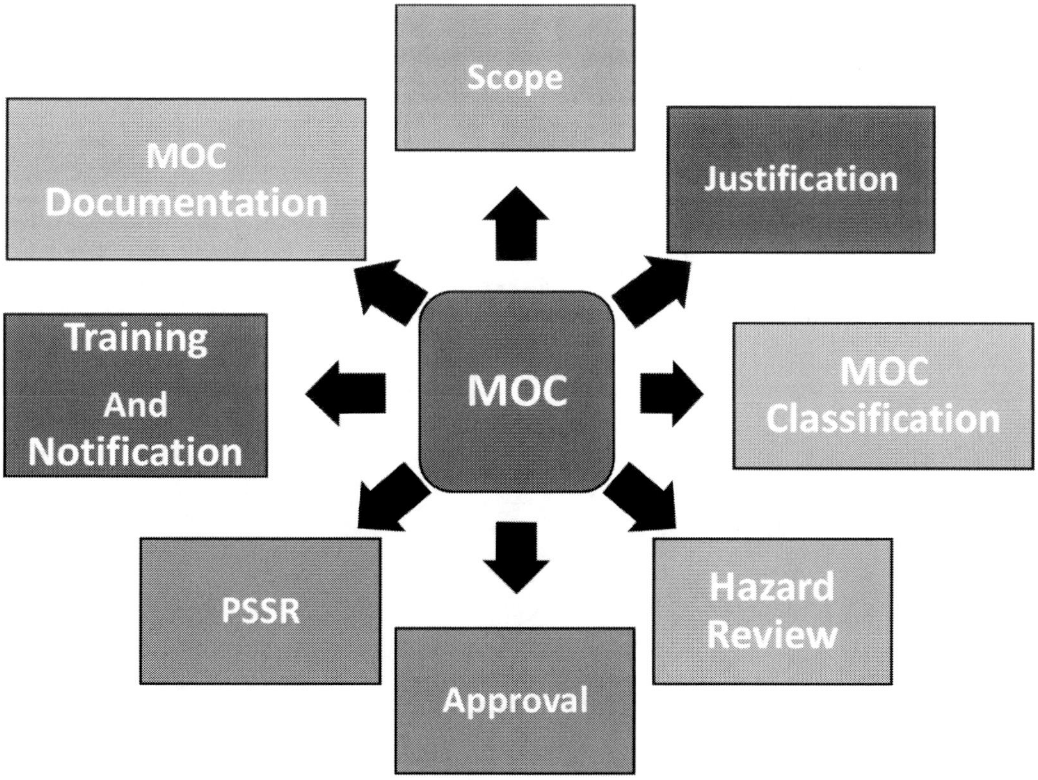

Figure 2- Fundamental Building Blocks of Management of Change

The fundamental building blocks of a Management of Change process are shown in the figure above. All the fundamental building blocks will be clearly explained in the management of change implementation section.

Management of Change - Exception

A management of change is changes related to Facilities, Procedures or Organization except for Replacement in Kind or identical changes. A Replacement in Kind (RIK) change can be an item (equipment, chemical, procedure, etc.) that meets the design specification of the item it is replacing. This can be an identical replacement or any other design alternative specifically provided for in the design specification, as long as the alternative does not in any way adversely affect the use of the item or associated items.

Management of Change – Why?

Some of the benefits of an effective Management of Change process are:

- It minimizes unplanned adverse impacts on system integrity, security, stability, and reliability for the business process being altered or added.
- It maximizes the productivity and efficiency of staff planning, coordinating, and implementing the changes.
- It provides a stable production environment.
- It ensures the proper level of technical completeness, accuracy of modifications, and testing of systems before implementation.
- It provides an appropriate level of management approval and involvement.

The safety program of a facility typically consists of process hazard analyses (PHAs), a mechanical integrity (MI) program, personnel training, PSSRs, PSI and MOC. MOC plays a central role in this program because:

- It provides updates to PSI
- It identifies when a PSSR is needed
- It ensures that employees are trained to carry out new procedures
- It adds new equipment to mechanical integrity test and inspection schedules

The bottom line is that Management of Change helps prevent incidents.

Implementing a Management of Change – How?

A management of change is a procedure for safely implementing a change to a facility and/or procedure by identifying the impact of the change. The figure below illustrates a simple management of change work flow which includes all the different building blocks for an effective and robust management of change program. A more detailed management of change work process with key players are shown in figure 4. The main objective of this simple work flow is to enunciate flow of information between different building blocks of the management of change program.

An effective MOC process consists of following basic steps:

1. Identify the need for change (MOC) - Initiation
2. Review and approval – Initial Review
3. Evaluate the Risk (PHA)
4. Review and Approval Process – Final Review
5. Construct/pre-commissioning
6. Effective Training and Communication
7. Pre-Startup Safety Review (PSSR)
8. MOC Closure

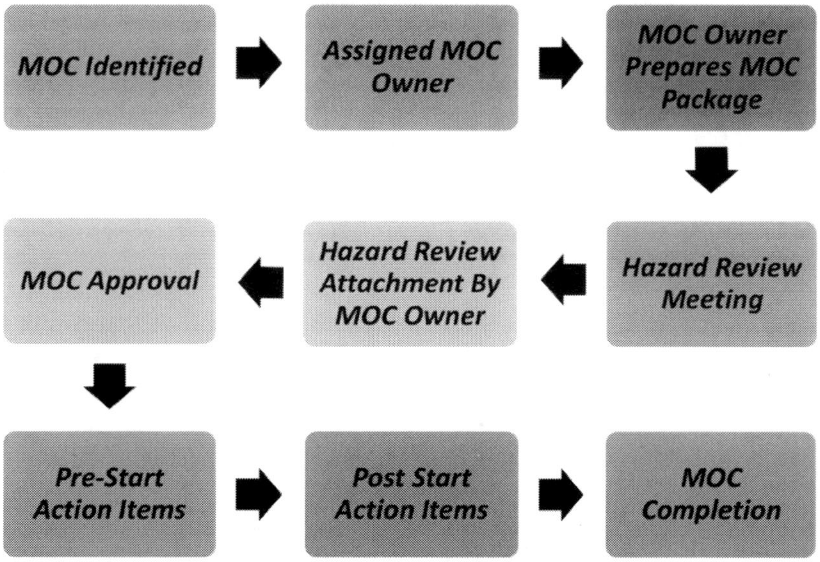

Figure 3- A Simple Management of Change Workflow

Identify the need for change: In the initiation step, any employee should be able to propose a change to a procedure, equipment or technology of the process/facility. The MOC forms should allow the initiator to document the what, why and when of the proposed change. The initiator should request the change in writing. This should include a description of the change, the technical basis for the change, the potential impact on safety and required PSI updates. The initiator then forwards the information to a member of the facility management team for review and approval.

Review and Approval – Initial Review: The preliminary review consists of a "go/no go" decision based on the information supplied by the initiator. This may include discussions between the initiator and the reviewer to clarify the details of the proposed change and alignment with the facility's needs. If the decision is made to proceed with the change, the preliminary reviewer should, as much as s/he is able, document the technical basis for the change, impact on safety and health and the timeframe to accomplish the change. If the change is temporary, the preliminary reviewer should also ensure that the date that the temporary change will be reverted to normal is included in the MOC documentation.

Evaluate the Risk (PHA): The risk assessment/technical review step is one of the most important activities within the MOC process. It is critical to involve the facility engineer, technical authority or someone with a deep understanding of the process in this step. The risk assessment should match the complexity of the change. For minor changes, it may be okay to ask the question "What if?" More complex changes require tools like process hazard analysis or failure mode-effect analysis. Some companies find it useful to include a risk assessment checklist in their MOC process. The checklist includes questions like "Does this change involve a shutdown or safety system?" and "Does this change increase or decrease the pressure, temperature or flow rates within the process?" The completed checklist can be the risk assessment or it can point the user toward a more sophisticated risk assessment tool or additional review by safety, environmental or technical experts. Regardless of the risk assessment method, it is important to document what could reasonably go wrong and the steps required to prevent it within the MOC prior to obtaining approval to design/construct.

Review and Approval – Final Review: The approval to design/construct step is typically granted by the facility superintendent or manager. This person is committing resources toward completing the change and agreeing that the change can safely be made based on the information provided. The final review is based on the availability of all design documentation along with a process hazard analysis to ensure proper safeguards.

Construct/ Pre-Commissioning: Most of the "real work" is accomplished in the design/construct/precommission step. In this step, final designs are completed, piping and equipment constructed, procedures updated, redline drawings created and equipment installation verified. If the design changes during this step, another round of risk assessment and approval is critical to ensure that risks associated with the new design are properly managed

Training and Communication: This step is part of the pre-startup safely review (PSSR) and involves critical activities that should be completed prior to putting the change into service. Notification involves informing all affected persons and departments that the change is about to occur. Training by a knowledgeable instructor is conducted (as needed) for all personnel who will be operating the process after the change has been made.

Pre-Startup Safety Review: Field verification is an inspection to ensure that what was done in the field matches the specifications in the engineering design package. The PSSR step consists of the set of verifications and assurances that each portion of the work associated with the change has been properly completed. The PSSR items are typically assigned to multiple people, including facility engineers, operations leaders and maintenance leaders. A PSSR must include the following items:

- equipment meets the design specifications and has been properly installed;
- adequate safety, operating, maintenance and emergency procedures have been developed or existing procedures modified accordingly;
- process hazard analyses have been completed for new facilities and recommendations have been resolved or implemented before startup;
- employees involved in operating the PSM-covered process have been trained on or notified of the change as appropriate;
- safe work practices have been revised accordingly;
- process safety information used by employees (e.g., piping and instrumentation diagrams, equipment spec sheets) has been updated accordingly;
- Items that will be needed in the future have been assigned for completion. This typically includes items like coordinating spare parts inventory with purchasing.

MOC Closure: Once all the post startup action items are completed then the MOC is closed. It is of utmost importance to ensure that all the process safety information is updated before the MOC is closed.

Conclusion

A simple, yet robust, MOC process can help ensure that risks associated with change are understood and properly managed, employees are trained on how the change affects them, accurate procedures are in place for maintenance and operations, equipment files are up to date and spare parts are available when they are needed. When these elements are in place, a facility can meet their goal of safe, reliable operations. Thus is could be concluded that a well-designed MOC process is an essential loss prevention tool for any business. It is not just for certain hazardous industries. The process applies to any company that wishes to avoid future losses resulting from today's changes. MOC does not have to be overwhelming or so difficult to use that it inhibits change. It cannot effortlessly compensate for past omissions, only reduce your future risk. So the time to start developing and implementing an effective MOC process is today.

Disclaimer

The information contained in this document represents the current view of the author at the time of publication. Process safety management is complex and this document cannot embody all possible scenarios or solutions related to compliance. This document is for informational purposes only. Oxy makes no warranties, express or implied, in this paper.

References

[1] "API RP 75, Recommended Practice for Development of a Safety and Environmental Management Program for Offshore Operations and Facilities", third edition. 2004. Washington, DC: API

[2] "Oil and Gas and Sulphur Operations in the Outer Continental Shelf--Safety and Environmental Management Systems", Bureau of Ocean Energy Management, Regulation and Enforcement Implementation (BOEMRE), *Final Rule*, 30 CFR Part 250, Volume 74, Number 199, (October 15, 2010)

[3] "Petroleum Refinery Process Safety Management National Emphasis Program", U.S. Occupational Safety and Health Administration (US OSHA), Directive CPL 03-00-004, (June 7, 2007)

[4] "PSM Covered Chemical Facilities National Emphasis Program", U.S. Occupational Safety and Health Administration (US OSHA), Directive CPL 09-06-002, (July 27, 2009)

[5] "Survey for API RP 75", Bureau of Ocean Energy Management, Regulation and Enforcement Implementation (BOEMRE), (September. 17, 2010)

[6] Occupational Safety and Health Administration. *29 CFR 1910.119, Process Safety Management of Highly Hazardous Chemicals*. 1992.

[7] "Precise Process Safety Information – A Key for Successful Process Safety Management Program", S. Laskar, *Global Process Safety Conference*, (April. 29, 2013)

Improving Human Factors Review in PHA and LOPA

Dave Grattan, P.E., CFSE
aeSolutions
10375 Richmond Ave. Suite 800
Houston, TX 77042
dave.grattan@aesolns.com

Prepared for Presentation at
American Institute of Chemical Engineers
2017 Spring Meeting and 13th Global Congress on Process Safety
San Antonio, TX
March 26 – 29, 2017

Improving Human Factors Review in PHA and LOPA

Dave Grattan, P.E., CFSE
aeSolutions
10375 Richmond Ave. Suite 800
Houston, TX 77042
dave.grattan@aesolns.com

Keywords: Human Factors, Human Error, PHA, LOPA, Facilitator, Task Analysis, Bias, Cognitive Psychology

Abstract

Human Reliability practitioners utilize a variety of tools in their work that could improve the facilitation of PHA-LOPA related to identifying and evaluating scenarios with a significant human factors component. These tools are derived from human factors engineering and cognitive psychology and include, (1) task analysis, (2) procedures and checklists, (3) human error rates, (4) systematic bias, and (5) Barrier effectiveness using Bow-tie.

Human error is not random, although the absent minded slips we all experience seem to come out of nowhere. Instead, human error is often predictable based on situations created external or internal to the mind. Human error is part of the human condition (part of being a human) and as such cannot be eliminated completely.

For example, a task performed at high frequency (e.g., daily or weekly) develops a highly-skilled operator with an expectation of a low error probability for that task. However, as the operator' skill increases, their reliance on procedures decreases, leaving them open to memory lapses caused by internal or external distractions. The fact that a skilled operator becomes less dependent on procedures is not a conscious decision. It is part of the human condition. Forcing a skilled operator to read the procedure while performing the task they are skilled at, is like asking you to think about what your feet are doing as you walk down a flight of stairs. In both cases a loss of adroitness will occur.

A large portion of this paper will be to describe with practical examples the five tools mentioned above. Task analysis is a talk-through and walk-through exercise of a task (typically focusing on one or two critical steps of a procedure) that is used to identify error likely situations (ELS). Quantitative human error rates can be attached to the ELS depending on if the error is associated with skill, rule, or knowledge (SRK) based performance. Systematic biases produced by Type 1 (fast) thinking cause judgment and diagnosis errors related to response to abnormal situations.

Having a working knowledge of these five tools will improve a PHA-LOPA facilitator's awareness and ability to better evaluate human error related scenarios and Barrier failure. In addition the facilitator will feel confident about recommending the need for a more detailed follow-up study such as an HRA (Human Reliability Analysis).

1 Introduction

This paper is written for facilitators of PHA and LOPA and their managers. The message is simple: Improve human factors review in PHA and LOPA by applying the already established tools of human factors and human reliability. How can we improve something we do not practice ourselves as facilitators? There is a caveat however. Much of the heavy lifting has to be done outside of PHA and LOPA studies. The argument for this is presented below in **Section 2.2**.

The LOPA method itself presents challenges and opportunities for using human factors and human reliability. Any LOPA scenario handling human error as initiating cause, human response, or administrative control barriers is a candidate for a safety critical task analysis. There is ample low-hanging fruit with regard to human factor based scenarios already coming out of PHA and LOPA. One of the tenets of this paper is that any credit given for human response to a critical alarm or other human barrier, or any human error initiating a catastrophic scenario, should undergo a safety critical task analysis, regardless if the gap is closed using generic credits (see **Section 2.4.5** for why).

The following tasks have been successfully treated with the qualitative and quantitative methods and concepts discussed in this paper. It is important to note that each of these tasks were well-defined, bounded, primarily skill and procedurally based tasks involving one or more personnel.

- Specific task performed during normal shutdown
- Specific task performed during start-up
- Safing for maintenance work
- Safing for N2 back-up in Instrument Air
- Response to critical alarm (including diagnosis)
- Restoration of critical equipment following ITPM (inspection, test, preventive maintenance)
- Catalyst drum change-out
- Rupture disk Barrier effectiveness
- Tanker Truck Unloading

Organizational factors are recognized as important human factors. Organizational factors such as social and institutional influences are discussed in this paper qualitatively. The quantitative treatment of organizational factors as they pertain to predicting incidents is complex (some would argue currently intractable) with large uncertainty. This paper is primarily written addressing skill or rule based tasks performed by one or more operators.

2 Why Human Factors and Human Reliability?

"Most industrial accidents are caused by human error: estimates range between 75 and 95 percent. How is it that so many people are so incompetent? Answer: They aren't. It's a design problem. If the number of accidents blamed upon human error were 1 to 5 percent, I might believe that people were at fault." Don Norman [1]

"Saying most accidents are due to human error is a bit like saying falls are due to gravity." Trevor Kletz [2]

2.1 We are Under-predicting Process Safety Incidents

There is published evidence [3] stating that current methods (HAZOP, LOPA, QRA) are predicting only about one-third of actual process safety incidents. The under-prediction is coming from the frequency (or likelihood) part of the risk equation. Identifying hazards and estimating consequences has been comparatively more robust.

Could this observed gap be coming from Human Factors impact?

2.2 The Case for Why PHA and LOPA are Not Enough

PHA and LOPA, and their derivatives including Procedural HAZOP, generic Human Factors checklists (see **Section 9** for a comparison with a Task Analysis checklist), etc. are targeted to evaluating states and conditions (e.g., is the alarm configured as priority, or, what-if this step in the procedure is omitted?) as opposed to behavior (e.g., how will the console operator interpret the alarm, or, how likely is the operator to skip that step and why?). States and conditions are important to validate, however behavior is just as important. The existing tools available to evaluate human (and organizational) behavior are task analysis and human reliability analysis. PHA and LOPA methodology cannot get you there.

Another reason is related to Barrier effectiveness. Barriers are listed as providing protection in a LOPA study. The hardware reliability of a Barrier can be and is typically estimated (or assumed to be a generic value). So to can the human impact on a Barrier be estimated, except it's typically not. Sometimes it's assumed the generic data value for hardware failure includes systematic (human) failures. See **Section 2.4.5** for why this is a myth. Human and organizational factors are at least as important as random hardware failures on a barrier, and they are not being accounted for. Methods outside of PHA and LOPA are needed.

2.3 Human Factors

2.3.1 <u>What is it</u>

A human factor is anything that affects human performance, good or bad. The goal of human factors is to match the task environment and system interface(s) to the abilities of the operator. If the physical or mental requirements of a task exceed the operator's ability, errors will occur.

2.3.2 Affordances, Signifiers, Forcing Functions, and Mapping

When performing a task analysis the facilitator must be able to spot deficiencies in the task environment. This can only come with practice. The following tools are used by human factors specialists [1]:

Affordances are used to afford a location to perform work. A simple example is a valve handle. It provides a place for the operator to place their hands. Affordances can be used improperly, for example, an operator may have to stand on said valve handle (or the piping) to reach a manual valve above him.

Signifiers tell us how to interact with the affordance. A round valve handle turns. A levered valve handle rotates. The best signifiers are intuitive (don't require conscious thought, i.e., Type 2 thinking – see **6.2**). Signs require effort and conscious thought (to see, read, and comprehend), and are therefore not the best use for signifiers (this is not to suggest getting rid of signs, but to understand the limitation). Populational stereotypes are learned signifiers, for example, "righty-tighty, lefty-loosey." Designs that violate populational stereotypes are considered error-likely.

Forcing functions force desired behavior. For example, car-seals, chain locks, and captive key systems placed on valves are recognized as reducing errors of omission and commission related to valve manipulation.

And, **mapping** is used to spatially connect controls with affordances. For example, procedures that use labels and drawings to map said valve to the actual valve in the field.

These concepts are only useful in the context of the operator performing the task, and so task observation of the operator with the equipment and procedures is necessary to be complete a successful analysis.

2.3.3 Human Nature

Slips and lapses (defined in **Section 3.4**) that sometimes occur are part of human nature. This type of human error cannot be trained, punished, rewarded, designed, or otherwise removed out of humans. It is part of being human. Policy, procedures, and good human factors help, but we should still expect the occasional human error. We should design systems to be tolerant of human error or allow for their recovery. If an operator commits an error, we should not look at the operator, but instead look at the task, procedures, environment, equipment interfaces, etc. the operator has to work with, because it is more likely those are the root cause (not the human).

2.3.4 Procedures

Different types of procedures are in use, including,

- Routine Operating
- Preventive Maintenance (e.g., for critical instrumentation)
- Response to an abnormal situation (critical alarm based response)

Each type of procedure is amenable to human factors and human reliability analysis via task analysis.

2.4 Human Reliability

Human reliability study is the quantification of human factors. Human reliability puts numbers to (i.e., attempts to quantify) human factors influences on tasks and can be used to help "sharpen the pencil" for error estimates to use in PHA-LOPA or QRA. Human reliability has a long history (see for example reference [4]) and is part of the renaissance of quantitative methods in the process industries [5]. Any number that is produced has an uncertainty associated with it (see **2.4.2** and **Section 8**).

2.4.1 <u>Tractability</u>

How can we predict human error? Humans are unpredictable (i.e., often irrational).

There are several ways to answer this question. The easiest and most relevant is the fact that there is so much low-hanging fruit associated with improving human (and organizational) factors related to hazard scenarios and Barrier effectiveness that it would take a generation worth of work in this field applied to the Process Industries to overtake the capabilities of the current methods.

The current methods to evaluate human behavior that include the SRK (skill, rule, knowledge) model and many human reliability models (e.g., THERP (Technique for Human Error Rate Prediction) – see ref [12]) have been validated as effective at predicting human error. The issue of human irrationality in the context of plant accidents even has a name: local rationality (what is in the head of the operator at the time they made that catastrophic error) [9].

The catastrophic accidents that post to the CSB (Chemical Safety Board) website at a steady rate, do not involve unpredictable human error.

One of the most incredible human factor case studies is the Browns Ferry fire of 1975. Browns Ferry nuclear plant is located in Alabama. The maintenance team had just installed some new cabling into a control room, and were leak testing the cable seals by looking for air flow in the duct, using the naked flame of a candle. The foam insulation caught fire, and burned through safety and control cables (common duct), and the operations team narrowly avoided a complete meltdown. One might ask how something like this could have been predicted a priori. It would be nearly impossible. But this and other extraneous acts often lie in wait for a number of years before causing an incident. The candle flame method to leak test had been in use for several years and was even recognized as a hazard.

2.4.2 <u>Model Error</u>

As mentioned above, the process industry has seen a "renaissance" in use of quantitative methods including SIL (safety integrity level) calculations, Fault Tree Analysis for likelihood of hazard, Event Trees for distribution of outcomes, and Bow-Ties (combined fault tree and event tree) as a means of communicating hazards and identified barriers to Operations staff, and identifying process safety leading indicators.

In any quantitative model of system performance, the following are typically recognized as sources of error:

- Equipment data (a function of equipment type, service, quality of maintenance, etc.)
- Incorrect model (incorrect logic)
- Human reliability impact
- Wrong assumptions (about plugging for example)
- Unforeseen hazards or ways in which hazards can occur (see **Section 8.1**).

The field of human reliability attempts to account for the uncertainty introduced by the impact of human factors on system performance. This may represent the most significant source of model error, next to bad assumptions.

2.4.3 HRA Methodologies

There are a myriad of HRA methodologies, some public, some proprietary, as evidenced by the veritable alphabet soup of acronyms associated with the methodologies.

The Health and Safety Executive (HSE) in the UK considered that it would be useful to be up to date with developments in the field of quantitative HRA methods and to have knowledge of the capability of the tools and an understanding of their strengths and weaknesses, to improve consistency, and determine acceptability of their use [6].

To sort this out, in 2009 the HSE published a review of 72 HRA methods that were "potentially relevant to HSE major hazard directorates." Of the 72 potential HRA tools, 17 were considered useful for major hazard directorates [6].

THERP (Technique for Human Error Rate Prediction) [12] is one of the 17 methods considered to be useful by the HSE.

2.4.4 THERP

THERP is considered a first generation human reliability model (in the public domain the models only go up to second generation). ATHEANA (A Technique for Human Error Analysis) is the second generation model developed for the USNRC (US Nuclear Regulatory Commission) that is intended to improve THERP, especially related to cognitive behavioral mechanisms, organization, and safety culture, in the context of response to abnormal events. But ATHEANA would benefit from further development and ultimately more widespread industry acceptance [6].

THERP is a perfectly acceptable methodology, especially for the low-hanging fruit coming out of current LOPA's. Some of the benefits of THERP:

- It's easy for someone with a PHA-LOPA-QRA-facilitator background to use
- It includes its own extensive human error data set
- It incorporates Task Analysis to uncover error mechanisms

- It has a very nice dependency model built-in
- It uses "recovery factors" such that unrecovered human error is the metric of interest
- It flanges up nicely with LOPA, QRA, and Bow-tie

THERP does take practice to get good at, and it does require an eye for human factors, and some study of cognitive behavior.

2.4.5 The Myth that Generic Equipment Failure Data includes Human Error

There is a belief that generic failure data (e.g., generic data tables for initiating events and protection layers) already includes data on human error. (It can be argued that *all* failure is ultimately human induced, but this is not the point I'm making.) This myth, however, is partly true. For example, failure associated with the quality of maintenance certainly contributes to the failure data of generic equipment. However, the following types of human errors are not included in generic equipment failure data [4]:

- Insertion of an incorrect component or device
- Wrong specification
- Miscalibration
- Left in wrong mode (manual, test, inhibit, etc.)
- Failure to restore system following maintenance (block valves)
- Inadvertent bypass during normal operation

But even if the above myth were true, it misses the point entirely. Hardware failure is considered random (statistical). Human error is systematic (i.e., deterministic). The two types of error differ only by the amount of information known about each. Behavioral models and task analysis provide information of why human error occurs or might occur. We can use this information to improve work processes, work situations, policy, competency, staffing, communication, contractor roles, etc. so that we might prevent the error from occurring. If we bury human error as a generic random variable, we miss this opportunity. Claiming that a generic value of human error is 1/10 and that it is conservative, is misleading and dangerous, because of the nature of determinism, the error likelihood (for an identified cause) is either 1 or 0. The same point could be made when discussing the numbers in **Section 5.2**. The way out of this problem is to *always* couple the quantitative analysis with the qualitative Task Analysis.

3 Tool 1: Task Analysis

3.1 Introduction

Human factors and human reliability share a technique called 'safety critical task analysis' or 'task analysis' for short. As the name implies, task analysis is a systematic analysis of a task (or barrier) that can help identify problems and reduce the likelihood of human error associated with that task (or barrier). Task analysis looks at all the ways that humans touch a system, how failure can occur, and ways to improve it. Task analysis is used to ferret out potential human errors related to a specific task, and make recommendations for improvement as well as estimate likelihood of occurrence.

Task analysis could look at the critical steps of a procedure (usually one or two steps of a large procedure), responding to a priority alarm, or restoring a critical system following maintenance. Task analysis can also be conducted on the escalation factors and escalation factor controls related to a Barrier. Task analysis is conducted as a team exercise, and includes talk-through and walk-through activities. Typically one task (or barrier) takes one day of team time to perform the task analysis, with additional follow-up reporting and some iteration.

You don't have to go far to find a task to analyze. They are right there in your PHA-LOPA study. Any catastrophic hazard scenario that has a significant human factors component including human barriers is a candidate for task analysis.

Task analysis can be picked up quickly by an experienced PHA-LOPA facilitator. The caveat is the facilitator needs to have an eye for human factors issues and an understanding of cognitive bias (see **Section 6.1**). It is recommended that a facilitator experienced in the methodology be engaged as a mentor.

The Functional Safety Assessment required by ANSI/ISA 84 as an audit of the Safety Instrumented System is really just a task analysis.

See **Section 9** for sample Task Analysis checklists.

3.2 Error-Likely Situations

A primary goal of task analysis is to identify error likely situations. This is partly accomplished by talking to operators that perform the task, reviewing procedures, and actually observing (real or a simulated walk-through) the task being performed. Here is where an eye for human factors is necessary. Much of human factors is common sense. If you talk to an operator who has been on the job for 30 years and he tells you this is the first time someone has ever come out to look at how he works, the fruit will be lying on the ground.

A lot of error likely situations are obvious, for example,

- Reading errors caused by missing valve handles or indicators (valve position), or instrumentation that is not back-lit.
- Omission errors caused by equipment that is not accessible.
- Commission errors caused by equipment layout.
- Non-response errors because alarms are buried in a DCS graphic, or the audible alarm volume is turned down too low.
- Interpretation error (of a signal) caused by failure to trend that signal on a DCS graphic (or use of process switches versus transmitters).
- Interpretation error of the state of equipment because there is not enough information for the Operator.
- Not believing an alarm is real because of so many other spurious alarms.
- Not knowing how to respond to an alarm because there is no training or documentation.
- Reliance on short-term memory because of deficient design.
- Spatial disorientation caused by layout (e.g., stairs) of identical equipment trains.

There are other error likely situations related to deficient procedures, work practices, communications, staffing, task load, competency, etc.

To be successful at identifying error likely situations takes practice. But a basic understanding of the types of human error possible is the way to begin.

3.3 Human Error is not a 'Cause'

Human error does not occur randomly (in a statistical sense). Human error is deterministic (i.e., it can be predicted using an appropriate behavioral model and task analysis). Examples of when human error will occur:

- When a person is rushing, distracted, fatigued, or complacent. (These are the Big 4.)
- Spatial disorientation (e.g., working on the wrong but identical equipment)
- Tasks that rely on short term memory (short term memory can hold approximately 5 items at most under ideal conditions)
- Interpreting a signal as false
- Knowledge based tasks (a novel situation) has higher error rates

We can reduce (but not eliminate) the likelihood of human error by evaluating the systems under which it occurs.

3.4 Human Error Classification

The following classification of error types has been established by engineering psychologists (a.k.a. human factors specialists) and has been used for many years [13].

3.4.1 Slips and Lapses

A 'slip' is an error of commission (a task done incorrectly).

A (memory) 'lapse' is an error of omission (forgetting a task).

Slips and memory lapses primarily occur during Type 1 thinking (see **Section 6.2**).

3.4.2 Mistakes

Mistakes are cognitive, that is they occur primarily during Type 2 thinking (see **Section 6.2**).

Examples of mistakes include:

- PHA errors
- Design errors
- Errors made writing procedures (not typos)
- Mis-diagnosing an abnormal situation

Mistakes primarily occur when faced with a new, ad hoc, or novel situation.

3.4.3 Craft Errors

Lack of adroitness related to tasks performed by maintenance, operations, pipe-fitters, etc. For a skilled craftsmen, these error likelihoods are lower than slips, lapses, and mistakes.

3.4.4 Social and Institutional Influences

These types of errors are the most insidious and can produce the most catastrophic incidents. Coupled with extended time and economic pressures, some hazard scenarios are especially susceptible to extreme outcomes. The added time a scenario may play out (e.g., days) allows multiple people (and therefore personalities) and conflicting objectives (safety v. economic) to creep-in. De facto Company policy, culture, remuneration, and the lines of communication are the organizational factors that are important.

Social and institutional factors can influence specific skill, rule, and knowledge based tasks. Task analysis has the potential to uncover some of these issues (e.g., by asking about non-compliance to a procedure), and can make these issues more discoverable.

3.4.5 Non-Compliance

A violation of a procedure or certain steps of a procedure. Not working to the procedure. Often times, the procedure is wrong, or the operator has found a better way of working. If there is a work process and culture to support operators writing their own procedures with appropriate checks of course, these types of non-compliance will not be an issue. On the other hand, ad hoc or unauthorized bypassing can be a serious non-compliance issue.

3.4.6 Summary of Error Classification

Table 1. Human Error Classification

Type of Error	Example	Most Likely Cause	Recovery Factors
Slip	Opening the wrong valve	Distraction	Feedback (usually immediate)
Memory Lapse	Forgetting a step in a procedure	Distraction	Checklist (see **4.1.5**)
Mistakes	Ignoring a real signal	Bias	Mistakes are hard to detect and correct
Craft	Loose Cam-lock fitting	Maladroitness	Feedback (usually immediate)
Social and Institutional	We've done it this way for many years and never had an accident.	Turkey Trap bias (see **8.1**)	Culture change
Non-compliance	Not following procedure	Poor procedures	Administrative
Knowledge based	Abnormal situation	Novel situation	Planning, Trial and error

4 Tool 2: Using Procedures and Checklists

4.1 How Routine Operating Procedures are Used

"*We have an opportunity before us, not just in medicine but in virtually any endeavor. Even the most expert among us can gain from searching out the patters of mistakes and failures and putting a few checks in place. But will we do it? Are we ready to grab onto the idea? It is far from clear.*" Atul Gawande [7]

Procedures are needed for safe operations and reliable maintenance. Working without reference to a procedure creates an error likely situation. Procedures are an important human factors tool.

4.1.1 Referencing Procedures In PHA-LOPA

In a PHA-LOPA, references to specific procedures where human action is involved in the study should be made. For example, operator error as an initiating cause, or operator response to a critical alarm. A laptop should be open in the study and should be able to pull up and reference the specific procedures as needed. At times, a quick read through of the procedure with the team is warranted. If a human action shows up in a PHA-LOPA without a procedural reference and none exists, consideration should be given for the need of one.

4.1.2 Using Procedures to Operate the Plant

Step-by-step routine procedures used by operations to run the plant are a critical administrative control system. Written procedures establish minimum requirements and expectations, which should be communicated via initial operator qualification and periodic refresher training thereafter.

Procedures used in plant practice have been successful at keeping human task error rates lower compared to other fields that do not use any procedures, for example, the error rates of the medical profession as described in reference [7].

In practice, procedures are used or referenced (or not used) in a variety of ways.

- Procedures may not be accessible, or don't exist
- Procedures may be printed and read before a job
- Procedures may be carried in the truck or in the back pocket
- Procedures may be carried by a scribe as part of a checklist
- Procedures may be posted at the location the task is performed
- Procedures may be placed at the location where data readings are taken to record
- Procedures are committed to long term memory
- Operators don't have enough hands to carry and read a procedure while performing the actual task at the same time (walking, bending, carrying tools, turning valves, wrenches, pushing buttons, etc.)

4.1.3 Working from Memory versus Procedure

Experienced operators that perform regular tasks are working primarily from long term memory. New operators rely heavily on written procedures. New operators will often read the procedure just prior to the task (for which they are qualified), while experienced operators will refresh their long term memories on a task performed infrequently. Experienced operators may carry the procedure in their back pocket for reference. New operators will shadow senior operators and be mentored in a task (to be qualified). In all cases procedures should be made available. Procedures should be reviewed annually and refresher training provided per regulatory requirements.

An experienced operator will not perform a task by reading a step in a procedure, performing that step, then reading the next step and performing that step, and so on. Experienced operators work from long term memory as mentioned above. Besides, procedures are insufficiently detailed to be used in that manner, nor should they be made to be. An operator will quickly lose interest in a procedure that spells out every mundane step. All procedures could be improved. Some procedures contain mistakes or critical omissions. If operators actually followed these procedures to the letter, errors would be made every time they were used.

As a skill is perfected over time, Human nature directs us to use Type 1 thinking (see **Section 6.2**), and the tasks we perform are largely done on auto-pilot (i.e., without conscious thought). Potential for slips and memory lapses creep in. This is often why newly qualified operators can often be more reliable than experienced ones (new operators may make more mistakes, but experienced operators have more potential to make the critical errors). New operators tend to spend more time in Type 2 thinking during a task.

4.1.4 Checklist v. Procedure

A checklist can be an effective forcing function (something that forces human behavior in a desired way). The caveat is, a checklist will not be effective if it is not used properly. A checklist will not be used properly if for example it is poorly crafted (e.g., too long) or not used in the correct situation (e.g., where long term memory is good enough). In these cases, checklists become equivalent to procedures, with no added benefit. Many checklists are "pencil-whipped" because it is difficult to check boxes at the same time as holding tools and turning valves. In these cases where a checklist is deemed absolutely necessary, a scribe should be used to initial the steps as they are performed by a different person. The next section discusses improving the use of checklists.

4.1.5 Improving use of a Checklist

Checklists are less useful to experienced operators because much of what an experienced operator can do comes from long term memory, and therefore long checklists are redundant to procedures, tedious, and a waste of good time. Let an experienced operator use their training, experience, intuition, judgment, competence, etc. to perform a task, but supplement this with short (60 to 90 second) "pause points" at which juncture a short check-list is reviewed for the most critical steps.

5 Tool 3: Human Error Rates

Quantification of human error rates provides several benefits beyond having a number to use in LOPA or QRA.

- It forces decomposition of the task (which is more accurate than guessing the whole) see **Section 5.2**
- It forces a deeper dive to understand qualitative factors

There are caveats to using numbers. At times too much effort is spent on the numbers. The uncertainty is too large to justify it. Without qualitative understanding, the numbers are meaningless.

5.1 Failure Modes Distribution

The following is an approximate distribution of error likelihoods for different error types (the more likely being listed first).

- Knowledge based errors. Those that require planning and trial and error.
- Selection errors or slips coupled in time and space (e.g., on the flight deck, selection errors on a DCS graphic, or the myriad of small electronic gadgets that pack multiple modes and functions into a small space). In our personal lives these types of errors may occur *daily*.
- Memory Lapses most typically caused by external interruptions (being pulled to another task) or internal interruptions (what we call 'absent-mindedness', our mind is occupied with something other than the task at hand). The Big 4 that can lead to absent-mindedness are: rushing, distraction, fatigue, or complacency.
- Selection errors decoupled in time and space (e.g., two identical valves 10 feet apart operated 60 seconds or more apart).
- Skill-based errors or craft error based on physical skill.

5.2 Using Human Reliability Data

"*If we break problems down into component questions (What is the chance of A happening? What is the chance of B happening? What is the chance that a man will be in the area when both happen? What is the chance of a fatality if he is there?) answering them with facts where possible and with opinions only when there are no facts available we are more likely to get a correct answer than if we try and guess the answer to the whole problem.*" Trevor Kletz [8]

5.2.1 Quantifying Human Error

The formula for calculating human error probability (HEP) is simple:

$$HEP = \frac{\text{Number of errors occurred}}{\text{Number of opportunities for error}}$$

The application of the equation is difficult for several reasons, and is why there exists a sparsity of human performance data.

Firstly, the numerator is hard to get because self-reporting of errors or reporting someone else's error has obvious problematic connotations.

Second, the denominator is hard to get because often we just don't know an accurate estimate of the number of opportunities for a given error.

5.2.2 Being better than 1/10

We are conditioned by LOPA that 1/10 is a typical and acceptable human performance value to use as a generic credit. That's fine (assuming we ignore the deterministic nature of human error, see **Section 2.4.5**). But human factors psychologists have recommended for us to consider error rates in our own personal lives that are as good as 1/1,000 and even under certain circumstances as good as 1/10,000. If you do this, you will convince yourself very easily that better than 1/10 is entirely possible and reasonable.

How good can humans be? Recognized and published human performance limiting values can be found in the literature, for example in Kirwan [4] as follows,

- Single operator carrying out task(s) on plant .. 1e-4
- Operators carrying out task on plant.. 1e-4 to 1e-5
- Control room based team.. 1e-5

These are HEP (human error probabilities). These are performance limits. Often human error rates for a specific task are found to be much higher than this.

To put this in perspective, one would need data for three years performing a daily task error free to demonstrate a HEP of 1e-3. One would need 30 years of data performing a daily task error free to demonstrate a HEP of 1e-4. For this reason, as a human reliability practitioner, one would be hesitant claiming as low as 1e-4 for a task performed by the same person even with good (self) recovery factors and good human factors.

For typical slips and memory lapses under Type 1 thinking (see **Section 6.2**) human error probabilities between 1e-3 and 1e-4 (for example, per step of a procedure) are perfectly reasonable, depending of course on the human factors.

5.2.3 The Relationship between Task Frequency and Human Error Rate

Operator tasks that are performed frequently (e.g., every shift) are especially problematic to handle quantitatively because the opportunity for error is so great, yet slips and lapses will occur at the same rate (there is no data to say they get lower). To convert the HEP (human error probability) into a frequency, one simply multiplies by the number of opportunities per time. Still, this issue probably has more to do with the fact that the HEP assumed are on the conservative side. Use of recovery factors, human redundancy, and inherent redundancy in the procedure itself are ways to handle this issue.

6 Tool 4: Understanding Systematic Bias

6.1 Bias and Heuristics

The following are some ways a bias can be described:

- A trap
- A distortion (the difference between what is seen and what is there)
- A systematic error (not statistical)
- Has the potential to be corrected for if it can be identified

A heuristic is a shortcut that bypasses cognition (i.e., thinking). Different biases provide the short-cut mechanism. We expect cognitive thinking to be less prone to bias (see **Section 6.2**).

Some examples of bias that have relevance to predicting and controlling major accident hazards are:

Confirmation bias and belief perseverance – we naturally look for corroboration for what we already know or believe, and we hold on to it. Rarely do people look to disconfirm. In the context of major accident hazards, what we don't know is far more relevant than what we do know. Why do we struggle with distinguishing between severity x or y when the hazard is already identified? Is the issue of taking multiple BPCS credits really that controversial when every barrier is already coupled by biased thinking (and we don't even realize it)? Are the consequences of summing versus not summing LOPA causes meaningful when accident sequences never play out in a linear fashion? I'm not arguing to ignore these issues, but to understand the context and prioritize. We already know these things, and the uncertainty is small. The uncertainty is very large for what we do not know, and worse if we don't look for it.

Hindsight bias – our inability to correctly remember or interpret past events, and our need to find linear causality (simplification) for explaining past events, produces a powerful distortion. If you've ever watched an accident video and thought, "How could they let that happen," you've fallen victim to hindsight bias. If you've ever read multiple accident reports from different sources on the same accident, and each report states a different cause or causes, this demonstrates hindsight bias at work.

Information bias – more information produces better results. Not always. What more information does do is make us more (falsely) confident in our results. This bias runs rampant in quantitative analysis. We drill down and sharpen certain variables in a calculation only to ignore other variables whose uncertainty far exceeds what we are attempting to calculate.

Tunneling bias – ignoring or missing sources of uncertainty. This bias may be the umbrella for the others. Any number or calculation produced should have an uncertainty associated with it.

Commitment bias and capture error – these manifest themselves in many accident sequences where an operator has gotten on the wrong equipment, or followed the wrong procedure or action.

Independence trap – simply becoming conscious of an IPL (independent protection layer) or barrier results in a loss of independence of that barrier. This is the most insidious of all traps and has been written about by McLeod [9] and Hopkins [10]. How does this trap work? Say I have two IPLs, x and y, and I am aware of both. Any decision made for one will be influenced by the other. For example, if x is bypassed we still have y. If x fails we still have y. If x works we don't need y. And, it would be nearly impossible for x and y to fail at the exact same time. While this reasoning may be technically and logically true, it can produce a gradual and subtle increase in risk and worst-case it can defeat all of the safety barriers for a scenario.

Normalization of deviation – There have been many examples of this since the term was first coined, and is sometimes why accidents take years to manifest themselves.

Group-think – the presumption that when decisions are made in small groups, everyone will (or must) agree with the decision. This trap has been extensively studied, and can result in bad decisions.

The Turkey trap – see **Section 8.1**.

6.2 How Humans Think

Psychologists, notably Daniel Kahneman [14], have written extensively about how humans think. Two distinct styles are recognized:

- Type 1 – Fast, intuitive, efficient. Not consciously thinking. Zoned out. Prone to bias and irrationality. If you've ever driven to work and don't remember doing it, you've experienced Type 1 thinking. Actually, humans spend the vast majority of their day in Type 1 thinking, while performing tasks, etc. most of the time very reliably.
- Type 2 – Slow and inefficient. Hard. Careful and rational. Conscious thought. Taking exams, writing procedures, performing calculations, require Type 2 thinking. Try to maintain Type 2 thinking *continuously* for more than a few minutes. It can become exhausting.

Type 2 thinking is not error free (see the definition of mistake in **Section 3.4.2**) however, in certain contexts it can be superior to Type 1 thinking. Obviously, because your brain is turned ON, which we assume is better for most people when performing tasks and making decisions.

It is impractical to expect any one to spend their day in Type 2 thinking. The so far elusive goal of Human factors Engineering is this:

> To jolt the operator out of Type 1 thinking into Type 2 thinking, just before he is about to make a catastrophic mistake.

For example, say an operator is staring blankly at a sign that states 'Reactor Valve A' getting ready to open that valve, but that the operator intended to be on 'Reactor Valve B'. As the operator reaches for the handle a small jolt of electricity or heat or something shocks him, and an audio output declares "You are on the wrong valve!"

To achieve this takes some serious human factors engineering and technologies that probably do not exist yet, or not available to general consumer.

7 Tool 5: Improving Barrier Effectiveness using Bow-tie

Bow-tie drawings (the left side) may be used as an effective means to communicate how barriers may fail due to human or organizational factors. Task analysis would be used to identify escalation factors (something that can cause a barrier to fail) as well as escalation factor controls (that prevent interference of the escalation factor). The same analysis can identify failures that could impact multiple barriers.

These escalation factors on the bow-tie drawings represent leading indicators of process safety and could be monitored for failure. The bow-tie can be quantified to determine barrier probability of failure. The bow-tie is the best way to show how barriers can fail and the controls to prevent failure.

Common escalation factors related to Barrier failure:

- Errors made by Stores
- Lack of technical competence
- Contractors (where they touch the barrier)
- Poor Maintenance procedures and practices
- Improper restoration following maintenance
- Lack of Change Management
- Poor Work processes
- Lack of Communications
- Wrong specification

8 Handling Uncertainty

Typically we treat uncertainty in a quantitative way by performing something like a sensitivity analysis (what-if this input value is changed to this) or by acknowledging data uncertainty intervals (i.e., confidence limits) for the data we use. Typically this handles uncertainty associated with random hardware failure.

There are at least two other types of uncertainty (see **Section 2.4.2** for others),

- That which we know about but do not account for, i.e., systematic human error, the subject of this paper
- That which we don't even know about (i.e., referred to as a Black Swan) addressed below

8.1 Black Swan Theory

I won't attempt to explain or even summarize Black Swan theory [11]. I can't do it justice. Further, I would not listen to or read any second hand explanation of it, other than from the book itself, reference [11]. It is an important book and should be read by every practitioner of process safety risk management.

Psychologists have recognized a fundamental flaw (a bias) in humans thinking that we can project (predict) the future based on the past. This is obviously a trap that PHA teams still fall into. Many of the most infamous accidents in Industry only occurred after years of performing the same task or operation "safely." These incidents could look something like **Figure 1**, adopted from Ref [11]. From the perspective of the turkey, the probability of a catastrophic incident is 1e-3 (looking backwards) the day before it strikes. In hindsight, the catastrophic event seems readily predictable (Americans eat turkeys on Thanksgiving), and the events leading up to it can be described in a nice linear fashion (feed, water, repeat, later kill). This is hindsight bias at work.

The best we can do to handle this kind of uncertainty is to maintain a sense of vulnerability and continue to use work processes (e.g., hazard identification, change management, preventive maintenance, etc.) and tools (e.g., databases, procedures, task analysis, reliability analysis, etc.), to manage what we can predict, in hopes of effecting what we cannot predict.

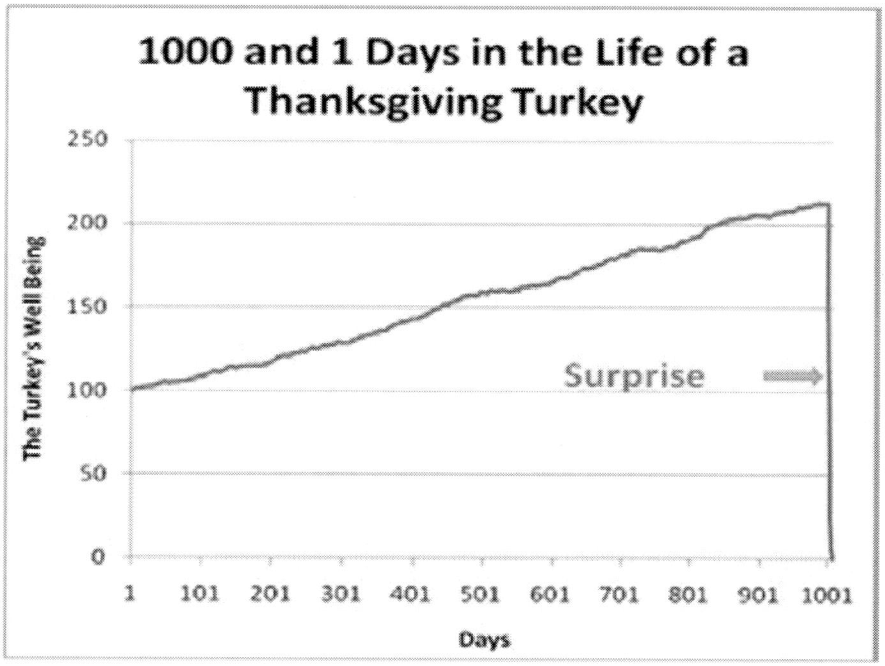

Figure 1. 1e-3 Probability of Catastrophe based on Past History

9 A Short Task Analysis Checklist

The following checklists represents a sample of the types of questions asked during a Task Analysis. A follow-up walk-thru to actually observe the task is always necessary. It is useful to observe how different the below checklists are compared to the standard human factors checklists used in PHA.

For any scenario with a human error initiating event that goes to LOPA, ask the team to answer the following questions:

- What is the cause of the error ("The operator." Why? "Wasn't paying attention." Why? "Gets pulled away during task." Why? etc.)
- What are the administrative controls in place (immediate feedback, procedure, permit, tagging, LOTO, human redundancy, housekeeping audits, etc.) that could allow recovery from the error?
- Is the task skill/ rule/ or knowledge based? Often tasks will be a combinations of the 3.
- How frequent is task performed? How long does the task take to complete?
- What is the qualification for the job?
- What Tools are required for the job?
- Can the task cross shift-change?
- What role do Contractors play in the task? Ask the same set of questions focusing on the Contractor.

For any scenario with a human response to abnormal condition (e.g., critical alarm barrier) that goes to LOPA, ask the team to answer the following questions (some of these can be observed directly during the walk-through):

Focus on the big 4: perceive, interpret, diagnose, respond (the first 3 represent the degree of situational awareness)

- Perceive: How does the Operator become aware of the alarm? Is the alarm prioritized? Is a dedicated alarm screen used? Are gray-scale graphics used?
- Interpret: What are the alarm metrics (how many alarms per hour, false alarms, etc.)? Do the metrics conform to good engineering practice? Are instrument signals trended? Is there a policy that every critical alarm that comes in must be treated as real? Are there periodic (e.g., quarterly) talk-throughs or drills for the alarm?
- Diagnose: Is the scenario known or novel? How many potential causes? How long does each cause take to diagnose? What actions are required?
- Respond: see checklist above.

For any scenario with a human barrier used in LOPA, ask the team to answer the following questions (some of these can be observed directly during the walk-through):

- What are all the human touch points and how can failure occur? E.g.,
    - Supplier
    - Purchasing

  - Stores
  - Maintenance
  - Quality Control
  - Engineering
  - Operations
  - Management
- What are the Contractor roles in the barrier?
- Has the barrier Specification (sizing, set-point, response time, etc.) been independently validated?

For any barrier or IPL that requires good situation awareness (SA*) (e.g., response to abnormal situation - AS), the following items should be looked at:

- Individual factors
  - How good is the operator's 'mental model' of the plant related to the AS (process conditions, equipment status, maintenance activities, etc.)?
  - What is the level of 'Automaticity' (how much cognition does the operator have to engage to resolve the AS)?
  - How well are the goals defined to resolve the AS?
- System factors
  - Interface design (alarm rationalization, use of gray-scale graphics, etc.)
  - System complexity (How many potential causes of the AS are there?)
  - Level of automation (Is there so much automation the operator has lost their hands-on operating skills?)

*SA is the difference between what is actually happening in the plant and what the operator thinks is happening.

These short checklists give you an idea of what task analysis is for each.

10 Summary

Improving Human factors in PHA-LOPA requires a diverse toolkit of methods and theory. Identification, prediction, and control of the human factors role in major accident hazards is currently soft. There is not a quick easy answer, however, improvement especially related to low hanging fruit is readily achievable.

PHA-LOPA facilitators need to begin practicing the methods of human factors and human reliability to improve said practice as it relates to major accident hazards. The skills a facilitator gains by practicing human factors and reliability will directly transfer to all of PHA-LOPA to improve decision making, the ability to write human factors scenarios, and better awareness of when further study is needed.

11 References

[1] Norman, Don. *The Design of Everyday Things*, Revised and Expanded Edition, Basic Books, New York, 2013.

[2] Chemical Safety Board Website. http://www.csb.gov/videos/csb-video-excerpts-from-dr-trevor-kletz/

[3] J. Robert Taylor. *Human Error in Process Plant Design and Operations – A Practitioner's Guide*, CRC Press, Taylor & Francis Group, Boca Raton, 2016.

[4] Kirwan, Barry. *A Guide to Practical Human Reliability Assessment*, CRC Press, Taylor & Francis Group, Boca Raton, 1994.

[5] CCPS. *Guidelines for Engineering Design for Process Safety* 2nd Edition, 2012.

[6] HSE. Review of Human Reliability Assessment Methods, 2009.

[7] Gawande, Atul. *The Checklist Manifesto: How to Get Things Right*, Metropolitan Books, 2009.

[8] Kletz, Trevor. *Hazard Analysis – A Quantitative Approach to Safety*, I.Chem.E. Symposium Series, No. 34, 1971.

[9] McLeod, Ronald W. *Designing for Human Reliability – Human Factors Engineering in the Oil, Gas, and Process Industries*, Gulf Professional Publishing, 2015.

[10] Hopkins, Andrew. *Disastrous Decisions – The Human and Organizational Causes of the Gulf of Mexico Blowout*, CCH, Australia, 2012.

[11] Taleb, Nassim Nicholas. *The Black Swan* 2nd Edition, Random House, New York, 2010.

[12] Swain and Guttmann. NUREG/ CR – 1278. *Handbook of Human Reliability Analysis*. 1983.

[13] Reason, James. A Life in Error – From Little Slips to Big Disasters, CRC Press, Taylor & Francis Group, Boca Raton, 2013.

[14] Kahneman, Daniel. *Thinking Fast and Slow*, Farrar, Straus & Giroux, New York, 2011.

13TH GLOBAL CONGRESS ON PROCESS SAFETY

Waste Solvents to Trash Haulers: Lessons Learned from Hazardous Waste Accidents

Delmar "Trey" Morrison, Ph.D., P.E.
Exponent, Inc.
4580 Weaver Parkway, Suite 100, Warrenville, IL 60555, USA
tmorrison@exponent.com

Michael Stern, Ph.D., P.E.
Exponent, Inc.
Natick, MA, USA
mstern@exponent.com

Carmen H. Osorio-Amado, Ph.D.
Exponent, Inc.
Los Angeles, CA, USA
cosorio@exponent.com

Prepared for Presentation at
American Institute of Chemical Engineers
2017 Spring Meeting and 13th Global Congress on Process Safety
San Antonio, TX
March 26 – 29, 2017

Waste Solvents to Trash Haulers: Lessons Learned from Hazardous Waste Accidents

Delmar "Trey" Morrison, Ph.D., P.E.
Exponent, Inc.
4580 Weaver Parkway, Suite 100, Warrenville, IL 60555, USA
tmorrison@exponent.com

Michael Stern, Ph.D., P.E.
Exponent, Inc.

Carmen H. Osorio, Ph.D.
Exponent, Inc.

Keywords: Hazardous waste, incident investigation, chemical reactivity

Abstract

Process safety management (PSM) of hazardous chemicals is a core tenet of the chemical, oil, and gas industries. Organizations in these industries have increasingly woven the elements of PSM into their strategic vision, day-to-day operations, and interactions with contractors and vendors. This cultural evolution has, in part, been driven by major accidents in these fields. In these industries, the chemicals, their hazardous properties, and their potential interactions, are usually either well-defined or capable of being well-defined. Control and mitigation strategies for those chemicals can be developed based on this available knowledge. In contrast, companies that handle wastes have experienced catastrophic accidents due to the ill-defined nature of the hazards. This industry handles the cast-offs, dregs, and refuse, which makes identification of the hazards and prevention of incidents more challenging. How do you implement a PSM program when your input streams can vary due to processes at a plant you do not control or monitor? Does the EPA (Environmental Protection Agency) Resource Conservation and Recovery Act (RCRA) provide sufficient guidance of safely handling of wastes? The overriding theme in the incidents is that PSM is more complex to apply to unknown, mixed, and uncontrolled chemicals, both solid and liquid. To assist companies that handle wastes, this paper provides several case studies of fires and explosions involving multiple waste material handling processes at different facilities. The incidents involved numerous matrices from solid industrial wastes, EPA RCRA hazardous wastes, to waste solvents. An analysis of the lessons learned will be presented and used to provide realistic guidelines to reduce safety risks based on a pragmatic implementation of process safety management principles. The paper will also compare OSHA (Occupation Safety and Health Administration) PSM with EPA RCRA approaches to characterizing wastes versus managing the hazards, and will emphasize those elements that played critical roles in the accident case studies.

1 Introduction

This paper discusses several case histories of process incidents occurring at waste handling or processing facilities and compiles lessons learned from those accidents. Depending upon the source, the industry, and the regulatory classification of the waste, industrial waste may pass through a municipal waste transfer station to a landfill or be collected, aggregated and then incinerated by a RCRA (Resource Conservation and Recovery Act) [1] permitted hazardous waste Treatment, Storage, and Disposal Facility (TSDF). Managing waste is inherently more challenging than managing typical chemical process hazards where the identity and properties of the process chemicals can be effectively defined. While many types of hazardous waste facilities test for the properties of incoming waste streams, these tests can be challenging if the waste is phase separated, if the waste is still reacting and changing over time, or if the waste is incompletely identified by the generator. Similarly, investigations of accidents are complicated by the fact that wastes involved in the accident could have contained unknown chemicals with unknown properties.

Several case studies are presented in this paper. The accident locations range from waste transfer stations, waste storage facilities, waste treatment facilities, and a cement manufacturing facility using waste as a fuel source. The wastes involved in the accidents ranged from municipal waste, unregulated solid and liquid wastes, to regulated hazardous solid and liquid wastes. Significant fires and explosions occurred in these incidents. The overarching causes of these incidents were due to the gap between the true properties and the recognized properties of the wastes. While the task of characterizing waste streams may be challenging, the overall process safety and incident investigation principles remain similar. In this paper, we start with a typical decision tree analysis for screening chemical reactivity hazards and modify it for use for waste processing facilities. We will then discuss several case studies involving fires and explosions in waste handling and treatment and provide lessons learned.

2 Background

Application of Process Safety Management (PSM) principles in the chemical industry has been codified for highly hazardous chemicals above threshold quantities through Title 29 Part 1910.119 of the Code of Federal Regulation since 1992 [2]. The Center for Chemical Process Safety (CCPS) subsequently expanded on the process safety model presented by the regulations to develop twenty elements of risk-based process safety [3]. The overlap between the CCPS process safety model and the OSHA PSM regulatory elements (bolded) is represented in the list below (Table 1).

Table 1. Overlap between OSHA (bold) and CCPS PSM elements

1. Process Safety Culture	2. Contractor Management **(h) Contractors**
3. Compliance with Standards	4. Training and Performance Assurance **(g) Training**
5. Process Safety Competency	6. Management of Change **(l) Management of change**
7. Workforce Involvement **(c) Employee participation**	8. Operational Readiness
9. Stakeholder Outreach	10. Conduct of Operations
11. Process Knowledge Management **(d) Process safety information**	12. Emergency Management **(n) Emergency planning and response**
13. Hazard Identification and Risk Analysis **(e) Process hazard analysis**	14. Incident Investigation **(m) Incident investigation**
15. Operating Procedures **(f) Operating procedures**	16. Measurement and Metrics
17. Safe Work Practices **(k) Hot work permit**	18. Auditing
19. Asset Integrity and Reliability **(j) Mechanical integrity**	20. Management Review & Continuous Improvement

Deficiencies in many of these elements can be traced back to the root causes of hazardous materials incidents, as will be illustrated in the case studies in this paper. However, it is uncommon for TSDFs and other waste handling facilities to formally apply PSM to their processes. Usually, facilities follow a prescriptive approach to waste characterization and handling that is based upon the EPA (Environmental Protection Agency) regulations and relies heavily upon the waste generator's characterization of the waste properties [4]. This reliance is not unusual when considering the quantity of distinct and unique waste streams that an individual municipal facility or TSDF may handle and dispose.

The primary hazard identified in the case studies is delayed reaction leading to pressure generation, heat generation, and unintended ignition of packaging or the waste material itself. This is commonly referred to as spontaneous ignition or spontaneous combustion since a piloted ignition source is not necessary. Trash and municipal waste handling facilities are frequently susceptible to delayed fires due to spontaneous combustion. Such ignition events may be caused by unintentional mixing of incompatible chemicals or exothermic oxidation of self-heating materials, such as vegetable-derived oils. For example, the National Fire Protection Association (NFPA) estimated that 4,460 annual fires involving spontaneous combustion of trash occurred between 2005 and 2009 [5].

Recent studies have identified several trends in hazardous waste TSDF fires and explosions [6], [7], [8]. Incidents can occur in any stage of the process from initial storage, waste repackaging, or transfer to the final treatment stage (i.e., incinerator, calciner, or kiln). The consequences of a fire or explosion can accelerate rapidly, depending upon the nature of the initial waste and its proximity to surrounding waste materials, surrounding waste materials' properties, and the adequacy of fire protection and emergency response actions. The causes of fires and explosions typically derive from unknown or unrecognized chemical reactivity issues with the waste material in question.

2.1 Waste generation in the United States

According to the US EPA, 26,237 generators produced 33.6 million tons of hazardous waste, of which over 6 million tons were non-wastewater [9]. Of the 6 million tons, 26% is combusted for energy recovery, 19% is put into landfills, and 13% is incinerated. Over 76% of the non-wastewater hazardous waste is treated off-site by a TSDF.

The EPA statistics also show that 258.5 million tons of non-hazardous Municipal Solid Waste (MSW) was generated in the United States in 2013. While the per capita waste has slightly decreased since the year 2000, the total annual generation continues to climb due to population growth [10]. Approximately 89.4 million tons (35%) of the MSW were recycled in 2013. Another 13% of the total MSW was combusted with some amount of energy recovery. The remaining 52.6% was placed into landfills.

2.2 Resource Conservation and Recovery Act

A hazardous waste can be any waste material with a potential to harm human health or the environment, either by itself or through interaction with other factors. However, for regulatory purposes in the U.S., a waste material must meet specific criteria to be considered as a hazardous waste. These conditions, summarized in Figure 1, are defined by the EPA under the RCRA in Title 40 of the Code of Federal Regulations [1].

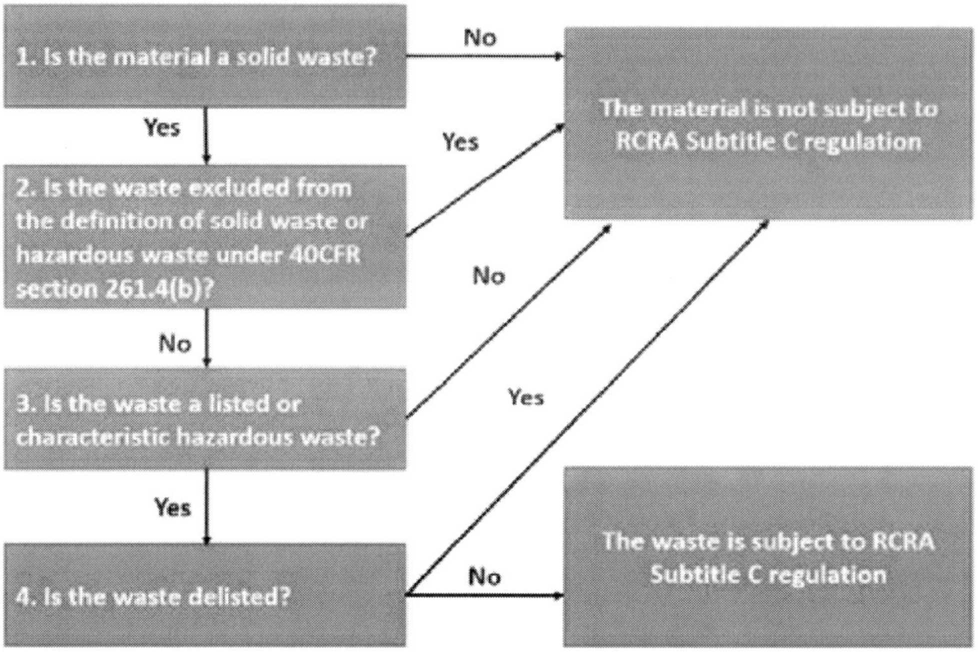

Figure 1. EPA's Hazardous waste identification process [11]

Per the EPA, the first condition of a hazardous waste is to be a solid waste, which paradoxically includes not only wastes that are physically solid, but also wastes that are liquid, semi-solid, or contained gaseous material. A hazardous waste is also required to meet certain characteristics based on its toxicity, ignitability, reactivity, and corrosivity, and needs to be listed in either the F, K, P, or U lists found at the 40CFR section 261 [12].

Some solid wastes are specifically excluded from RCRA regulatory compliance due to different reasons such as impracticability, economy impact, lack of information to justify applying the regulation, regulation by other laws, or they were removed (delisted) from this category. These exclusions refer to conditions (2) and (4) from the hazardous waste identification process shown in Figure 1. For example 40 CFR 261.4(b) excludes household waste and solid waste from commercial or industrial sources (that don't contain solid hazardous waste). It also excludes fly ash waste. These types of waste material may pose physical and chemical hazards, but they are not regulated under RCRA.

2.3 RCRA chemical compatibility testing

As part of the RCRA regulations, hazardous waste generators are required to classify their waste. When TSDFs receive waste, they must perform pre-acceptance analysis to confirm the waste's characteristics and identify incompatible wastes [13]. The EPA states "Compatibility testing may include determining water reactivity (e.g., foaming, heat generating, explosivity), oxidation-reduction potential, spot tests for cyanide and sulfide, among others" [14]. The EPA provides a list of some common potentially incompatible wastes in Appendix V to Part 264. When analysis is required, the EPA has published guidance on how to qualitatively evaluate the compatibility of wastes [15]. The guidance document uses a two-part approach: step-by-step compatibility analysis procedures and then a compatibility chart that considers 41 reactivity groupings. To simplify this analysis when chemical testing is not required, free computer software packages are available for performing quick assessments of chemical compatibility [16].

2.4 Guidelines for evaluating chemical reactivity hazards

According to the U.S. Chemical Safety Board (CSB), a chemical reactive hazard can be defined as "*a sudden event involving an uncontrolled chemical reaction with significant increases in temperature, pressure, or gas evolution that has the potential to, or has caused serious harm to people, property or the environment*" [17]. In order to decide whether a chemical can be considered a reactive hazard or not, the CCPS published a preliminary screening test based on a series of questions summarized in Figure 2 [18].

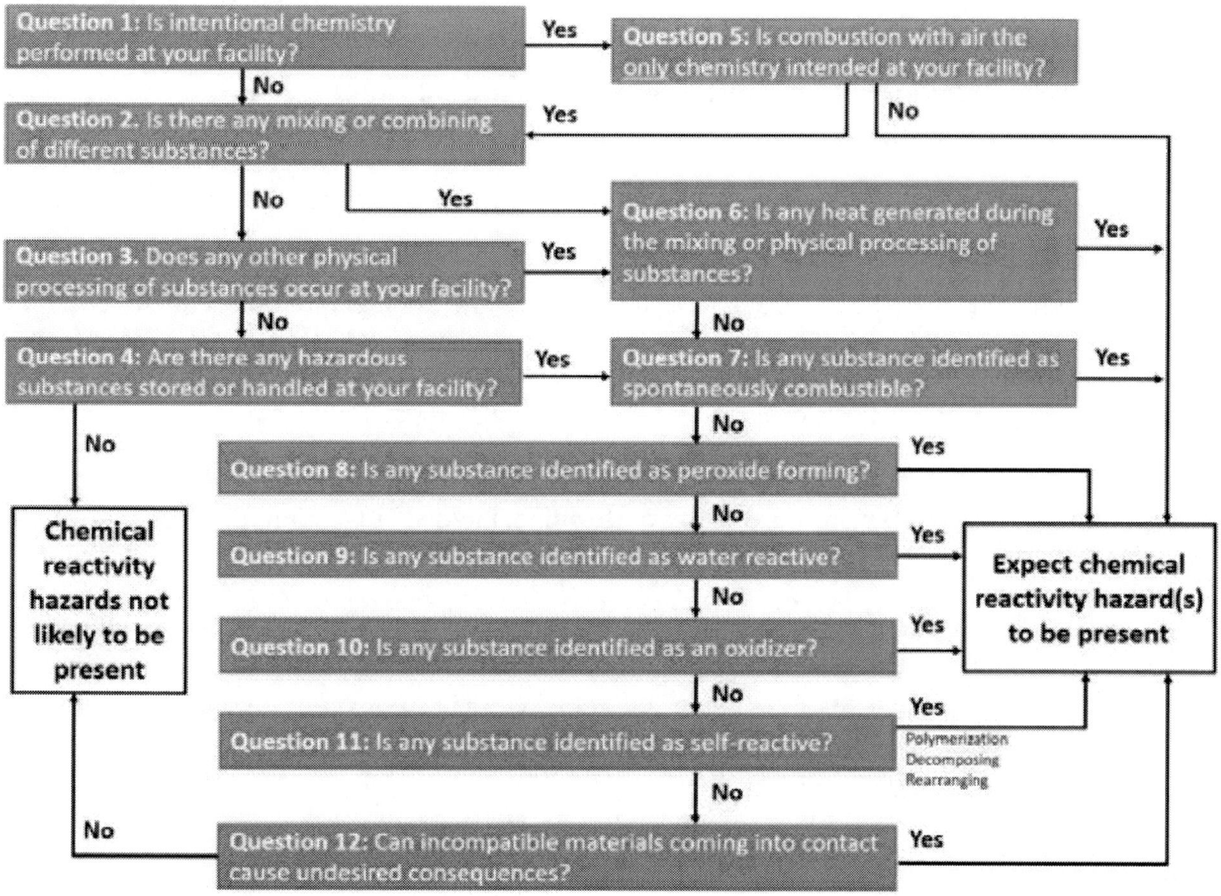

Figure 2. Preliminary screening for chemical reactivity hazards [18].

This test should be conducted on an ongoing basis, in particular when there have been changes in the process/conditions or a new chemical is introduced into the facility. If a chemical reactive hazard is expected to be present at the facility, the CCPS also suggests the implementation of a chemical reactivity management program structured consistent in Figure 3.

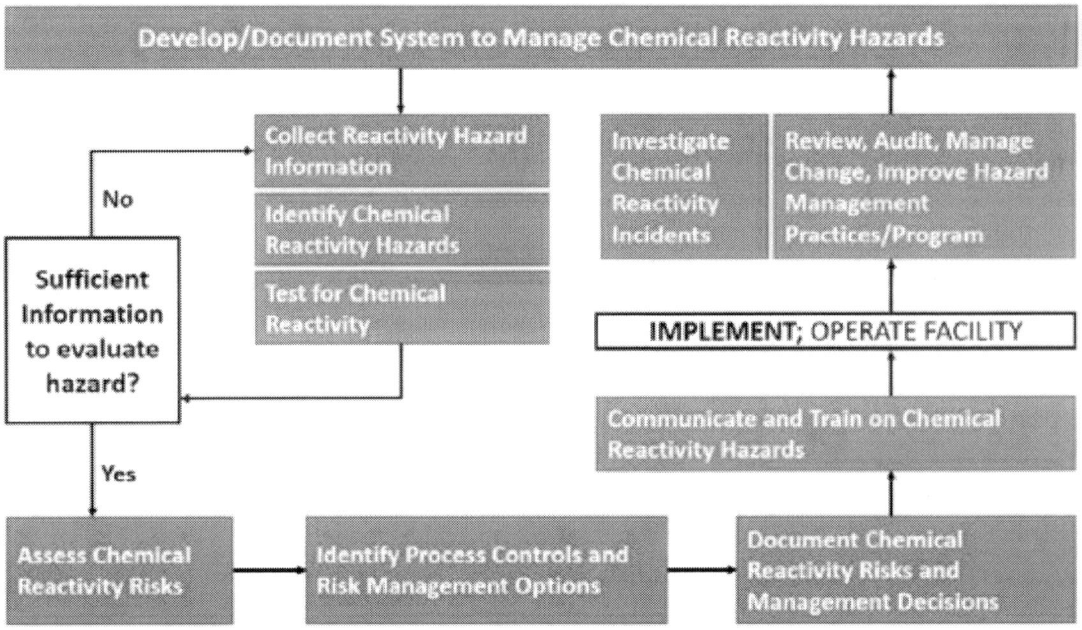

Figure 3. Flowchart for implementing chemical reactivity hazard management [18].

2.5 Why and how an event happened? Protocols for investigating reactive chemical incidents

Given the nature of fires and explosions, especially when involving hazardous waste materials, necessary evidence may be destroyed or lost during the incident or subsequent emergency response. Thus, conclusive determination of the chemistry and proximate causes may be impossible. The investigation process must rely upon the scientific method to develop hypothetical causes, test those causes, and determine the most likely causes and contributing factors.

Two commonly relied upon resources for performing incident investigations per the scientific method are NFPA 921 *Guide for Fire and Explosion* Investigations [19] and the CCPS publication Guidelines for Investigating Chemical Process Incidents [20]. Figure 4 shows the progression of an investigation as described in Chapter 4 of NFPA 921. Two essential concepts, inductive reasoning and deductive reasoning, are used to develop and test hypotheses, respectively. Inductive reasoning is using the analyzed data to develop a consistent hypothesis. Deductive reasoning is used to test the hypothesis by comparing it to all known facts of the accident and scientific literature. NFPA 921 cautions that "[t]he investigator does not have a valid or reliable conclusion unless the hypothesis can stand the test of a careful and serious challenge." Recent editions of NFPA 921 have also added cautionary language regarding the use of *negative corpus* also known as the process of elimination. Investigators should not confuse the lack of an alternative hypothesis for affirmative proof of a hypothesis with no supporting evidence (but that cannot be ruled out with the available information).

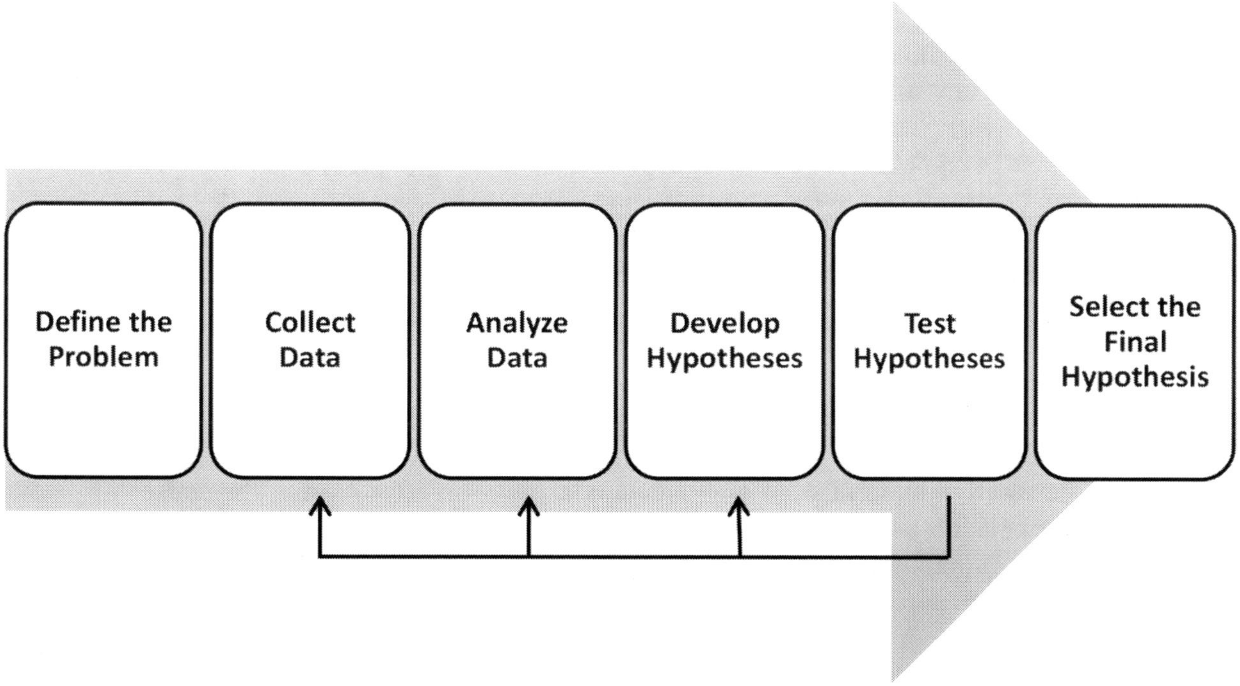

Figure 4. The scientific method of investigation.

The CCPS Guidelines text provides a similar scientific methodology for accident investigation but also offers information specific to industrial accident investigation. For example, information for designing incident management systems within a corporation is provided. Specific investigation and root cause analysis tools commonly employed for investigations, organized by the level of structure, are shown in Figure 5.

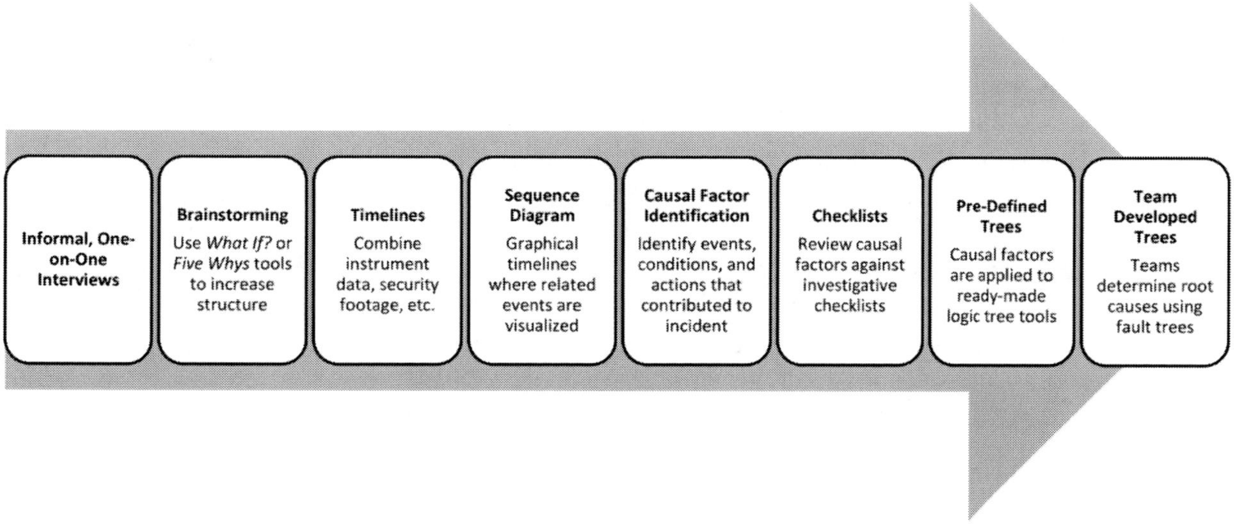

Figure 5. Investigation tools for process incident investigations.

The case studies below describe the causes and lessons learned from MSW handling, solvent reclamation, and TSDF incidents. All of the incidents were characterized by fires or explosions, which caused significant damage and posed potential for catastrophic injuries or fatalities to onsite personnel. Many of the incidents were investigated by external public agencies in addition to the company investigations. The authors also participated firsthand in the investigations of the incidents. Where possible, lessons learned from the investigation process as one element of PSM are shared.

3 MSW Transfer Facility Fires

Two fires, separated by three days, occurred at a MSW transfer station. The first fire involved a trailer of waste awaiting transfer to the landfill, and the second fire involved a different trailer and a large portion of the building and waste contained within. The facility was unsophisticated when considering the element of process safety management, which is understandable because it did not handle reactive materials. As a consequence, early investigation of the causes of the fires relied upon supposition and poor science to conclude that one waste stream was the cause of the fire. The incorrect conclusion adopted by the facility meant that the real cause went unaddressed until a later, thorough engineering investigation occurred.

The waste transfer station was a large, open warehouse structure with two sunken loading bays. The facility handled trash classified as municipal solid waste (MSW) and construction and demolition (C&D) debris. Waste materials were piled on the concrete and moved around using a front end loader. Different materials were commingled by the facility operators, both in piles and in the open trailers located in the loading bays. The commingled waste was periodically hauled to a municipal landfill. The facility did not accept RCRA hazardous waste or liquid wastes. The local regulations and the permit for the facility both acknowledged the reasonable possibility of "hot loads" - in other words, the receipt of loads of waste material that are undergoing self-heating or are already smoldering.

The first fire involved the contents and the rear of a trailer (Figure 6), which was located in a loading bay overnight. The trailer contained a mixture of bottom ash from a fluidized bed coal combustor and other unknown MSW and C&D materials. No formal investigation was conducted to attempt to determine the cause of the trailer fire. The waste material in the trailer was not fully extinguished or overhauled by the fire department, and the facility located the trailer away from the building so that it could eventually burn itself out. The fire witness statements indicated that the contents of the trailer continued to burn for the next several days. The fire damage patterns to the trailer are consistent with a fire originating at the rear wheels, perhaps due to overheated brakes. Three days after the initial fire, a second fire was discovered after a different trailer was loaded with similar waste materials from within the building and held overnight in the undamaged loading bay. The second fire burned to completion and destroyed the building.

Figure 6. Photograph of the trailer damage resulting from the first fire.

An early investigation of the second fire concluded that the combustor bottom ash was the equivalent of pure lime (CaO), caused both fires due to exothermic heat of hydration, and should have been classified as a RCRA hazardous waste. Pure CaO is referred to as "lime" or "quick lime." The reaction of CaO with water vapor results in solid calcium hydroxide, $Ca(OH)_2$, while mixing with excess water creates an aqueous solution, according to the chemical reaction:

$$CaO_{(s)} + H_2O \rightarrow Ca(OH)_2$$

This reaction is exothermic, meaning that it generates heat, with a heat of reaction of -65.27 kJ/mole of CaO reacted [21]. The quantity of heat released is directly proportional to the quantity of CaO and H_2O undergoing the reaction.

The transfer station's incorrect initial conclusion was based on the description of the waste by the generator as "agricultural lime," which is commonly used to describe calcium carbonate ($CaCO_3$), hydroxide ($Ca(OH)_2$), and oxide (CaO) used in farming. Further, the early investigation concluded that the mischaracterization of the waste was a direct cause of the fires. A subsequent engineering investigation analyzed that conclusion and also identified several plausible causes including an electrical fire, spontaneous combustion of MSW or C&D, or re-kindling of the smoldering remains of the first fire after being transported across the facility.

Chemical analysis of the combustor bottom ash and a review of the literature indicated that the ash did not contain sufficient CaO to cause spontaneous combustion through hydration and was excluded under 40 CFR 261.4(b). The residual CaO in combustor tram will undergo hydration by reacting with water (H_2O) in either a vapor (e.g., atmospheric) or liquid form. The hydration reaction for residual CaO would run largely to completion within a few days when exposed to the outdoor atmosphere [22]. Comparison samples of fresh combustor bottom ash revealed that it contained approximately 10% CaO by weight.

If concentrated CaO is mixed with liquid water, the hydration reaction can lead to very high temperatures capable of igniting combustible materials. However, it is unlikely that these conditions existed in the current matter. The maximum temperature reached by the combustor ash via this reaction, if any unreacted CaO is present, will be tempered by the presence of liquid water and surrounding solid material (i.e., $CaCO_3$, $Ca(OH)_2$, $CaSO_4$, sand, and onsite waste materials). The water will absorb the released heat and act to cool the material as the water evaporates; the other solid species will act to absorb the liberated heat and increase in temperature.

The false conclusion that the ash waste was CaO, thus causing the fires, highlights the importance of precise descriptions for waste streams even if they are not bound to the RCRA regulation. Effective description can help investigators, who may be unaware of the precise waste compositions, more effectively assess the important properties of the waste stream after an accident. Similarly, investigators must be careful when examining the properties of non-RCRA waste streams as their descriptions may be more colloquial. Conclusions should not be drawn until a more detailed composition of the waste can be obtained through chemical analysis or discussion with the generator.

4 Solvent Recovery Plant Vapor Cloud Explosion

In 2009, a release of ignitable liquid and vapors, followed by catastrophic vapor cloud explosion and subsequent fires, took place in a solvent recovery plant. The explosion severely injured two employees, destroyed part of the plant (see Figure 7), and broke windows in neighboring facilities and residences. The explosion was investigated by the Chemical Safety and Hazard Investigation Board (CSB) [23].

The solvent recovery plant (Figure 8) was located at a facility that handled hazardous and non-hazardous waste materials that were received via tanker truck, drums, totes, and lab packs. The facility performed fuel blending, solvent recovery, and waste consolidation. Incoming dirty solvents were unloaded into tanks in the dirty tank farm. Clean solvent was recovered from the dirty solvents via one of three distillation units (Unit 1, Unit 3, and Unit 4). The recovery process involved the circulation of distillation bottoms back into the source tank in the dirty tank farm, while purified "clean" solvent was condensed and transferred to the clean solvent tank farm. Final waste liquids and sludge were shipped offsite to a waste incinerator as blended fuels.

At the time of the incident, the three Processing Units were being run by a single outside operator. Unit 1 was used to recycle a paint thinner material whose primary compound was acetone, while Unit 3 was processing a tetrahydrofuran (THF)/water mixture. Unit 4 worked as a closed circuit reflux-distillation unit where a mixture of THF and water near the azeotropic concentration was processed.

Figure 7. Aerial view of the solvent recovery plant after the incident from the CSB report [23].

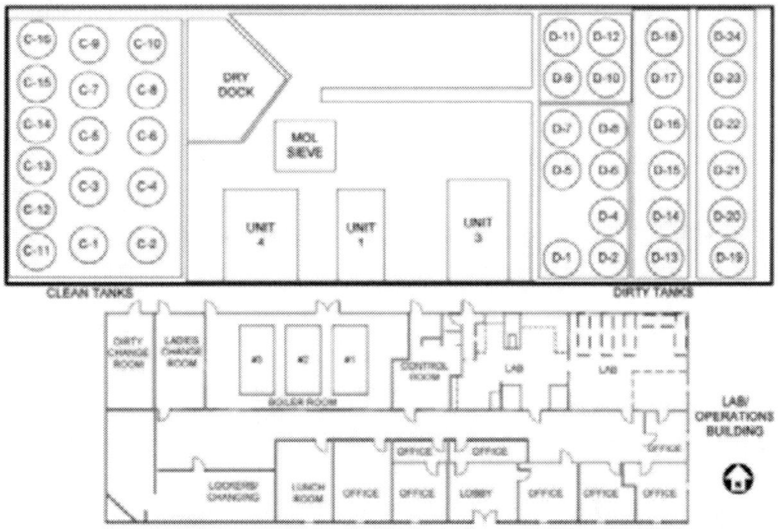

Figure 8. Facility layout (not to scale) [23].

Three of the six workers onsite reported hearing a loud noise like a pressurized release, accompanied by a strong solvent odor and a visible aerosol/or vapor cloud coming from the Dirty Tank Farm prior to ignition. Security camera footage from onsite and a nearby business supported the conclusion of a pressurized release, likely of Tetrahydrofuran (THF) solvent mixture; however, the exact composition and source of the flammable material released could not be determined.

The incident investigation concluded that the vapor cloud spread out from the Dirty Tank Farm, entered the laboratory/Operations Building, and ignited inside the Boiler Room causing, an initial explosion. Consequently, additional flammable material within the Dirty Tank Farm was also ignited causing a blast pressure wave (vapor cloud explosion) and subsequent pool fire that burned for several hours.

The CSB's findings from this incident included the following facility siting concerns:

- Lack of hazard analysis to determine the risks associated to the proximity of occupied buildings to processes: The most critical injured employees were at the Laboratory/operations building at the time of the incident. The nearby building was occupied with non-essential personnel during the day shift.
- Inappropriate area classification: Two boilers were in service in the Laboratory/Operation Building, where the ignition source was identified. The flammable vapor cloud was ingested and ignited by one of the boilers.

Although the facility was covered by the OSHA PSM standard, the CSB found that the TSDFs did not have industry-specific guidance for fire safety and facility siting.

5 Fires and Explosions in Storage Operations at TSDFs

Several fire incidents have occurred during storage of hazardous materials at TSDFs. The ultimate fate of hazardous materials is often destruction through incineration. The type of incinerator and local environmental emission limits dictate handling activities with the hazardous waste. To manage the limits on the heat of combustion for the incinerator, heavy metal emissions, and other factors, TSDFs commonly repackage wastes by either combining similar waste streams in the same container or by breaking up larger containers into smaller container sizes. For example, if the bulk heat of combustion of the 55-gallon drum of waste received by a TSDF is too high for the incinerator's operating limit, then the 55-gallon drum must either be metered (if a liquid) into the incinerator, or the drum must be subdivided into smaller containers (e.g., 30-gallon drums).

Thus, hazardous materials often pass through the following stages at a TSDF:

1. Receipt and characterization
2. Temporary storage
3. Repackaging – either bulk up or break down
4. Storage awaiting disposal
5. Final disposal

The waste generator must provide specific information on the hazardous waste that they are providing to the TSDF. The TSDF oftentimes works with the generator to help characterize the waste stream as part of the initial contracting process. The TSDF typically samples and performs basic characterization of individual wastes as they are received. However, this process is imperfect because the TSDF cannot be guaranteed that a set of samples from distinct locations in a waste container represent the bulk of the waste in that container. The potential deviation of actual properties from the results of this sample may grow significantly for larger-sized containers (e.g., a liquid tote or a 1 m^3 box), wastes that are heterogeneous, wastes with layers or pockets of different composition, or large volume wastes that are not completely sampled (e.g., sampling a subset of containers).

After initial contracting and waste specification, a TSDF will develop a plan for handling that waste stream, which will include sampling, characterization, and verification that as-received waste meets the contract specification. This process falls under Process Knowledge Management in the 20 elements of CCPS PSM. Deviations of as-received waste from the specification should be addressed under a Management of Change element, but the TSDF may not have a robust change management policy. Additionally, changes from the specification may not be identified during the characterization process as discussed above.

Incident investigation after a hazardous materials incident at a TSDF can be very challenging. A TSDF storage area may contain hundreds or even thousands of individual waste containers (see Figure 9). A fire or explosion in the storage area will destroy or damage the waste containers, displace waste containers, and mix the residue from involved containers. All these aspects make determination of the cause of the incident difficult. The likelihood of determination is usually directly related to whether the incipient stages of the fire or explosion were witnessed. The following three case studies provide incident summary and investigation results for a progressively larger TSDF fire incidents in order to characterize the difficulty in cause determination and lessons learned.

Figure 9. Example of TSDF storage warehouse.

5.1 Delayed Ignition - No Repackaging – Cause Undetermined

An investigation was performed on a fire that occurred in a small TSDF's storage bay. The facility handled a wide variety of wastes including flammables materials and acids. The facility included storage bays, laboratories, treatment equipment and office areas. In the storage bay where the fire occurred, several different types of containers were present including:

- Plastic buckets
- Plastic, steel, and paper fiber drums
- Steel encaged plastic totes known as intermediate bulk containers (IBCs)

The 55-gallon drums were placed on wooden pallets in sets of four. The pallets of 55-gallon drums and IBCs were stacked two or three levels high. A photograph of the bay after the fire is shown in Figure 10.

Figure 10. Photograph of storage area after fire.

Based on witness statements, the fire in the storage bay was preceded by a loud hissing sound and a pressurized release of a reddish dust/mist from a container near the center row of the image. The witness did not see the exact container where the release originated, but was able to place it within a few pallet stacks in the area. Immediately after the pressurized release, a fire ensued in the area. Due to fire suppression efforts, the fire damage was contained to the storage bay where it originated. The investigation focused on the adjacent rows containing the waste drums that were damaged. The exact location of each waste container in the neighboring pallet stacks was reconstructed based upon storage records, remaining containers, damage patterns, and witness statements about container placement. The origin was confirmed to be one pallet stack in the center row.

The containers in the center row had not been repackaged or disturbed from the time of receipt until the incident; therefore, commingling of incompatible wastes or mixing of a reactive pocket within the source drum were refuted as potential causes. Delayed decomposition of waste material in one of the 55-gallon drums is the likely source of the pressurization. Delayed decomposition may be the result of chemical incompatibility or thermal instability leading to a temperature rise, gas generating reaction, or both. Eventually, the pressure increased to the level that the source container ruptured. It was most likely that this release occurred from either a steel drum or a plastic drum given their ability to contain some degree of pressurization compared to a fiber drum. After the initial pressurized release and fire, several other containers were damaged and ruptured as a result of the growing fire.

Based on the photo-documentation of the scene, it was possible to identify several potential drums that showed evidence of pressurization and/or rupture. However, due to the extent of the damage, it was not possible to identify the serial number on all of the containers, and therefore, their origin and contents. Examination of the fire damage showed evidence of burned away paint, burned pallets, and melting of non-metal containers. Based on the damage patterns, it was possible to rule out some waste containers including a set of drums containing phosphoric acid. A ground-level area containing four drums was identified as the likely area of origin, but further specificity was not possible due to the extent of the damage and several other drums that could not be conclusively eliminated. In a regular facility, the containers of all the stored containers would be known, providing valuable information to incident investigators. In a waste handling facility, significantly less certainty is available about the exact contents of each container. The ability for many of the waste container to have been the source of the pressure release significantly complicated this specific investigation. The final determination as to which container was the source could not be determined given the degree of damage.

The extent of damage does not always preclude determination of the likely cause of a fire in storage. A similar TSDF facility experienced a fire in a storage bay, which consumed all of the hazardous waste and combustible containers in the area (see Figure 11). The point of origin of the initial fire could be more precisely placed than the incident just discussed because a nearby operator observed the location of the incipient fire at ground level at a specific pallet stack. The source of the fire was most likely a leak of fuming red nitric acid that spilled onto a wood pallet holding the drums. The drums were stacked on combustible wood pallets, and fuming nitric acid is a strong acid capable of igniting such materials. The facility did not identify the potential consequences of a spill of the acid; thus, a noncombustible pallet (i.e., steel) was not selected. Use of a what-if strategy for packaging and storage selection may have prevented the fire (but not necessarily the leak).

Figure 11. Another TSDF storage bay after extensive fire damage.

5.2 EQ Industrial Services Warehouse

A hazardous waste TSDF was complete destroyed due to a fire for which the initial cause was undetermined (see Figure 12). The magnitude of this incident forced thousands of local residents to be evacuated in two days. The CSB investigated the cause of the incident, and identified several hypothetical causes. Their recommendations focused on eliminating those hypothetical causes.

Figure 12. Facility after the incident [24].

Based on witness statements, a strong chlorine odor and plume of smoke came from the TSDF on the night of the incident. Initial firefighters responding to the scene reported that a small fire started in one of the hazardous waste storage bays and rapidly spread to the other locations of the facility, including a bay containing many 55-gallon drums of flammable solvents. The incident investigation revealed that the fire originated in a storage bay where containers of solid chlorine-based pool chemicals and unspent aircraft oxygen generators awaiting for final disposal.

Many incidents have occurred involving storage of chlorine-based pool chemicals (see for example [25]-[28]), and around the time of the incident, the NFPA was pursuing research into fire protection considerations for pool chemicals in warehouse settings [29]. Concentrated calcium hypochlorite or sodium dichloroisocyanurate containing pool chemicals pose potential thermal runaway concerns if contaminated with organic material or if stored at elevated temperature. Given their nature as solid oxidizers, pool chemicals would also serve to accelerate fire growth. The witness observations of a chlorine-odor associated with the smoke likely indicate that these chemicals were present near the fire origin, but the investigators could not conclude that these were cause of the fire.

Although the first material involved could not be determined due to the extent of damage, the CSB investigation findings focused on chemical oxygen generators [30]. The CSB concluded that chemical oxygen generators most likely contributed to the intensification of the fire, which quickly spread to an adjacent bay where flammable materials were stored. The investigation found that the unspent oxygen generators, coming from an aircraft maintenance facility, were not properly packaged by the aircraft maintenance personnel, and were misidentified as general oxidizer waste on shipping paperwork. A lesson learned from this incident was to consider oxygen generators that have exceed their service life as still posing a risk of heat and oxygen generation if accidentally actuated. Therefore, the CSB recommended that unspent oxygen generators should be actuated before they are transported [30].

5.3 Teris El Dorado Warehouse

This incident occurred in a warehouse at a TSDF that contained 4000 to 5000 drums of different wastes. The facility operated two rotary kiln incinerators for solid disposal and thermal oxidation for liquid incineration [31]. The incident was witnessed by an operator who saw a drum rupture, which quickly spread fire throughout the warehouse contents (see Figure 13). The entire contents of the warehouse were consumed, damaged, and/or released during the incident. No injuries or fatalities were reported; however, nearby residents were evacuated during the emergency response.

The investigation found that the fire was caused by spontaneous combustion of sodium chlorate solid waste recovered from a rail car accident site. This material was brought to the facility in $1 m^3$ boxes to be incinerated shortly before the incident. However, the waste generator incorrectly profiled this material as containing only sodium chlorate in soil; the waste actually contained vegetative matter, rail car components, and portions of railroad ties. The organic contaminants are incompatible with the strong oxidizer and would act as fuel sources, allowing spontaneous ignition of the mixture. The material was repackaged into smaller drums and a portion was moved into the warehouse prior to the fire. The remainder of the material had not been relocated at the time of the fire. This residual material was isolated in a trailer onsite. The trailer containing the residual material underwent spontaneous ignition several days after the original fire incident, which supported the conclusion that the sodium chlorate waste material was the cause of the fire.

Figure 13. Aerial image during fire event [31].

Given the non-homogeneous nature of the waste material received, grab sampling of a single container was insufficient to fully characterize the mixed material. In contrast to liquid sampling, a thorough visual examination of the mixed, non-homogeneous waste material may have identified the presence of organic materials in the oxidizer and soil waste. Further, the use of "soil" in the waste analysis plan submitted by the generator may have triggered a further review by the TSDF since soil typically contains a significant organic component. It is likely that the repackaging activity mixed reactive sodium chlorate with fresh organic fuel materials that led to the delayed spontaneous combustion event. After the incident, the facility enhanced their material analysis and handling plans for wastes containing solid oxidizers.

6 Fires & Explosions during Incineration Processes

6.1 Explosion during Solid Waste Homogenization

An explosion occurred at an incineration facility that processed hazardous and non-hazardous solid and liquid wastes through thermal destruction in a rotary kiln. Drums of waste were homogenized in a nitrogen-inerted shredder. Drums were scheduled to be incinerated and assembled onto pallets based upon the waste characteristics. The system automatically moved individual pallets into an airlock, purged the airlock with nitrogen, then dumped the pallets into the already inerted shredder chamber. The shredder was used to shred steel drums as well as plastic drums. After shredding, the homogenized waste was conveyed to the rotary kiln via a screw conveyor.

The pallets, packaging materials, and waste materials are combustible materials, and flammable materials may also be processed based on the incinerator waste schedule. Due to the potentially flammable nature of the waste and the likelihood of mechanical sparks and heat generation from the shredder, the system was inerted with nitrogen gas. The nitrogen inerting function was performed by using a fixed timing of nitrogen addition. Although the system did not use an oxygen measurement control loop to dictate the duration of purging, it did use oxygen measurement as a permissive control for the automated conveyor movements and door opening between chambers in the system.

A central incinerator control room at the facility monitored several operating parameters, including the oxygen concentration and temperature within the inerted system. Since one of the primary safety mechanisms for the unit was a reduced oxygen atmosphere, the unit was not intended for processing oxidizing materials including solid oxidizers, oxygen containers/cylinders, explosives, and/or water reactive materials. During one waste load, an explosion occurred in the shredder chamber, which damaged the unit and caused explosion panels to open on the shredder chamber and airlock.

An approximation of the oxygen concentration data and shredder chamber temperature is provided in Figure 14. There was a lag noted in the oxygen sensor response due to the delay for gas to traverse the sampling leg to the sensor. The temperature measurement had less of a response lag; thus, the sudden temperature rise in the shredder chamber actually occurred coincidentally with the explosion and the opening of the explosion vents. On the night of the incident, an additional waste load was fed into the unit, and shredding commenced less than one minute after it was added. Approximately one minute after shredding began, the oxygen concentration started to increase from 2% to over 13% over a period of 20 seconds. According to the recorded data, the temperature in the shredder chamber suddenly increased at approximately the same time as the oxygen concentration increased. This temperature increase is consistent with the heat generated by the explosion within the shredder chamber. An engineering analysis concluded that the most likely source of the rapid oxygen influx was from an item in the waste stream that released oxygen when shredded. All potential sources of air via external sources were examined and rejected.

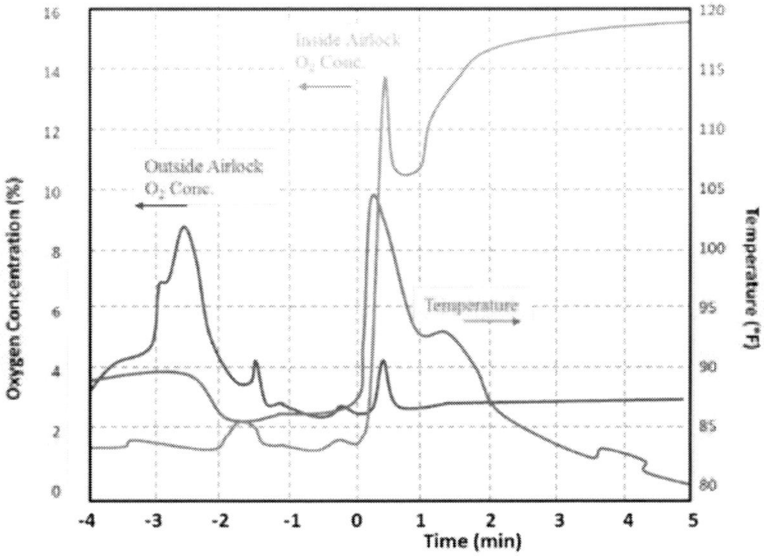

Figure 14. Process oxygen concentration and temperature during incident.

The explosion in the shredding system demonstrated the serious consequences of processing waste with unintended properties. In this case, a system that was intended to operate while inerted inadvertently contained compressed oxygen, resulting in a damaging explosion. While the waste was screened prior being fed into the shredder, the screening process did not identify and divert the oxygen-containing waste. In this case, the explosion mitigation system worked as intended by relieving the explosion pressure through explosion vents. The system was damaged, but the damage was not catastrophic, and there were no injuries.

6.2 Spontaneous Combustion at a Calciner

A fire occurred at a cement production facility that used an engineered solid fuel composited from MSW and other industrial wastes to co-fire in the calciner. The engineered fuel was a blended mixture of solid wastes including paper, plastic, and sawdust as well as non-hazardous waste liquids, slurries, and sludge. Because the waste streams possessed variable energy content, a qualifications process (including laboratory analysis) was used to determine the appropriate mixture for creating a consistent engineered feed. A block flow diagram of the shredding, blending, and mixing process is shown in Figure 15. The engineered fuel had the potential for self-heating due to containing oxidizing materials such as polymerizing agents and vegetable-derived oils. Material handlers inspected the stockpiles for external signs of high temperature due to self-heating and excavated the piles to identify internal high temperatures. Infrared pyrometers were used for surface temperature measurements. The temperatures were monitored with the handheld pyrometers and logged prior to the loading the material into trucks for transport to the cement production facility.

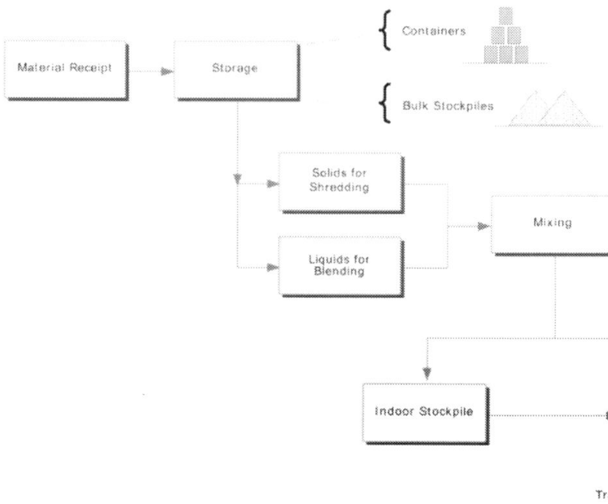

Figure 15. Block flow diagram of the engineered solids feed preparation process.

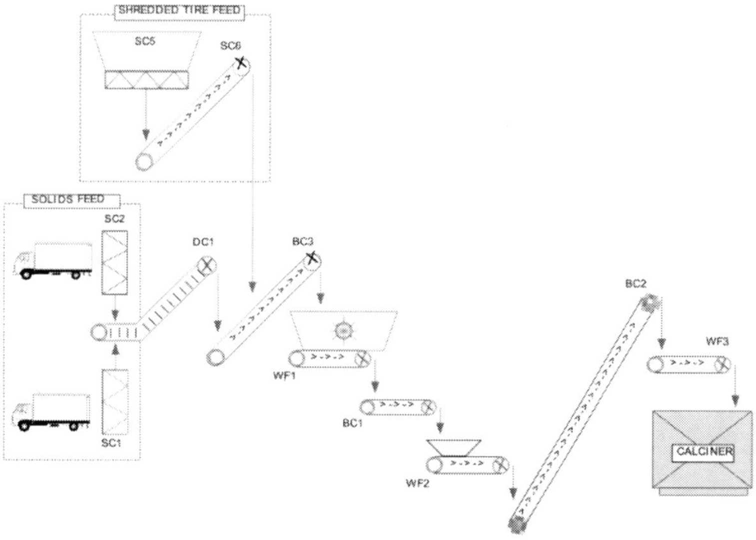

Figure 16. Conveyor flow diagram between truck delivery and the calciner.

The engineered fuel was transported to the cement production facility in tractor trailers for eventual feed into the calciner unit. A conveyor flow diagram is provided in Figure 16. The engineered fuel was unloaded into enclosed hoppers that discharged into horizontal screw conveyors. The horizontal screw conveyors discharged into a central drag conveyor that delivers the fuel to a belt conveyor which also received a separate shredded tire feed. The combined feed then traveled through an additional series of hoppers and conveyors before being deposited on a 1,600 feet long enclosed belt conveyor that climbed from ground level to the upper levels of the calciner structure. The belt of the long conveyor was originally a flame-resistant belt material but had been subsequently patched with non-flame resistant material. The total travel time of the engineered fuel was estimated to be 20-30 minutes by the cement facility's personnel.

A fire was discovered overnight soon after the conveyor system was restarted after a two-day outage. During that two-day period, the belts were left loaded with the engineered fuel and combined feed. When the kiln was restarted, the system was operated for approximately one hour then shut down due to an upstream clog during which a fire was discovered near the head section of the long conveyor. The belt failed and broke as a result of the fire.

An investigation into the fire cause determined that, while various friction-based ignition sources (belts, bearing, etc.) did exist within the conveyor system, the most likely cause of the fire was spontaneous ignition of the engineered feed. This was substantiated because three days after the fire, a stockpile of the engineered feed that had been fed to conveyor prior to the fire enflamed at the storage warehouse. Although staff had monitored stockpiles at the feed mixing facility for external signs of ignition, external methods can be insufficient to detect self-heating inside the pile. After exposing the self-heating pile to oxygen during transport and handling, self-heating may accelerate and lead to spontaneous ignition. The facility implemented a strategy that avoided leaving materials on conveyors for long periods of time. Through the use of conveyor interlock strategies and external dump locations, the system can be operated to empty conveyors in the event of an unintended shutdown.

6.3 TDI Waste Rail Tank Car

An explosion occurred at another TSDF cement plant in the liquid fuels rail tank car unloading area [6], [32]. As consequence, the tank ruptured and was propelled about 750 feet away. Other tank cars, as well as storage tanks located nearby, were also heavily damaged (Figure 17). No injuries or fatalities were reported; however, total property losses were estimated at nearly $8.2 million.

This facility incinerated liquid hazardous waste streams as supplementary fuel in rotary kilns, and used to receive several tank cars containing column bottoms waste from a toluene diisocyanate (TDI) manufacturing process. TDI is a monomer feedstock used in the manufacture of polyurethane polymers. The waste generator had stored tank cars containing this type of waste material for several years, and reprocessed them as the opportunity arose. Prior to the incident, several tank cars were shipped to the TSDF for incineration, and had been successfully emptied, with contents incinerated in the kilns. Despite the TDI waste was partially polymerized in most tank cars, the car's contents were non-homogeneous and did not have identical properties between individual tank cars.

Figure 17.Locations (1) and (2) indicate positions of the tank car's vessel before and after the explosion, respectively [33].

In order to decrease the viscosity of the polymerized TDI, so that it was pumpable to the kilns, the facility had successfully used external steam heating through the built-in steam jackets on the tank cars.. A schematic of the approximately 30,000 gallon tank car is provided in Figure 18. While attempting to unload the incident TDI material, the tank car vessel was steam-heated in two opportunities to ensure the waste mixture was sufficiently fluid to be pumped. After three unsuccessful unloading attempts, the shipment was rejected, but the tank car was left in the unloading location. The explosion incident took place two days after the last heating attempt.

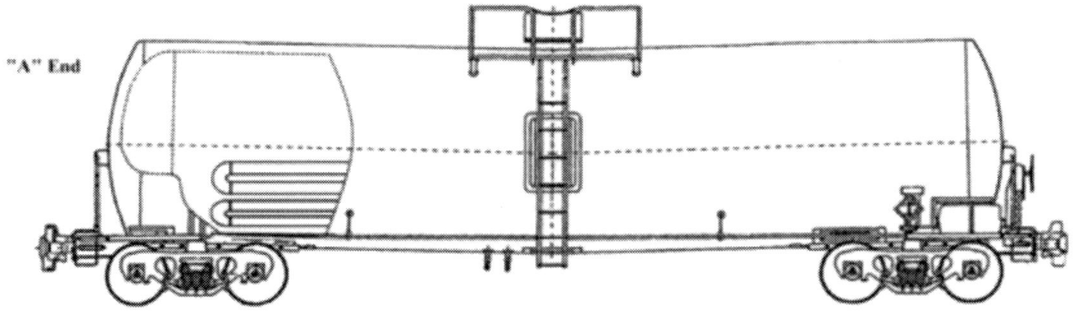

Figure 18.TDI tank car schematic with cutaway showing the external steam heating coils. [20]

Retained samples of the original material and residue ejected during the incident were analyzed, and the presence of self-polymerized TDI, TDI monomer, and ureas were determined. Thermal analysis of the materials indicated that the polymerized TDI and TDI monomer exhibited an onset temperature for thermal runaway that was much higher than possible with the steam heating system. A chemical hazard assessment revealed that the cause of the explosion was the runaway thermal decomposition of a contaminant mixed with the TDI waste. The TSDF operator relied on the waste generator for information regarding the waste composition and safe heating procedures. In this case, the waste generator supplied incomplete waste characterization data and inadequate instructions and warnings for heating and handling the TDI waste.

Four critical factors and lessons learned were identified from this incident [6] [20]:

(1) Use of inappropriate tank car design that was unsuitable for this service, creating difficulties for unloading of TDI solid waste.

(2) Storing of TDI waste for several years in a warm climate

(3) Providing inadequate instructions and warnings for unloading the TDI waste.

(4) Commingling an unknown contaminant with the TDI waste; this last factor was the most serious error in this incident.

7 Discussion and Conclusions

In the case studies presented above, hazardous properties that were not known to exist led to accidents. In several cases, the accidents were the result of a chemical reaction, leading to a release of heat or pressure that caused a storage vessel to fail. Chemical reactions due to the mixing of incompatible chemicals, storage of unstable chemicals, or heating of thermally unstable chemicals occurred in several of the case studies described. While characterization is a significant preventative safeguard for these types of failures, it may not always be possible to determine if a waste stream is completely stable based on testing; therefore, facilities may still need to consider the unlikely event of an exothermic or pressure-generating reaction within their waste streams. This may be accomplished by using separation distance or barriers between certain waste streams, fire protection systems, or fast turnaround from receipt through final processing.

In several incidents, oxidizers including oxygen tank, oxygen generators/tanks and sodium chlorate, are involved in the initiation of the incident or contribute to the severity of the incident. Because many waste streams contain flammable or combustible materials, enriched oxygen atmospheres or the presence of chemical oxidizers increases the likelihood and severity of a fire or explosion. While in many process operations, inerting atmospheres may be an effective and robust protective measure, due to the possibility of oxidizer ingress in mis-identified waste, additional protections such as explosion vents may be required. The following flowchart was created based on the cases presented in this paper, and may serve as a guideline for investigating suspected waste streams and their role on initiating a fire or explosion event.

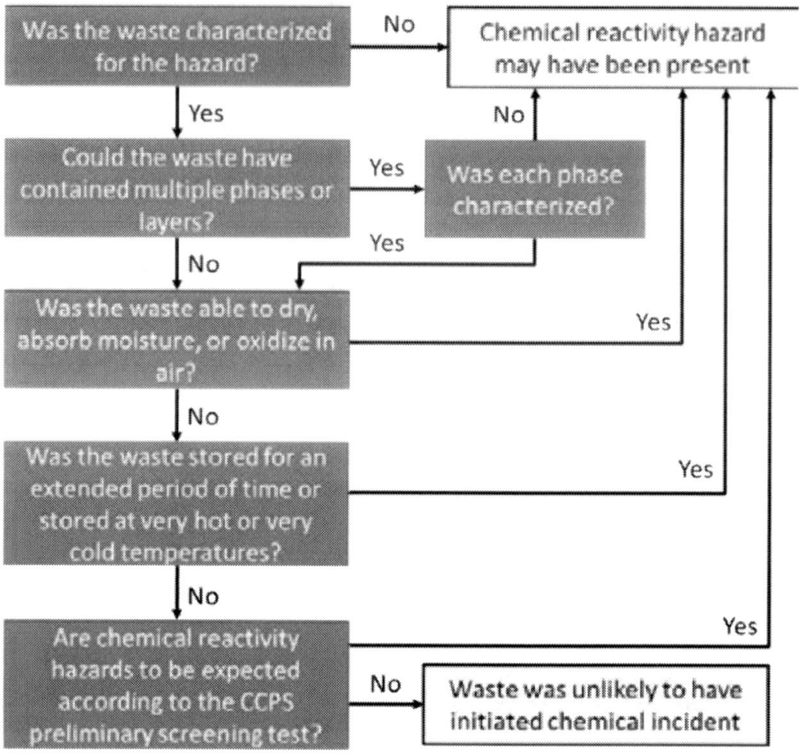

Figure 19. Role of suspected waste stream on initiating a fire or explosion.

8 References

[1] U.S. Environmental Protection Agency (EPA) Resource Conservation and Recovery Act, Title 40 of the Code of Federal Regulations.

[2] 29 CFR §§1910.119

[3] Center for Chemical Process Safety (CCPS). Guidelines for Risk Based Process Safety. Center for Chemical Process Safety, American Institute of Chemical Engineers, New York, NY, 2007.

[4] U.S. Environmental Protection Agency: Resource Conservation and Recovery Act (RCRA) Laws and Regulations, available at https://www.epa.gov/rcra. Accessed on February 15, 2017.

[5] Evarts, B. "Fires Caused by Spontaneous Combustion or Chemical Reactions," National Fire Protection Association, November 2011.

[6] Cox BL, Carpenter AR, Ogle RA. Lessons learned from case studies of hazardous waste/chemical reactivity incidents. Process Safety Progress 2014; 33(4):395–398.

[7] E.C. Buc. "Hazardous Waste Treatment, Storage, and Disposal Facility Fire Code Gap Analysis, Final Report," The Fire Protection Research Foundation, 2015.

[8] Environmental Protection Agency. Report on Emergency Incidents at Hazardous Waste Combustion Facilities and Other Treatment, Storage, and Disposal Facilities (TSDFs). EPA 1999

[9] U.S. Environmental Protection Agency (EPA): 2015 Biennial Report Summary, available at: https://rcrainfo.epa.gov/rcrainfoweb/action/modules/br/national/view. Accessed January 15, 2017.

[10] U.S. Environmental Protection Agency: Advancing Sustainable Materials Management: 2014 Fact Sheet. US EPA, November 2016.

[11] U.S. Environmental Protection Agency, Available at https://www.epa.gov/hw. Accessed on February 15, 2017.

[12] 40 CFR §§261

[13] 40 CFR §§264/265.17

[14] U.S. Environmental Protection Agency. Waste Analysis at Facilities that Generate, Treat, Store, and Dispose of Hazardous Wastes – Final: Guidance Manual (EPA 530-R-12-001). April 2015.

[15] Hatayama H.K., J. J. Chen, E. R. de Vera, R. D. Stephens, and D. L. Storm. A Method for Determining the Compatibility of Hazardous Wastes. U.S. Environmental Protection Agency, Washington, D.C., EPA/600/2-80/076 (NTIS PB80221005), 1980.

[16] National Oceanic and Atmospheric Administration (NOAA), "Chemical Reactive Worksheet," NOAA, available at http://www.aiche.org/ccps/resources/chemical-reactivity-worksheet-40. Accessed January 15, 2017.

[17] Davis E.M., D. Silva, and J. Murphy. The CCPS Chemical Reactivity Evaluation Tool. 8th Global Congress on Process Safety. April 2012.

[18] Center for Chemical Process Safety (CCPS). Essential Practices for Managing Chemical Reactivity Hazards. Center for Chemical Process Safety, American Institute of Chemical Engineers, New York, NY, 2013.

[19] National Fire Protection Agency. NFPA 921: Guide for Fire and Explosion Investigations

[20] Center for Chemical Process Safety (CCPS). Guidelines for Investigating Chemical Process Incidents. 2nd Ed. 2003

[21] Anthony, EJ, L Jia, M Caris, F Preto, and S Burwell, "An examination of the exothermic nature of fluidized bed combustion (FBC) residues," Waste Management Journal, 19 (1999).

[22] Serris, E, L Favergeon, M Pijolat, M Soustelle, P. Nortier, RS Gartner, T Chopin. "Study of the hydration of CaO powder by gas–solid reaction," Cement and Concrete Research 41(2011) 1078-1084.

[23] U.S. Chemical Safety Board (CSB). Veolia ES Technical Solutions, L.L.C. Case Study. Report 2009-10-I-OH. Washington, D.C. 2010

[24] U.S. Chemical Safety Board (CSB). EQ North Carolina Case Study. Report 2007-01-I-NC. Washington, D. C. 2008.

[25] Gray B.F. and B.W. Halliburton. 2000. "The thermal decomposition of hydrated calcium hypochlorite (UN 2880)." *Fire Safety Journal*. 35.3: 223-239.

[26] Gray B.F. 2002. Spontaneous Combustion and Self-Heating. In SFPE Handbook of Fire Protection Engineering, eds. P.J. DiNenno, D. Drysdale, C.L. Beyler, W.D. Walton, R.L.P. Custer, J.R. Hall and J.M. Watts, Quincy, Massachusetts: National Fire Protection Association, 211-228.

[27] Ogle R. A, Carpenter A. R. Toxic gas release caused by the thermal decomposition of a bulk powder blend containing sodium dichloroisocyanurate. Process Safety Progress 2003; 22:75–82.

[28] Ogle RA, Morrison DR, Carpenter AR, Su YS. Missed opportunities in reactive chemical hazard evaluations. Process Safety Progress, March, 2006, 25(1):2-7.

[29] Buc E.C. and D.J. Hoffman, "Development of an Enhanced Hazard Classification System for Oxidizers Research Project," The Fire Protection Research Foundation, One Batterymarch Park, Quincy, Massachusetts, April, 2006

[30] U.S. Chemical Safety Board (CSB). Safety Advisory Dangers of Unspent Aircraft Oxygen Generators (2007-I-NC-01-SA). 2007

[31] U.S. EPA, "Teris El Dorado Explosion," downloaded on March 13, 2017, https://response.epa.gov/site/site_profile.aspx?site_id=NRC745957.

[32] Ogle R.A, Carpenter AR, Morrison DR. Explosion of a railcar containing toluene diisocyanate waste. Process Safety Progress 2004 Dec; 23(4):316-320.

[33] National Transportation Safety Board. Rupture of a Railroad Tank Car Containing Hazardous Waste Near Clymers, Indiana. Hazardous Materials Accident Report NTSB/HZM-01/01. Washington, DC. 2001

479039 ExxonMobil Torrance Refinery FCC ESP Explosion

Abstract:

On February 18, 2015, an explosion occurred in the electrostatic precipitator (ESP) at the ExxonMobil Refinery in Torrance, California near Los Angeles. The explosion injured four workers caused significant property damage to multiple process units within the refinery, and resulted in an off-site accidental release of catalyst dust. As a result of the incident, the FCC unit was down for over one year. The explosion occurred when hydrocarbons from the FCC unit reached the energized ESP while the unit was idling to perform maintenance on equipment that had failed. Multiple failures contributed to the incident. During the ESP explosion, there was also a near miss release of hydrofluoric acid when a large piece of ESP debris fell within feet of a storage vessel storing thousands of gallons of modified hydrofluoric acid in the neighboring alkylation unit. If the storage vessel had failed due to impact from ESP debris following the explosion, hydrofluoric acid would have been released. This presentation will detail key findings from the CSB investigation into the incident.

Overfill to Overpressure – Investigation Findings and Contributing Events to the CAPECO Explosion

Scott Davis
GexCon US
Bethesda, MD
sgdavis@gexcon.com

Tom DeBold
Claudio Marsegan
Bryant Hendrickson
GexCon US
Bethesda, MD

Prepared for Presentation at
American Institute of Chemical Engineers
2017 Spring Meeting
13th Global Congress on Process Safety
San Antonio, Texas
March 26-29, 2017

Overfill to Overpressure – Investigation Findings and Contributing Events to the CAPECO Explosion

Scott Davis
GexCon US
Bethesda, MD
sgdavis@gexcon.com

Tom DeBold
Claudio Marsegan
Bryant Hendrickson
GexCon US
Bethesda, MD

Keywords: CAPECO, petroleum, overfill, bulk storage, explosion

Abstract

On October 23, 2009, a vapor cloud explosion occurred after an above ground gasoline storage tank was overfilled at the Caribbean Petroleum Corporation (CAPECO) tank farm facility in Bayamón, Puerto Rico. The blast registered a 2.9 on the Richter scale and sparked several fires throughout the facility that burned for another three days and damaged a total of 21 tanks before emergency responders were able to extinguish the flames.

Operations on the day of the incident included a ship transfer of unleaded gasoline to the CAPECO facility and the transfer was scheduled for four smaller capacity tanks. At approximately 10:00 pm, all of the unleaded gasoline was being transferred to tank 409 and a visible vapor cloud was noted during the walk around at 12:00 am on October 23, 2009. It was determined that tank 409 was overfilled by approximately 5,000 barrels (210,000 gallons) of gasoline, which caused gasoline to flow out of the tank vents near the top of the tank and cascade down the side forming a vapor cloud and liquid pool. This is equivalent to overfilling the tank by over 20 gasoline tanker trucks, which if 10% of the liquid formed the vapor cloud it would cover an area larger than 30 American football fields of a rich flammable mixture. The gasoline vapors mixed with the air forming an ignitable vapor cloud spanning a vast portion of the facility, which subsequently ignited neighboring tanks that burned for three days.

US regulations do not consider CAPECO and other bulk petroleum storage terminals as high hazard facilities, yet a large vapor cloud formed. This paper will present investigation findings and key lessons learned into the massive explosion and fire. This includes: effects of confinement and congestion in a seemingly open space that contributed to the incident; ship offloading and tank filling procedures that revealed significant operational deficiencies; failures to secure the confined drainage system; and level controls.

1 Introduction

On October 23, 2009, a vapor cloud explosion occurred after an above ground gasoline storage tank was overfilled at the Caribbean Petroleum Corporation (CAPECO) tank farm facility in Bayamón, Puerto Rico. The blast registered a 2.9 on the Richter scale and sparked several fires throughout the facility that burned for another three days and damaged a total of 21 tanks before emergency responders were able to extinguish the flames. The explosion and ensuing fires did not cause any fatalities; only a few minor injuries were reported. However, approximately 300 homes and business within 1.25 miles of the site reported damages [1].

GexCon was retained to perform an independent and objective investigation into the cause and origin of the explosion. Due to the nature of the investigation, some information provided to GexCon was confidential and could not be provided. Therefore, information provided in the reports of other agencies were used in the present analysis, and include: the ATF Investigation 764045-10-0011 Report [1]; U.S. Chemical Safety and Hazard Investigation Board Final Investigation Report – Caribbean Petroleum Tank Terminal Explosion and Multiple Tank Fires [2]; and the Puerto Rico Fire Department Fire Marshal Report [3] . This paper will present investigation findings and key lessons learned into the massive explosion and fire that occurred at the CAPECO facility. This includes: effects of confinement and congestion in a seemingly open space that contributed to the incident; ship offloading and tank filling procedures that revealed significant operational deficiencies; failures to secure the confined drainage system; and level controls.

We will first present a brief description of the facility, events leading up to the incident, and details of the explosion. Next, we will evaluate the conditions and contributing factors that lead to the explosion.

CAPECO Facility

The Caribbean Petroleum Corporation (CAPECO) facility was a petroleum storage terminal located in Bayamón, Puerto Rico. At the time of the incident, the petroleum storage terminal employed approximately 65 people. Prior to the gasoline overfill and ensuing explosion, the CAPECO facility (Figure 1.1) consisted of a tank farm of above ground storage tanks, offices, decommissioned refinery, waste water treatment plant and distribution piping that serviced the San Juan Bay. The CAPECO facility stored gasoline, diesel, fuel oil, and jet fuel at the time of the incident. The tank locations along with what fuel was stored in each can be found in Figure 1.2.

The tank farm was split by roads into three areas; the northern, central, and southern tank farms. The tanks were separated by dikes or berms designed to hold 110% of the largest tank capacity in that berm (see Figure 1.2) [3]. The southern tank farm contained eleven tanks, which included diesel and fuel oil tanks, divided by six berms. Jet fuel, diesel, and fuel oil were contained in the central tank farm. The northern tank farm held gasoline, diesel, and fuel oil. CAPECO estimated that 613,816 barrels (25,780,272 gallons) of gasoline, 428,260 barrels (17,986,920 gallons) of diesel, 360,496 barrels (15,140,832 gallons) of fuel oil, and 132,724 barrels (5,574,408) of jet fuel were present at the time of the incident [4].

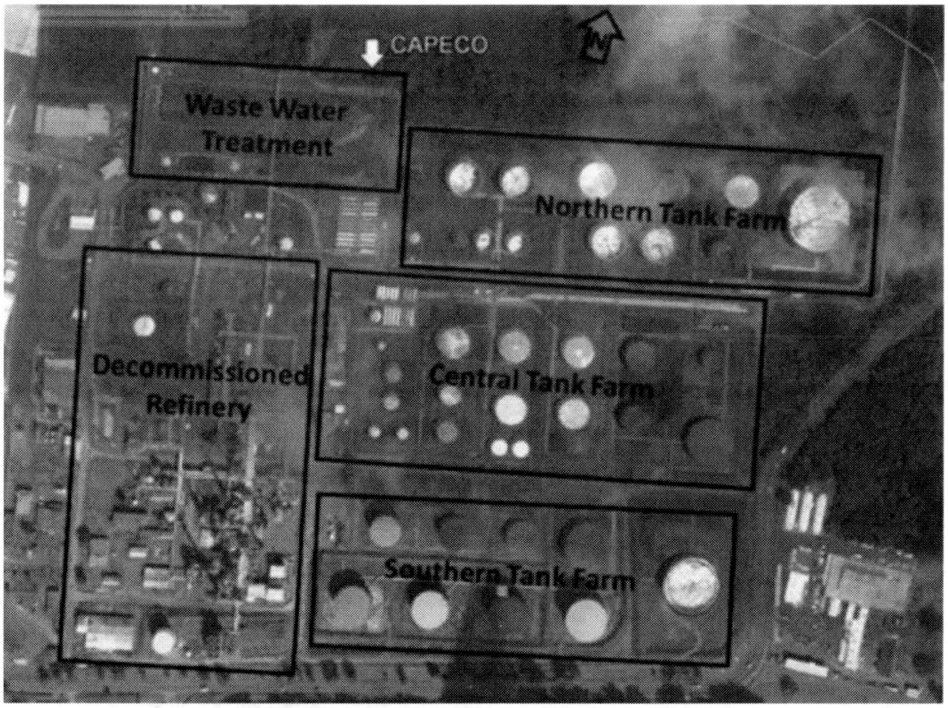

Figure 1.1: Facility Layout.

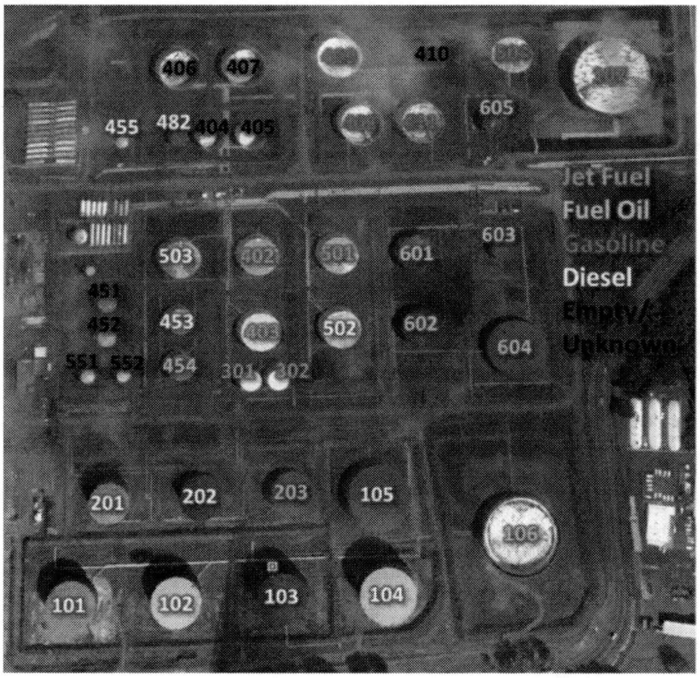

Figure 1.2: CAPECO tank location and inventory.

During fuel transfer operations, vessels would load/unload through the facility's 2 ½ mile pipeline which ended at a loading dock located in the San Juan Bay [2]. The terminal dock could accommodate one tanker of up to 75,000 tons. The transfers were made through six pipelines that connected to the facility transferring crude oil, FCC stock, fuel oil, gasoline, diesel, jet fuel and kerosene. The pipelines were equipped with remote operating valves that shutdown through: an electrical actuator; manually closing the valve; thermal shut-off switch set to 221°F (105°C); five push buttons in the operator's office, guards office, stairway to the main platform, fire pump building and Chiksan platform [5].

Events Prior to the Incident

Operations the day of the incident included: 1) A transfer of diesel fuel to the Puerto Rico Electric Power Authority Plant in Puerto Nuevo from tank 202; 2) A transfer of fuel oil to the Puerto Rico Electric Power Authority Plant in Palo Seco from tank 105; and 3) A transfer of gasoline from the Cape Bruny vessel to the CAPECO facility. Due to CAPECO's largest tank 107 (capacity 500,000 bbls – 21,000,000 gal) being used for a different quality gasoline, the transfer was scheduled for four smaller capacity tanks (see Figure 1.3) [6].

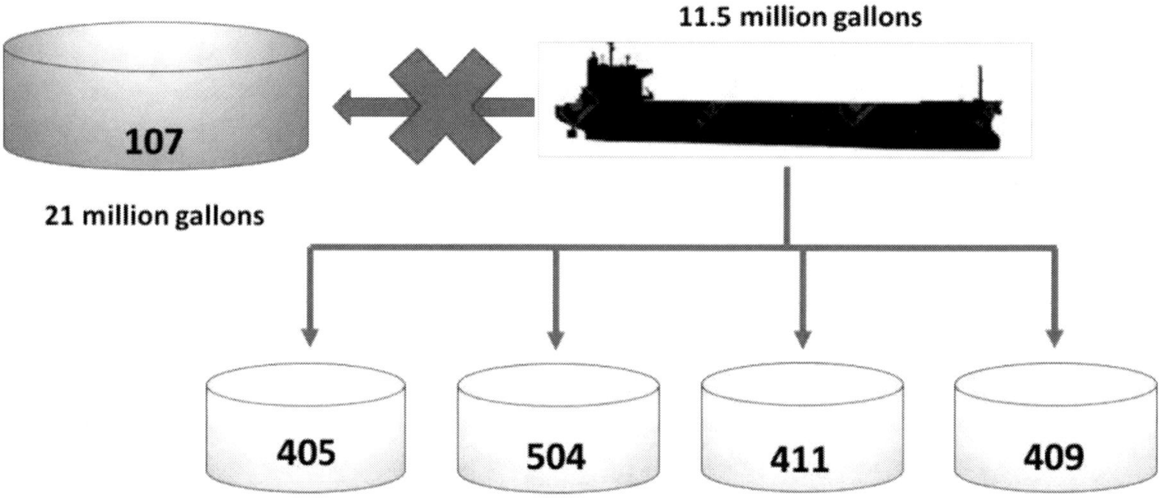

Figure 1.3: Cape Bruny transfer plan.

Pumping from Cape Bruny to the CAPECO facility began on October 21, 2009. Four tanks (Tank 405, 504. 409, and 411) in the northern tank farm received gasoline during the transfer. Initial pumping started at approximately 8:47 p.m. A sequence of operations can be found in Figure 1.4 [2].

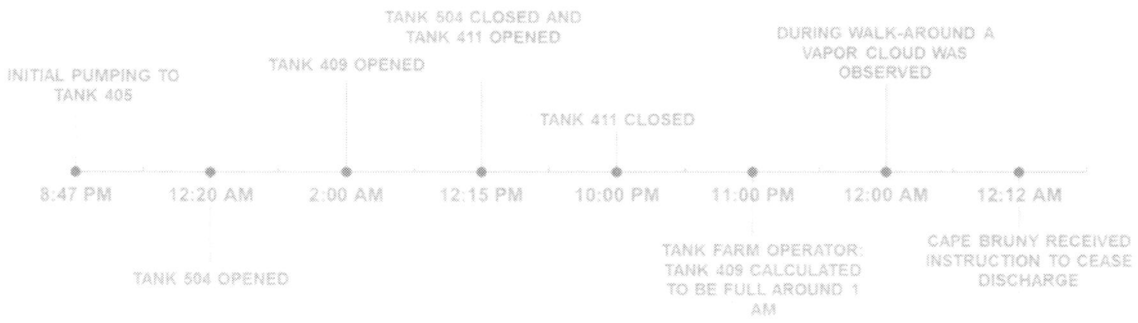

Figure 1.4: Transfer Operations [6].

Shortly after 11:00 p.m., the tank farm operator advised the supervisor in charge of the operations that night that the inspector found a difference between the side gauge, monitor console reading and the physical status of Tank 411 (already closed), and advised that the discrepancy was to be investigated [1].

ATF interviews determined that numerous operators and employees observed a vapor cloud starting at midnight. Tank 409 began to overfill between the 11:00 p.m. and 12:00 a.m. tank checks while the average pumping rate was approximately 11,134 bbls/hr (467,628 gal/hr) [7]. The overfilling of tank 409 caused gasoline to flow out of the tank vents near the top of the tank and cascade down the side forming a vapor cloud and liquid pool (as will be discussed in detail later). Tank 409 was contained in a berm with Tank 410 having a capacity of approximately 132,000 barrels (5,544,000 gallons) based on the 110% capacity requirement [3] . As the overfilling continued, gasoline (liquid and vapor) began to fill the containment area and expand towards the waste water treatment facility resulting in the white mist or vapor cloud that was seen in the facility by employees [6]. It was determined that approximately 5,082 barrels (213,444 gallons) of gasoline overflowed from tank 409 [7]. This is equivalent to overfilling the tank by over 20 gasoline tanker trucks, which if 10% of the liquid formed the vapor cloud it would cover an area larger than 30 American football fields of a rich flammable mixture.

The employees saw a white cloud extending approximately two to three feet above the ground by tank 404, and the white vapor cloud could be seen throughout the area between tanks 453 and 301, and near tanks 201 and 409 (Figure 1.5). The vapor cloud could only be seen in areas that were illuminated and extended approximately 200 yards wide by 400 yards long (183 m by 366 m) [1].

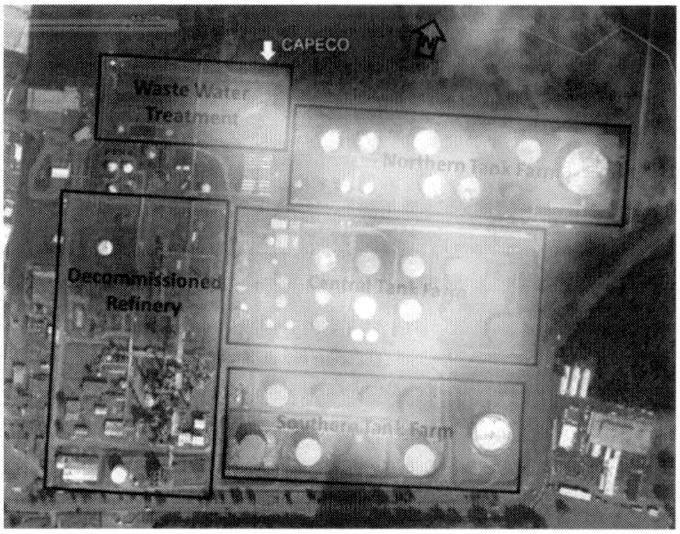

Figure 1.5: Approximate migration area of the vapor cloud.

Incident

At approximately 12:23 a.m. on October 23, 2009, the resulting vapor cloud from the overfill of gasoline ignited on the western side of the facility, away from the tank farm area. As the flame propagated (Figure 1.6 and Figure 1.8) towards the above ground storage tanks, a large explosion (Figure 1.8) occurred in the tank farm area that registered 2.9 on the Richter scale. Due to ignition of the large vapor cloud and ensuing explosion, many of the facility storage tanks were consumed by fire, which burned for another three days before emergency responders extinguished the flames [3].

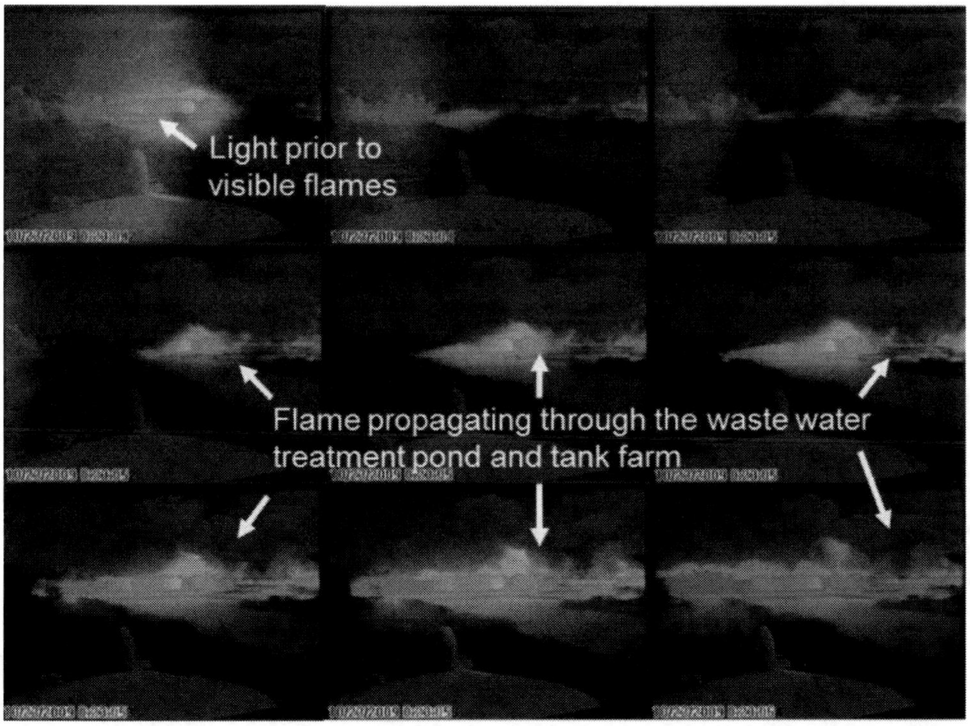

Figure 1.6: CAPECO security footage from camera 4. Visible light is seen prior to the flame propagation through the waste water treatment pond and tank farm.

Figure 1.7: Approximate propagation of the flame front prior to the explosion in the underground drainage as indicated by arrows. Camera locations and angles (red-camera 3 and blue-camera 4) at the first sign of ignition.

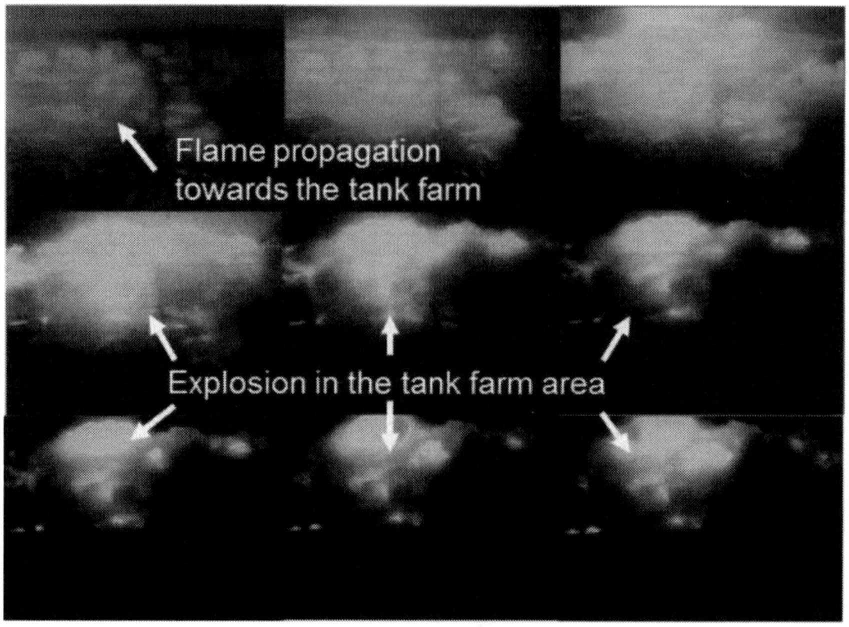

Figure 1.8: CAPECO security footage from camera 3 as the flash fire approached the tank farm from the west, leading to an explosion within the tank farm area (top-left to bottom-right).

Post Incident and Blast Damage

Emergency crews were alerted to the incident when a nearby citizen heard a large explosion and saw flames at the CAPECO facility [1]. At approximately 12:30 a.m. on October 23, 2009 the Bayamón and Cataño Fire Departments arrived at the CAPECO facility where they waited for the site fire chief to arrive approximately 45 minutes later. Eleven of the CAPECO tanks were damaged by the initial fire and explosion.

Firefighting efforts, as shown in Figure 1.9, lasted almost three days until approximately October 25, 2009 at 11:40 a.m. when the fires were extinguished [6]. Additional tanks were damaged by ensuing fires (see Figure 1.10 and Figure 1.11) [3] and when the fire was completely extinguished, the damage caused by the large gasoline vapor cloud explosion and fire included 21 tanks onsite. Both blast and thermal damage were observed on the tanks at the CAPECO facility (see Figure 1.12), consistent with the 11 tanks initially damaged by the large vapor cloud explosion and ensuing gasoline fire, and the others falling victim to the resultant fire [1].

Figure 1.9: Firefighting efforts after the incident [8].

Figure 1.10: Aerial view of the CAPECO facility after the initial gasoline vapor cloud explosion.

Figure 1.11: Secondary tank fires caused by the initial vapor cloud explosion and fire from the gasoline vapors.

Figure 1.12: Post incident damage to the tank farm.

2 Physics of vapor cloud explosions

Prior to the explosion at the CAPECO facility, tank 409 was overfilled and gasoline was being discharged from the overflow vents at the top of the tank. Recall that ignition and burning of liquid hydrocarbons occurs when the hydrocarbon is in the vapor state. A large vapor cloud developed resulting in the 'white cloud' described by CAPECO employees. As the vapor cloud grew from the continuous overflow, the gasoline vapors continued to mix with air forming a flammable mixture throughout the facility. This cloud found a competent ignition source in the western side of the facility and flames propagated towards the tank farm leading up to the main explosion event. This phenomenon is referred to as a vapor cloud explosion (VCE). The specifics on how the vapor cloud was formed and on the VCE will be discussed next.

Formation of vapor clouds

When evaluating the potential for large vapor cloud formation, which may lead to a massive explosion event, it is important to consider the mechanism by which a vapor and vapor clouds are formed. Recent experimental and modeling research has been performed to explore the generation of vapor clouds from overfilling gasoline storage tanks [9], [10]. The research determined that during a tank overfill, the gasoline would flow out of the ventilation holes at the top of the tank and over the deflector plate. The deflector plate channeled some of the escaped fuel onto the tank wall, but the rest ran over the top of the plate, fragmenting into droplets that cascaded through the air. Most of the fuel running down the wall hit a wind girder (a structural stiffening ring) and detached from the tank wall, creating a second cascade of droplets.

These conditions promoted the evaporation of the lighter components of gasoline (e.g., butanes, pentanes and hexanes). The free-fall of droplets leads to entrainment of air and mixing between the air and fuel vapor, and the formation of a rich fuel/air mixture. In addition, the wind conditions at the time of the overfill were calm, and had the wind speeds been higher, the evaporating gasoline vapor would be quickly diluted as it escaped the berm. Therefore, the overfill led to conditions that increased the break-up of gasoline into droplets, which significantly increased the evaporation rate of the lighter components of gasoline as compared to that from the pooled gasoline. Therefore, the overfill rate coupled with low wind speeds generated very large vapor clouds capable of filling a significant region of the CAPECO tank farm as seen in modeling shown in Figure 2.1.

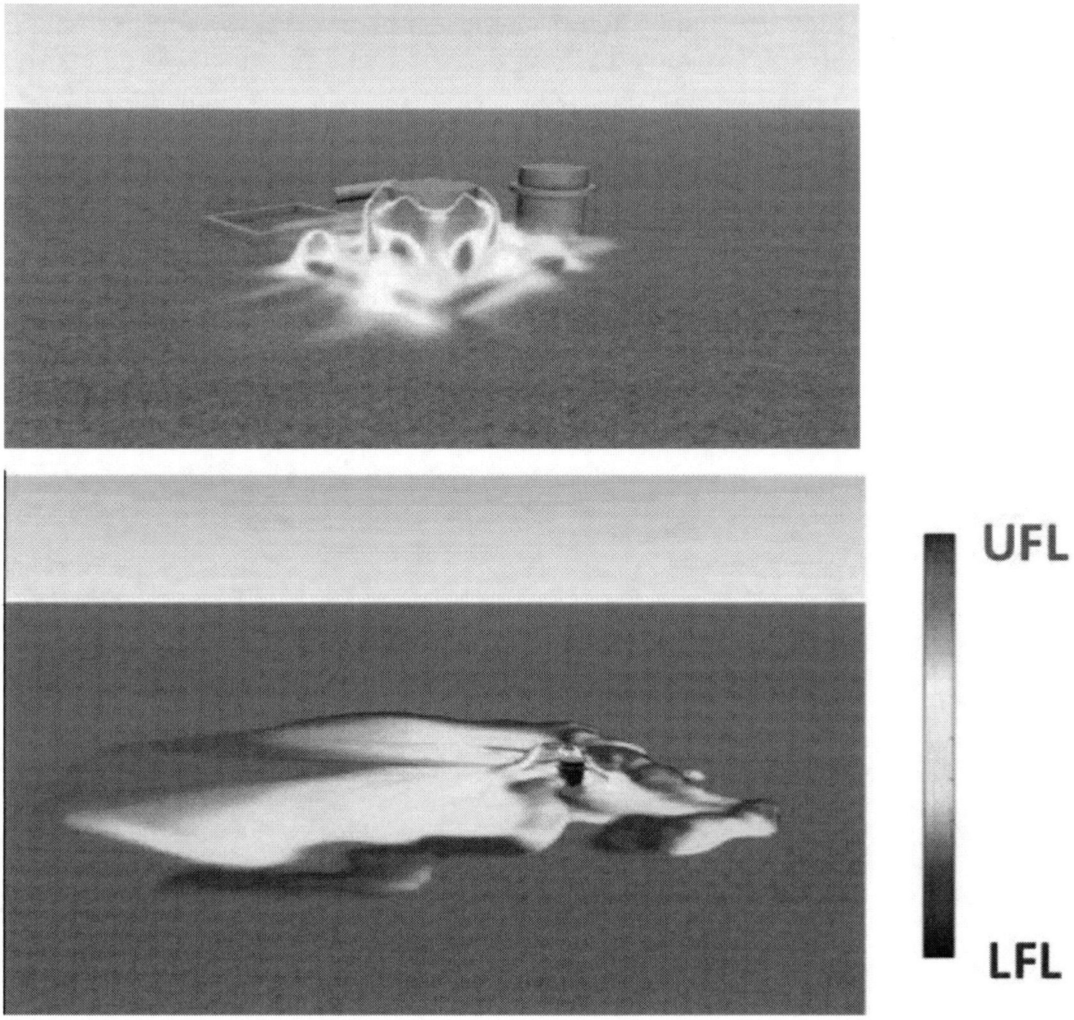

Figure 2.1: Vapor cloud formation and growth as gasoline initially cascades down the side of a storage tank and is dispersed.

Vapor cloud explosions

Vapor cloud explosions (VCEs) are oftentimes understood as "confined explosions", where a flammable mixture of gas (or vapor) and air is ignited within a confined structure or vessel. The heat released from the burning mixture causes the gases to expand and the pressure within the confinement increases above normal atmospheric pressure; this is referred to as "overpressure". This occurs until the envelope of the structure ultimately fails, similar to an ignition of a flammable mixture within an empty tank causing the roof to blow off. The sudden failure of the structure/vessel causes a shock or blast wave to form. However, conditions at CAPECO were mostly unconfined and open, except for the underground drainage system. An open and unconfined environment presents little to no resistance of an expanding flame front when a mixture is ignited; therefore, an overpressure is not expected if the gases are allowed to freely expand. The formation of a shock or blast wave in an open area cannot be explained in the same manner, and in fact, requires a unique set of circumstances.

Overpressures can indeed be created for unconfined vapor cloud explosions and require mechanisms that accelerate the combustion or burning process. The combustion rate for burning vapor/air mixtures is measured by how fast the flame propagates (flame speed) relative to the unburned mixture. For quiescent fuel/air mixtures the flame propagates at approximately 0.5 m/s and minimal overpressure is observed as the hot gases can expand in the unconfined environment. However, the flame propagation rate is extremely sensitive to turbulence in the flammable mixture into which the flame is burning. This turbulence can be caused by repetitive obstacles, (e.g. pipes, equipment, trees) within the vapor cloud itself. This results in a feedback loop, which has been extensively studied by Gexcon (see Figure 2.2), whereby the turbulence causes the flame to burn faster, which in turn pushes the unburned gases past the obstacles even faster creating higher turbulence levels, which in turn causes the flame to burn even faster. This results in the flame accelerating and can lead to very high propagation rates.

Figure 2.2: Flammable fuel air experiments for "unconfined" vapor clouds with no obstacles (top) and with trees (bottom). Clouds ignited on the left and maximum flame speeds with trees are six times larger than without.

The flame may propagate at such high speeds that the expanding gas ahead of the flame is actually compressed, which can ultimately result in a shock or blast wave. The strength of the blast wave will depend on the flame speeds achieved during the acceleration process.

If the congestion in the cloud is low, the flame speeds will remain low and will not generate a shock or blast wave (i.e., a "flash" or "whoosh" type sound), while relatively higher flame speeds can occur in high congestion areas, resulting in shock wave and "boom" type sound. Flame speeds between 100-150 m/s can generate significant overpressures, and those above 400 m/s can generate pressures well in excess of 1 barg, and can even transition to detonations. This phenomenon was precisely what occurred at the Buncefield explosion in 2005, whereby the flame accelerated into highly congested tree lines that extended over 100 meters. No confinement was found at the Buncefield site and the flame accelerations, and extremely high overpressures, were exclusively caused by the tree density [11].

In order to generate the massive explosion observed at the CAPECO facility, the resulting large vapor cloud needs to migrate into a congested (as described above) or confined region to accelerate the flame. Similar to Buncefield, the CAPECO facility had congestion that included numerous low-lying pipelines connecting the storage tanks, and vegetation to the north and east of the facility. During the post-incident scene investigation, thermal damage was found to extend into the tree line approximately 50 yards past the east property fence [1]. Unlike Buncefield, the CAPECO facility had additional long channels of confined drainage pipes that were not properly closed. Recall that the vapor cloud was ignited western side of the plant propagating towards the waste water treatment (see Figure 1.7) and it propagated east towards the tank farm and vegetation. Video analysis confirmed (see Figure 2.3) that while the main flame propagated above ground through the tank farm, a faster flame front traveled via the underground drainage pipes where a fireball exited the sump pump encasement at the northeast corner of the central tank farm well before the above ground flame.

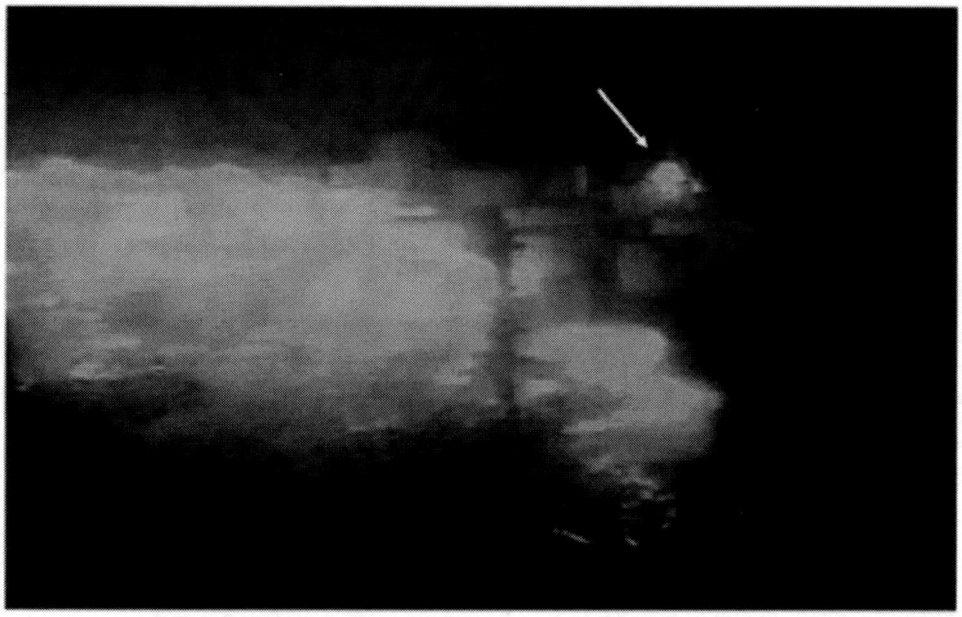

Figure 2.3: View from camera 3 showing the flame front in the underground drainage pipes reaching the northeast corner of the central tank farm before the above ground flame front. Ignition occurred on the western side of the facility resulting in flame propagation from left to right in the image above.

In fact, the 10 ft x 10 ft concrete lid on the sump pump, weighing about 12,000 pounds, was found over five feet from its original location [1]. Furthermore, blast damage to the above ground storage tanks was generally in the direction of the sump pump encasement and eastern vegetation and not the western side of the facility where the cloud was first ignited [1]. This confirms that the explosion pressures were exacerbated due to the confinement caused by the underground drainage pipes, which resulted in higher overpressures and faster accelerating flames that also entered the congested region. To evaluate the effect of the underground drainage system, FLACS simulation were performed to evaluate the resulting overpressures and blast damage both with and without the presence of the drainage system.

The CAPECO tank farm area (both with and without the drainage system) was modeled in FLACS, a computational fluid dynamics (CFD) tool predominantly used in the Oil & Gas and chemical industries to predict the consequences of hydrocarbon deflagrations. A large vapor cloud (e.g., stoichiometric cloud 4 meters tall) was placed throughout the tank farm and water treatment areas, and ignited in the water treatment area. Figure 2.4 shows the tank farm with the large vapor cloud and drainage system.

The flame propagation rates were evaluated both through the above ground cloud and through the drainage system. Figure 2.5 shows that the vapor cloud entering the drainage system resulted in flame speeds of over 400 m/s, significantly higher than above ground flame propagation. In fact, the confinement and acceleration of the flame in the drainage system resulted in overpressures that exceeded 2 barg (see Figure 2.6), which could possibly result in condition where the exiting deflagration transitions to a detonation. In contrast, overpressures generated by the flame burning through the above ground portion of the cloud did not exceed 0.03 barg. Therefore, the confined underground drainage system resulted in flames burning faster than the speed of sound and with pressures exceeding 2 barg, which ignited the remaining unburned vapor cloud in the tank farm. This high-pressure, supersonic flame front resulted in a significantly more devastating explosion than the "flash" or "whoosh" that would have been observed had the drainage system been properly closed prior to the overfill.

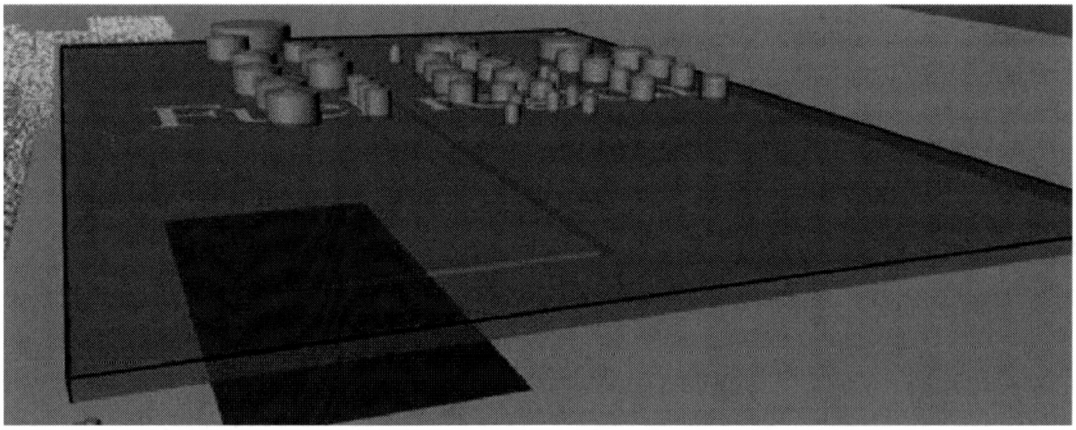

Figure 2.4: FLACS model of the CAPECO underground drainage system.

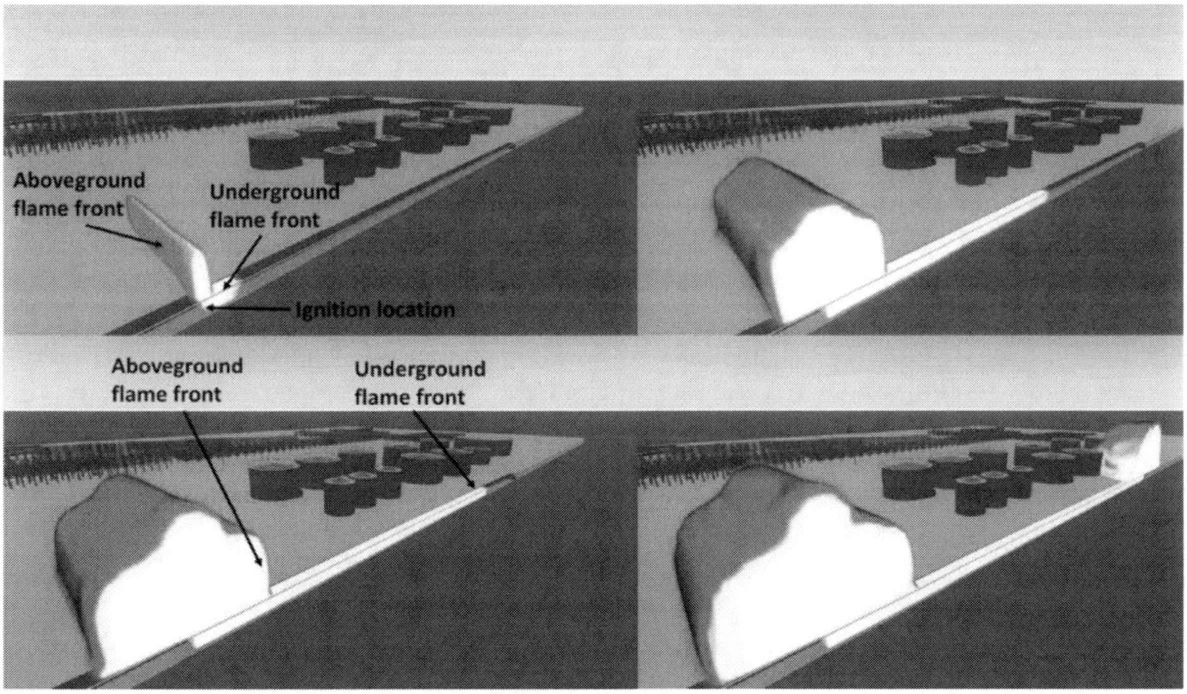

Figure 2.5: Flame front above and underground.

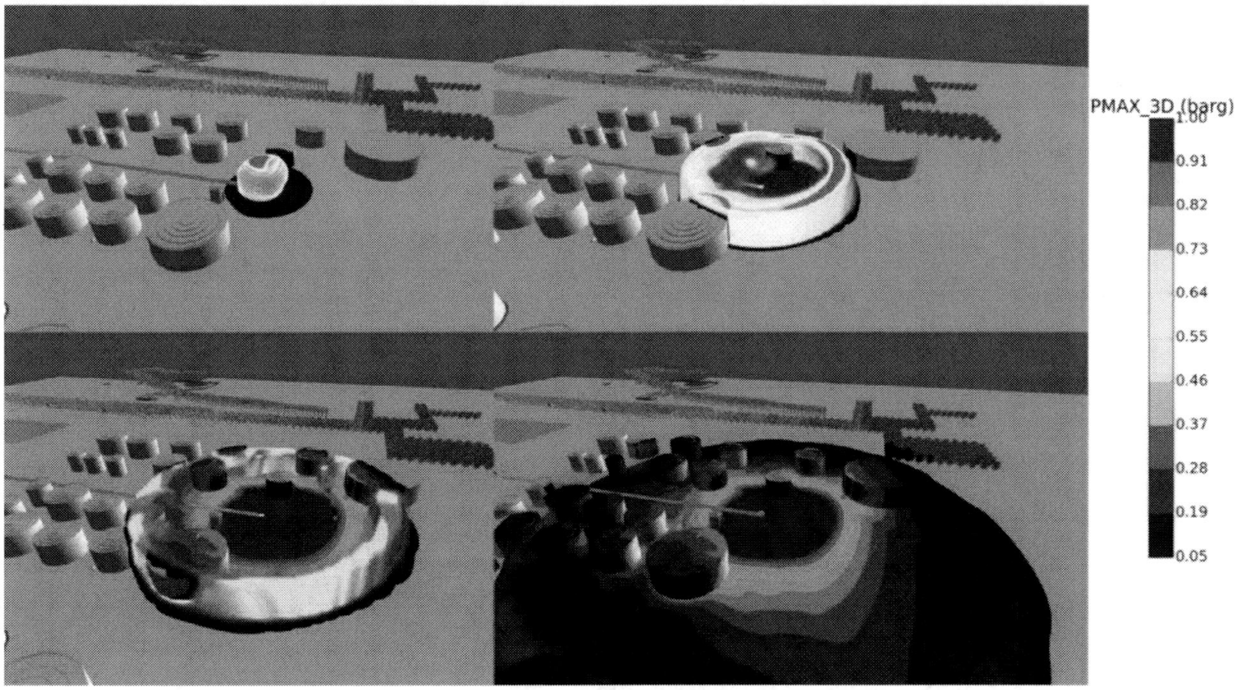

Figure 2.6: Pressure generated from the explosion in the underground drainage system.

3 Investigation findings and key lessons learned

Many of the contributors to the explosive atmosphere present at the CAPECO facility during the early morning hours on October 23, 2009 could have been avoided if the proper precautions were followed. While previous investigation reports have outlined certain contributing factors related to the incident, other major findings and lessons learned were not identified. For example, other groups have determined that the facility did not have automatic level alarms and that the automatic tank gauge system at the facility was not operational, which contributed to the event by not providing key indicators automatically and relying solely on manual measurements. However, key contributors to the incident (in addition to the automatic gauge system) were not identified and included: a failure to accurately monitor and verify the rate at which the gasoline storage tanks were being filled during the vessel offloading; and a failure to ensure "inherent safety" at the facility by not securing the dike drainage systems, which resulted high-pressure, supersonic flame fronts capable of causing the massive explosion as opposed to a "flash" or "whoosh" had the vapor cloud remained above ground. These factors will be discussed next.

Operational deficiencies that led to the Tank 409 Overfill

Recall that automatic controls were down and operations were manually monitored. Using the vessel discharge rates and tank strapping tables, it was estimated that when tanks 504 and 409 were being filled from 4:00 a.m. to approximately 12:00 noon on October 23, 2009, the transfer to each tank was approximately equal (see Figure 3.1). However, when 411 was brought online from 12:00 noon to 10:00 p.m. on October 23, 2009, it was determined that approximately 66% of the fuel was being transferred to tank 411 (see Figure 3.2), with the remaining fuel going to tank 409. This is consistent with operators' observations when preparing to take tank 411 offline: the valve to tank 409 was noted to be only open a little, confirming that tank 411 was receiving a much higher flow rate than tank 409 while they were being filled simultaneously [1].

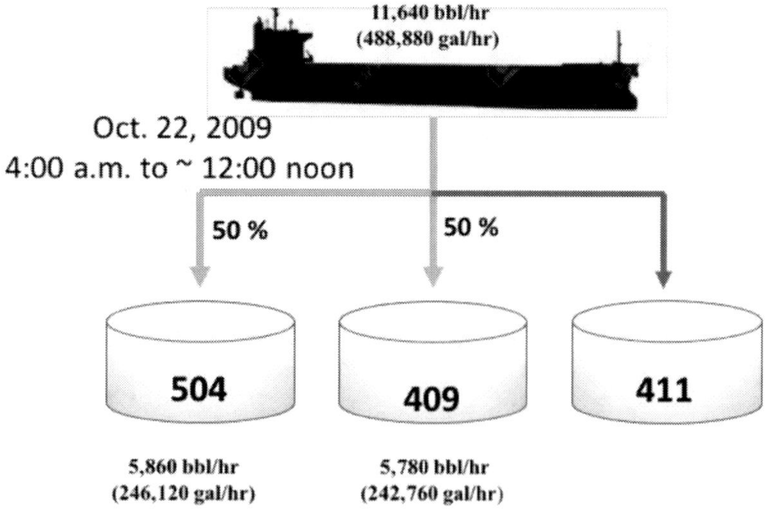

Figure 3.1: Transfer from the Cape Bruny to tanks 504 and 409 from 4:00 a.m. to approximately 12:00 noon on October 22, 2009.

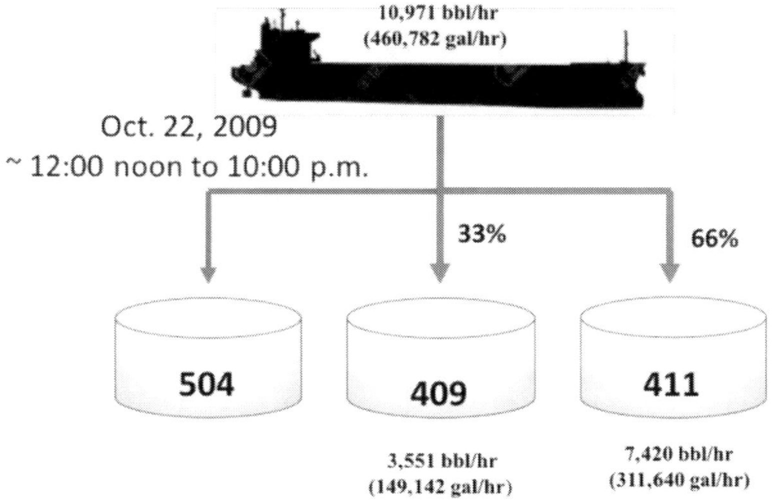

Figure 3.2: .Transfer from the Cape Bruny to tanks 411 and 409 from approximately 12:00 noon to 10:00 p.m. on October 22, 2009.

At 10:00 p.m. on October 22nd, tank 411 was closed and tank 409 was fully opened. Tank 409 contained approximately 98,716 gross barrels (4,146,072 gallons) of gasoline (overfill occurs at approximately 121,062 barrels).[1] The supervisor incorrectly estimated that the tank would be full at 1:00 a.m. by assuming only 33% of the discharge rate was going to tank 409. However, with the full vessel discharge of approximately 11,134 bbl/hr [7] going into tank 409, approximately 5,082 barrels (213,444 gallons) of gasoline overflowed from tank 409 contributing to the generation of a large flammable vapor cloud. This means that overflow occurred for approximately 27 minutes, beginning at approximately 11:45 p.m. on October 22nd until 12:12 a.m. the next morning.

Table 3.1: Estimated amounts of gasoline added to each tank and overflown at the CAPECO facility.

Tank	Estimated gasoline added Gross Barrels (Gallons)
405	4,411 (185,262)
504	62,984 (2,645,328)
411	74,198 (3,116,316)
409	114,325 (4,801,650)
Total	255,918 (255,918)
Total from Cape Bruny	261,000 (10,962,000)
Overfill	**5,082 (213,444)**

Therefore, one major cause of the incident was due to an incorrectly estimated fill time of 1:00 a.m. for tank 409, likely caused by ignorance of the increased flow rate to tank 409 after closing tank 411. Therefore, failure to compare vessel discharge rates to tank fill rates and verify that tank 409 had sufficient space to receive the scheduled amount of product resulted in tank overflow which generated a large flammable vapor cloud [12].

Drainage system

In the post incident investigation, the drainage valve in the berm surrounding tank 409 was found to be in the open position. The open valve drained to the waste water treatment facility, creating a pathway for the liquid gasoline to spread, increasing vapor generation and allowing for flammable gasoline vapor to fill the underground drainage system. The dike drainage valves should be closed during transfer operations so that in the event of an overfill, pool spread is limited. In this case, failure to close the drainage valve led to flammable vapors building up within the underground drainage system, and after ignition, resulted in a deflagration propagating within a completely confined drain and causing significant flame accelerations and overpressures. CFD analyses confirmed that these flame accelerations led to high pressure, supersonic flame fronts capable of causing the massive explosion as opposed to a "flash" or "whoosh" had the vapor cloud remained above ground. Keeping the drainage valves closed during a fuel transfer would have vastly limited the consequences and ensured inherent safety at the facility.

Level controls

The CAPECO facility has an automatic tank gauge system that transmitted the tank liquid depth to the operator's office. It would calculate the fill rate, as well as plot the flow rate and tank liquid depth versus time. Due to issues with the automatic tank gauge system (typical float and tape gauge), manual inspection of the tanks and calculation of the fill time had to occur through hourly readings.

4 Conclusions

Although US regulations do not consider bulk petroleum storage terminals as high hazards, an explosion at the CAPECO facility and ensuing fire destroyed a vast majority of the tank farm facility. This explosion was caused by a vapor cloud resulting from the overfill of a gasoline bulk storage tank, and this vapor cloud entering an improperly closed drainage system. The two main lessons learned from this incident (and previous incidents like Buncefield) include: (1) having systems and procedures in place to limit the likelihood of having such an overfill condition, and (2) providing inherent safety to such facilities to vastly limit the explosion consequences in the event such an overfill occurs (e.g. limiting confinement and congestion).

Regarding reducing the likelihood of such an event, the CAPECO facility was operating as a fully-attended facility and was in compliance with federal regulations at the time of the incident. As of today, fully-attended tank farm facilities do not require independent liquid level alarms and automatic tank gauging which allows other facilities to operate in the same fashion as CAPECO when transferring and storing flammable liquids such as gasoline. Therefore, such facilities should consider redundant controls and procedures to ensure: fill and transfer rates are constantly updated and verified throughout the transfer; and that the automatic tank gauge system is operational and complements manual measurements made at the facilities; and areas of high confinement like dike drainage systems are secured and closed prior to transfer of volatile hydrocarbons.

Inherent safety should also be evaluated at such facilities to ensure significantly lower consequences, even in the event of a significant release. CFD simulations have confirmed that physical layout of both Buncefield and CAPECO resulted in significantly higher consequences that could have resulted for identical overfill conditions. In fact, the densely coppiced tree lines at Buncefield and high confinement of the improperly opened drainage system at CAPECO resulted in explosion pressures and consequences orders of magnitude higher had such congestion/confinement not been present. Inherent safety of such facilities can be enhanced if areas of confinement and congestion are identified and mitigated in standard risk assessments. In fact, the CAPECO and Buncefield incidents would have likely resulted in large flash fires instead of the more devastating VCEs had these congestion and confinement areas been mitigated.

References

[1] ATF Investigation 764045-10-0011 Report.

[2] U.S. Chemical Safety and Hazard Investigation Board Final Investigation Report – Caribbean Petroleum Tank Terminal Explosion and Multiple Tank Fires.

[3] Puerto Rico Fire Department Fire Marshal Report.

[4] Inventory Logs.

[5] Dock Operating Manual.

[6] ATF Investigation 764045-10-0011 Interviews.

[7] Ship Transfer Logs.

[8] Caribnewspr https://www.flickr.com/photos/43578995@N03/.

[9] Davis, SG, Hinze, PC, Hansen, OR, van Wingerden, K (2010). 2005 Buncefield Vapor Cloud Explosion: Unraveling the Mystery of the Blast. Interflame 2010, Nottingham, UK..

[10] Coldrick, S., Atkinson, G.T., and Gant, S.E., Health and Safety Laboratory, Hazards XXII, 2011, pp. 560- 569..

[11] Van Wingerden, M. and Wilkins, B. (2008) Experimental investigation of the effect of flexible obstructions on flame propagation in vapour cloud explosions, GexCon report no. GexCon-08-F44110-RA-1, rev 01.

[12] API RP 2350, Appdx B.3.3 (2005).

Produced Water Tank Explosion Case Study

Presented by:

J. Kelly Thomas, Ph.D.
Baker Engineering and Risk Consultants, Inc.

2017 AIChE
Global Congress on Process Safety

BAKERRISK®

BAKERRISK

Outline

- Purpose
- Background
- Facility Description
- Incident
  - Overview
  - Fuel
  - Oxidant
  - Ignition
- Related Incidents
- Lessons Learned
- Potential Prevention and Mitigation Approaches

2

Purpose

- Highlight commonly encountered hazard
  - Produced water tank with flammable atmosphere
  - Similar hazard associated with waste (slop) water tanks
    - Incidents involving waste water tanks in industry have occurred
- Identify lessons learned
- Identify potential prevention and mitigation approaches
  - Examples, not exhaustive compilation
  - Other approaches may provide similar level of risk reduction

3

Background

- Produced water (brine water)
  - Water separated from oil / condensate / gas streams
  - Collected in tanks and treated as waste
  - Head space may be in flammable range depending on operating parameters and separator performance (e.g., condensate carryover)
- Not typically viewed as posing significant explosion risk
  - Reasonable on a relative basis if other explosion hazards present which pose much more severe consequences and/or risk
  - Reasonable from a building siting perspective
  - Hazard is local, although debris may be thrown considerable distance
  - However, produced water tanks can, and do, explode
- Should be considered in terms of hazard/risk to personnel

4

Facility Description (1 of 6)

- Incident occurred at gas production site (West Virginia, 2013)
- 6 gas wells (no oil or condensate produced), each with own separator
  - Separators are 2.5 ft diameter by 8 ft long, and operate at 675 psig
  - Gas composition (97.2% CH_4, 2.1% C_2H_6, 0.1% C_3H_8, 0.3 CO_2, 0.3% N_2)
- Produced water tanks
  - Tanks in two groups of 3 (i.e., 6 tanks)
  - Group of 3 tanks share common overflow (3 in.) and vent (4 in.) lines
    - Common vent line not supplied by tank manufacturer
    - No flame arresters on these lines
  - Fixed roof API 12F style, 12 ft. (3.7 m) diameter, 20 feet (6.1 m) tall
  - 400 barrel (16,800 gallons, 64 m^3) & hydrotested at 1.5 psig (0.10 barg)
  - 2 access hatches (Enardo ES-660 and Jayco JT-8); relieve at 0.5 psig

5

Facility Description (2 of 6)

6

Facility Description (3 of 6)

- Fill line and access hatches

7

Facility Description (4 of 6)

- Common vent and overflow lines

- Common vent line terminates with open end at opposite end of 3-tank group

Facility Description (5 of 6)

- Sign at entrance to pad

9

Facility Description (6 of 6)

- Warning sign on each produced water tank
- Flammability rating of 3 would be indicative of gasoline

10

Incident (Overview, 1 of 7)

- Approx. 100 bbl removed from each tank every 3 or 4 days
  - Driver goes up to top of tank via catwalk, opens hatch, and gauges using wooden rod before pumping out desired liquid volume
- Driver arrived between 5:30 and 5:45 am, hooked truck up to tank manifold, and (at some point) ascended catwalk to tanks
- Explosion occurs prior to 6:45 am (February, sunrise at 7:17)
- One tank fails at shell-floor joint, ejected, and thrown 110 feet (34 m)
- One tank fails at shell-roof joint, roof thrown 165 feet (50 m)
- Catwalk damaged and displaced, throwing driver from catwalk and outside tank battery, approximately 30 feet (9 m); fatality.

11

Incident (Overview, 2 of 7)

12

BAKERRISK

Incident (Overview, 3 of 7)

13

BAKERRISK

Incident (Overview, 4 of 7)

- Ejected tank

14

Incident (Overview, 5 of 7)

- Failed tank roof
- Shell-floor joint shows uplift (was approaching joint failure)

15

Incident (Overview, 6 of 7)

- Catwalk displacement (catwalk supported by tank attachment)

16

Incident (Overview, 7 of 7)

- Tank structural capacity roughly 10 to 15 psig (gradually applied internal pressurization) based on past experience
- Trajectory analysis indicates in range of 5 to 15 psig required to achieve observed tank and tank roof throw distances
- Indicates weakly flammable mixture was present in two tanks which failed, and mixture in "middle" tank was either not flammable or barely flammable
- Post-incident flammable gas measurements
  - Tank in adjacent set ≈ 65% LFL days after incident
  - Truck ≈ 15% LFL 2 months after incident
  - Consistent with elevated fuel concentration in tanks at time of incident

17

Incident (Fuel, 1 of 3)

- Separator does not remove "all gas" from the fluid, even when functioning properly and as designed
- Produced water contains gas since at elevated pressure (675 psig), which is released when pressure dropped to atmospheric
  - Gas content function of water pressure, temperature and composition
- Can have gas "carry" over or with stuck/jammed valve in separator
  - Separators were tested post-event and found to work properly
  - Concluded gas source was solely content of produced water transferred to storage tanks from separators

18

Incident (Fuel, 2 of 3)

- Methane solubility data from Duan & Mao (2006) Geochemica, 70: 2269-86.
  - Approximately 0.043 moles/kg at relevant conditions
- See vapor space fuel concentration plot on following slide:
  - Low level to high level of 50 to 150 barrels (1/4 tank capacity)
  - Conditions present (separator pressure, fluid composition, etc.)
  - No air in-leakage (only draw in air when drop liquid level)
- Can easily exceed UFL
  - Fuel concentration would be lower at lower separator pressure
  - Minimum fuel concentration would decrease if either high level or liquid volume extracted were increased (vs. values noted above)
  - > UFL is not a "universal condition" for produced water tanks

19

Incident (Fuel, 3 of 3)

- Fuel concentration for series of 100 bbl fill / extract cycles

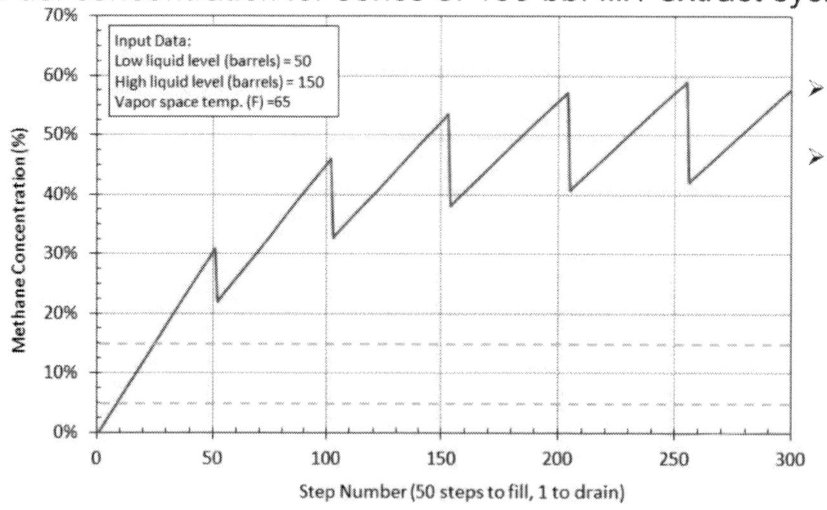

- Can easily exceed the UFL
- Issue is how get enough oxidant to be in flammable range

20

Incident (Oxidant, 1 of 4)

- Need additional air to enter flammable range (i.e., < UFL)
- Additional air can be drawn into tank via natural convection
  - Gas within tank lighter than outside air (i.e., unless outside air is hotter)
  - Light gas can flow out common vent line, pulling air in through leak points (e.g., access hatch seals)
    - Analysis subject to uncertainty since inlet flow path not well defined, but leakage rate in range of 125 scfd (2.5 lpm, 22 bbl/day) reasonable
    - Acts to limit maximum fuel concentration
    - Steady decrease in fuel concentration when tank is "stagnant" (i.e., no produced water additions or removals)
  - Much higher air in-leakage rate when hatch is open (2 to 5 scfm)
- Tanks stagnant for period before explosion & hatched opened
- See illustrate plots on following page

21

Incident (Oxidant, 2 of 4)

- Fuel concentration for series of 100 bbl fill / extract cycles, followed by stagnant period (125 scfd air in-leakage)

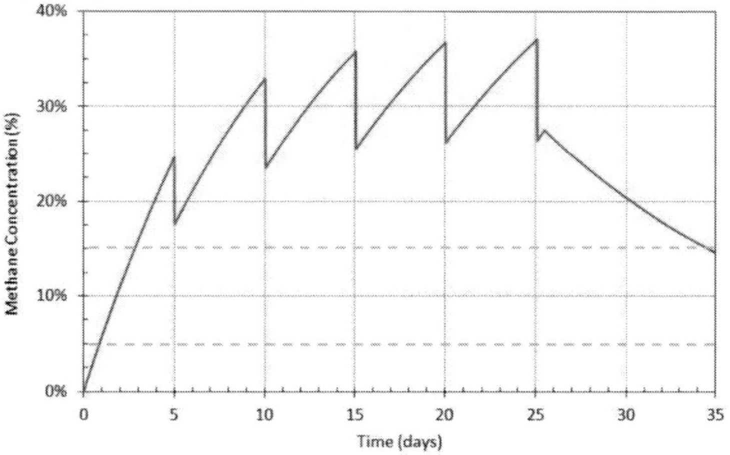

- Above UFL until enter stagnant period
- Air in-leakage during stagnant period allows fuel concentration to drop below UFL and enter flammable range
- Reach flammable condition in about 2 weeks

22

Incident (Oxidant, 3 of 4)

- Fuel concentration for series of 100 bbl fill / extract cycles, followed by stagnant period with hatch open (2 scfm air in-leakage)

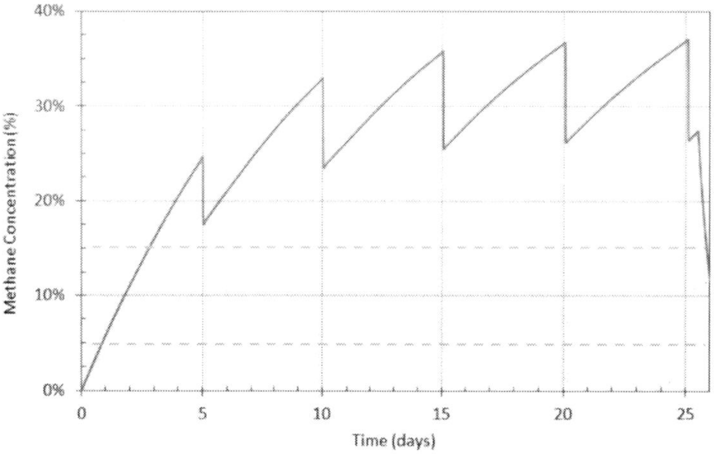

- Above UFL until hatch open for extended period
- Air in-leakage with hatch open for extended period allows fuel concentration to drop below UFL and enter flammable range
- Reach flammable condition in about 8 hours

23

Incident (Oxidant, 4 of 4)

- Natural convection limits maximum fuel concentration and can bring tank vapor space into flammable range, particularly with a tank which is stagnant for an extended period or with a hatch left open
- Note should <u>not</u> conclude that require natural convection in order to form a flammable mixture with a generic produced water tank
  - Lower separator pressure would mean less gas and hence tank could be flammable for portion of time regardless of natural convection
  - Larger volume of produced water withdrawn from tank or higher tank levels could put the tank vapor space into the flammable range
- Should assume a flammable mixture can result if the fuel concentration can exceed the LFL

24

Incident (Ignition Source)

- Driver was performing operation before sunrise
  - No headlamp or flashlight recovered from scene
  - Lighter found 25 feet from body (in same direction as thrown)
  - Driver was a smoker and one glove was off
  - Opened cigarette pack present in truck
  - Reasonable to assume driver used lighter to check liquid level
  - Driver not equipped with an LFL meter
- Static charge cannot be completely ruled out
- Flame propagation to adjacent tanks via connecting lines
- Ignition potential recognized (e.g., API RP 505)

25

Related Incidents

- Produced water tank explosions have happened in the past
- Cursory literature search
  - Texas (2013): Set of six 200 bbl produced water tanks, 2 people smoking on top of tank, tank exploded
  - Canada: Welding on tank which had contained produced water resulted in explosion
- CSB investigation report (2011-H-1):
  - "Public Safety at Oil & Gas Storage Facilities"
  - 26 incidents over 27 year period (44 fatalities and 25 injuries)
  - Includes at least one produced water tank
- Produced water SDS packages vary in flammability warnings (as does the signage on the produced water trucks)

26

Lessons Learned

- Produced water tank explosions can and do happen
  - Lesson already learned in prior incidents .....
- Hazard awareness may not be adequate among all personnel working at oil and gas production sites
- Development of flammable atmosphere in tank depends on separator pressure, tank arrangement and tank ventilation
  - Should assume tank vapor space may be flammable
  - Can also have separator malfunction leading to increased fuel concentration

27

Potential Prevention and Mitigation Approaches

- Increase hazard awareness for all personnel working at site
- Measure level at grade to eliminate routine access to most hazardous location (i.e., tank roof)
- Inert tanks using natural gas
  - Take feed from product gas line
  - Eliminate open vent paths and institute pressure control system (flow as required to maintain positive pressure and preclude air in-leakage),
  - Will result in increased emission of natural gas from facility
- Use a "lead tank" as a secondary separator
  - Inert only this tank with natural gas (limit use of natural gas)
  - Pump or gravity flow from "lead tank" to other produced water tanks
- Monitor tank vapor space

28

Questions

?

29

481704 Milestone Company Incidents and Their Impact on Air Products Process Safety Journey

Abstract:

Air Products has experienced several significant Process Safety events in our history that have helped define the Process Safety culture that exists at the company today. This presentation describes these events and their root causes, as well as the changes that they launched in the company's approach to Process Safety. The first incident was an explosion at a Nitrous Oxide plant in 1976 that resulted in four on-site fatalities. In the aftermath of this event the need for a formal Hazard Review Process was identified. The second event occurred in 1978 at an air separation plant in which a cryogenic Liquid Oxygen storage tank was ruptured and 800 tons of Liquid Oxygen was released. The Liquid Oxygen vaporized and drifted off-site, resulting in five off-site fatalities. This incident highlighted the necessity of Risk Quantification. While these milestone events served as the impetus for Air Products Process Safety journey, that journey continues to current day, and the presentation will close with a brief update on future direction.

481925 Plymouth Incident

Abstract:

The Williams Company LNG facility on their Northwest Pipeline system located in the state of Washington experience an explosion March 30th 2014. The incident occurred during a routine start-up of the liquefaction process. Directly after the facility was deemed safe the site preservation process was initiated and the investigation began. The various regulatory agencies were involved during the investigation to help determine the cause and to provide recommendations to prevent like kind occurrences elsewhere within the industry. Additional 3rd party firms were called in to review potential structural deficiencies, design flaws, metallurgy testing, and auto ignition testing.

AUTHOR INDEX

AUTHOR INDEX